Water Resources and Coastal Management

Managing the Environment for Sustainable Development

Series Editors: R. Kerry Turner
Professor of Environmental Sciences at the School of Environmental Sciences
and Director of CSERGE and the Centre for Environmental Decisionmaking, University of East Anglia, UK

Ian J. Bateman
Reader in Environmental Sciences at the School of Environmental Sciences
and Senior Research Fellow of CSERGE and the Centre for Environmental Decisionmaking, University of East Anglia, UK

1. Urban Planning and Management
   Kenneth G. Willis, R. Kerry Turner and Ian J. Bateman

2. Rural Planning and Management
   Joe Morris, Alison Bailey, R. Kerry Turner and Ian J. Bateman

3. Water Resources and Coastal Management
   R. Kerry Turner and Ian J. Bateman

Future titles will include:

Environmental Risk Planning and Management
Simon Gerrard, R. Kerry Turner and Ian J. Bateman

Waste Management and Planning
Jane C. Powell, R. Kerry Turner and Ian J. Bateman

Environmental Ethics and Philosophy
John O'Neill, R. Kerry Turner and Ian J. Bateman

Wherever possible, the articles in these volumes have been reproduced as originally published using facsimile reproduction, inclusive of footnotes and pagination to facilitate ease of reference.

For a list of all Edward Elgar published titles visit our site on the World Wide Web at
http://www.e-elgar.co.uk

Water Resources and Coastal Management

Edited by

R. Kerry Turner

Professor of Environmental Sciences, School of Environmenal Sciences and Director, CSERGE and the Centre for Environmental Decisionmaking, University of East Anglia, UK

and

Ian J. Bateman

Reader in Environmental Sciences, School of Environmental Sciences and Senior Research Fellow, CSERGE and the Centre for Environmental Decisionmaking, University of East Anglia, UK

MANAGING THE ENVIRONMENT FOR SUSTAINABLE DEVELOPMENT

An Elgar Reference Collection
Cheltenham, UK • Northampton, MA, USA

Published by
Edward Elgar Publishing Limited
Glensanda House
Montpellier Parade
Cheltenham
Glos GL50 1UA
UK

Edward Elgar Publishing, Inc.
136 West Street
Suite 202
Northampton
Massachusetts 01060
USA

A catalogue record for this book
is available from the British Library

Library of Congress Cataloguing in Publication Data
Water resources and coastal management / edited by R. Kerry Turner and Ian J. Bateman.
    p. cm. —(Managing the environment for sustainable development ; 3)
Includes bibliographical references and index.
    1. Coastal zone management. 2. Coastal engineering—Environmental aspects. 3. Water resources development. 4. Shore protection. I. Turner, R. Kerry. II. Bateman, Ian. III. Series.

HT391 .W33 2001
333.91—dc21

2001040114

ISBN 1 84064 222 X

Printed and bound in Great Britain by MPG Books Ltd, Bodmin, Cornwall

Contents

PART III   INTEGRATED COASTAL MANAGEMENT

PART IV   VALUATION OF COASTAL RESOURCES

Acknowledgements

The editors and publishers wish to thank the authors and the following publishers who have kindly given permission for the use of copyright material.

Academic Press for articles: J.C. Doornkamp (1998), 'Coastal Flooding, Global Warming and Environmental Management', *Journal of Environmental Management*, **52**, 327–33; C.A. Davos (1998), 'Sustaining Co-operation for Coastal Sustainability', *Journal of Environmental Management*, **52**, 379–87.

American Institute of Biological Sciences for article: John H. Steele (1991), 'Marine Functional Diversity: Ocean and Land Ecosystems May Have Different Time Scales For Their Responses to Change', *BioScience*, **41** (7), July/August, 470–4.

Blackwell Publishers Ltd for article: Stephen J. Essex and Graham P. Brown (1997), 'The Emergence of Post-Suburban Landscapes on the North Coast of New South Wales: A Case Study of Contested Space', *International Journal of Urban and Regional Research*, **21** (2), June, 259–85.

Coastal Education and Research Foundation, Inc. for articles: Stephen P. Leatherman and Robert J. Nicholls (1995), 'Accelerated Sea-Level Rise and Developing Countries: An Overview', *Journal of Coastal Research*, **14**, Special Issue, Winter, 1–14; Amalia Moriki, Harry Coccossis and Michael Karydis (1996), 'Multicriteria Evaluation in Coastal Management', *Journal of Coastal Research*, **12** (1), Winter, 171–8.

Helen Dwight Reid Educational Foundation (published by Heldref Publications) for articles: V.M. Kotlyakov (1991), 'The Aral Sea Basin: A Critical Environmental Zone', *Environment*, **33** (1), January/February, 4–9, 36–8; Janusz Kindler and Stephen F. Lintner (1993), 'An Action Plan to Clean Up the Baltic', *Environment*, **35** (8), October, 7–15, 28–31.

Elsevier Science for articles: Edward D. Goldberg (1995), 'Emerging Problems in the Coastal Zone for the Twenty-First Century', *Marine Pollution Bulletin*, **31** (4–12), 152–8; R.J. Nicholls and F.M.J. Hoozemans (1996), 'The Mediterranean: Vulnerability to Coastal Implications of Climate Change', *Ocean and Coastal Management*, **31** (2–3), 105–32; Stephen Olsen, James Tobey and Meg Kerr (1997), 'A Common Framework for Learning from ICM Experience', *Ocean and Coastal Management*, **37** (2), 155–74; Timothy O'Riordan and Rosie Ward (1997), 'Building Trust in Shoreline Management: Creating Participatory Consultation in Shoreline Management Plans', *Land Use Policy*, **14** (4), 257–76; Jörg Köhn (1998), 'An Approach to Baltic Sea Sustainability', *Ecological Economics*, **27** (1), October, 13–28; Blair T. Bower and R. Kerry Turner (1998), 'Characterising and Analysing Benefits from Integrated Coastal Management (ICM)', *Ocean and Coastal Management*, **38**, 41–66; Nguyen Hoang Tri,

W.N. Adger and P.M. Kelly (1998), 'Natural Resource Management in Mitigating Climate Impacts: The Example of Mangrove Restoration in Vietnam', *Global Environmental Change*, **8** (1), April, 49–61; Russell S. Arthurton (1998), 'Marine-Related Physical Natural Hazards Affecting Coastal Megacities of the Asia–Pacific Region – Awareness and Mitigation', *Ocean and Coastal Management*, **40**, 65–85; Staff of the National Oceanic and Atmospheric Administration, US Environmental Protection Agency, and US Geological Survey (1999), 'The Ocean's Role in Climate Variability and Change and the Resulting Impacts on Coasts', *Natural Resources Forum*, **23**, 123–34; R.K. Turner, W.N. Adger, S. Crooks, I. Lorenzoni and L. Ledoux (1999), 'Sustainable Coastal Resources Management: Principles and Practice', *Natural Resources Forum*, **23**, 275–86; R. Kerry Turner, Stavros Georgiou, Ing-Marie Gren, Fredric Wulff, Scott Barrett, Tore Söderqvist, Ian J. Bateman, Carl Folke, Sindre Langaas, Tomasz Zylicz, Karl-Göran Mäler and Agnieszka Markowska (1999), 'Managing Nutrient Fluxes and Pollution in the Baltic: An Interdisciplinary Simulation Study', *Ecological Economics*, **30**, 333–52.

Helsinki Commission for material included in article: Janusz Kindler and Stephen F. Lintner (1993), 'An Action Plan to Clean Up the Baltic', *Environment*, **35** (8), October, 7–15, 28–31.

Wolfgang Kaehler (www.wkaehlerphoto.com) for photograph included in article: Janusz Kindler and Stephen F. Lintner (1993), 'An Action Plan to Clean Up the Baltic', *Environment*, **35** (8), October, 7–15, 28–31.

Kluwer Academic Publishers for article: Edward B. Barbier and Ivar Strand (1998), 'Valuing Mangrove–Fishery Linkages: A Case Study of Campeche, Mexico', *Environmental and Resource Economics*, **12** (2), September, 151–66.

V.M. Kotlyakov for material included in his article: (1991), 'The Aral Sea Basin: A Critical Environmental Zone', *Environment*, **33** (1), January/February, 4–9, 36–8.

Evelyn Letfuss for photographs included in article: Janusz Kindler and Stephen F. Lintner (1993), 'An Action Plan to Clean Up the Baltic', *Environment*, **35** (8), October, 7–15, 28–31.

Marine Resource Economics for articles: John C. Whitehead (1993), 'Total Economic Values for Coastal and Marine Wildlife: Specification, Validity, and Valuation Issues', *Marine Resource Economics*, **8** (2), Summer, 119–32; John B. Loomis and Douglas M. Larson (1994), 'Total Economic Values of Increasing Gray Whale Populations: Results from a Contingent Valuation Survey of Visitors and Households', *Marine Resource Economics*, **9**, 275–86.

Opulus Press for article: Donald F. Boesch (1996), 'Science and Management in Four U.S. Coastal Ecosystems Dominated by Land–Ocean Interactions', *Journal of Coastal Conservation*, **2**, 103–14.

Pearson Education Ltd for excerpt: Keith Clayton and Timothy O'Riordan (1995), 'Coastal Processes and Management', in Timothy O'Riordan (ed.), *Environmental Science for Environmental Management*, Chapter 8, 151–64.

Risk, Decision and Policy for article: Stavros Georgiou, Ian J. Bateman, Ian H. Langford and Rosemary J. Day (2000), 'Coastal Bathing Water Health Risks: Developing Means of Assessing the Adequacy of Proposals to Amend the 1976 EC Directive', *Risk, Decision and Policy*, **5**, 49–68.

Routledge, Inc. (Taylor and Francis Group) for articles: Rutherford H. Platt (1994), 'Evolution of Coastal Hazards Policies in the United States', *Coastal Management*, **22**, 265–84; Robert J. Nicholls and Stephen P. Leatherman (1996), 'Adapting to Sea-Level Rise: Relative Sea-Level Trends to 2100 for the United States', *Coastal Management*, **24**, 301–24.

Royal Swedish Academy of Sciences for articles: Carl Gustaf Lundin and Olof Lindén (1993), 'Coastal Ecosystems: Attempts to Manage a Threatened Resource', *Ambio*, **22** (7), November, 468–73; Kerstin Lindahl Kiessling (1998), 'Conference on the Aral Sea – Women, Children, Health and Environment', *Ambio*, **27** (7), November, 560–4.

Springer-Verlag for articles and excerpts: R. Kerry Turner and Jan Brooke (1988), 'Management and Valuation of an Environmentally Sensitive Area: Norfolk Broadland, England, Case Study', *Environmental Management*, **12** (2), March, 193–207; Jonas Larsson, Carl Folke and Nils Kautsky (1994), 'Ecological Limitations and Appropriation of Ecosystem Support by Shrimp Farming in Colombia', *Environmental Management*, **18** (5), September/October, 663–76; Henning Karup (1999), 'Fixed Link Projects in Denmark and Ecological Monitoring of the Øresund Fixed Link', in Manfred Vollmer and Henning Grann (eds), *Large-Scale Constructions in Coastal Environments: Conflict Resolution Strategies*, Chapter 10, 105–16; A.R.D. Stebbing and R.I. Willows (1999), 'Quality Status, Appropriate Monitoring and Legislation of the North Sea in Relation to its Assimilative Capacity', in Wim Salomons, R. Kerry Turner, Luiz Drude de Lacerda and S. Ramachandran (eds), *Perspectives on Integrated Coastal Zone Management*, Chapter 8, 145–82.

Every effort has been made to trace all the copyright holders but if any have been inadvertently overlooked the publishers will be pleased to make the necessary arrangement at the first opportunity.

In addition the publishers wish to thank the Library of the London School of Economics and Political Science, the Marshall Library of Economics, Cambridge University and the Library of Indiana University at Bloomington, USA, for their assistance in obtaining these articles.

Towards Integrated Coastal Management

R. Kerry Turner and Ian J. Bateman

Coastal zones and their hydrologically linked catchment areas have come under heavy environmental pressure in recent decades. A number of megacities (> 10 million inhabitants), for example, are located on the coast. This pressure is expected to intensify further over the next 20 to 30 years as coastal urbanization and inward population migration increase. Multiple and complex feedback effects have been spawned as socio-economic systems and the supporting ecological systems interact and coevolve. The resilience of the coastal-catchment system, in terms of its ability to cope with stress and shock, is being threatened by pollution and resource over-utilization. The degree of environmental risk present is also increased by the threat of climate change-induced sea-level rise and storm intensification. Cumulative stress pressure can build up through unexpected environmental change events, disrupting the livelihoods of millions of people. It is probably the case that more than half the coastal zones in most regions around the globe are at moderate to high risk (FAO, 2000). All countries with a coastline have an interest in the sustainable management of coastal resource systems. The core objective of such management is the sustainable utilization of the many goods and services produced by coastal ecosystems, together with the equitable distribution of the consequent welfare gains and losses. The management challenge is made more problematic by the natural variability present in coastal systems and by the multiple stakeholder interests and competing resource uses and values typically found in such zones.

An important concept in the sustainability context is functional diversity and its social science counterpart, functional value diversity. The basis notion is that ecosystems through their continued functioning provide society with an array of valuable goods and services, including socio-cultural, historical and symbolic assets. A strategy designed to maximize or maintain as much functional diversity as is feasible in coastal-catchment areas would seek to maintain ecosystem integrity; that is the maintenance of system components, interactions among them and the resultant behaviour or dynamic of the system. Safeguarding functional diversity is compatible with enhancing the resilience of systems to better adapt to change and therefore to retain the capacity for secure wealth creation over time.

Before examining these issues in more detail, the five-Part organizational framework for this volume as a whole will be outlined. The contents of the volume have been set out in the following way.

Part I contains articles which provide the reader with a basic introduction to, and overview of, marine and coastal science. The Steele paper (Chapter 1) provides a concise definition and explanation of the functional diversity notion; that is, the variety of responses to environmental change with a particular focus on the range of spatial and temporal scales with which organisms react to each other and to the environment. He points out that marine and terrestrial ecosystems differ significantly in their functional response to environmental change and that this will have practical implications for management strategies. It is the case that although marine systems may be much more sensitive to changes in their environment, they may also be much more resilient.

Coastal zones comprise a continuum of aquatic systems, including the network of rivers, the estuaries, the coastal fringe of the sea, and the continental shelf and its slope. Regional seas and their linked drainage basins cover even more extensive scales. The functional value diversity concept encourages analysts to take such a wider perspective and examine changes in large-scale ecological processes, together with the relevant environmental and socio-economic driving forces. At the global scale, while climate has fluctuated throughout time, a global warming scenario could lead to accelerated sea-level rise, changes in rainfall patterns and storm frequency or intensity, and increased siltation (see the paper by NOAA, USEPA and USGS, Chapter 2). The consequences might include shoreline erosion and associated loss of habitats, such as saltmarshes, mangroves and mudflats. An economic multiplier effect would then be generated, leading to, for example, losses in tourism income and fisheries productivity, together with increased costs of water supply and biodiversity conservation. The paper by Clayton and O'Riordan (Chapter 3) surveys the important coastal processes and links them to a series of coastal management issues. The paper by Boesch (Chapter 4) continues the discussion of the interface between science and management and looks at four US case study areas: Chesapeake Bay, San Francisco Bay, the Mississippi Delta and Florida Bay. Boesch concludes that both science and management have been challenged by the spatial, functional and temporal scale mismatches inherent in the watershed–coastal ecosystem relationship. These scale mismatches underlie the difficulties management agencies have had in recognizing, assigning causes to and effectively managing large-scale modifications of coastal ecosystems associated with changes in the inputs of fresh water, sediments and nutrients from the land.

The final paper, by Goldberg (Chapter 5), in this Part highlights the fact that threats to coastal ecosystems and their resilience are multiple and cumulative. They are characterized by long residence times, relatively slow accumulation rates, increasing fluxes with time and large spatial distribution. Among the most significant contemporary problems are:

- hydrologic disruption in the catchment and eutrophication impacts (i.e. environmental changes stimulated by an oversupply of nutrients (nitrogen and phosphorus));
- contamination of sediments and the water column via sewage, plastics and environmental oestrogens;
- the introduction of non-indigenous species through the discharge of ship-ballast waters; and
- climate change-induced risks.

Part II explores the various facets of the problem of variability in coastal processes and systems and the implications for human communities seeking to live in and utilize the resources provided by coastal areas. Under natural conditions the form of a coastline is an optimal, but ephemeral, morphodynamic response to changing sea level and the impact of wave and tidal energy. Platt (Chapter 6) has examined how society has sought to adapt to the dynamic characteristics of life in a coastal zone. He argues, using the USA as an example, that the range of societal responses to hazards such as flooding and erosion has broadened and evolved during the last one hundred years. Research has also helped to refine public understanding of physical coastal processes. Initial reliance on hard engineering protection has been supplemented by beach nourishment, flood insurance, building and land regulations, and coastal zone planning. He concludes, however, that even in the USA, with all its scientific and financial resources, there remains much conflict between federal programmes to mitigate coastal hazards and other policies that stimulate new coastal construction.

Doornkamp (Chapter 7) goes further and argues that past management decisions about socio-economic activity in the coastal zone have had an impact on relative land and sea levels and have done more to increase the risk of coastal flooding than can be assigned so far to global warming. The placement of fixed engineering structures (for resource exploitation, sea defence and coastal erosion protection reasons) within this constantly changing system has in many cases reduced the resilience of coastlines to respond to the stresses and shocks of environmental change. Arthurton (Chapter 8) focuses on the plight of the fast-growing coastal megacities of the Asia-Pacific region. He distinguishes between high-risk hazard scenarios (e.g. extreme events such as storm surges) and long-term incremental hazards (e.g. global sea-level rise). Mitigation strategies should be adapted so as to recognize this basic distinction. High-risk hazards should be countered by effective warning networks and emergency planning while long-term hazards can be addressed with a strategic planning approach, involving relocation and selective hard engineering works.

Nicholls and Leatherman (Chapter 9) explore three relative sea-level rise scenarios for the USA and similar projected ranges for a variety of developing countries (Chapter 10). Given the uncertainties which surround these long-run (up to 2100) predictions, the authors argue that policymakers and scientists must identify both low-cost, no-regret responses, which would maintain choices for future coastal managers, and sectors where reactive adaptation would be relatively costly and where allowance for future sea-level rise can be considered a worthwhile 'insurance policy'. But coastal systems provide for multiple uses, and as urbanization and development pressures continue to build up conflict between different interest groups it is becoming increasingly difficult to mitigate through traditional planning and other means. Essex and Brown (Chapter 11) examine these value conflicts in the context of the North Coast of New South Wales in Australia, where population growth and its associated infrastructure is putting heavy pressure on important environmental assets.

The final three chapters in Part II (Chapter 12 by Karup, Chapter 13 by Tri, Adger and Kelly, and Chapter 14 by Larsson, Folke and Kautsky) deal with specific coastal resource management issues; that is, the impact of large-scale constructions such as bridges on coastal environments and the role of natural ecosystems such as mangrove wetlands and their conservation versus development/replacement value in developing economies. Tri, Adger and Kelly show that mangrove rehabilitation in Vietnam makes economic sense even if only the direct benefits derived by local communities are appraised. But the restoration schemes are even more economically efficient when wider mangrove use values, such as storm protection, are also included. Larsson, Folke and Kautsky include so-called 'ecological footprint' analysis in their evaluation of shrimp farming in mangrove areas in coastal Colombia. They argue that a semi-intensive shrimp farm needs a spatial ecosystem support – the ecological footprint – that is

35–190 times larger than the surface area of the farm. They conclude that shrimp farming (compared with other forms of aquaculture) ranks as one of the most resource-intensive food production systems. While this type of analysis provides some useful data it is not in itself a sufficient method for policy option evaluation (Black *et al.*, 1997; van den Bergh and Verbruggen, 1999; Hanley *et al.*, 1999). The footprint approach fails to take sufficient account of the multifunctionality of ecosystems, or the workings of the international economic system and technological innovations. As such it cannot provide any precise sustainability rank ordering (or, for that matter, efficiency or equity rankings) of alternative policy options.

Part III of this volume is concerned with the issues that serve to advance or hinder progress towards a more integrated approach to coastal resources management (ICM). Chapter 15, by Turner *et al.*, adopts a scoping framework known as the 'Driving forces–pressure–state–impact–response' (DP–S–I–R) approach in order to analyse the coastal zone and its future prospects. Given the pressure on coastal areas, their socio-economic and cultural value significance and the competition among stakeholders seeking to utilize and/or conserve coastal resources, many analysts have been advocating a much more integrated and holistic approach to coastal management (Salomons *et al.*, 1999).

ICM (as a future goal) is a continuous, adaptive, day-to-day process which consists of a set of tasks, typically carried out by several or many public and private entities (see Chapter 16 by Bower and Turner). The tasks together produce a mix of products, services and other gains/losses of socio-cultural significance from the available coastal resources. In principle, the core objective of coastal zone management is the production of a 'socially desirable' mix of coastal environmental system states, products and services. In practice, this mix is subject to intense stakeholder debate and is likely to change over time with changing demands, changing knowledge and changing pressures.

A future, more integrated coastal management process should include:

- integration of programmes and plans for economic development, environmental quality management and ICM;
- integration of ICM with programmes for such sectors as fisheries, energy, transportation, water resources management, disposal of wastes, tourism and natural hazards management;
- integration of responsibilities for various tasks of ICM among the levels of government – local, state/provincial, regional, national, international and between the public and private sectors;
- integration of all elements of management, from planning and design, to implementation; that is, construction and installation, operation and maintenance, monitoring and feedback and evaluation overtime;
- integration among disciplines; for example, ecology, geomorphology, marine biology, economics, engineering, political science, law; and
- integration of the management resources of the agencies and entities involved.

The ICM process should aim to unite government and the community, science and management, and sectoral and public interests. It should *inter alia* improve the quality of life of human communities who depend on coastal resources while maintaining the biological diversity and productivity of coastal ecosystems (Gesamp, 1996). Clearly this is a formidable task, especially for developing countries, and one that will only be achieved incrementally over time (see Chapter 17 by Lundin and Lindén).

Some analysts have questioned the whole rationale for ICM. According to Nichols (1999), ICM is actually an attempt to reorganize coastal spaces and political systems for the purpose of facilitating investment penetration by governments and/or transnational corporations. The consequence (particularly in developing countries) is the political and spatial marginalization of pre-existing resource users. In order to address this equity issue and others, ICM has to be more than just a process by which efficient utilization of coastal resources is promoted. In Chapter 19, Olsen, Tobey and Kerr strongly argue that the fundamental challenge of coastal management is one of governance (objective, process and structures) and not of technology transfer or refined scientific knowledge. They recommend a learning-based approach to coastal management, which assumes that such intervention is a young endeavour inevitably beset by uncertainties, instability and rapid rates of change. It follows that progress towards effective coastal management and sustainable forms of coastal development will only come incrementally, through analysis and experience over decades. A learning-based approach calls for framing coastal management initiatives as experiments and subjecting them to formal scientific testing and analysis.

Progress through the ICM cycle will also be conditioned by the degree to which 'accountability' and 'trust' issues are successfully tackled. No process of ICM can produce legitimate answers (and effective solutions) to the challenges posed without meaningful public participation. The public will must be incorporated in a proactive, participatory and conflict-minimizing fashion. Davos (Chapter 18) believes that if ICM is crucially dependent on the voluntary co-operation of stakeholders, this raises doubts about the value of positivistic or normative ICM prescriptions in the absence of consensus. He argues that the alternative is to pursue a 'co-operative CZM' approach, which would have as its defining property a reliance on social discourse. Such discourse also needs a guiding framework to facilitate the achievement of co-operative collective decisions. There is a need to establish 'windows of opportunity' where policy, politics and participants can operate together to set the sustainable resource utilization agenda and to implement it effectively. The final chapter in Part III, by O'Riordan and Ward (Chapter 20), speculates on the type of participatory management process that is required in the future if a sustainable coastal resources utilization strategy is to become a practical reality.

Part IV deals with the prospects for more precisely valuing a range of coastal resources. One key to valuing a change in an ecosystem function is establishing the link between that function and some service flow valued by people. If that link can be established, then the concept of derived demand can be applied. The value of a change in an ecosystem function can be derived from the change in the value of the ecosystem service flow it supports.

The main problem when including the range of biodiversity services in economic choices is that many of these services are not valued on markets. There is a gap between market valuation and the economic value of biodiversity. To fill these gaps the non-marketed services must first be identified and then, where possible, monetized. In the case of biodiversity the identification of economically relevant services is of special importance, since over time those perceived benefits not allocated by the market have increased in importance.

The mainstream economic approach to valuation takes an instrumental (usage-based) approach (as opposed to an intrinsic value which resides in the object itself) and seeks to combine various components of value into an aggregate measure of resource value labelled total economic value (TEV). This TEV can be usefully broken down into a number of categories.

The initial distinction is between use value and non-use value. Use value involves some interaction with the resource), either directly or indirectly:

1. Indirect use value derives from services provided by the ecosystem, such as, for example, a wetland. This might, for instance, include the removal of nutrients, providing cleaner water to those downstream, or the prevention of downstream flooding.
2. Direct use value, on the other hand, involves interaction with the ecosystem itself rather than via the services it provides. It may be consumptive use, such as the harvesting of reeds or fish, or it may be non-consumptive, such as with some recreational and educational activities. There is also the possibility of deriving value from 'distant use' through media such as television or magazines, although whether or not this type of value is actually a use value, and to what extent it can be attributed to the ecosystem involved, is unclear.

Non-use value is associated with benefits derived simply from the knowledge that a resource, such as an individual species or an entire ecosystem, is maintained. It is by definition not associated with any use of the resource or tangible benefit derived from it, although users of a resource might also attribute non-use value to it.

In Chapter 21, Turner and Brooke investigate the management and valuation problems related to an important coastal wetland area in England. Barbier and Strand (Chapter 22) focus on a specific ecosystem function provided by a mangrove wetland in Mexico and assign economic values to it in terms of a production function relationship. In Chapter 23, Georgiou, Bateman, Langford and Day are concerned with a different type of coastal service – recreational bathing at the coast – and utilize a mixed quantitative/qualitative methodology in a risk–benefit analysis. Chapters 24 and 25, by Loomis and Larson, and Whitehead respectively, adopt the TEV typology in order to deploy a survey-based (contingent valuation) appraisal method to value marine/coastal biodiversity. Finally, Moriki, Coccossis and Karydis deploy a multi-criteria evaluation method to analyse a land use conflict situation on the Greek island of Rhodes (Chapter 26).

Part V of the volume brings together a number of papers which set out to examine the more spatially extensive environmental pressure and policy options associated with regional seas – the Baltic Sea, Aral Sea, North Sea and Mediterranean; see Chapters 27 (Kindler and Lintner), 28 (Köhn), 29 (Turner *et al.*), 30 (Kotlyakov), 31 (Kiessling), 32 (Stebbing and Willows) and 33 (Nicholls and Hoozemans).

Scoping Framework

The DP–S–I–R framework is a useful device for the scoping of coastal science and management issues and problems (Turner *et al.*, 1998). The objective in this approach is to clarify multisectoral inter-relationships and to highlight the dynamic characteristics of ecosystem and socio-economic changes. The DP–S–I–R framework provides a way of identifying the key issues, questions, data/information availability, land use patterns, proposed developments and existing institutional frameworks, as well as timing and spatial considerations. The framework reflects the hypothesis that anthropogenically induced change in coastal ecosystems follows a consistent, and therefore predictable, sequence. Speed and historical timing of change, however, vary significantly.

For any given coastal area (defined to encompass the entire drainage network) there will exist a spatial distribution of socio-economic activities and related land uses – urban, industrial mining, agriculture/forestry/aquaculture and fisheries, commercial and transportation. This spatial distribution of human activities reflects the final demand for a variety of goods and services within the defined area and from outside the area. Environmental pressure builds up via these socio-economic driving forces and is augmented by natural systems variability, which stimulates changes in environmental systems states.

The production and consumption activities result in different types and quantities of residuals, as well as goods and services measured in terms of gross national product. Thus the concern might be, for example, the risks posed by contaminants (artificial substances) released into the environment and the role and extent of changes in Carbon, Nitrogen, Phosphorus and sediment fluxes as a result of land use change and other activities. Conceptually, what we have are a multiplicity of input–output (I–O) relationships, with the outputs being joint products (combinations of goods and services and non-product outputs or residuals, which if not recycled become waste emitted/discharged into the ambient environment). I–O relationships will operate at the individual industrial process/plant level, through population settlements, agricultural cropping regimes/practices, and up to regional drainage basin scale. These residuals estimates will then serve as the input to the natural science models, such as nutrient budgets and ecotoxicological assessments. Environmental processes will transform the time and spatial pattern of the discharged/emitted residuals into a consequent short-run and long-run time and spatial ambient environmental quality pattern.

These state environmental changes impact on human and non-human receptors, resulting in a number of perceived social welfare changes (benefits and costs). Such welfare changes provide the stimulus for management action which depends on the institutional structure, culture/value system and competing demands for scarce resources and for other goods and services in the coastal zone. An integrated (modelling) approach will need to encompass within its analytical framework the socio-economic and biophysical drivers that generate the spatially distributed economic activities and related ambient environmental quality, in order to provide information on future environmental states.

In summary, the scoping/audit phase should raise, among others, the following fundamental issues/questions/problems:

- the need for, and feasibility of, a basic characterization of the study area encompassing both natural science and social science (socio-economic activity patterns and drivers) data;
- the extent of scale, particularly the system boundaries for the proposed study;
- the modelling/analysis goals, the need for, and feasibility of, some predictive power in the analysis to be adopted, for example via environmental change scenarios or management strategies;
- the contribution the chosen study can make to any scaling-up process.

Functional Value Diversity

From an anthropocentric viewpoint all ecosystems can be classified in terms of their structural and functional aspects. Ecosystem structure is defined as the tangible items, such as plants, animals, soil, air and water, of which it is composed. Thus wetland structural benefits (of instrumental value to humans) include fish, waterfowl, peat, timber, reed and fur harvests as well as non-consumptive use benefits such as recreation, research and education. By contrast, ecosystem processes are encompassed by the dynamics of exchange of means of energy. The processes are subsequently responsible for the services – life support services, such as assimilation of pollutants, cycling of nutrients and maintenance of the balance of gases in the air.

The terminology used regards processes and functions as relationships within and between natural systems; uses refer to actual use, potential use and non-use interactions between human and natural systems; and values refer to assessment of human preferences for a range of natural or non-natural 'objects', services and attributes. A management strategy based on the sustainable utilization of coastal resources principle should have at its core the objective of ecosystem integrity maintenance; that is, the maintenance of system components, interactions among them and the resultant behaviour or dynamic of the system.

Evaluating Change in Coastal Zones

The assessment of the impact of changes in the coastal zone on human use of resources (wealth creation) and habitation (quality of life aspects) requires the application of socio-economic research methods and techniques in the context of coastal resource assessment and management. A particular contribution of socio-economic research is the incorporation of evaluation methods and techniques which can be applied to specific resource damage and utilization situations (projects, policies or courses of action which change land use/cover, alter or modify residuals from point and non-point sources, and so on). Most of these valuation studies will be at a local/regional level and there may therefore be a scaling-up problem if transboundary resource management issues are to be addressed, as will be the case in the regional seas context. However, the transfer of economic valuation estimates (known as benefits transfer) across time and geographical and cultural space is controversial (Costanza, 1998).

The last 20 to 30 years have seen the gradual evolution of a strategy aimed at an integrated assessment of environmental science, technology and policy problems. A multi-disciplinary tool kit has been presented which, for example, global climatic change researchers have tapped into (Schneider, 1997). The integrated assessment framework must include coupled or integrated models (biogeochemical and socio-economic) but it is not limited to this. According to Rotmans and van Asselt (1996), integrated assessment is 'an interdisciplinary and participatory process of combining, interpreting and communicating knowledge from diverse scientific disciplines to achieve a better understanding of complex phenomena'. The critical importance of making value-laden assumptions highly transparent in both natural and social scientific components of integrated assessment models (IAMs) needs to be highlighted. Practitioners now argue that incorporating decision makers and other stakeholders into the early design of IAMs greatly facilitates this process. Valuation in this process is more than the assignment of monetary values and requires, among others, multi-criteria assessment methods and techniques in order to identify practicable trade-offs.

The economic concept of TEV can be traced back to two fundamental dimensions of value, production and individual values. Production values of biodiversity are arguments in the production and cost functions of market-allocated goods. These production inputs affect individual welfare via changing prices of goods or other inputs, for example, use of ecosystems for aquaculture production. Individual values, on the other hand, are a direct argument of individual utility functions. These include recreational and aesthetic values, as well as passive use, non-use or existence values. These two categories of value can, however, be supplemented with another category, which considers the ecological importance of biodiversity by describing the ecological–functional role of biodiversity in natural systems. Included here are those services of biological resources that stabilize the ecological system and perform a protective and supportive function for the economic system. The recently developed approaches for considering the ecological functions of biodiversity include the following (again somewhat overlapping) categories of values under this umbrella (Fromm, 2000):

- *inherent value* – describes those services without which there would not be the goods and services provided by the system;
- *contributory value* – considers the economic–ecological importance of species diversity, such that even species not useful to humans are important since they contribute to increases in diversity, which contributes to the generation of more species (Norton, 1986);
- *indirect use value* – is related to the support and protection provided to economic activity by regulatory environmental services (Barbier, 1993);
- *primary value* – incorporates the fact that the existence of the wetland structure is prior to the range of function/service values (secondary values); and that the system holds everything together in a healthy functioning state – a 'glue' value contribution (Turner and Pearce, 1993);
- *infrastructure value* – relates to a minimum level of ecosystem 'infrastructure' as a contributor to its total value (Costanza *et al.*, 1997).

The economic relevance of including the structures and functions of ecosystems in total economic value follows from their input functions for production and individual values and the protection services for human capital, man-made capital and natural capital (as output). The conventional total economic value restricted to the individual and production values of biodiversity contains no value component that gives credit to the contributory, infrastructure and glue value dimension of ecosystems. It therefore remains incomplete.

The social value of an ecosystem (total system value) therefore may not be equivalent to the aggregate private total economic value of that same system's components, because of the following factors (Turner, Button and Nijkamp, 1999):

1. The full complexity and coverage of the underpinning 'life support' functions of healthy evolving ecosystems are currently not precisely known in scientific terms. A number of indirect use values within systems therefore remain to be discovered and valued (quasi-option value; that is, conditional expected value of information).
2. Because the range of secondary values (use and non-use) that can be instrumentally derived from an ecosystem is contingent on the prior existence of such a healthy and evolving system, there is in a philosophical sense a 'prior value' that could be ascribed to the system itself. Such a value would not, however, be measurable in conventional economic terms and is non-commensurate with the economic (secondary) values of the system.

3.   Following on, the continued functioning of a healthy ecosystem is more than the sum of its individual components. There is a sense in which the operating system yields or possesses 'glue' value; that is, value related to the structure and functioning properties of the system which hold everything together.

4.   A healthy ecosystem also contains a redundancy reserve, a pool of latent keystone species/processes which are required for system maintenance in the face of stress and shock. This is what the quasi-option value concept seeks to capture and here it is interpreted to mean more than an ex-post option value judgement.

In environmental economics, an individual preference-based value system operates in which the benefits of environmental gain (or the damages from environmental loss) are measured by social opportunity cost (i.e. cost of foregone options) or TEV. The assumption is that the functioning of ecosystems provides society with a vast number of environmental goods and services which are of instrumental value to the extent that some individual is willing to pay for the satisfaction of a preference. It is taken as axiomatic that individuals almost always make choices (express their preferences), subject to an income budget constraint, which benefit (directly or indirectly) themselves or enhance their welfare. Households are assumed to maximize well-being from different sources of value subject to an income constraint. Their private willingness-to-pay (their valuation) is a function of prices, income and household tasks (including environmental attitudes) together with conditioning variables such as household size. The social value of environmental resource committed to some use is then defined as the aggregation of private values. Nature conservation benefits should be valued and compared with the relevant costs. Conservation measures should only be adopted if it can be demonstrated that they generate net economic benefits.

The economic component of assessment consists of the identification and economic valuation of positive and negative effects; that is, the costs and benefits associated with any proposed management option and the comparison with the 'do nothing' approach. The difference is the incremental net benefit arising from the project investment. Cost–benefit analysis is one of the evaluation tools developed by economists to determine whether a policy, project or action is economically efficient. Its principal feature is that all the pros and cons of a project, if technically possible, including social and socio-cultural and historical contexts that surround particular value gain/loss, are translated into monetary terms. As a rule, a project is efficient if total benefits exceed total costs.

Other environmental analysts, on the other hand, either claim that nature has non-anthropocentric intrinsic value and non-human species possess moral interests or rights; or that while all values are anthropocentric and usually (but not always) instrumental, the economic approach to valuation is only a partial approach. The socio-cultural and historical contexts in which environmental assets exist provide for alternative dimensions of environmental value which may not be captured by the market paradigm. Heritage coastlines, floodplain landscapes, river corridors and other assets can possess local and national symbolic value which cannot be expressed adequately in monetary terms. These environmentalist positions lead to the advocacy of environmental sustainability standards or constraints, which to some extent obviate the need for valuation of specific components of the environment. It is still necessary, however, to quantify the opportunity costs of such standards; or to quantify the costs of current and prospective environmental protection and maintenance measures. Nevertheless, for some people

it is feasible and desirable to manage the environment without prices. According to O'Neill (1997), for example, conflicts of values in forestry and biodiversity management issues in the UK are resolved through pragmatic methods of argument between botanists, ornithologists, zoologists, landscape managers, members of a local community, farmers, and so on.

In summary, the main generic approaches which can form the methodological basis for strategic socio-economic options appraisal are:

- stakeholder analysis;
- cost-effectiveness analysis;
- cost–benefit analysis;
- multi-criteria analysis.

There will be circumstances where one approach is preferable to the others. Where there are clear and commonly agreed objectives or targets to be reached then the most appropriate approach may be to look at cost-effective options. On the other hand, when targets cannot be pre-defined but must be determined within the assessment exercise, and all or most of the impacts can be expressed in money terms, then cost–benefit analysis will be favoured. In contrast, if impacts cannot be monetized, but are instead expressed via a variety of measurement units, then multi-criteria analysis may be most appropriate.

In any multiple resource use problem context, it will be necessary to identify the complete range of stakeholders present and their pressure impacts and influences. Multiple stakeholders translate into multiple worldviews and potential values conflict. The stakeholder/revenue conflict situations that may be identified in any given coastal zone could be assessed and evaluated via multi-criteria evaluation methods which encompass both monetary and non-monetary valuation procedures (Janssen, 1994). One way of conceptualizing this value conflict problem over time is via the formulation and analysis of environmental change scenarios. For this approach to produce meaningful results, a trend scenario (i.e. the implications of current trends remaining substantially unaltered until some chosen terminal date in the future) needs to be contrasted with the results derived from one or more alternative futures scenarios.

Usually all of the above approaches will involve some form of stakeholder analysis – in that they can involve stakeholders at a number of different points within the appraisal process. Stakeholders could, for example, be involved in the setting of management objectives or in the determination of values. Deciding how stakeholders should be involved is a key issue.

A stakeholder analysis focuses primarily on the people who have some kind of interest in the area and who will be positively or negatively affected in welfare terms by a change in the area's management regime. Examples of stakeholders are farmers, households dependent on the resources of the ecosystem or holidaymakers enjoying the beauty of the area. Each of these interest groups may exert a certain pressure on the ecosystem involved. Together, these pressures may directly or indirectly, in the long run, impair the various functions the system provides to each stakeholder group. In other words, the various interests and uses of the ecosystem may be conflicting.

Suggested changes in management practices in the area, arising within national and international environmental regulation, may reduce or reinforce conflicts between the various interests involved. Trying to satisfy all interest groups will often be difficult From a policy-making point of view, processes that can provide effective and efficient consensus outcomes are key future goals. In order to be able to meet these goals, insight is needed into what the various interests in the area are, who the stakeholders are and what the distribution of the positive and negative effects of changes in management regimes will be.

Benefits Transfer (Meta-analysis)

A study by Costanza *et al.* (1997) has engaged environmental scientists and policy makers, but the global, biome scale, economic value calculations contained in this study risk criticism from both scientists and economists. On the basis of the data and methods cited in the article, and supporting inventory, the conclusion that the value of the biosphere services is, on average, US$ 33 trillion per year, is not to be taken literally. Apart from raising policy maker, scientist and citizen awareness about the environment's economic value and the possible significance of the loss of that value over time, the global value calculations do not serve to advance meaningful policy debate in efficiency and equity terms, in practical conservation versus development contexts. Such calculations, with their 'single number' outcomes, shroud a number of fundamental 'scaling' problems to do with valuation contexts; that is, the temporal, spatial and cultural specificity of economic value estimates. Such values can also only meaningfully be assigned to relatively small ('marginal') changes in ecosystem capabilities (functions/services) (see Table 1). The practical problem is that determining precisely what is and what is not a discrete and marginal change in complex ecological systems is not straightforward.

Table 1   Composition of value elements for selected ecosystems

Coral reefs		Mangroves $ per hectare per year	
Coastal protection	2750	Coastal protection	1839
Waste treatment	58	Nutrient cycling	6696
Food production/biological control	259	Food production/biological control	797
Recreation	3008	Recreation	658
Total	**6075**		**9990**

Source: Derived from Costanza *et al.* (1997).

The issues of site, context, cultural and historical value specificity are generic and serve to constrain the transfer of site-based function and system services economic values across time and geographical and cultural space. It is not being argued that all such benefits transfer is invalid, but we do believe that such procedures must be handled with extreme caution and have real limits. Many value estimates will not be amenable to legitimate aggregation beyond local to 'regional' (defined biogeographically and including cross-national boundaries where necessary) scales. Further research to define these limits more precisely and to formulate a robust validity and reliability testing protocol is an urgent requirement (Brouwer, 1998; Costanza, 1998).

References

Barbier, E.B. (1993), 'Sustainable use of wetlands. Valuing tropical wetland benefits: economic methodologies and applications'. *The Geographical Journal* 159(1): 22–32.

Black, E.A., Gowen, R., Rosenthal, H., Roth, E., Stechy, D. and Taylor, F.J.R. (1997), 'The cost of eutrophication from salmon farming – a comment'. *Journal of Environmental Management* 50: 105–9.

Brouwer, R. (1998), 'Future research priorities for valid and reliable environmental value transfer'. Global environmental change working paper 98-28, Centre for Social and Economic Research on the Global Environment (CSERGE), University of East Anglia and University College London.

Costanza, R. (ed.) (1998), Special Issue, 'Ecosystem services value, research needs and policy reference', *Ecological Economics* 25.

Costanza, R., d'Arge, R, de Groot, R., Farber, S., Grasso, M., Hannon, B., Limburg, K., Naeem, S., O'Neill, R.V., Paruelo, J., Raskin, R.G., Sutton, P. and van den Belt, M. (1997), 'The value of the world's ecosystem services and natural capital'. *Nature* 387: 253–60.

Food and Agriculture Organisation (2000), FAO Web Site Data Base: www.fao.org, Rome.

Fromm, O. (2000), 'Ecological structure and functions of biodiversity as elements of its total economic value'. *Environmental and Resource Economics* 16(3) : 303–28.

Gesamp (1996), *The Contributions of Science to Integrated Coastal Management*, Report 61, FAO, Rome.

Hanley, N., Moffatt, I., Faichney, R. and Wilson, M. (1999), 'Measuring sustainability: a time series of alternative indicators for Scotland'. *Ecological Economics* 29.

Janssen, R. (1994), *Multiobjective Decision Support for Environmental Management*, Kluwer, Dordrecht.

Nichols, K. (1999), 'Coming to terms with integrated coastal management: problems of meaning and method in a new area of reserve regulation'. *Professional Geographer* 51: 388–99.

Norton, B.G. (1986), *Towards Unity Among Environmentalists*. Oxford University Press, Oxford.

O'Neill, J. (1997), 'Managing without prices: the monetary valuation of biodiversity'. *Ambio* 26: 546–50.

Rotmans, J. and van Asselt, M. (1996), 'Integrated assessment: a growing child on its way to maturity'. *Climate Change* 34: 327–36.

Salomons, W. *et al.* (1999), *Perspectives on Integrated Coastal Zone Management*. Springer, Berlin.

Schneider, S.H. (1997), 'Integrated assessment modelling of global climate change: transparent rational tool for policy making or opaque screen hiding value-laden assumptions?'. *Journal of Environmental Modelling and Assessment* 2: 229–49.

Turner, R.K. and Pearce, D.W. (1993), 'Sustainable economic development: economic and ethical principles', in Barbier, E.B. (ed.), *Economics and Ecology: New Frontiers and Sustainable Development*. Chapman and Hall, London.

Turner, R.K., Button, K. and Nijkamp, P. (eds) (1999), *Ecosystems and Nature: Economics, Science and Policy*. Edward Elgar, Cheltenham.

Turner, R.K. *et al.* (1998), 'Coastal management for sustainable development'. *Geographic Journal* 164: 269–81.

van den Bergh, J. and Verbruggen, H. (1999), 'Spatial sustainability, trade and indicators: an evaluation of the technological footprint'. *Ecological Economics* 29: 61–72.

Part I
Marine and Coastal Science

[1]

Marine Functional Divers

Ocean and land ecosystems may have different time scales for their responses to change

John H. Steele

The term *biological diversity* has the virtue of containing many different interpretations, scientifically and emotionally. A recent report (OTA 1987) suggested three technical meanings: genetic, species, and ecological diversity. I suggest a fourth category that needs to be considered, which I term *functional diversity*—the variety of different responses to environmental change, especially the diverse space and time scales with which organisms react to each other and to the environment.

The public is becoming increasingly aware of biological diversity losses imposed by human activity directly and through possible climatic change. Change as such is not new, and it is an integral element in systems. Concerns arise from the large scales of these changes and particularly from the rapid rates at which they may occur—rates that are much greater than any in the historical or even the geological past. Thus scientific and social interest focuses on the relative rates of change of so-called natural processes and of human interventions. In turn, this interest focuses on ecological aspects. How do ecological systems change through time? How diverse are these rates of change? How rapidly or slowly do communities respond to changes in their physical and chemical environments? In particular,

John H. Steele is a senior scientist at Woods Hole Oceanographic Institution, Woods Hole, MA 02543. © 1991 American Institute of Biological Sciences

Changes can have major economic consequences without being ecological disasters

I consider the response of marine communities to natural and human stresses, but in the context of current views of changes in terrestrial systems.

Marine and terrestrial ecosystems differ in their functional responses to environmental change. The current attention on climate focuses on time scales of several years to decades. I hypothesize that at these time scales marine systems are more closely coupled to their physical environment than are terrestrial systems (Steele 1985). I conclude that, in consequence, many open-sea systems respond to rapid changes in their overall environment, and they are adaptable in an ecological sense.

Comparison of land and marine systems

Trees, which are the longest-living component of terrestrial ecosystems, show changing geographic distributions with time scales usually measured in millenia (Delcourt et al. 1983). These changes are generally related to long-term large-scale climate trends such as the retreat of ice after the last Ice Age, 18,000 years

ago (Figure 1). But the probable rate of dispersion of forests is on the order of tens of kilometers per century (Davis 1989). Future geographical distributions can be predicted on the hypothetical rapid climate changes deduced from the large computer simulations. A 2–4° C increase in a century implies a northward movement of isotherms on the order of several hundred kilometers. Forests may not be able to keep up with this rapid climate change, and a disequilibrium in the overall ecological system is expected to result.

In the open sea, the longest-lived organisms are at the top of the food webs—just the opposite of the terrestrial system, where the oldest organisms are trees, which are primary producers (Steele 1985). For fish populations on the continental shelf, where we have the most data, we see changes in distribution where the spatial scales are comparable with forest patterns (Figure 1) but are much more rapid (Cushing 1982). A typical example is the sand lance (*Ammodytes* spp.), a benthic species of fish with pelagic larvae that lives on the continental shelf of the eastern United States (Figure 2). Sand lance distribution changed in three years as much as the boundaries of the hardwood forest moved in 5000 years.

Until recently, the major changes in fish populations were considered to result from fishing activities, which are certainly a factor. But other elements, particularly climate, contribute to these changes and may be dominant in the largest, longest-term fluctuations. These large variations in abundance of fish stocks can have severe economic consequences. But are they ecologically deleterious? Do they upset the entire food web? Should we expect a breakdown in the overall marine ecological system similar to that predicted for the land?

Marine scales of response

Data from Georges Bank, describing the relative abundance of different species in the catches, shows that species changes can result in severe economic disturbance without any evidence of overall ecological deterioration. Figures 2 and 3 illustrate that the decline in the larger pelagic species such as herring and mackerel (Figure 4a) coincided with the increase in the sand lance. The North Sea shows a similar picture (Figure 4b), with the decline in the same large and commercially valuable pelagic species being balanced initially by an increase in smaller species, such as sand lance, that are used predominantly for animal feed. However, the most interesting feature of the changeover was the dramatic increase in the stocks of demersal fish species, such as haddock and whiting, that feed mainly on the sea bed. Are the simultaneous changes in the different species abundances shown in Figure 4 coincidental? Or are there causal links between the increase in one species and the decrease in another?

It is not clear whether succession—the slow and predictable sequence of dominant species—seen in terrestrial systems, occurs in the ocean. A focus of current research is to untangle the diverse functional interrelations that a population of one species has with its physical environment and with the other members of its biotic community. Some possible interactions can be seen in the population fluctuations of North Sea herring. The marked decline during the 1970s (Figure 4a) was associated with a decrease in recruitment of juveniles to the adult stock (Corten 1986). Over-fishing was blamed. But in the 1980s, there was a marked increase in numbers of larvae and juveniles before the sexually mature stock itself increased. This finding made it apparent that the increasing recruitment of larvae could not be ascribed solely to an increase in the adult stock (Rothschild 1986). Thus attention has focused on the

earlier life stages.

The striking feature of this development is the close correspondence between ocean currents and the life-cycle patterns of herring stocks (Figure 5; see Bailey and Steele in press and Corten 1986 for full discussion). The comparison illustrates not only the dependence on the average ocean-water circulation but the consequences of departures from this average. Data on interannual larval distributions (Corten 1986) suggest variable transport from the spawning grounds to the nursery areas. Numerical simulations of the wind-driven currents within the North Sea provide a possible mechanism (Backaus 1985). But inflow of water from the Atlantic is a dominant

feature (Figure 5) and can have effects lasting from several years to a decade (Dooley et al. 1984). Thus there may be several links between specific physical current patterns and different scales of climate variation. At the population level, such links are explained by a much closer interaction between the reproductive processes and the particular patterns of ocean currents and mixing.

This example may help to explain the importance of recruitment, the process by which populations of immature marine organisms settle into their adult habitats. Scientists are now examining the egg and larval stages of marine populations for factors affecting population density, es-

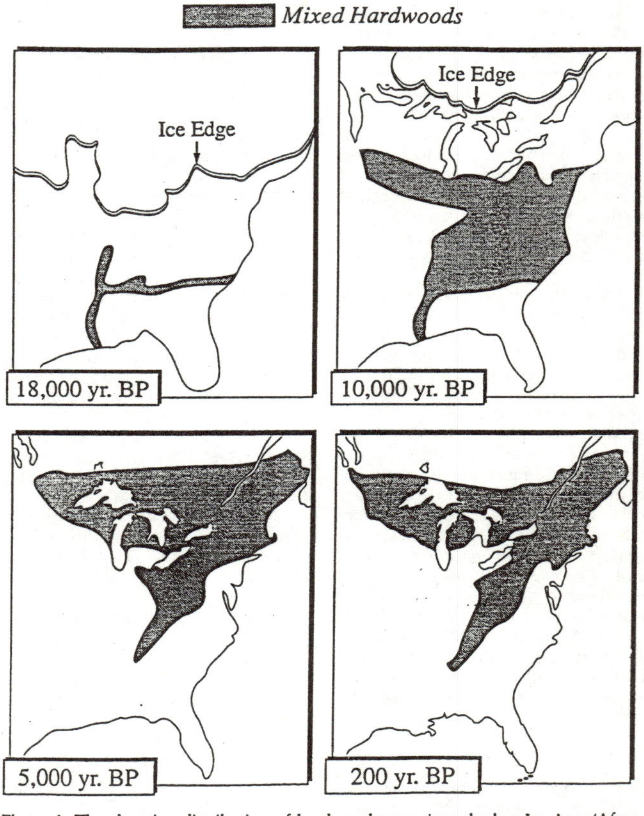

Figure 1. The changing distribution of hardwood trees since the last Ice Age. (After Delcourt et al. 1983.)

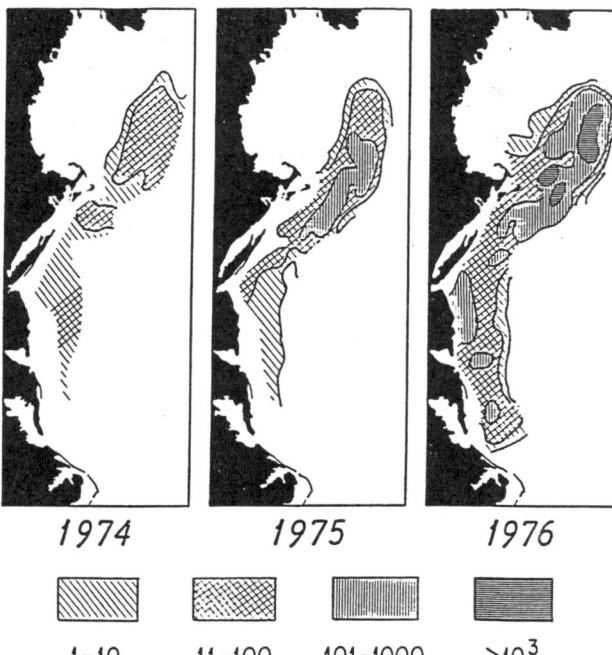

1974 1975 1976

1-10   11-100   101-1000   >10^3

Figure 2. The changing distribution and density of the sand lance off the US East Coast from Cape Hatteras to Nova Scotia. (From Sherman et al. 1981.)

pecially variations in the physical dynamics at longer periods that are superimposed on the annual cycles. As in herring, in barnacle and reef fish populations (Roughgarden et al. 1988, Sale 1982), the effects of water movement on larval recruitment rather than adult competition for space may determine population densities.

Yet community interactions cannot be ignored. For example, the North Sea herring is part of a complex food web (Hardy 1924). In the last three decades, there have been changes in the relative abundances of herring and some of the planktonic organisms it consumes. These changes may be related to climate trends (Aebischer et al. 1990), but there have also been changes in demersal fish species (Figure 4). The food of demersal fish, such as haddock, is predominantly from the sea bed, so these fish occupy a different sector of the food web than herring

(Figure 6). But the collapse of the pelagic fish stocks in the North Sea during the 1970s corresponded to a marked increase in several demersal species (Figure 4). Even more striking, the subsequent return of the herring a decade later was accompanied by a severe decline in haddock. This reversal might be seen as a diversion of the energy flow from basic primary production between the pelagic and benthic components, but it has not proved possible to explain the switches in terms of particular causal processes in this web.

There are many other examples of striking switches in community structure; the Russell cycle in the English Channel (Southward 1980) consists of marked changes in the herring and pilchard populations during the last century. Primary explanations for each species are sought in terms of changes in the physical dynamics, but

those changes are insufficient to explain the community interactions where there are marked changes in balance and even in composition.

Three conclusions about changes in marine communities can be made:

- Large switches in open-sea marine communities can last several decades. These changes can have major economic consequences but cannot be considered as ecological disasters or even as being deleterious in any way within the marine systems.
- Such switches have occurred without human involvement, for example before heavy fishing was a factor, but they may be increased in frequency or amplitude by human actions.
- The effects of physical processes on particular populations during the early life stages can be explained, but observed changes at the community level have not been explained.

These conclusions differ greatly from the usual views of terrestrial changes. The focus in studies of terrestrial systems is generally on community interactions as the explanation for changing patterns in observations of species compositions.

Changes in the environment

The effects of fluctuations in the physical environment are usually treated as noise, depending on the time scales involved. Fluctuations are greater in terrestrial systems than in the ocean. For example, on top of the regular diurnal and seasonal cycles of temperature, there is a much larger unpredictable temperature variability on land than in the ocean. I have speculated that such variability has tended to result in the evolution of terrestrial adaptations that can eliminate or smooth out the consequences of this random component (Steele 1985).

On land, the coupling of ecosystems changes with large-scale trends in climate is observed at time-scales of centuries to millennia (Delcourt et al. 1983). But within the ocean it appears that responses to physical variability are found at decadal periods and may be closely coupled to variations in ocean dynamics.

Comparing fish and trees

When we consider climate at the same time scales—from interannual to decadal—we see very different responses in marine and terrestrial systems, one illustration of functional diversity. These functional responses to changing physical environments depend on relative rates of growth and mortality, and these rates are known to be related to body size (Bonner 1965). Some elements of this diversity can be seen in the usual relations of size to trophic status when the pelagic marine and the terrestrial species are separated. Trees are the longest-lived organisms on land, whereas phytoplankton cells, the primary producers in the ocean, have lifetimes measured in days. There are several obvious biophysical reasons for these differences: among others, gravity and heat conduction. These differences in time scales must have an effect on the way the marine and terrestrial communities are structured.

The comparisons between trees and fish represent the extremes of pelagic marine communities and forests. What about the intermediate ecosystems? Should coral reefs and rocky shores be viewed as marine or terrestrial? It may be beneficial to use a combination of the classical terrestrial mechanism—competition for space—and the classical marine mechanism—larvae carried by currents.

The major difference between land and sea communities appears to be that marine systems have evolved to exploit regularities in the physical dynamics of the environment as part of their reproductive processes (Denman and Powell 1984) as well as using diffusive dispersal to counteract the longer-term consequences of variability in the physics. It has been pointed out by several ecologists, including Roughgarden (1988), that larval dispersal in marine communities provides for open systems, so that recruitment to areas where adults are not present can occur. This method of dispersal may allow populations in one area to live dangerously if there is always the possibility of recruiting larvae from elsewhere. Terrestrial systems can also recruit, but on much longer time scales, on the order of

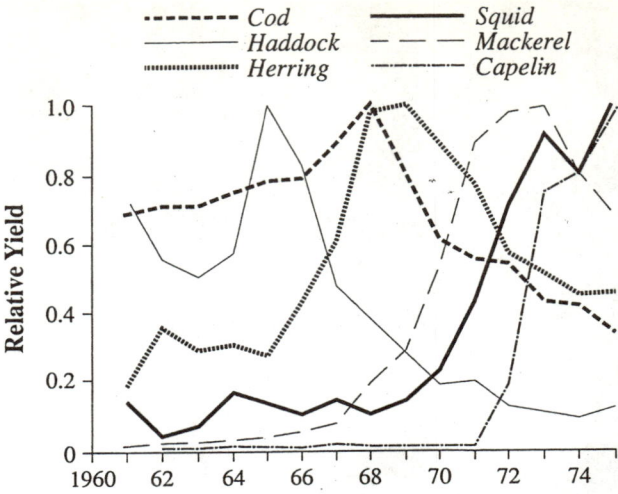

Figure 3. Relative yield of different fish and squid stocks off the East Coast of North America. (From Horwood 1981.)

centuries or millenia for forest spread.

The replacement of natural communities of trees or perennial grasses by annual crops can be viewed as a change in ecological time scales: from long to short, that is, annual or less. Agricultural systems often do not include any buffering mechanisms against environmental variability at time scales (years to decades) longer

than those of agricultural practice (one or a few years) but shorter than those of the natural perennial communities. A few years of drought can be devastating to an agricultural system.

Conclusions

Marine and terrestrial ecosystems differ significantly in their functional re-

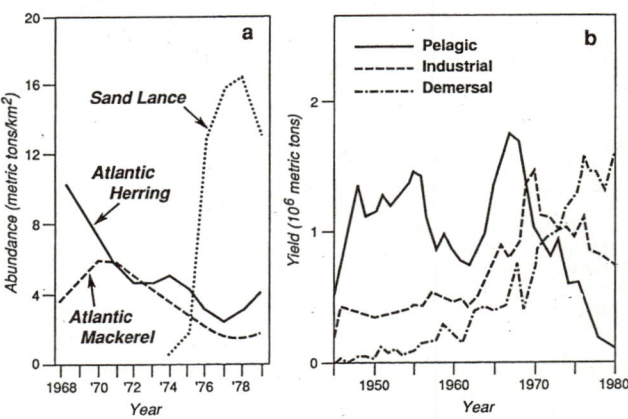

Figure 4. Changes in fish stocks (a) for the US East Coast (from Sherman et al. 1981) and (b) for the North Sea (from Bailey and Steele in press).

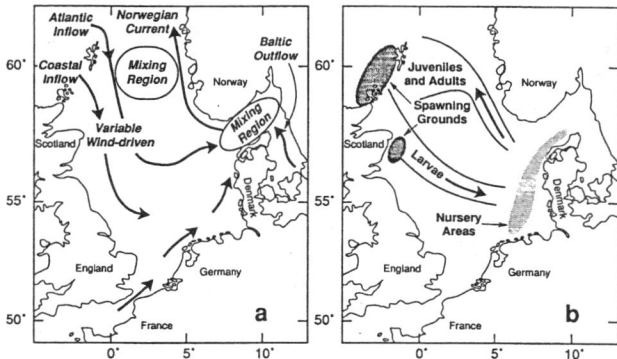

Figure 5. Schematic presentation for North Sea of (a) the main current systems (Dooley 1983) and (b) the life cycle and movements of herring stocks (Corten 1986, Cushing 1982). The larvae move by wind-driven currents, and the juveniles and adults move by the Norwegian current.

sponses to environmental change. Three principal functional relations exemplify this diversity.

• Size, life span, and trophic status. In the ocean and on land, the larger organisms are usually more long-lived, but they differ in trophic status—their places in the food web. Forest trees and open-ocean plankton represent the extremes.

• Relationship of organisms with physical processes. Marine reproductive cycles, particularly in the larval and juvenile phases, are closely linked to water movements. Terrestrial animals tend to evolve reproductive patterns that

eliminate the effects of interannual or decadal variability in the physical environment.

• Differences in processes on the land and in the sea, observed on the same time scale. Some of this contrast can be eliminated by using different time scales; the same responses can occur in different systems at different time scales. The rates, rather than the magnitudes, of environmental change may have more effect on the disturbance of community structure.

The practical implication of land-sea functional diversity is that it would be inappropriate to apply terrestrial perspectives to marine communities, particularly in the context of management or conservation. Although marine systems may be much more sensitive to alterations in their environments, they may also be much more adaptable.

Acknowledgment

This article is WHOI contribution no. 7684.

References cited

Aebischer, N. J., J. C. Coulson, and J. M. Colebrook. 1990. Parallel long term trends

across four marine trophic levels and weather. *Nature* 347: 753–755.

Backhaus, J. O. 1985. A three-dimensional model of the simulation of shelf sea dynamics. *Deutsche Hydrographische Zeitschrift* 38: 165–187.

Bailey, R. S., and J. H. Steele. In press. North Sea herring fluctuations. In M. H. Glantz, ed. *Climate Variability, Climate Change and Fisheries.*

Bonner, J. T. 1965. *Size and Cycle: An Essay on the Structure of Biology.* Princeton University Press, Princeton, NJ.

Corten, A. 1986. On the cause of recruitment failure of herrings in the central and northern North Sea in the years 1972–78. *Journal du Conseil pour Exploration de la Mer* 42: 281–294.

Cushing, D. H. 1982. *Climate and Fisheries.* Academic Press, New York.

Davis, M. B. 1989. Lags in vegetation response to greenhouse warming. *Clim. Change* 15: 75–82.

Delcourt, H. R., P. A. Delcourt, and T. Webb. 1983. Dynamic plant ecology: the spectrum of vegetational change in space and time. *Quaternary Science Review* 1: 153–175.

Denman, K. L., and T. M. Powell. 1984. Effects of physical processes on planktonic ecosystems in the coastal ocean. *Oceanogr. Mar. Biol. Annu. Rev.* 22: 125–168.

Dooley, H. D., J. H. A. Martin, and D. J. Ellettm. 1984. Abnormal hydrographic conditions in the Northeast Atlantic during the 1970s. Rapports et Process-Verbaux des Reunions. *Cons. Int. Explor. Mer* 185: 179–187.

Hardy, A. C. 1924. The herring in relation to its animate environment. *Fisheries Investigations London* 7: 1–53.

Horwood, J. W. 1981. Management and models of marine multispecies complexes. Pages 339–360 in C. W. Fowler and T. D. Smith, eds. *Dynamics of Large Mammal Populations.* John Wiley & Sons, New York.

Office of Technology Assessment (OTA). 1987. *Technologies to Maintain Biological Diversity.* OTA-F-330, US Govt. Printing Office, Washington, DC.

Rothschild, B. J. 1986. *Dynamics of Marine Fish Populations.* Harvard University Press, Cambridge, MA.

Roughgarden, J., S. Gaines, and H. Possingham. 1988. Recruitment dynamics in complex life cycles. *Science* 241: 1460–1466.

Sale, P. F. 1982. Stock-recruit relations and regional coexistance in a lottery competitive system: a simulation study. *Am. Nat.* 120: 139–159.

Sherman, K., C. Jones, L. Sullivan, W. Smith, P. Berrien, and L. Ejsymont. 1981. Congruent shifts in sand eel abundance in western and eastern North Atlantic ecosystems. *Nature* 291: 486–489.

Southward, A. J. 1980. The western English Channel—an inconstant ecosystem. *Nature* 285: 361–366.

Steele, J. H. 1974. *The Structure of Marine Ecosystems.* Harvard University Press, Cambridge, MA.

———. 1985. Comparison of marine and terrestrial ecological systems. *Nature* 313: 355–358.

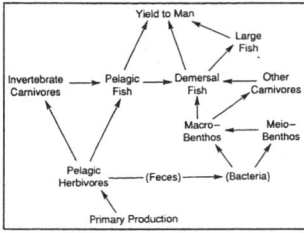

Figure 6. A general food web for the North Sea, showing the pelagic and benthic pathways to the two main fish categories. (From Steele 1974.)

[2]

PERGAMON

Natural Resources Forum 23 (1999) 123–134

Natural Resources
FORUM

The ocean's role in climate variability and change and the resulting impacts on coasts

Staff of the National Oceanic and Atmospheric Administration, US Environmental Protection Agency, and US Geological Survey[1]

Abstract

This article describes the oceans' influence on weather and climate and identifies selected global climate change impacts on coastal areas. It is divided into three parts: seasonal to inter-annual climate impacts; decadal to centennial climate impacts; and coastal global climate change impacts.

The article describes how the weather and climate are driven by the redistribution of heat. The major source of heat to the surface of the earth is the sun, principally through incoming visible radiation. Most of it is absorbed by the earth's surface. This radiation is redistributed by the ocean and the atmosphere and the excess is radiated back into space as longer wavelength, infrared radiation. Clouds and other gases, primarily water vapour and carbon dioxide, absorb the infrared radiation emitted by the earth's surface and reemit their own heat at much lower temperatures. This "traps" the earth's radiation and makes the earth much warmer than it would be otherwise.

Most of the incoming solar radiation is received in tropical regions, while very little is received in polar regions especially during winter months. Over time, energy absorbed near the equator spreads to the colder regions of the globe, carried by winds in the atmosphere and by currents in the oceans. Compared to the atmosphere, the ocean is much denser and has a much greater ability to store heat. The oceans also move much more slowly than the atmosphere. Thus, the oceans and the atmosphere interact on different time scales. The ocean moderates seasonal variations; it stores and transports, via ocean currents, large amounts of heat around the globe, resulting in changing weather patterns.

While global climate has fluctuated throughout time, a global warming scenario could speed climate change, possibly causing accelerated sea-level rise, alterations of rainfall patterns and storm frequency or intensity, and increased siltation. This in turn could result in the increased erosion of shores and associated habitat, increased salinity of estuaries and freshwater aquifers, altered tidal ranges in rivers and bays, changes in sediment and nutrient transport, a change in the pattern of chemical and biological contamination in coastal areas, and increased coastal flooding.

Some coastal ecosystems are particularly at risk, including saltwater marshes, coastal wetlands, coral reefs, coral atolls, and river deltas. Other critical coastal resources, such as mangroves and seagrass beds, submerged systems and mudflats, are at risk from climate change impacts, exacerbated by anthropogenic factors. Changes in these ecosystems could have major negative effects on tourism, freshwater supplies, fisheries, and biodiversity that could make coastal impacts an important economic concern. Coastal structures, including homes would also be more vulnerable to increased sea-levels. A number of alternative management strategies are discussed, as are some of the difficulties of convincing the public of the need for action. © 1999 United Nations. Published by Elsevier Science Ltd. All rights reserved.

Keywords: Climate change impacts; Ocean–climate interactions; Coastal ecosystems

1. Introduction to ocean–climate interactions

Understanding the ocean's role in weather and climate

variability requires first an understanding of the earth's heat budget, i.e., how the energy from incoming solar radiation is redistributed around the globe. The major source of heat to the surface of the earth is the sun, principally through incoming visible radiation. Heat is also generated in the earth's interior, for example through the decay of radioisotopes. However, this contribution to the heat balance of the surface of the earth, where most life exists, is small

[1] Corresponding author: C. Mason, Senior Coastal Oceanographer, Senior Scientist's Office, National Ocean Service, National Oceanic and Atmospheric Administration, Silver Spring, MD, USA.
E-mail address: curt.mason@noaa.gov (C. Mason)

124 *C. Mason / Natural Resources Forum 23 (1999) 123–134*

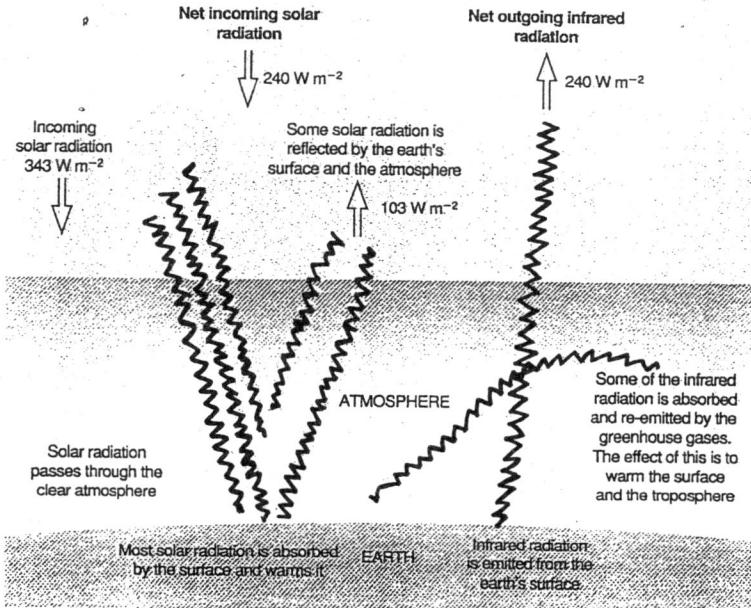

Fig. 1. A simplified diagram illustrating the global long-term radiative balance of the atmosphere.

compared to the solar flux. Clouds, ice caps and snow make the earth a relatively bright planet, such that about 30% of the earth's incoming solar radiation is reflected back into space. A very small fraction is absorbed directly by gases and aerosols as it passes through the atmosphere. Most of it is absorbed by the earth's surface, about 75% of which is water. Thus, most of the radiation is redistributed by the ocean and the atmosphere, and the excess is radiated back into space as longer wavelength, infrared radiation.

If our atmosphere consisted of just nitrogen and oxygen, the global average surface temperature of the planet would be about 33°C colder than it is now and the earth would be a frozen wasteland. This is not the case because of the presence of clouds and very small quantities of other gases, primarily water vapour and carbon dioxide, that absorb the infrared radiation emitted by the earth's surface and reemit their own heat at much lower temperatures. This "traps" the earth's radiation and is the mechanism for planetary warming, "the greenhouse effect" (Fig. 1).

Aside from gases in the atmosphere, clouds also play a major role in climate. By reflecting solar radiation away

from earth, some clouds act to cool the planet while other types of clouds warm the earth by trapping heat near the surface. For years, scientists could not tell whether clouds warmed or cooled the planet. Recent satellite measurements have proved that clouds exert a powerful cooling effect on the earth as a whole. In some areas, however, such as the tropics, heavy clouds may markedly warm the regional climate.

Clouds and greenhouse gases fit into a global radiation budget, a budget that must balance itself. Most of the incoming solar radiation is received in tropical regions, while very little is received in polar regions, especially during winter months. Over time, energy absorbed near the equator spreads to the colder regions of the globe, carried by winds in the atmosphere and by ocean currents such as the Gulf Stream. The small amount of energy retained in the atmosphere is redistributed by winds. The time it takes for the atmosphere to mix around the globe is around one month. Compared to the atmosphere, the ocean is much denser and has a much greater ability to store heat. The oceans also move much more slowly than the atmosphere. They therefore distribute heat at a much slower rate. Because the oceans cover nearly three-fourths of the

Normal Conditions

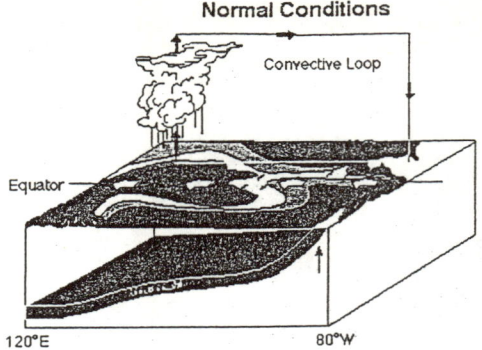

El Niño Conditions

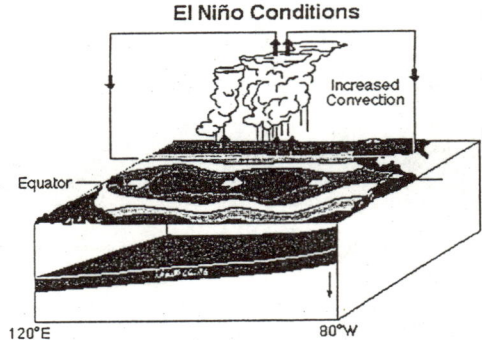

Fig. 2. Normal (top) and El Niño (bottom) conditions. El Niño conditions include a reduction in the trade winds. an increase in SSTs, and an eastward shift of the precipitation.

surface of the earth, the combined effect of the oceans' coverage of the earth's surface and its heat capacity means that most of the heat received from the sun is stored in the oceans.

The oceans and the atmosphere interact on different time scales. As the time scales change from weather time scales, minutes to weeks, to the longer time scales of climate, the interaction between the ocean and the atmosphere changes, as more of the ocean becomes involved. Thus, on weather time scales generally only sea surface temperatures (SSTs) are involved. At time scales of seasons to years, the upper layers of the ocean (a few hundred metres) have an influence, while at time scales of decades and longer the entire ocean plays a role. The transport of heat by surface ocean currents, for example, modifies mid-latitude temperatures across ocean basins so that land areas on the eastern boundary are generally warmer than areas at the same latitude on the western boundary.

2. Seasonal to interannual impacts

2.1. Fundamental processes

The ocean's influence on climate can be considered in two parts: (1) a "normal" seasonal climate cycle; and (2) departures from normal, also known as climate change and variability. Until now, commerce, agriculture and industry have all evolved to function optimally within the framework of the "normal" seasonal cycle. However, changes from the seasonal normal, for example floods and droughts, can lead to economic disruptions and human suffering. New and more accurate predictions of climate anomalies on seasonal and longer time scales can help in developing more effective management strategies for dealing with climate change and variability, and thus can be of great importance to society.

The best understood and most robust inter-annual air-sea climate signal comes from the El Niño Southern Oscillation

WARM EPISODE RELATIONSHIPS DECEMBER - FEBRUARY

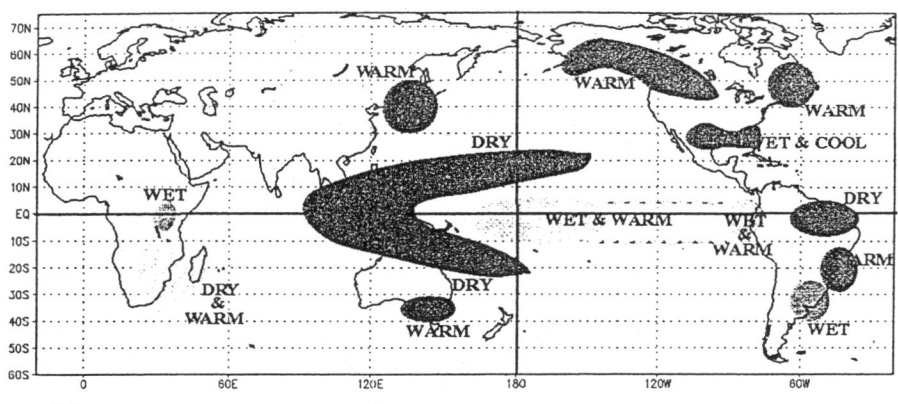

WARM EPISODE RELATIONSHIPS JUNE - AUGUST

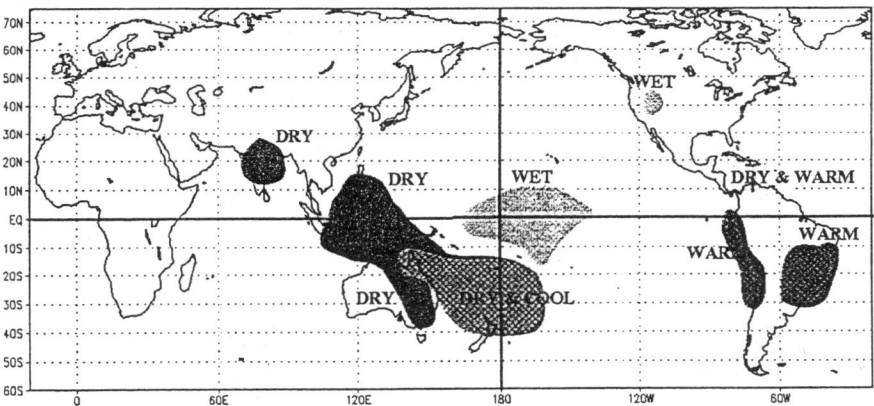

Fig. 3. Expected changes in temperature and precipitation during a warm episode (El Niño) during December through February (top) and June through August (bottom).

(ENSO) which originates in the tropical Pacific. Under normal conditions, the prevailing trade winds blow from east to west which contributes to higher ocean temperatures in the west. Associated with these temperatures are a higher sea-level and deeper thermocline in the west than the east. (The thermocline is the boundary between warmer surface waters and the colder water below.) In addition, convective rainfall is located in the far western Pacific over the warmer

sea surface temperatures (see the upper panel of Fig. 2). In a warm episode (El Niño), the trade winds weaken and warmer waters expand eastward, carrying with them portions of the precipitation. This change includes a reduction of the sea-level and thermocline depth in the west and an increase in the east (see the lower panel of Fig. 2). There are also cold episodes which are generally the inverses of the warm episode shown in the figure. The term ENSO

C. Mason / Natural Resources Forum 23 (1999) 123–134　　　　　127

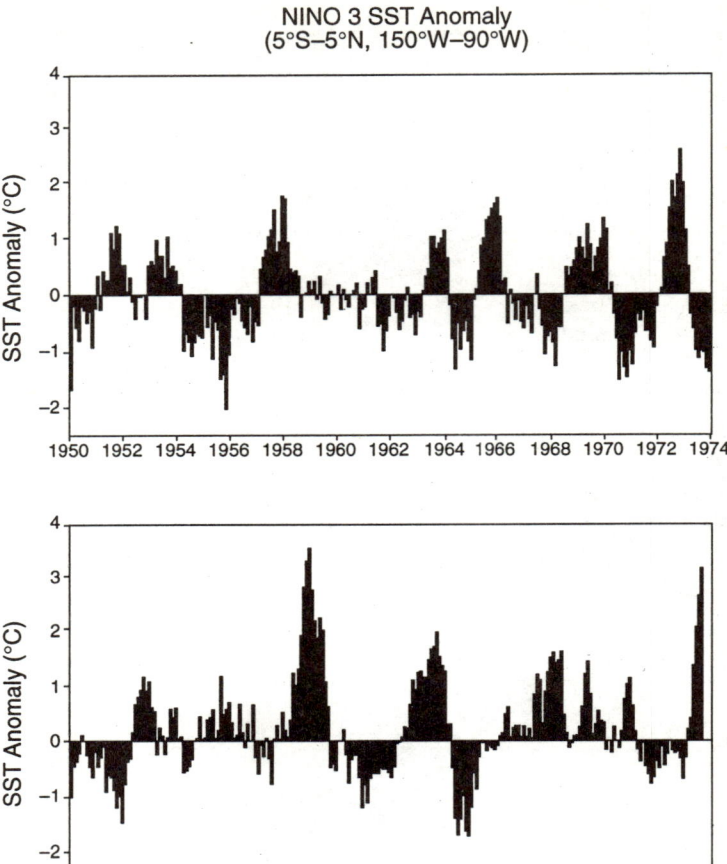

Fig. 4. Monthly SST anomalies (departure from climatological normal) in °C. The climatological normal period was 1950–1970.

refers to both warm and cold episodes; El Niño refers to the warm episode and La Niña to the cold episode.

The shift in the distribution of winds, surface temperatures, and tropical convection leads to changes in the atmospheric circulation with the possibility of regional droughts, floods, and temperature changes in areas well beyond the tropical Pacific. The typical dependence for El Niño is shown in Fig. 3 for Northern Hemisphere winter and summer. In northern middle latitudes, the strongest relationship occurs in the Northern Hemisphere winter when the atmospheric circulation is strongest. The figure shows that El Niño tends to cause warmer than normal winter temperatures in the US Pacific northwest and higher than normal winter precipitation along the US Gulf Coast. La Niña generally impacts the same areas as El Niño but with opposite effects. Although the exact nature of the cause of the ENSO is unknown, the cycle is rooted in the instability of the coupled atmosphere-ocean system. The instability produces repetition of an irregular, quasi-periodic cycle which varies between three and seven years.

Understanding the ENSO and developing appropriate response strategies requires studies of the evolution of SSTs as part of the oceanic response to atmospheric forcing and meteorological studies on regional and large-scale air–sea interactions. Thus, monitoring and prediction of sea surface temperatures is an important part of monitoring and predicting ENSO. SST anomalies for 1950 to present are shown for a region with strong ENSO variability ($10°N$–$10°S$, $150°W$–$90°W$) in Fig. 4. The anomalous SSTs shown here are computed as the difference between measured monthly SSTs and the normal expected monthly SSTs for the period 1950–1979. The figure shows positive and negative SST anomalies. Although the distinction between normal, El Niño, and La Niña is not rigorous, anomalies which persist for at least six months above roughly $0.75°C$ can be considered to indicate El Niño, while those that persist below $-0.75°C$ can be considered to indicate La Niña. The figure also shows an overall warming of the tropical ocean by $0.5°C$ in the decades of the 1980s and 1990s with stronger El Niño episodes occurring in the latter part of the record.

2.2. Economic impacts of climate variability

ENSO (El Niño and La Niña) episodes cause changes in the normal global atmospheric circulation. The changes lead to changes in precipitation and temperature which strongly depend on season and location as shown in Fig. 3. Areas that are strongly impacted during Northern Hemisphere fall and winter are the south of Africa, Australia, South America, and the US. The occurrence of El Niño or La Niña does not guarantee a specific precipitation or temperature response, but only increases the likelihood that a deviation from normal will occur.

Although crop yields depend on many factors, rainfall during part of the growth cycle is often critical. Despite the uncertainties, links between both El Niño and La Niña and crop yields have been established in a number of regions. As an example, winter crop yields in Texas, Oklahoma, Kansas and Colorado show that the presence of El Niño, with its likelihood of increased rainfall, can increase yields by 15%, while La Niña, with its likelihood of decreased rainfall, can decrease yields by 15%.

During normal years, winds along the equator and along the coast of Peru and the west coast of North America, lead to upwelling of nutrient rich colder waters. This results in well-established plankton populations and the fish, birds, and marine mammals which feed on them. During El Niño periods, the winds change direction, the upwelling decreases, and the normal plankton population decreases. The animals which feed on the plankton either move elsewhere or die. For example, the 1972–1973 El Niño coupled with overfishing caused a collapse of the anchovy fishery off Peru. In addition, in the Pacific northwest, El Niño leads to changes in the salmon fisheries. This is linked to more

northward migration of mackerel which prey on juvenile salmon.

ENSO episodes can now be predicted with a level of skill and with enough lead time that hundreds of millions of dollars a year can be saved. A recent interdisciplinary study estimated the value of improved ENSO forecasts to the US agriculture sector to be between US$240 and US$325 million per year. A draft study estimates the benefits of a perfect forecast in crop storage to be US$240 million annually for corn alone. Advanced knowledge of ENSO will allow farmers to make decisions to maximize agriculture yields. ENSO forecasts also have the potential to improve fisheries management because ENSO episodes strongly influence marine catches from Chile to Alaska. In addition, ENSO induced changes in precipitation can lead to increases in the threat of mosquito borne diseases such as malaria. The Center for Disease Control and Prevention and the World Health Organization are building programmes to utilize climate forecasts for enhanced surveillance and health early-warning systems.

Benefits to the water resources and energy sectors are potentially large. The availability of fresh water for irrigation and household use is fundamental to economic well being and varies drastically during ENSO episodes. ENSO episodes have been linked to regional droughts and an increase in the number of forest fires due to decreased precipitation. Decisions on the purchase and distribution of fuels could be made more cost effective, or estimates of fuel demand based on anticipated climate trends could contribute to more efficient decisions regarding options for purchasing different energy supplies.

Many developing countries are strongly affected by ENSO episodes because their economies are dependent upon agriculture sectors as a major source of food supply, employment, and foreign exports. In these countries, droughts predicted up to several months in advance, coupled with the response of local farmers, have already contributed to maintenance of food supplies. For example in the Brazilian state of Ceara, agriculture officials used the predictions of the 1991–1992 El Niño to change timing and types of crops planted. That year the state had harvests at near normal levels compared to the massive crop failures during the 1986–1987 El Niño.

2.3. Education and human resources

A major focus of climate science programmes is the development of an informed and responsible citizenry who are knowledgeable about climate variability. This includes assessing the impacts of climate variability on human activity and economic potential, and improving public education so climate forecasts are understood and used by resource managers.

Societies from around the world could benefit by

C. Mason / Natural Resources Forum 23 (1999) 123–134   129

participating in a shared multinational mechanism to maintain and enhance predictability and to learn to incorporate the information into decision-making for broad based environmental and economic gain. Activities such as agriculture, fishing, water management and fuel distribution take into account the climatological mean annual cycle: crops are planted in anticipation of the optimal growing season; fishing vessels in Peru and Oregon are readied for the seasons when wind-driven upwelling provides nutrients for the food chain; reservoir levels are lowered in anticipation of spring flooding; and fuel oil is distributed in anticipation of wintertime heating needs.

3. Decadal to centennial climate impacts

3.1. Fundamental processes

Both the atmosphere and the oceans act together as a giant heat engine with the oceans also playing the role of a flywheel in the system. It takes approximately four years for the surface currents of the world's oceans to circulate around the globe. As they do so, they give up their heat to their surroundings and cool. These relatively warm currents also tend to have a slightly higher salt content than the waters they circulate through. This is due to increased evaporation at low latitudes, resulting in a small increase in the concentration of salt. In certain parts of the globe these waters can cool sufficiently such that the cold temperatures combined with the higher salt content make them denser than the surrounding waters. When this happens, the cold, dense water sinks and enters into the circulation of the deep ocean. The deep currents of the oceans eventually surface, primarily in the North Pacific, where they enter into the surface circulation again. It takes on the order of 700–1000 years to complete a single circuit of the oceans.

The oceans are not only immense reservoirs of heat and water, but also of carbon dioxide (CO_2). On geological time scales, marine biological processes take place through the uptake of dissolved CO_2 (photosynthesis) and its conversion to inorganic carbonate that is precipitated as carbonate rock. These processes are the major control on CO_2 distributions in the earth's biogeochemical system. On time scales measured in years, marine biological systems, as with faster growing terrestrial systems, equilibrate fairly rapidly with carbon dioxide in the atmosphere. On longer time scales, the transfer of CO_2 to woody vegetation, soils, and to the deep ocean removes CO_2 from the atmospheric system. The oceanic sink of CO_2 is considerably larger than the terrestrial sink. While 10–20% of the CO_2 emitted to the atmosphere by man's activities is utilized by terrestrial processes, between 30 and 60% of the total CO_2 emitted by man has been removed from the atmosphere by long-term oceanic processes.

3.2. Limitations and uncertainties

Increased confidence in understanding climate variability, and potential impacts by man on climate can only be obtained through improved representation of ocean climate processes in models and systematic collection of long-term instrumental observations of climate system variables in the oceans. Key uncertainties limit our ability to detect and project future climate change. In particular, the Intergovernmental Panel on Climate Change (IPCC, 1995) report lists the following as priority topics:

- Representation of climate processes in models, especially feedbacks associated with clouds, oceans, sea ice and vegetation, in order to improve projections of rates and regional patterns of climate change.
- Systematic collection of longterm instrumental and proxy observations of climate system variables (e.g., solar output. atmospheric energy balance components, hydrological cycles, ocean characteristics, and ecosystem changes) for the purpose of model testing, assessment of temporal and regional variability, and for detection and attribution studies.

These priorities recognize that predicting climate change resulting from emissions of greenhouse gases and formulating future decisions on the possible regulation of these emissions require more accurate models, models that have been adequately tested against a well designed network of observations. Observations also serve other purposes. Of paramount importance are long-term observations which are needed to detect and quantify climate change. In addition, observations can provide increased understanding of climatically important ocean processes. Finally, chemical and physical oceanographic observations are needed to separate anthropogenic from natural variability.

The present state of models and sparse observations are factors that lead to uncertainties in estimates of oceanic transport, uptake and sequestration. For example, the current coupled general circulation models (GCMs) used to simulate climate change fail to produce long-term trends in Pacific seasurface temperature. Observationally, three estimates of oceanic heat flux at 24°N in the Atlantic show a monotonic rise from 1957 through 1992. Available data cannot resolve whether these changes represent a natural or an anthropogenic induced trend and/or whether they are biased by an unresolved annual signal. Furthermore, the uncertainty in measurement of ocean dissolved inorganic carbon (DIC) uptake is 40% of the total (2 billion metric tons of carbon per year). Such high uncertainties must be reduced if we are to increase confidence in CO_2 warming scenarios.

3.3. Education and human resources

The earth's climate system is extremely intricate. Clouds, ocean currents, solar radiation and other elements interact in

C. Mason / Natural Resources Forum 23 (1999) 123–134

Based on data from:
"Sea Level Variations for the United States
1855-1986", National Ocean Service, NOAA, 1988

Fig. 5. Long term sea-level trends for the United States.

a complex way to determine our climate. Mathematical models allow us to study parts of the climate system and how those parts interact. These models include many aspects of the climate system (air, oceans, land, biology) partitioned into many small grid boxes. Even though they may require weeks of powerful computer time to run, they are relatively simple when compared to the natural system. The models indicate that temperature could rise considerably over some areas of the globe due to increased emissions of greenhouse species. At issue is the accuracy of these predictions. We know that, despite their complexity, these models do not adequately represent the roles of oceans and clouds in the climate system. These models do provide useful insights into the climate system. When attempting to make conclusions regarding long-term climate variability, these models require decades of precise observations for verification. Only now are we close to a sufficiently long record of accurate data to make some preliminary assessments of model predictions. Unfortunately, the long periods of time involved in the oceanic response and the large inertia of the oceans mean that any action taken to reverse impacts of man's activities in long-term climate patterns will require decades to centuries before significant results could occur.

4. Global climate change impacts on coastal areas

4.1. Fundamental processes

Throughout time, climate change has affected the coastal environment and will continue to do so in the future. However, human activities and alterations have rendered coastal resources more vulnerable to climate change-induced processes, such as accelerated sea-level rise, alterations of rainfall patterns and storm frequency or intensity, and increased siltation. Climate change and a rise in sea-level or changes in storms or storm surges could result in the increased erosion of shores and associated habitat, increased salinity of estuaries and freshwater aquifers, altered tidal ranges in rivers and bays, changes in sediment and nutrient transport, a change in the pattern of chemical and microbiological contamination in coastal areas, and increased coastal flooding. Some coastal ecosystems are particularly at risk, including saltwater marshes, coastal wetlands, coral reefs, coral atolls, and river deltas. Other critical coastal resources, such as mangroves and sea-grass beds, submerged systems including submerged aquatic vegetation (SAV), and mudflats, are at risk from climate change impacts, exacerbated by anthropogenic factors. Changes in these ecosystems could have major negative effects on

C. Mason / Natural Resources Forum 23 (1999) 123–134   131

tourism, freshwater supplies, fisheries, and biodiversity that could make coastal impacts an important economic concern. These impacts would further modify the functioning of coastal ocean zones and inland waters that has already resulted from pollution, physical modification, and material inputs related to human activities. Secondary impacts associated with climate change, such as inundation of waste disposal sites and landfills that will reintroduce toxic materials and increased siltation into the environment also pose threats to the health of coastal populations and ecosystems.

Global sea-levels have been rising since the conclusion of the last ice age approximately 15 000 years ago. During the last 100 years, sea-level rise has occurred at approximately 1–2.5 mm/year. Fig. 5 shows long-term sea-level rise in the United States. This figure represents eustatic sea-level (the absolute elevation of the earth's ocean) that has been determined from tidal stations around the globe. However, there are large regional variations due to: subsidence, isostatic (glacial) rebound, tectonic uplift, etc. that contribute to a 'relative' sea-level rise. For example, within the US, portions of the Gulf Coast are experiencing a relative sea-level rise of 10 mm/year. Concurrently, the coast of Alaska is experiencing a negative relative sea-level rise of up to 8 mm/year; i.e., sea-level is receding. Fig. 5 illustrates the change in sea-level along the US coasts as determined from historical tidal data. If this historical rate of sea-level rise is projected to 2100, sea-levels would rise 10–27 cm globally.

Over the next 100 years, global warming could further raise sea-levels by expanding ocean water, melting alpine and other small glaciers. The most recent IPCC assessment (1995) forecasts a rise in global sea-level of 5 mm/year, within a range of uncertainty of 2–9 mm/year, which represents a rate of sea-level rise that is still about two to five times the rate experienced over the last 100 years. Although still debated and somewhat speculative, global warming might also cause parts of the large ice sheets of Greenland and Antarctica to melt or slide into the oceans, contributing significantly to sea-level rise.

4.2. Economic impacts on developed coasts

Coastal development, including buildings, transportation infrastructure, and recreational and agricultural areas, are vulnerable to inundation and increased erosion as a result of climate change. All lowlands are threatened by a rise in sea-level. Estuaries are also threatened by potential hydrologic changes that could increase the range of saltwater intrusion as well as alter the amount of freshwater reaching an estuary. If a one-metre rise in sea-level occurs during the next century, the worst-case IPCC scenario, thousands of square miles of US coastal lands could be lost, particularly in low-lying areas such as the Mississippi delta, where the subsidence rate is about one metre per century.

Assessing total economic impacts from sea-level rise on coastal areas on a national or global scale is still somewhat speculative. The potential social, economic, and environmental impacts to a particular country will vary greatly depending on its vulnerability to accelerated sea-level rise. Small islands and atolls may have to be abandoned, deltas may be severely degraded or lost, and low-lying areas may need the equivalent of the dikes of The Netherlands, the seawalls of Tokyo, or constant renourishment of beaches. Natural systems, like tidal wetlands, can provide flood control, storm protection, and waste recycling and may have significant economic value.

Coastal erosion is already a widespread problem in much of the world. For example, in Oahu, Hawaii, over the past 50 years a quarter of the beaches have been lost or significantly degraded due to causes that are poorly understood. Heightened storm surge could increase the rate of erosion. The highest-risk areas are those currently experiencing rapid erosion rates and with very low relief, such as the southeastern United States and the Gulf Coast. Coastal areas would also be more vulnerable to hurricanes, as well as to increased or decreased freshwater and sediment flux from river systems.

Rising sea-level will increase storm surge flooding by a comparable amount, exacerbating the impacts of every coastal storm. However, some areas will experience dramatic changes, ranging from no flooding to extensive flooding. Many coastal features, such as levees, seawalls, and naturally occurring sand dunes and ridge lines effectively block storm surges for most storms. Whenever one of these features is overtopped by storm surge from either a hurricane or an extratropical storm, the areas inland will flood. Numerical modeling has shown that large amounts of water can move over such barriers, flooding the marshland or bay behind the barrier, and sweeping over mainland areas.

4.3. Impacts on coastal living marine resources

Coastal wetlands are already eroding in many parts of the world. For example, in the US, Louisiana's coastal area lost an estimated 3950 km^2 of wetlands from 1930 to 1990. This loss of wetlands resulted, for the most part, from inundation or erosion of wetlands due to the development of canals in support of oil and gas activities, rather than from the draining or filling that is characteristic of many wetland losses elsewhere. In addition, large areas of brackish and freshwater wetlands have become progressively more saline, as salt water has increasingly invaded the deteriorating coastal zone. As 40% of US coastal wetlands are found in Louisiana, this loss constitutes about 80% of the total national coastal wetland loss. Louisiana coastal wetlands are exceptionally valuable in terms of coastal fisheries and migratory waterfowl, protection of low-lying population centers from hurricanes and other storms, and oil and gas production. Further, the greatly accelerated rates of coastal wetland loss appear to be the unintended result of massive human disturbances of these wetlands and intervention (for purposes of flood protection, water supply, maritime commerce, energy production, and wildlife management) in the processes that sustain coastal wetlands.

Wetlands require a delicate balance of sediment, fresh and salt water and are particularly vulnerable to inundation and erosion as a result of sea-level rise. Coastal wetlands are also vulnerable to changes in the source or decreased flux of fresh water and sediment, if upstream areas become more arid. Wetlands naturally migrate as land subsides and sediment supply changes, but migration has been limited in several areas by the encroachment of urban areas with sea walls and other protective structures. In addition, the possible rate of sea-level rise predicted by some climate change models is more rapid than the natural rate of wetland migration. Thus wetland losses will likely increase.

Climate change also has the potential to significantly affect coastal biological diversity. It could cause changes in the population sizes and distributions of species, alter the species composition and geographical extent of habitats and ecosystems and increase the rate of species extinction. Fragile systems, such as coral reefs, are highly susceptible to temperature increases. Short-term increases in water temperatures of the order of only 1–2°C, combined with other environmental stresses (such as pollution or siltation from human activities), can cause "bleaching," leading to significant reef destruction. Reefs in many parts of the world, including the US, have undergone episodes of bleaching, particularly in the 1980s. Sustained increases of 3–4°C above long-term average seasonal maximum can cause significant coral mortality. Biologists suggest that full regeneration of these coral communities could require several centuries.

Fisheries in estuaries and the coastal ocean are also vulnerable to changes in water temperature and freshwater inflow. The loss of coastal wetlands has already been implicated in the decline of shrimp harvests in Louisiana, and would also likely reduce yields of crab and menhaden. Projections of general circulation models suggest that freshwater discharge from the Mississippi River to the coastal ocean will increase 20% if atmospheric carbon dioxide concentrations double. This is likely to affect water column stability, surface productivity, and global oxygen cycling in the northern Gulf of Mexico, which is already suffering from persistent hypoxia. In the open ocean, increased temperatures could result in a shifting of the geographical distribution of certain species. Decreasing freshwater flow, when combined with rising sea-level could result in the encroachment of saltwater species into typically freshwater habitats. For example, in estuaries, decreased freshwater inflow would result in increased salinity and, in turn, a replacement of some freshwater species by saltwater species.

4.4. Climate change and integrated coastal management

Management strategies in coastal areas can be divided into three categories:

- "Accommodate," where vulnerable areas continue to be occupied, accepting the greater degree of negative effects, e.g., flooding, saltwater intrusion, and erosion. In these areas, advanced coastal management, including improved early-warning of catastrophic events, and building codes modified to strengthen structures, could be used to avoid the worst impacts;
- "Protect," where vulnerable areas, particularly population centres, high-value economic activities, and critical natural resources are defended by, for example, sea walls, bulkheads or saltwater intrusion barriers. Protective "soft" structural options such as periodic beach renourishment, dune maintenance or restoration, and wetlands creation can also be introduced where other infrastructure investments are made.
- "Retreat," where existing structures and infrastructure in vulnerable areas are abandoned, inhabitants resettled, government subsidies withdrawn, and new development is required to be set back specific distances from the shore, as appropriate.

The cost-effectiveness of these adaptation strategies would be enhanced to the extent that they are planned and implemented in the context of integrated coastal management (ICM) programmes carried out at various levels of government. ICM is a continuous, iterative, adaptive, and consensus-building process comprised of a set of related tasks, all of which must be carried out to achieve a set of goals for the sustainable use of coastal areas, including adapting to the effects of climate change. The dimensions of ICM include:

- Integration of policies and programmes across and among sectors of the economy, e.g., economic development, transportation, recreation, and agriculture.
- Integration among agencies involved in coastal management at all levels of government, including both vertical (national, subnational, and local) and horizontal (across the same level of government) integration.
- Integration between public- and private-sector management activities.
- Integration between management actions that affect the land and water environments of coastal areas, and areas upstream and upwind of coastal areas.
- Integration among the disciplines of coastal management, including ecology, economics, engineering, and political science.

In the US, state and local government responses to future accelerated sea-level rise scenarios include:

- taking accelerated sea-level rise into account when filling wetlands for water-dependent facilities and ports which requires an applicant to raise the fill level so they will not have to abandon facilities during the 50–100 year life expectancy of the facility;
- raising the elevation of new facilities such as sewage treatment plants to protect the integrity of facility use;

- incorporating accelerated sea-level rise into setback laws for larger facilities which goes beyond the normal incorporation of only historical rates of erosion;
- passing legislation in support of a retreat policy that normally prohibits sea wall construction to allow backward migration of a beach.

' In February 1997, the US Country Studies Programme and NOAA sponsored an international workshop that developed guidelines for integrating coastal management and climate change adaptation strategies (Cicin-Sain et al., 1997). They can serve as a guide for coastal nations to implement or strengthen an ICM programme, and simultaneously meet the obligations of international agreements that have a special focus on the management of coastal areas and the related implications of climate change.

4.5. Research needs and scientific uncertainties

Continued investments in research and monitoring at the national and international levels are needed to improve the information base for adapting to climate change. For example, coastal wetlands naturally migrate in response to changes in sediment supply and relative sea-level. However, it is unknown if the rate at which wetlands can naturally migrate is sufficient for the possible rates of sea-level rise that would be caused by climate change. Establishing locations for wetlands to migrate to by expanding reserves and protected areas adjacent to current coastal wetlands can facilitate adaptation. Creation or restoration of wetlands is another adaptive strategy that requires the development of effective methods for restoring coastal wetlands and for measuring the effectiveness of those restoration efforts.

Two major international climate change assessments (1990 and 1995) have been conducted by the IPCC. These cover the available science, the magnitude of human-induced climate change and appropriate response options. Some areas that require further scientific research and data include scaling down of current general circulation models to obtain local climate change estimates, as the current GCM resolution is too low. In addition, sea-level rise alone is not an exclusive feature of climate change in coastal areas. Therefore, climate change studies need to be broader, combining the effects of sea-level rise, storminess, atmospheric circulation change, precipitation, etc.

The IPCC 1995 assessment also indicates that more work is necessary for quantifying the social costs of climate change. Net climate change damages include both market and non-market impacts as far as they can be quantified at present and, in some cases, adaptation costs. However, the non-market damages (e.g. human health, risk of human mortality and damage to ecosystems, etc.) are highly speculative and not comprehensive, and therefore are a source of major uncertainty in assessing the implications of global climate change for human welfare.

4.6. Global framework for ICM

In 1994, many nations signed the United Nations Framework Convention on Climate Change (FCCC). The major objective of this effort is to achieve the stabilization of greenhouse gas emissions and to identify national adaptation strategies. Article 4 of the Convention commits nations to, among other things, develop integrated plans for coastal zone management as part of their adaptation strategies. In response to potential commitments and obligations under the convention, many nations, including the US, are preparing national climate change action plans that identify management strategies to reduce greenhouse gas emissions and adapt to the potential impacts of long-term climate change.

The successful implementation of these plans and their management strategies within individual countries will depend to a large measure on the extent of their integration into the implementation of other national and sectoral management plans, including coastal management plans. For example, the US Coastal Zone Management Act as amended authorizes assessments of management implications of sea-level rise scenarios. Individual state responses have varied according to the circumstances of their vulnerability, political situations, and availability of resources.

4.7. Education and human resource issues

Education, training and outreach that target all sectors of society are essential components in the successful implementation of ICM and climate change adaptation strategies. Well-designed public education programmes should use target specific clientele—including elected officials, user groups, women's groups, school children, and the general public—to develop support for ICM and climate change action plans. Public education ought to include informal education programmes that will reach all segments of the community, including the illiterate, who may form a significant segment of the stakeholder population.

Despite the scientific evidence that sea-level is rising globally as a result of climate change, many individual homeowners are undeterred from purchasing beach-front property that is threatened even now by beach erosion resulting from human interventions (construction of jetties, sea walls, sand mining, etc which obstruct natural sand replenishment) and natural causes (storms and hurricanes). Even though rising sea-levels will exacerbate these current problems, construction in these areas continues.

4.8. Managing in the face of uncertainty

While climate change impacts, such as accelerated sea-level rise are a potentially major concern, many countries and most state and local governments in the US have hesitated to address the issues effectively because of the relatively long-term nature of the problem, scientific uncertainties, and large investments required. Because of

134 *C. Mason / Natural Resources Forum 23 (1999) 123–134*

scientific debate and uncertainties, few are willing to develop potentially costly management strategies for which the benefits will not be realized for years to come. Consequently, management responses to climate change impacts have been gradual since the early 1980s, but pioneering efforts have been important for US policy makers.

Nevertheless, a number of state and local governments in the US have included the consequences of potential sea-level rise in their planning efforts. Maine, Rhode Island, South Carolina, and Massachusetts have implemented some form of rolling easement policy to ensure that wetlands and beaches can migrate inland as sea-level rises. Several other states require counties to consider sea-level rise in their comprehensive coastal management plans. A few communities have enacted ordinances requiring building lots or new structures to be elevated an additional foot or two.

Adapting to sea-level rise and other effects of climate change will involve important trade-offs that weigh environmental, economic, social, and cultural values. Effects will depend not only on the local patterns and intensity of climate change, but also on the nature of the local coastal environment, on the human, ecological, and physical responsiveness of the affected coastal system, and on actions in other sectors of the coastal and national economies. Given the long time frames involved in reducing the magnitude of global warming, it is vital that steps be taken now to manage the impacts that almost certainly will occur in islands and low-lying coastal areas.

Acknowledgements

The present article is a synopsis of a paper by Richard W. Reynolds. NOAA National Weather Service, Ben Mieremet. NOAA Office of Sustainable Development & Intergovernmental Affairs, Steve Morrison, NOAA National Ocean Service. Steve Piotrowicz, NOAA Office of Oceanic and Atmospheric Research, Ellen Prager, Kathryn Ries, NOAA National Ocean Service, James G. Titus, US Environmental Protection Agency, and S. Jeffress Williams. US Geological Survey, published in the "Year of the Ocean Discussion Papers," US Department of Commerce, Washington, DC, March 1998. It was adapted for the *Natural Resources Forum* by Curt Mason with the advice of the other authors.

References

Cicin-Sain, B. et al., 1997. Guidelines for Integrating Coastal Management Programs and National Climate Change Action Plans. Taipei.
Intergovernmental Panel on Climate Change (IPCC) 1995. Second Assessment Report. Climate Change 1995. World Meteorological Organization. Geneva. Switzerland.

[3]

CHAPTER 8

Coastal processes and management

KEITH CLAYTON AND TIMOTHY O'RIORDAN

The coast is an amazingly awkward zone to manage. Yet it is of crucial significance for economic activity, leisure and natural restoration processes. Over two-thirds of marine biological activity takes place at or near the coast, and most especially in estuaries. In crucially significant nutrient-rich zones such as the Waddensee in the eastern North Sea, well over half the commercial fish population spawn and develop in their formative life cycle. To lose that area would be a major blow to North Sea marine ecology. Mangrove swamps in the coastal tropics play indispensable roles in regulating tides and floods, trapping nutrient-rich sediment and providing refuges for fish and invertebrates. They are under threat, as are the equally valuable coral reefs, from coastal mismanagement – everything from poaching of coral for commercial sale, to polluting discharges from hotels and coastal residences to altering marine currents as a result of port, road or marina developments.

The coastal zone is awkward to manage because it covers three 'territories', three administrative 'regions' and three 'styles' of management. The territories are:

- the offshore waters, beyond low tide and within national territorial jurisdiction;
- the coastal margin between low and high water tides and including estuaries;
- the littoral landward zone, including headlands, beaches and coastal settlement.

Normally, though not in every country, these zones are administered by different agencies reporting to a variety of government departments. In the UK, for example, the offshore is run by the agriculture departments, the shore by the crown and environment departments, and the littoral area by local authorities answerable to environment departments. Coastal protection is the responsibility of local government with all their heavy budget restrictions imposed by threats of rate 'capping' by central government, while flood defence is the responsibility of the National Rivers Authority answerable to the agriculture departments.

In other countries the mix is different, but the potential for conflict between competing administrations is ever-present. In the chapter that follows a case is made for *shared governance* of this zone, the sophisticated administrative marriage of local, regional and national authorities, coupled with regulatory agencies for planning, coastal defence, offshore minerals and fisheries management. In no country yet has there been a successful union of the triple alliance of land-use planning controls, offshore management controls and integrated coastal protection utilizing ecological-geomorphological principles.

This takes us to the third difficulty facing any much-needed reform in the management of the coastal zone. The management styles are so very different. Offshore, the long reach of formal planning regulation and full environmental impact assessment is often just beyond grasp. Marine gravel extraction, offshore pipelines and port developments, marinas beyond the coastline – all of these can elude the gambit of both strategic and local planning. Management here tends to be more pragmatic, alleviating problems of extraction and development by fairly aggressive measures according to tradition and influence.

On the actual coast, as the chapter that follows reveals, there is a persistent tension between the 'eco' orientated management measures, which are pro-active and exploratory, and the 'techno' style of build and be damned. This tends to be reactive, waiting for evidence of crisis or failure in protective structures, carried by the howl of protest from residents and property owners who have been allowed to move into vulnerable zones. They believe they are entitled to protection by a publicly financed agency whose remit is to safeguard. Similarly, on the landward part of the coastal triad of zones, the management style is primarily facilitative, allowing development to proceed with regard to flooding and possible sea-level rise, but by no means fully taking into account the special demands of the 'eco' protective approach of safeguarding sacrificial headlands, permitting coastal retreat of salt marshes into areas that should be specifically zoned against developments and designing new dunes in locations where development should be sympathetic and tourism appropriately limited. The tools are there in the planning kitboxes: as yet they have not been deployed as sensitively as they should.

This raises the perennial issue in environmental management, namely that of *policy integration*. Despite years of interdepartmental, interstate and intersectoral co-ordination the coastal zone is still a policy battleground in the USA. Subsidies encourage the overuse of chemicals that provide the coastal arena with nutrients and toxins, destroying shellfish and creating tourist-repelling algae. Flood insurance schemes promote insensitive waterside development, that is subsequently backed out by federal subsidies when the next hurricane strikes. In South Carolina, for example, Hurricane Hugo struck Folly Island, off the coast of Charleston, destroying or damaging 89 of 290 properties. Despite legislation to limit rebuilding, the act was changed to enable reconstruction in zones known to be subject to erosion and inundation. The flood insurance programme will cover such property. Federal tax provisions do not take into account environmentally sensitive economics, so it pays to develop wetlands, and receive a tax deduction in so doing, while the social value of the marsh is lost, because there is no tax subsidy for its survival.

These policy conflicts are very troublesome. They are rooted in the administrative history of the managing agencies, they create beneficiaries who lobby hard and successfully to protect their interests, they establish precedent that can be supported by the courts on technical rather than moral grounds, and they weaken the resolve of would-be reformers because the hurdles are not just high, but are also unpredictable. The chapter that follows offers a number of ways forward, building on the principles that guide this text as a whole.

Be aware: there is an administrative revolution taking place on the coast. For example, in the UK about half the coastal district authorities now co-ordinate planning and development in the coastal zone. Some have detailed inter-agency strategies backed by project officers and participatory steering groups. As yet, all have insufficient powers and resources, but their presence is influential.

The prospect of sea-level rise is likely to stimulate innovation and adaptation even more. Humans, like all organisms, are capable of responding to environmental change. But unlike their non-human comrades in the evolutionary game, humans make a fuss about it.

Further reading

R. Platt, T. Beatley and H.C. Miller (1991) The folly at Folly Beach and other failings of US coastal erosion policy. *Environment*, **33**(9), 6–9, 25–32.

Coastal zones are both vulnerable and populated. More than half the world's inhabitants live within 60 km of the coast. By 2020 this figure could be 70 per cent, many of whom will be concentrated in cities or densely packed agricultural areas. The rise of international tourism and the permanent lure of the beach has also created a new style of settlement and breed of resident in coastal areas, packed into resorts and expensive homes. The coast is both attractive and dangerous. It is constantly undergoing physical and biological alteration, yet habitation continues almost as if the zone was peaceful and stable. Coasts also contribute about 25 per cent of total biological productivity, supporting over two-thirds of the world's fisheries and nine-tenths of the shellfish industry.

The coastal zone has always proved a problem to manage. In the introductions both to this section and to this chapter the point is made that both institutions and policies usually fail to coincide for compatible and anticipatory management of this vulnerable and unstable area. The trouble is that the coast is too much in demand for it to be left to nature to determine its fate. Sea-level rise presents more of a formidable problem than is generally realized because it threatens wealth, property protectionism, powerful economic interests, it can swing political votes and it further reinforces the status of an engineering fraternity that does not like to be beaten. It is no wonder that the coast is such an administrative battleground, full of agency jealousies, ill-matched budgets and suspicious tiers of government.

In this relative state of managerial anarchy, powerful interests can gain ground. Engineers of the 'old school' delight when crisis creates a serious threat, even if they are very concerned about the suffering and the distress thereby caused. In the wake of a coastal disaster, the tendency is to reassert an engineering-dominant approach, almost as if it is a mission in itself. The true adjusters are those prepared to relocate, to identify and enforce strict zones of managed retreat, which should have considerable recreational and wildlife advantage, and to establish sound budgets for compensation for those deliberately left vulnerable in the face of a managed retreat management regime. All this will take great vision, much public involvement, steadiness of nerve and extremely effective communication. These are all qualities of good interdisciplinarity in applied environmental science.

Coastal processes

Coasts are formed by waves and currents. Waves in turn are generated by wind, which converts its kinetic energy to the sea surface. The longer the fetch or distance of wind-disturbed sea, the more powerful the waves. Surfing is always most popular on open ocean coasts. As waves approach shallow water they dissipate their energy in sediment disturbance, erosive power and broken water. The shape of the coast influences the pattern of energy. Converging orthogonals, the force of energy at right angles to the wave crest, focus wave power, while diverging orthogonals disperse it. With satellite technology it is possible to map potential wave energy for different wind directions, and thus to explain the pattern of erosion and accretion zones for subsequent planning purposes. By adding an understanding of the processes at work to historical records of coastal change, this allows the recognition of persistently hazardous coastal sectors. How far that will in time be converted into planning restrictions will depend on the credibility of the science and the political influence of planning and management agencies.

As waves usually approach any beach at an angle, part of their energy is converted into lateral movement of sediment. Even where wind patterns are diverse, the importance of fetch will ensure that this drift will be predictable and often significant, for example beach sediment moves southward along the California coast at a rate of about 300 000 m³ each year. Severe storms, maybe occurring with a probability of once in 10 or 20 years, will account for an extraordinary amount of sediment disturbance. Whole beaches can form or disappear overnight. This episodic convulsion not only gives rise to much property loss and human suffering; it is the very essence of coastal formation. Figure 8.1 shows how the coast of England and Wales has prograded and eroded over the past 100 years.

In addition to wave-generated sediment movement, tidal currents sort sediment and help to shape the geomorphology of the coastline. This is dominant in deeper areas offshore and especially important in estuarine areas protected by offshore bars or spits. From a management point of view, estuarine tidal-induced sediment movement will offset dredging and can move toxic material buried below the surface. In densely populated areas where river waters may carry sediment from polluting industrial zones, such as the Rotterdam harbour at the mouth of the Rhine, this can be a severe problem – so severe that the Dutch government is

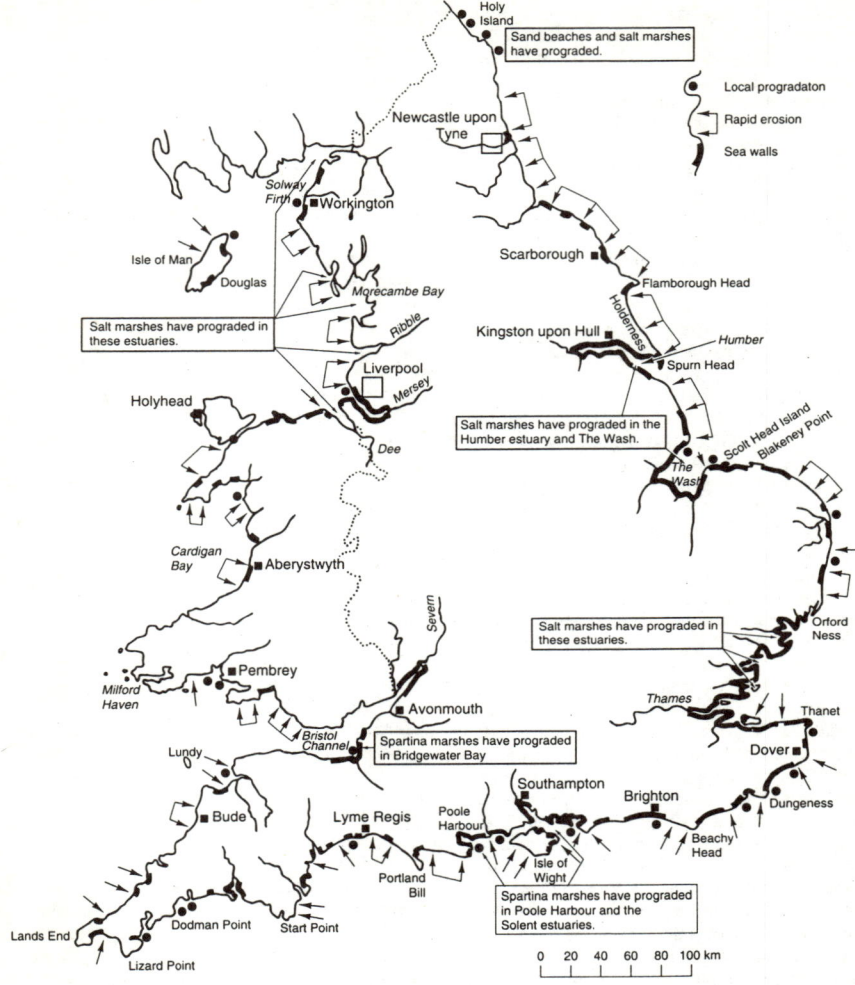

Holy
Island

Sand beaches and salt marshes
have prograded.

Local progradaton

Rapid erosion

Sea walls

Newcastle upon
Tyne

Solway
Firth ■Workington

Scarborough

Flamborough Head

Isle of Man

Douglas

Morecambe Bay

Kingston upon Hull

Humber

Salt marshes have prograded in
these estuaries.

Ribble

Spurn Head

Liverpool

Mersey

Holyhead

Dee

Salt marshes have prograded in the
Humber estuary and The Wash.

Scolt Head Island
Blakeney Point

The
Wash

Cardigan
Bay

Aberystwyth

Severn

Salt marshes have prograded in
these estuaries.

Orford
Ness

Pembrey

Milford
Haven

Avonmouth

Thames

Thanet

Bristol
Channel

Spartina marshes have prograded
in Bridgewater Bay

Dover ■

Lundy

Southampton

Brighton

Dungeness

Bude

Lyme Regis

Poole
Harbour

Beachy
Head

Portland
Bill

Isle of
Wight

Spartina marshes have prograded
in Poole Harbour and the
Solent estuaries.

Lands End

Dodman Point   Start Point

0   20 40 60 80 100 km

Lizard Point

Fig 8.1 Comparison of late 19th century maps with the 1980 outline of England and Wales has identified the pattern of coastline changes shown here.

contemplating an ingenious variant of a tradeable permit. The proposal is to establish a property right in effluent discharge by issuing a covenant to the polluting firm. This is a form of voluntary payment by the identified polluter to pay for sediment removal, or stabilization, in lieu of court proceedings and the possibility of a heavy fine. The scheme relies on proof of effluent discharge and sediment transport. 'Finger printing' by coding indicator chemicals of particular effluents together with sound modelling can provide this service.

Beaches are forever readjusting. Their prime function is to buffer the coast from wave and tidal energy. Without this protection, soft coastlines would disappear very rapidly. Beaches form from a 'sediment store' collected from the sea floor, from eroded headlands or from riverborne materials. Shingle tends to be found where wave energy is concentrated, and sand or mud where it is more dispersed. Figure 8.2 describes coastal terminology. Dunes are especially important in this arrangement for they feed beaches when they erode, but re-establish naturally during periods of relative calm. Dunes may be stabilized by long-rooted grasses such as marram. This not only binds the sand, forming a thin organic soil on the surface, it also lowers wind

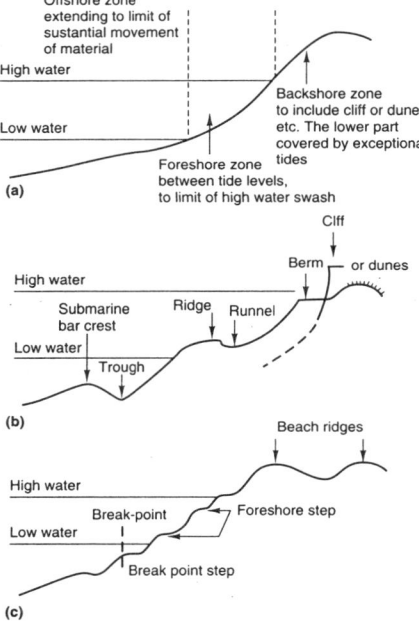

Fig 8.2 Coastal terminology: (a) The beach in profile; (b) composite sand beach profile; (c) composite shingle beach

speed, thereby encouraging greater sand settlement.

Dunes can be damaged more by the trampling of visitors than by natural factors: hence the need for tough dune management schemes to keep people off vulnerable or damaged areas. More often than not wardens and public information programmes are necessary to ensure that trampling is kept to a minimum. Better public understanding of the ecological and geomorphological role of dunes in beach formation and coastal protection has to be actively promoted. This will become even more of an issue as dune reconstruction and new dune construction is undertaken as part of the strategic response to sea-level rise. In the Netherlands, for example, dune protection is a major component in coastal management. All existing dunes are safeguarded by law and by sympathetic management practice. Dune restoration zones have also been identified. The cost benefit justification of this programme includes estimates of the recreational value of the dunes using the travel cost method as outlined in Chapter 3, and economic-ecological approaches as presented in Chapter 2. Without these techniques, it would have been a little more difficult to invest in the restoration programme. They are even used as sources of fresh groundwater supply.

Salt marshes are found in sheltered coastal regions where wave energy is dissipated by offshore bars or barrier beaches or alongside estuaries. The marshes themselves are efficiently flooded and drained by an intricate web of tidal creeks, bordered by vegetation that traps the mud and builds the marsh slowly above to the level of the highest tides. The marshes themselves are important converters of sulphur, and sequesterers of nutrients and toxic sediment. Their bio-geo-chemical role in the ocean–atmosphere interface is nowadays a focus of much scientific research. Early findings suggest a self-regulating ability to maintain carbon, sulphur and methane generation. Not only do such marshes protect soft coasts and flood banks from wave action, they may have an important role to play in accommodating the rising fluxes of carbon and sulphur. In addition, salt marsh habitats are extremely valuable for bird-life. Wading birds can be found in great numbers, as for example on the Waddensee and the North Norfolk Coast around the North Sea and the Carolina–Delaware estuaries of the eastern United States. Many salt marshes are protected zones for conservation and amenity purposes, with a social value so large as to justify their extension by carefully managed controlled retreat, i.e. the removal of flood banks from in front of reclaimed former marshes.

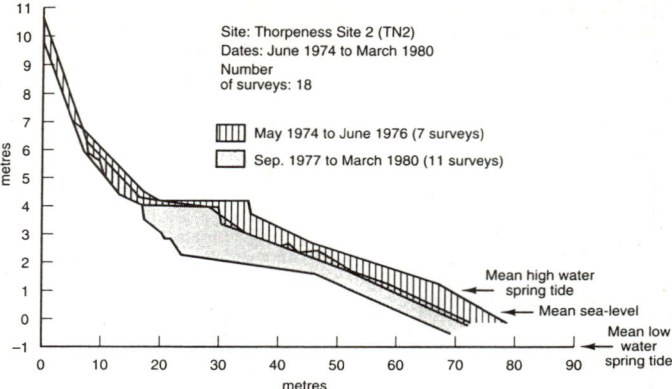

Fig 8.3 Cross-section illustrating the sweep zones.

Figure 8.3 illustrates a typical set of beach profiles. The profile has an enormous influence on the erosion potential of a beach, as well as on the likely shape of the beach during a period of persistent sea-level rise. Under what is known as the Bruun Rule, as devised by a Norwegian engineer:

$$\text{shoreline erosion}\,(R) = \frac{\text{Profile width} \times \text{sea-level rise}(s')}{\text{profile depth}\,(z)}$$

as illustrated in Figure 8.4.

The rate of erosion can be modelled as follows:

- record wind direction and speeds over as long a time as possible;
- model the relationship between wind speed and direction and wave height from both observation and wave tank experimentation;
- in turn, model wave diffraction on the basis of the beach profile, using Snell's Law of wave refraction.

Persistently eroding coastal cliffs provide an important supply of beach material. They involve an erosive offshore zone and at the coast an efficient littoral drift. Offshore sand and shingle banks in areas of relative stability act as sediment stores and can prove invaluable as buffers for exceptionally high tides. Zones which are relatively self-regulating in this regard are sometimes referred to as coastal cells. These are forever changing, but can be used to measure sediment erosion, transport and deposition. Figure 8.5 shows how this technique can be used

to reveal the sand budget of the East Anglian coast. The role of 'feeder cliffs' as sources of continual replenishment is of the utmost importance, especially in lowlands where rivers do not deliver sand and shingle to the coast. Such areas should be protected from any form of development and deliberately allowed to continue to erode.

Sea defence and coastal protection

In the UK the responsibility for managing the coast falls to two ministries and two political jurisdictions (see introduction to this chapter). The coast itself is the responsibility of the Ministry of Agriculture, Fisheries

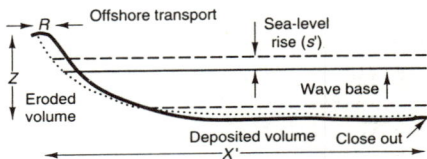

Fig 8.4 The Bruun Rule: in 1962, Bruun described a concept that attempted to explain the development of an equilibrium coastal profile during landward movement of the shoreline as a result of sea-level rise. Bruun assumed that there were no long-term changes in energy input and sediment volume.

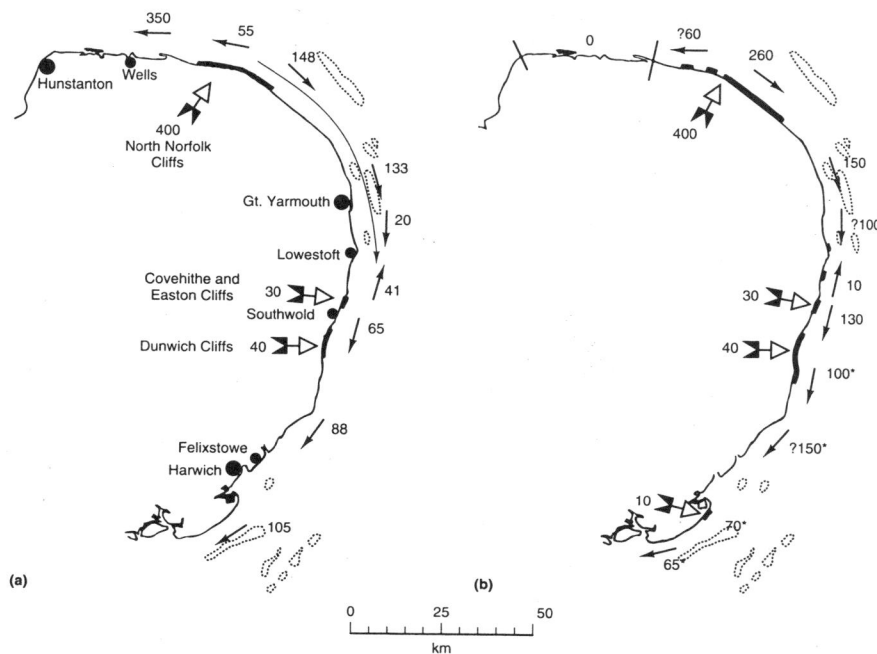

Fig 8.5 The East Anglian sand budget. Cliff inputs and littoral drift values in 1000 m³/yr. (a) Computed net longshore sand transport values over 13 years (1964–76). (b) Most probable values derived from examination of gross values and net values for longshore transport over varying periods of time, using both computed values and those calculated from wave observer observations, 1974–79. These are regarded as relatively reliable values (over a period of 20 years or more), but for those marked ? the value remains uncertain, although the direction is certain. The asterisks note theoretical values not reached due to lack of sand and/or lack of exposed beach at all states of tide. The main offshore banks are indicated on both maps.

and Food (MAFF) and its agency, the National Rivers Authority (NRA). The NRA is divided into ten regions, and directed in its work by local flood defence committees. These are composed, for the most part, of landowners in flood-prone areas and local authority councillors, more often than not, representing coastal constituencies. The coastal zone landward of the sea is the responsibility of the Department of the Environment (DoE), and through it, the district councils bordering the coast. They aim to control erosion and to stabilize the coastal zone. They receive grants for approved coastal protection schemes from the MAFF, but generally have to meet the cost of repairs themselves.

Cash bottlenecks are also common for flood protection schemes known as sea defence. In the UK the county councils must contribute about a quarter of the NRA sea defence capital works programme. That money eventually comes back from the DoE a year later. In the meantime the county councils must levy their council taxpayers for the capital when they also have to raise revenue for a host of other requirements, such as the police, social services and roads. The formula on which county spending is assessed does not take into account specific sea defence needs. Consequently, the counties are regularly faced with a bill that exceeds their actual spending allocation. Even though

Box 8.1 The Venice Lagoon

By any standards Venice is both priceless and imperilled. Its maritime dominance in the 13th to 16th centuries endowed it with the stability and wealth to construct a magnificent city on a series of islands lying within a lagoon, feebly protected by an offshore bar. Over the centuries the bar has been breached for commercial shipping, the lagoon has been dredged for access and landfill, and much of the 'barriers' or ecologically rich mudflats has disappeared, or has been marginalized. Yet Venice is subject to periodic storm surges, bursts of exceptionally high tides pushed by strong winds on the eastern flank of a southerly moving alpine depression. On 4 November 1986 the tide reached 194 cm compared with an average high tide of 50–60 cm. The famous St Mark's Square was flooded to over a metre, and over 70% of the city was under water. Nowadays, tides of 100 cm occur five or six times per year: a sea-level rise of 30 cm would mean that St Mark's Square would flood nearly every day. These tides are not only a great nuisance to the tourist and to commerce generally: they corrode the foundation of the buildings and increase both odour and mosquito problems.

In 1992 the Italian government allocated $1–2 billion to the preparation of a plan to place hydraulic barriers across the four entrances to the Lagoon. These structures will consist of 79 flap gates, each one of which is designed to rise to an angle of 50 ° against the rising sea. The gates can operate independently, allowing tides of a modest height to enter through one gate so as to increase the flushing ca-

pacity of the Lagoon itself, which is heavily polluted with nutrient and agricultural run-off. In addition, the Lagoon will be manipulated by four schemes designed to restore its capacity to absorb high waters:

1. Reconstruction of the coastal marshlands, presently only half their original extent of 100 km^2. This is being undertaken via a series of experimental schemes of salt marsh reconstruction.
2. Maintenance of the complex hydraulic network of the Lagoon, presently altered by erosion and dredging, to permit greater water circulation to coastal marshes.
3. Replanting the marsh vegetation in shallow areas both to stabilize the muds and colonize the mudflats for ecological reasons.
4. The construction of sand bypasses to recapture part of the sediment carried by the marine currents, presently deflected by the inlet jetties.

The Venice Lagoon Project is well analysed on paper, but a long way from actual completion. Part of the problem lies in the multiple jurisdictions between local, regional and national governments. But a major difficulty is the obtaining of agreement amongst local governments to fund sewage treatment works, removal of nutrients and reduction of agricultural fertilizer application. Ironically the success of the physical barrier to control tides could seriously damage the morphology and ecology of the Lagoon unless the Project is designed and managed as an integrated whole.

---

the money is eventually returned, this is of little consequence if the county is restricted in its total spend by central government allocations based on theoretical spending assessments.

There is thus a tug of war between the coastal counties and the local land drainage committees, in which they are well represented, and the specific needs of the NRA. Engineering works rarely receive funding to the degree to which finances are required, and each year of short-fall builds up a capital spend backlog that can prove politically embarrassing.

Behind the facade of busy engineering activity there is much political and economic turmoil. In no country is there administrative cohesion over managing the coast. New Zealand passed the Resource Conservation Act that provides for a single coastal agency: but the landward position is still administered by two tiers of local authorities. In the USA conflicts abound between federal, state and local agencies, especially over land-

use planning controls and the protection of vulnerable beach-funding areas and wetlands. The Venice Lagoon project has been static for many years because of an inability to co-ordinate regional and national priorities for spending, despite the increasing threat of '*aqua-alta*' (high tides) and an administrative committee designed to co-ordinate the various responsible parties (see Box 8.1).

Clearly it is in the best interests of all to build durable and cost-effective coastal protection works. The question is, what is durable and what is cost effective when it comes to coastal defence?

The commonest form of defence is the shingle or clay bank, or seawall. These are usually designed to withstand a high tide of 1 in 20 years' recurrence for small areas of farmland, 1 in 50 years' recurrence for roads and rail communications, and 1 in 200 years' recurrence for commercial and residential property. These structures rarely last more than 40 years, and

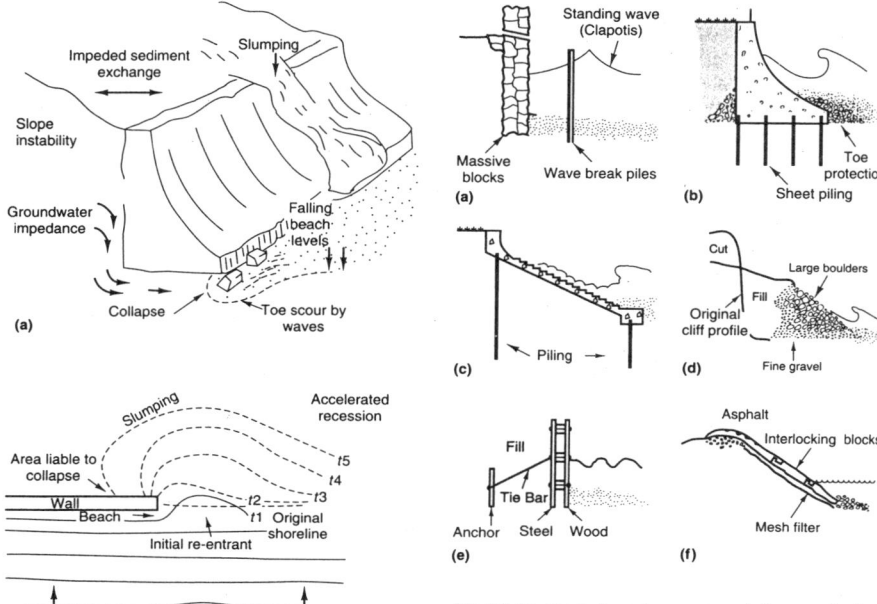

Fig 8.6 Some environmental problems relating to seawalls: (a) falling beach levels, poor drainage and impeded sediment exchange lead to both undercutting and overslumping; (b) at the end of the wall 'flanking' is common, with an erosional re-entrant forming and eventually scouring behind the wall causing collapse. t_1 to t_5 represent progradation over time.

Fig 8.7 An energy-based sequence of shore protection designs (high to low, a–f): (a) vertical seawall constructed of resistant interlocking blocks; (b) curved seawall with toe protection; (c) curved and stepped wall secured by piling; (d) rubble-mound armouring plus regrading of the coastal slope; (e) bulkhead of wood or steel; (f) revetment made of armour blocks, gabions or asphalt.

some fail quite soon after construction if the coastal geomorphology is not fully understood. Figure 8.6 shows how inadequate drainage protection can undermine a cliff floor wall, partly because the cliff is not allowed to erode. This causes a loss of feed to the beach. Also, where a seawall or groyne system ends there may be additional erosion on what might otherwise have been a relatively stable coast.

Figure 8.7 shows more clearly how various coastal defences operate. The main problem with most of these designs is that they cause wave reflection, and hence greater erosion of the beach, thereby exposing the foundations to more erosive force and seepage. Revetments, or wood structures designed to dissipate wave energy, are equally damaging if fully planked. The better design should be to allow some seawater to pass through with shore or cliff sediment to a controlled extent. Similarly groynes, or wood structures built at right angles to the coast, should be widely spaced to allow for the accumulation of sediment. This means that beach material should rest between the upper and lower sections of adjacent groynes, rather like a continually filling bathtub with a controlled outlet. The correct spacing is a matter of trial and error, knowledge of wave patterns and intensity, and calculations as to trapping ability and groyne effectiveness. Figure 8.8 describes the sequence of calculations. It is also essential that sand or shingle is placed between newly constructed groynes and not 'stolen' from the natural littoral drift.

Integrated coastal zone management

Integrated coastal zone management is the ideal arrangement for which everybody strives, but few attain. Essentially it is the recognition of four principles:

1. That natural processes of defence and protection should be encouraged, costed properly and fully incorporated into any plan or management scheme.

2. That natural zones essential to this purpose, such as headlands, dunes, salt marshes and wetlands, should be adequately protected by law, cleared of existing settlement, with compensation if necessary, and carefully monitored for their continuing role.

3. That coastal defence works should always be designed sympathetically and encourage the retention of a natural beach, and that cost benefit analyses should recognize the essential linkage between the two.

4. That land-use planning formally take into account the vulnerable areas of coast subject to sea-level rise and increased storminess, so that no new settlement or economic activity is permitted in such areas, and, where possible, existing buildings are left unprotected, again with compensation where necessary.

These are tough and controversial requirements. One can readily see why they would not easily be met even in cohesive administrations. But in the case of jealously protected governmental levels, where cost benefit has not yet come of age in terms of the ideas promoted in Chapters 2 and 3, it is easy to understand that the politics of coastal protection confound integrated management.

Box 8.2 The Broads flood alleviation strategy

The Norfolk and Suffolk Broads (lakes) in eastern England are one of the most important wetlands in North Western Europe. The 20 000 ha region is essentially the drained and undrained valleys of three rivers which converge at Great Yarmouth. The whole area is below the level of high tides, and is very vulnerable to storm surge tides carried by northerly winds on a depression running south-east across the North Sea. At present, the coastal defences are relatively sound, having been built and rebuilt since the great storm surge of 1 February 1953. Ten years ago, the justification of any scheme to upgrade the slowly sinking river walls and to block the rising sea by a barrier at Yarmouth required some heroic calculations of improved agricultural output in the drained marshes. Since cereals were both heavily subsidized and in surplus, this justification simply would not be made in the face of criticism by environmental groups anxious to retain the historical grazing marsh landscapes and the diverse aquatic plantago of the drainage dykes.

Nowadays the whole region is designated as an environmentally sensitive area. This means that the UK Ministry of Agriculture pays about £2.5 million annually to farmers to retain unprofitable grazing and to recreate the ecologically diverse drainage systems. This investment provides the basis for an ecologically biased cost benefit analysis along the lines discussed in Chapter 2. The conservation interests in the region believe that a combination of a barrier on the Bure river along with washlands to absorb tidal surges on the southern rivers would provide adequate levels of protection and a much more diverse ecology. The local land drainage committee prefers a single barrier on the inner Yare with no washlands but with a much higher level of protection for all the valleys. Current government guidelines to give priority to the protection of urban areas suggest an outer barrier at the Yare mouth which would protect Great Yarmouth as well as the Broads. For navigation-related reasons, however, this option has been ruled out. The matter is still to be resolved, but it seems likely that neither a barrier on the Yare nor a smaller barrier on the ecologically sensitive Bure will prove cost-effective. This is the case even with liberal use and contingent valuation studies in the case of the Yare barrier. Therefore a program of bank strengthening along with planned localized washlands, coupled to land set-aside schemes for surplus agricultural land for local tidal relief schemes is being prepared, along with a scheme to augment low river flows by freshwater injection from boreholes. This should keep saltwater intrusion to the less ecologically sensitive lower reaches.

The Broads story reveals the intricate politics of land versus soft engineering styles, the changing role of cost benefit analysis as agricultural profitability and land-use decline, and the scope for extending a biodiversity component of agricultural extensification as part of a total approach to flood protection. This is a case of integrated coastal management, permitting the scope for controlled retreat work on protective scheme that is both desirable and flexible. But the compromise took three years of intense argument and counter argument to achieve. Antagonistic attitudes of mind were not noticeably altered by the process.

Integrated coastal management is not a single-minded outcome. It is a process with many possible pathways, depending on the physical, institutional and political circumstances of coastal protection. It would be most unwise to lay down a blueprint planning guide. Three criteria can be used to assist the evaluation of good performance in integrated coastal management:

1. *Optimization of multiple objectives.* Multiple objections cannot be met without some losing while others gain. The aim of integrated coastal management should be to create structures of administration and public involvement geared to achieving agreement. This may mean sub-optimal realization of objectives for each special interest group, but at least respect for and support towards a common goal. It also means flexible and adaptive management geared to shifts in information, scientific experimentation and shifting public interest demands as all this innovation proceeds. Clearly this also indicates the need for imaginative and innovative cost benefit analysis backed by informed public support (see Box 8.2).

2. *Maintenance of life-support processes.* A well-managed coastal project should demonstrably enhance the life-support systems of the coastal zone, by empathetic design, by creating and protecting habitats, and by ensuring that there is room for ignorance because of the necessarily limited understanding of ecological resilience. This requires a sound scientific basis of the workings of the coast, good ecological and hydrological modelling as far

as knowledge allows, and honest acceptance of the limits of knowledge. Uncertainty is not an ogre if it is properly taken into account.

3. *Responsive management.* The 'cost' of any coastal zone scheme must be seen in terms of its success in balancing inherently contradictory objectives and in guaranteeing ecological viability. Where these objectives are evident, the 'benefits' are partly measured by the procedures that achieve such success. Responsive management translates into the satisfaction of trust in a scheme that is well grounded, well monitored and well communicated to those who have a stake in its outcome. Participation is not a democratic symbol: it should be a process of guidance and continual readjustment to a changing set of optimized objectives that become compatible only by the revelation of shared interest (see Box 8.3).

Still, progress is being made. There has been a steady shift amongst all engineering organizations in favour of a mix of beach nourishment and the concept of controlled retreat along with conventional revetments, seawalls and groynes. There is much talk of 'soft engineering' and of 'working with nature'. This is an important step forward, for the cost benefit analyses have to incorporate some measure of recreational benefit (using the travel cost method) and amenity provision (using contingent valuation).

Secondly, the precautionary principle is now being applied more readily to such aspects as safeguarding coastal wetlands, controlling polluting emissions, notably nutrients, and placing more of the burden of an

---

Box 8.3 The Australian coastal zone inquiry

Throughout 1992 to 1994 the Resource Assessment Commission of the Australian Commonwealth government held hearings and developed position papers on the management of the whole Australian coast. The Commission concluded along the following lines:

- *Coastal zone management.* There is a need for much better co-ordination of commonwealth, state and local government, with pooling of both objectives and budgets, and a full acceptance of the principle of shared governance. This means establishing legally supported administrative co-operatives headed by project co-ordinators on a sub-regional basis for structures of waste.
- *Precaution in pollution and land-use planning.* Coastal pollution is already excessive and wor-

rying. The precautionary approach, coupled to ecological economics, should be applied to sewage treatment clean-up, control of agricultural wastes and estuary clean-up programmes. Tourist developments should be strictly controlled, with safeguards for key habitats, and the polluter pays principle should be extended to a tourist impact tax to fund sustainable tourism projects.

- *Involvement of indigenous peoples.* In many parts of northern and western Australia indigenous peoples have *de facto* property rights in fishing, tourism development, marine parks management and the protection of cultural and sacred sites. These requirements should be built into the coastal management co-operatives.

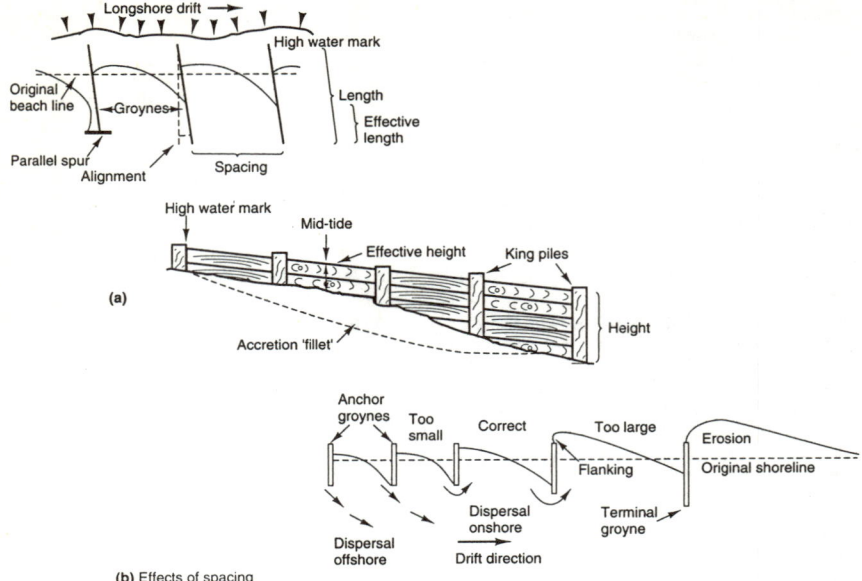

(b) Effects of spacing

Fig 8.8 Groyne design and spacing. (a) Groynes should be designed to trap and retain sediment, but all too often this is not accomplished. Trapping ability and groyne effectiveness can be assessed by knowing the amounts of material moving in and out of the groyne field. The accretion 'fillet' in the middle diagram may occupy only a small proportion of the inter-groyne volume. (b) Correct spacing of groynes is probably a function of wave parameters. Relatively small spacings promote flanking. Terminal groynes are often responsible for downdrift erosion. Trapping ability (%) = volume retained × '100'/(length × spacing × height); groyne effectiveness (%) = (volume input − output (downdrift)) × 100/volume input (updrift).

acceptable environmental assessment on the shoulders of developers. In addition, strategic environmental impact assessment is becoming in vogue. This is the meshing of coastal management policies over nature conservation, pollution control, tourism restraint and water use limitation with long-term plans to shape the coast to accommodate sea-level rise. To date, achievements in this area have been modest, because policy integration is still a distant vision. But at least the matter is getting a hearing in this period of local level Agenda 21s, natural sustainable development strategies, and increasing pressure on governments to incorporate environmental thinking into all aspects of resource management policy.

Sadly all this will come too late for many populated and vulnerable coasts. The low-lying islands of the Pacific and Caribbean are extremely vulnerable even to

modest sea-level rise, yet they cannot obtain insurance cover for flooding to existing property, including resort hotels, let alone any compensation for property removal. The consequences for future economic development are potentially severe, yet no compensation is in the offing. In every country battles are looming over the denial of taxpayer-financed protection for property in the wave firing line, especially after a flurry of storm damage. The politics of property wealth coupled to post-disaster sympathy usually combine to protect such areas just when controlled retreat would prove a better option. Despite all the huffing and puffing, EIAs are not particularly effective in safeguarding wetlands and other vital national zones on the coast. Nor is the revamped cost benefit analysis a particular safeguard.

As the IPCC gathers its strength in identifying more comprehensive approaches to coastal zone management, so one hopes that some of these deficiencies will be overcome. The omens are not good, when seawater is lapping at the door, and the cry is to protect rather than to retreat. One possible breakthrough might come with the strengthening of national strategic coastal management plans, on the back of the Agenda 21 exercise, and cash for compensating the removal of protection of property on coasts that should be returned to natural processes. That money could come, in part, from a carbon tax. After all, sea-level rise is arguably 60 per cent due to carbon dioxide. This opens up the Pandora's box of directional revenue from environmental taxation. When all these jigsaw pieces fit into place, integrated coastal management might just come of age.

Further reading

A good start can be made in R.W.G. Carter (1988) *Coastal environments: an introduction to the physical, ecological and cultural systems of coastlines* (Academic Press, London). For the geomorphological story see K.M. Clayton (1979) *Coastal geomorphology* (Macmillan Educational Books, London). A useful perspective on how various countries approach this topic is given in Organization of Economic Cooperation and Development (1993) *Coastal zone management: integrated policies* (OECD, Paris). For a perspective on ecological planning see R.V. Salm and J.R. Clark (1984) *Marine and coastal protected areas: a guide for planners and managers* (International Union for the Conservation of Nature, Gland, Switzerland).

[4]

Journal of Coastal Conservation 2: 103-114, 1996
© *EUCC; Opulus Press Uppsala. Printed in Sweden*

Science and management in four U.S. coastal ecosystems dominated by land-ocean interactions

Boesch, Donald F.

Center for Environmental and Estuarine Studies, University of Maryland, P.O. Box 775,
Cambridge, MD 21613 USA; Tel. +1 410 228 9250; Fax +1 410 228 3843; E-mail boesch@co.cees.edu

Abstract. The influence of science in the recognition of the effects of landscape changes on coastal ecosystems and in the development of effective policy for managing and restoring these ecosystems is examined through four case studies: Chesapeake Bay, San Francisco Bay, the Mississippi Delta, and Florida Bay. These ecosystems have undergone major alterations as a result of changes in the delivery of water, sediments and nutrients from their watersheds. Both science and management have been challenged by the spatial, functional and temporal scale mismatches inherent in the watershed-coastal ecosystem relationship. Key factors affecting the influence of science on management include (1) sustained scientific investigation, responsive to but not totally defined by managers; (2) clear evidence of change, the scale of the change and the causes of the change; (3) consensus among the scientific communities associated with various interests; (4) the development of models to guide management actions; (5) identification of effective and feasible solutions to the problems.

Keywords: Chesapeake Bay; Environmental policy; Estuaries; Florida Bay; Mississippi Delta; Nutrients; River discharge; Salinity; San Francisco Bay.

Introduction

The emerging, widespread environmental threats confronting coastal ecosystems around the world, such as eutrophication, hydrologic disruption, introduction of non-indigenous species, and global climate change, pose new challenges to environmental policy, management and science. Meeting these challenges requires different approaches from those used to manage traditional problems such as point-source discharges of industrial and municipal effluents, coastal land use, direct habitat destruction, and oil spills (Anon. 1994). In particular, there is a growing appreciation that coastal ecosystems are heavily influenced by human activities on the land–often hundreds of kilometers from the coast–as well as by activities in the coastal zone itself. This is well exemplified in the United States, with its large continental land mass and rivers which drain into coastal

ecosystems important in terms of their economic value and natural heritage.

In this paper I will relate the role science has played in understanding these important connections between activities on the land and the coastal zone and guiding effective management solutions based on experiences in four ecosystems: Chesapeake Bay, San Francisco Bay, the Mississippi Delta, and Florida Bay (an ecosystem which has recently undergone major changes which have just begun to attract concerted scientific appraisal) (see Fig. 1). My analysis of the history of the influence of science is based in part on the published perspectives of others and in part on my own experiences as a research scientist, scientific administrator, or scientific advisor within each of the four coastal ecosystems (see also Boesch 1995). After reviewing these four case studies, I will then attempt to draw some generalities and to develop recommendations for improving the effectiveness of science in guiding the formulation of effective policies and the implementation of these policies.

Chesapeake Bay

The largest estuary in the United States, the Chesapeake Bay has been the site of extensive scientific research and is the subject of what is probably the world's most ambitious effort to manage and restore a coastal ecosystem. Despite the great attention the Chesapeake has received by scientists and environmental management, the dimensions of environmental change that has taken place in the Chesapeake Bay and its tributary sub-estuaries and their relationship to changes on the land have not been appreciated until relatively recently. The most pervasive and consequential environmental change has been an increase in nutrients reaching the bay. Nutrient enrichment has caused changes in plankton communities, productivity, the extent of bottom-water hypoxia, and increased turbidity. Increased hypoxia is responsible for reductions in some living resources and increased turbidity has caused great reductions in submersed vascular vegetation.

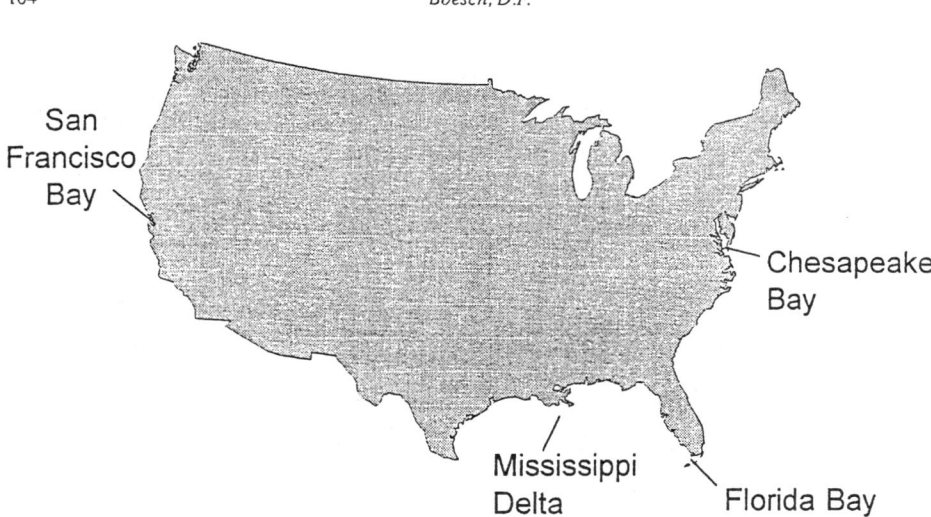

Fig. 1. Conterminous United States showing the location of the four coastal ecosystems discussed.

Through paleontological and chemical analyses of cores of sediment from the depositional deep trough of the bay, Cooper (Cooper & Brush 1991; Cooper 1995) demonstrated that the sedimentation rate, primary production (reflected in the accumulation of carbon and biogenic silica), and extent and severity of anoxia (reflected in the formation of iron pyrite from reduced sulphur) began to change dramatically around 1760 following widespread land clearing for agriculture by European colonists (Fig. 2). In addition, the diatom community shifted from a diverse community with nearly equal representation by centric and pennate forms to a less diverse community heavily dominated by centric diatoms as the enriched and more turbid system became dominated by planktonic rather than benthic primary production. Although, according to Cooper & Brush (1991), intense seasonal anoxia is mainly a phenomenon of the late twentieth century, the eutrophication of the Chesapeake started with European settlement and, in particular, the proliferation of agriculture.

The combination of a large watershed (166 000 km^2), significant volume in relation to freshwater input and tidal exchange, and partial stratification disposes the Chesapeake Bay to nutrient retention and recycling. Therefore, this ecosystem is particularly sensitive to nutrient inputs from the watershed (Boynton et al. 1995). Yet, for many years both the scientific and management community focused on smaller scale human impacts

associated with activities directly affecting portions of the bay while assuming that, with the exception of widespread overfishing of certain species, the bay was in good health overall.

Malone et al. (1993) examined in considerable detail how this view began to change in the early 1970s and how the importance of nutrient loadings to the Chesapeake Bay was embraced by scientists, managers and policy makers. By the 1960s the upper end of the Potomac River sub-estuary below Washington, DC had become obviously over-enriched as evidenced by massive algal blooms and depleted oxygen (Jaworski 1990). Following the recent successes in addressing over-enrichment problems in Lake Erie, one of the North American Great Lakes, major federal investments were made in 1972 to provide advanced treatment, in particular phosphorus removal, of municipal wastewaters of metropolitan Washington. This was after all the nation's capital and these were the days of the Great Society when government was thought to accomplish what it set out to do. The results were dramatic: water quality greatly improved, nuisance algal blooms retreated, and fish returned to the upper Potomac (Jaworski 1990). This experience had the effect of instilling confidence in regional environmental managers that commitment to waste treatment would yield positive results, but it also focused attention on point sources of pollutants, obscuring the effect of non-point sources on the bay.

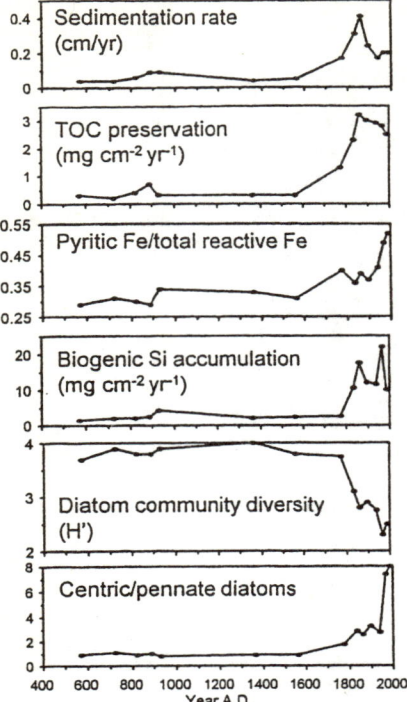

Fig. 2. The history of eutrophication of the Chesapeake Bay as revealed in a sediment core (R4-50) from the central channel (Cooper & Brush 1991; Cooper 1995). Sedimentation increased greatly following extensive land clearing for agriculture around 1760. The increased deposition of total organic carbon (TOC) and biogenic silica reflect the significant enrichment of the estuary by nutrients following this landscape change. A reduction of diatom community diversity and an increase in centric diatoms reflect a shift from a benthic-dominated to a plankton-dominated, light-limited system. Seasonal hypoxia, reflected by an increase of pyritic iron, has intensified in the latter half of this century.

Also in 1972, the entire Chesapeake watershed was affected by record floods associated with the passage of the tropical storm Agnes. The resulting freshet had profound effects on the Chesapeake Bay and its river sub-estuaries in terms of circulation, sedimentation, chemical inputs, biotic changes, and declines in important fisheries. It forced the scientific community and

some of the management community to begin to think of the bay not as a vast arm of the sea, but as an estuarine ecosystem heavily influenced by its watershed (Malone et al. 1993). Following Agnes, concern about large scale changes in the bay, such as the declines of submersed vascular plants in both the upper bay and lower bay, stimulated Congressional pressures on the Federal government to study and fix the problems afflicting this ecosystem. The resulting multi-year Chesapeake Bay study began in 1978 with heavy financial support by the U.S. Environmental Protection Agency. It focused on aquatic vegetation, toxic materials and nutrient enrichment. Interestingly, the studies of nutrient enrichment never advanced to the stage of field research, involving instead a series of workshops and conferences. Although this did help coalesce opinions about the importance of non-point sources, for example, it was left to the studies of submersed vegetation to identify widespread nutrient over-enrichment as the primary culprit in the disappearance of these bay grasses.

Meanwhile, controversies developed over the effects of population growth and expanding wastewater discharges into the Patuxent River sub-estuary, just to the north of the Potomac. On one side of the controversy were estuarine scientists, who had been studying the Patuxent and had become alarmed at signs of over-enrichment, and their allies, officials of the rural local governments at the lower end of the estuary. On the other were state and federal environmental officials who were planning the wastewater treatment plants to handle the population growth in suburban upstream areas. A particular bone of contention was the need to remove nitrogen as well as phosphorus in wastewater treatment. State and federal managers and their engineering consultants, borrowing from the experience in the upper Potomac sub-estuary, held that phosphorus removal was all that was required. But the upper Potomac is freshwater, and the estuarine scientists pointed to literature which indicated that N rather than P tended to be the limiting nutrient for marine phytoplankton. The phytoplankton of the mesohaline lower Patuxent, they argued, was likely to be N-limited, therefore costly nitrogen removal would be required in the new treatment plants to avoid further degradation in the estuary. A lawsuit ensued, with estuarine scientists appearing for the plaintiff in opposition to the very agencies which supported their research. The matter was ultimately settled in 1981 by a 'charette' in which the parties committed to hammer out a consensus during a time-constrained meeting. The agreement to remove N was a milestone in the scientific influence on nutrient management policies.

With the conclusion of the five-year Chesapeake Bay study, the three states (Virginia, Maryland and Pennsylvania) which occupy most of the bay's watershed, the District of Columbia which includes the nation's capital, and the federal government represented by the Environmental Protection Agency endorsed the first Chesapeake Bay Agreement in 1983, thus launching the intergovernmental Chesapeake Bay Program for restoration of the bay. A Scientific and Technical Advisory Committee was formed and in 1986 it released a report which presented clear and compelling evidence that both P and N removal would be required to improve water quality in the bay and its tributaries and, very importantly, that cost efficient technologies were available for the combined removal of these nutrients. This scientific consensus provided the rationale and credibility for the bold action of the Second Chesapeake Bay Agreement in 1987 which committed the signatories to achieving a 40 % reduction in controllable inputs of N and P by the year 2000.

To the scientists involved in these debates this adaptation to scientific understanding may seem to have been painfully slow, but, because I was away from the Chesapeake scientific community during the 1980s, I was impressed by the speed of the effect of science on the management paradigm. Hennessey (1994) reviewed the Chesapeake Bay Program and observed that the evolution and refinement of its objectives and its use of monitoring and scientific information evidenced an effective application of adaptive management. But, in the sense of the originators of this concept, adaptive management involves a more structured approach to environmental management in the face of high uncertainty which emphasizes learning and pursuing multiple options (Walters 1986). While this may be true of the Chesapeake Bay Program viewed from a distance or over several decades, adaptive management as a concerted process requires even tighter linkages and shorter time steps.

Two technical tools have been of central importance in the Chesapeake Bay Program since its inception: modeling and monitoring. Over ten years of monitoring of water quality and living resources has made continuous and otherwise unattainable environmental data available to regional scientists for use in·extending their research. Furthermore, many researchers actually perform some of the monitoring. There has likewise been a mutualistic relationship between researchers and modelers. An array of linked and coupled models has been developed to predict water quality and ecological conditions in the bay in response to inputs of energy, water and nutrients from the atmosphere, watershed, and coastal ocean. The water quality model of the main stem of the bay started as a hydrodynamic model of the · type used in sanitary engineering analyses of the effects of biological oxygen demand of wastes on oxygen conditions. Through the creative tension between scientific

critics of the model's assumptions and the practically minded engineers, many new discoveries about biological and chemical processes in the bay have been incorporated into the water quality model (e.g. regarding factors affecting nutrient limitation of primary production, effects of animals on nutrient flux at the seabed, and grazing and settling rates of different forms of phytoplankton) such that today it is one of the most realistic and effective coastal ecosystem models that exist. So much have managers come to rely, and perhaps over-rely, on this model that many will only believe something when it is 'confirmed' by the model. A case in point is the announcement in 1994 that the model has demonstrated that phytoplankton production in the bay is N-limited – this eight years after results of mesocosm experiments were published which clearly showed this to be the case!

Now, the principal weak links in the models concern the watershed rather than the estuary. The watershed model is an adaptation of generic streamflow modeling and is not built on as rich an understanding of how this particular ecosystem works as the bay water quality model. There have not been comparable investments in advancing hydrology, geochemistry and ecology in the watershed and the terrestrial and freshwater scientists are not well linked with their estuarine counterparts. This is unfortunate because, as· the Chesapeake Bay Program emphasizes non-point source control within the tens of hydrologic units which comprise the Chesapeake watershed through what is known as Tributary Strategies, models well grounded in scientific understanding will be essential in guiding local communities to the most cost-effective targets for reducing the nutrients which actually reach the bay.

San Francisco Bay

The San Francisco Bay estuary, including the large tidal delta at the confluence of the Sacramento and San Joaquin rivers, is perhaps the major U.S. estuary most modified by human activity (Nichols et al. 1986). Although the Spanish settled in the area in 1769, the bay remained little affected until the discovery of gold in the Sierra Nevada foothills in 1848. Hydraulic mining of ore resulted in massive downstream sedimentation in the upper bay. Virtually all of the freshwater marshes of the delta were reclaimed for agriculture and salt marshes were filled or diked. Of the original 2200 km^2 of tidal marsh, only 125 km^2 of undiked marshes remained in 1986. Once abundant populations of commercial fishery species have been over-harvested and otherwise affected by habitat degradation to the point that only herring and anchovies are harvested today. Many non-indigenous species of invertebrates and fishes were introduced either purposefully or inadvertently with transplanted oysters or via ship ballast. Many of these exotics have established populations and, in fact, now constitute the dominant biota in the bay (Nichols et al. 1986; Carlton & Geller 1993).

Presently, the most significant management issue for the San Francisco Bay estuary is the consumption and diversion of fresh water for agriculture and for urban uses in central and southern California. To serve these needs the world's largest human-made water system removes about 40 % of the historic flow of the Sacramento-San Joaquin river system for local consumption upstream and in the delta, and exports another 24 % in aqueducts for agricultural and municipal consumption elsewhere. By the mid-1980s the flow actually reaching the estuary had decreased to less than 40 % of historic levels, and was projected to decline below 30 % by the year 2000 (Nichols et al. 1986).

The consequences of reduced river inflow to the estuarine ecosystem have been profound and include interference with migrations of anadromous fish species (i.e. fishes that migrate into fresh water to spawn); changes in estuarine circulation and increased residence time; upstream movement of isohalines and the null zone where sediments and phytoplankton accumulate; and suppression of the pelagic food web (Nichols et al. 1986). However, the lack of clear consensus among scientific and technical experts representing different interests on these effects and relationships led to the continued low priority given to the estuary in water allocation decisions, particularly in the face of strong and tangible interests of the other water users. The lack of definable and achievable objectives for the estuary led to what Kimmerer & Schubel (1994) referred to as 'regulatory gridlock'. The lack of technical consensus may be attributed to the difficulties in bridging the advocacy coalitions (Sabatier 1994) of agency representatives, resource users and scientists which form around specific resources or concerns, e.g. agricultural and municipal water supplies, fisheries, wildlife, or water quality.

A major breakthrough recently occurred as a result of a concerted effort to forge consensus on what management criteria should be used to guide water allocation to the estuary. This was effected through a series of workshops (Kimmerer & Schubel 1994) and an innovative statistical analysis relating the position of isohalines in the estuary to key ecosystem variables, including several directly related to important living resources (Jassby et al. 1995). It was shown that the longitudinal position in the estuary of the two practical salinity units (psu) isohaline measured 1 m off the bottom was directly related to such variables as total

input of organic carbon including *in situ* production, biomass of molluscs, survival from egg to young-of-the year and year class strength of striped bass, survival of salmon smolts passing through the delta, and the abundance of several important prey species. Freshwater inflow can thus be regulated by managing releases from upstream dams and withdrawals from the delta to maintain the desired position of the 2 psu isohaline. After years of impasse and protracted negotiation between the Federal and State governments, this management guideline has now been included in the 1994 agreement for water allocation.

Mississippi Delta

Even in comparison to such expansive catchment areas as those of the Chesapeake and San Francisco bays, the catchment of the Mississippi River is vast, over 3.3 million km^2, including 41 % of the conterminous United States. In contrast to the other coastal ecosystems considered here, the coastal area receiving water of the Mississippi is not a semi-enclosed embayment, but a distributary delta and the open continental shelf. Like the Chesapeake Bay and San Francisco Bay watersheds and river tributaries, the Mississippi watershed has also been greatly changed by agricultural conversion and damming (Meade 1995). But the flow of the Mississippi has also been greatly affected by channel deepening and straightening for navigation and by an extensive flood-control system of earthwork levees, revetments, weirs and dredged channels that has isolated most riverine wetlands in the flood plain (Turner & Rabalais 1991).

At the mouth of the Mississippi a vast distributary deltaic plain has been constructed by fluvial and marine processes during the past 7000 years, following Holocene transgression of sea level (Boesch et al. 1994). This deltaic plain includes extensive tidal (estuarine) wetlands and lagoons between the active or abandoned distributaries of the Mississippi Delta. Normally, the river and its distributaries — several of which were active at the same time — would overtop their banks during spring floods, bringing fresh water and sediments to the extensive wetlands of the interdistributary lagoon basins. However, soon after colonization of New Orleans by the French in 1719, construction of flood protection levees and closure of minor distributary channels began, culminating in an unbroken barrier of levees extending to the hub of the distributaries of the present active delta, a rather modest delta precariously perched at the edge of the continental shelf. Following catastrophic floods in 1927 a controlled diversion of 30% of the flow of the Mississippi and Red rivers was made into the Atchafalaya River in order to relieve the hydrologic inefficiency that resulted from so constraining a long channel to the sea. Thus, today the mighty Mississippi has two effective mouths, the deep-water birdsfoot delta and the entry of the Atchafalaya into a large shallow embayment.

Major changes have taken place in the Mississippi deltaic plain and in the offshore waters of the continental shelf during the latter half of the twentieth century. Best documented is the accelerated rate of coastal wetland loss from marshes and swamps to open water (Fig. 3), and conversion to more salt tolerant species of wetland vegetation. By the late 1960s, approximately 73 km^2 of vegetated wetlands were being lost per year (Boesch et al. 1994). The factors responsible for this massive wetland change are multiple, complex and difficult to apportion

but are related to widespread channelization of the wetlands for navigation and oil and gas extraction, increasing salinity, and a deficit in the aggradation of soil (mineral sediments and peat) in the rapidly subsiding marshes. The high rate of relative sea level rise resulting from the rapid regional subsidence offers a model for forecasting the effects of accelerated eustatic sea level rise on coastal environments (Day & Templet 1989). Prevention by the levees of the introduction of fresh water and sediments into the wetlands and lagoons was certainly a factor in wetland loss, but it appears that channelization was largely responsible for the great increase in wetland loss between 1950 and 1980 (Boesch et al. 1994). Nonetheless, it is widely held that the long-

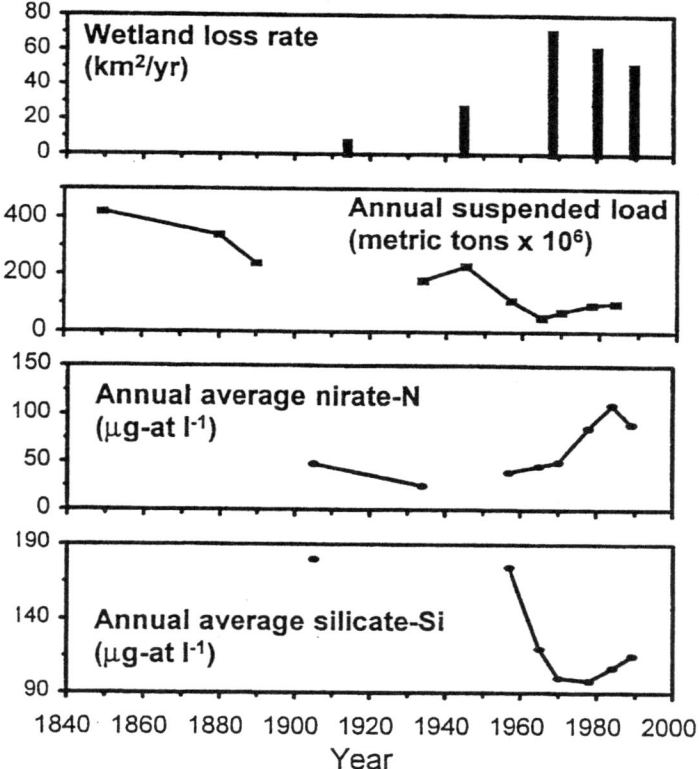

Fig. 3. Rapid changes in the Mississippi River Delta and its effluent have taken place in the latter half of the twentieth century. Coastal wetland loss rates are from data summarized by Boesch et al. (1994), suspended load estimates are from Kesel (1989), and average nitrate and silicate concentrations are taken from Turner & Rabalais (1994b).

term survival of wetlands in the deltaic plain must depend on the re-introduction of river flow into the interdistributary basins in order to stem salinity intrusion and supply sediments for wetland aggradation and progradation.

Significant changes in the composition of the Mississippi's effluent have also been well documented (Fig. 3). Average annual suspended sediment load in the lower river has declined by at least one-half since the late 19th century, ostensibly as a result of improved soil conservation practices and, particularly, the construction in the 1950s of dams which trap sediments (Kesel 1989). The concentration of dissolved nutrients has also changed from the 1950s; nitrate concentrations more than doubled and silicate concentrations have declined by 40% or more. Turner & Rabalais (1991) presented evidence suggesting that the increase in nitrate was coincident with the increase in use of chemical fertilizers in the U.S. In studies of nutrient concentrations throughout the length of the river, Antweiler et al. (1995) demonstrated that the source of most of this nitrate is the heavily agricultural, upper Mississippi basin over 1500 km upstream. Smith et al. (1987) suggested that increased atmospheric deposition of N in the industrialized Ohio River basin may also be a contributing factor; but Howarth et al. (1996) estimated that atmospheric deposition would account for no more than 25% of the anthropogenic N-loading of the Mississippi River system. The decline in concentrations of silicate are presumably related to reductions in suspended sediment loadings and to the biodeposition of removal of silicate by diatoms, the production of which has been enhanced by phosphorus enrichment and the construction of reservoirs and navigation pools which reduce turbidity and light limitation.

The consequences of the changes in nutrient delivery to continental shelf waters are incompletely known. Although hypoxic bottom waters reflective of eutrophication were occasionally reported in the literature, it was not until 1985 that the spatial and temporal extent of shelf hypoxia was surveyed. A region of up to 18 200 km^2 of the inner continental shelf east of the mouth of the Mississippi River and extending to the west from the Atchafalaya River occasionally to Texas, has been found to have bottom water oxygen concentrations too low (<2 mg/l) to sustain fishes and decapod crustaceans during the long summer season (Rabalais et al. 1991; N. Rabalais pers. comm.).

The key question of whether shelf hypoxia has spread or become more intense as a result of increased nitrate loading is difficult to answer because there were very few observations prior to 1985. In contrast to the wetlands of the Mississippi deltaic plain which have been studied intensively beginning in the 1960s, the continental shelf

off Louisiana and Texas – an area that produces virtually all of the offshore oil and gas and a large portion of the fishery landings in the U.S. – remained a *mare incognito*. Nonetheless, analyses of sediment cores from the region of chronic summer hypoxia have shown changes in benthic foraminifera microfossils consistent with worsening hypoxia (Rabalais et al. 1996); an increase in the accumulation of biogenic silica (Turner & Rabalais 1994a); increased concentration of organic carbon; and C and N isotopic signatures (Eadie et al. 1994) consistent with increased productivity since the 1950s. In addition, the changed ratios of N, P and Si in the river discharge may have shifted the nutritional conditions for phytoplankton growth, perhaps favoring flagellates over diatoms (Turner & Rabalais 1994b; Rabalais et al. 1996).

Science has played an increasing role in environmental management of the wetlands of the Mississippi deltaic plain. The publication in 1981 of the first comprehensive measurements of land change rates stimulated much concern among resource managers, the general public and, eventually, political leaders. During the 1980s extensive research was conducted which greatly increased understanding of the causes of wetland loss. By 1990, the U.S. Congress had enacted the Coastal Wetlands, Planning, Protection and Restoration Act (CWPPRA) which established a process of planning and active restoration focused on Louisiana's coastal wetlands (Boesch et al. 1994).

Controversies among scientists and managers still rage on the effectiveness of various restoration and management techniques, including river diversions, barrier island restoration, and structural control of water levels. Planning and implementation of restoration projects within the inter-distributary basins are progressing with varying levels of involvement of the scientific community. An independent assessment of the process by a group of scientists emphasized the need for greater involvement of the scientific community in planning and monitoring and for a more holistic approach within the entire Mississippi deltaic plain as well as within each of its constituent interdistributary basins (Boesch et al. 1994). One factor that must be taken into account is the decline in suspended sediment loads, and thus basic building material, in the river as a result of dams put into place upstream.

On the other hand, policy-makers and managers have not yet developed any mitigative responses to the eutrophication of shelf waters, such as reduction of point and non-point sources as in the Chesapeake Bay. There are several potential reasons for this: the evidence for worsening eutrophication, although very strong, is new and not yet widely understood and accepted; the consequences of eutrophication to important resources, although potentially major, have not been well documented; hypoxia occurs offshore, out of view and without obvious massive fish kills; and the idea of trying to control nutrient discharges throughout the huge Mississippi watershed has been just too daunting for many to contemplate. In January 1995 a coalition of environment organizations petitioned the U.S. Environmental Protection Agency (EPA) to take action to control nutrient pollution of the Mississippi River under federal law which allows federal intervention when pollutants discharged on one state affect another downstream. So far the response of the EPA has been to initiate an assessment of the evidence, hold a workshop, and begin to discuss the problem with agricultural and other upstream interests.

There is an under-appreciated interaction between wetland restoration initiatives, particularly river diversions, and the eutrophication of coastal waters. Diffusing the rivers effluent more broadly over coastal marshes rather injecting it onto the continental shelf may effect some N removal but probably a small part of the total loading, particularly during high spring flow. On the other hand, introducing nutrient-rich river waters into an interdistributary bay or to the east of the river's mouth may stimulate excess phytoplankton growth in areas not now experiencing hypoxia. Large scale diversions, such as that proposed to divert flow from the birdsfoot delta to the east, could result in eutrophication of portions of the continental shelf not now experiencing hypoxia and affect stratification and buoyancy-driven circulation over a large scale.

Florida Bay

Florida Bay is a large (about 2200 km^2 in surface area), very shallow (average depth less than 1 m) lagoon bordered on the north by the Florida mainland and on the south by the Florida Keys (McIvor et al. 1994). In contrast to the three other coastal ecosystems considered here it is tropical, frequently hypersaline, contains carbonate sediments and outcroppings, and has extensive seagrass meadows and mangroves.

Florida Bay has received relatively little attention and was thought to be little affected by human activities. Beginning in 1987, seagrass meadows, which had covered as much as 80 % of the bay's bottom began to die with the area of die-off as large as 18 % of the total area of the bay (Roblee et al. 1991; Boesch et al. 1993; McIvor et al. 1994). Blooms of cyanobacteria and other phytoplankton, which were first noticed as early as 1979, began to occur with increasing frequency and intensity, turning the once clear waters a turbid green. Populations of water birds, forage fish and juveniles of

game fish seem to be reduced in the upper end of the bay, coincident with increased salinities. Many large sponges along the Florida Keys margin of the bay died, potentially threatening a significant decline in the catch of spiny lobsters, the juveniles of which use the sponges as critical habitats.

A number of scientists who were investigating these changes argued that most were related –one causing another– and have as a root cause changes in freshwater flow through the Everglades into Florida Bay (McIvor et al. 1994). A conceptual model of these cascading effects is presented in Fig. 4. Florida Bay lies at the distal end of the 28 000 km^2 Kissimmee-Okeechobee-Everglades drainage basin. In order to reclaim wetlands for agriculture, furnish irrigation water, and provide flood protection for the sprawling population of the Miami region which encroaches on this watershed, an extensive series of canals has been dug which have the effect of moving water outside of this drainage basin to discharge sites along the Atlantic coast. As little as one-fifth of the historic surface water flow into the northeastern corner of the bay may now escape such diversion under the present water management regime. However, other scientists suggested that the changes may have been the manifestations of natural cycles, including the frequency of hurricanes, related to the filling in of the Florida Keys and occlusion of water exchange, or were caused by greater infusion of plant nutrients from the mainland watershed (Boesch et al. 1993).

In order to assess the evidence associated with the intense differences of opinion among scientists, which had become publicized by the news media, the Author was asked by the Assistant Secretary of the U.S. Department of the Interior to chair a panel of outside scientists. Although our report was unable to resolve these scientific controversies, it did indicate that there was enough evidence that the deteriorating conditions in the northeastern part of the bay were related to increased salinities and that rediversion of freshwater flows was needed to restore this part of the bay. Further, the panel crafted a set of hypothesis-driven research, monitoring and modeling needs to address the questions critical to effective management. The structure of questions and scientific priorities developed by the panel are now being used as guidance of a greatly expanded ($7 million per year) research program coordinated by a Federal-state inter-agency committee. Although this scientific evaluation started with a more limited base of knowledge about this coastal ecosystem than in the case of the Chesapeake Bay, San Francisco Bay or Mississippi Delta, it now has the opportunity to resolve the unknowns through a more strategic and coordinated approach. It will be interesting to observe whether this

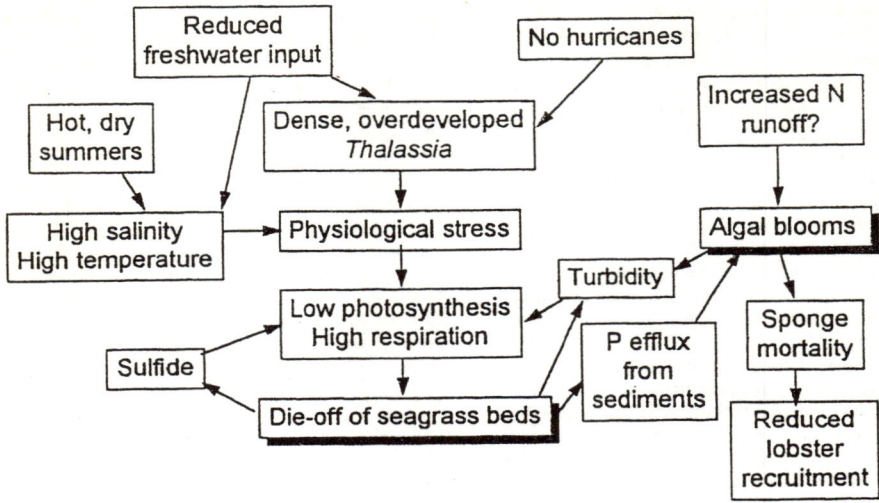

Fig. 4. Hypothetical model relating the potential causes of massive die-off of seagrasses and algal blooms in Florida Bay, based on McIvor et al. (1994) and Boesch et al. (1995).

speeds up the process by which effective management solutions are identified and implemented.

Discussion

Scale mismatches underlie the difficulties in recognizing, assigning causes to and effectively managing large-scale modifications of coastal ecosystems associated with changes in the delivery of fresh water, sediments and nutrients from the land. As Lee (1993) notes, "when human responsibility does not match the spatial, temporal or functional scale of natural phenomena, unsustainable use of resources is likely, and it will persist until the mismatch of scales is cured." In the case studies considered here, spatial scale mismatches occurred because of the lack of awareness of the consequences of the clearing of land, use of agricultural fertilizers, dam construction, or flood protection on the flux of materials into coastal ecosystems far removed from these activities. Functional scale mismatches occurred, for example, in the allocation of water among users when the interests of the estuary as a user was not considered or not considered important. Temporal scale mismatches occurred when the longer-term effects of actions were not understood or considered, for example the eventual unsustainability of Mississippi Delta wetlands deprived

of periodic sediment subsidies from floods.

Both science and management have been challenged by the spatial, functional and temporal scale mismatches inherent in the watershed-coastal ecosystem relationship. Based on both the successes and problems identified in the four U.S. case studies considered here, five key factors seem to affect the degree and timeliness of influence of science on environmental policy and management:

1. *Sustained scientific investigation, responsive to but not totally defined by managers.* Scientific investigations and institutions were sustained over long periods in both the Chesapeake Bay and San Francisco Bay. In the Chesapeake this took place primarily in academic research institutions which were established and have been supported to provide scientific information to the states of Virginia (Virginia Institute of Marine Science) and Maryland (Center for Environmental and Estuarine Studies). In San Francisco Bay, despite the proximity to prestigious universities, this role was played largely by governmental agency scientists from the U.S. Geological Survey and the California Department of Fish and Game. In the Mississippi Delta, there had been sustained research on wetlands by university scientists, particularly through the Sea Grant Program, but little effort on the continental shelf until the late 1980s. In

contrast, Florida Bay has been the subject of little sustained research which could have detected and understood changes earlier. Now, there is new, intense scientific effort, but without the benefit of much corporate experience. It has also proved important that the research be responsive to the management issues, but not totally prescribed by environmental and resource managers who are more prone to support science based on present understanding than to invest in the potentially heretical investigation that have led to paradigm shifts.

2. *Clear evidence of change, the scale of the change and the causes of the change.* Because of the scale of changes and the confounding effects of other human activities and natural phenomena, it has frequently proved difficult to communicate in a clear and convincing way that the coastal ecosystem has indeed been affected by changes in the watershed. This is particularly so where, as is usually the case, historical data are lacking or sketchy. The Chesapeake Bay and Mississippi Delta examples provided here (Figs. 2 and 3) illustrate the power of historical analyses, but comparative ecosystem analyses must become increasingly applied to extend understanding from better studied coastal ecosystems to those potentially experiencing emerging problems (Anon. 1994). Also, more attention should be devoted to the process of making the connections necessary to promote the appropriate use of science in policy-making (Anon. 1995).

3. *Some level of consensus among the scientific communities associated with various interests.* The power of scientific consensus is illustrated by the Chesapeake Bay Scientific and Technical Committee report on nutrient controls and the recent San Francisco Bay workshop proposals for managing for optimal location of the 2 psu isohaline. High-level governmental agreements were concluded quickly after the articulation of each consensus. Surely management actions were being considered parallel with these processes, but broad consensus within the scientific community, including those scientists associated with different sectors such as fisheries, water supply, water quality, waste treatment and agriculture make difficult political decisions more acceptable or less risky. In a similar vein, Haas (1990) points out the critical importance of the international scientific community– an epistemic community in the terms of political scientists– in obtaining multinational agreement for the Mediterranean Action Plan. Tempering the reliance on consensus, however, is Walters' (1986) suggestion that adaptive management requires embracing alternatives rather than promoting consensus.

4. *The development of models to guide management actions.* Models are particularly important in making complex relationships understandable, defining management indicators, pointing to effective solutions, and assessing progress. These models may be process models or statistical models and range from descriptive to highly quantitative. They provide a means of articulating scientific information in a way that can be understood and used by managers. Of course, an oversimplified model or one based on false premises can be dangerous in the hands of these very same managers. For that reason, modeling must be a process which actively engages scientists as well as modelers and managers and must be closely coupled with monitoring as part of an adaptive management approach (Walters 1986).

5. *Identification of effective and feasible solutions to the problems.* In the cases studied here, management became engaged in responding to the problems only after effective and feasible solutions were identified (e.g. biological nutrient removal in the Chesapeake Bay, managed river diversions in the Mississippi Delta, and management for isohaline position in upper San Francisco Bay). Coastal environmental scientists are more oriented to uncovering problems than in identifying solutions, particularly when those solutions must be implemented far from the coast. Toward this end, much better communication and integration must take place among the scientific and engineering communities working on coastal ecosystems, watershed processes, agricultural practices, and waste treatment.

Acknowledgements. My understanding of how science is used in environmental policy in general and in the four case studies in particular has benefited greatly from interactions over the years with Jerry Schubel, Thomas Malone, Wayne Bell, Nancy Rabalais, Denise Reed, James Cloern, John Day, Biliana Cicin-Sain, William Eichbaum, Brock Bernstein, Peter Douglas, John Teal, Michael Orbach, John Costlow, Edward Goldberg, Frederick Holland, Clifford Randall, Christopher D'Elia, Sherrie Cooper, Michael Haire, Robert Perciascepe, Donald Scavia, and Bill Matuszeski.

References

Anon. 1994. *Priorities for coastal ecosystem science.* National Research Council. National Academy Press, Washington, DC.

Anon. 1995. *Science, policy, and the coast: Improving decision-making.* National Research Council. National Academy Press, Washington, DC.

Antweiler, R.C., Goolsby, D.A. & Taylor, H.E. 1995. Nitrates in the Mississippi River. In: Meade, R.H. (ed.) *Contaminants in the Mississippi River, 1987-92*, pp. 73-86. U.S. Geological Survey Circular 1133, Denver, CO.

Boesch, D.F. 1995. Coastal ecosystem management: Challenges for science. In: *Improving interactions between coastal science and policy, Proceedings of the Gulf of Maine Symposium.* National Academy Press, Washington, DC.

Boesch, D.F., Armstrong, N.E., D'Elia, C.F., Maynard, N.G., Paerl, H.W. & Williams, S.L. 1993. *Deterioration of the Florida Bay ecosystem: An evaluation of the scientific evidence.* National Fish and Wildlife Foundation, Washington, DC.

Boesch, D.F., Josselyn, M.N., Mehta, A.J., Morris, J.T., Nuttle, W.K., Simenstad, C.A. & Swift, D.J.P. 1994. Scientific assessment of coastal wetland loss, restoration and management in Louisiana. *J. Coastal Res.* Special Issue 20: 1-103.

Boynton, W.R., Garber, J.H., Summers, R. & Kemp, W.M. 1995. Inputs, transformations, and transport of nitrogen and phosphorus in Chesapeake Bay and selected tributaries. *Estuaries* 18: 285-314.

Carlton, J.T. & Geller, J.B. 1993. Ecological roulette: The global transport of nonindigenous marine organisms. *Science* 261: 78-82.

Cooper, S.R. 1995. Chesapeake Bay watershed historical land use: Impact on water quality and diatom communities. *Ecol. Appl.* 5: 703-723.

Cooper, S.R. & Brush, G.S. 1991. Long-term history of Chesapeake Bay anoxia. *Science* 254: 992-996.

Day, J.W., Jr. & Templet, P.H. 1989. Consequences of sea level rise: Implications from the Mississippi Delta. *Coastal Manage.* 17: 241-257.

D'Elia, C.F., Sanders, J.G. & Boynton, W.R. 1986. Nutrient enrichment studies in a coastal plain estuary: Phytoplankton growth in large-scale, continuous cultures. *Can. J. Fish. Aquacul. Sci.* 43: 397-406.

Eadie, B.J., McKee, B.A., Lansing, M.B., Robbins, J.A., Metz, S. & Trefry, J.H. 1994. Records of nutrient-enhanced coastal ocean productivity in sediments from the Louisiana continental shelf. *Estuaries* 17: 754-765.

Haas, P. 1990. *Saving the Mediterranean: The politics of international cooperation.* Columbia University Press, New York, NY.

Hennessey, T.M. 1994. Governance and adaptive management for estuarine ecosystems: The case of Chesapeake Bay. *Coastal Manage.* 22: 119-145.

Howarth, R.W., Billen, G., Swaney, D., Townsend, A., Jaworski, N., Lajtha, K., Downing, J.A., Elmgren, R., Caraco, N., Jordan, T., Berendse, F., Freney, J., Kudeyarov,

V., Murdoch, P. & Zhu, Z.-L. 1996. Regional nitrogen budgets and the riverine N & P fluxes for the drainages to the North Atlantic Ocean: Natural and human influences. *Biogeochemistry.*

Jassby, A.D., Kimmerer, W.J., Monismith, S.G., Armor, C., Cloern, J.E., Powell, T.M., Schubel, J.R. & Vendlinski, T.J. 1995. Isohaline position as a habitat indicator for estuarine populations. *Ecol. Appl.* 5: 272-289.

Jaworski, N.A. 1990. Retrospective of the water quality issues of the upper Potomac estuary. *Aquatic Sc.* 3: 11-40.

Kesel, R.H. 1989. The role of the Mississippi River in wetland loss in southeastern Louisiana, U.S.A. *Envir. Geol. Water Sci.* 13: 183-193.

Kimmerer, W.J. & Schubel, J.R. 1994. Managing freshwater flows into San Francisco Bay using a salinity standard: results of a workshop. In: Dyer, K.R. & Orth, R.J. (eds.) *Changes in fluxes in estuaries: Implications from science to management*, pp. 411-416. Olsen & Olsen, Fredensborg.

Lee, K.N. 1993. Greed, scale mismatch, and learning. *Ecol. Appl.* 3: 560-564.

Malone, T.C., Boynton, W., Horton, T. & Stevenson, C. 1993. Nutrient loading to surface waters: Chesapeake Bay case study. In: Uman, M.F. (ed.) *Keeping pace with science and engineering*, pp. 8-38. National Academy Press, Washington, DC.

McIvor, C.C., Ley, L.A. & Bjork, R.D. 1994. Changes in freshwater inflow from the Everglades to Florida Bay including effects on biotic and biotic processing: A review. In: Davis, S.M. & Ogden, J.C. (eds.) *Everglades: The ecosystem and its restoration*, pp. 117-146. St. Lucie Press, Delray Beach, FL.

Meade, R.H. 1995. Setting: Geology, hydrology, sediments, and engineering of the Mississippi River. In: Meade, R.H. (ed.) *Contaminants in the Mississippi River, 1987-92*, pp. 13-28. U.S. Geological Survey Circular 1133, Denver, CO.

Nichols, F.H., Cloern, J.E., Luoma, S.N. & Peterson, D.H. 1986. The modification of an estuary. *Science* 231: 567-573.

Rabalais, N.N., Turner, R.E., Wiseman, W.J., Jr. & Boesch, D.F. 1991. A brief summary of hypoxia on the northern Gulf of Mexico continental shelf: 1985-1988. In: Tyson, R.V. & Pearson, T.H. (eds.) *Modern and ancient continental shelf anoxia*, pp. 35-47. Geological Society, London Special Publication No. 58.

Rabalais, N.N., Turner, R.E., Justi, D., Dortch, Q., Wiseman, W.J., Jr. & Sen Gupta, B.K. 1996. Nutrient changes in the Mississippi River and system responses on the adjacent continental shelf. *Estuaries* 19: 386-407.

Roblee, M.B., Barber, T.B., Carlson, P.R., Jr., Durako, M. J., Fourqurean, J.W., Muehstein, L.M., Porter, D., Yabro, L.A., Zieman, R.T. & Zieman, J.C. 1991. Mass mortality of the tropical seagrass *Thalassia testudinum* in Florida Bay (USA). *Mar. Ecol. Prog. Ser.* 71: 297-299.

Sabatier, P.A. 1994. Alternative models of the role of science in public policy: Applications to coastal zone management. In: *Improving interactions between coastal science and policy: Problems and opportunities, Proceedings of the California Symposium*, pp. 69-80. National Academy

Press, Washington, DC.

Smith, R.A., Alexander, R.B. & Wolman, M.G. 1987. Water-quality trends in the nation's rivers. *Science* 235: 1605-1615.

Turner, R.E. & Rabalais, N.N. 1991. Changes in Mississippi River water quality this century. *BioScience* 41: 140-147.

Turner, R.E. & Rabalais, N.N. 1994a. Coastal eutrophication near the Mississippi river delta. *Nature* 368: 619-621.

Turner, R.E. & Rabalais, N.N. 1994b. Changes in the Mississippi River nutrient supply and offshore silicate-based phytoplankton community responses. In: Dyer, K.R. & Orth, R.J. (eds.) *Changes in fluxes in estuaries: Implications from science to management*, pp. 147-150. Olsen & Olsen, Fredensborg.

Walters, C. 1986. *Adaptive ecosystem management of renewable natural resources*. MacMillan, New York, NY.

Received 15 September 1995;
Revision received 3 July 1996;
Accepted 23 August 1996.

 Pergamon

0025–326X(95)00102–6

Marine Pollution Bulletin, Vol. 31, Nos 4–12, pp. 152–158, 1995
Copyright © 1995 Elsevier Science Ltd
Printed in Great Britain. All rights reserved
0025–326X/95 $9.50 + 0.00

Emerging Problems in the Coastal Zone for the Twenty-First Century

EDWARD D. GOLDBERG
Scripps Institution of Oceanography, La Jolla, CA 92093-0220, USA

The continued availability of some marine resources is threatened by the increased fluxes to the oceans of identifiable and measurable collections of pollutants, which include plant nutrients, plastics, environmental oestrogens, and organisms contained in ship-ballast waters. Characteristic of these societal discards that will guide research progress are long residence times; slow accumulation rates; increasing fluxes with time; and dissemination over large areas. The resolution of these problems will require data collections over decadal time-scales. Finally, some classical and some perceived marine pollution problems, such as those involving specific metals, can now be discontinued in the face of the absence of unacceptable impacts on living organisms.

Maintaining or improving the health of the coastal zone brings about an increased quality of life and improved economy for a nation's citizenry. For about the past half century scientists and engineers have recognized that the activities of human societies can significantly alter the nature of the marine environment, sometimes negatively affecting public health, the well-being of marine organisms and such resources as aesthetics and transportation. The first serious challenge to the integrity of the oceans came about in the early 1950s through the potential discharge of artificial radio-activities from nuclear power plants. Scientists and engineers in many countries recognized the possible impact upon public health through the ingestion of marine foods or through exposure to contaminated waters. They formulated protocols to protect the potentially most exposed individuals through the regulation of releases of the produced radionuclides to the environment. Their actions were most successful.

Subsequently, the morbidities and mortalities of marine organisms through uptake of chlorinated hydrocarbon pesticides directed efforts to protect the integrity of ecosystems through regulation in the uses of these chemicals. Both catastrophic events and scientific intuition revealed unacceptable activities of societies. Monitoring programmes for pollutants were initiated to measure ocean quality for life in the sea. The current status was well documented in *The State of the Marine Environment* (GESAMP, 1990):

'While no areas of the ocean and none of its principal resources appear to be irrevocably damaged, while there are encouraging signs that in some areas marine contamination is decreasing, we are concerned that too little is being done to correct or anticipate situations that call for action, that not enough consideration is being given to the consequences for the oceans of coastal development and that activities on land continue with little regard to their effects in coastal waters.'

With these words as guidance, I will now put forth what I consider to be some of the important marine pollution problems facing us as we enter the twenty-first century.

Identifying Potential Pollution Problems

The primary factor driving coastal ocean pollution is the increase in populations that will be more affluent and hence use more energy and material resources. With improper management their wastes can enter the oceans in unacceptable amounts. Certain characteristics of the discards will guide scientists in their investigations: *1.* long residence times or persistence in the marine environment; *2.* slow accumulation; *3.* increasing flux with time; and *4.* dissemination over wide areas. The following polluting substances will be considered: plastics, plant nutrients, environmental oestrogens, pathogens, ballast waters and algal toxins.

Eutrophication

Perhaps the most studied marine pollution problem involves the consequences of the over-fertilization of coastal surface waters. The entry of plant nutrients such as phosphates, nitrates and silicates leads to excessive biomass production in the waters and in the sediments. This is followed by a transfer of organic materials to the deeper waters where they can be oxidized by dissolved oxygen gas. Hypoxia and anoxia can then develop. Declines in fishery and shellfishery yields, exotic and toxic algal blooms, alterations of community structure in coastal ecosystems and decreases in water quality are also attributed to the nutrient enrichments.

Understanding the course of eutrophication in a given water body requires a long-term commitment to a monitoring programme. The responses of the coastal ocean to over-fertilization can well be seen in changes in the dissolved gaseous oxygen in the water column. Oxygen measurements by the Winkler method have

Volume 31 'Numbers 4–12 'April–December 1995

been essentially unchanged since the beginning of the century and, on this basis, they are claimed to be reliable. There can be over-saturation of oxygen in the surface waters from its production in the photosynthetic process and depletion in deeper waters from its involvement in respiration. Both effects were seen in the Northern Adriatic for the period 1911–1984 by Justic *et al.* (1987). The increase in oxygen content in surface waters was 0.013 cm^3 O$_2$ dm^{-3} yr^{-1}, while there was a corresponding decrease in the bottom waters of 0.015 cm^3 O$_2$ dm^{-3} yr^{-1} (Fig. 1). There were no significant changes in the overall oxygen content of the waters. The more evident changes between surface and bottom water oxygen levels occurred during the summer, when vertical and horizontal water exchange was small. A similar situation was observed in Swedish coastal waters by Rosenberg (1990). The annual minimum oxygen concentrations were measured at 14 stations (Fig. 2). Twelve of the stations showed drops in oxygen concentration over the last few decades. The trend is also recorded in the sediments where the anoxia, measured by hydrogen sulphide, and the initiation of varving, first appeared in the period

1956–1960s. Also, significant decreases in the benthic biomass were recognized. Such programmes establish that the time-frames of eutrophication involve periods of decades for significant and measureable chemical changes in the water column. Also, the geographical extent of eutrophication can be extensive, as has been the case in the Northern Adriatic, where pollutants, primarily from the Po River drainage area, have affected the coastal regions of Italy, Slovenia and Croatia.

Systematic programmes to assess the general problem of eutrophication have yet to appear. Such an activity will require measurements to be made over large areas and over a time-scale of decades. In order that relevant and essential data can be gathered in economically rational ways, novel methodologies will be needed, perhaps involving the use of satellites and buoys. The development of instruments that can continuously monitor the relevant chemical and physical parameters is crucial.

Because of the uncertainties in the factors controlling eutrophication in a given coastal zone, a large set of measurements has been proposed by a Committee of

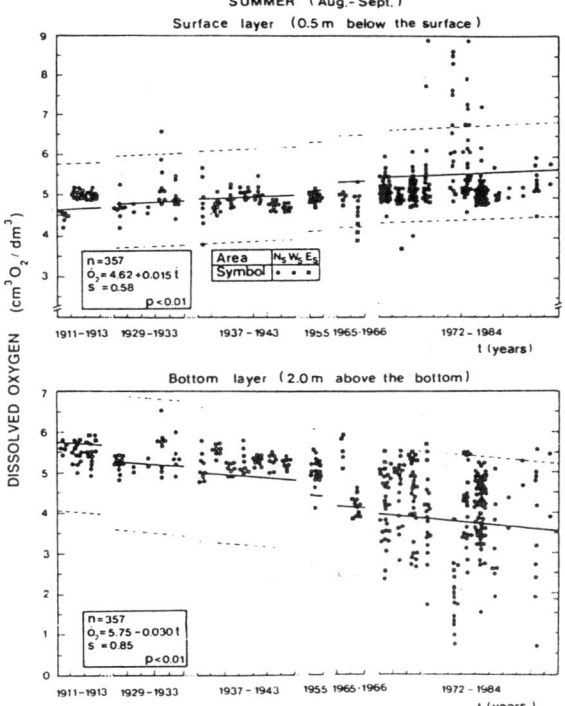

Fig. 1 Oxygen contents in the northern Adriatic Sea during August and September 1911 and 1984. From Justic *et al.* (1987).

Marine Pollution Bulletin

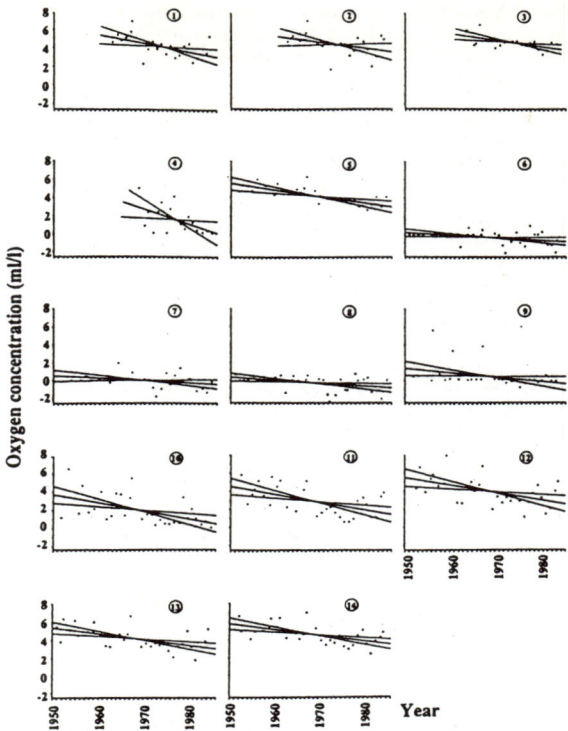

Fig. 2 Temporal trends in annual minimum bottom oxygen in Swedish coastal waters. The regression line is shown together with the 95% significance levels for the slope of that line. From Rosenberg (1990).

the Marine Board of the US National Research Council.

• Chemical measurements: nutrients, dissolved oxygen, chlorophyll, vertical particle flux.
• Biological measurements: primary productivity, phytoplankton species composition, bacterial mass and production, benthic respiration, water column respiration.
• Physical measurements: current speeds and tidal direction, density fields light field.

Eutrophication programmes will thus be expensive to carry out. But for most countries only a few areas need to be involved—those with very high inputs of plant nutrients. For example, the National Research Council Committee recommended that only three sites in the US should be considered as high priority regions: New York Bight/Long Island Sound, Chesapeake Bay and the Northern Gulf of Mexico which receives the Mississippi–Atchafalaya outflow. Since these areas all have limited studies of the eutrophication, the

implementation of a larger plan could take place with a firm foundation.

Exotic Algal Blooms

Related to the chemical and physical factors that cause eutrophication, perhaps, is the occurrence of exotic algal blooms, often called red tides. These massive growths of phytoplankton, often dinoflagellates, may contain highly toxic chemicals that can cause illness, and even death, to marine organisms and humans. Outbreaks of these organisms can result in red discolorations of the waters—although some can remain colourless and others may be green, yellow or other hues. Some scientists argue that their frequency and geographic extent are increasing, possibly reflecting greater inputs of polluting chemicals with time. Along with the direct ingestion of pathogens, the algal diseases are important causes of human mortality from exposure to the components of the marine environment. Recent

Volume 31/Numbers 4–12/April–December 1995

events of algal blooms are chronicled in *Harmful Algae News* published periodically by the Intergovernmental Oceanographic Commission of UNESCO.

There have been four types of human illnesses associated with algal toxins so far identified: PSP (paralytic shellfish poisoning) causes numbness and can result in human deaths, DSP (diarrhetic shellfish poisoning) brings about diarrhoea and nausea and NSP (neurotoxic shellfish poisoning) causes diarrhoea, vomiting, abdominal pain, among other symptoms. A novel illness, ASP (amnesic shellfish poisoning), was first identified in Canada in 1987 (Anderson, 1994) and resulted in memory loss.

The toxins can enter the human food chain through the ingestion of phytoplankton by filter-feeding organisms, followed by their consumption by humans. For example, Canadians who had eaten mussels from Prince Edward Island in 1987 suffered three deaths in 105 cases of ASP.

Poisoning episodes occur in the wild and in mariculture pens where the organisms cannot escape the toxic organisms. A major shellfish poisoning (NSP) occurred in North Island, New Zealand (Chang, 1994), for example. Over 180 citizens who consumed shellfish suffered from numbness of lips, tingling fingers, reversed sensation to hot and cold, muscular weakness, nausea, diarrhoea and vomiting. The organism responsible was a dinoflagellate, a *Gymondinium breve*-type species. Curiously, this toxic algal event coincided with a spell of unusual climatic conditions, attributed to the El Niño.

Red tides of *Chattonella antiqua* have caused massive killing of farmed fish, mostly yellowtail, in the Seto Inland Sea of Japan (IOC, 1987). A similar event occurred in Antifer, France (near Le Havre), where the entire stock of a fish farm perished after a red tide, dominated by *Exuvialelola* sp. producing a PSP toxin (Jenkenson, 1987). Clearly, the increasing intensities and frequencies of red tides can bring about untold economic losses to mariculturers. Perhaps one necessary activity in the future will be the formation of monitoring systems to assist in the prediction of algal blooms and the development of preventive measures for farmers.

Economics will drive the scientific community to a better understanding of the red tide phenomenon. Because undesirable effects from red tides can involve two or more nations, multilateral activities are clearly in the offing.

Plastics

A visitor to a well-used beach can immediately sense the plastics problem. A layer of debris, mostly plastic, will be seen. Bottles, sheets, dining ware and cigarette filters will be all too evident. The eventual fate of these materials generally involves burial in the adjacent sediments. Although freshly introduced plastics are buoyant, they quickly accumulate organic coatings which sorb shells, sand and other debris and sink to the bottom with the increased density. They are virtually indestructible unless combusted. Besides the coastal deposits the only other sinks involve land burial and incineration.

Plastic materials can provide habitat to opportunistic organisms. But, probably more serious, is their inhibition of gas exchange between the overlying waters and the pore waters of the sediments. Anoxia and hypoxia can come about near the water–sediment interface. Such effects may seriously interfere with the normal functioning of ecosystems and may alter the make-up of life on the sea floor (Goldberg, 1994).

Clearly, monitoring programmes to ascertain whether or not the coverage of the coastal sea floor is increasing to the extent that life processes are threatened should be initiated. So far there are few studies in sea floor debris. Galgani *et al.* (1994a,b) have surveyed the north-western Mediterranean Sea, the Bay of Biscay and Seine Bay. In the latter case, specimens were obtained using a 55 mm mesh bottom trawl with a mean opening of 14 m. The tows were conducted for 120–150 m and covered an area of 0.155 km^2. Plastics were the most abundant type of debris, with an average of 80% in Seine Bay and up to 95% for some stations in the Bay of Biscay. In the north-western Mediterranean, plastics constituted most of the debris, at an average of about 77%. Considerable geographic variations were noted, with peak plastic abundances found in deposits off large cities. Bags constituted more than 90% of the total plastic debris.

The framers of the International Convention for the Prevention of Pollution from Ships (MARPOL) through its Annex V were most far-sighted in proposing a total prohibition of any discharge of plastic material to seawaters. Now it is up to sovereign nations to ascertain the plastic coverage of the sediments of their heavily used beaches and its changes with time. Such data will allow the benthic scientists to ascertain whether or not any disruptions to the ecosystems have taken place or are imminent. Monitoring programmes might utilize photographic studies, divers, trawls or submersibles. Trawl surveys appear to be the least expensive and perhaps can provide the statistically most satisfying results. The results of monitoring can be used as a springboard for ecological studies. It is worth noting that many plastic bottles and containers have embossed on their surfaces the country of manufacture, date of manufacture and the contained materials. Such data are most useful in reconstructing the plastic invasion of a given area. In many areas the plastics are transported many thousands of kilometres from their site of entry.

Environmental Oestrogens

One of the most contentious issues in environmental toxicology involves the widespread occurrence of endocrine-disrupting chemicals (Stone, 1994). Much of the attention has been directed to environmental oestrogens and especially the risks to humans through exposure. Many of the compounds act as anti-oestrogens by interfering with the activity of the oestrogen receptors or by reducing the number of receptors in the organisms.

Marine scientists have been involved in the actions of some of these chemicals since the early 1960s, with the impact of the DDT family of chemicals on the reproductive success of fish-eating birds. Later, tri-

Marine Pollution Bulletin

butyltin, the anti-fouling agent employed in marine paints, was shown to cause sexual changes in gastropods. Females, upon exposures to levels in the parts per trillion range, developed male characteristics, such as penises, the so-called imposex problem, and populations of gastropods were devastated.

The number of potential endocrine-disrupting chemicals have been tabulated by Colborn *et al.* (1993) and are displayed in Table 1. Most are very weak, but the disturbing concern is whether or not they can act in concert. Colborn *et al.* (1993) emphasize the lack of knowledge as to which of the chemicals present in the environment might be responsible for endocrine-disrupting events. They also argue that the cumulative effects of these chemicals might not be recognized until young adulthood, when dysfunction in the reproductive systems might become apparent.

Lake Apopka, studied by Colborn and her colleagues, provides one of the most important recent studies on the effect of environmental oestrogens upon wildlife. The high levels of DDT introduced by an accident spill resulted in a 90% fall-off in the birthrate of alligators and reduced penis size in many of the young alligators (Stone, 1994).

The potential effect of environmental oestrogens upon human health stirs up intense debate. The sceptics argue that normal levels of oestradiol bind to oestrogen receptors thousands of times more strongly than environmental oestrogens. On the other hand, extremely low levels of dioxin in the human body from the environment can interfere with the body's response to oestrogen (Stone, 1994). The impact of environmental oestrogens upon human health awaits the results of further research.

Non-indigenous Organisms in Ballast Waters

An unusual collection of pollutants involves the non-indigenous organisms moved over long distances in the ballast waters and on the outer surfaces of ships. If they take hold in a guest port of the carrying ship, they can displace indigenous species, some of which may have commercial value, reduce species diversity and change the normal functioning of ecosystems. The make-up of the natural population of San Francisco Bay has been so altered by invasions that Hedgepeth (1993) argues that the flora and fauna are impossible to describe before the entry of the Europeans. He points out that there were at least 255 foreign invertebrates there as of 1979.

The ballast water pathway, rather than fouling organisms, appears today to be the more important vector. The reduction in attached organisms through higher vessel speeds and the use of anti-fouling agents has directed major studies to the transported waters.

Carlton & Geller (1993) characterize the problem as 'ecological roulette'. They found that plankton samples from ballast waters on Japanese ships contained 367 taxa, any one of which might be successful in establishing a colony. The waters were subsequently discharged in ports in Oregon, USA. Most taxa that had a planktonic phase in their life-cycle were found in the ballast waters. In addition, all major marine habitats and trophic groups were found. Certain taxa prevail: five phyla accounted for 80% of the taxa: crustaceans (31%), polychaete annelids (18%), turbellarian flatworms (14%), cnidarians (11%) and molluscs (8%).

Ballast can be taken from oceanic, fresh or brackish waters or from combinations of the three. Likewise, discharge can take place in the three zones. The possibility of survival of the transported organisms is probably higher where the chemical and physical characteristics of the receiving waters are similar to those of the site from which they were drawn. However, with the large number of organisms in ballast waters and the few successful invasions, other factors must come into play.

Even more threatening to the quality of coastal waters is the importation of toxic organisms. Of the 80 cargo vessels that entered Australian ports in a year, six were found to contain the cysts of the toxic dinoflagellates *Alexandrium cutinella* and *A. tamarense* (Hallegraff & Bolch, 1991). Blooms of these organisms are now

TABLE 1

Chemicals with widespread distribution in the environment reported to have reproductive and endocrine-disrupting effects. From Colborn *et al.* (1993).

Biocides				Industrial chemicals
Herbicides	Fungicides	Insecticides	Nematocides	Industrial chemicals
2,4-D	Benomyl	β-HCH	Aldicarb	Cadmium
2,4,5-T	Hexachlorobenzene	Carbaryl	DBCP	Dioxin (2,3,7,8-TCDD)
Alachlor	Mancozeb	Chlordane		Lead
Amitrole	Maneb	Dicofol		Mercury
Atrazine	Metiram-complex	Dieldrin		PBBs
Metribuzin	Tributyltin	DDT and metabolites		PCBs
Nitrofen	Zineb	Endosulfan		Pentachlorophenol (PCP)
Trifluralin	Ziram	Heptachlor and H-epoxide		Penta- to nonylphenols
		Lindane (γ-HCH)		Phthalates
		Methomyl		Styrenes
		Methoxychlor		
		Mirex		
		Oxychlordane		
		Parathion		
		Synthetic pyrethroids		
		Toxaphene		
		Transnonachlor		

Volume 31/Numbers 4–12/April–December 1995

occurring in places where they had not previously been found.

Measures to minimize the import of toxic organisms clearly must be implemented by local authorities. These might include *1.* certification that, at the ports visited, the waters were free of toxic organisms; *2.* proof of reballasting at sea; *3.* treatment of ballast waters to remove toxic organisms by heat or by chemicals such as chlorine or ozone; *4.* discharge of ballast waters in safe areas; *5.* the use of mechanisms to keep sediments containing the organisms out of the ballast tanks; and *6.* prohibition of the discharge of ballast waters in the port of entry (Hallegraff & Bolch, 1991).

Carlton & Geller (1993) argue that bays, estuaries and inland waters with deep-water ports are the most threatened ecosystems on our planet as a consequence of invasions of organisms in ballast waters. In fact, past invasions may be unrecognized; some invaders may now be considered a component of native populations.

The zoogeography of many port areas may be totally confounded by the invasions of ballast water-borne creatures. Carlton & Geller (1993) provide us with food for thought in a tabulation of recent invasions mediated by ballast water (see Table 2).

Pathogens

The increasing population of the coastal zone introduces more and more enteric bacteria, viruses and fungi into adjacent waters as waste discharge. The organisms can jeopardize public health through the consumption of seafoods or through exposure. This form of pollution is probably responsible for the greatest number of human morbidities and mortalities as a consequence of involvement with the marine environment. The consumption of poorly cooked or uncooked seafood, primarily filter-feeding organisms, which can accumulate the pathogens, and the exposure of wounds are the primary causes of infection. This is often a consequence of working with seafoods.

Direct body exposure to contaminated waters appears to be unimportant. Godfree *et al.* (1990) describe it in the following way: '[direct exposure] has received considerable research attention during the last decade

TABLE 2
Examples of recent invasions probably mediated by ballast water. From Carlton & Geller (1993).

Higher taxon	Taxon	Species	Native distribution	Introduced to
Dinoflagellata		*Alexandrium catenella*	Japan	Australia
		Alexandrium minutum	Europe?	Australia
		Gymnodinium catenatum	Japan	Australia
Cnidaria	Scyphozoa	*Phyllorhiza punctata*	Indo-Pacific	California
	Hydrozoa	*Cladonema uchidai*	Japan, China	California
Ctenophora		*Mnemiopsis leidyi*	Western Atlantic	Black Sea
Annelida	Oligochaeta	*Teneridrilus mastix*	China	California
	Polychaeta	*Desdemona ornata*	South Africa, Australia	Italy
		Marenzelleria viridis	US Atlantic	Germany
Crustacea	Cladocera	*Bythotrephes cederstroemi*	Europe	Great Lakes
	Mysidacea	*Rhopalophthalmus tattersallae*	Indian Ocean	Kuwait
		Neomysis japonica	Japan	Australia
		Neomysis americana	US Atlantic	Argentina, Uruguay
	Cumacea	*Nippoleucon hinumensis*	Japan	California, Oregon
	Copepoda	*Pseudodioptomus inopinus*	Asia	Columbia River
		Pseudodiaptomus marinus	Japan	California
		Pseudodioptomus forbesi	China	California
		Sinocalanus doerrii	China	California
		Oithona davisae	Asia	California, Chile
		Limnoithona sinensis	China	California
		Centropages abdominalis	Japan	Chile
		Centropages typicus	US Atlantic	Texas
		Acartia omorii	Japan	Chile
	Decapoda: Brachyura	*Hermigrapsus sanguineus*	Asia	New Jersey
		Charybdis helleri	Indo-Pacific, Israel	Columbia (Caribbean)
	Decapoda: Caridea	*Salmoneus gracilipes*	Japan, Micronesia	California
		Hippolyte zostericola	Western Atlantic	Columbia (Atlantic)
		Exopalaemon styliferus	Indonesia, India	Iraq, Kuwait
Mollusca	Gastropoda	*Tritonia plebela*	Europe	Massachusetts
	Bivalvia	*Potamocorbula amurensis*	Asia	California
		Dreissena polymorpha	Eurasia	Great Lakes
		Dreissena sp.	Eurasia	Great Lakes
		Rangia cuneata	Southern US	New York
		Theora fragilis	Asia	California
		Musculista senhousia	Japan, Australia	New Zealand
		Enis americanus	US Atlantic	Germany
Ectoprocta		*Membranipora membranacea*	Europe	New Hampshire, Maine
Pisces		*Gymnocephalus cernuus*	Europe	Great Lakes
		Proterorhinus marmoratus	Black Sea	Great Lakes
		Neogobius melanostomus	Mediterranean	Great Lakes
		Butis koilomatodon	Indo-west Pacific	Nigeria, Cameroon, Panama Canal
		Rhinogobius brunneus	Japan	Arabian Gulf
		Mugiligobius sp.	Taiwan, Philippines	Hawaii
		Sparidentex hasta	Arabian Sea	Australia
		Parablennius thysanius	Philippines, Indian Ocean	Hawaii

Marine Pollution Bulletin

with little concrete progress to facilitate the definition of appropriate standards for the management and control of perceived risks'. They conclude with the argument that there is little risk from direct exposure unless the waters are so polluted with sewage as to be aesthetically revolting.

On the other hand, there are many reports of wound infections by the marine bacterium *Vibrio vulnificus* through fish bites, seafood handling or wounds exposed to seawater (Chuang *et al.*, 1992). Seafood workers do have a significantly elevated antibody response to certain phases of the bacterium (Lefkowitz *et al.*, 1992). Finally, a very unusual incident involved the development of endometritis in a previously healthy woman following underwater copulation in seawaters containing *V. vulnificus* (Tison & Kelly, 1984).

Overview

There are certain commonalities in some of the potential pollution problems outlined. First of all, long time periods of monitoring, perhaps of decades, will be required to firmly establish any impact upon the biosphere. This has been established in the case of eutrophication. Clearly, the effects upon the functioning of benthic communities by the coating of the coastal sea floor with plastics will require evidence for population changes in the benthos that may only be established over similar periods. Secondly, accurate measurement of the pollutant fluxes must be determined, in many cases by techniques yet to be established. This is especially true in the cases of oestrogen imitators and plastics. Such measurements will be necessary to evaluate whether or not regulatory mechanisms are working. Also, only a few of the affected organisms are yet to be identified. In the case of the oestrogen imitators, individual compounds or collectives of similar substances are known to have had unacceptable effects: tributyltin on the reproductive successes of gastropods; DDT and its metabolites on egg-shell thinning of the eggs of fish-eating birds. But are there other species of organisms, either poorly studied or providing less obvious effects, that are suffering dysfunction by the invasion of these alien chemicals?

Then there is the problem of identifying the factors resulting in a pollution event. For sample, what conditions cause exotic algal blooms, especially those that give rise to toxic chemicals? Then there are the haunting questions about the known hormone disruptors acting in concert or antagonistically.

The appropriate study of these lacunae in knowledge will require extensive field and laboratory studies. There will probably be inadequate financial and scientific resources to address all of these informational needs. How can priorities be established? I would submit that serious consideration be given to those pollution problems that endanger human life such as the algal toxins and the pathogens. What strategies are available to reduce mortalities and morbidities from the consumption of seafoods tainted with pathogens? Obviously, the education of those potentially affected on how to cook the fish or shellfish properly is necessary. Improved handling of the products before they reach the consumer is also called for. Many scientists have proposed the continuous monitoring of commercial shellfish for disease-causing pathogens. This may be of limited value since many of the consumed shellfish are not caught commercially.

Finally, are there monitoring programmes being carried out now, and presumably dedicated to pollution problems, that might be abandoned because of irrelevance? I have pointed out that only three of the dozen or so metals that are routinely monitored have actually been involved in well-established pollution episodes: tributyltin, methyl mercury and organically bound copper (Goldberg, 1992). Clearly, until some causality is established between current levels of other metals in a given area and the loss of a resource there is no reason to continue their marine assays.

Anderson, D. M. (1994). Red tides. *Sci. Amer.* August, 62–68.

Carlton, J. T. & Geller, J. B. (1993). Ecological roulette: the global transport of nonindigenous marine organisms. *Science* 261, 78–82.

Chuang, Y.-C., Yuan, C.-Y., Liu, C.-Y., Lan, C.-K. & Huang, A. H.-M. (1992). *Vibrio vulnificus* infection in Taiwan: report of 28 cases and review of clinical manifestations and treatment. *Clin. Infect. Dis.* 15, 271–276.

Colborn, T., Vom Saal, F. S. & Soto, A. M. (1993). Developmental effects of endocrine-disrupting chemicals in wildlife and humans. *Environ. Health Perspect.* 101, 378–384.

GESAMP (1990). The State of the Marine Environment. UNEP Regional Seas Reports and Studies No. 115.

Godfree, A., Jones, F. & Kay, D. (1990). Recreational water quality. The management of environmental health risks associated with sewage discharges. *Mar. Pollut. Bull.* 21, 414–422.

Goldberg, E. D. (1992). Marine metal pollutants: a small set. *Mar. Pollut. Bull.* 25, 1–4.

Goldberg, E. D. (1994). Diamonds and plastics are forever? *Mar. Pollut. Bull.* 28, 466.

Hallegraff, G. M. & Bolch, C. J. (1991). Transport of toxic dinoflagellate cysts via ship ballast water. *Mar. Pollut. Bull.* 22, 27–30.

Hedgepeth, J. W. (1993). Foreign invaders. *Science* 261, 34–35.

IOC (1987). IOC Workshop on International Cooperation in the Study of Red Tides and Ocean Blooms. Workshop Report No. 57.

Jenkenson, I. R. (1987). Red Tides and Toxic Phytoplankton on the North and West Coasts of France. IOC Workshop Report No. 57. p. 19.

Justic, D., Legovic, T. & Rottini-Sandrini, L. (1987). Trends in oxygen content 1911–1984 and the occurrence of benthic mortality in the Northern Adriatic Sea. *Est. Coastal Shelf Sci.* 25, 435–445.

Lefkowitz, A., Fout, G. S., Losonsky, G., Wasserman, S. S., Israel, E. & Morris, J. G. (1992). A serosurvey of pathogens associated with shellfish: prevalence of antibodies to *Vibrio* species and Norwalk Virus in the Chesapeake Bay Region. *Am. J. Epidemiol.* 135, 369–380.

Rosenberg, R. (1990). Negative oxygen trends in Swedish coastal bottom waters. *Mar. Pollut. Bull.* 21, 335–339.

Stone, R. (1994). Environmental oestrogens stir debate. *Science* 265, 308–310.

Tison, D. L. & Kelly, M. T. (1984). *Vibrio vulnificus* endometritis. *J. Clin. Microbiol.* 185–186.

Part II
Human Activities and Coastal Variability

[6]

Coastal Management, Volume 22, pp. 265–284
Printed in the UK. All rights reserved.

Evolution of Coastal Hazards Policies in the United States

RUTHERFORD H. PLATT

University of Massachusetts
Amherst, Massachusetts, USA

The ocean and Great Lakes coasts of the United States are experiencing widespread economic and environmental damage from coastal flooding and erosion. During this century, public response to such coastal hazards has evolved haphazardly in response to particular disasters. Over time, however, the range of response has broadened as research has helped to refine public understanding of physical coastal processes, and specific disasters have been studied before longer term forms of institutional response have been formulated. Earlier reliance on engineered shoreline protection has been supplemented by beach nourishment, flood insurance, building and land use regulations, coastal zone planning, and other approaches. This article interprets the evolution of such public policy innovations in terms of a model that depicts the interaction of spatially differentiated systems of physical, legal, and cultural phenomena in the coastal context.

Keywords coastal, erosion, federal policy, flood insurance

At the foot of this cliff a great ocean beach runs north and south unbroken, mile lengthening into mile. Solitary and elemental, unsullied and remote, visited and possessed by the outer sea, these sands might be the end or the beginning of a world. Age by age, the sea here gives battle to the land; age by age, the earth struggles for her own, calling to her defence her energies and her creations, bidding her plants steal down upon the beach, and holding the frontier sands in a net of grass and roots which the storms wash free. The great rhythms of nature, today so dully disregarded, wounded even, have here their spacious and primeval liberty; cloud and shadow of cloud, wind and tide, tremor of night and day. Journeying birds alight here and fly away again all unseen, schools of great fish move beneath the waves, the surf flings its spray against the sun.

Henry Beston, The Outermost House, 1928

But look! here come more crowds, pacing straight for the water, and seemingly bound for a dive. Strange! Nothing will content them but the extremist limit of the land, . . . They must get just as nigh the water as they possibly can without falling in.

Herman Melville, Moby Dick, 1851

Received 10 August 1993; accepted 1 March 1994.
This article is partly based on research conducted under a grant from the National Science Foundation.
Address correspondence to Rutherford H. Platt, Department of Geology and Geography, Morrill Science Center, University of Massachusetts, Amherst, MA 01003, USA.

The ocean coast, as reflected in these quotations, is an arena of incessant change, tension, competition, and conflict. Values and interests that clash along coasts include biotic habitat, sport and commercial fisheries, recreation, personal dwellings, rental properties, economic development, navigation, tourism, aesthetics, privacy, and solitude. Conflicts abound at all scales between adjoining beach users, neighboring property owners, and political jurisdictions, as well as between public and private interests, and between natural and human phenomena.

A substantial proportion of our nation's population and economic activities are crowded near coasts. Nine of our ten largest metropolitan areas, inhabited by more than one-fourth (71.9 million) of the U.S. population, are on oceans or the Great Lakes. Counties within 50 miles of coasts gained 39.6% in population between 1960 and 1990, and now contain more than half (129 million) of the American population. Those within 50 miles of the Pacific Ocean and the Gulf of Mexico have grown by 78 and 88% in population, respectively since 1960 (U.S. Bureau of the Census, 1991, Table 34). These figures overstate the "coastal population" since they include large numbers of people who have little or no contact with the actual coast. But they reflect the relatively high densities and rates of growth of population potentially competing for scarce coastal resources. For instance, the Anaheim–Santa Ana Metropolitan Statistical Area, encompassing affluent Orange County, California, on the Pacific coast just south of Los Angeles, grew from 1.4 million to 2.4 million between 1970 and 1990, a gain of 71%.

Inhabited coasts have widely lost their natural qualities and have become increasingly artificial and hazardous. Nordstrom et al. (1986, 6) describe the New Jersey shore, the archetype of coastal transformation, as follows:

> As the development of New Jersey's barrier islands proceeded, little consideration was given to the importance of either the protective or aesthetic values of the beach and dune system. Dunes were systematically removed from the shorefront to be replaced with boardwalks, roads, or houses. The essential requirement was the view out to sea, and dunes were in the way. . . .
> Efforts to provide protection to the building led to the installation of engineering structures, such as bulkheads and seawalls, to block the storm waves. Building a seawall and maintaining a beach are not compatible objectives as seen at Cape May and Sea Bright.

Ecological damage has been an inevitable result of coastal development in the past. Between the mid 1950s and the mid 1970s, some 372,000 acres of estuarine saltmarshes were lost nationally out of a total of 5.5 million acres (U.S. Fish and Wildlife Service, 1983). This process has been slowed by wetlands protection laws, but degradation and encroachment continues to impair the functions of wetlands as buffers against erosion and as nurturing habitat for marine biota. Protective dune fields and the shorebird habitat that they provide have often disappeared through development and loss of sand supply, leading to demands for emergency shore protection projects to extend the life of threatened structures (Figure 1).

Hurricanes and winter storms have inflicted billions of dollars of damage on coastal communities on both the east and west coasts. Hurricane Hugo in 1989 alone cost $1.6 billion in federal disaster assistance, with total insured losses of $4.2 billion (including both coastal and inland damage) (U.S. General Accounting Office, 1991). Three years later, Hurricane Andrew narrowly missed major coastal metropolitan areas but nevertheless caused over $20 billion in damage to inland areas of South Florida and Louisiana.

Figure 1. Emergency beach scraping at Beach Haven, NJ, a month after the December 1992 northeaster. Note unfinished luxury home being protected at public expense. Photo by the author.

The possibility of accelerated sea level rise due to global warming would worsen coastal flooding and erosion along developed shorelines (Barth & Titus, 1984).

Coastal Erosion

The coastline of the United States (excluding Alaska and Hawaii) extends approximately 53,677 miles, plus 4500 miles along the American shorelines of the Great Lakes and their connecting waterways (National Oceanic and Atmospheric Administration, 1975). The physical nature of these shorelines is diverse. According to the National Research Council (1990, 45), principal types of natural shorelines include the following:

- Crystalline bedrock (e.g., central and eastern Maine coast)
- Eroding bluffs and cliffs (e.g., Cape Cod; parts of Long Island, NY; Great Lakes)
- Pocket beaches between headlands (e.g., Southern New England; California; Oregon)
- Strandplain beaches (e.g., Myrtle Beach, SC; Holly Beach, LA)
- Barrier beaches (e.g., generally along Atlantic and Gulf of Mexico ocean coasts)
- Coral reef and mangrove (e.g., South Florida)
- Coastal wetlands (e.g., Southern Louisiana; elsewhere generally landward of barrier beaches)

All of these shoreline types except crystalline bedrock and certain types of coral formations are potentially subject to measurable erosion. Erosion may be defined as the landward displacement of the shoreline (mean high water line) in response to a variety of causative factors. In the case of beach coasts, Kaufman and Pilkey (1979, 15) ascribe the process of shoreline change (involving either recession or accretion) to a state of "dynamic equilibrium" among four variables: (1) supply and loss of *beach material,* including sand, shell fragments, coral, silt, and flotsam; (2) *energy* imparted by wind, waves, currents, and tides; (3) *beach profile*—slope and width; and (4) *relative sea level,* including both eustatic sea level and local land subsidence or uplift. It is the interaction of these physical variables, including their possible modification by deliberate or inadvertent human intervention, that determines the position of the water's edge over time. Coastal erosion is not a "hazard" in the absence of human investment at risk. Structural efforts to protect such investments may compound the problem by accelerating shoreline recession or by "borrowing from Peter to pay Paul" in terms of intercepting sand to protect one location while depriving another location of its natural sand supply.

Sand may be lost to the local sand sharing systems in several ways. It may be carried by ocean currents too far seaward into deep water to be available for restoration to the beach. Storm waves may transport sand over or through the dune line to create outwash deposits further landward. (This sand may blow back to the dune and beach system.) Human activities may transport sand out to sea or truck it away for construction and other uses. Or coastal engineering projects, described below, may lock up sand in bluffs or other sources, thus interrupting the flow of sand in littoral currents and cutting off the supply of beach material to portions of the shoreline. All of these may cause beaches to recede landward. If prevented from readjusting to an equilibrium profile, due to armoring, beachscraping (Figure 1), or other forms of intervention, beaches become narrower and steeper and eventually disappear. Sea level rise causes most shorelines to recede in the long term and human efforts to deter this process usually make matters worse (Kaufman & Pilkey, 1979; Leatherman, 1979).

While the average rate of erosion must be determined locally through historical shoreline records or shoreline modeling (National Research Council, 1990), erosion is known to be a widespread but not universal phenomenon along the U.S. coastlines. In 1971, the U.S. Army Corps of Engineers, pursuant to Congressional mandate, published its National Shoreline Study, to date the only nationwide survey of erosion that has been conducted. This report estimated that 20,500 miles were experiencing "significant erosion," of which 2700 miles were subject to "critical erosion" that justified counteractive measures (presumably by the Corps). The report was criticized by the General Accounting Office (1975) for failing to precisely define the terms "significant" and "critical"[1] and for tolerating inconsistent estimation procedures among Corps district offices. Also the National Shoreline Study has been criticized as reflecting the Corps' political agenda to justify federal funding of shoreline protection projects.[2] Many states have prepared more detailed studies of coastal erosion rates under Coastal Zone Management auspices (e.g., Benoit, 1989).

The spatial incidence and magnitude of coastal erosion has been more recently assessed by researchers at the University of Virginia in collaboration with the U.S. Geological Survey and the Federal Emergency Management Agency. Geographers Robert Dolan, Bruce Hayden, and their colleagues over 20 years have created a computer-based Coastal Erosion Information System (CEIS). This system assembles available geographical data on shoreline position and erosion rates for the entire coastline of the United States, including the Great Lakes (Dolan et al., 1983, 287). The CEIS data of course vary widely in resolution and reliability since they are derived from diverse sources using a variety of mapping methodologies. Subject to these limitations, CEIS nevertheless represents a considerable improvement over the methodology of the 1971 National Shoreline Study. It provides a more statistical, albeit inexact, picture of the spatial incidence of erosion. A recent map (Dolan & Peatross, 1992) updating the CEIS for the Middle Atlantic coast (Sandy Hook, NJ to Cape Fear, NC) depicts available erosion data by three-mile segments. In the aggregate, 46% of the Middle Atlantic coast is reported to be eroding at more than a half meter per year, 17% is accreting at more than a half meter per year, and 37% is relatively stable (Table 1).

Table 1
Estimated Erosion and Accretion Rates for the Middle Atlantic Coast

Rates of Erosion and Accretion (m/year)	Percentage of Shoreline
−25.00 to −2.01	17.6
−2.00 to −1.51	4.3
−1.50 to −1.01	6.9
−1.00 to −0.51	17.1
−0.50 to 0.50	37.1
0.51 to 1.00	4.5
1.01 to 1.50	4.5
1.51 to 2.00	2.1
2.01 to 25.00	5.9

Source: Dolan and Peatross (1992).

These broad-scale estimates, however, mask the extreme variability of local erosion rates indicated on the Dolan–Peatross map that Table 1 summarizes. According to Williams et al. (1990, 10):

> While the same dynamic processes cause continuous change on every coast, coasts do not all respond in the same way. Interactions among the different processes and the degree to which a particular process controls change depend upon local factors. They include the coast's proximity to sediment-laden rivers and tectonic activity, the topography and composition of the land, the prevailing wind and weather patterns, and the configuration of the coastline and nearshore geometry.

The economic impacts of coastal erosion are difficult to quantify and are often intermingled with losses also attributable to coastal flooding and wind damage. Billions of dollars of coastal investment are at risk nationally. The North Carolina Coastal Commission, for instance, has identified nearly 5000 buildings in that state at risk of erosion within the next 60 years, of which 777 will be threatened within 10 years (Platt et al., 1992, 94). In Michigan, high lake levels between 1985 and 1987 caused about $222 million in damage to property bordering the Great Lakes, mostly due to shore erosion and bluff recession (Platt et al., 1992, 64). In California, winter storms in 1982–83 were estimated by the National Research Council (1984) to have caused "several hundred million dollars" in erosion and structural damage.

Dwellings elevated on substantial pilings may resist collapse, although the shoreline may recede beneath them, leaving them stranded in the surf zone (Figure 2). Failure of on-site septic systems renders structures uninhabitable even if they are still physically accessible. Erosion also plays havoc with public infrastructure such as roads, sewers, water lines, parking lots, bathhouses, restrooms, and boardwalks (Figure 3).

Land itself is an expensive casualty of erosion (which cannot be insured under the National Flood Insurance Program). In January 1987, a new inlet broke through a barrier spit protecting the fashionable mainland community of Chatham, Massachusetts, at the

Figure 2. House on pilings over the intertidal zone at Folly Island, South Carolina, a year after Hurricane Hugo. Note individual attempts at rip-rap in the background. Photo by the author.

Figure 3. Remains of a public boardwalk and beach snack bar at Bradley Beach, NJ, after the December 1992 northeaster. Dunes have migrated landward of the boardwalk. Photo by the author.

elbow of Cape Cod. The inlet widened rapidly to a mile across. Ocean surf pounded the erodible glacial shoreline of North Chatham, which receded 75 ft in one year, resulting in the loss of several houses (Giese et al., 1989). Further loss of residential real estate occurred in the October 1991 "Halloween Storm" and during the winter of 1992–93, which undermined the town's only beachfront parking lot (Figure 4). Further up the coast, the bluffs at the Cape Cod National Seashore experience relentless erosion. Their recession rate was estimated by Henry David Thoreau in the 1850s at about 6 ft per year (Thoreau, 1951, 149), remarkably close to a recent estimate of 4 ft annually (Benoit, 1989, 10).

Evolution of Public Response

The United States has no comprehensive national policy in response to coastal hazards. Development of such a policy is obstructed by two prevailing dogmas of the American political system:

1. *Privatism*, whereby owners are presumed to be entitled to use their land as they wish, subject only to reasonable constraints to protect the public interest. This was the subject of the recent U.S. Supreme Court decision in *Lucas v. South Carolina Coastal Council* (discussed below).
2. *Localism*, whereby land planning and management is considered a local governmental prerogative under the supervision of the state in which the federal government is not supposed to interfere (Platt, 1991). This contrasts starkly with the prevailing expectation that disaster assistance should be primarily a federal responsibility.

Figure 4. Undermined parking lot near Coast Guard Lighthouse, Chatham, Massachusetts, after "Halloween Storm" of October 1991. Photo by the author.

Additional obstacles to the formulation of a national policy on coastal hazards and erosion in particular include the following (Platt et al., 1992, 18–21):

- Spatial and temporal variation in both the processes and rates of coastal erosion
- The interdependence of coastal erosion and coastal flooding
- Fragmentation of authority over coasts (federal, state, local, private)
- Intermingling of public and private benefits from shoreline protection projects
- The problem of intergovernmental cost sharing of hazard mitigation projects
- Fear of the "taking issue" on the part of public officials
- Conflict between economic and environmental objectives in coastal management
- The problem of downdrift impacts from shoreline stabilization measures
- The circumstances of political representation in Congress for coastal areas

Response to coastal disasters thus is a hodgepodge of private and public efforts, usually in response to a recent disaster or an immediate threat (Table 2). When the political process has demanded federal attention to coastal problems, Congress has lurched from one approach to another according to the conventional wisdom of the moment and the political power of the legislators from the area in question. Such political attempts to manage coastal hazards have often lacked a coherent scientific or economic basis.

Learning from Experience

A closer look at Table 2, however, suggests that response to coastal disasters has not been entirely haphazard. A societal learning process has been at work and continues today. As with other physical and social risks, this learning process has resulted from improved perception and understanding of the hazard and its costs, often triggered by a destructive event or threat. Figure 5 represents a simple model of coastal management that relates three systems of spatial data or "geographies" that interact in the coastal context. These include, respectively, the physical geography of the coastal environment, including land, water, and biotic phenomena (circle 1); the political geography of governmental and private legal authority over coastal resources (circle 2); and the human or cultural geography of land and water use (circle 3) that results from the decisions and policies, or lack thereof, of the managers in circle 2.

The circles are linked by vectors representing impacts or information flows. The lure of the coast (initially driven by perception of amenity but not risk) stimulates private and public coastal managers to construct homes and communities on the coast. These modify the natural coastal environment ("environmental impact") and are in turn affected positively by amenity and negatively by risk of damage. Over time, with experience of coastal disasters and ongoing environmental harm, coastal managers at various levels in the legal hierarchy theoretically respond to better understanding of physical/biotic systems ("environmental and risk perception") and feedback on damage to human communities and activities ("socioeconomic data") by revising the laws, mitigation approaches, and building practices along coasts ("coastal management").

The model depicted in Figure 5 may be applied to the empirical data in Table 2 to better understand the evolution of public coastal policies over time. From the Galveston hurricane in 1900 through the "Ash Wednesday Storm" of March 1962, governmental response to coastal hazards primarily took the form of engineered approaches to control the effects of erosion and wave damage on private and public coastal investment. The model as applied to this period (Figure 6) depicts coastal management as limited to "shore protection" projects constructed in response to actual or imminent damage. Scores

Table 2
Policy Responses to Selected U.S. Coastal Disasters

Disasters	Response(s)
1900 Galveston Hurricane	Galveston Seawall constructed
1938 "Great Northeast Hurricane"	Structural Protection
1954–55 Six hurricanes—N.E.	P.L. 84-71 (Structural Protection)
1962 "Ash Wednesday Storm"	P.L. 87-874 (Structural Protection)
1965 Hurricane Betsy	P.L. 89-339 (charged Dept. of HUD to study flood insurance) Clawson and White Reports—1966 National Flood Insurance Act (NFIA)—1968

1972
Tropical Storm Agnes
	Flood Disaster Protection Act of 1973—P.L. 93-234
	    (strengthened NFIA and added erosion as insurable
	    hazard)

1979–80 Hurricanes Frederic, David	Sheaffer and Roland Report—1981 Coastal Barrier Resources Act of 1982—P.L. 97-348
1985–86 Great Lakes coastal flooding NC, SC coastal erosion	Michigan Home Moving Program Upton-Jones Amendment to NFIA—P.L. 100-242, sec. 544 SC Beachfront Management Comm. Report—1987 SC Beachfront Management Act—1988

1989
Hurricane Hugo
	National Research Council (NRC) Erosion Report—1990
	H.R. 1236 adopted NRC erosion recommendations—
	    1991 (S. 1650 died in committee after U.S. Supreme
	    Court decision in *Lucas v. SC Coastal Council*—1992)

1992 Hurricane Andrew Winter Storms—Northeast	Coastal amendments to NFIP reintroduced (H.R. 62)— 1993

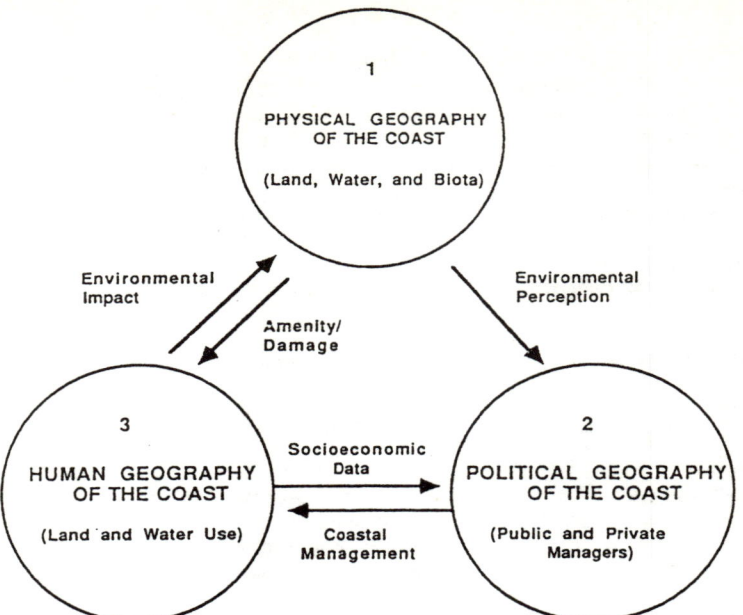

Figure 5. A basic model of coastal management.

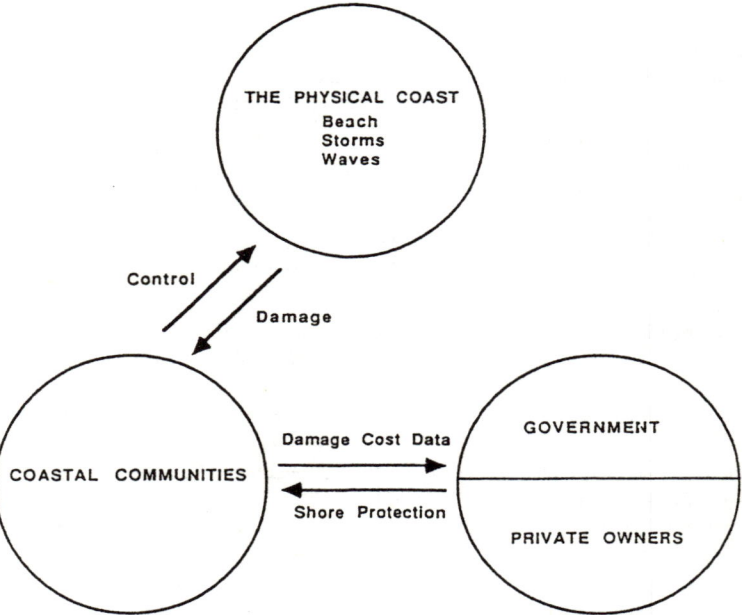

Figure 6. Response to coastal disasters: 1900–1960s.

of projects were authorized by a long series of federal public laws of which a few are listed in Table 2. During this period, the physical coast was viewed simplistically in terms of beaches, waves, and storms with little perception of the broader physical and biotic coastal processes. Shorelines developed before the 1970s (e.g., New Jersey) frequently were "armored" with concrete or sheet steel seawalls, rock revetments, bulkheads, or other structures intended to protect landward private and public facilities from destruction due to erosion (Nordstrom et al., 1986). Such devices often postpone the loss of valuable real estate but generally at the cost of the oceanfront beach (Williams et al., 1990, 16). An alternative approach involved the use of sand-trapping structures such as groins and jetties extending seaward from the shoreline into the surf zone to reduce wave energy and capture sand on their updrift side (thus often depriving downdrift shorelines of their natural sand supply).

With the establishment of the U.S. Army Corps of Engineers Coastal Engineering Research Center (CERC) in 1963 and the National Shoreline Study of 1971, the shore protection activities of the Corps were increasingly informed by longer term research on coastal processes (Figure 7). Beach nourishment, sometimes accompanied by dune restoration was added to the arsenal of Corps-sponsored approaches, and in the 1990s accounts for most of the Corps' shore protection activity. A 10-mile, $65 million Corps-sponsored beach restoration project at Miami Beach in the early 1980s is the most famous example of this approach. So far, Miami Beach has been spared direct hits by hurricanes so the project is untested by violent conditions. Beach restoration projects in more stormy latitudes, such as Ocean City, Maryland, Cape May, New Jersey, and the south shore of Long Island, New York, have experienced greater difficulty in retaining

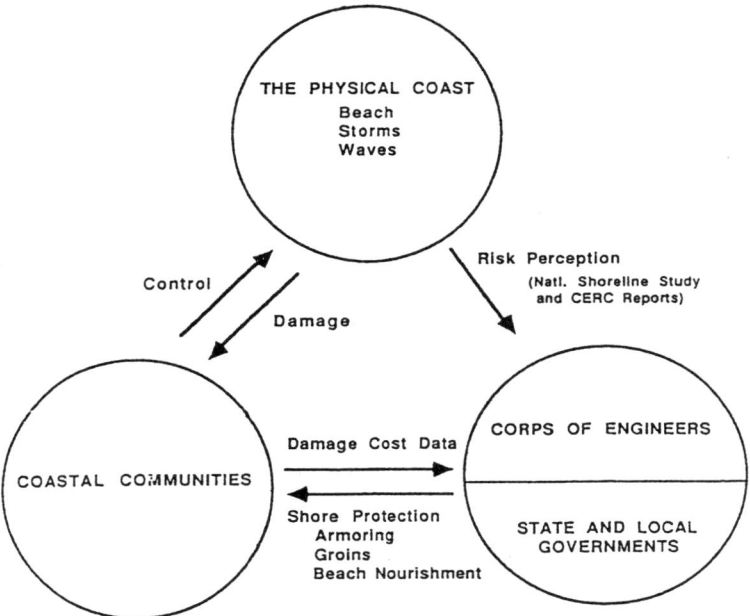

Figure 7. COE model of response to coastal disasters: 1970s.

sand over time, and sometimes require emergency beach scraping (Figure 1) and expensive renourishment.

Before the mid 1960s, Congress tended to respond directly to coastal disasters by enacting remedial legislation, usually authorizing structural shore protection projects. However, two months after Hurricane Betsy inflicted $2 billion in damage along the Gulf of Mexico shoreline, Congress passed the Southeast Hurricane Disaster Relief Act of 1965 (P.L. 89-339), which, among other provisions, called for a study of "alternative programs which could be established to help provide financial assistance to those suffering property losses in flood and other natural disasters, including alternative methods of Federal disaster insurance." The ensuing report by resource economist Marion Clawson (U.S. Congress, 1966a), and a parallel report by a Bureau of the Budget Task Force chaired by geographer Gilbert F. White (U.S. Congress, 1966b), jointly influenced the adoption in 1968 of the National Flood Insurance Act (NFIA) (P.L. 90-448; 42 U.S.C. 4001-4128, as amended). Thus, in contrast to its usual "shoot from the hip" approach, Congress sought to expand its understanding of the problem of flood hazards (enhancing its "environmental perception") before legislating a response to a disaster. The National Flood Insurance Act marked a clear departure from the longstanding reliance on purely structural approaches to coastal and riverine hazard reduction. The National Flood Insurance Program (NFIP) established by the act offers federally backed insurance at affordable rates[3] against riverine and coastal flood losses in communities that meet federal standards for floodplain management.

A key contribution of the NFIP to the perception of flood hazards by public and private land-use managers has been the mapping of areas subject to a 1% chance of flooding in any year (also known misleadingly as the "100-year flood"). This program expanded upon earlier flood hazard mapping efforts by the Army Corps of Engineers, the Soil Conservation Service, and the Tennessee Valley Authority. The NFIP has mapped flood hazard areas (A zones) along rivers and coasts in some 18,000 communities. In coastal communities the program also identifies "coastal high hazard areas" (V zones) along the open coast in which special rules for flood loss reduction apply. Within mapped hazard areas, local communities must adopt and enforce floodplain management measures established by the Federal Emergency Management Agency (FEMA) in order for property owners to be eligible for insurance coverage under the program. The FEMA approach to managing coastal hazards is depicted in Figure 8. (A weakness of the program is its reliance on local communities to enforce their own floodplain management restrictions.)

The NFIP has been amended several times in response to further disaster experience and hazards research (Table 2). Tropical Storm Agnes in 1972 stimulated the adoption of the Flood Disaster Protection Act of 1973 (P.L. 93-234), which required purchase of flood insurance as a condition of all federally related financing of flood-prone structures, thus vastly expanding the program's coverage. In 1993, total NFIP coverage stood at $240 billion, of which about 75% was in coastal communities or counties.

The Flood Disaster Protection Act of 1973 also added erosion as an insurable hazard in response to high levels of erosion losses along the Great Lakes, to the extent described as follows:

> The term "flood" shall also include the collapse or subsidence of land along the shore of a lake or other body of water as a result of erosion or undermining caused by waves or currents of water exceeding anticipated cyclical levels. (42 USCA 4121)

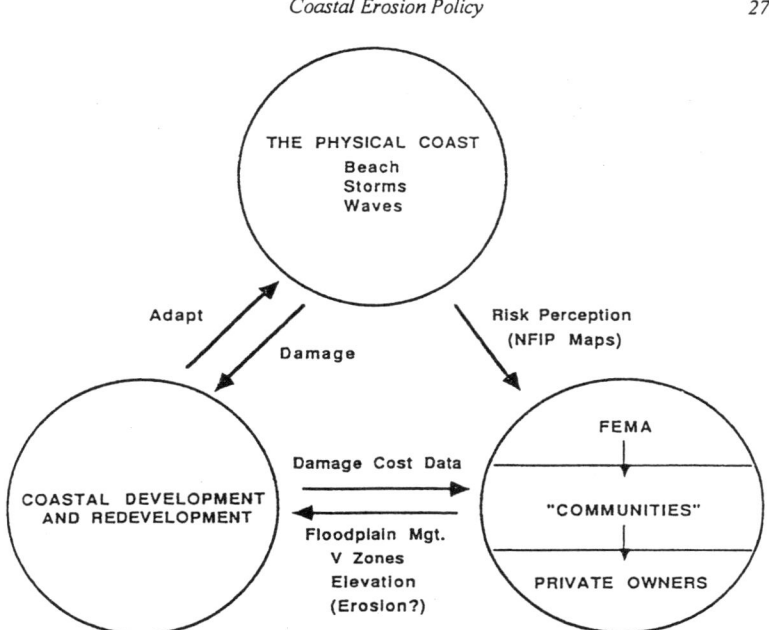

Figure 8. FEMA model of coastal floodplain management.

This definition, obviously written with the Great Lakes in mind, has been difficult to apply in the ocean coastal context. In practice, erosion losses have been combined with flood losses under the NFIP and to date no separate management standards have been adopted regarding erosion hazards, as discussed below.

The Coastal Zone Management Act of 1972 (P.L. 92-583; 16 U.S.C. 1451 et seq., as amended) (CZMA) reflected a broader federal approach to the problems of the coast. As proposed by the Commission on Marine Science, Engineering, and Resources (the "Stratton Commission") (1969), the CZMA established the federal Coastal Zone Management Program (CZMP) under the National Oceanic and Atmospheric Administration (NOAA). The commission report and the 1972 act each primarily addressed ecological and economic issues in coastal management, but were silent on coastal hazards. With the aid of further research (Office of Coastal Zone Management, 1976), the CZMA was amended in 1980 (P.L. 96-464) to include as an additional statutory purpose:

> . . . the management of coastal development to minimize the loss of life and property caused by improper development in flood-prone, storm surge, geological hazard, and erosion-prone areas and in areas likely to be affected by or vulnerable to sea level rise . . . and by the destruction of natural protective features such as beaches, dunes, wetlands, and barrier islands. (CZMA, Sec. 303(2)(B)

In contrast to both the Corps of Engineers and FEMA, the approach of the CZMP is to fund eligible coastal states to develop and administer their own coastal programs, which must address specified national concerns including coastal hazards. A variety of options are encouraged by NOAA, but the federal program contains no performance standards for hazard mitigation. State programs are encouraged to be comprehensive and interactive. Diverse state agencies responsible for economic development, recreation, natural resources, and natural hazards are expected to coordinate their coastal efforts under the umbrella of the state CZM plan. Furthermore, federal activities affecting the coastal zone are required to be "consistent to the maximum extent practicable" with approved state coastal plans [CZMA, Sec. 307(c)(1)(A)]. State studies of coastal hazards prepared under CZMP auspices presumably contribute to the perception of coastal flooding and erosion hazards by local governments and private owners, and may thereby influence their management policies and actions. The NOAA version of the basic model is set forth as Figure 9.

By the early 1980s, some researchers were questioning the wisdom of providing flood insurance and other forms of federal assistance for new development on hazardous coastal barriers. H. Crane Miller (1981) and Sheaffer and Roland, Inc. (1981), based on their analysis of barrier island damage in Hurricanes Frederic in 1979 and David in 1980, argued that it would be cheaper to buy coastal barrier islands outright than to subsidize their development and cover the resulting losses through federal disaster assistance. Miller's testimony to Congress directly contributed to the adoption of the Coastal Barrier Resources Act of 1982 (P.L. 97-348). The act banned flood insurance, shore protection, water projects, highway and bridge subsidies, and other federal incentives to development within certain undeveloped, nonprotected coastal barriers along the Atlantic and Gulf coasts (Platt et al., 1987). Congress thus "fine-tuned" several existing programs in response to research findings.

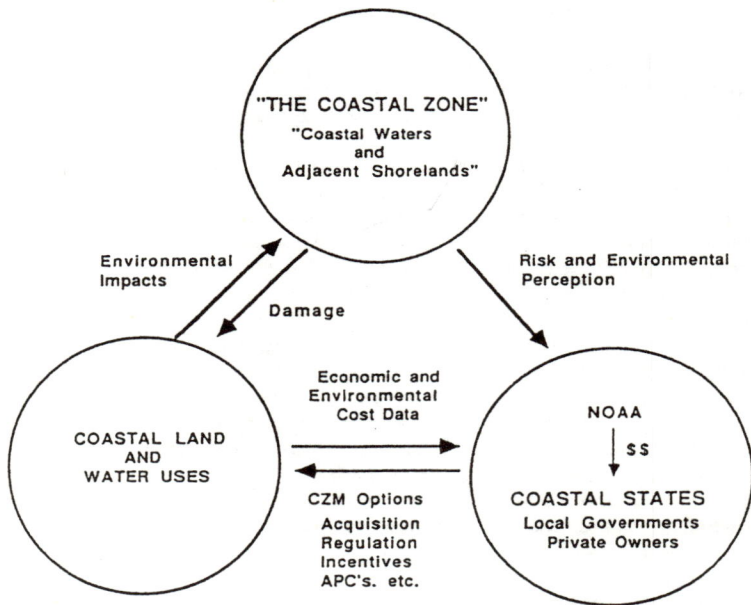

Figure 9. NOAA model of coastal zone management.

Broader public perception of coastal hazards was facilitated during the 1980s by several public education efforts undertaken by coastal researchers. The Duke University Press *Living with the Shore* series edited by Orrin H. Pilkey, Jr., with many collaborators (e.g., Nordstrom et al., 1986) has provided detailed summaries of coastal hazard issues for a number of specific state and regional coasts. An ad hoc group, Concerned Coastal Geologists (1981, 1985), twice convened by Pilkey, urged abandonment of in situ shore protection efforts and advocated a strategy of "managed retreat" landward away from eroding shores. The *Barrier Island Handbook* by Stephen P. Leatherman (1979) and various state handbooks prepared under the CZMP also have contributed to public awareness of coastal phenomena and hazards.

In 1987, Congress incorporated a limited version of Pilkey's "retreat" strategy into the NFIP. In response to erosion along the Great Lakes and the Southeast Atlantic coast (and the influence of legislators from Michigan and North Carolina, respectively, who invoked Pilkey's concept), the "Upton–Jones Amendment" (P.L. 100-242, Sec. 544) authorized the NFIP to fund the relocation or demolition of structures imminently endangered by erosion. The Upton–Jones approach has been little used, however, with only a few hundred claims nationally since it was adopted.

Several states have adopted their own coastal and erosion management laws (National Research Council, 1990; Platt et al., 1992). State erosion laws typically involve a minimum setback for new or rebuilt construction based on the local average annual erosion rate (AAER), as determined from historic shoreline data. The setback is measured landward from a specified reference feature, e.g., the vegetation line, dune crest, or mean high water line. State erosion management laws typically require setback of new or rebuilt structures landward of a line representing a certain number of years of erosion. The North Carolina Coastal Area Management Act of 1974, for instance, requires smaller structures to be set back 30 times the local AAER and larger structures 60 times, theoretically affording 30 and 60 years of erosion protection, respectively.

The Lucas Case

The evolution of public policy is not a linear process, however. Countervailing circumstances may override perception of hazards in persuading public decision-makers to reverse course and reject or rescind management initiatives. One type of countervailing circumstance is an adverse court decision resulting from challenge of hazard and erosion management regulations by stakeholders who are adversely affected. Such a case involving a challenge to the South Carolina Beachfront Management Act of 1988 was recently decided by the U.S. Supreme Court (*Lucas v. South Carolina Coastal Council* 112 S. Ct. 2886, 1992). The state law was adopted pursuant to a 1987 report of the South Carolina Blue Ribbon Committee on Beachfront Management, which informed the state legislature on problems of the coast. Under the act, no new construction was permitted seaward of baselines to be mapped along the state's coast. (Large structures were required to be set back an additional distance from the baseline equivalent to 40 times the AAER.) Lucas, the owner of two lots located entirely seaward of the baseline, challenged the law as a "taking" of the value of his property without compensation in violation of the 5th Amendment of the U.S. Constitution, which states, "nor shall private property be taken for public use without just compensation." Lucas claimed that denial of a permit to build on either of his oceanfront lots was a "taking" of the entire value of his property for "public use" without compensation in violation of the 5th Amendment.

The trial court agreed and awarded Lucas $1.2 million in damages. On appeal, the

South Carolina Supreme Court (404 S.E.2d 895, 1991) overturned that award and held the permit denial to be a reasonable exercise of the public power to regulate the use of hazardous land. Upon appeal by Lucas, the U.S. Supreme Court on 29 June 1992 overruled the state decision and required compensation for a "total taking" of the value of the land unless the state's erosion restriction would be justified under "background state common law." On remand, the state supreme court (424 S.E.2d 484, 1992) found that no such justification existed and held that Lucas was entitled to compensation for the period during which he could not obtain a building permit (about four years).

The Supreme Court in *Lucas* did not hold the state Beachfront Management Act to be unconstitutional per se and it therefore will remain in effect. Nevertheless, the decision was widely viewed as a victory for property owners who challenge government land use restrictions of many types. It will "hover like a black cloud over the environmental landscape" (Platt, 1992). Indeed, its political implications may outweigh its immediate influence in other court cases.

Lucas indirectly contributed to the defeat of proposed amendments to the NFIP in 1992 that would have required coastal communities to adopt erosion management, including minimum setbacks for new or rebuilt structures, similar to the North Carolina approach. The legislation was based on a report of the National Research Council Committee on Coastal Erosion Management (of which the present writer was a member). At the request of its agency sponsor, the Federal Emergency Management Agency, the committee urged limitation of new construction and availability of flood insurance within areas subject to erosion within the next 30 years. The House of Representatives in 1991 voted 388-18 in favor of a bill (H.R. 1236) that embodied most of the committee's recommendations. But the Senate version (S. 1650) was killed in committee in late 1992 due to intensive lobbying by the Fire Island Association and other coastal property owner associations newly empowered at least psychologically by the *Lucas* decision.[4]

The effect of *Lucas* on the basic model of coastal management is to affect the balance of power between public and private sectors within circle 2 of Figure 5. While constitutional challenges have always been available, public authorities have historically enjoyed a presumption of validity of public regulations. The burden of proof usually lies with the challenger to demonstrate that the regulation is unreasonable, arbitrary, or discriminatory (Platt, 1991, 195–199). At least where a "total taking" of property value is claimed, this burden of proof may be reversed with public regulators having to show that a restriction is consistent with "background common law" in the state in question. Although it may be read very narrowly, *Lucas* instilled new vigor into the "taking issue" as a legal and political limitation on the actions of governmental entities on the coast and elsewhere. The threat of being sued is a major constraint on hard-pressed municipalities and government agencies, even if they would ultimately win.

On the other hand, perception of the range of purposes and means of coastal management has broadened since the 1970s through improved recognition of coastal natural resource values, particularly of coastal wetlands. Beginning with the seminal *Life and Death of the Salt Marsh* (Teal & Teal, 1969), many publications and symposia have promoted public awareness of coastal wetlands as highly productive and diverse ecosystems, rather than merely as swamps and wastelands. While wetlands and natural hazards have seldom been viewed interactively, the recent report of the Federal Interagency Floodplain Management Task Force (1992) urges that the two objectives should be more closely linked (Kusler & Larson, 1993).

These recent inputs to the coastal management model can be assimilated into a new general model (Figure 10) that builds upon and incorporates elements of those versions

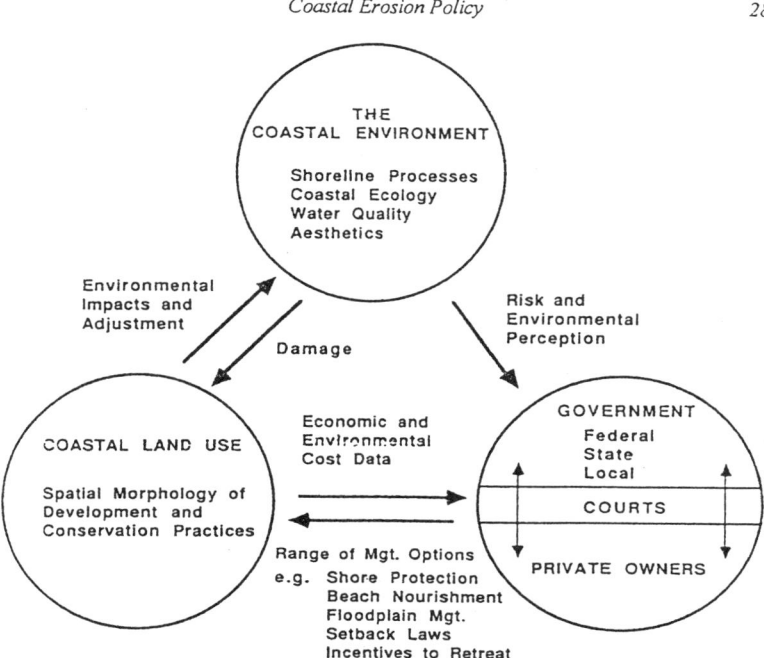

Figure 10. General model of response to coastal hazards.

discussed earlier. In place of the simplistic geographical and functional perspectives of earlier phases, this model envisions a broad ecosystem perspective, multiple objectives, and multiple means. It also reflects the increasingly contentious legal and political climate of the post-*Lucas* era and the conservatism toward private property rights that it reflects.

Conclusion

Two generalizations may be derived from Table 2 and the ensuing discussion. First, national policies to coastal erosion have evolved from reliance on in situ protection, especially involving engineering structures, toward a more flexible range of responses, including retreat from eroding coasts through setbacks and house relocation. While retreat is not yet reflected in the National Flood Insurance Program (other than the limited Upton–Jones Amendment), states such as North Carolina, Michigan, Rhode Island, and Florida have adopted setbacks through state coastal laws.

Second, coastal research has begun to play an increasingly important role in shaping legislative response by strengthening the environmental and socioeconomic perceptions of decision makers concerning the effects of disasters. The adoption of the Coastal Barrier Resources Act of 1982 in response to the findings by H. Crane Miller exemplify this linkage between applied research and policy revision. On the other hand, the impasse confronting efforts to address coastal erosion through the National Flood Insurance Program, especially in the aftermath of the *Lucas* decision, suggest that the pendulum swings both ways. Even clear and convincing evidence of the risks of oceanfront development, as was provided in the winter storms of 1992–93, has proved insufficient to rebut the assertions by property owner organizations that nature can be controlled through federally funded shore protection.

Until the 1970s, the dominance of in situ shore protection was based on several factors: (1) high property values of coastal real estate; (2) cultural preferences to build as close to the water as possible despite risks from floods, hurricanes, and erosion; (3) a prevalent faith in conquering nature with technology rather than adjusting to it; and (4) generous assistance from the federal government to construct shoreline protection projects. The influence of these factors, however, has been somewhat lessened by several countervailing considerations since the mid 1970s: (1) decreasing federal funding for shore protection and higher levels of nonfederal participation; (2) public interest in coastal and aquatic habitats; (3) judicial support for public regulatory programs despite reduction of private property values (albeit possibly undermined by the *Lucas* decision); and (4) predictions of higher rates of future sea level rise due to global warming (Barth & Titus, 1984).

Overall, national policy on coastal hazards in the United States has evolved during the 20th century from reliance on structural shoreline protection to a broad array of structural and nonstructural approaches. There remains much conflict, however, between federal programs to mitigate coastal hazards and other policies that stimulate new coastal construction, e.g., federal tax provisions granting various subsidies to investors in hazardous property, federal disaster assistance, federal infrastructure grants, and the availability of flood insurance for new construction up to the mean high water line. The Coastal Barrier Resources Act of 1982 was an unusual attempt to limit the mutual incompatibility of certain federal programs based on research findings. But the act applies only to a narrow subset of undeveloped coastal barriers. Comparable reexamination of federal policies toward hazardous coasts more broadly is urgently required.

Notes

1. The term "critical" as used in the National Shoreline Study was actually based on the level of coastal investment at risk rather than the rate of shoreline recession. Thus a segment of coast could theoretically shift from "significant" to "critical" status due to new coastal development rather than a change in natural processes.

2. This comment was made by an anonymous reviewer of this paper.

3. Originally, the NFIP offered a basic level of insurance for existing structures at a flat subsidized rate, while new structures (started after a community joined the NFIP) were eligible for higher levels of coverage but only at "actuarial rates" based on the level of risk for the structure in question. Now, 25 years since the program began, most NFIP coverage is "actuarial" rather than "subsidized." The Federal Insurance Administration (a component of the Federal Emergency Management Agency) has raised actuarial rates significantly and the NFIP has been self-supporting out of premium revenue since 1990. Nevertheless, its policy of insuring structures along eroding coasts even after they incur major damage (as at Fire Island, New York, after the 1992–93 winter storms) implicitly represents a subsidy to owners of expensive oceanfront homes unless the rates charged are truly actuarial (e.g., more than 10% of the insured value in the 10-year erosion zone). This issue is involved in efforts to amend the NFIP as discussed below.

4. New versions of the legislation were introduced in both the House (H.R. 62) and Senate (S. 1405) in 1993. As of February 1994, the House version has been reported out of committee that would impose a surcharge on NFIP policies in areas subject to erosion and a prohibition on new policies within 30-year erosion zones. As of the time of writing (February 1994) the prospects for passage of erosion amendments to the NFIP remain unclear (Kerry Kehoe, Coastal States Organization, personal communication, 17 February 1994).

References

Barth, M. C., and J. G. Titus, eds. 1984. *Greenhouse effect and sea level rise: A challenge for this generation.* New York: Van Nostrand Reinhold.

Benoit, J. R., ed. 1989. *Massachusetts Shoreline Change Project.* Boston: Coastal Zone Management Office.

Beston, H. 1928/1992. *The outermost house.* New York: Henry Holt Owl Books.

Commission on Marine Science, Engineering, and Resources. 1969. *Our nation and the sea.* Washington, DC: U.S. Government Printing Office.

Concerned Coastal Geologists. 1981. *Saving the American beach.* Savannah, GA: Skidaway Institute of Oceanography (mimeo).

Concerned Coastal Geologists. 1985. *National strategy for beach preservation.* Savannah, GA: Skidaway Institute of Oceanography (mimeo).

Dolan R., and J. Peatross. 1992. *Shoreline erosion and accretion of the Middle Atlantic Coast.* U.S. Geological Survey Open File Report 92-377. Reston, VA: U.S. Geological Survey.

Dolan, R., B. Hayden, and S. May. 1983. Erosion of the U.S. shorelines. In *CRC handbook of coastal processes and erosion,* ed. P. D. Komar, 285–290. Boca Raton, FL: CRC Press.

Federal Interagency Floodplain Management Task Force. 1992. *Floodplain management in the United States: An assessment report.* Washington, DC: Federal Emergency Management Agency.

Giese, G. S., D. G. Aubrey, and J. T. Liu, 1989. Development characteristics and effects of the New Chatham Harbor Inlet. Technical Report C.R.C. 89-4. Woods Hole, MA: Woods Hole Oceanographic Institute.

Kaufman, W., and O. H. Pilkey, Jr. 1979. *The beaches are moving.* Durham, NC: Duke University Press.

Kusler, J. A., and L. Larson. 1993. Beyond the ark: A new approach to U.S. floodplain management. *Environment* 35(5):6–11; 31–34.

Leatherman, S. P. 1979. *Barrier island handbook.* College Park, MD: Coastal Research Laboratory, University of Maryland.

Melville, H. 1851/1992. *Moby Dick.* New York: The Modern Library.

Miller, H. C. 1981. The barrier islands: A gamble with time and nature. *Environment* 23(9):6–11; 36–42.

National Oceanic and Atmospheric Administration. 1975. *The coastline of the United States.* Washington, DC: U.S. Government Printing Office.

National Research Council. 1984. *California coastal erosion and storm damage during the winter of 1982–83.* Washington, DC: National Academy Press.

National Research Council. 1990. *Managing coastal erosion.* Washington, DC: National Academy Press.

Nordstrom, K. F., et al. 1986. *Living with the New Jersey shore.* Durham, NC: Duke University Press.

Nummedal, D. 1983. Barrier islands. In *CRC handbook of coastal processes and erosion,* ed. P. D. Komar, 77–121. Boca Raton, FL: CRC Press.

Office of Coastal Zone Management. 1976. *Natural hazard management in coastal areas.* Washington, DC: OCZM.

Platt, R. H. 1991. *Land use control: Geography, law and public policy.* Englewood Cliffs, NJ: Prentice-Hall.

Platt, R. H. 1992. An eroding base. *The Environmental Forum* 9(5):10–15.

Platt, R. H., S. G. Pelczarski, and B. K. R. Burbank, eds. 1987. *Cities on the beach: Management issues of developed coastal barriers.* Research Paper No. 224. Chicago: University of Chicago Department of Geography.

Platt, R. H., T. Beatley, and H. C. Miller. 1991. The folly at Folly Beach and other failings of U.S. coastal erosion policy. *Environment* 33(9):6–9; 25–32.

Platt, R. H., et al. 1992. *Coastal erosion: Has retreat sounded?* Program on Environment and Behavior, Monograph No. 53. Boulder: University of Colorado Institute of Behavioral Science.

Sheaffer and Roland, Inc. 1981. *Barrier island development near four national seashores.* Washington, DC: Sheaffer and Roland.

Teal, J., and M. Teal. 1969. *Life and death of the salt marsh.* New York: Ballantine.

Thoreau, H. D., ed. 1951. *Cape Cod.* New York: Bramhall House.

U.S. Army Corps of Engineers. 1971. *Report on the National Shoreline Study.* Washington, DC: The Corps.

U.S. Bureau of the Census. 1991. *Statistical abstract of the United States—1991.* Washington, DC: U.S. Government Printing Office.

U.S. Congress. 1966a. *Insurance and other programs for financial assistance to flood victims.* Committee Print No. 43, 89th Cong., 2d sess. Washington, DC: U.S. Government Printing Office.

U.S. Congress. 1966b. *A unified national program for managing flood losses.* House Document 465, 89th Cong., 2d sess. Washington, DC: U.S. Government Printing Office.

U.S. Fish and Wildlife Service. 1983. *Status and trends of wetlands and deepwater habitats in the conterminous United States: 1950s to 1970s.* Washington, DC: U.S. Government Printing Office.

U.S. General Accounting Office. 1975. *National efforts to preserve the nation's beaches and shorelines: A continuing problem.* RED-075-364. Washington, DC: GAO.

U.S. General Accounting Office. 1991. *Disaster assistance: Federal, state, and local responses to natural disasters need improvement.* Washington, DC: GAO.

Williams, S. J., K. Dodd, and K. K. Gohn. 1990. *Coasts in crisis.* U.S. Geological Survey Circular 1075. Washington, DC: U.S. Government Printing Office.

[7]

Journal of Environmental Management (1998) **52**, 327–333

Article No. ev980188

Coastal flooding, global warming and environmental management

J. C. Doornkamp

A review of the difficulties associated with the definition of coastal flood frequencies and magnitudes leads to a recognition that there is considerable doubt in many parts of the world as to the precise nature of this particular hazard. Similarly, a review of the sea-level measurements that have been used to indicate a response to global warming shows that there is uncertainty about the amount of other controlling influences. What is clear, however, are that past management decisions about human endeavours in the coastal zone (including flood defences, occupance of flood-prone lands, extraction of ground water and natural gas) have had an impact on relative land and sea levels and have done more to increase the risk of coastal flooding than can be assigned so far to global warming. In addition, these changes induced by human activity may render inappropriate calculations of coastal-flood frequencies based on historical records since the latter relate to a period of time when the controls on flooding may have been very different.

© 1998 Academic Press Limited

Keywords: coastal flooding, global warming, tectonics, engineering, management, sea-level change, vulnerability, hazard, monitoring, prediction, planning, policy reactions.

Introduction

Environmental management always includes two components: (1) an understanding of environment; and (2) the inclusion of that understanding into a management system. The definition of 'environment' and of 'management systems' varies from one profession to another, and what may be a matter of concern to some may not be so to many others. In the case of the coastal environment, however, there are many concerned professions including engineers, planners, conservationists, port managers, those concerned with coastal navigation, those in the fishing industry, offshore dredging operators, insurers, re-insurers, members of the holiday and tourist industries, owners of real estate and ultimately Governments (whose very survival may be prejudiced by inappropriate policies regarding the occupance of coastal zones). Amongst each of these groups there is a concern about those coastal hazards that most directly affect their responsibilities.

The range of coastal hazards that are of concern includes: coastal erosion (including unstable cliffs), coastal siltation, the movement of sediment within estuaries and the near-shore area, flooding and wind storms. Of all of these hazards the ones that have the most impact are wind storms and flooding. Dynamically the two often occur together, and an analysis of flood hazard must include reference to wind storms.

Relative changes in the elevations of land and sea are important to the occupants of the coastal zone, and those whose commercial activities are linked to it. This is especially the case if those changes are rapid and cause a significant change in processes such as flooding, siltation and erosion that materially affect human activity.

Over the last decade, debate concerning the potential for future changes in sea-level associated with global warming has been considerable. This has raised the general level of awareness in the dynamic nature of the coastal zone, and caused coastal-zone managers considerable concern. Increased awareness has also been accompanied by

Department of Geography,
University of Nottingham,
Nottingham, NG7 2RD, UK

*Received 6 March 1997;
accepted 17 March 1998*

0301–4797/98/040327 + 07 $25.00/0

confusion. This paper is concerned to identify the key components of flooding in the coastal zone, under a global warming regime, which need to be considered by coastal-zone managers.

Fundamentals of hazard assessment

The two fundamental components of any natural hazard (peril) are 'magnitude' and 'frequency' (with an assumption that the higher the magnitude the lower its frequency). The 'spatial component' of a hazard and its 'temporal variability' (including the changes in applicable frequencies that accompany shifts in the climate) also need to be understood.

This account will concentrate on magnitudes and frequencies in respect of coastal flooding. This demands a close look at the historical perspective and at present-day tendencies. Neither can be understood, however, outside their spatial context. Concerns about an increase in the flood hazard (and storminess) in the context of projected changes in climate are widespread in the management groups listed above. Such concerns have a stronger focus if viewed against a historical perspective and the related spatial contexts.

Magnitudes

The magnitude of coastal flooding is usually measured in terms of elevation and inland extent. These two parameters are directly related to each other, and are controlled by the form and heights of the ground exposed to flooding.

Some of the greatest of the more recent floods include those of the North Sea in England and mainland Europe (e.g. 31st January–1st February, 1953). Storms of this magnitude are associated with specific atmospheric and storm conditions, especially if they coincide with exceptional high tides (e.g. Spring tides). A storm surge may also be involved [see Steers (1953) for an account of the 1953 flood]. In the North Sea such surges

Table 1. Return periods for water levels in the Tees and Middlesborough Docks, Tees Estuary, NE England (from Shennan and Sproxton, 1990)

Return period (years)	Tees Dock (m. OD)	Middlesborough Dock (m. OD)
1:10	3·65	3·82
1:100	4·00	4·25
1:1000	4·35	4·68
1:10000	4·70	5·11

OD, ordnance datum.

are associated with deep atmospheric depressions. Nineteen surges per annum, on average, have a magnitude of more than 0·6 m above 'normal' each winter. Maxima approaching 3·5 m were recorded in the Thames Estuary in 1921 and 1953 (Lee *et al.*, 1995), though the theoretical maximum surge is close to 4 m (Dugdale, 1990). Damaging surge events in the North Sea area occurred in 1825, 1894, 1897, 1906, 1916, 1921, 1928, 1936, 1942, 1943, 1949, 1953, 1969, 1976 and 1978 (Lee *et al.*, 1995).

A great problem in coastal flood-hazard management is that records of past magnitudes do not always exist. Obtaining historical records for some of the largest storms and associated floods is possible [as Lamb (1991) has done for a 500-year period, or so, for the North Sea area of Europe]. These may say very little about storms of lower magnitudes, which, nevertheless can be locally very damaging. A search through newspaper (which cover a shorter period of historical time) and other records may yield some information about flooding at lower magnitudes, but such information is unlikely ever to be complete.

The alternative, and usual, approach is to extrapolate extreme flood levels from measurements made over a shorter period. Shennan and Sproxton (1990) provide a set of predictions, based on measured water levels, for Tees Dock and Middlesborough Dock (NE England) (Table 1), but recognize that there may be doubts over a linear extrapolation of the records in order to determine high-magnitude flood frequencies. In addition the period over which the records were taken (1921–1983) may or may not have been typical of a longer period.

Magnitude and frequency of flooding are inextricably linked in such analyses.

Frequencies

Frequency analysis inevitably includes historical data where such are available. Such data may do more than just provide evidence of past flooding and the calculation (by linear extrapolation) of extreme flood events, they may also allow a statistical analysis that can also reveal other important aspects such as those shown by Eliasson (1996) for Reykjavik Harbour (Iceland). Eliasson showed that a probability integral could be defined which gives the expected tidal-surge level as a function of the return period when the latter falls in the range 30–100 years.

Two important constraints exist on the use of flood frequency analysis based on records. These are: (1) the assumption that even under natural conditions the climatic controls on flood frequencies and magnitudes have not changed over the centuries; and (2) that flood protection work will have changed the frequency of local flood experiences.

Indications exist that flood (and storm) frequency calculations cannot be applied to long-term records. The reason is that within one period of climatic conditions certain frequency–magnitude relationships will exist, while a change in the climatic conditions brings in a changed set of frequency–magnitude conditions. One indication is the discovery recorded by Lamb (1982) that the greatest number of storms, within the North Sea, occurs in the warmest centuries. Storm floods in the North Sea reached a maximum in the eleventh and thirteenth centuries, with severe floods in late Roman times and again in the twentieth century. The frequency of storm and flood events changes with climatic context. These may also change through time.

Engineering works have produced flood protection in respect of flooding of lower magnitudes. These works have removed many higher frequency (lower magnitude) floods from the system. What remains, therefore, is just two states: no flooding and catastrophic flooding (i.e. those flood levels capable of overtopping, breaching or bypassing flood defences). This has profound implications for coastal-zone management. The fundamental findings of Hewitt and Burton (1971), which experience has borne out, include the crucial hazard-management observation that as flood protection is provided an increased occupance of flood-prone lands takes place. This increases the vulnerability of the area to extreme floods. When a high magnitude event does occur, far more damage is done than would have been the case before the flood defence system was constructed and the floodable lands occupied. What appears to be forgotten (or ignored) is that the engineering design parameters are based on a calculated flood frequency (which may be based on limited records taken during an untypical period) of perhaps the 1 in 500 year event. When an event of greater magnitude occurs, as it will, a far greater level of vulnerability exists than was the case before the flood defence system was constructed.

Historical aspects

There has been a general rise in sea level since the last glacial maximum (and sea-level minimum) over at least the past 18 000 years, and perhaps even longer. This rise has not been uniform across the globe (Dugdale, 1990; Kidson, 1982). Indeed, on a larger scale, critical differences in the relative heights of land and sea may occur within one country. Thus, while, in general, the south of England is losing height relative to the sea, as is western France, parts of Scotland and Scandinavia are emerging (Jardine, 1982) (e.g. as a result of isostatic adjustment).

Coastal flooding has been documented over very different time-periods in different parts of the world. In some places it has been possible to use the geological record to extend the historical period farther back in time (e.g. Nio and Yang, 1991; Stanley and Warne, 1994). What emerges is that coastal flooding is a natural process by which adjustments occur (between land and sea, between water and sediments) and that human beings have invaded this dynamic space and, of necessity, have tried (by the introduction of rigid flood defences) to 'fix' a dynamic system whose natural tendency is to undergo continual change.

Spatial aspects

The potential for coastal flooding has a strong spatial component at a variety of scales. On the global scale such flooding is associated with large low-lying areas such as the largest deltas (e.g. the Brahmaputra, Bangladesh), the major river inlets (e.g. the Amazon), areas of dominant coastal wetlands (e.g. Everglades of Florida) and areas experiencing subsidence on a large scale (e.g. Venice, see Bandarin, 1994).

On a continental scale, such as within Europe, coastal flooding is associated with low-lying areas (e.g. the Rhone delta, the Rhine delta) and with areas that experience flooding because of on shore storms (e.g. the coastal towns of southern England). On a national scale one part of the coastline is unlikely to experience catastrophic flooding at the same time as another [e.g. major floods in East Anglia (on the east coast of England) have never been recorded at the same time as a catastrophic flood anywhere on the west coast].

There exists a flood hazard in all low-lying coastal areas. Sometimes, where the area is stable in geological terms, the historical frequency–magnitude relationships may continue to be applicable. In other areas, where geological subsidence is a part of the dynamic development of the area, the frequency of flooding at any (and every) magnitude may be increasing. In coastal Louisiana, for instance, the rise of sea-level relative to a subsiding land level is not only increasing the amount of flooding, it is also causing saline intrusion into the groundwater, an increase in the volume of sediment in estuaries and bays, as well as affecting the tidal range, changing the detailed position of the shoreline and creating new or extended wetlands.

Dugdale (1990) reports a consensus view that sea-level rise has an average rate of between 1·1 and 3·0 mm yr^{-1} though different rates apply in different oceanic regions. In addition, there appears to have been a tendency for rates to rise in more recent years (Barnett, 1984; Gornitz, 1995).

In some areas it is recognized that human activity has caused land subsidence relative to sea level (Belperio, 1993) and where this is in coastal areas, such as the Po Delta in Italy (Bondesan et al., 1995), this will continue to lead to greater risks from coastal flooding than would otherwise have been the case. Such subsidence may be related to the extraction of underground water-resources, as in Venice (Bandarin, 1994), or as a result of port development, land reclamation and/ or industrialization as in Port Adelaide (Belperio, 1993) and Manilla (Spencer and Woodworth, 1993).

The future

Future changes in flooding are usually discussed in terms of the potential for climate change. The predictions endorsed by the Inter-governmental Panel on Climate Change (IPCC) and published in May 1996 can be found in Houghton et al. (1996). Their implications for coastal flooding and storminess vary around the globe. It is accepted that it is the coastal zone which may suffer most, mainly through sea-level rise (Turner et al., 1995; Turner et al., 1996). However, conditions may become more temperamental with excited variations around the mean (Leinfelder and Seyfried, 1993) and this implies a greater frequency (in the short to medium term) of both coastal storms and associated floods. Increased intensity of atmospheric processes will tend to increase the magnitude and the frequency of high-magnitude floods and storms in many parts of the world (Berz, 1993).

Concern has also been expressed about the effect that predicted changes in climate and sea level will have on near-shore sedimentation (Healy, 1996). Along sand-dune coasts there is an increased threat of the erosion of coastal dunes with the potential for increased sedimentary deposition (of these eroded sands) in the near shore. This may have the secondary effect of extending embayments further inland and causing the additional flooding of wetlands.

Most of the realistic estimates of climate change accept an element of uncertainty (Bodansky, 1995; Chao, 1995; Leinfelder and Seyfried, 1993; Ungar, 1995) and some authors define this as 'considerable uncertainty'. This uncertainty also extends into any economic predictions concerning both the effects of climatic change and the impact of human management reactions on the nature of climate change (Turner *et al.*, 1996).

Although there are reports of a global sea-level rise within the past 100 years, there is considerable doubt about the reliability of some records, and no certain link to global warming as the cause. Reported rates of sea-level rise at various locations around the world lie in the range 0.3–3 mm yr^{-1}. This is a greater range than the consensus view of the global average (Dugdale, 1990). Local causes such as vertical earth movements and influences on specific water-level gauges are recognized as influencing many readings (Gornitz, 1995). However, for management purposes changes in sea level, whatever the reason, are an important consideration. The very rapid rates of change along the coast of Louisiana, which are about 10 times the global rate (Gornitz, 1995), are a cause for management concern. Delta areas are known to experience a more rapid sea-level rise than adjacent areas that are not a part of the deltaic system. This is because the earth's crust under deltas tends to subside more rapidly than that in adjacent areas (Day *et al.*, 1995). Under natural conditions the supply of sediment during flooding, within the river systems, allows the delta surface to build up at a rate which keeps pace with this rise in sea level. As flood control schemes have been established along many of the river systems leading into deltaic areas (e.g. Mississippi, Rhine) this natural supply of sediment has been withheld from the delta and its surface has been unable to adjust to higher sea levels. This inevitably leads to an increased flood risk in such deltaic areas.

Uncertainty in the climatic change context has been looked at by Shlyakhter *et al.* (1995) in terms of the risk management issues of sequential decision strategies, value of information and the problems of interregional and intergenerational equity. Permitted development of very low-lying ground now leads to the observation that some of these are under a greater risk from flooding than was previously the case (Hughes and Brundrit, 1995). The problem is that there are now so many such locations of increased flood risk around the world.

Flood risk management

In essence, flood risk management, like any other form of risk management faces dilemmas posed by uncertainty, ambiguity and a lack of consensus as to the 'best' course of action. The manager will attempt to establish the best available factual background (in this case information about the existing and predicted coastal flood hazard) and then consider the policy alternatives, with conclusions reached by consensus.

This review has identified some of the more significant aspects of the coastal flood hazard. It has identified the fact that the hazard is not going to go away, and in many cases is going to provide a greater threat in the future. This is partly because underlying controls, such as the geological subsidence of parts of the earth's crust and the encouragement of subsidence by human extraction of underground resources, and the increased occupance of flood-prone lands will make greater flooding and flood damage inevitable.

Conclusions

Those many professions concerned with the coastal zone and its occupants recognize the hazard posed by flooding. Their problem often lies in an inability to grasp the true scale at which flooding can occur, and the ways in which environmental change can increase either the magnitude or the frequency of the flood hazard.

This review has drawn on studies around the globe that show that changes in sea level will lead to changes in flood characteristics. It is also clear from these studies that the cause of such changes is not exclusively that of a response by sea level to global warming. It seems much more likely that present flood-related difficulties have more to do with natural geological subsidence (which has a variety of causes), man-induced subsidence, the control of inland flooding that leads to a reduction in the necessary supply of silt to depositional coastal areas (and thereby preventing them from maintaining a balance with changes in sea level) and above all with the fact that occupance of potentially floodable land has taken place leading to an increase in the people and properties vulnerable in the event of a catastrophic flood.

A presumption has been made that insurance is the route to coping with flood risks (Berz, 1993), but this may turn out not to be the case, and in any event avoids the moral dimension. The potential exists for a catastrophic event, or a sequence of catastrophic flood events such that insurers and their re-insurers may be rendered incapable of meeting the financial cost of claims. However, even to presume that insurance is a management option is to avoid the fact that the lives of people are involved, and in this insurance is an irrelevance.

When it is recognized that some of the causes of increasing catastrophic flood risk lie in the hands of those who manage the coastal zone it is inevitable that the agenda of management concerns as well as the policies of central Government should be re-examined. For example, should further occupance of areas liable to extreme floods be prohibited? Should a free flow of river sediment into deltaic washlands be restored? Should man-induced subsidence (e.g. through the extraction of natural gas, or the extraction of water, as has been done in Venice and Tokyo) cease? Should there be an attempt (as in California) to restore ground levels by pumping back into the ground? There are no easy answers to such questions, but it is clear that management issues may be as important in determining future coastal flooding as any changes that global warming may cause.

What needs to be decided is the nature of the flood hazard along the coastline of concern, and to identify the management decisions that need to be made. In this context a decision to do nothing becomes a positive decision. In each case it is worth asking four basic questions (see Hewitt and Burton, 1971):

(1) is there a technically feasible solution?
(2) is the solution economically justified?
(3) is the solution socially acceptable?
(4) is the solution environmentally sound?

The problem of coastal flooding is not going to go away. Management strategies are required to cope with this inevitability.

References

Bandarin, F. (1994). The Venice project – a challenge for modern engineering. *Proceedings of the Institution of Civil Engineers* **102**(4), 163–174.

Barnett, T. P. (1984). The estimation of "global" sea level change: a problem of uniqueness. *Journal of Geophysical Research* **89**, 7980–7988.

Belperio, A. P. (1993). Land subsidence and sea-level rise in the Port Adelaide estuary – implications for monitoring the greenhouse effect. *Australian Journal of Earth Sciences* **40**(4), 359–368.

Berz, G. A. (1993). Global warming and the insurance industry. *Interdisciplinary Science Reviews* **18**(2), 120–125.

Bodansky, D. M. (1995). The emerging climate-change regime. *Annual Review of Energy and the Environment* **20**, 425–261.

Bondesan, M., Castiglioni, G. B., Elmi, C., Gabbianelli, G., Marocco, R., Pirazzoli, P. A. and Tomasin, A. (1995). Coastal areas at risk from storm surges and sea-level rise in northeastern Italy. *Journal of Coastal Research* **11**(4), 1354–1379.

Chao, H. P. (1995). Managing the risk of global climate catastrophe – an uncertain analysis. *Risk Analysis* **15**(1), 69–78.

Day, J. W., Pont, D., Hensel, P. F. and Ibanez, C. (1995). Impacts of sea-level rise on deltas in the Gulf-of-Mexico and the Mediterranean – the importance of pulsing events to sustainability. *Estuaries* **18**(4), 636–647.

Dugdale, R. E. (1990). Global reactions of the oceans and seas. In *The Greenhouse Effect and Rising Sea Levels in the UK* (J. C. Doornkamp ed.), pp. 31–61. Long Eaton, England: M1 Press.

Eliasson, J. (1996). Probability of tidal surge levels in Reykjavik, Iceland. *Journal of Coastal Research* **12**(1), 368–374.

Gornitz, V. (1995). Sea-level rise – a review of recent past and near-future trends. *Earth Surface Processes and Landforms* **20**(1), 7–20.

Healy, T. (1996). Sea-level rise and impacts on nearshore sedimentation – an overview. *Geologische Rundschau* **85**(3), 546–553.

Hewitt, K. and Burton, I. (1971). *The Hazardousness of Place: a Regional Ecology of Damaging Events*. 154 pp. Toronto: University of Toronto Press.

Houghton, J. T., Meira Filho, L. G., Callander, B. A., Harris, N., Kattenberg, A. and Maskell, K. (eds) (1996). *Climate Change 1995: The Science of Climate Change*. pp. 572. Cambridge: Cambridge University Press.

Hughes, P. and Brundrit, G. B. (1995). Sea-level rise and coastal planning – a call for stricter control in river mouths. *Journal of Coastal Research* **11**(3), 887–898.

Jardine, W. G. (1982). Sea level changes in Scotland during the last 18,000 years. *Proceedings of the Geological Association* **93**(1), 25–41.

Kidson, C. (1982). Sea level changes in the Holocene. *Quaternary Science Reviews* **1**, 121–151.

Lamb, H. H. (1982). *Climate, History and the Modern World*, 387 pp. London: Methuen.

Lamb, H. H. (1991). *Historic Storms of the North Sea, British Isles and Northwest Europe*. Cambridge: Cambridge University Press.

Lee, E. M., Clark, A. R. and Doornkamp, J. C. (1995). *The Occurrence and Significance of Erosion, Deposition and Flooding in Great Britain*. 177 pp. London: HMSO.

Leinfelder, R. and Seyfried, H. (1993). Sea-level change – a philosophical approach. *Geologische Rundschau* **82(2)**, 159–172.

Nio, S. D. and Yang, C. S. (1991). Sea-level fluctuations and the geometric variability of tide-dominated sandbodies. *Sedimentary Geology* **70**, 161–193.

Shennan, I. and Sproxton, I. (1990). Possible impacts of sea-level rise – a case study from the Tees estuary, Cleveland County. In *The Greenhouse Effect and Rising Sea Levels in the UK* (J. C. Doornkamp ed.), pp. 109–133. Long Eaton, England: M1 Press.

Shlyakhter, A., Valverde, L. J. and Wilson, R. (1995). Integrated risk analysis of global climate-change. *Chemosphere* **30(8)**, 1585–1618.

Stanley, D. J. and Warne, A. G. (1994). Worldwide initiation of Holocene marine deltas by deceleration sea-level rise. *Science* **265(5169)**, 228–231.

Steers, J. A. (1953). The east coast floods January 31–1 February 1953. *Geographical Journal* **119**, 280–298.

Turner, R. K., Adger, N. and Doktor, P. (1995). Assessing the economic costs of sea-level rise. *Environment and Planning A* **27(11)**, 1777–1796.

Turner, R. K., Subak, S. and Adger, W. N. (1996). Pressures, trends, and impacts in coastal zones – interactions between socioeconomic and natural systems. *Environmental Management* **20(2)**, 159–173.

Ungar, S. (1995). Social scares and global warming – beyond the Rio convention. *Society and Natural Resources* **8(5)**, 443–456.

[8]

ELSEVIER

Ocean & Coastal Management 40 (1998) 65–85

Ocean &
Coastal
Management

Marine-related physical natural hazards affecting coastal megacities of the Asia–Pacific region – awareness and mitigation

Russell S. Arthurton*

British Geological Survey, Keyworth, Nottingham NG12 5GG, UK

Abstract

The fast-growing, coastal megacities of the Asia–Pacific region are expanding into areas that are vulnerable to marine-related physical natural hazards, or, because of physical environmental changes, will become increasingly vulnerable within the timescale of city planning. The hazards comprise those that are due to extreme events such as storm surge and tsunami which may be catastrophic in their impacts; and those that relate to continuing changes over the long-term, notably global sea-level rise, sedimentary consolidation and coastal erosion. The latter may be exacerbated by human activities such as the increasing production of 'greenhouse' gases and over-abstraction of groundwater, and, while not threatening catastrophic loss of life or destruction of property, do have important economic and social implications for the future.

There are two complementary approaches to hazard mitigation – constraining the hazard, and reducing vulnerability to the hazard. The contributions that science can make in the planning and implementation of sustainable adaptive measures are to improve the quantification of the incidence and severity of the various hazards, establishing realistic timescales of incidence, estimating return periods; and to establish the geographical limits of vulnerability to the hazards in a range of likely scenarios over timescales appropriate to the planning cycle. Contemporary, high risk, hazard scenarios for existing city developments demand an approach which focuses on effective warning networks and emergency planning; long-term, incremental hazards that are forecast to affect both developed and periurban areas can be addressed with a strategic planning approach, involving relocation and capital protective works. The selection of strategic measures demands the best possible predictive information on hazards and on vulnerability, including its full socio-economic evaluation so that the costs and benefits of the possible mitigation options can be realistically assessed. A predictive capacity, developed through modelling, requires the collection of reliable baseline and monitoring data relating

_____

* Tel.: 0115 936 3486; Fax: 0115 936 3460.

to the hazards over a range of timescales in local, regional and global perspectives. © 1998 Natural Environment Research Council. Elsevier Science Ltd.

1. Introduction

Coastal megacities have grown from historic port development. The attributes which favoured the use and development of such sites as ports were essentially physiographic — places on otherwise exposed coasts which afforded sheltered anchorage and wharfage for hinterland trade. While the port function generally remains a focus of economic activity, most coastal megacities have grown far beyond their original, protected port location. Cities have expanded rapidly, subject to physiographic constraints, not only along the waterfront and into the hinterland [1], but in many cases also seaward on land reclaimed from the sea. This trend of coastal urban growth is set to continue [2]. The geographic settings of individual coastal megacities have provided specific opportunities for urban development but they impose constraints to sustainable growth. Important among such constraints are those which are due partly or wholly to natural hazards. The understanding of these natural hazards, and a recognition of a city's vulnerability to them, are key elements in the planning and management of effective adaptive measures.

1.1. Natural hazards of the Asia–Pacific region

Coastal megacities of the Asia–Pacific region (Fig. 1) are subject to different suites and intensities of natural hazard. Depending on a range of physiographic and developmental factors, they differ greatly in their vulnerability. Some (Hong Kong, Manila) suffer, on an extensive scale, the extreme wind and rainfall effects of seasonal tropical cyclones (typhoons). Some (Karachi, Jakarta, Osaka) are located in regions prone to potentially damaging earthquakes, while Manila, e.g., lies within range of significant ash fall from nearby active volcanic centres [3].

In addition to the natural hazards described above, there are those which are specific to coastal megacities because of their maritime location. A major concern is the possibility of accelerated sea-level rise as a consequence of global climate change [4]. Most coastal countries in the Asia–Pacific region have long shorelines and major centres of population located in low-lying, coastal areas. Even under today's sea-level conditions, these coasts are prone to wave erosion or marine inundation, in particular that caused by storm surges [5]. The prospect of significantly higher global sea levels during the next century and the consequent exacerbation of these marine-related hazards have captured the attention of coastal scientists and managers alike (see, e.g., Ref. [6]).

Marine-related hazards occur over a wide range of timescales. They include catastrophic events, some of which may last perhaps only a few minutes, and incremental

R.S. Arthurton / Ocean & Coastal Management 40 (1998) 65–85   67

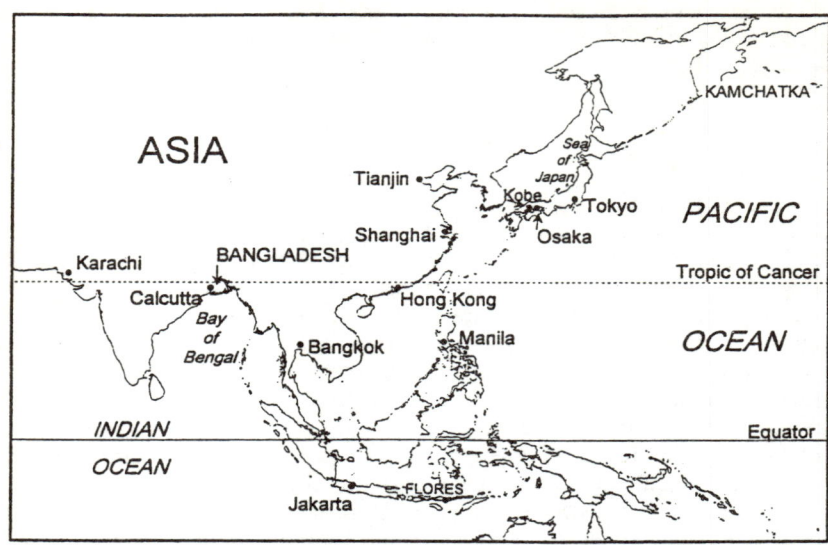

Fig. 1. The Asia–Pacific area showing principal locations referred to in the text.

physical changes, notably relative sea-level rise, that take place over much longer periods – from years to millennia and longer. These hazards differ greatly in their predictability. The ability of city managers and planners to set in place effective adaptive measures to cope with specific hazards depends on reliable information on the likely incidence and severity of hazard events. It is a role of science to improve knowledge and understanding of these natural hazards, and thus contribute to sound management and planning decisions. The priorities for action at local as well as regional and global levels need to be based on a sound and scientific assessment of the vulnerability of coastal areas to global change [7]. The research agenda is to set the spatial and temporal contexts of the hazards and the limits within which realistic assessments of vulnerability and risk can be made [8].

This paper reviews the marine-related, physical, natural hazards that pose risks [9] for the vulnerable elements of developing coastal megacities in the Asia–Pacific region. It considers the timescales relating to specific hazards and the geographical extents of vulnerability to those hazards. It identifies ways in which science can contribute to the appraisal of those risks by quantifying the hazards and assessing the context of vulnerability to them. It discusses how city management and the strategic planning of city development might respond to these risks by mitigation and adaptive measures implemented both through emergency procedures and by civil and social engineering over the medium and long terms.

2. Recognising vulnerability to marine-related hazards

Proximity to the sea is an obvious, necessary condition for the existence and growth of port facilities. However, this condition may be coincidental and irrelevant to the wider functions of the coastal megacity. While the megacity's waterfront may provide a recreational resource and sites for prestige property development and urban infrastructure, most of the city's inhabitants are likely to be indifferent to their maritime location. As a consequence, that population, whether as individuals or municipally, may be poorly aware of their vulnerability to hazards posed by the sea, and thus the risks that they might face in those respects. Vulnerability to hazards tends to occur where people lack the resources, awareness, knowledge, power or choices to mobilise defences against them [10]. In areas prone to such hazards people often appear ignorant of the potential for serious consequences, or, if aware of them, seem prepared to take unnecessary risks [11]. For some longer-term hazards and infrequent catastrophic events, they may simply be unaware of their vulnerability [12].

Important issues here are the city's expectations of the incidence and magnitude of hazard events. These are factors that science can address. They are described and discussed below in the context of specific hazards. Vulnerability to hazards differs greatly both between and within cities. It depends upon a range of physical, environmental, economic, social and cultural factors [13]. In particular, vulnerability depends on the spatial distributions of these factors (notably the extent of potentially floodable areas and their closely interrelated socio-economic systems) and their changes over time within the city's planning perspective. These are factors to be assessed by the application of the natural and social sciences, and also economics, because of the need in vulnerability assessment to quantify the economic activity (and infrastructure) at risk in hazardous coastal zones [14].

The pressures for urban growth are such that much of the development of coastal megacities takes place without the benefit of reliable information on natural hazards and vulnerability. Because of this, the need for mitigation measures, or the scale of such measures, may not be well understood by city managers. Even where plans and protocols are in place, they may be inappropriate to the real risk. The public's perception of risk today may be ill-informed and the scope of mitigation out of date, a consequence perhaps of unplanned development, or of changes in vulnerability due to changing physical and social conditions. Mitigation measures are unlikely to eliminate risk. Rather they should aim to reduce risks to levels that are acceptable within the limits of available resources.

The marine-related hazards which affect the coastal megacities of the region are of two main types, those caused by extreme physical events and those due to continuing changes over the long term. Extreme event hazards, which may be catastrophic in their impacts, are listed in Box 1. Hazards of the second type, listed in Box 2, impact incrementally over much longer periods. While they may not constitute a direct threat of catastrophic loss of life or property damage, they do have important economic and social implications over the long term.

R.S. Arthurton / Ocean & Coastal Management 40 (1998) 65–85   69

Box 1. Marine-related, physical extreme event hazards affecting coastal megacities

- *Severe waves* (high amplitude, storm-generated waves, may be cyclone-induced). These may cause local flooding of unprotected coastal lowlands and recession of erodible shores. They may also cause disruption to operations at exposed port terminals. Impacts tend to be most severe during high tidal states.
- *Storm surges* (cyclone-induced, abnormally high sea levels, commonly associated with high amplitude waves). These may cause the extensive inundation of unprotected coastal lowlands over periods ranging from a few hours to several days (events which may be exacerbated by landwater floods).
- *Tsunami* (high amplitude, long-period waves and run-up generated by near- or far-field submarine earthquakes and submarine 'land' slides). These may cause severe damage to waterfront land areas and associated coastal defences, with potential consequent loss of life and destruction of property and infrastructure over periods as short as only a few minutes.
- *Coastal earthquakes.* These may result in ground surface displacements in the coastal zone over seconds to days. The vertical component of displacement may induce flooding of coastal land or, depending on the sense of movement, emergence of the intertidal to shallow subtidal sea bed. These earthquakes may be accompanied by near-field tsunami.

Box 2. Marine-related, physical long-term hazards affecting coastal megacities

- *Relative sea-level change.* This is the increase (or decrease) at any given coastal location between mean sea level and the level of a reference point on the adjacent land surface or sea bed. Contributing factors include the possible human contribution to the forecast accelerated global sea-level rise as a predicted response to global warming. Continuing relative sea-level rise leads to an increasing frequency and severity of the marine inundation of low-lying coastal land, and, in the absence of engineering intervention, to long-term inundation.
- *Coastal erosion* and accretion (including siltation of navigation channels). These are physical manifestations of coastal change caused by a wide range of possible forcing factors, including some which are induced or exacerbated by human interventions, both local and regional. Consequences include progressive loss or gain of land and a possible enhanced need for maintenance dredging.
- *Saline intrusion* of coastal aquifers. This causes the progressive reduction or degradation of the coastal groundwater resource and is a likely consequence of relative sea-level rise. However, saline intrusion today being exacerbated by the over-abstraction of groundwater by coastal communities.

2.1. Extreme event hazards – knowing the risk

The extreme event, physical hazards are difficult to plan for. They may provide no warning of their incidence, or so little warning that, even if emergency procedures are in place, there may be insufficient time to implement them. They range from events that are unpredictable in their timing and location, notably earthquakes, to those of a seasonal nature that have relatively predictable return periods and severities, such as storm surges (Table 1). Scientific uncertainties continue to shroud the scale and significance of potential combined sea-level rise and storm event risks at the regional level [13].

2.1.1. Severe waves

Any unprotected coast exposed to an extensive, uninterrupted, marine fetch is prone to damage by local flooding and shoreline erosion by waves, notably during high tidal states and storm surge conditions. While port facilities have generally been developed in sheltered inlets or estuaries where wave energy is subdued, there are

Table 1
Conditions and likely coastal impacts of marine-related physical extreme event hazards

Extreme event	Main vulnerable locations	Likely return period	Likely impact	Critical tidal state
Severe waves	Coasts exposed to long oversea wind fetch	Storm-related: months to years	Inundation and erosion of unprotected water front	High
Storm surge	Areas prone to tropical cyclones	Storm-related: months to years	Extensive inundation of unprotected lowland	High
Tsunami	Coasts and critical inlets exposed to far-field events	Years to centuries	Inundation and severe damage to waterfront zone	High
Coastal earthquake	Any seismically active area	Unknown	Unknown	

many instances worldwide of urban expansion having taken place along less protected shores, resulting in vulnerable waterfronts. Wave damage is probably the commonest of the marine-related extreme event hazards. Because of this, its incidence is well documented in most urbanised coastal areas and the risks related to wave damage in existing, normal tidal conditions are usually well understood in respect of today's sea level and climatic conditions.

2.1.2. Storm surges

Storm surge (and related wave) inundation of coastal lowlands threatens human safety, and damages property and infrastructure. Surge levels are at their highest when these cyclone-related events coincide with high tidal states. Any unprotected land area at or below the surge sea level is vulnerable, as are its inhabitants and its susceptible service infrastructure, notably power and water supply, and sewerage. Landlocked harbours such as Tokyo and Osaka afford some protection from surges, though funnel-shaped coastal configurations and islands can accentuate the effect towards their closures [15]. Surges are normally associated with tropical cyclones, thus the incidence of this hazard varies greatly within the Asia–Pacific region. Tropical cyclones in the Bay of Bengal typically make landfall about three times a year [16]. Since 1882 Bangladesh has suffered significant marine flooding of coastal lowlands on average once in five years, although since 1950 less than once in two years [17]; such an event in 1991 killed 140 000 people.

2.1.3. Tsunami

Many coasts within the region are prone to tsunami impact. Elsewhere low-lying coasts exposed to the ocean and, notably, bays and estuaries are vulnerable, such embayments tending to enhance wave amplitude in run-up or bore effects. 'Tsunami'

R.S. Arthurton / Ocean & Coastal Management 40 (1998) 65–85   71

is the Japanese word for 'harbour waves' [18]. As with storm surge hazards, the existence of coastal defences may constrain some events, but the destructive power of tsunami waves in the shore zone may breach such defences and lead to significant loss of life and waterfront property before their energy is dissipated. The extent of the vulnerable zone depends on the wave magnitude and on local factors such as the nearshore bathymetry and shoreface profile. The impact of the 1960 Chilean far-field tsunami on the city of Hilo in Hawaii caused 61 deaths and carried blocks weighing more than 20 tonnes as much as 200 m inland [19]. The same tsunami claimed 119 lives and caused extensive damage to property and aquaculture in Japan [20].

The incidence of tsunami hazards, or tsunamigenic earthquakes, can be forecast and return periods estimated only in the most general way. Historical records provide some guidance in the prediction of impacts and the estimation of return periods [21], while the recent geological record at coastal sites may also provide evidence of past incidence [18]. In the Asia–Pacific region major tsunami affected many Pacific coasts in 1960 and 1964. The 1970s were generally uneventful but in 1983 a tsunami in the Sea of Japan killed 100 people [22], and in 1992 an event off Flores, Indonesia, also claimed many lives. Compared with the incidence of storm surges within the region, destructive tsunami are infrequent at any one site and, on many shores, e.g. western Kamchatka [23], they are regarded as the lesser of the two hazards.

2.1.4. Coastal earthquakes

Coasts in much of the Asia–Pacific region are earthquake-prone; these include the Indonesian archipelago and the islands of Japan. The 1995 earthquake off Kobe, Japan, provides a recent example of such devastation in a coastal urban area. Coastal lowlands are vulnerable to the marine-related impacts of coastal earthquakes in three ways. They may be affected directly by vertical ground displacement causing possible relative sea level rise or fall (e.g. Wellington, New Zealand, in the 1860s); or indirectly by marine inundation as a result of sediment consolidation triggered by the earthquake shock, or the impact of a near-field tsunami. Prediction of the incidence of catastrophic coastal earthquake events, as earthquakes in general, is imprecise. Information from the analysis and interpretation of the recent geological and historical records may provide the most realistic indication of the severity and return period of events which might affect coastal megacities. While such information may inform structural engineering and building regulations, its lack of precision offers little practical guidance to city planners in respect of the timing of significant events.

2.2. Understanding the risk from long-term, incremental hazards

Unlike extreme event hazards, the long-term, incremental hazards (Box 2) generally provide ample warning of their incidence. The processes causing the hazards can be analysed and the resulting changes monitored in space and time. The rates of change may be sufficiently slow and predictable to provide the opportunities for the planning and implementation of sustainable adaptive measures (Table 2).

Table 2
Components of relative sea-level change in coastal areas, with indications of operative timescales

Global and regional processes	Local effects– main processes	Contributing factors	Relative sea-level rise + fall −	Timescale
Global sea-level change			+ / −	Long term
Neotectonics	(Neotectonics)		+ / −	Long term (some extreme events)
	Consolidation	Natural loading	+	Decades/millennia
		Artificial loading	+	Months/decades
		Land drainage	+	Months/decades
		Weathering/ soil formation	+	Years/decades
		Groundwater abstraction	+	Months/decades
		Groundwater recharge	−	Months/decades
	Shrink-swell in clay		+ / −	Seasonal
	Floodland sedimentation	Fines from sus- pension/peats	−	Extreme events to long term
Isostatic adjustments		(Sea) water load- ing (or unloading)	+ / −	Centuries to millennia
		Sediment load- ing (or unloading)	+ / −	Centuries to millennia

2.2.1. Sea-level change

The assessment of relative sea-level change poses considerable problems in city planning. It is difficult to accurately assess the rate of this change at the local level with a resolution that is relevant to planning and management. The difficulty stems from the large number of separate contributory processes. Some components are due to global and regional processes, some to local. Relative sea-level change depends on a number of component factors of vertical displacement, some marine and some 'terrestrial' (though affecting the sea bed as well as the land), acting at different rates (Table 2) which themselves may be expected to change with time. While some of these factors, e.g. neotectonic crustal displacements in coastal areas, have strictly natural causes, most are either the result of human interventions of natural systems or of natural changes exacerbated by human activities. The net displacement of the land/sea-bed surface and the mean sea level at a given coastal location is the relative sea-level change at that site. In this review, relative sea-level change is described as a *natural* hazard, even though some of its contributing factors may be partly or wholly induced by human activities.

While the vulnerability (to sea-level change) of coastal megacities, their inhabitants and their economic infrastructure, may usually be geographically defined and quantified with some precision, the uncertain, multigenic, nature of this hazard makes its related risk difficult to assess. The likely incidence and severity of the hazard, in terms of the rates of net relative sea-level change and the changes of those rates with time, may be only poorly known. This presents a major challenge to science. Until the likely contributions from the various component factors are better understood and quantified at their respective scales, a precautionary approach should prevail in mitigation planning [24].

The difficulties implicit in relative sea-level change prediction at coastal locations arise from uncertainties at global, regional and local scales. At the global scale, forecasts by Warrick and Oerlemans [25] of accelerated *sea-level rise*, as a predicted – though yet unproven [26] – response to global climate warming, were 200 mm by the year 2030 and 660 mm by 2100. The best estimate given by Raper et al. [27] was that between 1990 and 2100 sea level would rise by 490 mm; taking uncertainties into account, estimates of the rise during this period were in the range 200–860 mm.

Regional and local circumstances may temper or exacerbate the effects of such a global change at the coastal site. For example, many major river outflows are sited in areas that have been subsiding tectonically over millions of years [28, 29]. While many coastal megacities in the Asia–Pacific region may be involved in similar regional tectonic subsidence, such *neotectonic effects* may be difficult to distinguish from other contributions. Notwithstanding this difficulty, subsidence of 1–2 mm/yr in Tianjin, China's third largest urban area, has been ascribed to neotectonics [30], a rate that is of concern in the timescale of coastal planning and management. Also acting at the regional scale are the processes of isostatic adjustment of the earth's crust. Perhaps of greatest significance in the context of coastal megacities in the region is the crustal loading effect of *sediment isostasy*, in which the crust responds to increasing sediment load by regional downwarping. This may be especially important in the vicinity of major deltas, where there is a substantial added crustal load due to the long-term accumulation of sediment, but distinguishing such subsidence from that due to neotectonics may not be feasible. The increase in crustal loading in coastal regions due to relative sea-level rise may also be a significant factor, this crustal deformation process referred to as *hydro-isostasy* (see, e.g., Ref.[26]). The post-Glacial global sea-level rise of more than 100 m [31] has differentially loaded what is now the inshore sea bed, the added load depending upon the pre-transgression relief. As with glacio-isostatic crustal deformation, both of these isostatic adjustments lag the loading in time, and are reversible.

At the local scale, natural physical processes and human interventions may lead to ground-level displacement over the short term, at rates considerably in excess of those predicted for global sea-level rise. The potential for *local displacement effects* depends mainly on the geology of the coastal area. While some cities, e.g. Hong Kong, are founded largely on bedrock (or weathered bedrock) or, like Karachi, on well consolidated sediments, many, including Bangkok, Shanghai, Greater Tianjin, Jakarta and Calcutta, are sited partly or wholly on poorly consolidated sedimentary formations, including marine muds and sands. Some of these sedimentary formations have been formed in (geologically) very recent times, or, as in the Yangtze River delta adjoining Shanghai [32, 33], are actively accreting. The land that these recent sedimentary deposits forms lies within only a few metres of mean sea level and may become increasingly vulnerable to relative sea-level rise within the planning timescale.

Largely irreversible land-surface subsidence due to the *consolidation* of these coastal and deltaic sediments can locally make an important contribution to relative sea-level rise. The potential for consolidation depends upon the types and thicknesses of sediments involved, some clay-rich muds reducing to about half of their deposited volume and peats to as little as little as one ninth [28, 29]. The rate of consolidation depends on natural and anthropogenic factors. Natural loading by superincumbent deposits and urban development enhances consolidation and consequent subsidence over the long term [34]. Lowering of the groundwater table, as a result, for example, of land drainage schemes, promotes consolidation of the superficial sediments and thus subsidence, while the weathering and oxidation of emergent associated peat deposits further enhances this effect [28]. The natural consolidation of coastal and deltaic muddy sediments may be catastrophically triggered by earthquakes. In part of Greater Tianjin, for example, abrupt land-surface subsidence of 500–600 mm occurred during the 1976 earthquake, in which some quarter of a million people are reported to have perished in Tangshan alone [30].

The over-abstraction of groundwater from aquifers within coastal and deltaic sediments is an especially important contributor to consolidation-related ground subsidence. There are several well documented instances of this problem within the region. In Bangkok between 1960 and 1988, up to 1.6 m of subsidence have been reported [35, 36], and in parts of Shanghai subsidence rates have exceeded 10 mm/yr [37, 38]. In the coastal plain of Tianjin, subsidence rates due to natural consolidation (1–3 mm/yr) are dwarfed by those induced by the extensive pumping of freshwater aquifers. Between 1960 and 1982 cumulative land subsidence of 1.5 m occurred [30]. When aquifers that have been over-pumped are recharged with water, there may be some expansion of the host sediments and thus some rebound of the land surface, though generally only a small proportion the original ground subsidence is recovered.

The monitoring of land-surface changes on clay land needs to take account of possible seasonal variations due to the *shrinking and swelling* effects of clay minerals within the soil and sub-soil. The magnitude of such changes – perhaps up to 100 mm within a seasonal cycle – may mask subtler changes due to consolidation or neotectonic effects.

The natural processes of *sedimentation* in coastal and deltaic areas are themselves significant contributors to relative sea-level change at the local scale as well as regionally. The natural flooding of undefended coastal lowlands by sediment-charged waters, whether of riverine or marine origin, leads to the progressive addition of sediment to the lowland surface. Many such low-lying coastal floodlands have been reclaimed for agricultural development during the last two millenia through the construction of protective dikes. Shanghai itself has grown on land that has been successively reclaimed as seaward sediment accretion has occurred (Tan, Q.X., 1987 [33]). The construction of coastal flood defences has protected the reclaimed land; but it has also effectively prevented the reclaimed land from being recharged with sediment

R.S. Arthurton / Ocean & Coastal Management 40 (1998) 65–85   75

during flood events. The natural sedimention history of reclaimed land has thus been interrupted by the reclamation, and floodland sedimentation has ceased, temporally, to be a factor contributing to relative sea-level change in such circumstances.

2.2.2. Coastal erosion

Coastal erosion may be a significant hazard for some coastal megacities in the region, though its incidence is seldom catastrophic. Of all the coastal hazards, erosion is perhaps the easiest to forecast. Indeed it is perhaps questionable whether coastal erosion should be classified as a hazard, but rather a predictable consequence of the impact of waves on physically vulnerable coasts.

Coastal erosion is the natural response of geological materials forming a shoreface to wave impact, or in some cases the impact of strong tidal currents. While there is widespread concern that the expected sea-level rise related to global warming will aggravate the effect [39], erosion can, and does, occur on coastlines where the sea level is stable or even falling [40]. Wherever and whenever sufficient wave energy reaches vulnerable coasts, erosion can take place. Vulnerability depends on (a) the degree of exposure to the waves, (b) the level of protection afforded by beach deposits and (c) the geological composition of the shoreface and its adjoining hinterland.

Erosion tends to be most severe when storm conditions coincide with high tidal states, particularly during surge events. In such conditions protective beach materials are drawn downshore, seawards from the backshore, exposing hinterland sediments or rocks to wave attack. Well lithified rocks forming the shoreface and hinterland resist erosion, while poorly lithified rocks and unlithified sediments may be readily eroded, with a consequent recession of the shoreline.

The protection afforded to erodible coasts by beach deposits depends on the maintenance of those deposits. Various factors in addition to the drawdown process may result in beaches becoming starved of sediment and thus ineffective in their protective role. In deltaic environments, the discharge of sediment which feeds beaches may change with time; climate variability may change the direction of net alongshore drift of beach sediment; or sediment may be removed from beaches, or its littoral transport impeded, by human interventions.

The process of coastal land loss by erosion has a counterpart process of coastal change – coastal land growth by sedimentary accretion. Such coastal accretion is commonly a feature of deltaic areas, where sediment-bearing rivers discharge to the sea. Coastal megacities sited in deltaic areas, e.g. Shanghai [33], may be wholly or partly founded upon sediments which have accreted to form new land within the last few millenia. The long-term stability of this type of land may be threatened by sea-level rise, but also by any significant reduction in the rates at which sediments are discharged by rivers to the sea [33]. It is therefore particularly important to understand the impacts at river mouths of dam construction schemes within catchments which might greatly reduce discharge.

2.2.3. Saline intrusion of coastal aquifers

For completeness, the process of saline intrusion is referred to in this review of marine-related natural hazards. In reality the effect is not a hazard in itself but rather a consequence of one or more of the following factors – relative sea level rise, the excessive pumped abstraction of groundwater from coastal aquifers and impeded flow in rivers due to human activities. The problem is now common in deltas and coastal urban areas [39].

3. Approaches to mitigation

This section considers the information requirements of city managers and planners to guide policies of mitigation to cope with marine-related natural hazards acting over time scales ranging from the immediate to the long term – say 50–100 years. Mitigation measures are those taken in advance of a hazard event aimed at decreasing or eliminating its impact on society and the environment [9]. It is a long-term investment in the welfare of all, and there is a growing need in the formulation of city development plans to emphasize mitigation rather than response [41].

The perspectives of *vulnerability* and *hazard* need careful definition. A hazard relates to vulnerability. Thus a storm surge is a real hazard in respect of unprotected coastal lowlands, but not a hazard in respect of (non-vulnerable) rock-cliffed coasts or the open ocean. Where a seawall provides protection, a surge becomes a hazard in respect of the lowlands only when the wall is in danger of being over-topped. From the perspective of vulnerable lowland inhabitants, the construction of the seawall constrains the hazard. The *vulnerability* of the coastal lowland relates to expected surge sea levels rather than the standard of protection afforded by seawalls. For example, much of the population and economic activity of The Netherlands lives, or takes place, below mean sea level and is therefore inherently vulnerable to flooding. However, because the standard of protection from storm surge (provided by seawalls and other tidal defences, and naturally by sand dunes) is high, the risk in respect of that hazard is low.

The problem in many coastal megacities in the Asia–Pacific region is that uncontrolled or poorly planned growth has created, or is creating, vulnerability (of people, services, economic activity) without the concurrent provision of an adequate standard of protection. The provision of adequate protection, e.g. by engineering intervention, may not be a realistic option, particularly if it is difficult to resource. Another possible adaptive option is the reduction of vulnerability, an approach which aims to safeguard people, economic activity and infrastructure by encouraging city development in areas removed from potential hazard impacts.

Thus there are two mutually complementary approaches to mitigation. One is to constrain the hazard, the other to reduce vulnerability to the hazard. Both contribute to a reduction in the level of risk. Planning and management strategies that aim to achieve a sustainable reduction in vulnerability should be designed to cope with the failure of protective systems [42]. Human intervention activities which are unsustainable contribute to vulnerability. In any event the approaches to mitigation must reflect vulnerability assessments and the different levels of immediacy attached to the various hazards. National and city governments are under an increasing obligation to assess risk and to ensure that mitigation strategies feature prominently in their development agendas [41].

R.S. Arthurton / Ocean & Coastal Management 40 (1998) 65–85   77

Contemporary high risk scenarios for existing city developments demand an approach which focuses on immediate emergency planning. Long-term scenarios of increasing marine-related hazard which may be forecast to affect both the developed and yet-to-be developed, periurban areas can be addressed with a considered, strategic planning approach, perhaps one involving capital protective works. Between these extreme scenarios, there is scope for tactical adaptive measures leading to a reduction in vulnerability over the shorter term. In practice, mitigation measures are likely to be pursued in all three modes – emergency, tactical and strategic.

In establishing a strategy for mitigation [41], a priority activity is the preliminary assessment of hazard, vulnerability and risk, using existing sources of information wherever possible (Box 3). Vulnerability is defined as being an estimate of the degree of loss resulting from a potentially damaging phenomenon [9]. The integration of the resulting assessments provides the basis for an initial quantification of risk and identifies the need for follow-up studies. With policy objectives in mind, a plan of action can then be developed by identifying the options for mitigation and the costs and benefits of these actions. The plan may identify the contributions that science can make to this quantification (Table 3).

3.1. Emergency measures

Recommendations for emergency procedures in the face of impending or actual extreme event hazard impacts have been well documented elsewhere [41], and are referred to only briefly in this review. The recommendations focus on short-term to immediate goals of reducing vulnerability. They include the development of public awareness and emergency warning systems, and the establishment of co-ordinated procedures for evacuation, rescue and rehabilitation. There may also be provision for the protection of vulnerable key service installations.

Warning networks for extreme marine-related events, other than coastal earthquakes, are generally well established in the Asia–Pacific region. The global forecasting of tropical cyclones through real-time observational and predictive modelling techniques [43] enables the issue of storm and related surge warnings, which may trigger the implementation of emergency procedures and provide an opportunity for the evacuation of vulnerable areas. The regional forecasting of tsunami impacts [21] involves a network of warning centres around the Pacific and uses data from specific recorded earthquake events or wave monitoring. Depending on the distance of the predicted impact from the source, the network can ideally provide vulnerable communities with up to a few hours' warning, and thus the opportunity of taking vital precautionary action. The effectiveness of the emergency procedures, however, may be hampered by a reluctance of the threatened community to take evasive action [11].

Box 3. Determinands for the preliminary assessment of city hazard vulnerability and risk [41]

> • The nature of potential hazards: their predicted frequency, intensity and duration.
> • The geographical areas of the city which are most vulnerable.
> • The communities, business sectors and infrastructure components which are most vulnerable.
> • The estimated losses which would result from hazard events of different magnitudes.

Table 3
Science activities at local, regional and global scales in respect of the quantification of marine-related physical hazards

Hazards	Analytical and monitoring activities		
	Local/city	Regional	Global
Extreme events			
Severe waves	Wave recording, monitoring	Weather/wave monitoring, forecasting	
Storm surge	Monitoring, modelling	Weather monitoring/forecasting	
Tsunami	Monitoring, modelling	Earthquake/tsunami warning network	Earthquake/tsunami warning network
Coastal earthquake	Monitoring	Monitoring network	
Long term			
Relative sea-level change	Tide-guage monitoring		
Global sea-level change		Tide-guage monitoring, sea surface altimetry	Tide-guage monitoring, sea surface altimetry
Neotectonic crustal displacement	Land surface altimetry monitoring	Land surface altimetry monitoring	
Isostatic crustal adjustment		Land surface monitoring/modelling	
Sediment consolidation	Land surface and sub-surface monitoring		
Floodland siltation	Land surface monitoring	Catchment/coastal sediment transport monitoring	
Coastal erosion	Shoreline/estuary monitoring	Catchment/coastal sediment transport monitoring	
Coastal progradation, channel siltation	Shoreline/estuary/ bathymetric monitoring		
Saline intrusion of coastal aquifers	Groundwater monitoring/modelling		

3.2. Tactical measures

Implementation of the dual approach of marine-related hazard constraint and vulnerability reduction, in the light of cost-benefit analysis, is appropriate to the management of established urban development (Table 4). The various regulatory and financial instruments (including insurance incentives) designed to divert or relocate people and economic activity from hazard-prone areas have been dealt with elsewhere [41]. Some possible mitigation options aimed at hazard constraint which may help to safeguard developed low-lying or waterfront urban areas are considered here. They

R.S. Arthurton / Ocean & Coastal Management 40 (1998) 65–85   79

Table 4
Mitigation and adaptive measures (excluding emergency measures) at local, regional and global scales in respect of the constraint of marine-related physical hazards

Hazards	Mitigation and adaptive options		
	Local/city	Regional	Global
Extreme events			
Severe waves	Construct/enhance maintain offshore breakwaters, seawalls. Promote/conserve saltmarsh		
Storm surge	Construct, enhance maintain seawalls. Promote/conserve sand dunes		
Tsunami	Construct/enhance maintain offshore breakwaters, seawalls. Conserve/promote saltmarsh. Restrict development on waterfront		
Coastal earthquake	(Take precautionary measures in construction and planning)		
Long term			
Relative sea-level change	Develop and relocate in low risk areas in response to rise, defend immovable assets		
Global sea-level change		Reduce emissions of "greenhouse" gases	
Neotectonic crustal displacement			
Isostatic crustal adjustment			
Sediment consolidation	Manage groundwater abstraction, regulate wetland drainage		
Floodland siltation	Consider planned siltation in periurban areas		
Coastal erosion	Restrict development in waterfront areas, regulate coastal defence interventions and beach/nearshore sand extraction	Assess impacts of climate and catchment land use change, river/coastal engineering affecting sediment discharge/ transport regimes	
Coastal progradation and channel siltation	Dredge to maintain/ enhance navigation/port function		
Saline intrusion of coastal aquifers	Manage groundwater abstraction		

include regulatory measures but not major capital works. They mostly aim to control the human activities which exacerbate the natural hazards of relative sea-level rise in vulnerable low-lying areas and coastal erosion by wave action.

Regulatory management of groundwater abstraction from aquifers underlying coastal cities can significantly reduce the rate of land surface subsidence. At Hangu, in Greater Tianjin, the aim has been to reduce pumping and to spread the abstraction over a wider area, reducing land-surface subsidence to about 20 mm/yr [30]. In the central part of Bangkok, where groundwater pumping is now prohibited, the rate of subsidence has decreased and in some places the ground level has shown a slight rebound [35]. Elsewhere, e.g. in Shanghai, programmes of pumped recharge of aquifers have led to some recovery of the land surface [29]. Land drainage schemes, carried out in advance of the development of periurban areas and which contribute to land-surface subsidence and the risk of inundation, need to be planned with a view to the provision of strategic protection works against relative sea-level rise.

Regulation of human interventions which impact on the urban and periurban shoreline may help to inhibit wave erosion of soft shorefaces and the consequent threat of recession of the city waterfront. The protective role of beach sediments can be enhanced by controlling or prohibiting the extraction of sand from the shoreface and adjoining sea bed. It may also be helped by avoidance of sea defence works that interrupt the natural transport of sediment in the littoral zone and result in beach sediment starvation, while such regulatory measures can be complemented by the periodic artificial replenishment of beach materials (see, e.g., Ref. [44]).

3.3. Strategic measures

Strategic mitigation measures concern long-term city planning and, if appropriate and affordable, major capital works arising. They address extreme event and long-term, incremental marine-related hazards. They have the dual aim of hazard constraint and vulnerability reduction, in part through hazard avoidance in urban development. They are concerned with the future development of periurban areas as well as with the management of existing urban areas. They should be considered, not only in a long (50–100 yr) time context, but in a wide geographical (e.g. catchment) context as well. Strategic mitigation measures above all demand the best possible information on hazards and vulnerability, as well as careful cost-benefit analysis. In view of the many uncertainties over hazard prediction over the long term, they should be implemented with due caution [24].

Long-term coastal city planning in respect of these hazards has three main strategic aims. These are (a) to encourage development outside areas vulnerable to marine-related hazards; (b) to relocate people, economic activities and key infrastructure to reduce vulnerability; and (c) to enhance the standard of protection from marine-related hazards where there is existing vulnerability but where relocation is not a feasible option (Table 4, Box 4).

Recognising the geographical extent, and assessing the scale, of vulnerability in the coastal urban and periurban areas under different hazard and developmental scenarios forms a basis for strategic planning (Box 3). Computer mapping techniques

R.S. Arthurton / Ocean & Coastal Management 40 (1998) 65–85   81

Box 4. Strategic aims for coastal city planning to take account of marine-related, physical hazards

- Promote new urban development away from areas vulnerable to marine-related hazards using financial incentives and regulatory constraints as appropriate.
- Relocate vulnerable urban population, economic activities and key infrastructure to areas of low hazard susceptibility.
- Enhance the standard of protection where there is existing vulnerability but where relocation is not a viable option.

provide the means of identifying areas that are unsuitable for development. These may involve hazard maps [41,45], or maps depicting vulnerability, based on social and economic factors; or composite risk maps, combining hazard information with vulnerability assessment. The analysis of possible development scenarios against the predictions of hazard impacts within the timescale of the planning cycle can inform the strategic planning process, including decisions on the implementation of capital works for hazard constraint.

Strategic, socio-economic adaptive measures, intended to encourage development or relocation in areas of low hazard risk within the planning guidelines, may include a range of regulatory and financial constraints and incentives [41]. Strategic coast and flood defence capital works, designed to constrain hazards in respect of existing, key development and infrastructure, include structures that are expected to withstand or dissipate extreme events and progressive long-term relative sea-level rise. These engineering interventions may include seawalls or other tidal defences; also offshore breakwaters, designed for extreme event wave, and especially tsunami, control. There may also be scope for the introduction of soft-engineered schemes, such as managed coastal 'setback' in periurban areas and the consequent promotion of saltmarsh wetland as a means of dissipating wave energy. Assessment of environmental impact over the long term is an essential part of any such engineering intervention.

4. Research priorities

Local research agendas relating to marine-related hazards as they affect coastal megacities in the region may be viewed as having three foci. One is the spatio-temporal prediction of the hazard impacts through the application of geological, oceanographic and climatological disciplines; this is considered briefly below. The second focus concerns the socio-economic issues relating to vulnerability in the context of predicted hazard scenarios, and aims to establish the geographical limits in vulnerability assessments and their predicted changes with time. Digital terrain mapping techniques, coupled with ground-level change data, provide a cost-effective geographical definition of vulnerability under likely hazard scenarios. The third focus is the translation of the scientific and socio-economic information to a format appropriate to the planning and integrated management of coastal megacities. In addition to this agenda, there are areas of generic research and development to be addressed by the international scientific community concerning, e.g., the development of marine-related hazard warning systems, or the optimization of groundwater abstraction in coastal urban environments.

Research needs for hazard impact prediction at the local or city-specific scale involve global, regional and local data (Table 3). Global and regional information, including global sea-level change predictions, tropical cyclone monitoring and tsunami modelling, must be taken into account when assessing potential local impacts. Complementary knowledge of the local physical conditions, in particular, the various effects that may be contributing to ground level change, is of paramount importance in the formulation of effective hazard mitigation strategies.

Overall, there is a need to put in place the means to collect these data, establishing baseline measurements against which future events and changes can be monitored and measured. The aim should be to produce predictive models which are soundly based on adequate observational information at the local to the global scale, and covering the range of timescales relevant to the quantification of extreme event and long-term incremental hazards.

5. Concluding summary

- Marine-related physical hazards affect, and will increasingly impact on, coastal megacities in the Asia–Pacific region.
- The hazards comprise those due to extreme events such as storm surges and tsunamis which may be catastrophic in their impacts; and those due to continuing changes over the long term, notably eustatic sea-level rise, sedimentary consolidation and coastal erosion.
- The long-term hazards may be exacerbated by human activities such as the increasing production of 'greenhouse' gases and over-abstraction of groundwater, and, while not threatening catastrophic loss of life or destruction of property, do have important economic and social implications for the future.
- There are two complementary approaches to mitigation: to constrain the hazard; and to reduce the vulnerability.
- The contributions that science can make to the planning of sustainable adaptive measures are to improve the prediction of the incidence and severity of the various hazards, and their impacts in space and time. A priority task is to establish the means to collect the essential baseline data for predictive modelling.
- Contemporary high risk hazard scenarios for existing city developments demand an approach that focuses on effective warning networks and emergency planning.
- Long-term, incremental hazards that are forecast to affect both developed and peri-urban areas can be addressed with a strategic planning approach, involving relocation and capital protective works.
- The selection of strategic measures demands the best possible predictive information on hazards and vulnerability, including its full socio-economic evaluation so that the costs and benefits of the mitigation options can be realistically assessed.

R.S. Arthurton / Ocean & Coastal Management 40 (1998) 65–85   83

Acknowledgements

The author is grateful to Chris Evans, Robin Wingfield, Martin Culshaw, Kerry Turner and Edmund Penning-Rowsell for their critical review of the manuscript. The paper is published with the approval of the Director, British Geological Survey (N.E.R.C.).

References

[1] Vallega A. Sea management. A theoretical approach. London: Elsevier Applied Science, 1992.

[2] Nicholls RJ. Coastal megacities and climate change. GeoJournal, 1995;37(3):369–79.

[3] Fernandez C, Gordon J. Natural disasters and their human consequences – overcoming the vacuum between humanitarian aid and long-term rehabilitation. In: Merriman PA, Browitt CWA, editors. Natural disasters: protecting vulnerable communities. London: Thomas Telford, 1993:432–46.

[4] Houghton JT, Jenkins GJ, Ephraums JJ, editors. Climate change: the IPCC scientific assessment. Intergovernmental Panel on Climate Change. Cambridge: Cambridge University Press, 1990.

[5] Amadore L, Bolhofer WC, Cruz RV, Feir RB, Freysinger CA, Guill S, Jalal KF, Inglesias A, Jose A, Leatherman S, Lenhart S, Mukherjee S, Smith JB, Wisniewski J. Climate change vulnerability and adaption in Asia and the Pacific: workshop summary. Water, Air and Soil Pollution, 1996;92:1–12.

[6] Carey JJ, Mieremet RB. Reducing vulnerability to sea level rise: international initiatives. Ocean & Coastal Management, 1992;18:161–77.

[7] Pernetta JC, Elder DL. Climate, sea level rise and the coastal zone: management and planning for global changes. Ocean and Coastal Management 1992;18:113–60.

[8] Arthurton RS. Physical environmental change and coastal zone management: estimation of economic consequences. In: Haq BU, Haq SM, Kullenberg G, Stel JH, editors. Coastal zone management imperative for maritime developing nations. Dordrecht: Kluwer Academic Publishers, 1997:93–8.

[9] UN-DHA. Hierarchy of disaster management terms. New York: United Nations Department of Humanitarian Affairs, 1992.

[10] Aysan YF. Vulnerability assessment. In: Merriman PA, Browitt CWA, editors. Natural disasters: protecting vulnerable communities. London: Thomas Telford, 1993:1–13.

[11] Horlick-Jones T, Jones DKC. Communicating risks to reduce vulnerability. In: Merriman PA, Browitt CWA, editors. Natural disasters: protecting vulnerable communities. London: Thomas Telford, 1993:25–37.

[12] Alley EE. Combatting the vulnerabilities of communities. In: Merriman PA, Browitt CWA, editors. Natural disasters: protecting vulnerable communities. London: Thomas Telford, 1993:67–77.

[13] Turner RK, Subak S, Adger WN. Pressures, trends and impacts in coastal zones: interactions between socioeconomics and natural systems. Environmental Management, 1996;20(2):159–173.

[14] Turner RK, Adger WN. Coastal zone resources assessment guidelines. LOICZ/R & S/96-4, LOICZ, Texel, The Netherlands, 1996.

[15] Haq BU. Regional and global oceanographic, climatic and geological factors in coastal zone planning. In: Haq BU, Haq SM, Kullenberg G, Stel JH, editors. Coastal zone management imperative for maritime developing nations. Dordrecht: Kluwer Academic Publishers, 1997:55–74.

[16] Hunt JCR. The contribution of meteorological science to wind hazard mitigation. In: Windstorm: coming to terms with Mankind's worst natural hazard. London: Royal Academy of Engineering, 1995:8–17.

[17] Madrell RJ. Alleviation of natural disasters in the south-western area of Bangladesh. In: Merriman PA, Browitt CWA, editors. Natural disasters: protecting vulnerable communities. London: Thomas Telford, 1993:51–66.

[18] Dawson AG. Shi S. Smith DE, Long D. Geological investigations of tsunami generation and long-term tsunami frequency. In: Merriman PA, Browitt CWA. editors. Natural disasters: protecting vulnerable communities. London: Thomas Telford, 1993:140–55.

[19] Yeh HH. Tsunami bore run-up. In: Bernard EN editor. Tsunami hazard: a practical guide for tsunami hazard reduction. Dordrecht: Kluwer Academic Publishers. 1991:209–20.

[20] Nagano O. Imamura F. Shuto N. A numerical model for far-field tsunamis and its application to predict damages done to aquaculture. In: Bernard EN, editor. Tsunami hazard: a practical guide for tsunami hazard reduction. Dordrecht: Kluwer Academic Publishers. 1991:235–55.

[21] Intergovernmental Oceanographic Commission, IOC workshop on the technical aspects of tsunami warning systems, tsunami analysis, preparedness, observation and instrumentation. IOC Workshop Report No. 58, UNESCO. Paris, 1991.

[22] Bernard EN. Fourteenth International Tsunami Symposium: opening address. In: Bernard EN editor. Tsunami hazard: a practical guide for tsunami hazard reduction. Dordrecht: Kluwer Academic Press, 1991:115–7.

[23] Kovalev PD, Rabinovich AB. Shevchenko GV. Investigation of long waves in the tsunami frequency band on the southern shelf of Kamchatka. In: Bernard EN, editor. Tsunami hazard: a practical guide for tsunami hazard reduction. Dordrecht: Kluwer Academic Press. 1991:141–59.

[24] Broadus JM. Economising human responses to subsidence and rising sea level. In: Milliman, JD, Haq BU, editors. Sea-level rise and coastal subsidence. Dordrecht: Kluwer Academic Publishers, 1996:313–25.

[25] Warrick R, Oerlemans J. Sea level rise. In: Houghton JT, Jenkins GJ, Ephraums JJ, editors. Climate change: the IPCC scientific assessment. Intergovernmental Panel on Climate Change. Cambridge: Cambridge University Press, Cambridge: 1990:257–81.

[26] Pirazzoli PA. Sea level changes: the last 20 000 years. Chichester: Wiley, 1996.

[27] Raper SCB, Warrick RA, Wigley TML. Global sea-level rise: past and future. In: Sea-level rise and coastal subsidence: causes, consequences and strategies. Dordrecht: Kluwer Academic Publishers, 1996:11–45.

[28] Jelgersma S, Van der Zijp M, Brinkman R. Sealevel rise and the coastal lowlands in the developing world. Journal of Coastal Research, 1993;9(4):958–72.

[29] Jelgersma S. Land subsidence in coastal lowlands. In: Milliman JD, Haq BU, editors. Sea level rise and coastal subsidence. Dordrecht: Kluwer Academic Publishers, 1996:47–62.

[30] Adams B, Grimble R, Shearer TR, Kitching R, Calow R, Chen DJ, Cui XD, Yu ZM. Aquifer overexploitation in the Hangu region of Tianjin, People's Republic of China. BGS Technical Report WC/94/42R, Keyworth: British Geological Survey, 1994.

[31] Fairbanks RG. A 17 000-year glacio-eustatic sea level record: influence of glacial melting rates on the Younger Dryas event and deep ocean circulation. Nature. 1989;342:637–42.

[32] Chen XQ. An integrated study of sediment discharge from the Changjiang River, China, and the delta development since the mid-Holocene. Journal of Coastal Research 1996;12(1):26–37.

[33] Ren M-e, Milliman JD. Effect of sea-level rise and human activity on the Yangtze delta, China. In: Milliman JD, Haq BU. editors. Sea level rise and coastal subsidence, Dordrecht: Kluwer Academic Publishers, 1996:205–14.

[34] Milliman JD, Haq BU. editors. Sea level rise and coastal subsidence. Dordrecht: Kluwer Academic Publishers. 1996.

[35] Nutalaya P, Yong RN. Chumnakit T, Buapeng S. Land subsidence in Bangkok during 1978–1988. In: Milliman JD, Haq BU. editors. Sea level rise and coastal subsidence. Dordrecht: Kluwer Academic Publishers, 1996:105–30.

[36] UNEP, Groundwater: a threatened resource. UNEP Environment Library No. 15, United Nations Environment Programme, Nairobi, 1996.

[37] Ren M-e. Relative sea level rise in China and its socio-economic implications. Marine Geodesy 1994;17:37–44.

[38] Wang Y. Sea-level changes, human impacts and coastal responses in China. Journal of Coastal Research 1998;14(1):31–6.

[39] Haq BU, Milliman JD. Coastal vulnerability: hazards and strategies. In: Milliman JD, Haq BU, editors. Sea level rise and coastal subsidence. Dordrecht: Kluwer Academic Publishers, 1996:357–64.

R.S. Arthurton   Ocean & Coastal Management 40 (1998) 65–85   85

[40] Bird ECF. Submerging coasts: the effects of a rising sea level on coastal environments. Chichester: Wiley, 1993.

[41] Institution of Civil Engineers. Megacities: reducing vulnerability to natural disasters. London: Thomas Telford, 1995.

[42] Green CH, Parker DJ, Penning-Rowsell EC. Designing for failure. In: Merriman PA, Browitt CWA, editors. Natural disasters: protecting vulnerable communities. London: Thomas Telford, 1993:78–91.

[43] Lyne WH, Radford AM. Prediction of exceptionally severe storms. In: Merriman PA, Browitt CWA, editors. Natural disasters: protecting vulnerable communities. London: Thomas Telford, 1993:84–193.

[44] Simm JD, Brampton AH, Beech NW, Brooke JS. Beach management manual. Construction Industry Research and Information Association, Report 153. London: CIRIA, 1996.

[45] Davis IR, Bickmore D. Data management for disaster planning. In: Merriman PA, Browitt CWA, editors. Natural disasters: protecting vulnerable communities. London: Thomas Telford, 1993:547–65.

[9]

Adapting to Sea-Level Rise: Relative Sea-Level Trends to 2100 for the United States

ROBERT J. NICHOLLS

School of Geography and Environmental Management
University of Middlesex
Queensway
Enfield, Middlesex, United Kingdom

STEPHEN P. LEATHERMAN

Laboratory for Coastal Research
Department of Geography
University of Maryland
College Park, Maryland, USA

Global sea levels have slowly risen during this century, and that rise is expected to accelerate in the coming century due to anthropogenic global warming. A total rise of up to 1 m is possible by the year 2100 (relative to 1990). To deal with this change, coastal managers require site-specific information on relative (i.e., local) changes in sea level to determine what might be threatened. Therefore as a first step, global sea-level rise scenarios need to be transformed into relative sea-level change scenarios which take account of local and regional factors, such as vertical land movements, in addition to global changes. Even present rates of relative sea-level rise have important long-term implications for coastal management—projecting existing trends predicts a relative sea-level rise from 1990 to 2100 of up to 0.4 m and 1.15 m for the Mid-Atlantic Region and Louisiana, respectively. Ignoring sea-level rise will lead to unwise decisions and increasing hazard with time.

This article adapts the Intergovernmental Panel on Climate Change (IPCC) global scenarios for sea-level rise (Warrick et al., 1996) to three relative sea-level rise scenarios for the contiguous United States. These scenarios cover the period 1990 to 2100 and provide a basis to assess possible proactive measures for sea-level rise. However, they are subject to the same uncertainties as the global scenarios as most of the sea-level rise will occur decades into the future. When considering what should be done now in response to future sea-level rise, given these large uncertainties, it is best to identify (1) low-cost, no regret responses which would maintain or enhance the choices available to tomorrow's coastal managers; and (2) sectors where reactive adaptation would have particularly high costs and where allowance for future sea-level rise can be considered a worthwhile "insurance policy." Sea-level rise will impact an evolving coastal landscape which already is experiencing a range of other pressures. Therefore, to be most effective, responses to sea-level rise need to be integrated with all other planning occurring in the coastal zone.

Keywords adaptation to sea-level rise, climate change, sea-level rise, subsidence

Received 1 January 1996; accepted 17 June 1996.

This research was made possible through a grant from the W. Alton Jones Foundation, Charlottesville, VA. The authors are indebted to Bruce Douglas who kindly provided unpublished sea-level data and trends up to 1991, and commented on earlier drafts of this manuscript. This article also benefited from the reviews of Jim Titus and two anonymous reviewers. Guan-hong Lee performed some of the initial calculations.

Address correspondence to Robert J. Nicholls, School of Geography and Environmental Management, University of Middlesex, Queensway, Enfield, Middlesex EN3 4SF, United Kingdom.

Coastal Management, 24:301–324, 1996
Copyright © 1996 Taylor & Francis
0892-0753/96 $12.00 + .00

A significant and growing proportion of the U.S. population lives within the coastal zone (Culliton et al., 1990), while about half of all residential and nonresidential construction between 1970 and 1989 has occurred in coastal areas (Culliton et al., 1992). Presently, many structures in the coastal zone are not adequately above existing flood levels and their survival and the safety of residents can be threatened by major storm activity. At the same time, global (or eustatic) sea levels have risen over the past century, and they are expected to rise more rapidly in the coming century with a possible additional rise of up to 1 m by 2100 (Warrick et al., 1996; Wigley & Raper, 1992). The general effects of such a rise in sea level on coastal lowlands are well known and include coastal erosion, inundation, saltwater intrusion, and higher water tables, as well as exacerbation of the existing storm flooding hazard as mentioned above (Leatherman et al., 1995; National Research Council, 1987). National assessments have suggested that a 1-m rise in global sea levels could have significant impacts, including the inundation of about 30,000 km^2 of land divided equally between wetlands and upland (Titus et al., 1991), while the coastal floodplain defined by a 100-year event could increase by 38% or at least 19,000 km^2 (Federal Emergency Management Agency [FEMA], 1991).

As a first step toward responding to these potential problems, coastal managers might ask: (1) What might be threatened by sea-level rise? (2) What responses are available? (3) How can the best option be selected? and (4) When should a response be implemented? The answers are conditioned by the fact that the impacts of accelerated sea-level rise will depend on both the magnitude of sea-level rise and the human responses to that change (Turner et al., 1996). Given the large and growing investment in the U.S. coastal zone, a "do nothing" policy will lead to unwise decisions, such as construction in the areas most vulnerable to sea-level rise. In contrast, a well-planned response that seeks to anticipate the physical impacts of sea-level rise in a timely fashion will minimize such problems and result in a wider range of options for future generations and lower costs for reactive responses (Nicholls & Leatherman, 1995; Office of Technology Assessment, 1993).

Coastal policymakers require knowledge of future changes in sea level to begin such planning. While scenarios for global change in sea level are widely available (e.g., Warrick et al., 1996; Wigley & Raper, 1992), they often are mistakenly applied directly at local scales. However, it is the relative (or local) rise in sea level which determines the magnitude of impacts at any site (Barth & Titus, 1984; National Research Council, 1987; Titus & Narayanan, 1995). Therefore, global sea-level rise scenarios must be transformed into relative sea-level rise scenarios for vulnerability assessment and coastal planning at specific sites.

Several detailed vulnerability assessments are available within the United States (Barth & Titus, 1984; Daniels, 1992; London & Volonté, 1991; Titus, 1991a), as well as regional and national studies including potential losses of coastal infrastructure (Yohe, 1990), protection costs and wetland losses (Smith & Tirpak, 1989; Titus et al., 1991), impacts on the National Flood Insurance Program (FEMA, 1991), and integrated assessment using a coastal vulnerability index approach (Gornitz et al., 1991, 1994). However, only a few studies provide guidance concerning local and state planning for sea-level rise (e.g., Craig, 1993; State of Maine, 1995).

Although transforming global sea-level rise scenarios to relative sea-level rise scenarios has been considered previously (e.g., National Research Council, 1987), a comprehensive national analysis for coastal management purposes has not been published. Further, the authors are concerned that despite the considerable attention given to sea-level rise in the scientific press, it still is not adequately addressed in U.S. coastal management and planning. To assist such efforts, this article presents scenarios of relative sea-level rise for the contiguous U.S. coast based on the recent Intergovernmental Panel on Climate Change (IPCC) Second Assessment Report scenarios for global sea-level rise (Warrick et al., 1996). These scenarios cover the time period 1990 to 2100. The article also considers the inherent uncertainties associated with these scenarios, hence their use in the evaluation and implementation of proactive adaptation measures for sea-level rise.

Relative Versus Global Sea-Level Rise

Over the timescale of interest (100 to 200 years), global sea-level rise is the result of an increase in global ocean volume. In addition to global changes, relative movements of sea level may occur due to vertical movements of the land. No location on the earth's surface can be considered stable, and vertical movements occur due to tectonic (e.g., plate movements), neotectonic (e.g., postglacial rebound [PGR]) and anthropogenic (e.g., groundwater extraction) causes (Emery & Aubrey, 1991; Parker, 1991). Relative sea-level rise (or fall) comprises the cumulative effect of all these local, regional, and global components, regardless of their cause; consequently its magnitude varies from place to place unlike global sea-level rise.

Tide gauges provide a direct measurement of relative sea-level rise as shown for New York City in Figure 1. There is a long-term rising trend, but with considerable scatter around this trend on an annual to decadal basis; hence in some periods sea level falls (e.g., 1920 to 1930 in Figure 1), while in others it rises more rapidly than the long-term trend (e.g., 1930 to 1950 in Figure 1). This is typical of relative sea-level observations. From a management perspective, the trend is of most interest as it is long-term rises in sea level which cause the most serious impacts. Based on the above discussion, the trend comprises two main components: (1) the global change in sea level, and (2) the local and regional vertical movement of the ground. Therefore, the vertical land component can be estimated if global sea-level rise is known.

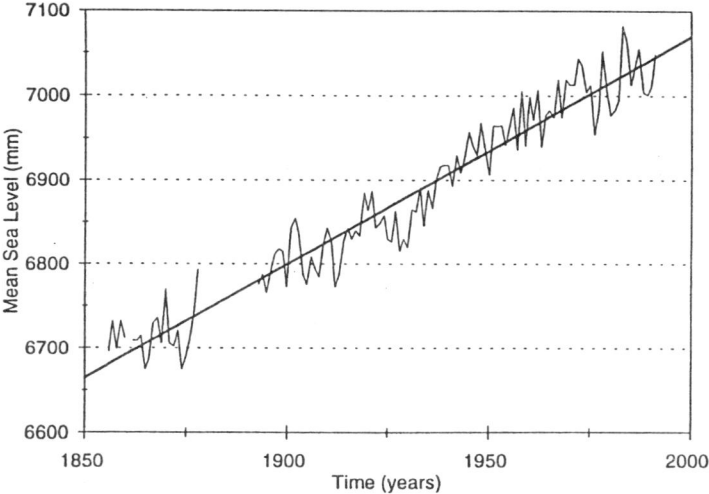

Figure 1. An example of relative sea-level rise data. Annual data for New York City (The Battery) from 1856 to 1991, with some early gaps, and the linear regression.

Based on the analysis of tide gauge records, it has been estimated that global sea-level rise has occurred over the past century (Warrick et al., 1996; Warrick & Oerlemans, 1990). This rise is primarily due to thermal expansion of seawater and the melting of land-based ice. Estimates of global sea-level rise range from 1 to 3 mm/yr, depending on the method employed (Douglas, 1995; Gornitz, 1995). In this study, Douglas's (1991, 1992) estimate of global sea-level rise from 1880 to 1980 was used—1.8 + 0.1 mm/yr with no evidence of any acceleration. This is one of the most comprehensive analyses to date, and it both is consistent with several other recent independent analyses (Peltier & Tushingham, 1989; Trupin & Wahr, 1990) and corresponds to the median IPCC estimate for sea-level rise over the past century (Warrick et al., 1996).

In the next century, global sea-level rise is expected to rise more rapidly due to anthropogenically produced global warming (Warrick & Oerlemans, 1990). This rise will result primarily from the continued thermal expansion of seawater and the melting of land-based ice in small glaciers. Greenland is likely to be a relatively small source of sea-level rise, while the ice mass stored in Antarctica is expected to increase given global warming, due to greater snowfall, partially offsetting the other sources of rise. The likely magnitude of these sources, and thus the estimates of future sea-level rise, have a large range of uncertainty. The IPCC has expressed this uncertainty by developing three scenarios for global sea-level rise from 1990 to 2100: low, best, and high (Warrick et al., 1996; Warrick and Oerlemans, 1990). Including the cooling effects of aerosols, these three scenarios are 8 cm, 20 cm, and 38 cm by 2050 and 20 cm, 49 cm, and 86 cm by 2100 (all relative to 1990). Probability-based scenarios for global sea-level rise recently have been published (Titus & Narayanan, 1995). These results further stress the uncertainty concerning future global sea levels. The 1 to 99 percentiles from 1990 to 2100 embrace a drop of 1 cm to a rise of 104 cm, with a median rise of 34 cm which is somewhat smaller than the IPCC best estimate scenario (49 cm). For this reason, the probability-based results are not used in this article.

At a regional scale, departures from these global estimates seem likely due to regional oceanographic factors. Ocean dynamics produce positive and negative anomalies on the ocean surface of up to 1 m. Therefore, changes in oceanic circulation, wind and pressure patterns, and ocean-water density also may cause local and regional sea-level change. Warrick et al. (1996) have estimated that net regional sea-level rise could be as much as two or three times the global rise, although no regional scenarios suitable for coastal planning have yet been available. This adds further uncertainty to future rates of relative sea-level rise. Given these substantial uncertainties, the authors have utilized the widely accepted IPCC scenarios in this article. However, in the long term, the authors favor the probability-based approach because it offers scenarios in a form more suitable for coastal management purposes.

Relative Sea-Level Change Analysis

In the United States, annual mean sea-level data are provided from water elevation measurements conducted at National Oceanic and Atmospheric Administration (NOAA) tide stations. The data to 1986 are summarized in Lyles et al. (1988). In four cases, these data extend back to the past century: San Francisco (1855), New York (1856), Fernandina Beach (1898), and Seattle (1899). While tidal data was originally collected for navigational and tidal prediction purposes, great care was taken to relate the measurements of these gauges to stable benchmarks on land. This has provided a data set which is suitable for precise analysis of sea-level change (Douglas, 1991; Parker, 1991). However, not all the NOAA data can be used in long-term analysis.

Short duration tide gauge records are of little use for determining the trend of relative sea level because of the large interannual fluctuations of sea level (Figure 1), and ideally at least 50 years of data are required (Douglas, 1995). This point is illustrated by comparing some sea-level trends on the East Coast between Eastport, Massachusetts, and Key West, Florida, for the longest period of record (>50 years) and from 1970 to 1991 (i.e., 21 years) (Figure 2). While the full record shows a rising trend at all the stations with a range from 1.6 to 4.2 mm/yr, the 1970 to 1991 results show a relative fall in sea level at 17 out of 22 stations considered, with a range from –4.7 to 1.0 mm/yr. Therefore, consideration of the 1970 to 1991 data alone would lead to the erroneous conclusion that sea level is falling along much of the East Coast.

Similarly, data gaps that are significant relative to the data record length may bias the derived trend (Douglas, 1991). In this study, only data which passed the following criteria were utilized: (1) those with a minimum record length of 37 years (or two 18.6-year lunar nodal cycles), (2) those with data gaps occupying less than 20% of the record length, and (3) those with statistically significant trends (at 95% confidence). Sea-level data from 39 stations satisfied these requirements. In addition, Crescent City, California, was also considered because there was lack of West Coast data and the regression was >94% confidence.

To maximize the record length, unpublished data from 1986 to 1991 was added to

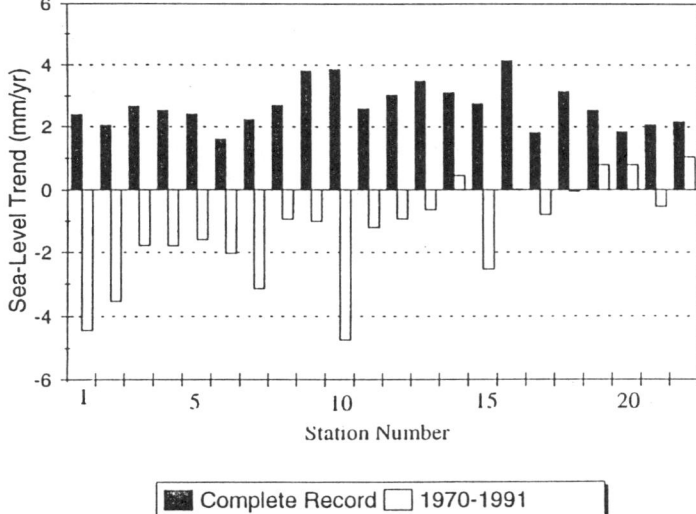

Complete Record □ 1970-1991

Figure 2. A comparison of relative sea-level change for 22 stations along the U.S. East Coast from Eastport, MA to Key West, FL, derived from the complete record period (>50 years) and 1970–91 (21 years). Stations are as follows: 1, Eastport, MA; 2, Portland, ME; 3, Boston, MA; 4, Woods Hole, MA; 5, Newport, RI; 6, New London, CT; 7, Willets Point, NY; 8, New York, NY; 9, Sandy Hook, NJ; 10, Atlantic City, NJ; 11, Philadelphia, PA; 12, Baltimore, MD; 13, Annapolis, MD; 14, Solomons Island, MD; 15, Washington, DC; 16, Hampton Roads, VA; 17, Wilmington, NC; 18, Charleston, SC; 19, Fort Pulaski, GA; 20, Fernandina Beach, FL; 21, Mayport, FL; 22, Key West, FL.

that contained in Lyles et al. (1988) for all the stations. Four records are shorter than the recommended 50-year criteria: St. Petersburg, Florida (44 years), Grand Isle, Louisiana (44 years), Port Isabel, Texas (47 years), and Port San Luis, California (45 years).

Variations in Sea-Level Change

Relative sea-level changes based on the 40 stations for the U.S. coast are presented in Figure 3. To facilitate visual interpretation, the estimates of relative sea-level change are grouped into five discrete classes with an inset for the Gulf Coast. Note the high density of data available between Portland, Maine, and Portsmouth, Virginia, as compared to other coastal areas, particularly the large gap between San Francisco and Crescent City, California, and again to Astoria, Oregon. Overall, sea level rose at 37 out of the 40 stations, showing that relative sea-level rise is already an important long-term issue for coastal management in the United States. Even those who are skeptical about the likelihood of global warming should expect the same rates of relative sea-level rise through the next century.

Regional Trends

The observed relative sea-level change shows regional variability which can be related to large-scale geological processes, such as plate tectonic setting and postglacial rebound (PGR) and hence, the rates of land surface uplift or subsidence (Emery and Aubrey, 1991; Inman & Nordstrom, 1971; Peltier, 1986). PGR is used to describe the isostatic readjustment of the crust due to changes in glacial loading over the last glaciation and deglaciation cycle (30,000 to 6,000 years ago) and can produce both an upward rebound and a downward subsidence. Contemporary vertical land movements due to PGR are often significant when compared to observed changes in relative sea level (Douglas, 1991, 1995), and therefore need to be considered in this type of analysis.

In Table 1 the observed relative sea-level change (RSL) is divided into global (GSL) and local components (LC). The LC is then compared to estimates of PGR from the model of Tushingham and Peltier (1991) and the residual sea-level change (RC) which is unexplained by PGR and GSL:

$$RSL = GSL + LC \qquad\qquad (1a)$$
$$LC = PGR + RC \qquad\qquad (1b)$$

Since subsidence of the land causes an increase in RSL, the sign convention adopted for PGR in Table 1 is positive when the land is sinking so that PGR values can be directly added to GSL. The residual changes are assumed to be due to other geological processes.

On the East and Gulf Coasts, relative sea-level rise has occurred at all stations. Considering the East Coast, from Virginia northward, PGR is causing significant subsidence (>1 mm/yr) (Douglas, 1991), and thus is contributing to relative sea-level rise (Table 1). The largest rise of between 3.0 and 4.5 mm/yr occurs between Sandy Hook, New Jersey, and Portsmouth, Virginia. To the north, including Philadelphia and New York, the high rates of subsidence predicted due to PGR appear to be partially offset by a regional uplift due to other causes. To the south, Wilmington, North Carolina, appears to be experiencing long-term geological uplift along the Cape Fear arch, while Charleston, South Carolina, is subsiding (Gornitz & Seeber, 1990).

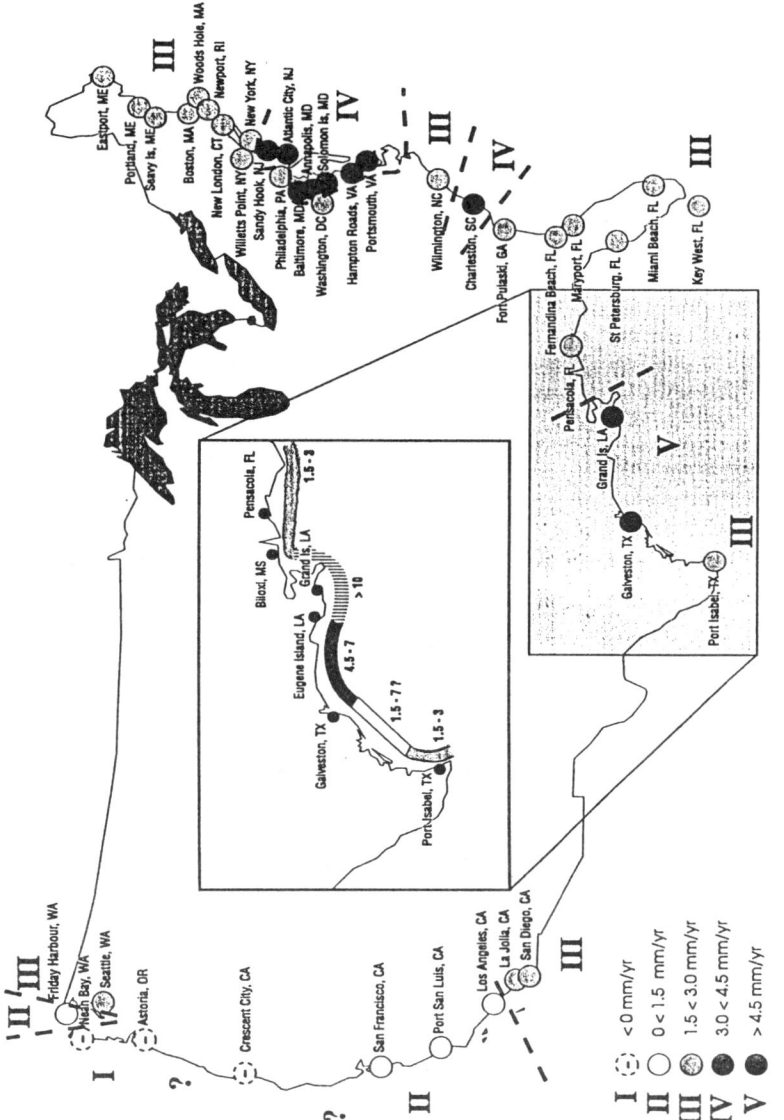

Figure 3. Relative sea-level trends around the United States divided into five classes. The inset for the Gulf Coast is adapted from Penland and Ramsey (1990).

Water Resources and Coastal Management

Table 1

Sea-level trends around the contiguous U.S. coast with relative sea-level components
(mm/yr) calculated using a global sea-level rise of 1.8 mm/yr

Station	RSL	LC	PGR	RC
Eastport, ME	2.4	0.6	1.8	−1.2
Portland, ME	2.1	0.3	1.0	−0.8
Seavey Island, ME	1.7	0.1	1.1	−1.2
Boston, MA	2.7	0.9	1.3	−0.4
Woods Hole, MA	2.5	0.7	1.8	−1.1
Newport, RI	2.4	0.6	1.6	−1.0
New London, CT	1.6	−0.2	1.5	−1.7
Willets Point, NY	2.3	0.5	1.5	−1.1
New York, NY	2.7	0.9	1.4	−0.5
Sandy Hook, NJ	3.8	2.0	1.4	0.6
Atlantic City, NJ	3.9	2.1	1.7	0.4
Philadelphia, PA	2.6	0.8	1.5	−0.7
Baltimore, MD	3.1	1.3	1.3	−0.1
Annapolis, MD	3.5	1.7	1.4	0.3
Solomons Island, MD	3.1	1.3	1.3	0.0
Washington, DC	2.8	1.0	1.3	−0.3
Hampton Roads, VA	4.2	2.4	1.2	1.2
Portsmouth, VA	3.8	2.0	1.2	0.8
Wilmington, NC	1.8	0.0	0.6	−0.6
Charleston, SC	3.2	1.4	0.2	1.2
Fort Pulaski, GA	2.6	0.8	0.0	0.8
Fernandina Beach, FL	1.8	0.0	−0.2	0.2
Mayport, FL	2.1	0.3	−0.2	0.5
Miami Beach, FL	2.4	0.6	−0.4	1.0
Key West, FL	2.2	0.4	−0.4	0.8
St. Petersburg, FL	2.4	0.6	−0.3	0.9
Pensacola, FL	2.2	0.4	−0.2	0.6
Grand Isle, LA	10.4	8.6	0.0	8.6
Galveston, TX	6.4	4.6	−0.1	4.7
Port Isabel, TX	2.7	0.9	−0.4	1.3
San Diego, CA	2.1	0.3	−0.6	0.9
La Jolla, CA	2.1	0.3	−0.6	0.9
Los Angeles, CA	0.7	−1.1	−0.6	−0.5
Port San Luis, CA	0.9	−0.9	−0.6	−0.3
San Francisco, CA	1.3	−0.5	−0.4	−0.1
Crescent City, CA	−1.0[a]	−2.8	0.1	−2.9
Astoria, OR	−0.5	−2.3	0.7	−3.0
Neah Bay, WA	−1.6	−3.4	0.0	−3.4
Seattle, WA	2.0	0.2	−0.7	0.9
Friday Harbour, WA	0.9	−0.9	−1.1	0.2

Postglacial rebound (PGR) is derived from the model of Tushingham and Peltier (1991).

Abbreviations: RSL, relative sea-level change (from tide gauge records); LC, local component; GSL, global component; PGR, postglacial rebound; RC, residual change, unexplained by GSL and PGR.

[a]A minus indicates a relative fall in sea level (or a relative rise in land level).

The Florida peninsula is experiencing relatively low rates of sea-level rise (1.5 to 3.0 mm/yr), but the neighboring areas of Louisiana and Texas are experiencing the highest rates of relative sea-level rise in the nation of up to 10.4 mm/yr at Grand Isle, Louisiana. At Galveston, Texas, sea-level rise is 6.4 mm/yr, declining to 2.7 mm/yr at Port Isabel, Texas. The high rates of subsidence in Louisiana are attributed to long-term geological processes, including the natural compaction of sedimentary deposits (Emery & Aubrey, 1991).

Penland and Ramsey (1990) used additional data from U.S. Army Corps of Engineers (USACE) tide gauges to examine sea-level change along the Gulf Coast (see inset in Figure 3). These records are typically 40+ years in duration. West of Eugene Island (the Chernier plain), rates of relative sea-level rise are in the range of 5 to 6 mm/yr, which is similar to Galveston. East of Eugene Island (the Mississippi River delta plain), rates of relative sea-level rise abruptly increase and generally exceed 10 mm/yr, showing that the results from Grand Isle are typical of this area. A USACE tide gauge at Biloxi, Michigan, suggests relative sea-level rise of 1.5 mm/yr, similar to Pensacola, Florida, although this record is only 27 years long (Penland & Ramsey, 1990; Turner, 1991). This pattern of relative sea-level rise can be related to the thickness of Holocene deposits. The greatest rate of relative sea-level rise is associated with the Mississippi River delta plain where the underlying Holocene section is thickest and the RC at Grand Isle indicates subsidence at 8.6 mm/yr (Table 1). Therefore, well-defined provinces exist which allow future changes to be predicted.

The large change in the rate of relative sea-level rise between Galveston and Port Isabel (nearly 4 mm/yr) is difficult to interpret without intermediate data. While the Port Isabel data provide a minimum rate of sea-level rise, the absolute magnitudes of relative sea-level rise remain more uncertain for much of the Texas coast than elsewhere along the Gulf and East Coasts (Figure 3).

Sea-level trends along the West Coast show smaller rises or even a fall (Table 1) and display a more spatially erratic behavior than the other coasts, largely because of tectonism along this active plate boundary (Douglas, 1991; Inman & Nordstrom, 1971). For instance, leveling surveys, as well as the sea-level measurements themselves, indicate a pattern of recent uplift along the coast of the Olympic peninsula, coupled with subsidence farther inland (Ando & Balazs, 1979). Thus, residual land movements at Neah Bay, Washington, and Astoria, Oregon, show uplift, whereas those at Friday Harbor and Seattle, Washington, show subsidence (Table 1). In management terms, care should be taken in interpolating and extrapolating the limited measurements of RSL along the West Coast because relative sea-level trends may vary significantly over small spatial areas. The large spatial gaps between some of the data points compound this problem.

The LC of relative sea-level rise alone (Table 2) emphasizes that many areas are subsiding, particularly the Mid-Atlantic Region and Louisiana and Texas, and these areas would experience relative sea-level rise without any global increase in sea level. For further discussions of the underlying geological processes which contribute to variations in relative sea-level rise, see Emery and Aubrey (1991), Gornitz and Lebedeff (1987), and Gornitz and Seeber (1990).

The Role of Human Induced Subsidence

The possible role of human induced subsidence due to the withdrawal of subsurface fluids needs to be considered, as this could contribute to the unexplained RC shown in

Table 2
The local component (LC) of relative sea-level rise (from Table 1)
projected from 1990 to 2050 and 1990 to 2100

Station	1990 to 2050 (cm)	1990 to 2100 (cm)
Eastport, ME	4	7
Portland, ME	1	3
Seavey Island, ME	0	−1
Boston, MA	5	10
Woods Hole, MA	4	8
Newport, RI	4	7
New London, CT	−1[a]	−2
Willets Point, NY	3	5
New York, NY	5	10
Sandy Hook, NJ	12	22
Atlantic City, NJ	12	23
Philadelphia, PA	5	9
Baltimore, MD	7	14
Annapolis, MD	10	18
Solomons Island, MD	8	15
Washington, DC	6	11
Hampton Roads, VA	14	26
Portsmouth, VA	12	22
Wilmington, NC	0	0
Charleston, SC	8	15
Fort Pulaski, GA	5	8
Fernandina Beach, FL	0	0
Mayport, FL	2	3
Miami Beach, FL	4	7
Key West, FL	2	4
St. Petersburg, FL	4	7
Pensacola, FL	2	4
Grand Isle, LA	52	95
Galveston, TX	27	50
Port Isabel, TX	6	10
San Diego, CA	2	3
La Jolla, CA	2	3
Los Angeles, CA	−7	−12
Port San Luis, CA	−6	−10
San Francisco, CA	−3	−5
Crescent City, CA	−17	−31
Astoria, OR	−14	−26
Neah Bay, WA	−20	−30
Seattle, WA	1	2
Friday Harbour, WA	−5	−10

[a] A minus indicates a relative fall in sea level (or a relative rise in land level).

Table 1. A number of U.S. coastal areas are known to have subsided during this century due to this cause, including localized areas in the Mississippi delta (Boesch et al., 1994); Savannah, Georgia; Houston, Texas, including Galveston Bay; Long Beach, California; and San Jose, California (Holzer, 1991; Holzer & Johnston, 1985). The maximum subsidence was 9 m at Long Beach (due to oil extraction), although subsidence in excess of 1 m affected only an area of about 30 km². In the Houston area, 13,500 km² has subsided more than 30 cm, necessitating coastal abandonment or increased protection around Galveston Bay (Holdahl et al., 1989; Holzer, 1991).

However, human-induced subsidence appears to be a reasonably local phenomenon which does not seem to have contributed significantly to the observed trends in relative sea-level rise shown in Figure 3. For instance, for the Galveston record, the possible contribution of subsidence in the Houston area as discussed above must be considered. The subsidence increased after 1945 due to excessive groundwater withdrawal, and subsequently decreased in the 1970s when policies to control the subsidence were successfully implemented (Holdahl et al., 1989). However, an unambiguous corresponding acceleration and deceleration of sea-level rise is not evident in the Galveston data (Figure 4), suggesting that Galveston has been subsiding due to other processes and the contribution to relative sea-level rise due to groundwater withdrawal is relatively minor (see also Turner, 1991). This supports using the Galveston record as a representative record for this region with the caveats noted earlier concerning the Texas coast. From a local perspective, the historic influence of subsidence on relative sea-level rise trends around Galveston Bay requires further investigation.

Around the Chesapeake Bay, it has been argued that subsidence from overpumping of surficial aquifers and sediment loading both are important regional contributors to relative sea-level rise (Gornitz & Seeber, 1987; Kearney & Stevenson, 1991). Davis (1987) reported considerable subsidence from large-scale groundwater withdrawal in the lower Chesapeake Bay (and broad areas of the Atlantic seaboard) since early in this century. However, if PGR is considered, the residual change around the Chesapeake

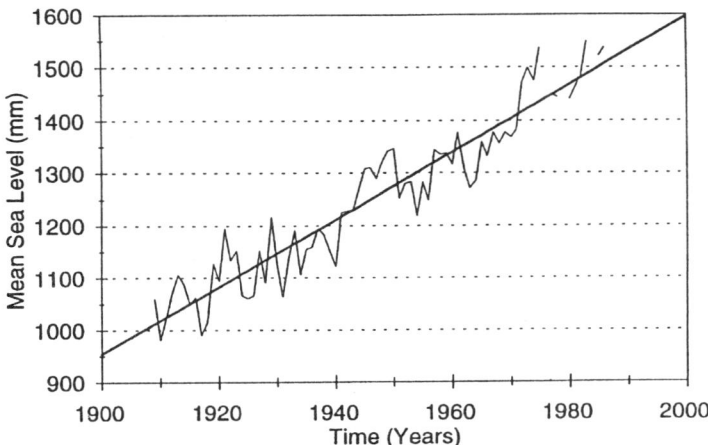

Figure 4. The relative sea-level curve for Galveston, TX, from 1909 to 1986, including the linear regression.

Bay is less than 1 mm/yr except at Hampton Roads, Virginia (see Table 1), and again, significant subsidence appears localized. Further, any subsidence is expected to continue and contribute to future changes in relative sea level.

Therefore, while human-induced subsidence locally is a major contributor to relative sea-level rise, it does not appear to contribute substantially to the patterns shown in Figure 3. More quantitative investigation of the magnitude of human-induced subsidence is also recommended (see below).

Future Trends of Relative Sea Level

Based on these observations of relative sea-level rise, any scenario of global sea-level rise can be transformed into a relative sea-level rise scenario for the U.S. coast. Using global sea-level rise observations (Douglas, 1991, 1992) and the IPCC global scenarios (Warrick et al., 1996), three global scenarios were selected. Following National Research Council (1987), they all are of the form (units in m):

$$GSL(t) = at + bt^2 \qquad (2)$$

where t is the time in years since 1990 and a and b are constant coefficients given in Table 3. This form allows prediction for any year of interest between 1990 and 2100. The low-rise scenario assumes no acceleration and simply continues the observed trend at each station, which is similar to the IPCC low scenario. The mid-rise scenario approximates the IPCC best estimate scenario and gives a total rise of 50 cm by 2100. The high-rise scenario approximates the IPCC high estimate scenario. This requires an almost immediate acceleration of global sea-level rise to 5 mm/yr and gives a total rise of 90 cm by 2100. Therefore, the resulting relative sea-level rise scenarios (units in m) are:

$$RSL(t) = at + bt^2 + (LC/1,000)t \qquad (3)$$

where LC is the local component in mm/yr derived from Table 1.

This analysis assumes that: (1) vertical land movement is responsible for all the deviation of relative sea-level rise from global sea-level rise, (2) the vertical land move-

Table 3
Global sea-level rise scenarios utilized in this study

Scenario		Coefficients for Equation 2		Rise (m) by	
This study	IPCC (Warrick et al., 1996)	a (m/yr)	b (m/yr²)	2050	2100
Low-rise	Low (or no acceleration)	0.0018	0.0	0.11	0.2
Mid-rise	Best	0.0018	0.000025	0.2	0.5
High-rise	High	0.005	0.000029	0.4	0.9

Abbreviation: IPCC, Intergovernmental Panel on climate change.

ments are regional in nature so extrapolation and interpolation is meaningful, and (3) they will continue unchanged until the year 2100. As discussed previously, the reasonably consistent regional patterns of RSL and LC support these assumptions for the East and Gulf Coasts (see Table 1).

The scenarios of relative sea-level rise for 2050 and 2100 are shown in Tables 4 and 5, respectively. For the low-rise scenario, by 2100 the predicted change in sea level ranges from a fall of 0.18 m at Neah Bay, Washington, to a rise of 1.15 m at Grand Isle, Louisiana—a cumulative national variation of about 1.3 m. In general, significant rises are forecast, reinforcing the message that existing rates of sea-level rise already are a long-term coastal management issue. Based simply on projecting existing trends from 1990 to 2100, 18 stations would experience a rise from 0.25 to 0.4 m, 4 stations in the Mid-Atlantic Region would experience a rise of about 0.4 m, with the greatest rises at Galveston (0.7 m) and at Grand Isle (1.15 m).

The mid-rise and high-rise scenarios raise the results for the low-rise scenario by only an equal amount (Tables 4 and 5). Therefore, the location of maximum and minimum rates of relative sea-level rise do not change. The three global scenarios (from Equation 2) are compared to the corresponding relative sea-level rise scenarios at Grand Isle, Neah Bay, and New York in Figure 5. Grand Isle and Neah Bay represent the stations with the largest and smallest predicted rises in relative sea level, respectively.

At many stations, the mid-rise scenario causes the same rise in sea level as the low-rise scenario, but 50 years earlier in 2050 rather than 2100. Considering the high-rise scenario, a rise of over 1.8 m (averaging 1.7 cm/yr) is predicted for Grand Isle from 1990 to 2100. Based on geological evidence this is close to a threshold rate of sea-level rise which has previously triggered significant land loss in the Mississippi delta plain (Boesch et al., 1994).

The relative sea-level scenarios from 1990 to 2050 for the East and Gulf Coasts are shown in Figure 6. The same pattern as Figure 1 is apparent, with high rates of relative sea-level rise in the Mid-Atlantic Region and in Louisiana and Texas. In the former region, relative sea-level rise scenarios are slightly smaller at inland stations such as Washington, DC, and Baltimore, Maryland, compared to those nearer the open coast.

Proactive Adaptation to Sea-Level Rise

The methodology presented in this article provides coastal managers with simple and robust guidelines for developing useful scenarios of relative sea-level rise for planning purposes. The coefficients *a* and *b* in Equation 3 can be adjusted easily to reflect new scientific knowledge, while in the extreme, other mathematical forms to represent relative sea-level rise over time are feasible, if necessary. The major deficiency of the approach is the absence of regional scenarios of sea-level change due to possible changes in oceanographic forcing. The solution to this problem is beyond our present understanding. Pragmatically, the scenarios presented here still are useful for coastal planning as the regional sea-level effect will be a perturbation on the global change, but this uncertainty sets the scenarios in context and emphasizes that their precision should not be overstated. In addition, any known localized subsidence also should be included where appropriate (e.g., Savannah, Georgia).

However, planning for sea-level rise requires a number of elements beyond relative sea-level rise scenarios, including the ability to handle the wide uncertainty concerning future rates of global, regional, and local sea-level rise.

Table 4

Relative sea-level rise scenarios from 1990 to 2050 for the low-rise (LR), mid-rise (MR), and high-rise (HR) global scenarios

Station	LR (cm)	MR (cm)	HR (cm)
Eastport, ME	14	24	44
Portland, ME	12	22	42
Seavey Island, ME	10	20	40
Boston, MA	16	25	45
Woods Hole, MA	15	24	44
Newport, RI	15	24	44
New London, CT	10	19	39
Willets Point, NY	14	23	43
New York, NY	16	25	45
Sandy Hook, NJ	23	32	52
Atlantic City, NJ	23	32	52
Philadelphia, PA	16	25	45
Baltimore, MD	18	28	48
Annapolis, MD	21	30	50
Solomons Island, MD	19	28	48
Washington, DC	17	26	46
Hampton Roads, VA	25	34	54
Portsmouth, VA	23	32	52
Wilmington, NC	11	20	40
Charleston, SC	19	28	48
Fort Pulaski, GA	15	25	45
Fernandina Beach, FL	11	20	40
Mayport, FL	12	22	42
Miami Beach, FL	14	24	44
Key West, FL	13	22	42
St. Petersburg, FL	15	24	44
Pensacola, FL	13	22	42
Grand Isle, LA	63	72	92
Galveston, TX	38	47	67
Port Isabel, TX	16	26	46
San Diego, CA	13	22	42
La Jolla, CA	13	22	42
Los Angeles, CA	4	13	33
Port San Luis, CA	5	14	34
San Francisco, CA	8	17	37
Crescent City, CA	−6[a]	3	24
Astoria, OR	−3	6	26
Neah Bay, WA	−10	0	20
Seattle, WA	12	21	41
Friday Harbour, WA	6	15	35

See Table 3 for global sea-level rise scenarios.
[a]A minus indicates a fall in relative sea level.

Table 5

Relative sea-level rise scenarios from 1990 to 2100 for the low-rise (LR), mid-rise (MR), and high-rise (HR) global scenarios

Station	LR (cm)	MR (cm)	HR (cm)
Eastport, ME	27	57	97
Portland, ME	23	53	93
Seavey Island, ME	19	49	89
Boston, MA	29	60	100
Woods Hole, MA	28	58	98
Newport, RI	27	57	97
New London, CT	18	48	88
Willets Point, NY	25	55	95
New York, NY	30	60	100
Sandy Hook, NJ	42	72	112
Atlantic City, NJ	42	73	113
Philadelphia, PA	29	59	99
Baltimore, MD	34	64	104
Annapolis, MD	38	68	108
Solomons Island, MD	34	65	105
Washington, DC	30	61	101
Hampton Roads, VA	46	76	116
Portsmouth, VA	41	72	112
Wilmington, NC	20	50	90
Charleston, SC	35	65	105
Fort Pulaski, GA	28	58	98
Fernandina Beach, FL	20	50	90
Mayport, FL	23	53	93
Miami Beach, FL	26	57	97
Key West, FL	24	54	94
St. Petersburg, FL	27	57	97
Pensacola, FL	24	54	94
Grand Isle, LA	115	145	185
Galveston, TX	70	100	140
Port Isabel, TX	30	60	100
San Diego, CA	23	53	93
La Jolla, CA	23	53	93
Los Angeles, CA	8	38	78
Port San Luis, CA	10	40	80
San Francisco, CA	14	45	85
Crescent City, CA	−11[a]	19	59
Astoria, OR	−6	24	64
Neah Bay, WA	−18	12	52
Seattle, WA	22	52	92
Friday Harbour, WA	10	40	80

See Table 3 for global sea-level rise scenarios.
[a]A minus indicates a fall in relative sea level.

Types of Response

All the planned responses to sea-level rise can be grouped under three, conceptually distinct categories (Bijlsma et al., 1996; Intergovernmental Panel on Climate Change, Coastal Zone Management Subgroup [IPCC CZMS], 1990; 1992):

(1) (Managed) retreat—planned abandonment of land in the face of progressive land loss and intensification of associated hazards with minimal loss of associated infrastructure;

(2) Accommodation—changing land use as water levels rise (e.g., raising build-

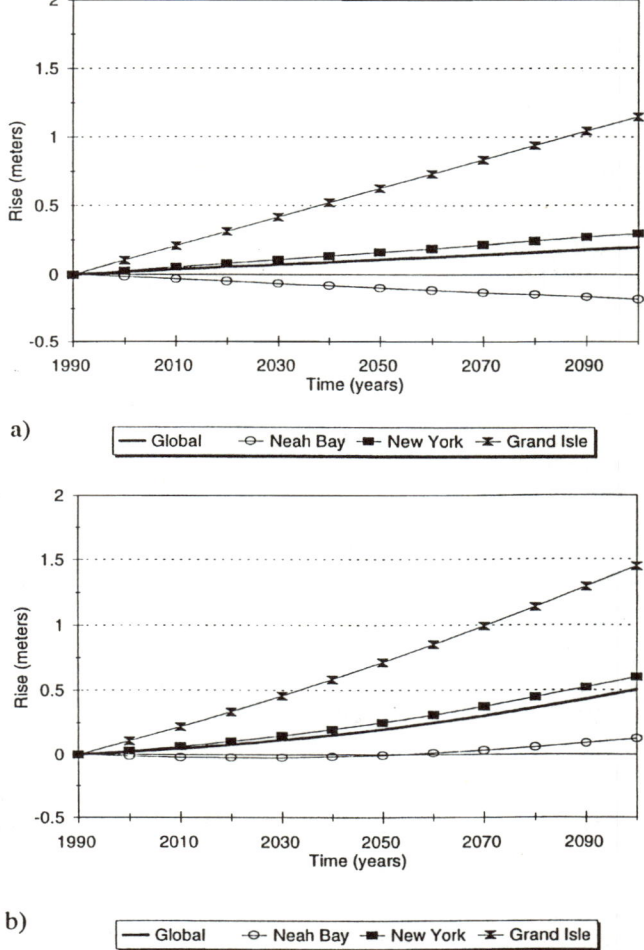

a)

b)

Figure 5. Relative sea-level scenarios for Grand Isle, LA, New York City, and Neah Bay, WA, and the global scenarios: (a) low-rise scenario, (b) mid-rise scenario.

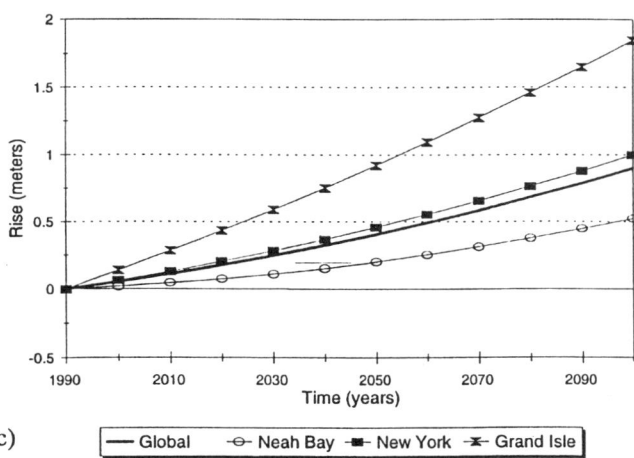

c)

Figure 5. Relative sea-level scenarios for Grand Isle, LA, New York City, and Neah Bay, WA, and the global scenarios: (*Continued*) (c) high-rise scenario.

ings on piles above the new flood levels, or converting to more salt tolerant crops);

(3) Protection—building dikes and levees, beach nourishment, and so forth.

The best management responses will vary from place to place and depend on local geomorphological and socioeconomic factors. For instance, the East and Gulf Coasts are more vulnerable to sea-level rise than the West Coast because the former areas have extensive low-lying coastal plains, while much of the West Coast is composed of cliffs (Gornitz et al., 1991; National Research Council, 1987). These regional characteristics are further fragmented by human use, such as large cities, resort towns, agriculture, and national seashores, making a range of responses appropriate.

Uncertainty and Coastal Planning

While sea levels are expected to rise, the magnitude of the change has a wide range of uncertainty which increases with time interval. Based on the scenarios used here, this uncertainty is 70 cm by 2100 (e.g., Figure 7) with additional uncertainties concerning regional sea-level change. This uncertainty is unlikely to be substantially reduced in the near future. The frequent and ongoing revisions to global sea-level rise scenarios often have less practical implications than first appearance suggests, as the range of the revised and previous scenarios have substantial overlap. Such uncertainty is often taken as a reason for inaction until the uncertainty is reduced. However, this course is the "do nothing" policy, which will lead to unwise decisions and increasing hazard with time. In contrast, a planned response needs to be effective despite these large uncertainties. When estimating possible impacts of sea-level rise, uncertainty can be handled by scenario-based vulnerability assessment (e.g., Titus et al., 1991; Yohe, 1990), but proactive responses should be effective for or easily adaptable to the full range of possible scenarios.

318 *R. J. Nicholls and S. P. Leatherman*

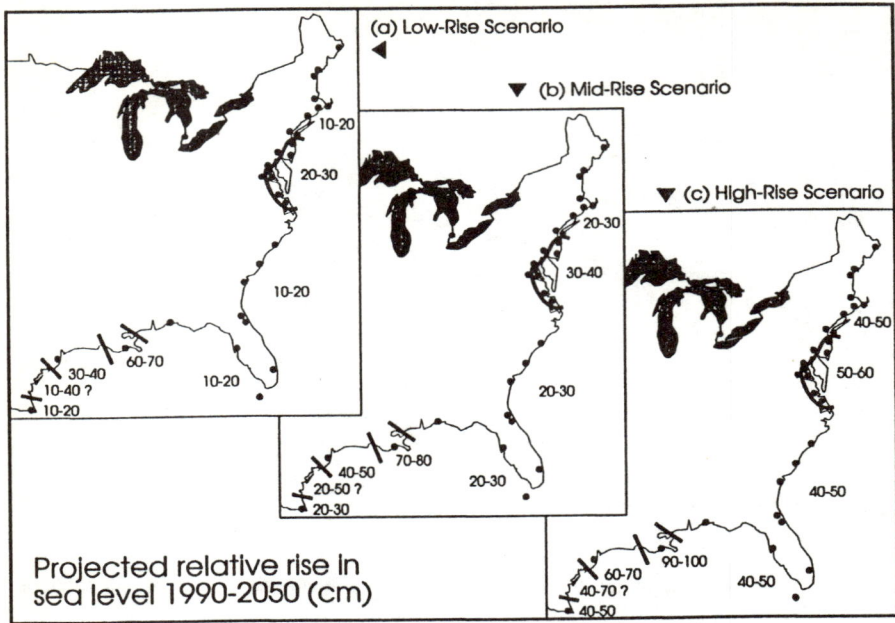

(a) Low-Rise Scenario

(b) Mid-Rise Scenario

(c) High-Rise Scenario

Projected relative rise in
sea level 1990-2050 (cm)

Figure 6. Relative sea-level rise scenarios for the East and Gulf Coasts from 1990 to 2050.

Given these large uncertainties, it is unwise to invest large sums of money today. Rather, attention should be focused on identifying (1) low-cost, no regret responses which would maintain and enhance the flexibility of tomorrow's coastal managers toward future impacts of sea-level rise; and (2) sectors where reactive adaptation would have particularly high costs and the additional costs of an allowance for future sea-level rise could be considered a worthwhile "insurance policy."

Examples of Proactive Responses

In the United States, it has not been customary to consider existing rates of sea-level rise in coastal management and engineering decisions, but elsewhere anticipating sea-level rise is not a new practice. For example, the Thames flood barrier in London anticipated the existing rate of rise of the high water level for 50 years (until 2030), amounting to an additional 0.4 m (Kelly, 1991). There is a small, but growing, number of examples of similar allowances for accelerated sea-level rise, including designing new infrastructure for higher sea levels and planning land use for increased coastal recession (Bijlsma et al., 1996; Nicholls & Leatherman, 1995). For instance, at Boston, Massachusetts, the Deer Island sewage treatment plant is built 0.46 m above static design levels. This will allow gravity-based flows under higher sea levels, avoiding the additional costs of pumping (Smith & Muller-Vollmer, 1993). Based on the relative sea-level rise scenarios in Tables 4 and 5, this design approach will remain effective at least until 2050.

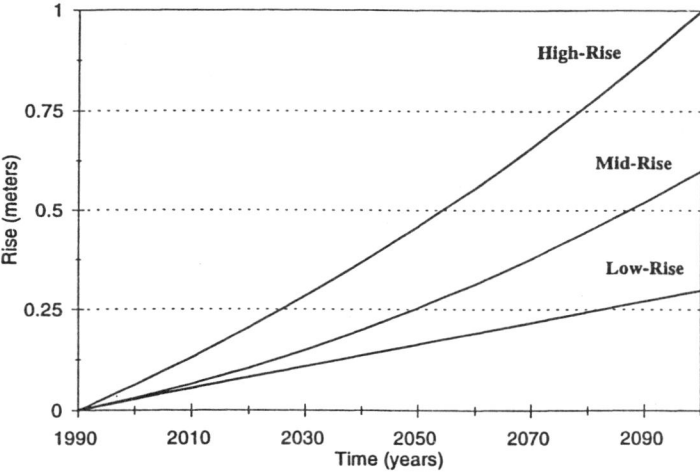

Figure 7. Relative sea-level scenarios from 1990 to 2100 for New York City.

Application to the U.S. Coast

The authors encourage state and local assessment of potential sea-level rise impacts and, where applicable, implementation of proactive adaptation strategies using the scenarios defined in Table 3. As an absolute minimum, the low-rise scenario should be considered in all coastal management decisions. More opportunities for implementing low-cost responses to sea-level rise will exist in areas which are being developed or redeveloped, or experiencing other land use changes. Selection of an appropriate sea-level rise scenario for planning and design purposes depends on a number of factors, including: (1) the associated time frame, (2) the additional costs of potential adaptation measures, and (3) the implications if the selected scenario is exceeded. Therefore, consideration of sea-level rise will be most important for enduring, high-value coastal infrastructure.

One approach is to allow for incremental upgrade or modification, which makes the future change in sea level of less importance, although even in these cases, the implications of the low-rise scenario still should be considered during the initial design. For instance, the design of coastal levees could allow progressive and routine raising of the crest elevation in response to sea-level rise. However, this incremental approach to design is not always practical. Coastal drainage infrastructure often has a long life and is costly to upgrade. Therefore, design standards for new installations in many low-lying coastal areas may usefully be raised to allow for future rises in sea level (Titus et al., 1987), as well as the likelihood of more intense rainfall events given global warming (Bijlsma et al., 1996). The results in Tables 4 and 5 can assist in the selection of appropriate design standards.

Appropriate land use planning is another important form of proactive adaptation. With or without climate change, one sensible policy is building setbacks, as most sandy shorelines are already eroding (National Research Council, 1990; Rogers, 1993). Maine and North Carolina have enacted legislation which encourages a retreat from the open coast by enacting building setbacks and prohibiting hard stabilization of the coast (National Research Council, 1987). Some Australian states have gone further and enlarged coastal building setbacks to include erosion expected due to accelerated sea-level rise (Caton & Eliot, 1993). The key decision is an appropriate setback distance which will integrate a number of factors, including likely recession rates and the life of coastal buildings. More radical concepts, such as "presumed mobility," which would allow coastal development but prohibit protection against sea-level rise, have been proposed as efficient planning mechanisms for sea-level rise (Titus, 1991b).

Losses and changes to natural systems also need to be considered because they have important implications for natural resources; coastal recreation and utilization; and, particularly when combined with increasing human pressure, coastal biodiversity (Reid & Trexler, 1991). While coastal wetlands are now protected from reclamation, they still are declining in areas of high rates of relative sea-level rise, such as the Chesapeake Bay (Downs et al., 1994; Stevenson et al., 1986) and Louisiana (Boesch et al., 1994). Therefore even existing trends suggest a progressive decline in wetland areas. Nationally, given a 50-cm rise in sea level, it is forecast that 38% to 61% of existing coastal wetlands would be lost. These losses might be partly offset by new wetland formation on former upland areas, although even under ideal circumstances not all the lost wetlands would be replaced, and hard protection of developed areas would further reduce the area available for such replacement (Titus et al., 1991).

Coastal policymakers need to address the implications of sea-level rise for coastal wetlands at a regional and local scale, embracing both losses and possible vegetative changes. Because some losses appear inevitable, the most realistic management goal in many cases is maximizing wetland survival. The relative sea-level rise scenarios presented here provide critical boundary conditions for this assessment, including defining the areas of upland which might be converted to wetlands as sea levels rise. In these areas, there likely will be conflict between those favoring coastal construction and those favoring wetlands formation. This is a difficult problem, but inaction will increase the likelihood of greater losses of wetlands (Titus, 1991b).

Finally, sea-level rise is only one of a number of long-term management pressures, and therefore, there is a need to integrate management policies for both a range of activities and a range of space- and timescales (National Research Council, 1995; World Coast Conference 1993, 1994). This is a great challenge, and robust administrative mechanisms are required to deal simultaneously with sea-level rise, other coastal impacts of climate change, and human pressures on the coastal zone. The more widespread utilization of adaptive management approaches has recently been recommended for the coastal zone (National Research Council, 1995). The uncertainty of long-term issues, such as sea-level rise and climate change in general, reinforces this conclusion. Better tools for assessing the overall merits of different response options within an integrated framework also are required: This calls for multidisciplinary research among the engineering, natural, and social sciences.

Further Developments

Coastal scientists and managers must work together to apply this knowledge to site-specific areas in order to evaluate what is threatened and define appropriate responses. Future studies need to address a number of issues so as to provide improved information on sea levels for coastal management purposes. First, there is a need to consider further the magnitude of human-induced subsidence and its possible contribution to relative sea-level rise, both at a regional scale (Davis, 1987) and at local scales in areas experiencing growing demand for groundwater (Holzer, 1991). This should be assessed on a case-by-case basis. In general, halting human-induced subsidence will have lower costs than adapting to the resulting relative sea-level rise and should be vigorously pursued where applicable (Holzer & Johnston, 1985; Nicholls, 1995).

There are a large number of data gaps, particularly on the West Coast (Figure 1), while higher spatial resolution of future sea-level rise is desirable elsewhere. As more tide gauge data are collected, so more stations will have a sufficient record length to be included in this type of analysis. Further, non-NOAA stations could be analyzed where available to increase spatial resolution (cf. Penland & Ramsey, 1990).

Finally, new data sources are becoming available (Warrick et al., 1996) and they may revolutionize our view of sea-level change and its relationship to coastal management. At ground level, improved geodetic measurements at tide gauges, such as global positioning systems and absolute gravity measurements, will allow the vertical land and sea surface changes in their measurements to be separated. In addition, satellite observations of the elevation of the sea surface are now routinely being collected with the Topex/Poseidon platform (Fu et al., 1996; Nerem, 1995). While the present time series is too short to reach conclusive results, in the longer term these data will allow unambiguous measurement of (1) regional and global sea-level trends, independent of vertical land movement; and (2) any acceleration of sea-level rise. However, these new data will not be a panacea, and the problem of planning, given uncertain future sea-level rise, appears likely to remain for the foreseeable future.

Conclusions

Relative sea level is rising along most of the U.S. coasts, and this rise is expected to accelerate due to global warming. By decoupling historic global sea-level rise and vertical land movement, relative sea-level rise scenarios have been developed for the contiguous U.S. coasts. "Hot spots" for relative sea-level rise include Texas and Louisiana and the Mid-Atlantic Region—by simply projecting present trends from 1990 to 2100, relative sea-level rise will range from 0.7 to 1.15 m and exceed 0.4 m in these two areas, respectively.

Local and state assessment of the implications of these scenarios is urgently needed. A "do nothing" policy will lead to unwise decisions and increasing hazard with time. However, such assessments must consider the inherent uncertainties of these scenarios, which are derived largely from the uncertainty of the global scenarios and the absence of regional scenarios of sea-level change due to changing oceanographic factors. Therefore, it is unwise to invest large sums of money today. Rather proactive measures should be focused on identifying (1) low-cost, no regret measures which would maintain and enhance long-term flexibility to respond to sea-level rise; or (2) sectors where reactive adaptation to sea-level rise would have particularly high costs, and therefore the additional response costs for sea-level rise could be considered a worthwhile "insurance policy." To be most effective, these responses need to be integrated with the management of other changes in the coastal zone.

References

Ando, M., and E. I. Balazs. 1979. Geodetic evidence for aseismic subduction of the Juan de Fuca plate. *Journal of Geophysical Research* 84:3023–3028.

Barth, M. C., and J. G. Titus. (Eds.). 1984. *Greenhouse effect and sea-level rise*. New York: Van Nostrand Reinhold.

322   *R. J. Nicholls and S. P. Leatherman*

Bijlsma, L., C. N. Ehler, R. J. T. Klein, S. M. Kulshrestha, R. F. McLean, N. Mimura, R. J. Nicholls, L. A. Nurse, H. Pérez Nieto, E. Z. Stakhiv, R. K. Turner, and R. A. Warrick. 1996. Coastal zones and small islands. In *Impacts, adaptations and mitigation of climate change: Scientific-technical analyses,* ed. R. T. Watson, M. C. Zinyowera, and R. H. Moss, 289–324. Cambridge: Cambridge University Press.

Boesch, D. F., M. N. Josselyn, A. J. Mehta, J. T. Morris, W. K. Nuttle, C. A. Simenstad, and D. J. P. Swift. 1994. Scientific assessment of coastal wetland loss, restoration and management in Louisiana. *Journal of Coastal Research* (Special Issue 20).

Caton, B., and I. Eliot. 1993. Coastal hazard policy development and the Australian federal system. In *Vulnerability Assessment to sea-level rise and coastal zone management, Proceedings of the Intergovernmental Panel on Climate Change, Coastal Zone Management Subgroup, Eastern Hemisphere Workshop,* ed. R. F. McLean and N. Mimura, 417–427. Tsukuba, Japan, August.

Craig, D. 1993. *Preliminary assessment of sea level rise in Olympia, Washington: Technical and policy implications.* Olympia, WA: Policy and Program Development Division, Olympia Public Works.

Culliton, T. J., M. A. Warren, T. R. Goodspeed, D. G. Remer, C. M. Blackwell, and J. J. McDonough III. 1990. *50 years of population change along the nation's coasts: 1960–2010.* Rockville, MD: National Ocean Service, National Oceanic and Atmospheric Administration (NOAA).

Culliton, T. J., J. J. McDonough III, D. G. Remer, and D. M. Lott. 1992. *Building along America's coasts: 20 years of building permits, 1970–1989.* Rockville, MD: National Ocean Service, NOAA.

Daniels, R. C. 1992. Sea-level rise on the South Carolina coast: Two case studies for 2100. *Journal of Coastal Research* 8:56–70.

Davis, G. H. 1987. Land subsidence and sea-level rise on the Atlantic coastal plain of the United States. *Environmental Geology and Water Science* 10:67–80.

Douglas, B. C. 1991. Global sea-level rise. *Journal of Geophysical Research* 96(C4):6981–6992.

Douglas, B. C. 1992. Global sea-level acceleration. *Journal of Geophysical Research* 97(C8): 12,699–12,706.

Douglas, B. C. 1995. Global sea level change: Determination and interpretation. In *Reviews of geophysics,* supplement (July), U.S. National Report to International Union of Geodesy and Geophysics 1991–1994:1425–1432.

Downs, L. L., R. J. Nicholls, S. P. Leatherman, and J. Hautzenroder. 1994. Historic evolution of a marsh island: Bloodsworth Island, Md. *Journal of Coastal Research* 10:1031–1044.

Emery, K. O., and D. G. Aubrey. 1991. *Sea levels, land levels and tide gauges.* New York: Springer Verlag.

Federal Emergency Management Agency (FEMA). 1991. Project impact of relative sea level rise on the National Flood Insurance Program. Unpublished report, Federal Emergency Management Agency, Washington DC.

Fu, L.-L., C. J. Koblinsky, J.-F. Minster, and J. Picaut. 1996. Reflecting on the first three years of TOPEX/POSEIDON. *EOS Transactions* 77:109–117.

Gornitz, V. 1995. Sea-level rise: A review of recent, past and near-future trends. *Earth Surface Processes and Landforms* 20:7–20.

Gornitz, V., R. C. Daniels, T. W. White, and K. R. Birdwell. 1994. The development of a coastal risk assessment database for the U.S. Southeast: Erosion and inundation from sea-level rise. *Journal of Coastal Research* (Special Issue 12):327–338.

Gornitz, V., and S. Lebedeff. 1987. Global sea-level changes during the past century. In *Sea-level fluctuation and coastal evolution,* ed. D. Nummedal, O. H. Pilkey, and J. D. Howard, 3–16. SEPM Special Publication No. 41.

Gornitz, V., and L. Seeber. 1990. Vertical crustal movements along the East coast, North America, from historic and late Holocene sea-level data. *Tectonophysics* 178:127–150.

Gornitz, V., T. W. White, and R. M. Cushman. 1991. Vulnerability of the U.S. to future sea-level rise. In *Proceedings of Coastal Zone '91*, 2354–2368. New York: American Society of Civil Engineers.

Holdahl, S. R., J. C. Holzschuh, and D. B. Zilkoski. 1989. *Subsidence at Houston, Texas, 1973–1987*. NOAA Technical Report NOS 131 NGS 44. Rockville, MD: NOAA.

Holzer, T. L. 1991. Neotectonic subsidence. In *The heritage of engineering geology: The first hundred years*, Centennial Special, Vol. 3, ed. G. A. Kiersch, 219–232. Boulder, CO: Geological Society of America.

Holzer, T. L. and A. I. Johnston. 1985. Land subsidence caused by ground water withdrawal in urban areas. *GeoJournal* 11:245–255.

Inman, D. L., and C. E. Nordstrom. 1971. On the tectonic and morphologic classification of coasts. *Journal of Geology* 79:1–21.

Intergovernmental Panel on Climate Change, Coastal Zone Management Subgroup (IPCC CZMS). 1990. *Strategies for adaption to sea level rise*. Report of the Coastal Zone Management Subgroup, Intergovernmental Panel on Climate Change, Working Group III. The Hague, the Netherlands: Rijkswaterstaat.

IPCC CZMS. 1992. *Global climate change and the rising challenge of the sea*. Report of the Coastal Zone Management Subgroup, Intergovernmental Panel on Climate Change, Working Group III. The Hague, the Netherlands: Rijkswaterstaat.

Kearney, M. S., and J. C. Stevenson. 1991. Island land loss and marsh vertical accretion rate evidence for historical sea-level changes in the Chesapeake Bay. *Journal of Coastal Research* 7:403–415.

Kelly, M. P. 1991. Global warming: Implications for the Thames barrier and associated defenses. In *Impact of sea level rise on cities and regions, Proceedings of the First International Meeting "Cities on Water,"* ed. R. Frassetto, 93–98. Venice, Italy: Marsilio Editori.

Leatherman, S. P., R. Chalfont, E. Pendleton, S. Funderburk, and T. McCandles. 1995. *Vanishing lands: Sea level, society and the Chesapeake Bay*. Annapolis, MD: U.S. Fish and Wildlife Service.

London, J. B., and C. R. Volonté. 1991. Land use implications of sea level rise: A case study at Myrtle Beach, South Carolina. *Coastal Management* 19:205–218.

Lyles, S. D., L. E. Hickman Jr., and H. A. Debaugh Jr. 1988. *Sea-level variations for the United States 1855–1986*. Rockville, MD: National Ocean Service, NOAA.

National Research Council. 1987. *Responding to changes in sea level: Engineering implications*. Washington, DC: National Academy Press.

National Research Council. 1990. *Managing coastal erosion*. Washington, DC: National Academy Press.

National Research Council. 1995. *Science, policy and the coast: Improving decisionmaking*. Washington, DC: National Academy Press.

Nerem, R. S. 1995. Global mean sea level variations from TOPEX/POSEIDON altimeter data. *Science* 268:708–710.

Nicholls, R. J. 1995. Coastal megacities and climate change. *GeoJournal* 37:369–379.

Nicholls, R. J., and S. P. Leatherman. 1995. Sea-level rise and coastal management. In *Geomorphology and land management in a changing environment*, ed. D. McGregor and D. Thompson, 229–244. Chichester, UK: John Wiley.

Office of Technology Assessment. 1993. *Preparing for an uncertain climate*. Washington, DC: U.S. Government Printing Office.

Parker, B. B. 1991. Sea level as an indicator of climate and global change. *Marine Technology Society Journal* 25(4):13–24.

Peltier, W. R. 1986. Deglaciation-induced vertical motion of the North American continent and transient lower mantle rheology. *Journal of Geophysical Research* 91:9099–9123.

Peltier, W. R., and A. M. Tushingham. 1989. Global sea level rise and the greenhouse effect: Might they be connected? *Science* 244:806–810.

Penland, S., and K. E. Ramsey. 1990, Relative sea-level rise in Louisiana and the Gulf of Mexico: 1908–1988. *Journal of Coastal Research* 6:323–342.

324   *R. J. Nicholls and S. P. Leatherman*

Reid, W. V., and M. C. Trexler. 1991. *Drowning the national heritage: Climate change and U.S. coastal biodiversity.* Washington, DC: World Resources Institute.

Rogers, S. M. 1993. Relocating erosion-threatened buildings: A study of North Carolina house-moving. In *Proceedings of Coastal Zone '93*, 1392–1405. New York: American Society of Civil Engineers.

Smith, J. B., and J. Muller-Vollmer. 1993. Setting priorities for adapting to climate change. Report prepared by RCG/Hagler, Bailly, Inc., Arlington, VA, for Office of Technology Assessment, Oceans and Environment Program. Contract No. 13-5935.0.

Smith, J. B., and D. Tirpak. (Eds.). 1989. *The potential effects of global climate change on the United States: Report to Congress.* Washington, DC: U.S. Environmental Protection Agency.

State of Maine. 1995. *Anticipatory planning for sea-level rise along the coast of Maine.* Augusta, ME: Maine State Planning Office.

Stevenson, J. C., L. G. Ward, and M. S. Kearney. 1986. Vertical accretion in marshes with varying rates of sea level rise. In *Estuarine variability*, ed. D. A. Wolfe, 241–258. New York: Academic Press.

Titus, J. G. 1991a. Greenhouse effect, sea level rise, and barrier islands: Case study of Long Beach Island, New Jersey. *Coastal Management* 18:65–90.

Titus, J. G. 1991b. Greenhouse effect and coastal wetland policy: How Americans could abandon an area the size of Massachusetts at minimum cost. *Environmental Management* 15:39–58.

Titus, J. G., C. Y. Kuo, M. J. Gibbs, T. B. LaRoche, M. K. Webb, and J. O. Waddell. 1987. Greenhouse effect, sea level rise and coastal drainage systems. *Journal of Water Resources Planning and Management* 14:146–171.

Titus, J. G., and V. K. Narayanan. 1995. *The probability of sea level rise.* Washington, DC: Environmental Protection Agency.

Titus, J. G., R. A. Park, S. P. Leatherman, J. R. Weggel, M. S. Green, P. W. Mausel, S. Brown, C. Gaunt, M. Trehan, and G. Yohe. 1991. Greenhouse effect and sea-level rise: Potential loss of land and the cost of holding back the sea. *Coastal Management* 19:171–204.

Trupin, A., and J. Wahr. 1990. Spectroscopic analysis of global tide gauge sea level data. *Geophysical Journal International* 100:441–453.

Turner, K. E. 1991. Tide gauge records, water level rise and subsidence in the northern Gulf of Mexico. *Estuaries* 14:139–147.

Turner, R. K., S. Subak, and W. N. Adger. 1996. Pressures, trends and impacts in the coastal zones: Interactions between socio-economic and natural systems. *Environmental Management* 20:159–173.

Tushingham, A. M., and W. R. Peltier. 1991. ICE-3G: A new global model of late-Pleistocene deglaciation based upon predictions of post-glacial relative sea-level change. *Journal of Geophysical Research* 96(B3):4497–4523.

Warrick, R. A., and H. Oerlemans. 1990. Sea-level rise. In *Climate change: The IPCC scientific assessment*, ed. J. T. Houghton, G. J. Jenkins, and J. J. Ephramus, 257–281. Cambridge: Cambridge University Press.

Warrick, R. A., J. Oerlemans, P. L. Woodworth, M. F. Meier, and C. le Provost. 1996. Changes in sea level. In *Climate change 1995: The science of climate change*, ed. J. T. Houghton, L. G. Meira Filho, and B. A. Callander. Cambridge: Cambridge University Press, in press.

Wigley, T. M. L., and S. C. B. Raper. 1992. Implications for climate and sea level of revised IPCC emissions scenarios. *Nature* 357:293–300.

World Coast Conference 1993. 1994. *Preparing to meet the coastal challenges of the 21st century.* Report of the World Coast Conference. Noordwijk, The Netherlands, November 1–5, 1993. The Hague, The Netherlands: Ministry of Transport, Public Works and Water Management.

Yohe, G. 1990. The cost of not holding back the sea. Toward a national sample of economic vulnerability. *Coastal Management* 18:403–431.

[10]

| Journal of Coastal Research | SI | 14 | 1-14 | Fort Lauderdale, Florida | Winter 1995 |

Accelerated Sea-Level Rise and Developing Countries: An Overview

Stephen P. Leatherman and Robert J. Nicholls[1]

Laboratory for Coastal Research
Department of Geography
University of Maryland
College Park, MD 20742 U.S.A.

INTRODUCTION

Accelerated sea-level rise is one of the more certain consequences of global warming. While detailed impact assessments have been undertaken in many industrial countries (e.g., GOEMANS, 1986; TITUS et al., 1991), much less research has been conducted in developing countries. This five year study was conducted to address this problem as part of the U.S. Environmental Protection Agency International Project, involving climate change impact analyses for agriculture, forestry, health, sea-level rise, and water resources (STRZEPEK and SMITH, 1994).

Sea-level rise scenarios from 0.2 to 2.0 meters by the year 2100 are considered with a 1-meter rise being the standard scenario. The physical effects of sea-level rise include: inundation of low-lying areas, erosion, salt-water intrusion into aquifers, higher water tables, and increased flooding and storm damage. We only considered the first two factors in this analysis: land loss due to inundation (submergence below high tide) and erosion (physical removal of sediment by waves).

This research consisted of three phases: (1) national overviews based on existing literature and available data, (2) case studies of specific areas based on new data, and (3) national assessments using aerial videotape-assisted vulnerability analysis (AVVA) to determine the coastal resources at risk and the cost of protection. These studies were conducted in concert with in-country researchers in order to obtain the best possible analysis and to build in-country expertise and presence in climate change issues. The principal collaborating scientists are listed in

Table 1. While our national overviews and case studies spanned the world's continents, the AVVA studies were restricted to the Atlantic coasts of South America and Africa (Figure 1).

In the first year we identified and initiated studies with scientists from eight developing countries. Initially we sent requests for proposals to over 25 scientists from 13 countries, allowing us to choose both the best proposals and the countries with the most interesting problems. In-country researchers produced a national overview, describing the impact of a 1-meter rise in sea level on their countries. An outline was prepared to assist the researchers in obtaining the

Table 1. *Principle collaborating scientists*

Dr. Jose Arismendi, Instituto de Ingenieria, Venezuela.
Dr. Virendra Asthana, Jawaharalal Nehru University, India.
Dr. Larry F. Awosika, Nigerian Institute for Oceanography and Marine Research, Nigeria.
Ms. Karen C. Dennis, University of Maryland, U.S.A.
Dr. Mohammed El-Raey, University of Alexandria, Egypt.
Mr. Gregory T. French, University of Maryland, U.S.A.
Dr. Mukang Han, Peking University, China.
Dr. Saleemul Huq, Bangladesh Centre for Advanced Studies, Bangladesh.
Dr. Say-Chong Lee, Drainage and Irrigation Department, Malaysia.
Dr. Zamali Midun, Drainage and Irrigation Department, Malaysia.
Dr. Dieter Muehe, Universidade Federal do Rio de Janeiro, Brazil.
Dr. Claudio F. Neves, Universidade Federal do Rio de Janeiro, Brazil.
Dr. Isabelle Niang-Diop, Universite Cheikh Anta Diop, Senegal.
Dr. Enrique Schnack, Laboratorio de Oceanografia Costera, Argentina.
Mr. Claudio R. Volonté, Canelones, Uruguay and University of Maryland, U.S.A.
Dr. Boacan Wang, East China Normal University, China.
Dr. Wyss Yim, University of Hong Kong, Hong Kong.

[1]Present Address: School of Geography and Environmental Management, University of Middlesex, Queensway, Enfield, Middlesex EN3 4SF, United Kingdom.
94185 received and accepted in revision 10 September 1994.

Figure 1.   Location Map: Sea-Level Rise Impact Studies. The countries studied with AVVA are highlighted.

type of information to be included in their national overviews (LEATHERMAN *et al.*, 1994).

The coastal scientists completed a draft version of their national overviews and presented them at a two-week workshop that was held at the University of Maryland in September 1989. Most of the national overviews were presented at the Intergovernmental Panel on Climate Change (IPCC) meetings in Miami, November 1989 (TITUS, 1990) and in Australia, February 1990, providing important input into the first IPCC assessment (IPCC, 1990).

During the second year more in-depth case studies were conducted in selected areas of high vulnerability to sea-level rise based on the information contained in the national overviews. The case studies followed a similar outline as the national overviews but concentrated on one site and included much more detailed information on the land area, population and resources at risk and potential responses. The researchers collected new data (including historic data) and analyzed the effects of a 1-meter sea-level rise, particularly for erosion and inundation at the site.

In addition, we established a network of people in developing countries who are likely to use

this information in formulating public policy. It is important that in-country scientists discuss with public officials the impacts of sea-level rise and possible resulting calamities as a first step towards sensible anticipatory adaptation (NICHOLLS and LEATHERMAN, 1994a).

In the last two years of this international project, we concentrated our efforts on a true national assessment of impacts and costs by using the aerial videotape-assisted vulnerability analysis (AVVA). This new approach was implemented in order to obtain quantitative data on a nationwide basis. While the national overviews proved useful in identifying critical problems and issues, most of the information was qualitative or anecdotal in nature. The case studies of selected areas provided numbers on population and real estate at risk, but only for a relatively small segment of the coast. Therefore, we had to devise a methodology that would allow us to quantify the amount of land loss and the cost of protecting that land.

We chose five countries along the Atlantic coast of South America and West Africa for the AVVA studies. As is true for most developing countries, there was a lack of basic topographic information; the best nationally-available maps only had 10 to 100-meter contours. This contour interval is not detailed enough to quantify land lost for any realistic sea-level rise scenario. We overcame this fundamental problem with a relatively simple and cost-effective methodology using aerial videotaping combine with limited ground surveys (see LEATHERMAN *et al.*, 1994). This approach also permitted us to work directly with our collaborators in the field in their own country, broadening experience and expertise for all team members.

NATIONAL OVERVIEWS

Sea-level rise is already much in evidence in terms of land loss along the world coastlines. Over 70% of the beaches worldwide are presently eroding (BIRD, 1985). There is much less documentation of loss of coastal wetlands, but it is already a major problem in some areas (*e.g.*, Chesapeake Bay, Maryland and coastal Louisiana U.S.A.), and relative sea-level rise is a principal causative factor. The following is an overview of sea-level rise impacts and responses in four countries: Bangladesh, Brazil, China, and Malaysia.

Bangladesh

Worldwide, this nation is often cited as a major loser to accelerated sea-level rise, and it is considered one of the most vulnerable countries to climate change. Its position on the delta of the Ganges-Brahmaputra-Megha River system makes Bangladesh extremely low-lying and naturally prone to flooding from both the rivers and the sea during tropical cyclones.

Assuming no responses, the impacts of a 1 m rise in sea level would be devastating. Over 25,000 km^2 (more than 17.5% of the existing land area) of the country will be underwater at high tide (Figure 2), which includes 85 cities and towns (HUQ *et al.*, 1994). More than 13 million people could potentially be displaced by this inundation without considering the rapid growth in population that is occurring today. Also, one-half of the country would be subject to storm flooding, and the amount of damage and loss of life during cyclones would increase significantly.

Bangladesh contains the Sunderbans, the second largest mangrove swamp in the world and one of the last refuges for the Bengal tiger. With a 1-meter rise in sea level, this swamp may be

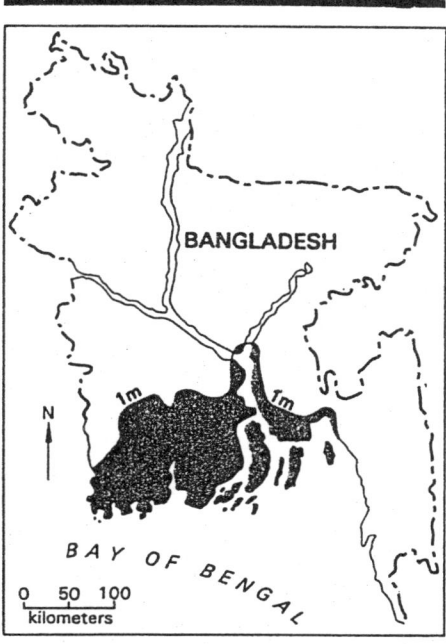

Figure 2. Land loss in Bangladesh in response to 1-meter sea-level rise (after Huq *et al.*, 1994).

lost, as would some of the best rice growing agricultural land.

One of the world's poorest countries, Bangladesh would find it difficult to adapt to such changes. One particular concern is the possibility that river-management structures might disable the natural sedimentation processes that otherwise help to mitigate the adverse consequences of sea-level rise (IPCC, 1990). All future human intervention in these rivers needs to be evaluated in terms of its effect on the national increase or decrease in vulnerability to climate change, particularly sea-level rise.

Brazil

While tide gauge records are limited by location and duration, the longest records show that relative sea-level rise is rising along the Coast of Brazil (MUEHE and NEVES, 1994). This finding refutes the widespread belief in Brazil that sea level is falling (based on geologic data for the last few millennia) and clearly indicates that the potential impacts of sea-level rise need to be evaluated for this the largest country in South America.

The length and variety of Brazil's coastline (7,400 km long) makes it difficult to summarize the impact of sea-level rise. Certainly, several major and expanding low-lying coastal cities would be threatened. In Recife, major development is proposed, including low-lying areas which would be inundated by sea-level rise. Portions of Rio de Janeiro are built on low-lying barrier beaches, and these areas will require expensive protection from increased erosion and flooding. The beaches around Rio (Figure 3), particular the world-famous Copacabana Beach (Figure 4), will require additional expenditures for beach nourishment (LEATHERMAN, 1986).

More generally, coastal flooding associated with heavy runoff is already a problem in parts of Brazil and would be exacerbated by sea-level rise.

MUEHE and NEVES (1994) conclude that municipal, state, and federal authorities should take a preventive approach when selecting coastal sites for urban expansion and location of industries. To avoid the high cost of protecting developed areas, this approach includes developing and enforcing coastal management programs, installing long-term tidal gauges, adopting flexible criteria for designing harbors and coastal structures which take into account sea-level rise, and formulating educational programs about environmental protection and global climatic effects.

Existing sea-level rise impact studies are already having some effect on public policy and coastal engineering. In the design of planned expansion of the port of Suape, an additional

Figure 3.   Brazilians are unfortunately emulating high-rise beachfront development practices along the U.S. Atlantic coast, such as Ocean City, MD. (Location: the Barra da Tijuca coastal plain, Rio de Janerio Province).

Figure 4. Typical use of Copacabana Beach in Rio de Janeiro, Brazil—the beach is slowly eroding and has already required nourishment.

0.25 m has been included to allow for sea-level rise over a 50 year period, based on the results of a study at nearby Recife (NEVES and MUEHE, 1994). This is the first documented example in South America where sea-level rise is being considered in design.

China

China has the longest coastline of the countries studied, particularly if the offshore islands are included. The major impacts of sea-level rise would be concentrated in the four low-lying and heavily populated alluvial plains: the Lower Liao River delta; the North China coastal plain; the East China coastal plain; and the Pearl River deltaic plain (Figure 5). If no measures were taken to hold back the sea, HAN *et al.* (1994a) estimate that over 73 million people, together with 70 counties and cities, would be living beneath the 100-year storm surge after a 1-meter rise. This includes major and ancient cities, such as Shanghai, large parts of which are already below high tide. This cannot be allowed to occur, and improvements to existing coastal defenses will be undertaken as such expenses are both affordable to local governments and economically very effective (HAN *et al.*, 1994a).

Dikes are the traditional Chinese approach to combat coastal flooding and relative sea-level rise (due to subsidence). Here dikes have been used long before the Dutch became the great dike-builders of Europe. The Chinese have been reclaiming land for over a thousand years for agriculture and, more recently, aquaculture. Most of the shrimp and an important quantity of the fish consumed by the Chinese come from these protein factories built on reclaimed land, protected by mud dikes. China cannot afford to lose any of this land and landward retreat is consid-

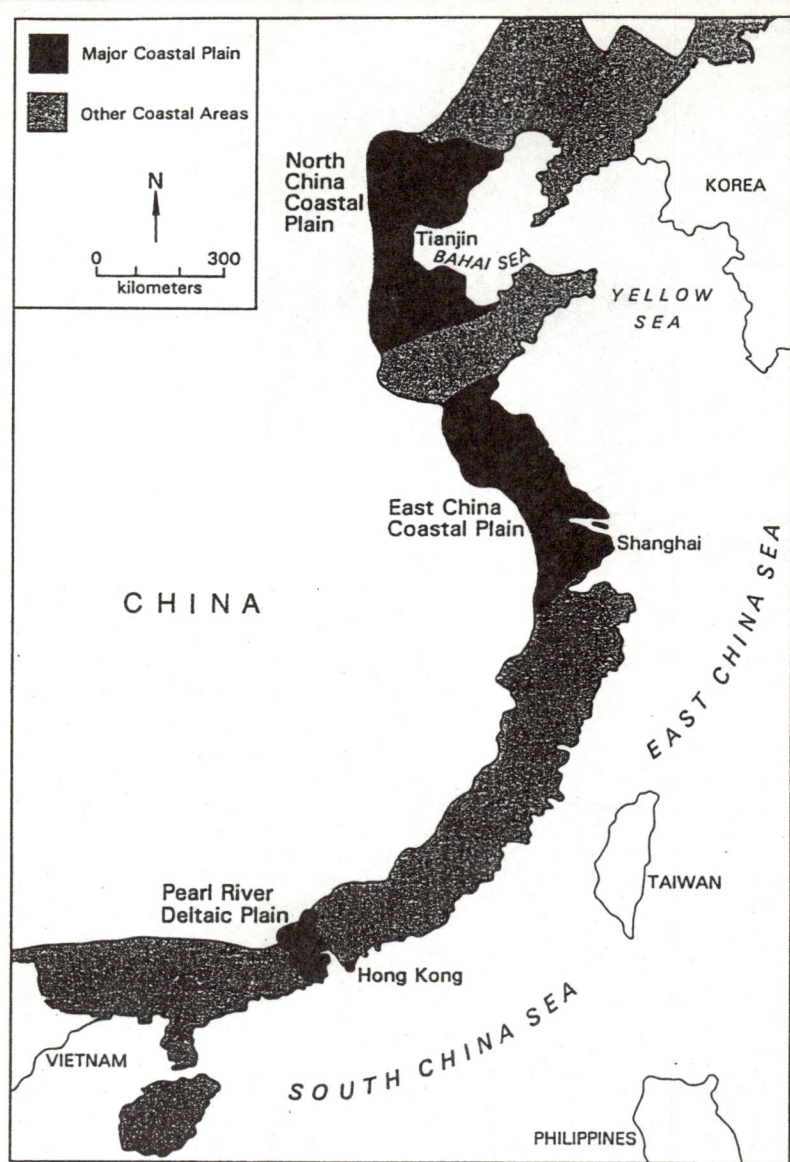

Figure 5.   The Chinese coast contains four major low-lying and heavily populated alluvial plains which are vulnerable to sea-level rise: the Lower Liao River Delta; the North China Coastal Plain (including Tianjin); the East China Coastal Plain (including Shanghai); and the Pearl River Deltaic Plain (after Han *et al.*, 1994a).

ered an impossible scenario. Therefore, upgrading and reconstruction measures will be required to prevent dike breaching in the face of accelerated sea-level rise (HAN *et al.*, 1994a).

While beaches are of much less importance economically, their ongoing erosion clearly indicates a present and continuing problem (Figure 6). The eastern end of the Great Wall of China is crumbling away as wave erosion has continued unabated in recent years.

Malaysia

Coastal erosion is expected to be the most severe impact, with further aggravation of existing erosion problems (MIDUN and LEE, 1994). Along Peninsular Malaysia the eastern shore is dominated by sandy beaches while the western shore is much more protected from oceanic waves

and is characterized by mangrove marshes. Already the famous Beach of Passionate Love is experiencing high erosion rates (several meters per year) (Figure 7), and coastal stabilization projects are underway in response to on-going coastal erosion. Along the western shore, sea-level rise impacts will be more severe as human reclamation of mangroves is already a serious and immediate problem.

The mangrove forests are considered as economic wastelands by most coastal developers, and an extensive network of bunds (*e.g.*, dikes) have been built for conversion of these wetlands to agriculture, coastal aquaculture, and industry/urbanism. In their natural state, mangroves can migrate landward up the coastal plain in response to a rise in sea level. Unfortunately, coastal development often occupies the areas the mangroves would colonize in the future. It is

Figure 6.   This pill box on a North China beach indicates the extent of erosion in the last 40 years (the pill box was built on the sand dunes).

Figure 7.   The Beach of Passionate Love, Malaysia—beach erosion is much in evidence. (Photograph by Zamili Midun).

estimated that over 50% of Malaysia's annual fishery income is related to mangrove areas (MIDUN and LEE, 1994) so that these wetlands are economically as well as ecologically valuable areas. In spite of this, reclamation is rapidly reducing the area of mangrove forest and extensive forests may disappear as soon as the year 2000 at present rates of reclamation. Therefore, it is critical that policymakers be made aware of the many critical coastal problems in Malaysia, including the likely impacts and economic consequences of accelerated sea-level rise.

CASE STUDIES

Five in-depth case studies were conducted by in-country scientists of particularly important areas to obtain better quantitative data on the extent of vulnerability and the cost of protection. In addition to the cities of Recife in Brazil

and Alexandria in Egypt, three Chinese areas were considered: North China coastal plain, Shanghai area, and Hong Kong.

Recife

Recife, known as the "Brazilian Venice," is the largest coastal city in the Northeast Region of Brazil. Much of the city is low-lying being built on reclaimed land at the expense of mangroves. Part of the downtown area is already flooded during exceptionally high tides (NEVES and MUEHE, 1994) so that accelerated sea-level rise will be a serious problem for many people in this city of two million inhabitants.

Beach erosion is also an existing problem which has been dealt with by the construction of seawalls, bulkheads, detached breakwaters, and groins. Unfortunately public officials are still allowing beachfront construction and a luxury hotel was recently built right on the beach in spite of the fact that the neighboring buildings already require rubble mound protection (NEVES and MUEHE, 1994). Protection and improved drainage seems to be the best option to deal with the present flooding and erosion problems and future sea-level rise impacts. In addition, coastal planning will be helpful to prevent further construction too close to the water's edge and in areas already or likely in the future to be subject to flooding.

North China Coastal Plain

This enormous coastal plain, which includes the major city of Tianjin, has an expanding industrial base and intensive agriculture and aquaculture. It is already subject to rapid land subsidence, saltwater intrusion, flooding, and erosion. Approximately 20,000 km^2 would be flooded given a 1 m rise in sea level, combined with a 100-year storm surge, assuming no upgrade of protection (HAN *et al.*, 1994b). This area presently contains 14 cities and a population exceeding 10 million people. Much of the land is at or near present sea level, and the existing dikes are often only sufficient to prevent flooding during normal tides.

Abandonment of this area and relocating the industries and cities is inconceivable. Protection has already been demonstrated to be physically and economically rational by recent practice and experience in this area. The local problems of land subsidence because of overpumping of groundwater must also be addressed. One approach is controlled flooding of land adjacent to the Yellow River to divert sediment-laden water onto the low-lying coastal plain. Similar innovative approaches are also being tested in Louisiana through controlled diversions of the Mississippi River.

Shanghai

This city of 13 million people is the largest economic center in China, being responsible for one-ninth of the gross national product (GNP). As is true of all the coastal plains, inundation and increased flooding would be the most serious consequences of sea-level rise. Protection through raising dikes that presently encircle the city is the only possible response.

Shanghai has undergone severe land subsidence this century due to groundwater withdrawal (WANG *et al.*, 1994). This practice has greatly increased the relative sea-level rise problem along this low-lying coastal plain, and significant land areas within the city are presently below sea level (termed "below zero level" cities by the Japanese who have similar situations in Tokyo and Osaka to name a few). Fortunately, the municipal government of Shanghai is addressing this issue by requiring more use of surface water and better management of pumping schemes to greatly reduce the rate of land subsidence.

Hong Kong

Hong Kong is perhaps the most densely populated area in the world, and new land must be created by chopping off the tops of the mountains and filling the adjacent sea with this fill. The cost of this land reclamation is extremely high as the nearshore waters are tens of meters deep; there is no wide intertidal coastal plain which can be diked to create dry land as is the custom on mainland China.

Storm surges are the greatest coastal hazard to the population living on these low-lying reclaimed lands, and severe typhoons, killing thousands of people in some events, have ravaged this area historically. The only realistic response to sea-level rise is protection by upgrading seawalls and other coastal defenses (YIM, 1994).

The government of Hong Kong is taking the sea-level rise issue quite seriously; the West Kowloon reclamation is being built 0.8 meters above previous practice to allow for accelerated sea-level rise. Future reclamations are expected to be similarly raised. While the additional cost will add millions of dollars to each project, the percentage increase in overall cost will be minimal for this "insurance policy" against climate change.

Alexandria

Alexandria, Egypt's second city after Cairo, is vulnerable to sea-level rise both through flooding and erosion. Much of the land area is low-lying; 35% of the land area in the Governorate of Alexandria is already below mean sea level (excluding Lake Mariut). A 1-meter rise in sea level could inundate about 1,200 km^2 or 68%

of the Governorate. This area contains most of the industrial and residential areas, and over 2 million people could be displaced from their homes by the rising sea (EL-RAEY, *et al.*, 1994). Given these severe impacts, protection will be essential.

Alexandria is the major tourist area for Egyptians trying to escape the summer heat of Cairo. Ever since its construction in 332 B.C. by Alexander the Great, this city on the Mediterranean coast has been popular for its sandy pocket beaches. However, erosion in response to rising sea level has already lead to the loss of three such beaches during this century. Six beaches in Alexandria already require nourishment, and a rise of only 0.5 meters could remove nearly all existing beaches. With all of Egypt's other immediate problems, especially a burgeoning population, it is important to educate the public and policymakers about environmental problems with lead times of several decades or more (EL-RAEY, *et al.*, 1994).

NATIONAL ASSESSMENTS

The national assessments were undertaken by utilizing the Aerial Videotape-Assisted Vulnerability Analysis (AVVA) approach (LEATHERMAN *et al.*, 1994) in conjunction with simple models (NICHOLLS *et al.*, 1994). The northern populated area of Argentina and the entire open-coast shorelines of Nigeria, Senegal, Uruguay, and most of Venezuela (excluding part of the almost uninhabited Orinoco delta) was flown and videotaped for this analysis. These five studies produced the most comprehensive data sets and hence lent themselves to the best quantitative analyses of populations and assets at risk and the cost of protection under various sea-level rise scenarios.

Argentina

This first quantitative national assessment of Argentina's vulnerability to sea-level rise showed that the major impacts will occur in the northern half of Buenos Aires Province, primarily because of the high coastal population in this area. A 1-meter rise in sea level could inundate at least 3,400 km^2 of land, which would threaten buildings valued at over U.S. $5 billion. At the same time only about 150 km or 2% of the shore-

line is presently medium to highly developed (DENNIS *et al.*, 1994a).

The major costs for protection will probably be port upgrade and beach nourishment. Beach erosion is already a problem at their most popular tourist beach, Mar del Plata, which boasts the world's single largest casino, located adjacent to the beach. Clearly this area will be protected along with the other tourist beaches.

Argentina is less vulnerable than many other countries in terms of land and value at risk and response costs (DENNIS *et al.*, 1994b). Given the large lengths of undeveloped shoreline, there is a great opportunity to anticipate the problems of sea-level rise. It is only hoped that the policymakers and planners will take advantage of the situation before heavy investment and development right at the water's edge locks them into an expensive protection strategy as is the case along much of the United States coast. A nationally mandated setback for new construction would be ideal as already established by the federal government along the neighboring Uruguayan shoreline (VOLONTE and NICHOLLS, 1994).

Nigeria

Much of Nigeria's population and economic activity is located along the low-lying coastline, within a few meters of existing sea levels (AWOSIKA *et al.*, 1990). A burgeoning population is encouraging development of low-lying barrier islands near the former capitol city of Lagos (Figure 8). Increased flooding, inundation and erosion will clearly be a problem in this area. A 1-meter rise would aggravate existing erosion, which is currently close to 25 m/yr at Victoria Island, Lagos due to a large updrift jetty. Increased flooding would be a particular problem for barrier islands, such as Victoria Island and Ikoyi Island, as well as industries and oil handling facilities in the Niger delta. A 1-meter rise in sea level could inundate as much as 18,000 km^2 of land, forcing as many as 3.2 million people to relocate their homes further inland assuming no human response (FRENCH *et al.*, 1994).

Even a small (*i.e.*, 30 cm) rise in sea level would affect mangroves and other wetlands with both saltwater intrusion and inundation. The loss of wetlands would threaten both subsistence fishing and the lumber industry.

Figure 8. Rapidly eroding beach on Victoria Island, Lagos caused by large harbor jetties. This beach is now being regularly nourished (Photograph by L.F. Awosika).

Traditional coastal engineering countermeasures may be too expensive for a developing country like Nigeria with huge external debts. Thus, responding to sea-level rise should consist largely of keeping some steps ahead of the projected rise by slowly 'disengaging' from the coast, including establishing and enforcing setback lines in areas of new development. In heavily built-up areas, protection will be necessary and low-cost, low-technology approaches need to be developed.

Senegal

A 1-meter rise in sea level could result in the loss of 6,000 km² of land (DENNIS *et al.*, 1994b). Estimates include land loss by both erosion and inundation, assuming that no measures are taken to counteract these impacts. Extensive inundation seems likely for the Saloum and Casamance estuaries and the Senegal delta, accounting for more than 95% of the land loss.

At the northern border, the Senegal River delta's main economic activities are growing rice and sugar cane, both of which would be threatened by sea-level rise. Moreover, two dams have been built on the Senegal River (NIANG, 1990). Reduced sediment supplies may impair the ability of this delta to keep pace with sea-level rise.

Along the sandy North Coast, erosion would only be a problem for a few coastal towns. The capitol city of Dakar is built on rocky cliffs and would not be significantly affected, except for saltwater intrusion into the underlying aquifer. Damage and loss of structures in this area would probably be the most pronounced along the South Coast (Dakar to the Saloum estuary), since this coast is highly developed with towns, industry, and international tourist resorts.

The Saloum and Casamance estuaries are located on either side of the Gambia. Sea-level rise threatens most of the remaining mangrove swamps, as well as endangering the spawning grounds of fish that depend on the mangrove swamps. This loss would reduce the numbers of fish available for both internal consumption and export. Agricultural interests, including rice paddies would also be affected.

The most southern section of coastline also includes active international tourism, which is a growing industry. Preventing the loss of the beaches in the tourist areas would require substantial beach nourishment as quantified by DENNIS *et al.* (1994b).

Uruguay

Tourism is by far the most important coastal industry in Uruguay which is essentially a fairly small country wedged between the two major countries in South America—Brazil and Argentina. The most famous resort is Punta del Este, which is built on a low-lying sandy coast including a fragile tombolo. Fortunately the national government has recently prohibited the construction of buildings within 250 m of the shoreline. This coastal buffer can serve as an effective building setback line and provides a great opportunity to adapt to sea-level rise impacts, notably increased rates of beach erosion. Therefore, there is the real possibility of a planned retreat along much of the coast as a wide coastal buffer zone already exists.

A 1 m rise in sea level is projected to erode or inundate 94 km^2 of land, which is some of the most valuable real estate in Uruguay (VOLONTE and NICHOLLS, 1994). Nourishing 543 km (or about 80%) of the coast to hold the present line on the sandy coast would cost U.S. $4.9 billion — a huge cost for such a small economy. Therefore, protection is likely to be less extensive. As mentioned above, the Uruguayans have bought some valuable time and will only have to expend limited funds to protect their coastal investments until the buffer zone is depleted. The importance of maintaining good beaches, however, should not be underestimated as beach recreation is a national pastime and a major part of the Uruguayan national character; also the buffer zone often contains a coast-parallel road whose loss may be unacceptable. Therefore, continued studies are necessary to develop an integrated coastal zone management plan (VOLONTE and NICHOLLS, 1994).

Venezuela

Venezuela has a relatively long (almost 3,000 km) and diverse shoreline. The original settlers entered the lagoonal coast and hence named the new land area "Venezuela" after the city of

Venice. Today oil extraction and processing is a major activity along this coast, and land subsidence due to withdrawal of subsurface fluids is already a problem in some coastal areas, similar to Venice, Italy.

Most of the coastal activities in Venezuela are presently concentrated in relatively small, intensely used areas, totalling only 385 km (or 13%) of the total coastline as directly measured by VOLONTE and ARISMENDI (1994). This allows the problems of sea-level rise to be anticipated for future developments. Response options will likely include the nourishment of recreational beaches and construction of seawalls along 200 km of developed shoreline, costing about U.S. $1 billion for a 1 m rise in sea level (VOLONTE and ARISMENDI, 1994).

Venezuela is expanding its international tourism, and Margarita Island is considered the new "jewel" in the Caribbean Sea. As this area continues to develop, increased vulnerability and increasing protection costs will be realized unless building setbacks are implemented as part of Venezuela's coastal zone management plan. As a consequence of the AVVA-developed data set, government officials are presently working with the U.S. Countries Studies Program to implement an integrated coastal zone management program that analyzes all the competing factors and potential problems, such as accelerated sea-level rise.

DISCUSSION AND CONCLUSIONS

These assessments show that a 1-meter rise in sea level would have a major impact on the coastal zone of developing countries (NICHOLLS and LEATHERMAN, 1994b). Lack of much of the needed basic data to make quantitative estimates resulted in development of the aerial videotape-assisted vulnerability analysis (AVVA).

Research carried out in the eleven countries, profiled in this overview paper, argues that coastal zones of the world are on a collision course! They are experiencing unprecedented and increasing human pressure due to the combination of coastal development and other impacts. These pressures will be greatly exacerbated by the likely impacts of climate change including accelerated sea-level rise. All the studies discussed here indicate that the developing world needs to carefully consider sea-level rise (NICHOLLS and

LEATHERMAN, 1994c). It is recommended that the coastal zone be treated as a finite resource through integrated coastal zone management. Whatever decisions are made, major changes are likely in the coastal landscape of the world in the coming decades.

Some of the key factors are as follows:

(1) Half of the world's rapidly-growing population and 13 out of 20 of the world's largest cities are situated in the coastal zone.

(2) Existing coastal hazards are often poorly understood.

(3) Over the last century global sea-level rise is estimated to have risen by 0.18 m (DOUGLAS, 1991), while by the year 2100 a further rise of 0.5 m is likely, with a rise of up to 1.0 m being possible (WIGLEY and RAPER, 1992).

(4) Coastal development is proceeding at rapid rates. Existing and future land use usually ignores present hazards, let alone the possible impacts of accelerated sea-level rise.

(5) For existing patterns and levels of coastal development, the financial cost of a 1-meter rise in sea level would be significant for many developing countries.

(6) Significant loss of wetlands at a global scale appears to be almost certain under the scenario of accelerated sea-level rise (illustrating limits to human adaptation to climate change).

Integrated coastal zone management is the only rational approach to deal with these problems. Such programs should commence with the collection of improved data on present coastal conditions and hazards. Coastal zone management will establish an institutional framework which should include the impacts of global change, especially accelerated sea-level rise. The issue of timing is also vital when considering sea-level rise: when and where will different policies be most appropriate?

ACKNOWLEDGMENTS

The research was funded by the Office of Policy Analysis, U.S. Environmental Protection Agency (Mr. Joel Smith, Project Officer; Mr. Jim Titus, Project Manager). The assistance of all our in-country collaborators is much appreciated, and this study would not have been possible without their full cooperation.

LITERATURE CITED

AWOSIKA, L.F.; IBE, A.C., and UDO-AKA, M.A., 1990. Impact of sea-level rise on the Nigerian coastal zone. *In:* TITUS, J.G., (ed.), *Changing Climate and the Coast,* Volume 2., Washington D.C.: U.S. Environmental Protection Agency, pp. 49–65.

BIRD, E.C.F., 1985. *Coastline Changes: A Global Review.* New York, New York: John Wiley & Sons, 219p.

DENNIS, K.C.; NIANG-DIOP, I., and NICHOLLS, R.J., 1994a. Sea-level rise and Senegal: Potential impacts and consequences, *Journal of Coastal Research,* Special Issue No. 14. This Volume.

DENNIS, K.C.; SCHNACK, E.J.; MOUZO, F.H., and ORONA, C.R., 1994b. Sea-level rise and Argentina: Potential impacts and consequences, *Journal of Coastal Research,* Special Issue No. 14. This Volume.

DOUGLAS, B.C., 1991. Global sea-level rise. *Journal of Geophysical Research,* 96(C4), 6981–6992.

EL-RAEY, M.; NASR, S.; FRIHY, O.; DESOUKI, S., and DEWIDAR, Kh., 1994. Potential impacts of accelerated sea-level rise on Alexandria Governorate, Egypt. *Journal of Coastal Research,* Special Issue No. 14. This Volume.

FRENCH, G.T.; AWOSIKA, L.F., and IBE, C.E., 1994. Sea-level rise and Nigeria: Potential impacts and consequences, *Journal of Coastal Research,* Special Issue No. 14. This Volume.

GOEMANS, T., 1986. The Sea Also Rises: The ongoing dialogue of the Dutch with the Sea. *In:* TITUS, J.G. (ed.), Effects of changes in stratospheric ozone and global change, *Sea-level Rise.* Volume 4. Washington, D.C.: United States Environmental Protection Agency and United Nations Environment Programme, pp. 165–189.

HAN, M.; HOU, J., and WU, L., 1994a. Potential impacts of sea-level rise on China's coastal environment and cities: A National assessment. *Journal of Coastal Research,* Special Issue No. 14. This Volume.

HAN, M.; WU, L.; HOU, J.; LIU, C.; ZHAO, G., and ZHANG, Z., 1994b. Sea-level rise and the North China coastal plain: A preliminary assessment. *Journal of Coastal Research,* Special Issue No. 14. This Volume.

HUQ, S.; ALI, S.I., and RAHMAN, A.A., 1994. Sea-Level Rise and Bangladesh: A Preliminary Analysis. *Journal of Coastal Research,* Special Issue No. 14, This Volume.

INTERGOVERNMENTAL PANEL ON CLIMATE CHANGE, 1990. Potential Impacts of Climate Change. *Report Prepared for IPCC by Working Group II.*

LEATHERMAN, S.P., 1986. Impacts of sea-level rise on the coasts of South Latin America. *In:* TITUS, J.G. (ed.), Impacts of changes in stratospheric ozone and global climate, *Sea-Level Rise.* Volume 4. Washington, D.C.: United States Environmental Protection Agency and United Nations Environment Programme, pp. 73–82.

LEATHERMAN, R.J.; NICHOLLS, R.J., and DENNIS, K.C., 1994. Aerial videotape-assisted vulnerability analysis: A cost-effective approach to assess sea-level rise impacts. *Journal of Coastal Research,* Special Issue No. 14, This Volume.

MIDUN, Z. and LEE, S-C., 1994. Implications of a greenhouse-induced sea-level rise: A national assessment for Malaysia. *Journal of Coastal Research*, Special Issue No. 14. This Volume.

MUEHE, D. and NEVES, C.F., 1994. The implications of sea-level rise on the Brazilian Coast: A preliminary assessment. *Journal of Coastal Research*, Special Issue No. 14. This Volume.

NEVES, C.F. and MUEHE, D., 1994. Potential impacts of sea-level rise on the Metropolitan Region of Recife, Brazil. *Journal of Coastal Research*, Special Issue No. 14, This Volume.

NIANG, I., 1990. Responses to the Impacts of Greenhouse Induced Sea-level Rise on Senegal. *In:* TITUS, J.G. (ed.), *Changing Climate and the Coast*, Volume 2. Washington D.C.: U.S. Environmental Protection Agency, pp. 67–87.

NICHOLLS, R.J. and LEATHERMAN, S.P., 1994a. Sea-level rise and coastal management. *In:* McGREGOR, D. and THOMPSON, D. (eds.), *Geomorphology and Land Management in a Changing Environment.* Chichester, U.K.: John Wiley, *in press.*

NICHOLLS, R.J. and LEATHERMAN, S.P., 1994b. Global sea-level rise. *In:* STRZEPEK, K. and SMITH, J.B. (eds.), *As Climate Changes: Potential Impacts and Implications.* Cambridge: Cambridge University Press, *in press.*

NICHOLLS, R.J. and LEATHERMAN, S.P. 1994c. The implications of accelerated sea-level rise and developing countries: A discussion. *Journal of Coastal Research*, Special Issue No. 14. This Volume.

NICHOLLS, R.J.; LEATHERMAN, S.P.; DENNIS, K.C., and VOLONTÉ, C.R., 1994. Impacts and responses to sea-level rise: Qualitative and quantitative assessments. *Journal of Coastal Research*, Special Issue No.14, This Volume.

STRZEPEK, K. and SMITH, J.B. (eds.), 1994. *As Climate Changes: Potential Impacts and Implications.* Cambridge: Cambridge University Press, *in press.*

TITUS, J., 1990. (ed.) *Changing Climate and the Coast*, Washington, D.C.: U.S. Environmental Protection Agency, Two Volumes, 396p. and 508p.

TITUS, J.G.; PARK, R.A.; LEATHERMAN S.P.; WEGGEL, J.R.; GREEN, M.S.; MAUSEL, P.W.; BROWN, S.; GAUNT, C.; TREHAN, M., and YOHE, G., 1991. Greenhouse effect and sea-level rise: Potential loss of land and the cost of holding back the sea. *Coastal Management*, 19, 171–204.

VOLONTÉ, C.R. and ARISMENDI, J., 1994. Sea-level rise in Venezuela: Potential impacts and responses, *Journal of Coastal Research*, Special Issue No. 14. This Volume.

VOLONTÉ, C.R. and NICHOLLS, R.J., 1994. Sea-level rise and Uruguay: Potential impacts and responses. *Journal of Coastal Research*, Special Issue No. 14, This Volume.

WANG, B.; CHEN, S.; ZHANG, K., and SHEN, J., 1994. Potential impacts of sea-level rise on the Shanghai Area. *Journal of Coastal Research*, Special Issue No. 14, This Volume.

WIGLEY, T.M.L. and RAPER, S.C.B., 1992. Implications for climate and sea level of revised IPCC emissions scenarios. *Nature*, 357, 293–300.

YIM, W., 1994. Implications of sea-level rise for Victoria Harbour, Hong Kong. *Journal of Coastal Research*, Special Issue No. 14, This Volume.

[11]

The Emergence of Post-Suburban Landscapes on the North Coast of New South Wales: A Case Study of Contested Space*

STEPHEN J. ESSEX AND GRAHAM P. BROWN

Introduction

Over the last 10–15 years, a restructuring of the Australian urban and regional system has begun to take place. The Parliament of the Commonwealth of Australia (1992) predicted major transformations to the urban pattern into the twenty-first century, including the emergence of a near-continuous line of urban development along the eastern seaboard from Cairns to Melbourne; a trend for cities to become multi-centred and suburban with a reduced role for the central cores; for state boundaries to become less important in growth regions (such as south-east Queensland, northern New South Wales (NSW) and the Australian Capital Territory/south-east NSW); and for isolated nodes of development in parts of Queensland and Western Australia to have stronger international links than with the state or nation. These transformations are attributed to economic restructuring and rationalization, technological change and the ageing of the population. As such, these expected patterns and processes of urban change bear resemblance to the new urban forms emerging in otherparts of the world, especially the post-suburban landscapes of California.

The principal aim of this paper is to consider recent changes to the urban landscape of Australia in light of the processes which have been shaping development elsewhere in the world, particularly as the Australian example appears to be in a period of transition. Focus is placed on the mechanics, outcomes and implications of change at the local scale and on the conflicts arising from contested development, whereby a host of different groups and interests compete for space. The difficulties of reconciling the aspirations of all interest groups, of exercising control by planning, of protecting natural environmental resources and of managing development in such a way that the scale of the new urbanization does not overwhelm the landscape represent the issues being faced by developers, planners and local communities as the urban restructuring of Australia takes place.

The paper will draw on evidence of urbanization on the North Coast of NSW and will provide a detailed examination of two shires in the region — Ballina and Byron — where

* The authors would like to thank Mark Brayshay, Clive Charlton and Mark Cleary for their constructive comments on previous drafts of this paper; Brian Rogers and Tim Absalom for the cartography; and Julie Shackleford, Kate Hopewell and Ros Bryant for the typing.

a study was carried out by the authors in 1994 and 1995. Both shires are areas of high environmental value and have been subject to substantial population growth since the 1970s. A particular feature of the social environment is the presence of vocal community groups which have formed to defend vigorously the area's qualities. The planning system is experiencing considerable operational problems by attempting to balance the demands for new development with claims for restricting further growth. The strength of local opinion has a number of implications for the likely success of planning intervention in this aspect of urban management.

The paper is divided into four sections. In the first section, the literature on post-suburban landscapes is reviewed; in the second section, previous research on the population and development trends in Australia are considered; in the third section, these patterns are related to NSW and its North Coast; and, in the fourth section, the implications of new development with reference to Ballina Shire and Byron Shire are discussed. In considering the emergence and impacts of post-suburban landscapes in this particular area of NSW, the paper explores the extent to which the notions and ideas of the postfordist concept may be applied. At the heart of the postfordism is the process of decentralization, increased local empowerment and a trend towards de-integration. Whether the model genuinely encapsulates the processes operating in Ballina Shire and Byron Shire clearly requires detailed discussion based on empirical investigation.

Post-suburban urban forms

A number of urban studies in the United States have identified a new and well-advanced pattern of urbanization. The established urban structure of metropolitan concentration based on industrial growth surrounded by economically inert, provincial and dormitory-based suburbs has begun to shift to increasingly economically active, polynucleated and amorphous suburbs with a less dominant central city (Kling *et al.*, 1991). Various terms have been coined to describe this new form of urbanization, including 'post-suburban' (Kling *et al.*, 1991); 'edge city' (Garreau, 1991); 'exopolis', 'technoburbs', 'silicon landscapes' (Soja, 1992); 'cyburbia' and 'ageographical city' (Sorkin, 1992). Key features of these extended urban forms are a homogenous landscape of low-rise buildings, often extending for hundreds of miles, with specialized residential, commercial and industrial zones (Gottdiener and Kephart, 1991: 34).

A unifying explanation of this new pattern of urbanization is postfordism, which encompasses a full sweep of related economic and social transformations such as the transition from industrial to post-industrial society and from modernism to postmodernism (Esser and Hirsch, 1989; Graham, 1992) (see Table 1). Postfordist technology, involving more flexible and specialized production, have meant that urban areas are no longer bound to mass production and large-scale assembly lines. New peripheral urbanization becomes possible to capture the new 'scope' economies of postfordist technology (Soja, 1992), allowing the deconcentration of ordinary labour intensive manufacturing (Gottdiener and Kephart, 1991). Also bound up with these changes has been the structural transformation to a post-industrial society, involving the replacement of an economic base dominated by manufacturing, with one dominated by service and information industries (Gottdiener and Kephart, 1991: 35). Kling *et al.* (1991: 15) have coined the term 'information capitalism' to describe occupations in which the processing and distribution of information are the central and time-consuming activities. These activities stimulate deconcentration of the urban form because of their use of mass communications and computer technology which frees them from central locations, and because such industries are more likely to require a greenfield site and a clean environment. Advanced technology and communications eliminate the importance of spatially defined communities (Winner, 1992: 54), and enable a shift from provincialism

Post-surburban landscapes on the North Coast of New South Wales 261

Table 1 *Representations of fordist and postfordist frameworks*

	Fordism	Postfordism
Economic relations		
Production	Mass production (large-scale assembly lines), deskilled labour, mass consumption.	Flexible production (specialized production units), skilled labour, market niches.
		Post-industrial: growth of service industries and 'information capitalism'.
Markets	Provincialism and regional markets.	Globalization/cosmopolitanism and international markets.
Management	Corporate management, centralized control, application of scientific principles ('taylorism').	Deconcentration, autonomous roles to section/division managers within/between organizations.
Political relations		
Role of state	Economic growth provided resources for state provision of infrastructure to promote production and improve productivity. Authoritarian regulation.	Economic instability eroded public expenditure and political legitimacy of state spending. Replaced by 'strong state, free economy'. Deregulation.
Local government	Large-scale, bureaucratic, management and control. Expert-driven intervention.	New flexible forms of corporate policy-making. New forms of local governance.
Social relations		
Social impacts of economic system	Worker alienation.	Two-tier workforce: 'core' (skilled and enhanced conditions) and 'peripheral' (low job security, part-time/ temporary, contracted, poor conditions, low wages).
Dominant culture	Welfarism/collectivism. Shared belief in infallibility of science and progress (modernism).	Enterprise culture/individualism. 'Safety net' welfare provision. Loss of faith in ideology of progress and science (post-modern). Growing environmentalism and development of alternative perspectives.
Patterns of urbanization		
Geography	Metropolitan concentration, economically inert and dormitory-based suburbs.	Decentralization. Less dominant central city, economically active, polynucleated and amorphous suburbs.

Table 1   (*continued*)

	Fordism	Postfordism
Form	Urban containment and concentration. Heterogeneous landscape. Functional architecture and design (modernist).	Urban sprawl and peripheral urbanization. Homogeneous landscape of low-rise buildings. Specialized zones. Aesthetic architecture and design (postmodernist).
Land use planning		
System	Emphasis on public sector regulation of land use.	Emphasis on enterprise culture/free market forces in more flexible allocation of land.
Dominant actors	Public sector agencies.	Private sector 'Growth networks'. Real estate developers.
Public involvement	Limited.	Increased public participation and empowerment. Environmentalism.
Outcome	Controlled development, maintaining 'status quo'.	Ad-hoc development. Conflict and contested space.

Source: Based and developed from Henry (1993, Table 7.1: 181).

to increasing globalization or cosmopolitanism (Kling *et al.*, 1992: 20). Local economies can be transformed from regionally-based markets to strong economies of international dimensions, where the importance of local space, linkages and control are reduced.

The homogeneity of post-suburban residential landscapes is attributed to private real estate developers, who are largely responsible for the construction of new properties. Developers tightly control residential developments by predetermined styles and designs, which venture too close to 'private-sector socialism' for some tastes (Soja, 1992: 114). The result is an urban landscape with little visual variety reflecting a standardized, quasi-global culture (Zukin, 1991: 20) and an expression of postmodernism shaped by aesthetic aims and principles (Harvey, 1989). Planning regulations play a role in stipulating the form and outcome of new urban development, although the underlying economic forces are beyond its control. Planning zoning laws usually favour relatively large plots and control building design, which act to keep property prices high (Zukin, 1991: 140). The regulation and intervention imposed by planning on private development under a fordist framework begins to be questioned as the economic instability of postfordism shifts the balance in favour of private enterprize. The formation of 'growth networks', consisting of coalitions of public officials (who are often significant owners of property), the real estate sector and representatives of finance and corporate capital, are other forces at work in the development process under postfordism (Olin, 1991: 224).

The implications of the post-suburban form are being increasingly recognized in the literature. Zukin (1991) focuses on the social inequalities emerging in these post-suburban areas. Kling *et al.* (1991: 23) discuss how post-suburban spatial organization may turn out to be far less than utopian as communities become increasingly segmented — spatially, economically, ethnically and socially. Within the globalization of the postfordist economy, Olin (1991: 223) highlights how powerful outside businesses begin to displace established local business interests, although not without opposition. Local businesses have the most to lose in such competition as they are tied to the growth of the particular region in which they are located. Conflicts can also emerge within local communities as new values, new cosmopolitan political agendas and new development are imposed on local populations. A feature of postmodern society is increased public participation in planning and heightened environmental sensitivity, which translates into an eagerness to defend the existing environment. Ironically, this empowerment becomes an obstacle to the emergence of alternative forms of urbanization and serves to perpetuate forms that were originally tied to fordist consumption patterns (Filion, 1996: 1654). Contested space therefore emerges as a central theme in the development of post-suburban landscapes and provides the focus for this study.

Restructuring of the Australian urban system

The population of Australia is not only highly urbanized, but is also concentrated in a small number of sprawling coastal cities. Moreover, Australian cities have traditionally been characterized by low population densities and high levels of suburbanization. This settlement pattern originates from the dominance of the colonial capital of each state (Logan *et al.*, 1981; Hamnett and Bunker, 1987), the clear, perhaps culturally ingrained, preference of Australians for low density living (Bunker and Houston, 1992) and the abundant availability of land. Since the 1980s, important changes have begun to modify this well-defined urban pattern. Hugo and Smailes (1985; 1992) have shown that during the 1970s there was a reversal in the established pattern of population concentration in large urban centres, as significant population increases began to be experienced in non-metropolitan areas. The non-metropolitan increase was spatially concentrated in the attractive areas of the south-east and east coast and the areas around the margins of the commuting zones of large cities. The growth involved the retention of established residents and net migration gain from major urban areas. Paris (1994) has suggested that, in the last 10–15 years, a restructuring of the Australian urban and regional system has begun to take place. He described this change as the development of a national urban system, involving a linkage between the three dominant sprawling conurbations (south-east Queensland, Newcastle-Sydney-Illawara and Melbourne) and a rapidly growing and extensive coastal zone of consumption centres and suburban development. Other research verifies Paris's findings. Mullins (1990; 1992; 1994) used the term 'tourism urbanization', to describe the rapidly developing cities and towns specially built for the 'consumption of pleasure'. The Gold Coast of south-east Queensland serves as the best example of this new form of urbanization in Australia. Murphy (1976; 1981; 1985) has demonstrated the importance of retirement settlement in the urbanization of the coast, especially in New South Wales.

For Sant and Simons (1993), recent shifts in the Australian urban settlement pattern were associated with processes of counterurbanization, where people's resources and desire for different (and sometimes new) lifestyles have demanded the creation of new urban spaces. Although the process of counterurbanization is rather nebulous in its definition, it is generally taken to mean the deconcentration of population from urban areas to remoter rural areas, which have previously been losing population (Berry, 1976; Vining & Pallone, 1982; Perry *et al.*, 1986; Champion, 1989). Attempts to explain the underlying motives which prompt counterurbanization have been multifaceted, ranging from a change in values and the rejection of urban lifestyles; the spatial freedom for populations and industries provided by technological advancements; the role of diseconomies of agglomeration in existing centres; and the exploitation of lower-waged and non-unionized labour in peripheral regions.

Sant (1993) has identified distinct forms of coastal development in Australia which can be related directly to the population growth. One of the most common forms of coastal development has been the extension (or infill) of existing settlements, although recent expansion of this sort has involved ribbons of development stretching along adjacent waterfronts. Existing freestanding, greenfield developments have also grown to become small urban centres. Marina villages and canal estates have been developed on low-lying land previously left vacant because of flood hazard, but which are now usable as a result of flood control engineering. Development on agricultural land, which has been subdivided into residential lots to achieve a higher return, has also become commonplace. These forms of coastal urban development have contributed to the spread of suburban landscapes in Australia, although they are in a much earlier phase of evolution than the post-suburban landscapes of California.

The implications of the new forms of urbanization are increasingly being recognized, especially in terms of being wasteful of productive rural land, destructive of environmental resources and a strain on local infrastructure. These concerns culminated in a government inquiry, which advocated the adoption of the concept of ecologically sustainable development in the management of Australia's coastal zone (Resource Assessment Commission, 1992). Paris (1994) has suggested that, as coastal urbanization becomes more dominant, significant political, economic and social implications will also take effect. The implications of continued coastal development will include increased local opposition to new construction, higher building densities to cope with the demand, and the gentrification of settlements, particularly in the more expensive coastal areas. Many of the new migrants, while contributing to this development process themselves, are well educated and effective in organizing and publicizing their opposition to new further development and their support for environmental protection. The behaviour of these activists in seeking to preserve their local environment appear to be universal and represent a common value system that contributes to a more homogeneous form of urban restructuring. Such groups can become extremely powerful due to the high profile they can achieve in what are still relatively small communities. The sociological changes and emerging conflicts involved in the population shifts associated with the restructuring of the Australian urban system present considerable implications and contradictions for the operation of planning and resource management, which this paper will explore further.

North Coast of New South Wales (NSW)

New South Wales (NSW) and its North Coast provides an appropriate case study to investigate these issues. The state has the country's largest population (c. six million people) and population growth. The North Coast possesses a highly valued environment and has experienced consistent population growth since 1976. Much of this growth has been associated with pressures from the bordering state (and from the Gold Coast conurbation of south-east Queensland in particular) rather than from pressures within NSW. At a very simplistic level, the North Coast of NSW represents a region in which the traditional dominance of the state capital is breaking down and being replaced by a developing national urban system. In this sense the processes of urban change in NSW bear the hallmarks of increasing postfordism whereby a trend towards decentralization and a multi-centred urban structure has evolved.

NSW covers 10.4% of the land surface of Australia and contains 34.1% of the country's population (Farrell, 1993). Sydney accounts for the largest proportion of the state's population (1989: 62.9%) and population growth (1981–89: 65.3%) (NSW Department of Planning, 1990) (see Figure 1). The metropolitan primacy of Sydney has been a long-established feature in NSW, often attributed to the city's role as the main gateway for international migration (Sant, 1993). Sydney is arguably Australia's only 'world city' (Berry and Rees, 1994). Outside the Sydney region, the majority of the population growth has occurred along the coast (especially the Illawarra, Hunter and

Post-surburban landscapes on the North Coast of New South Wales 265

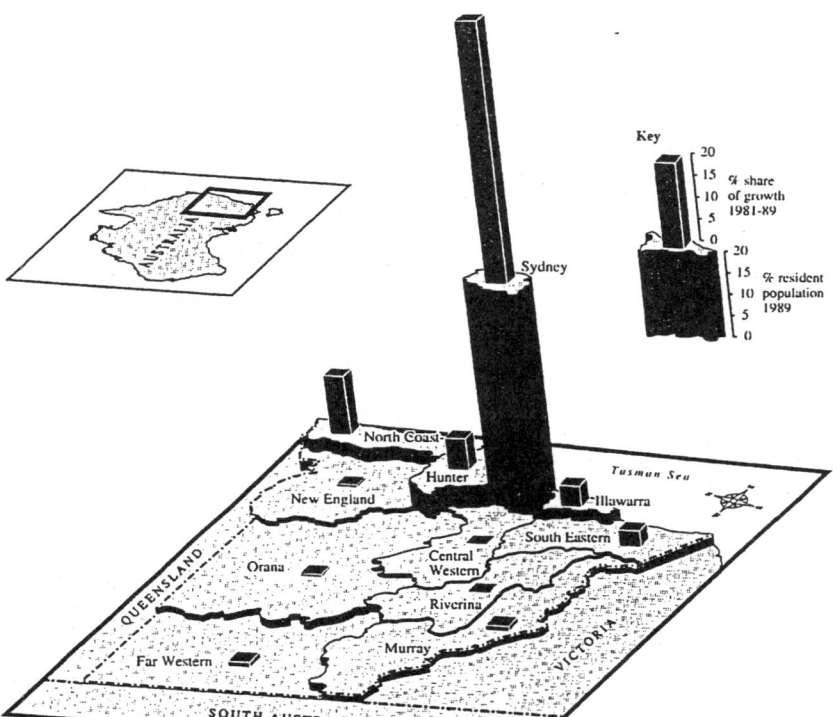

Figure 1 *Distribution of population (1989) and share of growth (1981–89) in New South Wales (source: NSW Department of Planning, 1990)*

North Coast regions). A total of 80% of the population of NSW is located in local government areas bordering the Pacific Ocean (Coastal Committee of NSW, 1994; Hugo, 1994).

The North Coast accounts for 6.8% of the population of NSW and 16.8% of the population growth (1981–89). Within the North Coast region, significant and consistent population growth has been evident since 1976, particularly in coastal shires such as Ballina, Byron, Hastings, MacLean, Nymboida, Tweed and Ulmarra (see Figure 2). In fact, the North Coast region is expected to increase its share of NSW's population from 6.5% in 1986 to between 9% and 11% by 2021, largely as a result of in-migration (NSW Department of Planning, 1993). The growth of population in this region has been rapid and significant: resulting in extensive urban development.

The extent and rate of development on the North Coast since the early 1980s is indicated by the key planning statistics produced by the state government, which relate to new dwelling approvals (NSW Department of Planning, 1994). Most shires reflect the pattern for the North Coast as a whole, which is characterized by cycles of boom (1981, 1983–84, 1986–89, 1992–93) and recession (1981–83, 1984–86, 1989–92) (see Figure 3). The overall trend has been for an increasing number of approvals per year and for a

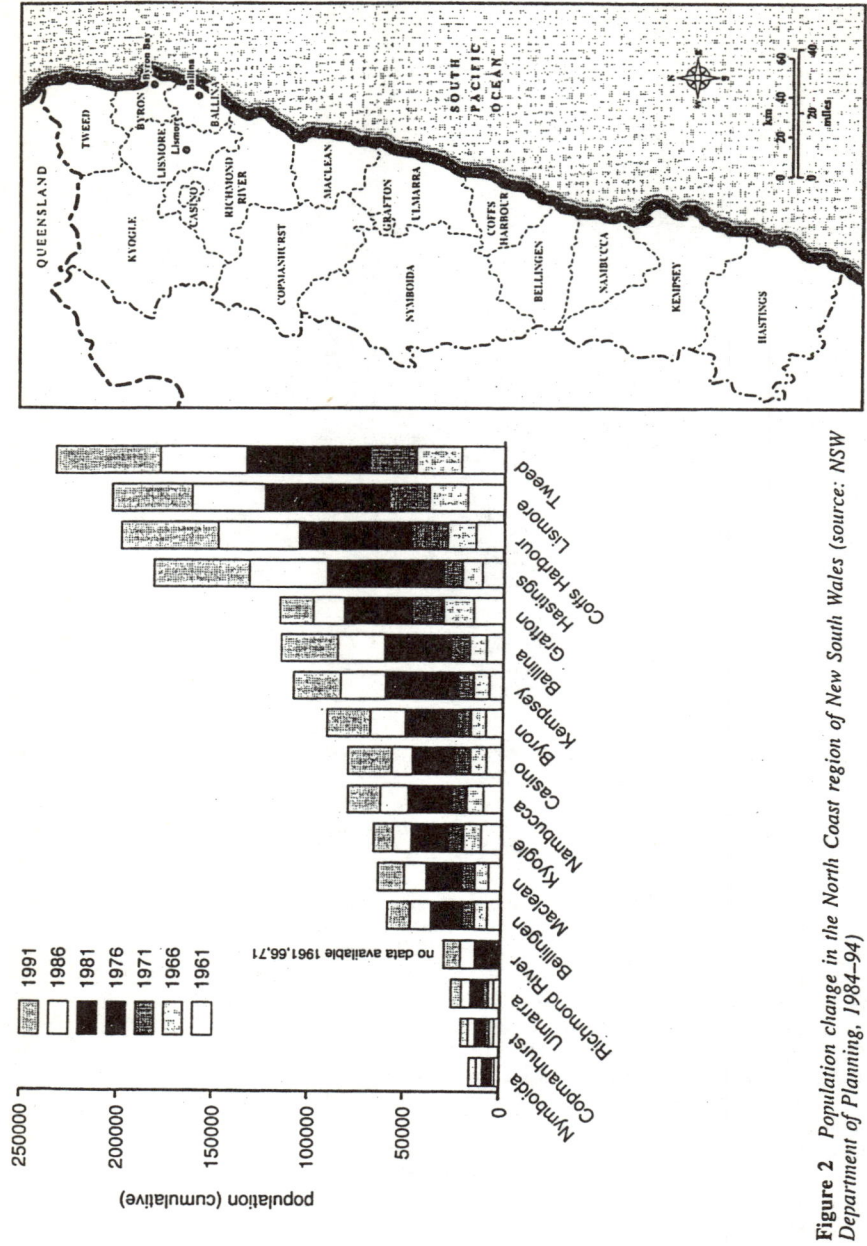

Figure 2 *Population change in the North Coast region of New South Wales (source: NSW Department of Planning, 1984–94)*

Post-surburban landscapes on the North Coast of New South Wales   267

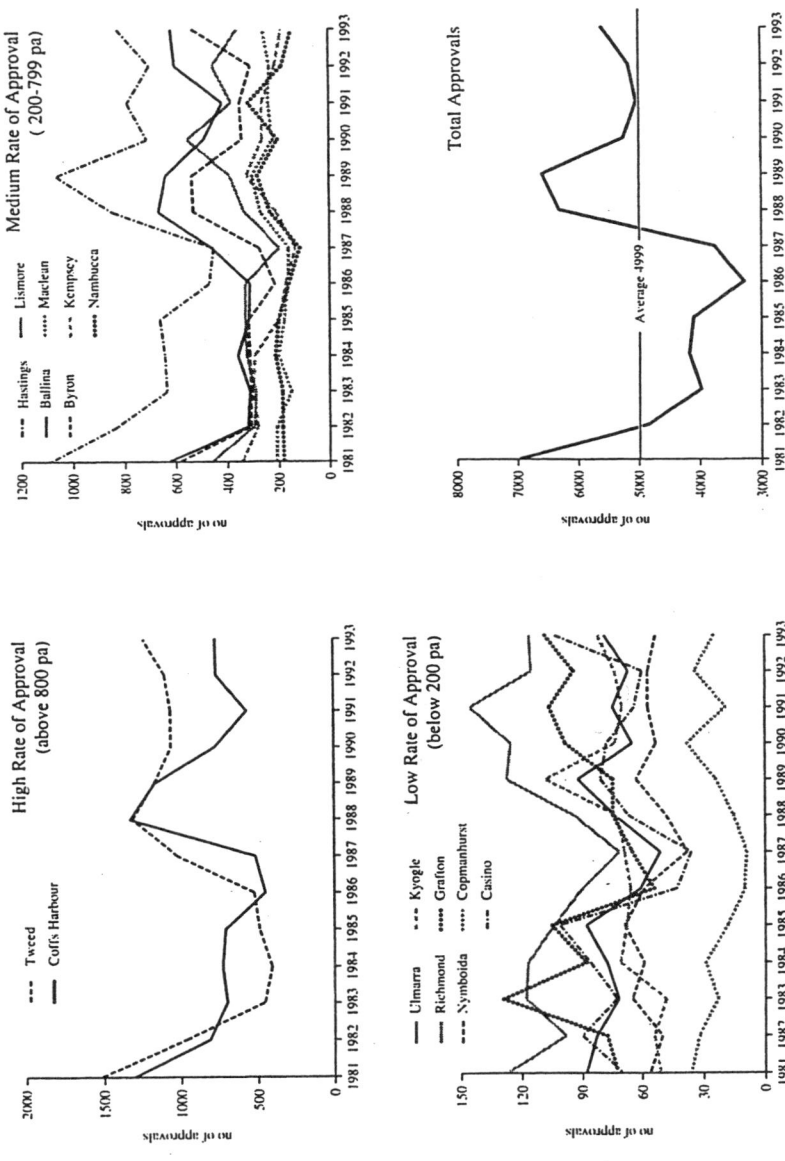

Figure 3 *New dwelling approvals in North Coast of New South Wales, 1981–93 (source: NSW Department of Planning, 1984–94)*

considerable amount of development since 1981 (64,989 approvals). Geographical variation is evident in the number of approvals for new dwellings over the period, 1981–1993. Consistently high rates of approvals (above 800 approvals per annum average) are concentrated in Tweed and Coffs Harbour, which represent the major coastal urban centres in the region. A small group of shires have approval rates of 300–799 per annum, including Ballina (470 per annum), Byron (378 per annum) and Lismore (361 per annum) forming a 'triangle' of growth in the far north of the area. Another group of shires have approval rates of 200–300 per annum (Kempsey, Maclean and Nambucca). The remaining seven shires have approval rates of less than 200 per annum, although these figures mask the real rate of development in some areas because of exemptions made by local planning regulations.

Ballina Shire and Byron Shire

The shires of Ballina and Byron present themselves as case study areas because of the high rates of development experienced in recent years and because they possess high landscape and ecological values. Both also exhibit a characteristic lifestyle which is cherished as an important element of contemporary Australia. Byron Shire, in particular, has evolved into a mecca for surfers and has achieved national recognition as a centre for people seeking an 'alternative' lifestyle. There is a strong desire among residents of Ballina and Byron Shires not to duplicate the type and scale of urban sprawl experienced on the neighbouring Gold Coast of south-east Queensland (Edols-Meeves and Knox, 1996).

The scale of population growth in the two shires since 1921 is shown in Figure 4. Both shires lost population prior to the 1970s, with Ballina Shire experiencing marginal losses between 1921 and 1947 and Byron Shire losing population in two periods (1921–33 and 1954–71). Even the gains in population experienced in Ballina Shire from 1947 and in Byron Shire between 1933 and 1954 were minimal, with growth never reaching more than 10% in an intercensal period. In contrast, however, the growth rates for both shires have increased markedly since 1976. These figures clearly show that there was a significant turnaround in population in the two shires after 1976.

As the population of the shires has increased, its composition has also changed. A comparison of the population pyramids for 1961 and 1991 for Ballina and Byron indicates an ageing of the community (see Figure 5), reflecting both national trends and the area's function as a destination for retirement. Slight increases are also evident in the age groups 30–44, pointing to an in-migration of people of working age. Population forecasts for both shires by the NSW Department of Planning indicate that Ballina Shire will increase from a recorded population of 30,192 in 1991 to 44,000 by the year 2000, while Byron Shire will grow from a recorded population of 22,629 in 1991 to 29,200 by the year 2001. Byron Shire Council have estimated a slightly higher population for its shire of 32,123 by the year 2000 (see Figure 4). Population pressures in the two shires are thus set to continue into the next century.

Given that statistics of building approvals provide only a partial and incomplete guide to the true scale of new development in Ballina and Byron, the manifestation of population growth on the rate of residential development in the two shires is difficult to illustrate in statistical terms. Figure 3 has already shown that Ballina Shire and Byron Shire have experienced an above average rate of building approvals (i.e. 300–799 per annum) over the period 1981–93 and that these rates broadly reflect trends on the North Coast as a whole. Further data on building approvals for Ballina Shire were available to provide a complete set of figures from 1966 to 1993 (see Figure 6).

In order to obtain a better understanding of the impact of new development in the two shires, a series of aerial photographs were examined. The spatial pattern of development was plotted, focusing on the growth of the area in and around the towns of Ballina and

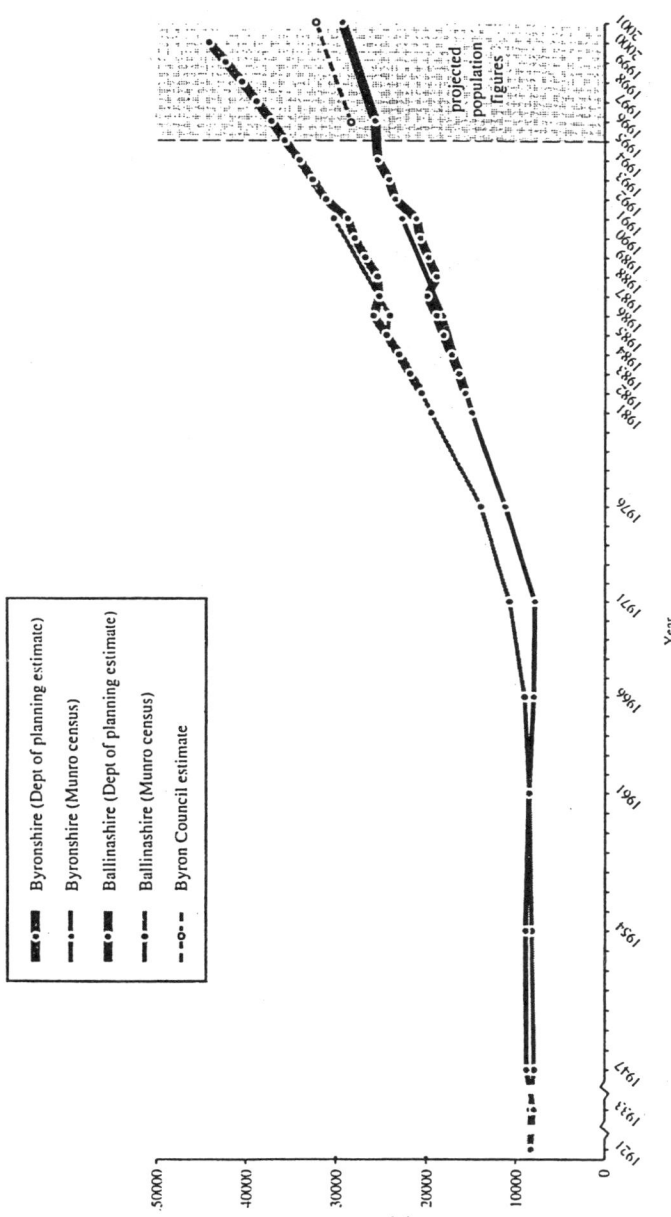

Figure 4 *Population change in Ballina Shire and Byron Shire, 1921–2001 (sources: Munro, 1976; NSW Department of Planning, 1984–94; Byron Shire Council, 1994)*

270 *Stephen J. Essex and Graham P. Brown*

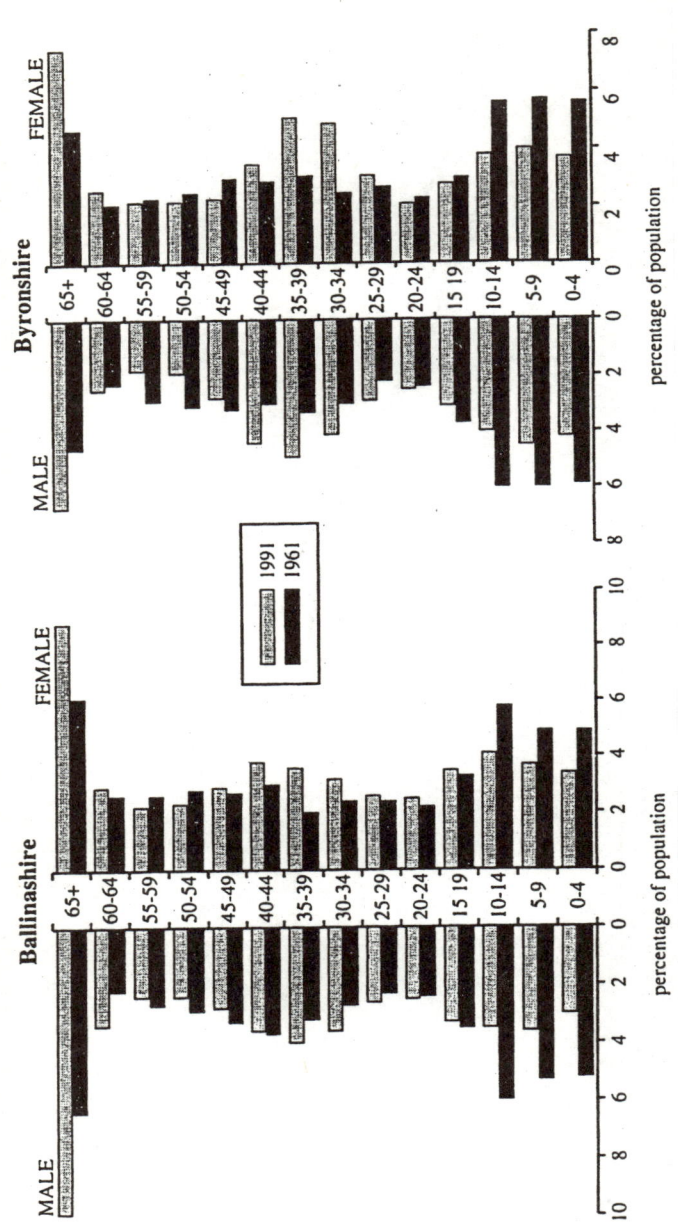

Figure 5 *Population pyramids for Ballina Shire and Byron Shire, 1961 and 1991 (sources: Munro, 1976; Ballina Shire Council, 1993; Byron Shire Council, 1994)*

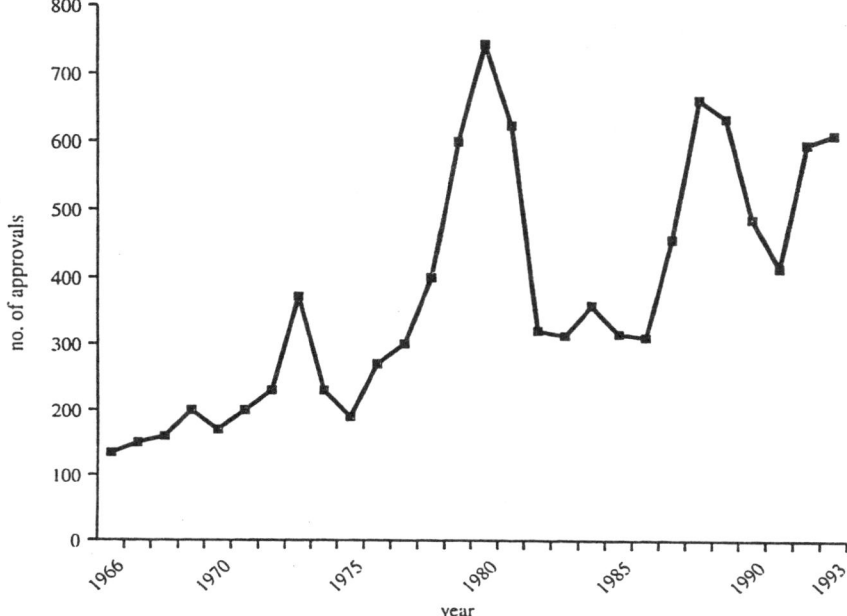

Figure 6 *Building approvals for Ballina Shire, 1966–93 (sources: Derrett, 1994; NSW Department of Planning, 1994)*

Byron Bay. The photographs provided a sequence of development for Ballina (1974–94) and for Byron Bay (1947–79), and was supported with reference to planning documents and field checking (1994).

The analysis revealed that, in the mid 1970s, urbanization in and around Ballina was concentrated in the town itself straddling both sides of North Creek (see Figure 7). A separate, beachside settlement to the north of Ballina, at Lennox Head, was starting to be developed. During the 1980s, the predominant form of urbanization occurred in the coastal strip between Ballina and Lennox Head thereby taking advantage of the seaward aspect. A marina development had also been established on the western side of Ballina on the Richmond River by the end of the 1980s. The coastal strip has continued to be developed in the 1990s and is being vigorously marketed as a 'lifestyle investment'. The subdivided plots are sold under slogans such as 'Life's too short not to live here!', which is a clear indication of the role of real estate developers in the emergence of these new urban forms and the emphasis placed on the 'quality of life' rather than 'functional' values of settlement.

The form of building development in Byron Bay apparent in 1947 indicates a low density 'village-like' township (see Figure 8). The seafront area of the town was not built up, enabling the sea and sand to encroach inland. There was also no development on Cape Byron, apart from the Lighthouse Keepers' Cottages. By the end of the 1970s, however, housing development on Watergoes, Cape Byron Headland and a suburban development to the south of Byron Bay at Suffolk Park was evident, together with a more formalized sea front in the settlement of Byron Bay itself. Field observations made in 1994 show

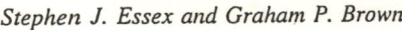

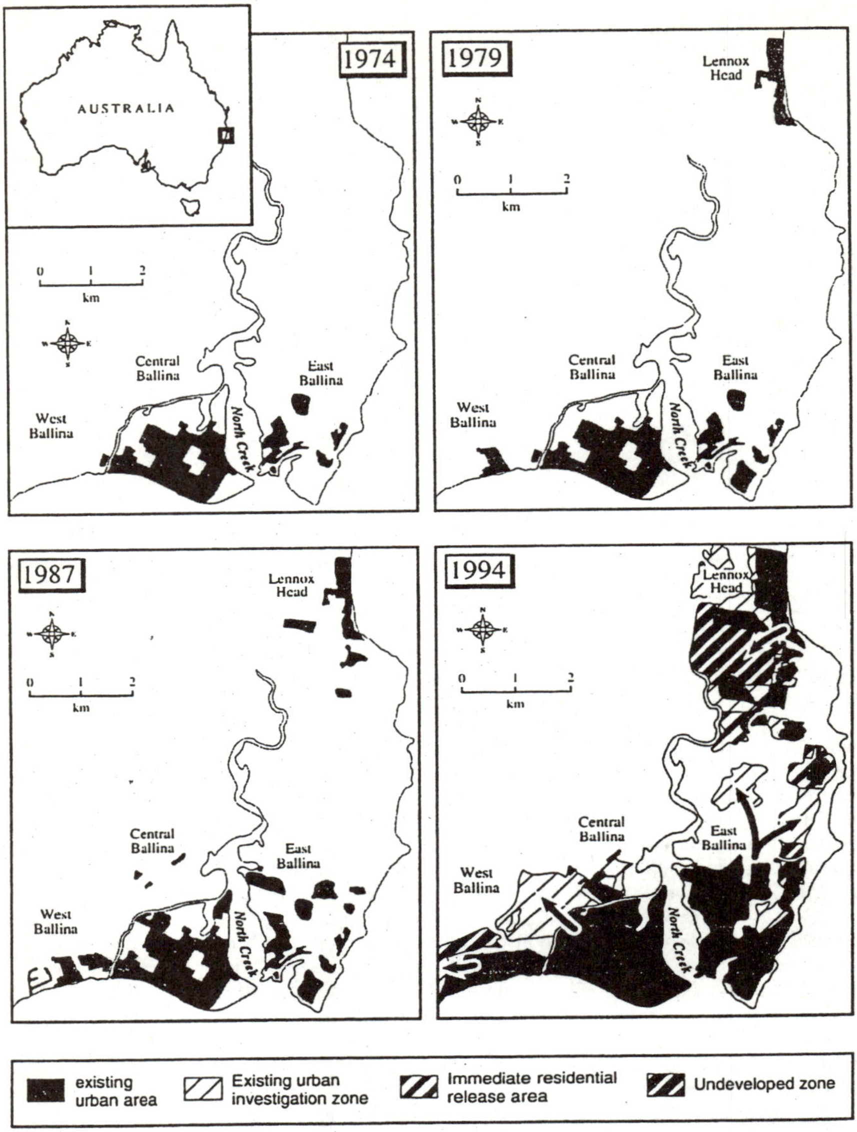

Figure 7 *Spatial development of Ballina, 1974–94 (sources: CMA Ballina 9640-3-N, 1:25,000 Second Edition 1984 Topographic map; Aerial photographs (14 March 1974, 3 April 1979, 1 August 1987, 14 March 1994); field checking in November 1994; Ballina Shire Council, 1993)*

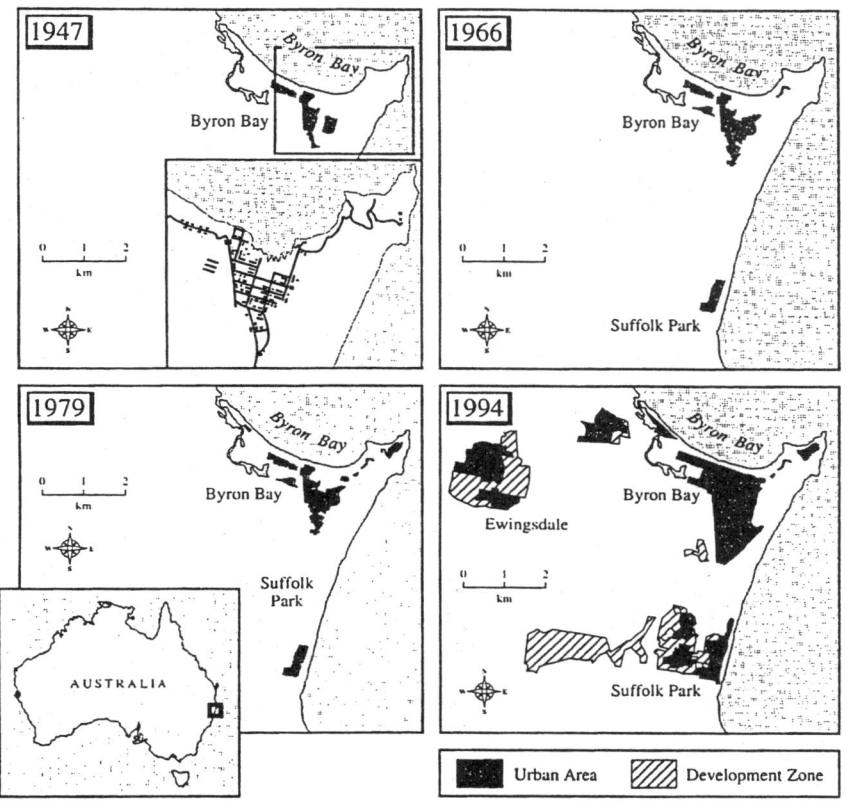

Figure 8 *Spatial development of Byron Bay, NSW, 1947–94 (sources: CMA Byron Bay 9640-4-S, 1:25,000 Second Edition 1984 Topographic map; Aerial photographs (27 May 1947, 22 September 1966, 3 April 1979); field checking in November 1994; Byron Shire Council, 1990)*

further extensions to the existing housing developments producing a much larger urban area, at a much higher density.

One consequence of population pressures in the area has been that land values in central Byron Bay have sharply increased. This change has encouraged some people living in the sought-after locations to rent their properties to visitors or short-term tenants and move to the suburbs, thus fuelling further development. Families remaining in Byron Bay have complained that many neighbourhoods now lack a sense of community spirit, representing a less tangible aspect of the wider experience of population and tourist pressure (Shantz, 1994).

The pattern of recent urban development around Ballina and Byron has not produced a sprawling amorphous suburb, bearing resemblance to the post-suburban landscapes of California, but perhaps represents an early stage in the transition to such an urban landscape. The scope for further growth of both Ballina and Byron Bay appears limited due to physical constraints. It was recognized in the North Coast Draft Urban Planning

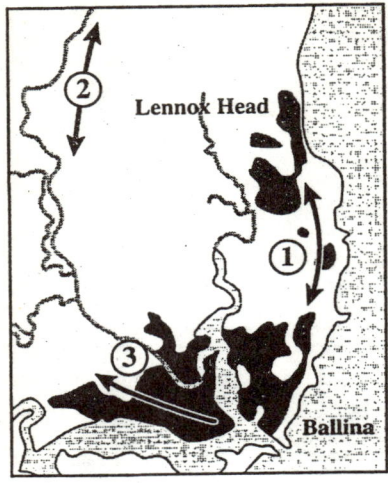

(1) Urbanization of the East Ballina-North Creek corridor is likely with increases in urban density on Ballina Island.

(2) Possible urban development in the Cumbalum-Sandy Creek area.

(3) Limited expansion of West Ballina if fill is available to raise land above flood heights.

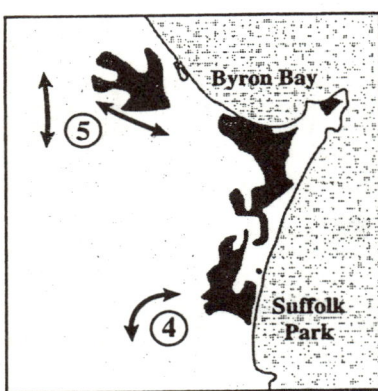

(4) Southern expansion of Byron Bay is limited by sensitive Broken Head habitat areas and buffers around extractive industries. Long-term expansion south-west of Suffolk Park may be possible.

(5) Infill of land to the west of Byron Bay is an option but may lead to the loss of good agricultural land. Important to consider a coastal habitat corridor based on remnant vegetation.

Figure 9 *Future expansion of Ballina and Byron Bay, NSW (source: NSW Department of Planning, 1993)*

Strategy (NSW Department of Planning, 1993) that Ballina is likely to expand northwards along the coastal corridor to Lennox Head with some additional expansion to the west (see Figure 9). The possibility of urban development in the Cumbalum–Sandy Creek area was recognized, to provide a new focus of development, but otherwise infill and increased densities were regarded as the only realistic options to accommodate further growth. These planning options represent another challenge to retaining the sense of community and low rise urban form valued by the local residents. The explanation for the growth of the local population and urbanization in the North Coast area is related to three factors: consumption-led factors, production-led factors and the inability of the planning system to regulate these processes.

Post-surburban landscapes on the North Coast of New South Wales 275

Consumption-led factors

The main cause of the population growth on the North Coast is seen as the redefinition and re-evaluation of the region's natural and lifestyle resources (Burnley, 1988; Murphy, 1992). The North Coast possesses rich land resources and a sub-tropical climate that has traditionally supported primary activities, such as agriculture, forestry and fishing. An appreciation of the area's scenic qualities was signalled by a growth of tourist numbers during the 1960s. Low land prices, relative to Sydney and other major cities, also contributed to the attraction of the area for settlement. Migrants of retirement age have been attracted to the area by its high environmental amenity, often having already purchased a second home in the region for recreational purposes. A survey of second home owners on the North Coast of NSW showed that 35% of buyers had future retirement as a purchase motive (Murphy, 1976). In the North Coast region, the proportion of retired population (over 65) is expected to increase from 14.3% in 1986 to 19.3% in 2016 (NSW Department of Planning, 1993). The migrants have also included younger people of working age who depend on business opportunities created by tourists and retirees. The staging of the Aquarius Festival at Nimbin in 1973 encouraged a large influx of people seeking a different way of life and an escape from the 'rat-race' of city life. These migrants, described as 'alternative' settlers, have been strongly motivated by lifestyle opportunities and the environmental quality of the region. Some of the migrants have professional backgrounds and have continued to pursue their careers in an environment which contrasts with their former residence. Others, who have moved onto large properties or have joined communes, have been able to adopt self-sufficient lifestyles. However, some are 'welfare' recipients (other than pensioners) who, while unemployed, benefit from the lifestyle, environment and lower cost of living of the region. Much of the population growth can therefore be considered as consumption-led, based on the residential choice of retirees, business operators and alternative lifestyle groups.

Questionnaire surveys in Byron Bay have indicated some of the main motivations for moving to the region. A consultancy study undertaken for Byron Shire Council in 1983 showed that the 'relaxing lifestyle' was the most valued feature about living in the area (Table 2) followed by the 'pleasant climate' and 'attractiveness of the coastal location' (Planning Workshop, 1983). Similar results were obtained from the responses to a question on the factors influencing the purchase of land in Byron Shire. The 'scenic beauty', 'existing character' and 'climate' were the main factors which were found to influence the purchase decision (see Table 3) (Planning Workshop, 1983). These results indicate the motivations of residents and migrants in the early 1980s, after the first decade of population growth, although studies undertaken in the 1990s have produced similar

Table 2 *Features enjoyed about living in Byron Shire*

	% of First Ranks	Mean of 3 Ranks
Relaxing lifestyle	19.6	18.0
Pleasant climate	18.1	18.7
Attractive coastal location	16.8	18.7
Low intensity and scale of development	11.6	2.6
Good place to raise a family	10.0	11.9
Attractive rural environment	9.4	9.7
Good farming land	4.8	3.8
Opportunity for growth and development	4.4	8.3
Good compromise between city and rural living	4.1	4.4
Proximity to Gold Coast/Brisbane	0.8	3.5
Availability of employment opportunities	0.4	0.4

Source: Planning Workshop (1983).

Table 3 *Factors influencing the purchase of land in Byron Shire*

	Strongly Influenced		Influenced		Not Influenced		Total	
	No	%	No	%	No	%	No	%
Climate	315	63.8	163	33.0	16	3.2	494	100.0
Availability of land/ housing	76	19.6	137	35.4	174	45.0	387	100.0
Competitive prices	62	16.9	128	34.6	180	48.6	370	100.0
Quality of beaches	231	54.5	139	32.8	54	12.9	424	100.0
Services provided	37	10.4	114	32.0	205	57.6	356	100.0
Scenic beauty	326	69.1	128	27.1	18	3.8	472	100.0
Anticipation of growth and development	117	28.8	125	30.8	164	40.4	406	100.0
The existing character of the area	281	61.6	139	30.5	36	7.9	456	100.0

Source: Planning Workshop (1983).

results. In 1993, a survey into reasons for living in Suffolk Park, a suburb of Byron Bay, showed that the 'attractive coastal location', the 'pleasant climate' and that the area was a 'good place to raise a family' were important to the residents (Shantz, 1994). The strength of 'environmental value' as a motivation for moving to the area gives an initial indication of the likely concern by the local community over further urban development.

Production-led factors
The influence of production-led factors in the population growth on the North Coast must also be highlighted (Sant and Simons, 1993). Although it should not be overemphasized, the growth of new employment sectors, such as tourism, has played an important role in attracting people to the area through the business opportunities created. Statistics, based on the three year averages of 1989, 1990 and 1991, indicate about 1.031 million domestic and 57,923 international tourists visit the upper North Coast region (Lane, 1993), with growth partly due to the 'spillover' effect of tourism on the Gold Coast, rather than to linkages with the rest of NSW (Hall, 1990). Accommodation statistics show a low level of serviced provision, with hotels and motels accounting for only 10% of visitor nights. Staying with friends and relatives accounts for 41% of stays and various types of budget accommodation account for the residue (backpackers hotels 20%, youth hostels 11%, camping and caravans 10%). However, the Tourism Commission of NSW (1987) has been encouraging the development of purpose-built resorts, large three-star hotels, holiday apartments and farm-stay accommodation as well as more motels and backpacker hostels. In many settlements, tourism development is concentrated in the more densely developed coastal fringe and orientated on the major north-south transport route of the Pacific Highway. It should also be noted that tourism can represent the initial stage in the resettlement process whereby satisfied tourists may ultimately return as part of a future wave of permanent working or retiree migrants (Burnley, 1988). This relationship highlights the significance of tourism in the urban restructuring of the North Coast of NSW, both as a form of new tourism-related development and as a stimulus to later residential development.

The influence of tourist-related development has increased in both Ballina and Byron shires. The number of visitor nights spent in Byron Shire increased by 94% between 1983 and 1993 and now stands at 1.5 million visitor nights. In Ballina Shire, the volume of visitor nights increased by 18% over the same period and now stands at 1.1 million visitor nights (Tourism Commission of NSW, 1986; 1989; 1994). The level of informal forms of accommodation provision, such as camping and caravan establishments, has changed very little in the shires, although Byron has experienced considerable growth in the number of private apartments — some of which have been able to command rents as holiday accommodation in excess of AUS$2000 per week. In addition, a number of the small, specialized hotels, which have recently been developed in Byron, charge room rates equivalent to those of five star hotels in Sydney. With the exception of a new three star beachside resort, Ballina continues to be dominated by two to three star motels which cater for the family holiday market and commercial travellers.

The shires are actively promoted by the Regional Tourism Organization using the slogan 'Tropical NSW' to evoke the natural and cultural attractions of the region. The North Coast of NSW is also receiving considerable federal assistance to promote and develop tourism in the region, through an eco-tourism strategy from the Commonwealth Department of Tourism (AUS$200,000) and a development strategy from the Department of Employment, Education and Training (AUS$2.5 million). These initiatives are also a significant influence on the restructuring of regional Australia, and are a rare example of public sector investment in postfordist times, which is usually characterized by a worsening fiscal crisis and growing sceptism concerning large-scale infrastructure projects (Filion, 1996: 1640).

Population growth on the North Coast has been facilitated by another production-led factor: the development and marketing of residential properties. Developers purchase land and then offer plots to prospective buyers. Buyers can have a considerable influence on the style and design of their property, usually by modifying one of the designs offered by the development company or builder. The end result is a housing estate of considerable local diversity but standardized in terms of building materials, architecture and quality — a situation noted by Zukin (1991). Many of the new estates on the North Coast of NSW offer residential environments that are difficult to distinguish from those of the suburbs of Sydney. The operation of the development control system in Australia is also important to understand as an influence on the style of new development and is now considered.

Public sector regulation of development
Numerous federal, state and local government agencies exist with duties to guide decisions about development (RAC, 1993) (see Figure 10). However, in terms of the day to day decisions over development applications, responsibility lies with the local shire councils. The Local Environment Plan (LEP) sets out the land zoning in the shire and therefore provides an indication to developers as to the suitability and location of different types of development (e.g. Ballina Shire Council, 1987). The Urban Land Release Strategy guides the release of undeveloped land for development according to forecasts of population growth and occupancy rates (e.g. Ballina Shire Council, 1990). The Strategy also determines the location, lot sizes and densities for new buildings and facilitates the planning of infrastructural requirements. These documents follow state guidelines on the release of land and design of development (NSW Department of Environment and Planning, 1988; NSW Department of Planning, 1989a; 1989b). This process assumes that a council's plan accords with private developers' views of the demand and location of new property. In cases where developers or landowners wish to utilize land not classified for urban uses in the LEP, commonly over the subdivision of land released from agricultural uses, a rezoning proposal can be considered. Decisions about rezoning proposals are based on the merits of the application and require a period of 'urban investigation', of about three to four months. During this time an Environmental Impact Assessment report is prepared and the public have a chance to comment on the proposed change of use. Rezoning of the LEP is demand led and is a constant and ongoing process. The

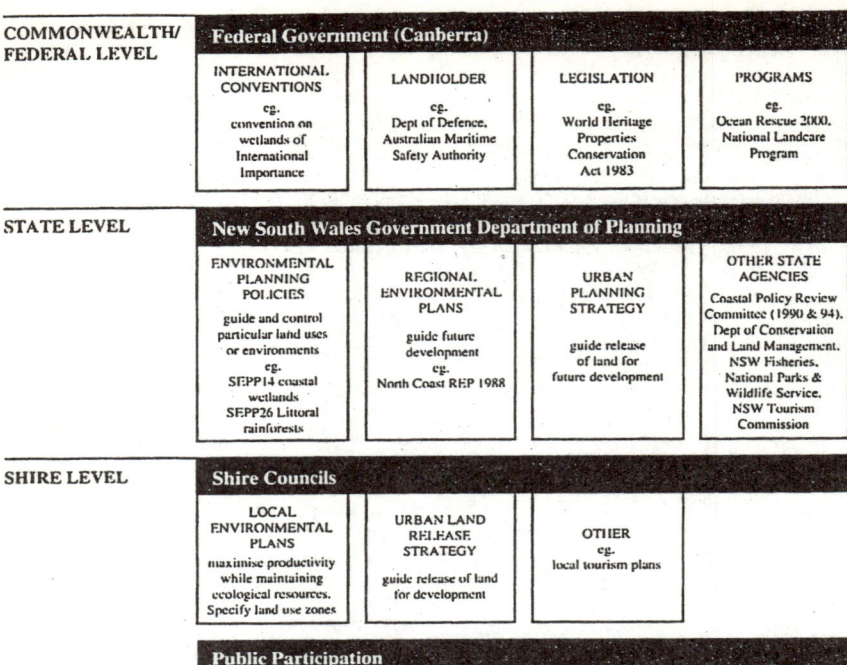

COMMONWEALTH/ FEDERAL LEVEL	Federal Government (Canberra)			
	INTERNATIONAL CONVENTIONS	LANDHOLDER	LEGISLATION	PROGRAMS
	eg. convention on wetlands of International Importance	eg. Dept of Defence, Australian Maritime Safety Authority	eg. World Heritage Properties Conservation Act 1983	eg. Ocean Rescue 2000, National Landcare Program

STATE LEVEL	New South Wales Government Department of Planning			
	ENVIRONMENTAL PLANNING POLICIES guide and control particular land uses or environments eg. SEPP14 coastal wetlands SEPP26 Littoral rainforests	REGIONAL ENVIRONMENTAL PLANS guide future development eg. North Coast REP 1988	URBAN PLANNING STRATEGY guide release of land for future development	OTHER STATE AGENCIES Coastal Policy Review Committee (1990 & 94), Dept of Conservation and Land Management, NSW Fisheries, National Parks & Wildlife Service, NSW Tourism Commission

SHIRE LEVEL	Shire Councils			
	LOCAL ENVIRONMENTAL PLANS maximise productivity while maintaining ecological resources. Specify land use zones	URBAN LAND RELEASE STRATEGY guide release of land for development	OTHER eg. local tourism plans	

| | Public Participation | | | |

Figure 10 *Levels of planning in the Australian Coastal Zone (based on Resource Assessment Commission, 1993)*

development process in Australia is therefore controlled by public sector regulation, but is responsive to private sector demands. However, as the role of planning is rooted in fordist and modernist patterns of production and consumption, involving rigid land use zoning, public regulation is experiencing increasing difficulties in adapting to the changing private sector pressures of postfordism, which require greater flexibility in land allocation decisions. It has been suggested that the planning system in Australia as a whole, and in NSW in particular, has three main weaknesses, which reflect these emerging conflicts.

First, the complexity of the planning system has been questioned. The Resource Assessment Commission's Inquiry on the Coastal Zone (RAC, 1992) took NSW as one of its case study areas and noted that existing practices, mechanisms and structures had not been effective in dealing with many issues that had arisen. The principal deficiencies of the system were noted as (RAC, 1992: 41):

- the fragmentary nature of decision-making between different levels of government;
- inadequate management mechanisms;
- inadequate public involvement;
- lack of a national approach to management;
- failure to implement procedures that ensured that decisions about the use of resources took account of the resources' real value to society;
- failure to integrate economic and environmental considerations into coastal zone management.

These issues highlighted the fact that the central problems with the planning system were related, not so much to the rate of population growth or to the absolute number of people wishing to live in coastal areas, but to the way in which competing demands for resources were being managed.

The second main weakness relates to the operation of planning in societies such as Australia, where emphasis is placed on private freehold ownership and minimum 'interference' in the right to develop land (including town and country planning) (McLoughlin, 1992). As a result of such cultural attitudes, common in Australia, planning laws tend to be subverted or even ignored by landowners and developers who feel they possess the development rights over their land irrespective of policy. The strength of these 'rights' is so strong that planners tend to acquiesce quietly or are politically impotent to resist (Foyel and Houston, 1992). In addition, development at the rural-urban fringe is often considered to be inevitable (Bunker and Houston, 1992). Traditional activities, such as agriculture and mining, often find their operations disrupted as residential areas encroach into the countryside and residents' complaints about smells, noise, agricultural spraying, harvesting and haulage increase (Ballina Shire Council, 1982). In some rural areas, the shadow of impending urbanization also raises land values beyond the ability of agriculture to remain commercially viable as a land use (Bunker and Houston, 1992). These circumstances, together with landowners' expectations of development rights irrespective of local planning policy, create a short-term approach to the investment, management and utilization of the land by landowners. This so-called 'impermanence syndrome' reinforces the inevitability of urbanization in the rural-urban fringe (Foyel and Houston, 1992). Planning systems seem unable to control such pressures.

A third weakness relates to concerns over corruption in the planning system, which have been raised in connection with the high level of involvement by the private sector, particularly in relation to rezoning proposals. Land development on the North Coast of NSW was investigated by an Independent Commission Against Corruption inquiry in 1990 (ICAC, 1990). The Commission established that improper payments to public officials and politicians had been made by consultants acting for landowners in return for favourable rezoning decisions. As a result of these circumstances, it is a common public perception that the planning system has been unable to control development pressures and has contributed to the strong public feelings and involvement in these issues. It might be argued that planning is part of the process of urban restructuring, but also appears to be unable to satisfactorily intervene in the operation of the free market.

Public concerns about development

The rate of new urban development on the North Coast of NSW has aroused considerable concern about the detrimental effects on the environment and society. The region has the second highest level of biodiversity in Australia, exceeded only by the wet tropics (National Parks and Wildlife Service, 1992, cited in NSW Department of Planning, 1993). It contains the majority of NSW rainforests, the largest number of eucalypt species and the habitat for many rare, endangered and migratory species. Aboriginal sites, buildings from the early period of European settlement and a vibrant community arts scene are evidence of the region's cultural heritage. The result is an area of great scenic, recreational and conservation value. Continued development may damage these environmental features and undermine the lifestyle so valued by its residents and prospective migrants.

In 1991, the National Parks and Wildlife Service calculated that 36% of the North Coast had already been cleared of natural vegetation and that the rate of population growth had outpaced the development of supporting physical infrastructure and human services (NSW Department of Planning, 1993). The Urban Development Strategy for the North Coast highlighted a number of locations where sewage treatment works were reaching capacity, where extra water supplies were required and where regional centres were not providing desirable levels of specialist medical, cultural and recreational facilities for the growing population.

Rural residential development, created by sub-division and producing a very dispersed settlement pattern, has aroused particular concern. Such development is considered to be extremely wasteful of land; is often uneconomic to service with basic infrastructure (roads, sewers, community facilities); can sterilize agricultural, geological and land resources; and can reduce the scenic and amenity value of the rural hinterland. The ad-hoc manner of such development can have significant implications for the environment as well as for the pattern of future urban growth. It is easy to appreciate why, in the late 1980s, there was a public perception that development along the NSW coast was uncontrolled (Coastal Committee of NSW, 1994).

The results of the 'Coast wise' project, undertaken by Southern Cross University to measure public opinion and awareness of coastal development on the North Coast of NSW in 1994, underlined some of these concerns (Dutton, 1994). The study showed that the three main threats perceived by over 70% of respondents were poor planning, lack of development control and overdevelopment. There was a great deal of public cynicism, made evident in the survey, about the will of government to control coastal urbanization.

Public contestation

Public opinion and involvement in planning issues can be very influential. Indeed it was acknowledged in the North Coast Draft Urban Planning Strategy that many settlements may grow more slowly in the future, not only as a result of natural constraints, but also due to the wishes of local communities (NSW Department of Planning, 1993). In many cases, public involvement in development issues can become a negative force. The RAC study in NSW reported that public groups often concentrated on technical or procedural aspects to prevent a development rather than on the merits of the proposals or decisions (RAC, 1993). Planners may also avoid making controversial decisions over applications and adopt a 'bunker mentality' for fear of causing public outcry. These issues represent potential barriers to the complete evolution of post-suburban landscapes in Australia, but also raise contradictions. While many of the local residents have contributed to the new patterns of growth and are willing to participate actively in planning decisions over new development, both of which are elements of postmodernist society, their opposition to change indicates an attachment to fordist consumption patterns (Filion, 1996: 1652).

Residential and tourism development in Ballina Shire and Byron Shire has received a vocal and sometimes hostile reaction from the local community. This situation is particularly the case in Byron Shire where, partly as a result of the characteristics of the migrants, a number of organized environmental groups have been established (e.g. Byron Environmental Centre and Byron Environmental and Conservation Organisation (BEACON)). These groups have a high profile, make their feelings known about planning and development issues and are critical of local government planners. For example, in 1993, BEACON held a conference which sought to create a future vision for the shire (BEACON, 1993). Considerable emphasis was placed on principles of ecological sustainability and public participation and provided tangible evidence of the level of commitment to environmentalism and empowerment in the community.

Specific tourism development proposals have faced vigorous local opposition. A Development Application (DA) submitted by Club Mediterranée (Club Med) to develop an existing tourist resort complex into an 800 bed resort in Byron Bay was narrowly approved by the Shire Council in 1994. The decision was successfully challenged in the Land and Environment Court by 'Byron Businesses for the Future', a group of local businesses which had been formed to oppose the DA. The reason for the court's decision was that the developers had not submitted a Fauna Impact Statement. This decision was regarded on the one hand as a moral victory by those who had opposed the development, but on the other hand, Club Med regarded it as a temporary delay, caused by a procedural technicality. The case is a good example of how local vested business interests successfully resisted powerful outside business interests (Olin, 1991; Mullins, 1994), in this instance a French multinational company. However, the means by which this outcome was achieved were not necessarily democratic or by a true representation of the facts.

The proposed Club Med development attracted considerable media attention. Local newspapers were used as a medium to promote a wide range of competing perspectives and the debate gained prominence following meetings which were held as part of the 'community consultation' process initiated by the developers. Interest, at the national level, was reflected by magazine articles and television documentary programmes. Opposition groups gained support from national figures, residing in Sydney, Melbourne and Canberra, for whom the preservation of the status quo in Byron Bay was seen as a *cause célèbre*. The role of the media was influential in the way in which views of particular interest groups, especially the opposition, were given prominence. The strength of opposition to the Club Med proposals portrayed in the media was contrary to previous research in Byron Bay which indicated that residents were in favour of tourism, particularly if there are benefits to them individually and to the community. Nearly two-thirds of a sample taken in Byron Bay supported the passive use of the shire's natural features as its major tourist attraction (Byron Shire Council, 1984). In a later survey, Brown (1992) showed that while there was concern about the impact of new tourist related development in Byron Bay, nearly 80% of the respondents perceived no threat from further development. The anomaly between the results of these previous surveys and the response to the Club Med proposal would seem to be a product of the ability of articulate, well-connected groups to promote their views in local and national media.

The rejection of the Club Med development does indicate that regional restructuring in Australia can be influenced by selective local participation in the development process. The role of the media in this process would appear to be critical. Local newspapers, in particular, are the usual source of information about a specific development issue. Journalists and editors of newspapers are therefore in a very influential position as they can sway local opinion depending upon the content, frequency, priority, debate and sensationalism attached to information presented on such issues. The ability of local interest groups to gain media attention to forward their views is also relevant to this process. Indeed, the role of the media in informing, stimulating and 'filtering' public participation in the development process would be a possible avenue of future research. The Club Med case would seem to suggest that the adoption of an open, inclusive form of communication may have negative consequences if community consultation is portrayed in the media in a way which serves to highlight the concerns of particular groups, whether local vested business interests or the vanguard of the environmental movement. This conclusion does not suggest that stakeholders should be excluded from the planning process but it does imply that conflict resolution may need to be attempted in a less widely publicized way. Public participation might then become more constructive and planning decisions reached for the benefit of the whole community and economy. A consequence of a community's myopic concentration on a single issue is that it allows other, less contentious, but sometimes significant, developments to proceed unopposed.

Conclusion

A new phase in the development of the Australian urban system is undoubtedly occurring, with an increase in population and development along the coastal zone. These movements are connected with aspects of the post-industrial society, such as the growth of service industries, tourism and leisure-related urbanization and retirement. The new patterns have obvious implications for the environment, in terms of the extent of built land, disruption to landscape and nature conservation and increased pollution, but they also have more subtle implications for the community and for cultural values, as illustrated in the case study of the North Coast of NSW. Both local residents and businesses have become motivated to protect their environment from further development, although the representation of the community in such opposition may be selective and may be biased from the use of the media to gain support and validation.

The maintenance of environmental and cultural values in light of development pressures has been particularly difficult to resolve by the existing planning structures. This situation exists not just because of the magnitude of the issue, but because of the strength of perceived development rights, the problems of establishing appropriate planning structures, the suspicion of corruption in development and planning and because of the often extreme positions taken in public participation in planning. In a few instances, strong public opinion and opposition to development has caused planning to work by restraint, rather than by policy and positive creation. Often developments have been halted by public participation on the basis of technical or procedural aspects rather than the merits of the proposals, as in the Club Med example. Under conditions of strong public resistance to local development, there is the potential that the trajectory of urban restructuring in Australia might be significantly modified in the future so as not to replicate the post-suburban forms of California and elsewhere. The post-suburban landscapes of Australia, while low-density, based on service industries (particularly tourism and the elderly) and incoming overseas investment, might be characterized by higher levels of environmental conservation and urban containment.

The case study, while only partial, does highlight a number of key issues in understanding the nature of urban development within a postfordist framework. The recent urban development on the North Coast of NSW has been rapid and extensive, but remains many stages behind the post-suburban landscapes of California. Nevertheless, some of the processes of urban restructuring in this region of Australia are similar to processes identified elsewhere, particularly in terms of the role of lifestyle aspirations, retirement, tourism and real estate development which are inextricably linked to postfordist trends. The contribution of spillover effects from the sprawl of the Gold Coast of south-east Queensland into the North Coast of NSW is an indication of the emergence of a national urban system, associated with the amorphous suburbanization of postfordism. The importance of tourism development to urban restructuring on the North Coast of NSW was particularly significant. Tourists, attracted to the area for holidays because of the local environment and culture and consequently stimulating tourism-related development, may later form a future wave of permanant working or retirement migrants, fuelling further residential development.

Within these strong postfordist processes, there were some remnant fordist forces at work, related to the involvement of the public sector. Some of the development and promotion of tourism in the region had been with the assistance of substantial public sector investment. Planning also contributed to the sprawling nature of the new urban development through regulations stipulating zoning, plot size and building design. The inability of the planning system to respond to the new flexibility in land allocation decisions required by postfordist developers has produced weaknesses in the system. Indeed, the effectiveness of public opposition to new urban development by both local residents and business interests have created barriers to further urbanization of this type. What the case study serves to highlight is an emerging contradiction in seeking to explain these trends with postfordist theories. On the one hand, the underlying forces of change and the empowerment of the public within the development process exhibit clear postmodernist tendencies. On the other hand, the most significant constraints on the development of a new pattern of urbanization are public opposition and planning regulations, which cling to fordist values and patterns of consumption. Clearly the postfordist concept provides a useful, but incomplete, explanatory framework or model within which the processes of change in Ballina and Byron Shires may be placed. Its failure to encapsulate entirely the trends observed in this area of NSW should not be seen as a critical flaw in the application of postfordist thinking in an empirical context. Instead, its strength lies in providing a set of ideas and perspectives which help to shape and sharpen analysis and discussion.

Stephen J. Essex, Department of Geographical Sciences, University of Plymouth, Drake Circus, Plymouth, Devon PL4 8AA, UK and **Graham P. Brown**, Centre for Tourism, Southern Cross University, PO Box 157, Lismore, NSW 2480, Australia.

References

Ballina Shire Council (1982) *Residential development: local environmental study.* Ballina, BC.
—— (1987) *Ballina local environmental plan 1987.* Ballina, BC.
—— (1990) *Urban land release strategy.* Ballina, BC.
—— (1993) *Ballina Shire facts and figures.* Ballina, BC.
Berry, B. (1976) *Urbanisation and counterurbanization.* Sage, London.
Berry, M. and G. Rees (1994) Australian urban and regional research: an introduction. *International Journal of Urban and Regional Research* 18.4, 549–54.
Brown, G. (1992) *Community attitudes toward tourism in Byron Shire.* Study conducted for Byron Shire Council, Centre for Tourism, University of New England - Northern Rivers, Lismore, NSW.
Bunker, R. and P. Houston (1992) Natural resource management meets metropolitan growth: contemporary rural-urban fringe planning in Australia. *Built Environment* 18.3, 221–33.
Burnley, I.H. (1988) Population turnaround and the peopling of the countryside? Migration from Sydney to country districts of NSW. *Australian Geographer* 19.2, 268–83.
Byron Environment and Conservation Organisation (1993) *A new approach to planning in Byron Shire: vision and policies.* BEACON, Byron Bay.
Byron Shire Council (1984) *Keeping Byron unique: a tourism strategy.* Byron Bay, BC.
—— (1993) *Byron residential development strategy.* Byron Bay, BC.
—— (1994) *Community profile.* Byron Bay, BC.
Champion, A.G. (1989) *Counterurbanization.* Edward Arnold, London.
Coastal Committee of NSW (1994) *Draft revised coastal policy for NSW.* CCNSW, Sydney.
Derrett, R. (1994) *Historical perspective report of Coastwise project.* Centres for Tourism and Coastal Management, Southern Cross University, Lismore, NSW.
Dutton, I. (1994) *Overview of Coastwise project and findings.* Conference paper presented at Coastwise Seminar, Invercauld House, Goonellabah, Southern Cross University, Lismore, NSW, 28 October 1994.
Edols-Meeves, M. and S. Knox (1996) Rural Residential Development. *Australian Planner* 33.1, 25–9.
Esser, J. and J. Hirsch (1989) The crisis of fordism and the dimensions of a 'postfordist' regional and urban structure. *International Journal of Urban and Regional Research* 13.3, 417–37.
Farrell, D. (1993) *New South Wales Yearbook no.73, 1993.* Australian Bureau of Statistics, Sydney.
Filion, P. (1996) Metropolitan planning objectives and implementation constraints: planning in a postfordist and postmodern age. *Environment and Planning A* 28.9, 1637–60.
Foyel, J. and P. Houston (1992) Planning in the rural-urban fringe: the challenge of perceived development rights and the significance of land use systems. *Australian Planner* 30.1, 45–50.

284   *Stephen J. Essex and Graham P. Brown*

Garreau, J. (1991) *Edge city: life on the new frontier*. Anchor Books/Doubleday, New York.
Gottdiener, M. and G. Kephart (1991) The multi-nucleated metropolitan region: a comparative analysis. In R. Kling, S. Olin and M. Poster (eds.), *Postsuburban California: the transformation of Orange County since World War II*, University of California Press, Berkeley, 31–54.
Graham, J. (1992) Post-fordism as politics: the political consequences of narratives on the left. *Environment and Society D: Society and Space* 10, 393–410.
Hall, C.M. (1990) From cottage to condominium: recreation, tourism and regional development in northern NSW. In D.J. Walmsley (ed.), *Change and adjustment in northern NSW*, Department of Geography and Planning, University of New England, Armidale.
Hamnett, S. and R. Bunker (eds.) (1987) *Urban Australia: planning issues and policies*. Mansell, London.
Harvey, D. (1989) *The condition of post modernity*. Basil Blackwell, Oxford.
Hugo, G. (1994) The turnaround in Australia: some first observations from the 1991 census. *Australian Geographer* 25.1, 1–17.
—— and R.J. Smailes (1992) Population dynamics in rural south Australia. *Journal of Rural Studies* 8.1, 29–51.
Independent Commission Against Corruption (ICAC) (1990) *Report on investigation into North Coast land development*. ICAC, Sydney.
Kling, R., S. Olin and M. Poster (eds.) (1991) *Postsuburban California: the transformation of orange county since World War II*. University of California Press, Berkeley.
Lane, S. (1993) *Tourism on the coast: northern NSW*. BTR Occasional paper No.15, Bureau of Tourism Research, Canberra.
Logan, M., J. Whitelaw and J. McKay (1981) *Urbanisation: the Australian experience*. Shillington House, Melbourne.
McLoughlin, J.B. (1992) The case for a legal basis for land-use planning systems. *Built Environment* 18.3, 214–20.
Mullins, P. (1990) Tourist cities as new cities: Australia's Gold Coast and Sunshine Coast. *Australian Planner* 28.3, 37–41.
—— (1992) Cities for pleasure: the emergence of tourism urbanisation in Australia. *Built Environment* 18.3, 187–98.
—— (1994) Class relations and tourism urbanization: the regeneration of the petite bourgeoisie and the emergence of a new urban form. *International Journal of Urban and Regional Research* 18.4, 591–608.
Munro, R.G. (1976) *North Coast region demographic analysis and projections 1921–2001*. Research Unit, Northern Rivers College of Advanced Education, Lismore.
Murphy, P. (1976) Residential resort land in NSW. *Australian Geographical Studies* 14.2, 103–15.
—— (1981) Patterns of coastal retirement migration. In A.L. Howe (ed.), *Towards an older Australia*, University of Queensland Press, St. Lucia.
—— (1985) Development of strata units in NSW North Coast resorts. *Australian Geographer* 16.4, 272–9.
—— (1992) Leisure and coastal development. *Australian Planner* 30.3, 145–51.
NSW Department of Environment and Planning (1988) *North Coast regional environmental plan*. Department of Environment and Planning, Sydney.
NSW Department of Planning (1989a) *North Coast: design guidelines*. Department of Planning, Sydney.
—— (1989b) *Tourism development near natural areas:guidelines for the North Coast*. Guidelines based on a study prepared by Ludwig Rieder and Associates Pty Ltd, NSW Department of Planning, Grafton.
—— (1990) *Major demographic trends in NSW: implications*. Department of Planning, Sydney.
—— (1993) *North Coast draft urban planning strategy*. Department of Planning, Grafton.
—— (1984–1994) *North Coast population and development monitor*. Nos 1-16, Department of Planning, Grafton.
Olin, S. (1991) Intraclass conflict and the politics of a fragmented region. In R. Kling, S. Olin and M. Poster (eds.), *Postsuburban California: the transformation of Orange County since World War II*, University of California Press, Berkeley, 223–53.
Paris, C. (1994) New patterns of urban and regional development in Australia: demographic restructuring and economic change. *International Journal of Urban and Regional Research* 18.4, 555–72.

Perry, R., K. Dean and B. Brown (1986) *Counterurbanization*. Geobooks, Norwich.

Planning Workshop Pty Ltd (1983) *Byron Shire environment study, working paper no 7: the community*. Planning Workshop, Sydney.

Resource Assessment Commission (1992) *Coastal Zone inquiry background paper*. Australian Government Publishing Service, Canberra.

—— (1993) *Coastal Zone inquiry: NSW case study*. Resource Assessment Commission, Canberra.

Sant, M. (1993) Coastal settlement systems and counterurbanization in NSW. *Australian Planner* 31.2, 108–13.

—— and P. Simons (1993) Counterurbanization and coastal development in NSW. *Geoforum* 24.3, 291–306.

Shantz, T. (1994) Personal communication.

Soja, E.W. (1992) Inside exopolis: scenes from Orange County. In M. Sorkin (ed.), *Variations on a theme park: the new American city and the end of public space*, Hill and Wang, New York, 94–122.

Sorkin, M. (ed.) (1992) *Variations on a theme park: the new American city and the end of public space*. Hill and Wang, New York.

The Parliament of the Commonwealth of Australia (1992) *Patterns of urban settlement: consolidating the future?* Report of the House of Representatives, Standing Committee for Long-Term Strategies, Australian Government Publishing Service, Canberra.

Tourism Commission of NSW (1986) *Regional tourism trends in NSW*. Tourism Commission of NSW, Sydney.

—— (1987) *North Coast region: tourism development strategy*. Tourism Commission of NSW, Sydney.

—— (1989) *Regional tourism trends in NSW*. Tourism Commission of NSW, Sydney.

—— (1994) *Regional tourism trends in NSW*. Tourism Commission of NSW, Sydney.

Vining, D.R. and R. Pallone (1982) Migration between core and peripheral regions: a description and tentative explanation of the patterns in 22 countries. *Geoforum* 13.4, 339–410.

Winner, L. (1992) Silican Valley mystey house. In M. Sorkin (ed.), *Variations on a theme park: the new American city and the end of public space*, Hill and Wang, New York, 31–60.

Zukin, S. (1991) *Landscapes of power: from Detroit to Disney World*. University of California Press, Berkeley.

[12]

CHAPTER 10

Fixed Link Projects in Denmark and Ecological Monitoring of the Øresund Fixed Link

Henning Karup

10.1
Introduction

Denmark consists of a large number of islands (Zealand and Funen are the two largest) and the mainland Jutland, which is the only part connected with the European continent. Fixed links between the various regions have been discussed during most of the twentieth century. In the 1930s the first bridge from Jutland to Funen and the bridge from Zealand to Falster Island were built. At the same time the first plans to construct a Great Belt link between Funen and Zealand and a fixed link from Denmark to Sweden were presented and, in the following years, new plans appeared in the public debate with regular intervals. In the 1970s the capacity of the bridges erected 40 years earlier became inadequate and new bridges were constructed from Jutland to Funen, and Sealand to Falster.

10.2
Great Belt Link

After a long political discussion and public debate the construction of the Great Belt Link started in 1989. The scheme consists of a four-lane motorway and a twin-track electrified railway. From Sealand to the small island of Sprogø in the middle of the Great Belt the railway is led through a tunnel and the motorway lies on a high bridge. From Sprogø Island to Funen both the railway and the motorway continue on a low bridge. In total, the Great Belt Link is 18 km from coast to coast. The link opened for trains 1 July 1997 and the motorway will be opened in June 1998.

In connection with the public debate on the Great Belt Link project, environmental queries were raised for the first time regarding a construction scheme of this type. The Great Belt Link is erected across the main route of water exchange with the Baltic Sea. It was therefore decided to adopt a so-called zero solution, meaning that the inflow and outflow of water to the Baltic Sea was not to be affected. The zero solution has been made by optimising the link design, and by the use of compensation dredging in the Great Belt.

A comprehensive hydrographic investigation programme was establised together with the control and monitoring programmes dealing with environmental impact on the area concerned. The programmes are conducted by the Great Belt Consortium.

10.3
The Øresund Link

In 1991 the Danish and Swedish governments decided to construct a fixed link from Denmark to Sweden, and the construction of the coast to coast facility started in 1995.

The link extends 430 m from the Danish coast at the airport of Copenhagen where an artificial peninsula has been establised. From there the link will continue under the Drogden navigation channel in an immersed tunnel with a length of 3510 m between the embankments. The immersed tunnel will emerge on a 4055-m-long artificial island south of the existing island of Saltholm. From the artificial island the link continues on a 7845-m-long two-level bridge across the Flinte and Trindel fairways and joins the Swedish mainland at Lernacken south of Malmø. The fixed link comprises a twin-track electrified railway and a four-lane motorway. Denmark and Sweden each on their part undertake to construct the necessary access facilities for railway and road traffic from the fixed link to the existing railway and road networks. The construction work is expected to be finalized in the year 2000.

On the basis of experience gained from the Great Belt Link, a heavy discussion of environmental questions took place during the decision-making process for the Øresund Link. Notably, the spill from the dredging activities in the Great Belt had had some impacts on the local environment. The main environmental questions in connection with the Øresund Link were therefore to balance the impact on the Baltic Sea caused by the blocking effect of the construction, and the impact of compensation dredging activities on the local environment including the shallow waters around the island of Saltholm which is an EU bird protection area. A zero solution for the Baltic Sea would necessitate an increase in dredging in the proximity of the link with a disturbance of the local environment as a consequence.

In order to obtain a balance between the Baltic Sea and the local environment, it is stipulated that the completed structure and the construction works must satisfy a number of environmental objectives and criteria imposed by the Danish and Swedish environmental authorities, and for all the dredging operations, spillage limits in time and space have been set up. In order to ensure that the spillage limits are not exceeded and the quality objectives not violated the authorities, Øresundskonsortiet and the contractors, have developed comprehensive control and monitoring programmes. The aim of the monitoring programme and the results will be described in further details below (see Sect. 10.5).

10.4
The Fehmarn Belt Link

As a part of the Danish-Swedish agreement on the Øresund Link, the Danish government accepted working for a fixed link across the Fehmarn Belt between Denmark and Germany. The experience from the Great Belt Link and the Øresund Link shows an increasing political and public concern for the environmental impact of fixed links on the coastal and marine environment. Seven different technical solution models have therefore been identified for the Fehmarn Belt Link, and comprehensive environmental investigation programmes have been started. The first phase of the investigations was finalized in August 1996 and it has been decided to continue the next phase of the investigations with five solutions models. The second phase of the investigations is expected to be finalized by the end of 1998. The Danish and German governments then have to decide whether they want to proceed with the project. If the two governments agree to do so, an Environmental Impact Assessment procedure will be initiated.

10.5
Ecological Monitoring Programme of the Øresund Link

10.5.1
Objectives

The Danish Public Works Act for the Øresund Link states that the link shall be executed with due considerations of what is ecologically motivated, technically feasible and financially reasonable in order to prevent detrimental effects on the environment. Ecological objectives for the environmental impact should be established together with a control and monitoring programme. The completed structure and the construction work shall satisfy two overall environmental objetives imposed by the authorities (Ministry of Transport and Ministry of Environment and Energy 1995):

- Far field
  - The Øresund Link must not affect the Baltic Sea in such a way that chemical/ physical and hence biological changes arise.
- Near field
  - The Øresund Link must only transiently cause conditions in the Øresund that are in conflict with the national plans for the coastal areas. More extensive effects can be accepted in an *inner impact zone* covering the area 500 m either side of the link trajectory measured from the north and south sides of the completed link, respectively, including areas where compensatory dredging is undertaken. Around the island of Saltholm, however, the zone must not extend closer than to a water depth of 1 m. Temporary effects can be accepted in an *outer impact zone* lying 7 km either side of the inner impact zone.
  - The permanent loss of areas as a result of the establishment of the artificial peninsula, artificial island and bridge piles, and permanent effects resulting from local changes in hydrographic conditions can be accepted.
  - Outside the outer impact zone the effects of the construction work must not hinder fulfilment of the objectives and criteria for coastal waters stipulated in the regional environmental plans. As far as the open parts of the Øresund are concerned, the construction work must not reduce the possibilities for establising an indigenous natural flora and fauna.

10.5.2
Criteria

More extensive effects can be accepted in the inner and outer impact zones, and special criteria for fulfilment of the objectives have therefore been drawn up for a number of selected aspects of nature and the environment (Ministry of Transport and Ministry of Environment and Energy 1995). The first objective implies that the link is to be establised in such a way that there will be no change in the throughflow of water in Øresund nor in the saltwater and oxygen input to the Baltic Sea. This is achieved by optimizing the design of the construction so the blocking effect from the construction is less then 0.5% and then carrying out compensation dredging until unchanged flow conditions have been established. The second objective is achived by careful planning and execution of the dredging operations in order to keep the spillage percentages below pre-scribed maximum limits in time and space.

10.5.3
Dredging Operations

The construction work for the Øresund Link involves dredging of approximately 7 million m³ material from the seabed. The overall requirement states that the total average spillage from dredging operations must not exceed 5% (Ministry of Transport and Ministry of Environment and Energy 1995). The dredging and reclamation work is divided into a number of sub-operations (Fig. 10.1). A dredging instruction and a spill budget for the work have to be approved by the environmental authorities before each sub-operation starts. The dredging instructions lay down the detailed guidelines for the extraction work.

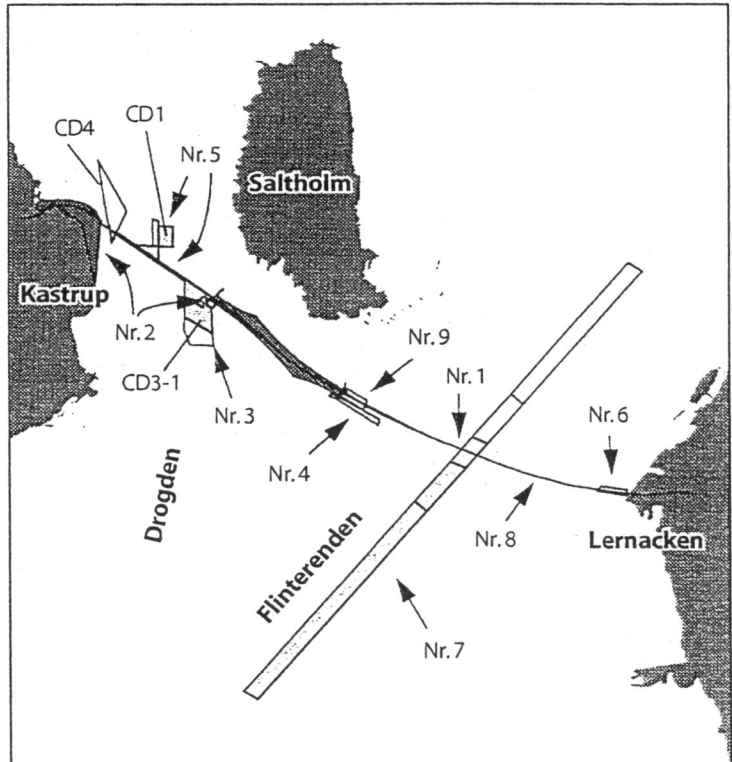

Fig. 10.1. Approved extraction areas as at 31 December 1996. Areas for individiual dredging instructions (Nr. 1–9) are indicated by arrows. *CD1, CD3-1* and *CD4* show position of compensation dredging areas. (Modified from Øresundskonsortiet 1997)

10.5.4
Monitoring Strategy

The Danish and Swedish environmental authorities and Øresundskonsortiet have developed control and monitoring programmes to ensure that the spillage limits are not exceeded and the quality objectives are not violated. The control and monitoring programmes are conducted at three different levels:

- The contractor is contractually obliged to ensure that total spillage limits are not exceeded, and that the requirements for spillage intensity in time and space are fulfilled. The spillage monitoring conducted by the contractor is supervised by Øresundskonsortiet and the environmental authorities in Denmark and Sweden.
- Øresundskonsortiet is responsible for a feedback monitoring programme in order to ensure that timely measures are taken to avoid any risk of violation of any of the authorities' requirements both to the marine environment and to the execution of the work.
- The authorities independently carry out monitoring and control of the environmental impact on water quality, bottom fauna, bottom vegetation, coastal morphology, herring migration, birds and bathing water quality.

The authorities in Denmark and Sweden have to report the results of the ecological control and monitoring programmes semi-anually to the governments.

10.5.5
The Contractors' Spill Monitoring Programme

Spillage measurements are carried out while sailing across the sediment plumes and in profiles in the sediment plumes themselves. In 1996 the measuring was done on a 24-h basis, but during the autumn of 1996 a special method was developed for determining spillage from dredging work done by dredgers equipped with a bucket. The new method determines the spillage percentage using a combination of concrete measurements and model calculations based on the experience gained so far. The aim is to limit measuring activities and so utilize the resources of the measuring vessels more effectively while satisfying the uncertainty requirements with regard to determining the spillage percentage. The new method for determining spillage was put into operation in spring 1997 (Øresundskonsortiet 1996a).

Monitoring of the bed load transport of sediment from dredging operations has been carried out in two stages. The first stage consists of visual observation of the seabed in and around the various dredging areas, and the second stage of more detailed investigation if spilled sediment in thicknesses in excess of 10 cm is found (Øresundskonsortiet 1996b). The non-quality-assured spill data shall within 48 h be stored in an Environmental Information System to which the authorities have direct access. On a weekly basis the contractor has to prepare a report with quality-assured spill data within 3 weeks from the end of the report period.

By the end of 1996 the overall average spillage percentage for all dredging operations was calculated as approx. 234 000 tons or approx. 4.4% of the total dredged amount (approx. 5 300 000 tons (Fig. 10.2); (Danish Ministry of the Environment and Energy, Danish Ministry of Transport, KSÖ 1997a)

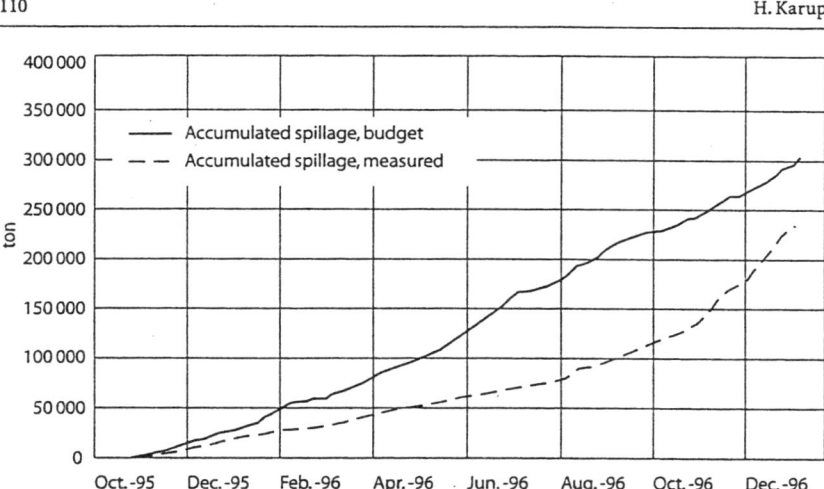

Fig. 10.2. Development over time of measured spillage from dredging and extraction work up to end of December 1996 in relation to budgeted spillage for the period. (Øresundskonsortiet 1997)

10.5.6
Calculation of the Zero Solution

According to the conditions set by the Danish and Swedish authorities, two independent three-dimensional hydrographic models have to be set up for use in calculating the blocking effect of the Øresund Fixed Link and the scope of the compensation dredging which will have to be carried out to reduce the blocking effect to zero. The models were adjusted and tested in the course of 1996 (Øresundskonsortiet 1997). Once final testing of the models is complete it will be possible to start new model calculations. An intensive measuring campaign is planned for 1997 to verify the model. It will be used to assess the effect on currents of building the artificial peninsula and artificial island.

10.5.7
Feedback Monitoring Programme

Øresundskonsortiet uses a feedback programme in connection with the individual dredging operations which involves frequent monitoring of individual, selected parameters in the marine environment around the dredging areas. The feedback programme has the aim of ensuring that early action is taken with regard to the execution of the construction work if there is a risk of the authorities' environmental requirements being exceeded. The programme includes monitoring of sediment, common mussels and eelgrass.

The results from the feedback programme are compared with forecasts made using models which have been updated with detailed information on the individual dredging operations. The models are also used for environmental optimization of the dredging work, with spillage in time and space and a biological evaluation of effects being described and processed. Planning is done with the assistance of a calculation model which includes a hydrodynamic model, a wave model, two dispersion models for calculating spilled sediment and an ecological model for calculating effects on eelgrass, phytoplankton and macroalgae.

Øresundskonsortiet has also set up so-called operational environmental criteria for sediment, eelgrass and common mussels which are stricter and more specific that those of the authorities. This enables corrections to be made for any developments which might otherwise result in an infringement of the authorities' criteria, as well as making it possible to document compliance with those criteria.

10.5.8
Effect Monitoring

As part of their supervision of the construction work the environmental authorities are running a general programme with a view to assessing whether the observed effects of the construction work lie within the frameworks of the expected effects decribed in the Environmental Impact Assessment (Øresundskonsortiet 1995). The programme includes monitoring within the following areas of the Øresund's environment: water quality, benthic vegetation and benthic fauna, fish, birds, bathing water quality and coastal morphology (Ministry of Transport and Ministry of Environment and Energy 1995).

The effect monitoring is long term and is primarily aimed at identifying effects and changes in the environment which are caused by the construction work. In this way the monitoring will form the basis for an assessment of the need for adjustments to the construction work in the longer term. The monitoring also addresses a broader cross-section of the ecosystem than the feedback monitoring, which focuses on fewer variables and the local area around the dredging operations.

10.5.8.1
Water Quality

The monitoring programme is being carried out at four stations in the north, middle and south of the Øresund (Semac 1997a). Each sample is analysed for salinity, temperature and visibility (Secchi) depth, and for concentrations of oxygen, chlorophyll and nutrients, as supporting parameters to help interpret the other sub-programmes. When assessing the oxygen content of the water, oxygen depletion is regarded as beginning at concentrations of less than 2.8 ml $O_2 l^{-1}$ (corresponding to 4 mg $O_2 l^{-1}$). Fish will leave areas with such a low oxygen content. Acute oxygen depletion with values below 1.4 ml $O_2 l^{-1}$ (corresponding to 2 mg $O_2 l^{-1}$) over an extended period can harm benthic fauna so seriously that it will die out (Danish Ministry of the Environment and Energy, Danish Ministry of Transport, KSÖ 1997a).

10.5.8.2
Benthic Flora and Fauna

The programme for benthic fauna includes monitoring of the shallow-water fauna and deep-water fauna together with a separate programme for common mussel. Shallow-water fauna is collected at a small number of transects and stations during the spring, while a more extended sampling takes place in the autumn. The samples are analysed for the occurrence of species, number and biomass. Samples of deep-water fauna are collected annually in April/May at 17 stations with a view to determining biomass, number and size distribution. Common mussel is sampled annually in October/November. Coverage is assessed and biomass, number and size distribution determined. The content of heavy metals in mussels at ten stations is also determined.

The vegetation programme includes monitoring and control of the incidence of tassel pondweed (*Ruppia*), eellgrass (*Zostera*) and sea tangle (*Laminaria*). The programme for tassel pond weed and eellgrass is carried out once a year in August/September and includes aerial photography of the distribution of the vegetation along the coast out to the 6-m-depth contour line. In addition, analyses are made of the distribution and biomass. The monitoring of sea tangle takes place at six localities in August/September using photography and surveying.

10.5.8.3
Fish

The cruises involve surveying the occurrence of herring by means of echo sounding along an extensive system of survey lines throughout the Øresund. The surveys are further supplemented by fishing to establish the occurrence of Rügen herring, and the North Sea herring which has the northern Øresund as the furthest reach of its area of distribution. At Drogden close to the alignment of the fixed link, a stationary sonar system for registering migrating herring shoals was being run in 1996. Echoes were found during the period which presumably represent migrating fish. Proper verification of the echoes with simultaneous fishing is planned.

10.5.8.4
Birds

The bird monitoring programme emphasizes monitoring of the breeding eider population at Saltholm and monitoring of moulting waterfowl (greylag geese and mute swan). More extensive monitoring is performed for occurrence of stageing migrants and wintering waterfowl, mainly tufted duck (National Environmental Research Institute 1994).

10.5.8.5
Bathing Water Quality

A supplementary bathing water programme is being carried out throughout the whole year at the Danish beaches close to the alignment by agreement with the local authorities. On the Swedish side, supplementary bathing water studies were carried out in July and August 1996 by agreement with the municipalities.

10.5.8.6
Coastal Morphology

A total of 65 profiles at right angles to the coast were surveyed on the Danish and Swedish sides in 1996. Thirty of the profiles are located on the coast along Copenhagen, fifteen are located on Saltholm and twenty are on the Swedish side of the Øresund (Semac 1997b). The surveying was done partly with a theodolite from the shore and partly by means of echo sounding from a vessel, with both methods being supplemented by aerial photography. Sediment samples were also taken from twenty four of the profiles for analysis of grain size distribution. Data from earlier studies is being compiled and evaluated in order to compare with the 1996 study to see if any changes of the coastline have occurred as a result of the construction work.

10.6
Discussion and Conclusions

Spillage from the dredging operations is the most important source of impacts on the marine environment during the contruction period of the fixed links. The sediment spill shades vegetation areas in shallow waters and settles on the vegetation and mussels beds. This has potential secondary effects on the feeding resources of breeding eiders and moulting mute swans and greylag geese on the island of Saltholm. The sediment plumes could also potentially prevent migration of Rügen herring through Øresund when the herring during the autumn and winter period migrate from the summer feeding areas in Kattegat/Skagerrak to the spring spawning areas along the German coast in the Baltic Sea.

At the end of December 1996 the Danish and Swedish authorities approved nine dredging instructions and approx. 2.7 million m³ seabed material was dredged with a total spill of 4.4% or approx. 118 000 m³ (Danish Ministry of the Environment and Energy, Danish Ministry of Transport, KSÖ 1997a). The results from the control and monitoring programme for the first year of the construction period show that the dredging was performed without exceeding the spill limits and without violation of the environmental quality objectives (Danish Ministry of the Environment and Energy, Danish Ministry of Transport, KSÖ 1997a). The oxygen content was relatively high in 1996 and there was no risk that the construction works caused or increased acute oxygen depletion (Danish Ministry of the Environment and Energy, Danish Ministry of Transport, KSÖ 1997a).

The concentration of nutrients above and below the pycnocline was generally lower in 1996 than in the previous 10-year period. The contribution of nutrients from the construction works did not lead to a measurable increase in the concentration of nutrients in the water in 1996 (Danish Ministry of the Environment and Energy, Danish Ministry of Transport, KSÖ 1997a). By the end of 1996 the changes in the fauna in the area as a whole did not show a significant reduction in relation to the baseline studies carried out in 1995, and there is no directional trend in relation to the construction work.

The investigation of the benthic vegetation in 1996 showed a decline in the accessible biomass of tassel pondweed in the area south-west and west of Saltholm with the most affected areas in shallow waters less then 2 m and in areas closest to the construction work. The effects in shallow waters indicate that the decline could be caused by the severe ice cover during the winter period 1995–1996 while the effect gradient towards the construction work indicates that the impact is caused by a combination of ice cover and sediment plumes. Provisonal investigations in 1997 show that the tassel pondweed has recovered. The accessible biomass of eelgrass in 1996 was comparable with the level observed in the baseline period of 1993–1995 (National Environmental Research Institute 1997).

The monitoring cruises for the 1995/1996 herring migration season showed concentrations of herring in the Øresund in October which corresponded to the concentrations which were found in the autumn months during the baseline studies in the period 1993–1995, while the monitoring cruises in March and April 1996 showed slightly higher concentrations than during the baseline studies in March and April 1994 and 1995. These higher concentrations in the spring of 1996 point to a later herring migration out of the Øresund owing to the very cold winter of 1995–1996. The concentrations still show, however, that the herring made their usual migration out of the Øresund (Danish Ministry of the Environment and Energy, Danish Ministry of Transport, KSÖ 1997a).

For the 1996/1997 herring migration season two cruises took place in September/October and November. Another two cruises are planned for the spring of 1997. The first cruise in autumn 1996 showed the concentrations of herring to be slightly different to what was expected in relation to the concentrations found during the baseline studies in 1993–1995 and monitoring in 1995. By the second cruise in the autumn of 1996 the concentrations were back to the expected levels, however.

In 1996, studies of overwintering and moulting mute swans were carried out. The highest number of swans in the moulting period, 1500 individuals, was recorded in early August, which means that the decline observed in the period 1993–1995 continued in 1996 (National Environmental Research Institute 1997a). The decline can presumably be attributed to high mortality in the period January–February 1996, which was probably caused by the harsh winter and ice sheet, and it is highly likely therefore that it was not caused by the construction work. It can be added that the number of moulting swans along the west coast of Scania in July/August was also lower than before, while the number of swans on Saltholm (approx. 2300) in March 1996 was the highest since 1993 (National Environmental Research Institute 1997).

Counts of resting migratory birds on Saltholm in September and October 1996 produced a smaller number of surface-feeding ducks in relation to previous years, presumably because of very dry conditions on the island. The number rose substantially following rain in late October and early November, when the numbers were very similar to those in the preliminary studies. The number of waders showed a similar development (National Environmental Research Institute 1997).

The analyses of the distribution of eiders, greylags and mute swans on Saltholm in 1996 showed that there had been changes in relation to previous years. In all cases there were fewer birds in the immediate area of the alignment and more birds in the areas east, north and north-west of Saltholm. It must therefore be judged very probable that this redistribution is due to disturbances caused by the construction work (Fig. 10.3). As far as all three species are concerned, however, there were still substantial numbers seeking food in and close to the immediate area itself. Subsequent studies showed that the eiders completed their breeding cycle without problems, and that the physical condition of the geese and swans at the end of the moulting period was no worse than in previous years. The birds were therefore able to find sufficient food by using alternative areas further away from the construction work (National Environmental Research Institute 1997).

The overall effects on the birds on Saltholm in 1996 can therefore be characterized as being at the lowest possible level. Thus, it can therefore be concluded that there were

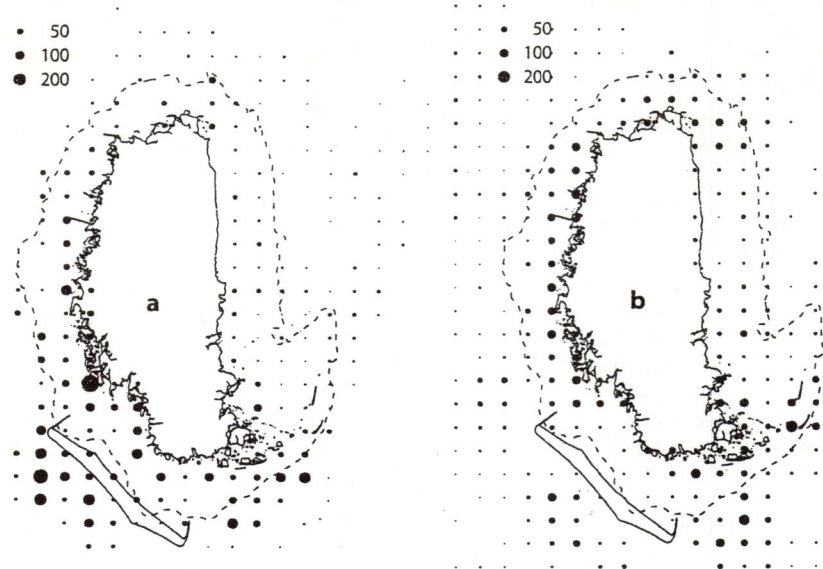

Fig. 10.3. Distribution of eiders around Saltholm on 30 March 1995 (a) and 3 April 1996 (b) charted on the basis of aerial photography. Charting only shows males. (National Environmental Research Institute 1997)

no infringements of the conditions for the construction works as regards breeding eiders, moulting greylags and mute swans on Saltholm (Danish Ministry of the Environment and Energy, Danish Ministry of Transport, KSÖ 1997).

The results of the bathing water studies carried out have only shown few occasions outside the normal bathing season with cloudy, turbid water with reduced visibility (Secchi) depth at the beach south of the Øresund Link on the Danish coast. No abnormal conditions were observed along the beach north of the Øresund Link and on the Swedish coast.

References

Danish Ministry of the Environment and Energy, Danish Ministry of Transport, KSÖ (1996) Second semi-annual report on the environment and the Øresund Fixed Link's coast-to-coast installation. 1st. half of 1996. 25 pp

Danish Ministry of the Environment and Energy, Danish Ministry of Transport, KSÖ (1997a) Third semi-annual report on the environment and the Øresund Fixed Link's coast-to-coast installation. 2nd half of 1996. 25 pp

Danish Ministry of the Environment and Energy, Danish Ministry of Transport, KSÖ (1997b) Fourth semi-annual report on the environment and the Øresund Fixed Link's coast-to-coast installation. 1st half of 1997

Ministry of Transport and Ministry of Environment and Energy (1995) Objectives and criteria and the environmental authorities' requirements for the overall control and monitoring programme for the Øresund Fixed Link coast-to-coast facility

National Environmental Research Institute (1994) Bird monitoring in relation to the establishment of a fixed link across Øresund. Proposal for the Programme. 35 pp

National Environmental Research Institute (1997) Monitoring of moulting mute swans around Saltholm, 1996. 40 pp

Øresundskonsortiet (1995) Supplementary assessment of the impacts on the marine environment of the Øresund Link. 193 pp

Øresundskonsortiet (1996a) Estimation methods for dredged material spill – mechanical dredging. 63 pp

Øresundskonsortiet (1996b) Bed load monitoring programme status report for dredging instruction no. 1–4, October 1995 to September 1996. 63 pp

Øresundskonsortiet (1997) Semi-annual environmental report, July–December 1996. 42 pp (In Danish)

SEMAC JV (1997a) Status report 1996. Water quality

SEMAC JV (1997b) Status report 1996. Coastal monitoring

[13]

Pergamon

PII: S0959-3780(97)00023-X

Global Environmental Change, Vol. 8, No. 1, pp. 49–61, 1998
© 1998 Elsevier Science Ltd. All rights reserved
Printed in Great Britain
0959-3780/98 $19.00 + 0.00

Natural resource management in mitigating climate impacts: the example of mangrove restoration in Vietnam

Nguyen Hoang Tri, WN Adger, PM Kelly

The risk that tropical storm occurrence may alter as a result of global warming presents coastal managers, particularly in vulnerable areas, with a serious challenge. Many countries are hard-pressed to protect their coastal resources against present-day hazards, let alone any increased threat in the future. Moreover, the threat posed by climate change is uncertain making the increased costs of protection difficult to justify. Here, we examine one management strategy, based on the rehabilitation of the mangrove ecosystem, which may provide a dual, "win-win" benefit in improving the livelihood of local resource users as well as enhancing sea defences. The strategy, therefore, represents a precautionary approach to climate impact mitigation. This paper quantifies the economic benefits of mangrove rehabilitation undertaken, *inter alia*, to enhance sea defence systems in three coastal Districts of northern Vietnam. The results of the analysis show that mangrove rehabilitation can be desirable from an economic perspective based solely on the direct use benefits by local communities. Such activities have even higher benefit cost ratios with the inclusion of the indirect benefits resulting from the avoided maintenance cost for the sea dike system which the mangrove stands protect from coastal storm surges. © 1998 Elsevier Science Ltd. All rights reserved

(continued on page 50)

Climate change, coastal protection and the mangrove ecosystem

Tropical cyclones can cause considerable damage along the 3 000 km coastline of Vietnam. One typhoon, in October 1985, was responsible for the loss of almost 900 lives, 3 300 boats were sunk and over half a million people were rendered homeless. While this was a particularly extreme impact, over 400 000 hectares of crops were lost in the coastal provinces of Vietnam as a result of tropical cyclone impacts over the ten-year period 1977–1986 (Thu, 1991). Protecting vulnerable coastal areas from typhoon impacts is, therefore, of high social and economic importance. Yet in a nation where resources are limited, affording adequate protection can prove difficult even in the present-day (Wickramanayake, 1994). From one to 12 typhoons a year have approached the Vietnamese coasts during recent decades (Kelly, 1996). In future decades, the characteristics of this risk may change as a result of global warming and there is concern that the frequency of occurrence may increase. According to the Second Assessment Report of the Intergovernmental Panel on Climate Change (IPCC), regional changes in cyclone frequency may occur as ocean temperatures rise and the atmospheric circulation alters (Lighthill *et al.*, 1994; Houghton *et al.*, 1996). There are, however, many uncertainties and the IPCC concludes that no firm assessment can be made.

This uncertainty leaves decision makers in a difficult position. No firm assessment does not mean that the risk is minimal, and it is in the nature of extreme events that unpreparedness itself increases vulnerability substantially. What can be done to plan for an uncertain future? What does a precautionary approach to reducing climate impacts entail? Kelly *et al.* (1994) argue that a precautionary approach to climate impact mitigation

49

and adaptation must involve identifying "win–win" situations in which action to reduce future risk also minimizes vulnerability in the present day: to climate change, to other environmental problems, or to social and economic threats. In this paper, we examine one such approach to coastal protection which takes advantage of a natural resource found along much of the coastline of Vietnam, the mangrove ecosystem.

Wetlands, such as the mangrove, cover 6% of the world's land surface and are found in all climates from arctic tundra to the tropics (Matthews and Fung, 1987). Wetlands provide humans directly and indirectly with ranges of goods and services, including staple food plants, fertile grazing land, support for coastal and inland fisheries, flood control, breeding grounds for numerous birds and fuel from peat. Despite being amongst the most productive ecosystems in the world, the global area of coastal mangrove forests has been decreasing through conversion for agriculture, forestry and urban uses, and due to extraction of timber for fuel, to the extent that many remaining significant areas are being protected under the Ramsar Convention. Mangrove swamps, dominated by 60 species of mangrove tree, are intertidal tropical and sub-tropical coastal wetlands (usually found between 25°N and 25°S). In Vietnam, large areas of mangroves have been converted to agriculture and, in particular, to shrimp aquaculture, causing ecological disturbance and enhancing instability in the coastal physical environment compared to the situation that prevailed under mature mangrove forests (Hong and San, 1993). Many aquaculture practices may be inherently unsustainable; see the critiques presented by Folke and Kautsky (1992) and Kelly (1996).

The various functions and services provided by mangrove areas have been documented and appraised (Lugo and Snedaker, 1974; Mitsch and Gosselink, 1993; Reimold, 1994). It has also been recognized in economic analysis that the functions and services provided by mangroves, and wetlands in general, have positive economic value and that these are often ignored in the ongoing process of mangrove conversion (Farber and Costanza, 1987; Barbier, 1993; Ruitenbeek, 1994; Swallow, 1994; Costanza *et al.*, 1997). Mangrove wetlands display the features of public goods in that their use is non-exclusive, and they are converted to other uses because these functions are undervalued. Often mangrove conversion takes place through overriding traditional common management of the resources (Walters, 1994). Identification of the functions and services and the incorporation of these into policy and the encouragement of appropriate property rights, whether communal or private, are, therefore, necessary first steps in promoting sustainable utilization of such resources.

Initiatives by local institutions in many parts of the world are reversing the dominant trend of wetland loss by undertaking restoration or rehabilitation. Natural wetland restoration activities are undertaken for diverse reasons, such as for wastewater and stormwater treatment (Kent, 1994) or for the supply of resources for local use. Critical issues in promoting the adoption of such schemes, and hence their ultimate sustainability, include the timing of the costs and benefits and the assignment of property rights to the various stakeholders in the restoration process. In the instance under analysis here, the benefit of reduction in maintenance of sea dikes is an important issue, and forms a central argument for schemes where mangroves are planted in front of existing sea defenses.

This paper documents the economic rationale behind mangrove rehabilitation in a case study of three coastal Districts of Nam Dinh Province

(continued from p. 49)

Nguyen Hoang Tri is with Mangrove Ecosystem Research Division, Centre for Natural Resources Management and Environmental Studies, Vietnam National University, Hanoi, Vietnam; W. N. Adger and P.M. Kelly are with Centre for Social and Economic Research on the Global Environment, University of East Anglia and University College London, UK; P. M. Kelly is also with Climatic Research Unit, University of East Anglia, Norwich, UK in northern Vietnam. In these areas, mangrove rehabilitation is subsidized by international development agencies through income generating projects (see, for example, Save the Children Fund Vietnam, 1992), based largely on an assumed benefit to local communities. The desirability of mangrove restoration is quantified in this paper using data on the costs and benefits of certain functions and services. The results provide an initial assessment of the likely effectiveness of this response to the risk of cyclone impacts in providing enhanced protection as well as improving local livelihoods.

Characteristics of Vietnam's coastal zones and risks to mangrove integrity

Nam Dinh Province is located in the southwest of the Red River delta in northern Vietnam (cc. 20°N, 106°E). The province includes three coastal administrative units – Xuan Thuy, Hai Hau and Nghia Hung Districts – and has a sea dike system to protect people, houses and crops. Freshwater reserves help mitigate against the impacts of saline intrusion, flood, storm and sea water rise. The total area of the three coastal Districts is approximately 72 000 ha. According to local hydro-meteorological stations, the annual mean temperature is close to 23°C with a maximum in the monthly means of around 28°C in July and a minimum of about 16°C during January–February. The annual mean monthly rainfall is close to 1 850 mm with a maximum of 330–350 mm during July–August and a minimum of about 25 mm during December–January. The rainy season extends from May to September. The tidal regime is daily with a mean amplitude of 3–3.5 m as far as the typical regime of Tonkin Bay is concerned. The area is affected by two main wind regimes: the northeast monsoon wind which occurs in winter and the south–southeast wind which occurs in summer (the rainy season). The area is towards the northern extreme of the mangrove range. At present, a belt of mangroves of approximately 8 410 ha (see Table 1) acts as a buffer for the sea dike system which has been built over the centuries to protect the intensively used agricultural land from coastal storm surges and floods.

The greatest number of typhoons and associated storm surges occurs in Tonkin Bay in September and October, the so-called 'months of the shifting season' when the monsoonal current changes direction. During these months, typhoons also land to the south on the central coast of Vietnam. At this time of the year, storm surges and sea level rise, as well as high waves and strong winds, may cause extensive damage to economic assets such as agriculture and aquaculture. Estimates of the magnitude of impacts in Nam Dinh Province from floods

Table 1 Mangrove area in Nam Dinh Province

District	Present mangrove areas (ha)	Land estimated to be available for planting (ha)
Xuan Thuy	3 000	7 640
Hai Hau	200	641
Nghia Hung	5 200	9 826
Total	8 400	18 107

Source: Nam Dinh Province data.

Table 2 Socio-economic characteristics of the coastal areas in the three Districts

| | District | | | |
	Xuan Thuy	Hai Hau	Nghia Hung	Total
Total area (ha)	16 246	12 985	9 006	38 237
Population (000 persons)	180.6	153.1	78.1	441.7
Number of households	35 374	39 380	19 363	94 117
Population density(people per km²)				1076
Labour force (000 labourers)	83.2	70.8	26.3	180.3
Average yield for food crops (tha)	5.0	4.0	3.7	4.4
Total food production (000 t rice equivalent)	47.0	29.0	16.4	92.4

Source: Nam Dinh Province data.

and typhoons for the 20 years between 1973 and 1992 show that there were more than 990 injured people, including fatalities, in total and over VND 470 billion damage (1993 constant prices) as a result of severe storms (VND = Vietnam Dong; US$1 = VND 11 000).

Within Nam Dinh Province, the impacts of severe storms are generally concentrated within the coastal Districts. The total population of the three coastal Districts of Xuan Thuy, Hai Hau and Nghia Hung is 445 000, with a population density of 1 076 per km² which is typical of the densely populated areas of the Red River delta plain. The economy of these Districts is primarily dependent on agriculture. Paddy cultivation, aquaculture and salt making are the major agricultural activities. Each of these activities is susceptible to, and differentially affected by, typhoon impacts. Other climatic extremes may also have significant consequences. Rainfall levels and sunshine hours, for example, affect the viability of salt-making. The socio-economic status of the coastal parts of the three Districts is summarized in Table 2.

Given the prevailing circumstances in the coastal Districts of Nam Dinh Province, and similar regions elsewhere, it is clear that mangrove rehabilitation can have a variety of benefits where the topography of the coastal shelf and other social, physical and ecological factors are appropriate. In such situations, mangrove rehabilitation can provide income where households are often severely constrained in cash income sources, as well as bringing about environmental benefits in terms of productive assets and reducing the impact of coastal storm surges. The following sections quantify an economic model of this form of natural resource management.

Economic framework for assessing mangrove rehabilitation and estimated costs and benefits

The economic framework

Economic analysis of resource use can be undertaken in order to assess the magnitude of benefits to local users of the resource. Some values of the goods and services can be assessed by observation of existing markets, but some of the functions and services of mangroves are indirect, or functional, benefits (see, for example, Pearce and Turner, 1990).

The crucial aspects of value for local decision-making, and for the differential impacts of global change, are the direct and indirect use benefits rather than option and existence values which often accrue at the global scale to those not associated with management decisions. It should be noted that some economic benefits of the mangrove resource will increase in value over time, while others will remain constant or decline. For example, as agricultural development intensifies, the potential economic losses from storm surges increases, so the value of the coastal protection function of the mangroves will rise accordingly. Exogenous environmental change associated with global climate change may increase the frequency and intensity of storm surges, and hence the value of this function of the mangroves will rise. Regardless, in this analysis, we have assumed a steady-state situation as our primary concern is with the present-day and near-term future.

The economic cost benefit analysis of mangrove rehabilitation schemes in this case is of the form

$$NPV = \sum_{t=1}^{\gamma} (B_t^{T} + B_t^{NT} + B_t^{P} - C_t)/(1 + r)^t,$$

where NPV is the net present value (VND per ha), B_t^{T} the net value of the timber products in year t (VND per ha), B_t^{NT} the net value of the non-timber products in year t (VND per ha), B_t^{P} the value of the protection of the sea defenses in year t (VND per ha), C_t the costs of planting, maintenance and thinning of mangrove stand in year t (VND per ha), r the rate of discount and γ the time horizon (25 yr rotation).

Costs

Estimates of the costs of establishing the rehabilitated mangrove stands are presented in Table 3. These costs are estimated primarily based on the cost of labour for the activities described. Survey research was carried out in 1994 with the cost for a work day in that year being typically 2.5 kg of rice or VND 5500. Planting of 1 hta of mangroves required 95 work days or VND 522000, as shown in Table 3. The estimates are averaged across the three Districts, with variations in costs dependent on where the

Table 3 Benefits and costs of mangrove rehabilitation in Vietnam and their valuation

Impact or asset valued	Method and assumptions for valuation	Timing of costs and benefits
Benefits		
Timber benefits	Market data: Thinning (VND 180 per tree); extraction mature trees (VND 5000)	Thinning and extraction from year 6 with 3 year rotation
Fish	Market data: Mean price of VND 12500 per kg; yield 50 kg per ha.	Fishing benefits from year 2 after planting
Honey	Market data: Potential yield estimated at 0.21 kg per ha.	Honey collected from year 5 after planting
Sea dike maintenance costs avoided	Morphological model: costs avoided = f (stand width, age, mean wavelength).	Benefits rising from year 1.
Costs		
Planting, capital and recurrent costs	Market and labour allocation data: Costs of seedlings and capital (VND 440000 per ha); Workdays valued at local wage in rice equivalent (VND 5500 per day).	Planting costs at year 1; thinning from year 6 on 3 year rotation

Note: US$1 = VND 11000.

seedlings were obtained. The planting and handling fees for seedlings obtained from forests in the area under rehabilitation are not significant compared to costs for collecting, handling and transportation for other areas which increase depending on the distance from the seedling source site to the planting site. The seed mortality rate between time of collection and time of planting adds an additional cost factor. For some mangrove species, such as *Sonneratia sp*, *Avicennia sp*, *Aegiceras sp* and others, planting directly onto mud flats is unsuccessful due to the exposure to strong wind and wave forces which wash away the seedlings. The cost of raising such species in a nursery and transplanting them at eight months old is relatively high, with fees for maintaining the nursery, care, protection and transportation adding to overall expenditure. The costs of establishing a stand, including planting, gapping and protection, occur mainly in the first year. Maintenance, from the second year on, incurs an estimated annual expenditure of VND 82 500 per hectare. The cost of thinning occurs in years 6, 9, 12, 15, 20 and 25.

Direct use benefits from rehabilitation

The benefits from wood and fuelwood sources from the processes of periodic thinning and extraction are derived from observations in local markets, and are shown in Table 3. It is assumed that the stands are managed in a sustainable fashion. The timber benefits represent wood for poles and fuelwood. The benefits from direct fishing sources were estimated on-site. Fishing activities in the three Districts are undertaken through the use of simple fishing nets, simple tools or even by hand. Aquatic products include fish, crabs, shrimps and shell fish. The yield is estimated at approximately 50 kg per hectare within mature mangrove stands annually for all types of aquatic products. The average unit price in 1994 was around VND 12 650 per kg averaged across the products. There is some evidence that present exploitation of mangrove aquatic products in the Red River delta in general may be leading to declines in fish stocks, so the estimated yield estimates may be not be sustainable, although they are considered conservative for the Districts surveyed.

Honey from bee-keeping is derived from the flowers of a number of mangrove species, though the season spans a limited number of months. The honey from mangroves is obtained during the first flowering season of *Kandelia candel* from January to March and from July to September for other mangrove species and the second flowering season of *Kandelia candel*. The potential yield from this bee-honey source was estimated to be an annual minimum of 0.21 kg per hectare. Honey production is possible from five years after planting, though some species of mangrove can flower after three to four years, and even after one and a half years, from planting.

Indirect use benefits from rehabilitation

The planting of mangroves on the seaward side of the extensive sea dike system provides a benefit of cost avoided in maintenance of these defenses. Such maintenance takes place on an annual basis in the coastal Districts of Vietnam through the obligatory labour of district inhabitants organized by the district committees and paid for through local land taxes. These commitments draw a heavy burden on labour-scarce households and are a source of conflict regarding the inter-district allocation of labour contracts (Adger, 1996). The rehabilitation of mangroves to reduce the costs of maintaining the dike system, however, is perceived to have high short term

Mangrove restoration in Vietnam: N.H. Tri et al

costs (Save the Children Fund Vietnam, 1992; Macintosh and Hong, 1995) as outlined above. Yet the benefits of rehabilitation have rarely been quantified and, when this has been attempted, it has often been without consideration of the indirect benefits.

The evaluation of the role of mangroves in protecting sea dikes is estimated from expenditure on their maintenance and repair in comparison with a case where no mangroves exist, with the control situation assumed to have similar morphological characteristics. In general terms, the greater the area of mangrove, the greater the benefit in terms of avoided maintenance costs. Establishing a precise set of relationships in order to estimate the benefits is not, however, a straightforward matter as the mechanisms by which mangroves protect the adjacent dike are complex. Mangrove stands provide a physical barrier, resulting in drag effects and the dissipation of wave energy. They also stabilize the sea floor, trapping sediment, and can affect the angle of slope of the sea bottom and again the dissipation of wave energy.

Studies in southern China have resulted in an empirical relationship through which the benefit, in terms of avoided cost (B_i^r), can be expressed as a function of the width of the mangrove stand as a proportion of the average wavelength of the ocean waves that the stand is exposed to and various parameters related to the age of stand (mangrove size and density) expressed as a buffer factor. The key parameters are illustrated in Fig. 1. The relationship was developed and tested in mangrove stands in southern China and has been calibrated in Vietnam through simulation (Vinh, 1995). We have used a simplified version of this relationship in estimating indirect use value in this study. Here, the buffer factor, α, is given by

$$\alpha = \frac{2\pi R^2}{1.73b^2},$$

where R is the mean radius of the canopy of an individual tree (m), which increases with age, and b is the typical distance between trees (m), which generally increases with time. As the stand matures, α increases from a minimum of around 0.1 to close to 1.0 as the stand presents a more and more effective obstacle.

Observations indicate that a mature stand will avert 25–30% of the costs of dike maintenance assuming a stand width at least comparable to the characteristic wavelength of the incident waves. The relationship

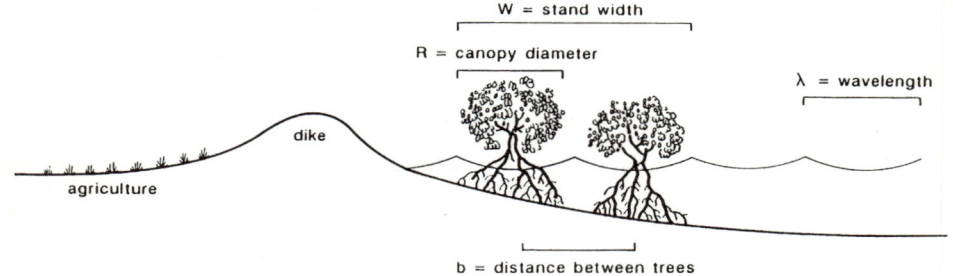

Figure 1. Profile of rehabilitated mangrove stands showing parameters for estimation of avoided maintenance costs.

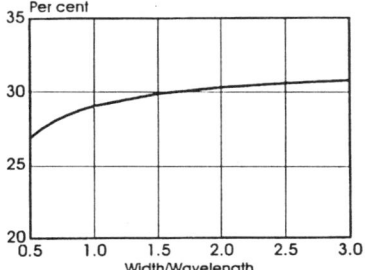

Figure 2. Relationship between percentage maintenance costs avoided and the ratio of stand width to the wavelength of the incident ocean waves (W/λ). Buffer factor, $\alpha = 0.9$ (i.e. mature stand).

between percentage costs avoided and the ratio between stand width, W, and wavelength, λ, for a mature stand over a realistic range of values is illustrated in Fig. 2. It can be seen that, beyond a certain point, increasing stand width results in decreasing gains in protection. Typical wavelengths would be between 25 and 75 m, suggesting stand width should be of the order of 50–150 m.

For the Nam Dinh example, the model was calibrated using survey data on the annual costs of maintenance of sea dikes in each of the three coastal Districts and data on mangrove productivity (growth in terms of mean annual increment, height, canopy density) for *Rhizophora apiculata* (Aksornkoae, 1993). The model was tested for its sensitivity to various parameters including the costs of maintenance in the Districts and the design of the protection schemes in terms of the width of the stand in front of the sea dikes.

The model used here must be regarded as a provisional attempt to estimate the benefits associated with reduced maintenance costs. In particular, we consider the model may be overestimating the benefits when the stand is not fully developed or the width of the stand is much less than the incident wavelength. Nevertheless, uncertainties in this area may not be critical for two reasons. First, as will be seen, the direct benefits from use of the resources are considerably more significant than this indirect use value and, second, the value estimated here is only part of the true storm protection value, which must also include broader damage avoidance benefits, and is, therefore, a lower bound figure.

The model of maintenance costs avoided was used for the three coastal Districts to derive the indirect benefit of mangrove rehabilitation. The baseline costs of maintenance are incurred by the District Committees which keep detailed records of work days and expenditure on annual maintenance. Recent estimates of the number of person-days a year spent on dike maintenance were used in the calculations. As the results represent the average situation, the impact of the most severe storm surges on both the cost of maintenance and repair of dikes is not accounted for. It should also be noted that this model does not account for other damage costs associated with storm occurrence, such as agricultural losses.

Comparing the costs and benefits of rehabilitation

The full results of the cost benefit analysis are presented in Table 4. This cost benefit analysis is of a partial nature, comparing establishment and

Table 4 Costs and benefits of direct and indirect use values of mangrove restoration compared. Stand width = 100 m; incident wavelength = 75 m.

Discount rate	Direct benefits (PV million VND per ha)	Indirect benefits (PV million VND per ha)	Costs (PV million VND per ha)	Overall B̄C ratio
3	18.26	1.40	3.45	5.69
6	12.08	1.04	2.51	5.22
10	7.72	0.75	1.82	4.65

Notes: US$1 = VND 11 000. B̄C ratio = NPV total benefits/NPV costs.

extraction costs with the direct benefits from extracted marketable products and with the indirect benefits of avoided maintenance of the sea dike system. It is assumed that present-day conditions continue to prevail with respect to storm frequency and so on. The results show a benefit to cost ratio in the range of four to five for a range of discount rates. The low relative changes in benefit cost ratios illustrates that most of the costs, as well as the benefits of rehabilitation, occur within a relatively short time frame, with even the reduced maintenance cost beginning to accrue within a few years of initial planting.

Figure 3 illustrates that the direct benefits from mangrove rehabilitation are more significant in economic terms than the indirect benefits associated with sea dike protection over a range of realistic parameter values. As might be expected, the greater the stand width, the more important the direct benefits in comparison to the avoidance of maintenance costs (Figure 3a). Yet even at the lower end of the range of realistic stand widths, offering the greatest return per hectare given suitable conditions, the direct benefits dominate (Figure 3b). As discussed above, the sea dike protection estimates do not include the benefits of reduced repair after serious storm damage, nor the potential losses of agricultural produce when flooding occurs. Flooding associated with severe tropical storms can lead to large economic losses, as well as to loss of life, and a reduced probability of flooding associated with the protection from the mangrove itself would be an additional indirect benefit. This benefit has not been estimated to date, though production function approaches to estimating such benefits for naturally occurring mangrove stands are presently underway.

In any event, it is clear from Figure 3 that the direct benefits from mangrove rehabilitation mean that this activity is economically desirable, as evidenced by the positive Net Present Values (NPV) at all discount rates considered. NPV increases at each rate of discount. The increase in NPV associated with mangrove planting resulting from including dike maintenance savings would promote the desirability of planting. The results presented in Table 4 show that this indirect joint-product benefit of mangrove rehabilitation is significant in further strengthening the economic case for such action in these locations.

Conclusions

This paper has quantified, in a preliminary fashion, various economic benefits of mangrove rehabilitation tied to sea defense systems in three coastal Districts in northern Vietnam. The results from the economic

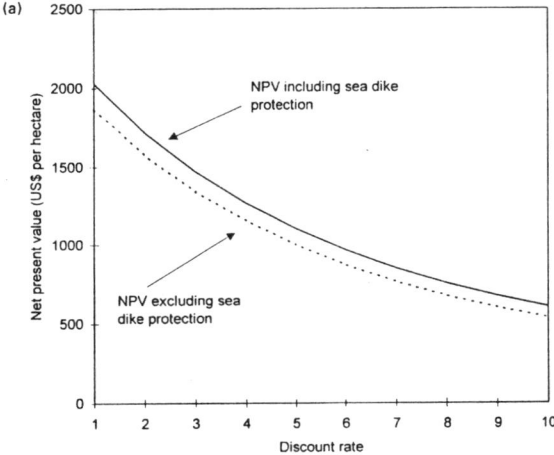

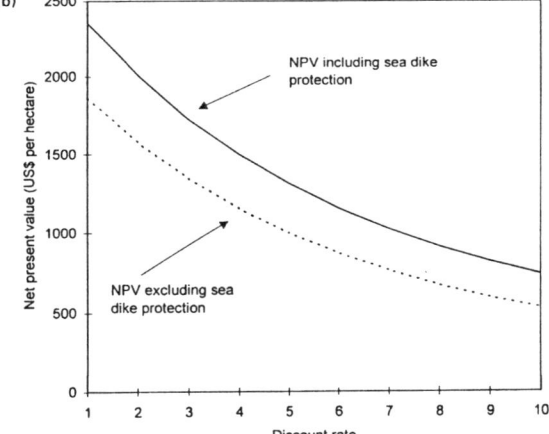

Figure 3. Net Present Value of mangrove rehabilitation, including value of sea dike protection, for two cases: (a) stand width = 100 m; incident wave-length = 75 m; (b) stand width = 33.3 m; incident wavelength = 25 m.

model show that rehabilitation is desirable from an economic perspective based solely on the direct benefits of use by local communities. The rehabilitation schemes have even higher benefit cost ratios when the indirect benefits of the avoided maintenance cost of the sea dike system, protected from coastal storm surges by the mangrove, are included. This analysis neglects a number of aspects which need to be incorporated in a full evaluation. In particular, no account has been taken of the economic benefits resulting from improved protection against storm damage, in terms of more sustainable agricultural yields and so on. There is also a need to test various assumptions, parameters and models used in the analysis. Nevertheless, these initial results do suggest that there is a strong case for mangrove rehabilitation as an important component of a sustainable coastal management strategy which is proof against future, as well as present-day, risk.

There is a broader lesson which can be drawn from this analysis regarding approaches to the problematic issue of adaptive responses to long-term environmental change. As noted in the introduction, decision makers face difficult decisions in assigning priorities when faced with an uncertain future in a resource-limited present. We would argue that this difficulty can be minimized by adopting a precautionary approach which focuses attention on present-day or near-future benefits which will accrue regardless of the nature and magnitude of the impact of environmental change. In the example presented in this paper, mangrove rehabilitation provides immediate economic benefits to the local people, those most vulnerable to storm impacts, while reducing the potential for storm damage over the near and long term.

It should be noted that the result derived from this analysis is dependent on the nature of the control over these resources by existing users and local institutions. The presently widely observed conversion of mangroves for use in aquaculture occurs because the calculus of economic loss and benefit is often undertaken in the context of overriding existing traditional property rights to mangrove products and services. If such constraints can be overcome, mangrove rehabilitation has the potential, under suitable social and physio-ecological conditions, to provide 'win–win' situations whereby the dichotomy between short- and long-term concerns is avoided. It can be contrasted with one alternative course of action, building higher sea dikes, which, although it may ultimately be necessary if the threat of increased storm impacts materializes, provides limited benefits in the short–term.

Acknowledgements

CSERGE is a designated research centre of the Economic and Social Research Council. Funding by the ESRC Global Environmental Change Programme, under the project 'Socio-economic and Physical Approaches to Vulnerability to Climate Change in Vietnam' (Award No. L320253240) is gratefully acknowledged. Assistance, in the form of resources and information, has also been given by Mr A. M. de Klock and Mr T. H. Dan of the World Food Programme; Mr N. H. Nuoi from the Ministry of Water Resources of Vietnam; Oxfam UK and Ireland; Save the Children Fund UK; Danish Red Cross; and Japanese Action for Mangrove Reforestation. Mr T. T. Vinh provided valuable information regarding the dike protection model. Finally, the valuable research recommendations and literature provided by Professor P. N. Hong of the Mangrove Ecosystem Research Division are gratefully acknowledged.

References

Adger, W.N. (1996) A theory of social vulnerability in coastal Vietnam. Presented at *'Designing Sustainability', Fourth Biennial Conference of the International Society for Ecological Economics*, Boston University, August 1996.

Aksornkoae, S. (1993) *Ecology and Management of Mangroves*. IUCN, Bangkok.

Barbier, E. (1993) Sustainable use of wetlands. Valuing tropical wetland benefits: economic methodologies and applications. *Geographical Journal* **159**, 22–32.

Costanza, R., d'Arge, R., de Groot, R., Farber, S., Grasso, M., Hannon, B., Limburg, K., Naeem, S., Oneill, R.V., Paruelo, J., Raskin, R.G., Sutton, P. and van den Belt, M. (1997) The value of the world's ecosystem services and natural capital. *Nature* **387**, 253–260.

Farber, S. and Costanza, R. (1987) The economic value of wetland systems. *Journal of Environmental Management* **24**, 41–51.

Folke, C. and Kautsky, N. (1992) Aquaculture with its environment: prospects for sustainability. *Ocean and Coastal Management* **17**, 5–24.

Hong, P.H. and San, H.T. (1993) *Mangroves of Vietnam*. IUCN, Bangkok.

Houghton, J.T., Meiro Filho, L.G., Callander, B.A., Harris. N., Kattenberg, A. and Maskell, K. (eds) (1996) *Climate Change 1995: The Science of Climate Change*. Cambridge University Press, Cambridge.

Kelly, P.F. (1996) Blue Revolution or red herring? fish farming and development discourse in the Philippines. *Asia Pacific Viewpoint* **37**, 39–57.

Kelly, P.M. (1996) *Tropical Cyclones on the Coast of Vietnam: Prospects for the Future*. Climatic Research Unit and CSERGE, University of East Anglia, Norwich.

Kelly, P.M., Granich, S.L.V. and Secrett, C.M. (1994) Global warming: responding to an uncertain future. *Asia Pacific Journal on Environment and Development* **1**, 28–45.

Kent, D.M. (ed) (1994) *Applied Wetlands Science and Technology*. Lewis, Boca Raton.

Lighthill, J., Holland, G.J., Gray, W.M., Landsea, C., Craig, G, Evans, J, Kuhihara, Y. and Guard, C.P. (1994) Global climate change and tropical cyclones. *Bulletin of the American Meteorological Society* **75**, 2147–2157.

Lugo, A.E. and Snedaker, S.C. (1974) The ecology of mangroves. *Annual Review of Ecology and Systematics* **5**, 39–64.

Macintosh, D.J. and Hong, P.H. (1995) Case study of replanting mangroves in Ha Tinh Province, Central Vietnam. Paper presented at *Workshop on Environment and Aquaculture Development*, Institute for Oceanographic Sciences, Haiphong, May.

Matthews, E. and Fung, I. (1987) Methane emissions from natural wetlands: global distribution and environmental characteristics of source. *Global Biogeochemical Cycles* **1**, 61–86.

Mitsch, W.J. and Gosselink, J.G. (1993) *Wetlands*, 2nd Edn. Van Nostrand Reinhold, New York.

Pearce, D.W. and Turner, R.K. (1990) *Economics of Natural Resources and the Environment*. Harvester Wheatsheaf, Hemel Hempstead.

Reimold, R.J. (1994) Wetlands functions and values. In *Applied Wetlands Science and Technology*, ed. D.M. Kent, pp. 55–78. Lewis, Boca Raton, FL.

Ruitenbeek, H.J. (1994) Modelling economy-ecology linkages in mangroves: economic evidence for promoting conservation in Bintuni Bay, Indonesia. *Ecological Economics* **10**, 233–247.

Save the Children Fund (1992) *Mangrove Planting Project Interim Evaluation*. December 1992, Save the Children Fund, Hanoi, Vietnam.

Swallow, S.K. (1994) Renewable and non-renewable resource theory applied to coastal agriculture, forest, wetland and fishery linkages. *Marine Resource Economics* **9**, 291–310.

Thu, T.V. (1991) Advances in forecast dissemination and community preparedness tactics in Vietnam. Paper presented at the *Second International Workshop on Tropical Cyclones*. Hydrometeorological Service, Hanoi.

Vinh, T.T. (1995) Tree planting measures to protect sea dike systems in the Central Provinces of Vietnam. Paper presented at the *Workshop on Mangrove Plantation for Sea Dike Protection*, 24–25 December 1995, Hatinh, Vietnam.

Walters, J.S. (1994) Coastal common property regimes in Southeast Asia. In *Ocean Yearbook 11*, ed. E.M. Borgese, N. Ginsburg and J.R. Morgan, pp. 304–327. University of Chicago Press, Chicago.

Wickramanayake, E. (1994) Flood mitigation problems in Vietnam. *Disasters* **18**, 81–86.

[14]

Ecological Limitations and Appropriation of Ecosystem Support by Shrimp Farming in Colombia

JONAS LARSSON[1,2,*]
CARL FOLKE[1,2]
NILS KAUTSKY[2]
[1]The Beijer International Institute of Ecological Economics
The Royal Swedish Academy of Sciences
Box 50005, S-104 05 Stockholm, Sweden
[2]Department of Systems Ecology
Stockholm University
S-106 91 Stockholm, Sweden

ABSTRACT / Shrimp farming in mangrove areas has grown dramatically in Asia and Latin America over the past decade. As a result, demand for resources required for farming, such as feed, seed, and clean water, has increased substantially. This study focuses on semiintensive shrimp culture as practiced on the Caribbean coast of Colombia. We estimated the spatial ecosystem support that is required to produce the food inputs, nursery areas, and clean water to the shrimp farms, as well as to process wastes. We also made an estimate of the natural and human-made resources necessary to run a typical semiintensive shrimp farm. The results show that a semiintensive shrimp farm needs a spatial ecosystem support—the ecological footprint—that is 35–190 times larger than the surface area of the farm. A typical such shrimp farm appropriates about 295 J of ecological work for each joule of edible shrimp protein produced. The corresponding figure for industrial energy is 40:1. More than 80% of the ecological primary production required to feed the shrimps is derived from external ecosystems. In 1990 an area of 874–2300 km^2 of mangrove was required to supply shrimp postlarvae to the farms in Colombia, corresponding to a total area equivalent to about 20–50% of the country's total mangrove area. The results were compared with similar estimates for other food production systems, particularly aquacultural ones. The comparison indicates that shrimp farming ranks as one of the most resource-intensive food production systems, characterizing it as an ecologically unsustainable throughput system. Based on the results, we discuss local, national, and regional appropriation of ecological support by the semiintensive shrimp farms. Suggestions are made for how shrimp farming could be transformed into a food production system that is less environmentally degrading and less dependent on external support areas.

KEY WORDS: Aquaculture; Shrimp farming; Colombia; Sustainability; Resource use; Life support system; Carrying capacity

*Author to whom correspondence should be addressed at the Beijer Institute.

Natural ecosystems play an important role in supporting human activities (Odum 1989, Folke 1991, de Groot 1992). However, they are still seldom explicitly recognized as such assets in environmental management, despite the fact that their continuous support is a prerequisite for economic activity. Degrading this support actually means that ecological limits for economic development are being approached more rapidly. Estimates of how close human societies are to such limits, i.e., ecological carrying capacity (Daily and Ehrlich 1992), indicate that humans already appropriate 40% of global net primary production (Vitousek and others 1986) and that current resource use patterns in Western society, if applied on a worldwide scale, would require a resource base greater than the present carrying capacity of the earth (Rees and Wackernagel 1994). The absurdity of such a notion is a vivid indication of the urgent need for a change towards ecologically sustainable economic development.

If economic development is to be sustainable in this sense, efforts must be directed towards human activities that make use of the natural environment without severely or irreversibly degrading it. Human production and consumption should require less resources, use them more efficiently, and emit wastes that do not exceed the assimilative capacity of the environment (e.g., Costanza and Daly 1992). A major goal of society should be to stimulate production and consumption that does not diminish the capacity of life-support systems to recovery after disturbance (Holling 1986) and that remains within the carrying capacity of the supporting ecosystems.

To achieve such goals, comprehensive indicators of ecosystem health and recovery in relation to human activity need to be developed (Kelly and Harwell 1990, Costanza and others 1992). One approach is to estimate the ecosystem area—the ecological foot-print—required to support human activities (Odum 1975). Rees and Wackernagel (1994) found that typical urban/industrial regions (>300 people/km^2) sequester or appropriate the biological production of at least 20 times more land (as land for cultivation, degraded ecosystems, and land required for energy production) than is usually contained within the regions themselves. Folke (1988) estimated that the production of salmon in a net pen in coastal waters requires the work of the marine food-web over an area 40,000–50,000 times larger that the area of the cages themselves. Similar estimates have been made for trade in fisheries resources (Hammer 1991), for endangered species (Wright 1990), and for entire sea areas (Folke and Kautsky 1989, Folke and others 1991).

Such estimates are helpful in indicating how big economic activities can grow in relation to their resource base or whether they might already have passed this limit. Furthermore, estimates of ecosystem support may serve as a guideline for policy formulation and may be especially useful at highlighting areas of overlap between conflicting activities, not all of which may be directly apparent.

Shrimp farming is, in this respect, a useful example. The environmental impact of shrimp aquaculture has been well documented as a result of the explosive growth of such operations in South East Asia and, to a lesser extent, in Latin America (CLIRSEN 1990, Snedaker and others 1986, Olsen 1990, Aiken 1990, Chua and Kungvankij 1990, Vesga F. 1990) and has also caused social impacts (Meltzoff and LiPuma 1986, Bailey 1988, 1991, Southgate 1992, Panvisavas and others 1991). The construction of shrimp ponds has led to widespread destruction of mangrove forests, depletion of groundwater, saltwater intrusion into groundwater tables, eutrophication and disease transmission in a number of places (e.g., Phillips and others 1990, Primavera 1991, 1993, Hamilton and Snedaker 1984, Rubino 1990). At the same time, shrimp farms require large amounts of clean, nutrient-rich water, wild shrimp fry from undisturbed mangrove lands (Naamin 1991, Paw and Chua 1991) and, in its intensive and semiintensive forms, fish and cereals for feed (in the form of pellets). There is clearly a conflict between the need for a healthy ecological support system and the effects of shrimp farming on the surrounding environment.

Critics have pointed out that most commercial shrimp farms, particularly the more intensive cultivation schemes, are ecologically unsustainable throughput systems. Throughput systems use large amounts

of fossil fuel energy directly and indirectly, in the form of feed and fertilizer, and require large ecological support areas to be sustained. In such systems resources are pumped in, used up, and pumped out in a linear fashion, rather than being recycled. This leads to accumulation of wastes in the recipient ecosystems, often causing severe and irreversible environmental problems (such as eutrophication and accumulation of pesticides in food chains). The leaky nature of these systems and their high resource demand also makes them very costly or impossible to sustain in the long run (Folke and Kautsky 1992). Although it may not be immediately apparent, throughput systems depend entirely on a resource base which, directly or indirectly, is linked to the very ecosystems they degrade. The failure of throughput systems to recognize and respond to these linkages makes them inherently liable to collapse, as the degradation of their support systems remains unnoticed.

The purpose of this study is to reveal the dependence of current farming practices on external systems and to what degree they are throughput systems. Specifically, the study focuses on semiintensive (see below) shrimp culture as practiced on the Caribbean coast of Colombia.

We estimate the spatial ecosystem support that is required to produce food inputs, nursery areas, and clean water to the shrimp farms, as well as to process wastes. We also make an estimate of the natural and human-made resources necessary to run a typical semiintensive shrimp farm. The results are compared with similar estimates for other food production systems, particularly aquacultural ones, and extrapolated on a nationwide level. Based on the results, we discuss local, national, and regional appropriation of ecological support by the semiintensive shrimp farms and its relation to other uses of this support. Finally, we offer some suggestions for how to transform shrimp farming into a food-producing system that is less environmentally degrading and less dependent on external support areas.

Study Area

The Bay of Barbacoas (10°10′N, 75°35′W) is located on the Caribbean coast of Colombia approximately 25 km south of the city of Cartagena. Fresh water from the Magdalena river is diverted into the bay through a number of branches of the Canal del Dique, which may be a factor conducive to the extensive mangrove formations south of the bay. A large number of shrimp farms are situated adjacent to the

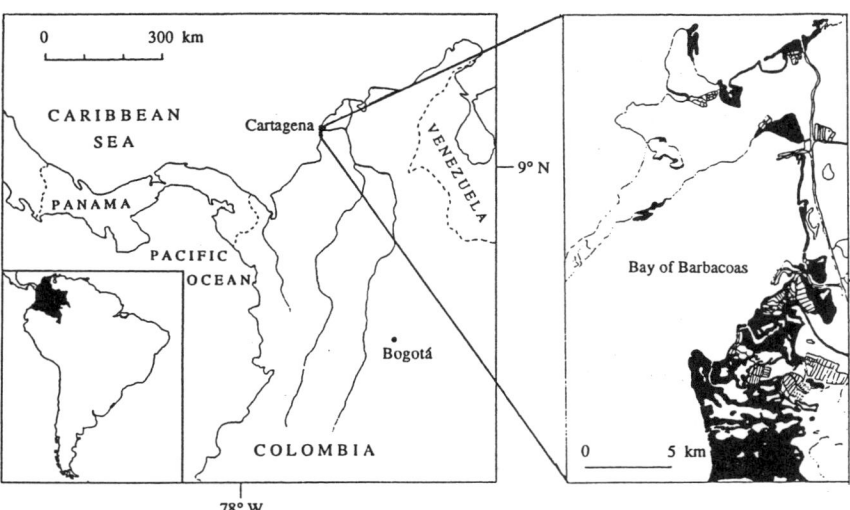

Figure 1. Map of the Bay of Barbacoas area, showing its location on the Colombian Caribbean coast and on the South American continent (inset). The shaded areas on the local area map are mangroves; striped polygonal areas are shrimp ponds.

mangrove swamps surrounding the bay, particularly at its southern end (Figure 1).

Colombian shrimp farms along the Caribbean coast use semiintensive cultivation practices; thus, shrimp receive nutrition both from natural pond productivity and added feed. Stocking densities are moderate to high (16–18 individuals/square meter) and diesel or electric pumps are used to circulate and aerate water at relatively high rates (5%–20% of pond water is renewed daily). Additional aeration is sometimes provided by paddlewheel aerators. Yields are close to 4000 kg/ha/yr, distributed over three harvests. Each growing season is about 100 days long. Ponds are on average 10 ha large and are normally constructed on salt pans or in shallow lagoons (Vesga F. 1990). There is no evidence that mangroves have been cut down to make way for ponds except on a very small scale (Larsson 1992), in contrast to other countries such as Ecuador (CLIRSEN 1990).

Only two species are cultivated on any scale, *Penaeus stylirostris* and *P. vannamei*, both natives of the Pacific coast. They are usually cocultivated in proportions that vary from 1:3 to 1:9. Shrimp postlarvae are bought from hatcheries that import wild postlarvae (mainly from Panama) in addition to rearing their own seed from gravid females caught at sea and, to a lesser extent, from captive breeding stocks. The proportions between postlarvae caught in the wild and those reared in laboratories vary greatly and depend largely on the time of the year (e.g., in January wild postlarvae are scarce and supplies uncertain) (Vesga F. 1990). If available, wild shrimp are much preferred by farmers, since artificially reared postlarvae grow slower and often have deformities. For our calculations we assume that 10%–50% of stocked postlarvae are caught in the wild, which agrees with estimates made by farmers themselves (H. Cárdenas Mahecha, personal communication).

The majority of Colombia's Caribbean shrimp farms are found in the Bay of Barbacoas area, and some of these are among the largest in the country. Most of the data for this study apply to all of the farms on the Caribbean coast, although certain data are from studies of farms in this region specifically. These data should nevertheless be applicable to most shrimp farms along Colombia's Caribbean coast, since farming methods and climatic factors are similar throughout this region.

Methods

All estimates were made on an annual basis, assuming three harvests and a 300-day growing season. They are divided into three parts: ecosystem support areas, natural capital inputs, and human-made capital inputs. Natural capital and human-made capital are measured in solar energy and industrial energy terms (e.g., Hall and others 1986), respectively. More detailed descriptions of the calculations can be found in the notes to the tables and figures in the article.

Ecosystem Support Areas

Ecosystem support areas were estimated for six different systems. The marine ecosystem support area was calculated as the sea surface area required to sustain a fish yield equivalent to the amount of fish meal in feed pellets for a 1-ha shrimp pond. The agricultural support area was calculated similarly, using known figures for the cereal content of feed pellets (Anon. 1986a), and the average agricultural yield per area for temperate crops under fertilization (Pimentel and Pimentel 1979). The mangrove postlarval nursery area required to produce sufficient amounts of shrimp postlarvae (Vance and others 1990, Turner 1977, 1986, von Prahl 1980, Robertson and Duke 1987, Kapetsky 1985, Pauly and Ingles 1986) was estimated using densities of wild shrimp fry (Pedini 1981), larval mortality, the proportion of seed reared or hatched artificially (H. Cárdenas Mahecha, Granjas Marinas Ararca, Cartagena, personal communication) and average stocking densities (Vesga F. 1990).

There is an annual pronounced drop in shrimp yield (30%–50%) in Colombian shrimp farms (Vesga F. 1990) during the dry season. The absence of terrestrial runoff means that no mangrove-derived litter is exported to the surrounding lagoons in this period. Tidal flushing, by contrast, plays a minor role in the transport of mangrove litter to these lagoons, since the amplitude of tides is very small along the Caribbean coast. The lagoons are the only source of water for shrimp ponds. Consequently, ponds receive sharply decreased amounts of mangrove litter during the dry season. The drop in yield may thus indicate a dependence on mangrove detritus as an important component of shrimp food under normal conditions, as has been suggested by a number of studies of the food requirements and mangrove association of estuarine shrimp (von Prahl 1980, Turner 1977, 1986, Lahmann and others 1987). On basis of this, we assumed that 30% of shrimp food in ponds is normally made up of mangrove-derived detritus (bacterial and fungal films on mangrove leaf detritus), which enabled us to calculate the mangrove support area necessary to produce this litter. Average local mangrove litterfall was estimated at 5 tons/ha (Larsson 1992) and bacterial conversion efficiency at 10% (Colinvaux 1986).

Using standard pumping rates (about 10% of pond volume exchanged daily for 300 days per year) and an estimated average depth of source lagoons of 5 m, an estimate for the size of the lagoon ecosystem providing clean water was made (since lagoons generally are shallower than 5 m this is probably an overestimate, but allows for the fact that some of the water is pumped directly from the sea where depths are greater). It is not known to what extent polluted water affects the shrimp ponds or how fast lagoon water is recycled, but the figure may serve as an indication of the extent of the support system that is necessary to provide water for the ponds and also to receive discharge (especially since some farms have their water intake close to another farm's outlet).

Finally, an attempt was made to estimate the ecosystem area needed to sequester the carbon dioxide released by industrial energy inputs (directly and indirectly) to shrimp farming (Rees and Wackernagel 1994, Krause and others 1990). This was done by calculating the CO_2 release directly from the fuel burned at the farming site, and by estimating the amount of CO_2 released from the use of indirect industrial energy to produce tools and other inputs to the farms (see inputs of human-made capital below). The latter was derived from the proportion of various energy sources (coal, petroleum, and natural gas) to the total energy consumption in Colombia in 1988 (Anon. 1991a, Hellsten 1992). For carbon sequestration a reconversion of tropical pasture to managed forest was assumed, with an assimilative capacity of 5 tons of carbon per hectare per year (t C/ha/yr) (Krause and others 1990). The method is described in more detail below in the legend notes to Figure 2.

Natural Capital Inputs

Natural resource demand was calculated as the food requirements, expressed in gigajoules of input per hectare pond (GJ/ha), of an average 1-ha shrimp pond for which yield and feed inputs were known.

Total food energy and indigenous net primary production (NPP) and gross primary production (GPP) were calculated by using growth efficiencies for shrimp (Kurmaly and others 1991) and zooplankton (Colinvaux 1986), combined with data for the energy content of shrimp (Odum and Arding 1991), zooplankton, and phytoplankton (Kurmaly and others 1991, assuming that phytoplankton carbon content is

50% by dry weight. The energy content of feed pellets was subtracted from total food requirements to give the indigenous food energy requirements, i.e., the proportion of shrimp food that is not derived from feed pellets.

By using trophic efficiencies assembled from a variety of sources, the NPP and GPP supporting this food web were calculated. For marine ecosystems, primary production was calculated as the average NPP per unit sea area multiplied by the marine ecosystem support area. A large proportion of the fish in the feed pellets used in Colombia is caught in South Eastern Pacific upwelling areas (Luis Martinez Silva, Grupo de Acuicultura Marina, INDERENA, Cartagena, personal communication), which enabled us to use standard figures for fish yield (Odum and Arding 1991) and primary productivity (Colinvaux 1986) in these areas, as a first approximation.

Agricultural NPP supporting the shrimp pond was calculated using known figures for the percentage of agricultural products contained in feed pellets and for the proportion of crops that is not harvested (Pimental and Pimental 1979, Zucchetto and Jansson 1985). These estimates are obviously less precise since a great variety of cereals and cereal products (such as wheat, corn, barley, molasses, etc.) are used in the manufacture of feed pellets. It is, however, hard to calculate accurately the exact NPP required to produce these components, since it is difficult to obtain detailed recipes for pellets from the manufacturers. This is further complicated by the various climatic and nutrient regimes under which these crops may be cultivated, since these regimes control where and when each crop is grown. Still, even though each cereal component is added to pellets in variable amounts, the total proportion of cereals and other agricultural products in pellets is relatively constant, according to the major manufacturer's own specifications (Raza Alimentos S. A., Bogotá). A rough estimate may, therefore, be equally plausible, in particular since agricultural NPP is a minor component of the total NPP required to farm shrimps.

Human-Made Capital Inputs

The human-made capital demand, expressed as industrial energy use (e.g., Hall and others 1986), was estimated by multiplying the dollar cost of different inputs [e.g., investments, postlarvae, miscellaneous expenses (Larsson 1992)] with the energy use/GNP ratio for Colombia (Anon. 1991a). The energy/GNP ratio is an imprecise way of estimating industrial energy use (Hall and others 1979, Cleveland 1992), but

was nevertheless chosen as a first rough estimate, since no data were available for more precise calculations. For some items, such as pellets and fuel, it was possible to calculate the industrial energy per unit commodity, using figures from Folke [embodied energy in feed pellets (1988)] and Odum and Arding [energy content in fuel (1991)].

Results

Our calculations suggest that a semiintensive shrimp farm requires a spatial ecosystem support, or ecological footprint, that is 35–190 times as large as the surface area of the farm. The shrimp farm appropriates about 295 J of ecological work for each joule of edible shrimp protein it yields. The corresponding figure for industrial energy is 40:1. The contribution from natural capital per joule of edible protein produced is thus about 7.5 times larger than the one from human-made capital. In 1990 an area of 874–2300 km^2 of mangrove was required to supply shrimp postlarvae to the farms, corresponding to about 20%–50% of Colombia's total mangrove area. The results are reported in detail below and in the tables and figures.

Ecosystem Support Areas

Figure 2 shows the estimated ecosystem support areas. By far the largest support system is the mangrove nursery for shrimp postlarvae, which may be as much as 160 times the size of the cultivation pond, if a large proportion (50%) of the postlarvae used are caught wild and postlarval density in the mangrove nursery is low (0.3 individuals/m^2). If, on the other hand, only 10% of postlarvae are caught wild and seed are more abundant (1/m^2), then the mangrove nursery need be no larger than 10 times the size of the pond area. This is on par with the sea area required to support the fish catch for pellet manufacture, which was estimated to be 14.5 ha/ha pond. Agricultural ecosystems, by contrast, only occupy 0.5 ha/ha shrimp farm area and are thus a very minor component.

The amount of water pumped yearly into the ponds is equivalent to a lagoon 5 m deep and about 7 ha in area for every hectare of pond. The adjacent mangrove area needed to produce a sufficient litterfall to fulfill the hypothetical 30% shrimp dietary requirement for mangrove detritus was estimated at about 4 ha/ha pond. The forest area necessary to sequester the carbon dioxide released directly and indirectly (by energy use in producing industrial inputs to the farms) by shrimp farming were estimated to be

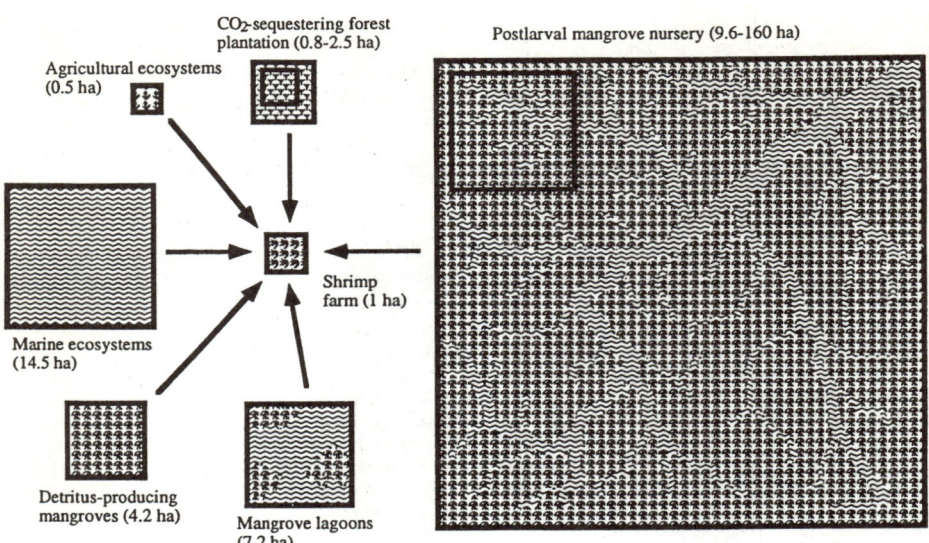

Figure 2. Ecosystem support areas required to sustain a semiintensive shrimp culture, ha/ha cultivated area. Marine up-welling ecosystems: fish yield 6.71 t C/km², 2.44 t fish used with a carbon content of 40% (Odum and Arding 1991); agricultural ecosystems yield 3.5 t/ha dry weight, 1.5 t used; postlarval density in mangroves 0.3–1 individual/m² (Pedini 1981), assuming a prestocking mortality of 50% and the proportion of postlarvae derived from wild fry (as opposed to hatchery-raised and/or reared from eggs from gravid females) assumed to be between 10 and 50% (H. Cárdenas Mahecha, Granjas Marinas Ararca, Cartagena, personal communication); mangrove area needed to yield enough litter (by carbon content) to provide 30% of feed for shrimp (as indicated by the 30–50% drop in productivity when no such litter is flushed into the ponds) estimated on basis of productivity measurements with an average mangrove litterfall of 5 t/ha and a 10% trophic efficiency in converting mangrove carbon into detrital organic matter available to shrimp (bacteria, fungi, etc.); lagoon water area calculated as the area of the pumped yearly volume (10% daily, ponds 1.2 m deep, 300 days/yr) assuming the source lagoon is on average 5 m deep (which may be an overestimate, in which case this figure should be higher). This water area is necessary for the replenishment of freshwater to the ponds and, although, of course, the same water may be recirculated, especially if one farm pumps its water from a point close to the discharge of another, the figure serves as an indication of the volume and area of the lagoon support system of the ponds. CO_2 sequestering area is calculated along the lines of Rees and Wackernagel (1994). Direct energy use (fuel) 6085 l/ha (Vesga F. 1990), corresponding to 14.9 t CO_2/ha (Hellsten 1992); maximum estimate includes indirect energy use (452 GJ/ha, Table 2) distributed according to national energy consumption in Colombia 1988 (Anon. 1991a). Thus 17.4% of this energy is obtained from natural gas, corresponding to 78.5 GJ/ha or 2344 m³, which in turn yields 4.1 t CO_2 (Hellsten 1992). Similarly for petroleum products (average for petrol, diesel, kerosene and oil) and coal: 54% of energy consumption is oil, or 244 GJ/ha, equivalent to 6.2 t, yielding 19.4 t CO_2/ha; 16.3% is coal-based energy, i.e., 73.6 GJ/ha, or 2.7 t coal per hectare, which yields 6.7 t CO_2/ha (Hellsten 1992). No correction was made for energy quality of different energy sources (Hall and others 1986), nor was carbon release from hydropower production estimated. Total CO_2 released is between 14.9 and 45.1 t CO_2/ha. In order to estimate the surface area needed for carbon sequestration (through reconversion of tropical pasture land to managed forest plantations), a carbon assimilation capacity of 5 t C/ha was assumed (Krause and others 1990), or 18.3 t CO_2/ha.

between less than 1 and 2.5 ha/ha pond. In all, external support ecosystems were estimated to be between 35 and 190 ha/ha shrimp pond.

The total pond surface area of Colombia's shrimp farms was 2913 ha in 1990 (Anon. 1991b), which im-plies that the size of the mangrove support system supplying postlarvae for Colombian shrimp farming was between 874 and 2300 km². This is equal to 20%–50% of the total mangrove area in the country [less than 4400 km² (Saenger and others 1983)].

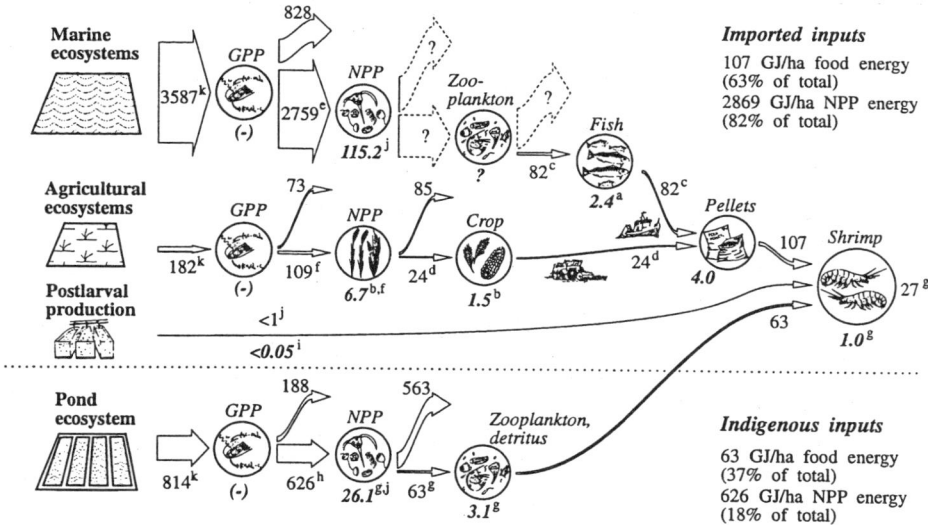

Figure 3. Minimum annual natural solar energy and resource inputs required to sustain a 1-ha semiintensively managed shrimp farm. Energy flows, GJ/ha (normal typeface); material resources, metric tonnes/ha (bold italics). (-) = not determined; ? = unknown or not used for calculations. Dashed lines represent unknown proportions. The width of arrows is roughly proportional to the size of the energy flow. Open-ended arrows represent system losses (e.g., through metabolism of waste). Figures may not add up due to rounding. Detailed explanations of calculations are as follows: a. Production costs US$20,969,000 for the Colombian Atlantic Coast (US$3.25/kg * 6,452,000 kg); pellet costs 25%, or US$5,242,350. With a cultivated area of 1631 ha and a pellet price of US$550/t this gives 5.83 t/ha feed or 5.07 t/ha dry weight (Anon. 1991b, Anders Alm, IOCARIBE, Cartagena, personal communication). 48% is fish meal (Anon. 1986a), or 2.44 t/ha.

b. 18% of feed consisting of cereals and 12% soybean meal (Anon. 1986a), equivalent to a total of 1.52 t/ha dry weight.

c. Energy content of fish used in fish meal 6.14 GJ/t wet weight (Anon. 1986a), wet weight = 5.5 * (dry weight) (quoted in Folke 1988).

d. Agricultural products 1.78 t/ha wet weight with a pellet moisture of 13%, cereal dry weight 83.5% of wet weight (Uhlin and others 1975) and soybean meal dry weight 88% of wet weight (Eriksson and others 1972), food energy 13.38 J/t (Anon. 1986b).

e. Fish yield of Southeast Pacific upwelling system 6.71 g C/m^2 (Odum and Arding 1991), primary productivity of ditto 225 g C/m^2 (Colinvaux 1986), 6.71/225 * 82.3.

f. Total energy content of cereal and soybean plants equal to 4.44 times the energy content of edible products (Pimentel and Pimentel 1979, Zucchetto and Jansson 1985).

g. Total food energy = (shrimp yield 3.956 t/ha) * [0.26 dry weight (Odum and Arding 1991)] * [energy content 25.953 GJ/t (Odum and Arding 1991)]/[shrimp growth efficiency 0.15766 J/J (for *P. japonicus*, from Kurmaly and others (1991)] = 169.3 GJ/ha; total imported food energy = (marine food energy 82.3 GJ/ha) + (agricultural energy 23.82 GJ/ha) = 106.7 GJ/ha; indigenous food energy = total food energy − imported food energy − postlarval food energy (see below) = 62.6 GJ/ha, energy content of zooplankton 20.057 GJ/t (Kurmaly and others 1991), 62.6/20.1.

h. Zooplankton growth efficiency 10% (Colinvaux 1986), food energy 62.6 GJ/ha, carbon-based energy content 47.88 GJ/t C (Kurmaly and others 1991), carbon content of phytoplankton assumed to be 50%.

i. 160,000 postlarvae/ha per growing period, three growing periods per year, postlarval size 0.025 g wet weight (Martinez Silva and others 1987).

j. 480,000 postlarvae/year, each one requiring 70 * 18.78 J/individual to raise from egg (Kurmaly and others 1991) (where 70 is a size correction factor to account for the smaller size of the *P. monodon* larvae used by Kurmaly and others (1991) compared to the size of the *P. vannamei* and *P. stylirostris* larvae used in Colombia), gives a total food energy of 0.63 GJ/ha. Assuming carbon content of feed (phytoplankton) to be 50%, this energy corresponds to 0.63/(47.88 GJ/t/2) = 0.0264 t. Assumed that the different species have similar resource requirements, since feed is similar. Note: larval mortality is not included, nor are energy inputs to egg production, since only approximate estimates were available for the former (~50%) and none for the latter. In any case, the effect of including mortality on the final result should be very minor, since the negligible mass of the postlarvae stocked make their contribution to total resource use very small.

k. Agricultural GPP = 1.67 * NPP (Odum 1983, Pimentel and Pimentel 1989, Zucchetto and Jansson 1985), marine GPP = 1.3 * NPP (Folke and Aneer 1988, Zucchetto and Jansson 1985).

Table 1. Industrial energy requirements for semiintensive shrimp culture, minimum–maximum (average) estimates (GJ/ha).

Investments[a]	Labor[b]	Pellets[c]	Postlarvae[d]	Fuel[e]	Miscellaneous[f]	Total
7.4–12.5	32.9–52.5	146–408	68.4–82.0	217	91.1	563–863
(10.1)	(42.7)	(233)	(75.3)			(669)

[a]Investments US$8427–14,045/ha (US$11,236), usable for 30 years (Odum and Arding 1991).

[b]Labor costs US$1236–1977/ha (US$1606).

[b]Pellet industrial energy input 25–70 GJ/t (40 GJ/t) (Folke 1988, Uhlin and others 1975, Pitcher 1977).

[d]Postlarvae cost US$2571–3085/ha (US$2828).

[e]Fuel 1582 gallons/ha, energy content 137 MJ/gallon (Odum and Arding 1991).

[f]Miscellaneous inputs include maintenance, harvest costs (other than fuel and labor), fertilizer, lime for pond preparation, etc. Annual expenses US$3427/ha, 0.0266 GJ/US$ (in Colombia) by total energy use/GNP (Anon. 1991a). The use of unit energy per unit GNP is necessarily imprecise but since data for a more detailed analysis were not available, this method may suffice for a first estimate. For comparison, calculating pellet industrial energy by the energy/GNP ratio yielded 68.4–102.7 (85.4) GJ/ha, which is rather less than the result obtained using the values from Folke (1988). This is what may be expected in the economy of a developing country, where a large proportion of GNP is derived from nonmechanized agriculture. As a result energy use by industries is proportionately greater relative to its share of GNP, compared to agriculture, than it would be in an industrialized country. Using energy/GNP ratios to calculate industrial energy inputs for relatively energy-intensive processes, such as pellet production, will therefore underestimate such inputs. Considering that many of the inputs to shrimp farming are rather energy- and technology-intensive (for instance, postlarval rearing and pond construction), the above figures should be seen as lower end estimates.

Natural Capital Inputs

Figure 3 lists the estimates of natural capital demands of a 1-ha semiintensive shrimp pond. Added feed, about 5.8 t/ha, amounts to 106.7 GJ/ha, or about 63% of total food requirements (169.3 GJ/ha). The feed energy is derived primarily (about 80%) from marine ecosystems (i.e., as fish caught to make the fish meal, which is the major component of fodder pellets). A smaller proportion, about 20%, is made up of agricultural products. The remaining 37% of total food requirements must be met by natural pond productivity or carried into the ponds by pumped lagoon water. Expressed in terms of the primary production that forms the base of the food chains of the various support ecosystems, the proportion derived from exogenous sources is even greater. As much as 82% of total required primary production is imported through feed, corresponding to 2869 GJ/ha, and only 18% is derived from the pond's own productivity. Total GPP required to sustain food needs was estimated at 4584 GJ/ha, or 1158 J/t.

Human-Made Capital Inputs

Estimates of the industrial energy consumed for producing the human-made capital used in shrimp farming are presented in Table 1. Pellets and fuel are the two largest single inputs and together make up about two thirds of the estimated total industrial energy use of 669 GJ/ha. Total industrial energy use is 142–218 GJ/t harvested shrimp, corresponding to an average of 43 J/J shrimp protein, or if labor is excluded, 40 J/J. This is about 13% of the natural (solar) energy appropriated per joule edible shrimp protein (295 J/J) (see Table 2 below).

Discussion

The results of this study show, not surprisingly, that shrimp farming is highly dependent on inputs from external sources and ecosystems, both human-made and natural. The former are due to the material investments both in constructing but especially in running the ponds, such as the large amounts of pumping needed, the energy-intensive methods for producing feed pellets and fertilizer, or for catching and/or rearing postlarvae. The latter are due to the many trophic levels between the shrimp and the pelagic primary productivity on which the fish in feed pellets depend. Since shrimp in its natural environment do not feed directly on phytoplankton or macrophytes (except for the young larval stages), shrimp farming is ecologically inherently less efficient than food production systems based on phytophagous organisms (e.g., grass carp, or cattle raising).

With the exception of the agricultural support area, the ecosystem support areas are all larger than the pond area, ranging from only a few times the size of the pond area to two orders of magnitude greater, as is the case for the maximum estimated postlarval nursery area. The implication of the size of the supporting mangrove nursery area becomes clearer when shrimp farming is viewed at a nation- and region-wide level. The size of the mangrove support system supplying postlarvae is between 874 and 2300 km². The total mangrove area in Colombia does not exceed 4400 km² (Saenger and others 1983) and far from all of this is suitable habitat for shrimp postlarvae. In fact, since only Pacific shrimp species are cultivated on any scale, the available mangrove area is appreciably smaller. The majority of postlarvae used in the country are imported, primarily from Panama (H. Cárdenas Mahecha, personal communication), which implies that the greater part of the postlarvae support system of Colombian shrimp farms is outside the country itself. The mangrove area needed to supply enough postlarvae is a sizable proportion of Panama's mangrove area. Given that Panama exports shrimp postlarvae to other countries as well, there is clearly a great pressure on the natural environment to produce sufficient seed. The same holds true for other seed-producing countries such as Ecuador.

It is possible that the postlarval population estimates used herein may underestimate the regenerative capability of shrimp and that there is a large natural surplus. However, as the highly fluctuating supply of shrimp postlarvae has shown, intense fishing for seed may result in shortages of postlarvae when natural productivity is low, for example, during the recurring El Niño episodes (Vesga F. 1990). This has been exacerbated by the widespread destruction of mangrove nurseries for shrimp pond construction, which is believed to be a major cause of the decline in catches of seed postlarvae in Ecuador (Primavera 1993, Lahmann and others 1987). There is also a conflict with offshore fisheries that catch adult shrimp. This puts additional pressure on wild stocks.

The largest support ecosystem, mangrove nurseries for postlarvae, extends far beyond the physical location of the shrimp farms and is a vulnerable link in the farming operation. Shrimp farming should, therefore, strive to reduce its impact and reliance upon external ecosystems for postlarvae production. Development of closed cycle rearing systems that can produce genetically variable shrimp fry of high quality would benefit not only the mangroves by relieving natural shrimp populations of overintensive harvesting pressure, but also the shrimp farming industry by making it more self-sufficient and the postlarval supply more predictable.

A more accurate estimate of ecosystem support areas would also have to take into account the sea area required to support sufficient numbers of adult parental shrimp. In practice, however, it is very difficult to estimate how many offspring each individual gives rise to that survive until the postlarval stage. Likewise it is difficult to assess shrimp population densities at sea.

Shrimp farming is particularly resource-intensive in terms of the large human-made investments and operation costs. These are proportionally greater than for salmon farming in cages since stocking densities are much higher in fish cages and the volume of water that needs aeration and management is proportionally considerably smaller. Thus every ton of shrimp harvested requires approximately 1.5 times as much industrial energy to rear as an equivalent amount of cage-cultured salmon [129–205 GJ/t compared with 97–107 GJ/t for salmon (Folke 1988), labor excluded]. It is interesting to note, for instance, that for each kilogram (wet weight) of shrimp produced about 1.5 liter of diesel fuel is required directly, mainly to power the pumping of freshwater into the cultivation ponds. The corresponding direct energy figure for farmed salmon is between 0.14 and 0.18 liters/kg (Folke 1988).

Conversely, if the total energy input needed to yield a given amount of protein is considered (Table 2), the picture is reversed. This is due to the higher protein content of shrimp than salmon. Thus shrimp requires 80% of the industrial energy needed to produce an equivalent amount of salmon (40 J and 50 J, respectively). This is still enough to rank shrimp farming as one of the most industrial energy-intensive means to produce protein, as is apparent when compared with other food production systems (Table 3).

By contrast, a much smaller amount of natural solar energy is required to raise shrimp than salmon. To produce 1 J of edible shrimp protein requires a GPP of 295 J, whereas 1 J of farmed salmon requires a solar energy subsidy as large as 1204 J (Folke 1988). This indicates that semiintensive shrimp farming, by its nature a throughput system where a great deal of feed and fertilizer is lost rather than recycled (Larsson 1992, Reymond and Lagardere 1990), still may be more efficient on a protein production basis than salmon farming in terms of natural and industrial resource demand. The reasons for this difference are discussed below.

The sea surface area supporting the fish meal production that is required to produce feed pellets is one order of magnitude greater than the area of the pond itself. It is, nevertheless, rather small if compared to equivalent values for salmon cage-farming. For this culture type, studies have shown that a sea area 40,000–50,000 times as large as the surface area of the cage farm is required to produce enough fish for feed pellets (Folke 1988, Folke and Kautsky 1989). Such a comparison is misleading, however, since cages are stocked at very much higher densities than shrimp ponds: a 0.1-ha cage farm may yield 40 t of fish per

Table 2. Natural and industrial energy inputs required to produce 1 J of cultured shrimp and cage-farmed salmon, average values (J/J)[a]

	Shrimp	Salmon
NPP/J biomass[b]	131	600
GPP/J biomass[c]	172	783
GPP/J edible protein[d]	295	1204
Industrial energy input per unit biomass produced, GJ/t[e]	158	103
Industrial energy input per J edible protein[f,g]	40.3	50.2
Direct fuel energy per J edible protein[f]	13.9	2.9
Natural and industrial energy inputs per J edible protein	338	1264
Natural/industrial energy input ratios[h]	7.5	24

[a]Values for salmon have been calculated from Folke (1988).

[b]Total NPP energy required/total harvested energy (from Figure 3).

[c]Total GPP energy required/total harvested energy (from Figure 3).

[d]Edible parts of shrimp 68.9% by weight (Martinez Silva and others 1987), protein 84.6% of energy content (Altman and Dittmer 1968).

[e]From Table 1 (energy input, labor excluded) and Figure 3 (yield).

[f]From Table 1.

[g]Excluding labor.

[h]GPP/J edible protein divided by industrial energy input (excluding labor) per J edible protein.

Table 3. Estimates of industrial energy inputs (excluding labor) per protein output for various food production systems[a]

Food type	Industrial energy input/ protein energy output, J/J
Gracilaria spp. culture, West Indies	1
Vegetable crops	2–4
Sea-ranching of Atlantic salmon using delayed release	7
Mussel culture	10
Sheep farming	10
Rangeland beef farming	10
Conventional sea-ranching of Atlantic salmon	12
Cod fisheries	20
Broiler farming	22
Cage-farming of rainbow trout	24
Atlantic salmon fisheries	29
Atlantic salmon fisheries based on delayed-release hatchery-reared smolts	33
Semiintensive shrimp farming	**40**
Cage-farming of Atlantic salmon	50
Atlantic salmon fisheries based on hatchery-reared smolts	52
Lobster fisheries	192
Shrimp fisheries	3–198

[a]**Bold type**—result from this study. Adapted from Folke and Kautsky (1992).

year, whereas the same shrimp yield requires a surface area of 10 ha. A more appropriate way to compare would be to calculate the sea surface area required to produce a given amount of product.

Yet even then shrimp farming emerges as considerably less dependent upon marine support ecosystems than salmon cage-farming. Each kilogram of edible shrimp protein sequesters the primary production of about 205 m² of sea, which is less than one sixth of the sea surface area required to produce an equal amount of salmon protein [1310–1385 m²/kg (calculated from Folke 1988)]. This corresponds with the higher primary production needed to sustain a salmon cage-farm than a semiintensive shrimp farm. In part, this is explained by the fact that shrimp derive only about two thirds (see also Figure 3) of their food requirements from marine feed, so that almost twice as much fish (as feed pellets) is fed to salmon for the same harvested yield.

However, this still leaves an appreciable difference in support area. This is partially a reflection of the higher productivity of the upwelling areas where fish for shrimp feed are caught, compared with the corresponding marine ecosystems on which salmon feed manufacture in Scandinavia depends (the North and

Barent seas): 225 g C/m² and 90–160 g C/m², respectively (Colinvaux 1986, Steele 1974, Elmgren 1984). The difference also results from the greater fish yield relative to primary productivity in the Pacific upwelling ecosystem compared with the North and Barent seas. However, since overfishing has occurred repeatedly in the South East Pacific upwelling system and the figure represents actual catches rather than a sustainable yield, it is prudent to assume that the marine support system in reality is larger.

The remaining support areas are minor in size, although of special importance since they must be located close to the farms (except the agricultural support system and the carbon sequestering system). It is notable that the calculated mangrove support area for detritus production (4 ha/ha) closely matches the area of existing mangroves draining into lagoons from which the Bay of Barbacoas shrimp farms draw their water (approximately 3.5 ha/ha). This suggests that further construction of ponds using the same water resource could cause a decline in productivity in the existing ones and that new farms could be better located elsewhere. Since transportation costs are a major expense, there is a great incentive to locate new farms as close to roads or existing farms as possible, because the necessary facilities, such as boats and jetties, already exist. This could lead to crowding and degradation of the necessary lagoon and mangrove support system and could also facilitate disease transfer from one farm to another, with potentially very serious results. The collapse of shrimp farming in Taiwan and the abandoning of shrimp farms in Thailand, Indonesia, and Japan, among other countries, after only a few years of operation was largely the result of such overcrowding and degradation of the local environment through acidification of soils and shortage of clean water (Primavera 1993).

In other words, the local environment in the Bay of Barbacoas area already appears to be utilized to its fullest, which is not immediately apparent from an inspection of the area. This is an example of how an analysis of spatial ecosystem support can yield important information about ecological constraints that may otherwise be overlooked in coastal zone management. Fortunately, judging by the attitudes of the shrimp farmers interviewed, there appears to exist an awareness of the need for healthy surroundings, which may prevent such a development.

It must be stressed that the support functions of the external ecosystems are not only appropriated by the shrimp farms. Other uses of these support systems may overlap and compete with the shrimp farms (e.g., clear-cutting of mangrove nurseries), others may not (e.g., small-scale logging and fishing), and others may even benefit (e.g., conservation of mangroves to stimulate ecotourism). If several activities that use a mutual ecological system are to be compared by a spatial ecosystem support analysis, it must be established to what extent they overlap, and, if so, whether they compete or supplement each other in order to avoid a misleading, mechanistic application of the method.

For example, cutting down mangroves for intensive shrimp farming may impact on the fish populations that utilize mangrove lagoons as breeding and nursery areas. This may affect fisheries, including the fishing industry providing feeds to the shrimp farms. Furthermore, catching shrimp seed may conflict with and lessen the capacity of mangroves to support pelagic shrimp fisheries, since postlarvae of noncultivated species are also caught (but not used) in the nets used for seed collection. The hypothetical forest plantation that would be needed to sequester carbon released by industrial processes and direct fuel use at the shrimp farms would certainly support a great many uses in addition to the intended function. Like-

wise, only part of the productivity of the detritus-producing mangrove area goes to the shrimp farms, and even small-scale logging might not interfere with its support function. If a significant proportion of mangrove litterfall is sequestered by shrimp ponds, the shrimp farm may decrease the productivity of lagoon fisheries. This could be offset by the enrichment of lagoon waters by shrimp pond discharge, supporting a larger fish yield than would be obtained in the absence of pond effluents.

In fact, mangrove ecosystems (as well as other support ecosystems) could benefit from a diversity of uses, which may enhance the ecological services derived from these systems. If mangroves are managed in such a way that noncompeting uses overlap maximally, there is a much greater chance of preserving sufficiently large areas for purposes that exclude other uses, such as wildlife and habitat preserves and, to a lesser extent, shrimp postlarval nurseries. A way to achieve this is to involve local communities and resource users in the management process. This has been a successful approach in other areas, for example, in St. Lucia (Smith and Berkes 1992). Preserving the natural mosaic of mangroves and lagoons would be one approach to maximize use and minimize impact, by retaining natural flow patterns and avoiding overcrowding of shrimp farms. The implication for management would be to locate new farms in existing salt pans spaced as far apart as possible (but without infringing upon biologically important wilderness areas).

Conclusion

Our results indicate clearly that semiintensive shrimp farming is highly dependent on external ecosystems. The existence of extensive healthy external support systems is a prerequisite for shrimp farming that is rarely recognized or accounted for in management. Great quantities of water, feed, and industrial energy are needed for the production of shrimp. In addition, the biological production of large land and sea areas is appropriated more or less exclusively for production of feed. Even larger areas, mostly mangroves, are needed to support the farms in ways that put pressure on and restrict the number of other uses of these areas.

The highly energy-intensive, throughput nature of shrimp farming makes it an exclusive method for food production that can be justified only because of the high price of the product. We would argue that if the environmental and social costs to society of the present farming technology were paid by the shrimp farmer (internalized in the language of economists), it would not be profitable to farm shrimps. This has been shown to be the case for intensive salmon farming (Folke and others 1994).

In terms of sustainability and environmental impact, shrimp farming methods could benefit considerably from novel approaches that lessen the dependence upon fossil fuels, reduce wastes, increase efficiency, and are more in tune with the processes of the surrounding and external ecosystems. Such approaches may include polyculture and biological water control, e.g., by filter feeders (Wang 1990, Rubino 1990, Shpigel and Blaylock 1991), improving artificial rearing methods, using renewable local energy sources (e.g., wind and/or solar-powered pumps), and less energy-intensive feeds. By using more of the ponds' natural productivity and formulating feeds derived from vegetable rather than animal products (Lim and Dominy 1990), dependence on external support areas could be reduced, which in addition would reduce the pressure on already intensively exploited marine resources. Such systems traditionally exist, for example, in Indonesia (Costa-Pierce 1988), and have been successful for long periods (Berkes and others 1994). Polyculture approaches are now being implemented on a small scale (e.g., Chua and Kungvankij 1990, Vesga F. 1990, Primavera 1993).

Clearly the present behavior of the shrimp farming industry is ecologically unsustainable. The Rio declaration on environment and development states that unsustainable patterns of production and consumption should be reduced and eliminated. At present, countries allow the shrimp farming market to base production on throughput monocultures in isolation from their ecological support systems. If shrimp farming is to be sustainable, production systems that are integrated with their environment have to be developed. Economic and social incentives for shrimp farmers to invest in such systems must be created to make a transition possible. The fundamental value of ecological support systems must be explicitly incorporated in such incentives.

Acknowledgments

We thank the three reviewers—John Cairns, Jr., David Gruber, and J. M. Marcus—as well as Cutler Cleveland and Charles Hall for their valuable comments and support. This work is based on a Minor Field Study (MFS) supported by grants from Swedmar (Swedish Centre for Coastal Development and Management of Aquatic Resources) and the Swedish International Development Authority (SIDA). Additional grants were provided by the Swedish Council for Forestry and Agricultural Research and the Swedish Agency for Research Cooperation with Developing Countries (SAREC).

Literature Cited

Aiken, D. 1990. Shrimp farming in Ecuador: Whither the future? *World Aquaculture* 21(4):26–30.

Altman, P. L. and D. S. Dittmer (eds.). 1968. Metabolism. Federation of American Societies for Experimental Biology. Bethesda, Maryland.

Anonymous. 1986a. Fiskeforkatalog (Fish food catalogue). FK-EWOS, Södertälje, Sweden (in Norwegian).

Anonymous. 1986b. Livsmedelstabeller (food tables). Statens livsmedelsverk, Uppsala, Sweden.

Anonymous. 1991a. Colombia. Pages 390–414 *in* G. T. Kurian (ed.), Encyclopedia of the third world, 4th ed. Facts On File, New York.

Anonymous. 1991b. Estadísticas y estimativos—industria camaricultora 1989–1991. Asociación Nacional de Acuicultores de Colombia (Acuanal), 5 pp.

Bailey, C. 1988. The social consequences of tropical shrimp mariculture development. *Ocean and Shoreline Management* 11:31–44.

Bailey, C. 1991. Social and environmental impacts of shrimp aquaculture in Indonesia. Paper presented at the 1991 meeting of the Rural Sociological Society, Columbus, Ohio (mimeograph), 23 pp.

Berkes, F., C. Folke, and M. Gadgil. 1994. Traditional ecological knowledge, biodiversity, resilience and sustainability. *In* C. Perrings, K.-G. Mäler, C. Folke, C. S. Holling, and B.-O. Jansson (eds.), Biodiversity conservation. Kluwer Academic Publishers, Dordrecht.

Chua, T.-E., and P. Kungvankij. 1990. Una evaluación del cultivo del camarón en el Ecuador y estrategía para su desarrollo y diversificación de la maricultura. Pages 21–56 *in* Programa de Manejo de Recursos Costeros, 3er serie estudios, Quito, Ecuador.

Cleveland, C. J. 1992. Energy quality and energy surplus in the extraction of fossil fuels in the U.S. *Ecological Economics* 6:139–162.

CLIRSEN (Centro de levantamientos integrados de recursos naturales por sensores remotos). 1990. Estudio multitemporal de los manglares, camaroneras y áreas salinas de la costa ecuatoriana mediante información de sensores remotos. Pages 57–93 *in* Programa de Manejo de Recursos Costeros, 3er serie estudios, Quito, Ecuador.

Colinvaux, P. 1986. Ecology. John Wiley & Sons, New York, 725 pp.

Costanza, R., and H. E. Daly. 1992. Natural capital and sustainable development. *Conservation Biology* 6:37–46.

Costanza, R., B. G. Norton, and B. D. Haskell. 1992. Ecosystem health: New goals for environmental management. Island Press, Washington, DC.

Costa-Pierce, B. A. 1988. Traditional fisheries and dualism in Indonesia. *Naga* 11(2):3–4.

Daily, G. C., and P. R. Ehrlich. 1992. Population, sustainability and Earth's carrying capacity. *Bioscience* 42:761–771.

de Groot, R. S. 1992. Functions of nature: evaluation of

nature in environmental planning, management and decision making. Wolters-Noordhoff, Amsterdam. 315 pp.

Elmgren, R. 1984. Trophic dynamics in the enclosed brackish Baltic Sea. *Rapports et Proces-verbeaux des Réunions, Conseil International pour l'Exploration de la Mer* 183:152–169.

Eriksson, S., S. Thomke, and S. Sanne. 1972. Fodermedlen. LT Publishers, Borås, Sweden.

Folke, C. 1988. Energy economy of salmon aquaculture in the Baltic Sea. *Environmental Management* 12:525–537.

Folke, C. 1991. Socio-economic dependence on the life-supporting environment. Pages 77–94 *in* C. Folke and T. Kåberger (eds.), Linking the natural environment and the economy: Essays from the eco-eco group. Kluwer Academic Publishers, Dordrecht.

Folke, C., and G. Aneer. 1988. Estimations of solar and fossil energy flows in Atlantic salmon (*Salmo salar*) aquaculture in the Baltic. Contributions from the Askö Laboratory No. 34, Askö Laboratory, University of Stockholm, Sweden.

Folke, C., and N. Kautsky. 1989. The role of ecosystems for s sustainable development of aquaculture. *Ambio* 18:234–243.

Folke, C., and N. Kautsky. 1992. Aquaculture with its environment: prospects for sustainability. *Ocean and Coastal Management* 17:5–24.

Folke, C., M. Hammer, and A.-M. Jansson. 1991. Life-support values of ecosystems: A case study of the Baltic Sea Region. *Ecological Economics* 3:123–137.

Folke, C., N. Kautsky, and M. Troell. 1994. The costs of eutrophication from salmon farming. *Journal of Environmental Management* (in press).

Hall, C. A. S., E. Kaufman, S. Walker, and D. Yen. 1979. Efficiency of energy delivery systems: II. Estimating energy costs of capital equipment. *Environmental Management* 3:505–510.

Hall, C. A. S., C. J. Cleveland, and R. Kaufmann. 1986. Energy and resource quality: The ecology of the economic process. John Wiley & Sons, New York.

Hamilton, L. S., and S. C. Snedaker. 1984. Handbook for mangrove area management. Environment and Policy Institute, East-West Center, International Union for Conservation of Nature (IUCN) and UNESCO, with additional support from UNEP, 123 pp.

Hammer, M. 1991. Marine ecosystem support to fisheries and fish trade. Pages 189–209 *in* C. Folke and T. Kåberger (eds.), Linking the natural environment and the economy: Essays from the eco-eco group. Kluwer Academic Publishers, Dordrecht.

Hellsten, G. 1992. Tabeller och diagram, energi- och kemiteknik. Almqvist och Wiksell, Uppsala, Sweden, 128 pp. (in Swedish).

Holling, C. A. 1986. The resilience of global ecosystems: local surprise and global change. Pages 292–320 *in* W. C. Clark and R. E. Munn (eds.), Sustainable development of the biosphere. International Institute for Applied Systems Analysis (IIASA)/Cambridge University Press, Cambridge.

Kapetsky, J. M. 1985. Mangroves, fisheries and aquaculture. Pages 17–36 *in* FAO Fisheries Report 338, Supplement. FAO, Rome.

Kelly, J. R., and M. A. Harwell. 1990. Indicators of ecosystem recovery. *Environmental Management* 14:527–545.

Krause, F., W. Bach, and J. Koomey (eds.). 1990. Energy policy in the greenhouse: From warming fate to warming limit. Earthscan Publications, London.

Kurmaly, K., A. B. Yule, and D. A. Jones. 1991. An energy budget for the larvae of *Penaeus monodon* Fabricius. *Aquaculture* 81:13–25.

Lahmann, E. J., S. C. Snedaker, and M. S. Brown. 1987. Structural comparisons of mangrove forests near shrimp ponds in southern Ecuador. *Interciencia* 12(5):240–243.

Larsson, J. 1992. An ecosystem analysis of shrimp farming and mangroves in the Bay of Barbacoas, Colombia. A minor field study. Fisheries Development Series 68. Swedmar (Swedish Centre for Coastal Development and Management of Aquatic Resources), Gothenburg, Sweden, 67 pp.

Lim, C., and W. Dominy. 1990. Evaluation of soybean meal as a replacement for marine animal protein in diets for shrimp (*Penaeus vannamei*). *Aquaculture* 87:53–63.

Martinez Silva, L. E., D. Osorio Dualiby, and M. J. Torres V. 1987. Estudio comparativo del comportamiento y desarrollo en el cultivo de camarones marinos del Pacífico y del Caribe colombiano, con enfasis en *Penaeus stylirostris* Simpson. INDERENA, Cartagena, Colombia.

Meltzoff, S. K., and E. Lipuma. 1986. The social and political economy of coastal zone management: shrimp mariculture in Ecuador. *Coastal Zone Management Journal* 14:349–380.

Naamin, N. 1991. The ecological and economic roles of Segara Anakan, Indonesia, as a nursery ground for shrimp. Pages 119–130 *in* L. M. Chou, T.-E. Chua, H. W. Koo, P. E. Lim, J. N. Paw, G. T. Silvestre, M. J. Valencia, A. T. White, and P. K. Wong (eds.), Towards an integrated management of tropical coastal resources. ICLARM conference proceedings 22, International Center for Living Aquatic Resources Management, Manila, Philippines.

Odum, E. P. 1975. Ecology: The link between the natural and social sciences, 2nd ed. Holt Saunders, New York, 244 pp.

Odum, E. P. 1983. Basic ecology. Holt-Saunders, New York, 613 pp.

Odum, E. P. 1989. Ecology and our endangered life-support systems. Sinauer Associates, Sunderland, Massachusetts.

Odum, H. T., and J. E. Arding. 1991. Emergy analysis of shrimp mariculture in Ecuador. Working paper, Coastal Resources Center, University of Rhode Island, Narragansett, Rhode Island, 113 pp.

Olsen, S. 1990. After the shrimp mariculture boom. CAMP Network 12, May 1990.

Panvisavas, S., P. Agamanon, T. Arthorn-Thurasook and K. Khatikarn. 1991. Mangrove deforestation and uses in Ban Don Bay, Thailand. Pages 223–230 *in* L. M. Chou, T.-E. Chua, H. W. Koo, P. E. Lim, J. N. Paw, G. T. Silvestre, M. J. Valencia, A. T. White, and P. K. Wong (eds.), Towards an integrated management of tropical coastal resources. ICLARM conference proceedings 22, International Center for Living Aquatic Resources Management, Manila, Philippines.

Pauly, D., and J. Ingles. 1986. The relationship between shrimp yields and intertidal vegetation (mangrove) areas: a reassessment. Pages 277–283 *in* IOC Workshop Report No. 44, Supplement. IOC, Paris.

Paw, J. N., and T.-E. Chua. 1991. An assessment of the ecological and economic impact of mangrove conversion in Southeast Asia. Pages 201–212 *in* L. M. Chou, T.-E. Chua, H. W. Koo, P. E. Lim, J. N. Paw, G. T. Silvestre, M. J. Valencia, A. T. White, and P. K. Wong (eds.), Towards an integrated management of tropical coastal resources. ICLARM, conference proceedings 22, International Center for Living Aquatic Resources Management, Manila, Philippines.

Pedini, M. 1981. Penaeid shrimp culture in tropical developing countries. FAO Fisheries Circular 732. FAO, Rome.

Phillips, M. J., C. Kwei Lin, and M. C. M. Beveridge. 1990. Shrimp culture and the environment—lessons from the world's most rapidly expanding warm water aquaculture sector. Paper presented at the Bellagio Conference on Environment and third world aquaculture development, 17–22 September 1990 (mimeograph), 51 pp.

Pimentel, D., and M. Pimentel. 1979. Food, energy, and society. John Wiley, New York, 165 pp.

Pitcher, T. J. 1977. An energy budget for a rainbow trout farm. *Environmental Conservation* 4:59–65.

Primavera, J. H. 1991. Intensive prawn farming in the Philippines: ecological, social and economic implications. *Ambio* 20:28–33.

Primavera, J. H. 1993. A critical review of shrimp pond culture in the Philippines. *Reviews in Fisheries Science* 1:151–201.

Rees, W. E., and M. Wackernagel. 1994. Appropriated carrying capacity: Measuring the natural capital requirements of the human economy. Pages 362–390 *in* A. M. Jansson, M. Hammer, C. Folke, and R. Costanza (eds.), Investing in natural capital: The ecological economic approach to sustainability. Island Press, Washington, DC.

Reymond, H., and J. P. Lagardere, 1990. Feeding rhythms and food of *Penaeus japonicus* Bate (Crustacea, Penaeidae) in salt shrimp ponds: Role of halophilic entomofauna. *Aquaculture* 84:125–143.

Robertson, A. I., and N. C. Duke. 1987. Mangroves as nursery sites: Comparisons of the abundance and species composition of fish and crustaceans in mangroves and other nearshore habitats in tropical Australia. *Marine Biology* 96:193–205.

Rubino, M. C. 1990. Sustainable development of shrimp aquaculture. Paper presented at the International Society for Ecological Economics conference: The ecological economics of sustainability. The World Bank, Washington, DC, 21–23 May 1990 (mimeograph), 10 pp.

Saenger, P., E. J. Hegel, and J. D. S. Davie. 1983. Global status of mangrove ecosystems. *IUCN/The Environmentalist* 3(Suppl. 3):88 pp.

Shpigel, M., and R. A. Blaylock. 1991. The Pacific oyster

Crassostrea gigas as a biological filter for a marine fish aquaculture pond. *Aquaculture* 92:187–197.

Smith, A. H., and F. Berkes. 1992. Community-based use of mangrove resources in St. Lucia. *International Journal of Environmental Studies* 43:123–131.

Snedaker, S. C., J. C. Dickinson, III, M. S. Brown, and E. J. Lahmann. 1986. Shrimp pond siting and management alternatives in mangrove ecosystems in Ecuador. (Ubicación de piscinas camaroneras y alternativas de manejo en ecosistemas de manglares en el Ecuador, Proyecto de Manejo de Recursos Costeros, Spanish translation 1988). PMRC, Quito, Ecuador.

Southgate, D. 1992. Shrimp mariculture development in Ecuador: Some resource policy issues. Working Paper No. 5. Environmental and natural resources policy and training project/Midwest universities consortium for international activities (EPAT/MUCIA). University of Wisconsin, Madison.

Steele, J. H. 1974. The structure of marine ecosystems. Blackwell, Oxford.

Turner, R. E. 1977. Intertidal vegetation and commercial yields of penaeid shrimp. *Transactions of the American Fisheries Society* 106:411–416.

Turner, R. E. 1986. Relationships between coastal wetlands, climate and penaeid shrimp yields. Pages 267–276 *in* IOC Workshop Report No. 44, Supplement. IOC, Paris.

Uhlin, H. E., E. Johansson, I. Lindström, P.-O. Nilsson, D. Myhrman, G. Möller, H. Petre, U. Renborg, H. Wiktorsson, and U. Wunsche. 1975. Resource flows in Swedish agriculture and forestry with emphasis on energy flows. Report from the Department of Economics and Statistics No. 64 and 65. Swedish university of Agricultural Sciences, Uppsala, Sweden, 127 pp. (in Swedish).

Vance, D. J., M. D. E. Haywood, and D. J. Staples. 1990. Use of a mangrove estuary as a nursery area by postlarval and juvenile banana prawns *Penaeus merguiensis* de Man, in northern Australia. *Estuarine, Coastal and Shelf Science* 31:689–701.

Vesga, F., R. 1990. Casos de éxito de desarrollo exportador en Colombia: Las exportaciones de la camaricultura. Asociatión Nacional de Acuicultores de Colombia (Acuanal); 121 pp.

Vitousek, P. M., P. R. Ehrlich, A. H. Ehrlich, and P. A. Matson. 1986. Human appropriation of the products of photosynthesis. *Bioscience* 36:368–373.

van Prahl, H. 1980. Importancia del manglar en la biologia de los camarones peneidos. Pages 341–343 *in* Memorias del seminario sobre el estudio científico e impacto humano en el ecosistema de manglares. UNESCO, Montevideo, Uruguay.

Wang, J.-K. 1990. Managing shrimp pond water to reduce discharge problems. *Aquacultural Engineering* 9:61–73.

Wright, D. H. 1990. Human impacts on energy flow through natural ecosystems, and implications for species endangerment. *Ambio* 19:189–194.

Zucchetto, J., and A. M. Jansson. 1985. Resources and society: A systems ecology study of the island of Gotland, Sweden. Ecological Studies 56, Springer-Verlag, Heidelberg, 246 pp.

Part III
Integrated Coastal Management

[15]

PERGAMON

Natural Resources Forum 23 (1999) 275–286

Natural Resources
FORUM

www.elsevier.com/locate/natresfor

Sustainable coastal resources management: principles and practice

R.K. Turner, W.N. Adger, S. Crooks, I. Lorenzoni, L. Ledoux

CSERGE, School of Environmental Sciences, University of East Anglia, Norwich, UK
and University College London, London, UK. E-mail: r.k.turner@uea.ac.uk

Abstract

Coastal zones are currently experiencing intense and sustained environmental pressures from a range of driving forces. Responsible agencies around the globe are seeking ways of better managing the causes and consequences of the environmental change process in coastal areas. This article discusses the basic principles underpinning a more integrated approach to coastal management, as well as the obstacles to its implementation in both developed and developing countries. The fulfilment of the goal of sustainable utilisation of coastal resources via integrated management is likely to prove to be difficult. Any successful strategy will have to encompass all the elements of management from planning and design through financing and implementation. An interdisciplinary analytical and operational approach is also required, combined with a more flexible and participatory institutional structure and emphasis to account for multiple stakeholders and resource demands. As historical and institutional perspectives as well as socio-economic and cultural contexts are also important, two case studies (based on UK and Vietnamese experiences) are presented in order to identify arguments and examine these aspects in more detail. © 1999 United Nations. Published by Elsevier Science Ltd. All rights reserved.

Keywords: Coastal management; Pressure-state-impact-response; Institutional change; United Kingdom; Vietnam

1. Introduction

With the relentless and cumulative process of global environmental change, driven by, among other things, population growth, urbanisation, industrial development, trade and capital flows, liberalisation of transnational corporation activity and changes in lifestyle and attitude, the world's coastal zones have come under increasingly severe pressure. The consequences of this process manifest themselves across a range of spatial and temporal scales, and now pose a significant threat to the environmental and socio-economic systems located in coastal areas. These zones contain a wide diversity of assets—human populations, physical as well as natural biological capital—with the capacity, when sustainably managed, to provide extensive opportunities for wealth creation and the maintenance and enhancement of the quality of life. The loss of biodiversity may impose negative impacts on the functioning and adaptability of ecological systems themselves and the provision of their goods and services. It also seems to be the case that the current rate of biodiversity loss is unprecedented. All coastal areas are facing an increasing range of stresses and

shocks, the scale of which now poses a threat to the resilience of both human and environmental coastal systems. The coastal resource systems are being stressed by growing and multiple usage demands, many of which are competing (Turner et al., 1996). A more integrated approach to coastal management is urgently required.

The present article addresses this issue through an analysis of both the principles and practice of coastal management. The article first examines the basic principles underpinning the process of integrated coastal management, as well as the obstacles to its implementation, in both developed and developing countries. Two case studies are then introduced (UK and Vietnam) to illustrate key arguments in more detail and to emphasise the importance of the historical, socio-economic and cultural contexts necessary for sustainable coastal resources management to operate.

2. Pressures, impacts and responses in the coastal zone

In order to scope the myriad of issues, problems and arguments surrounding the scientific analysis, valuation

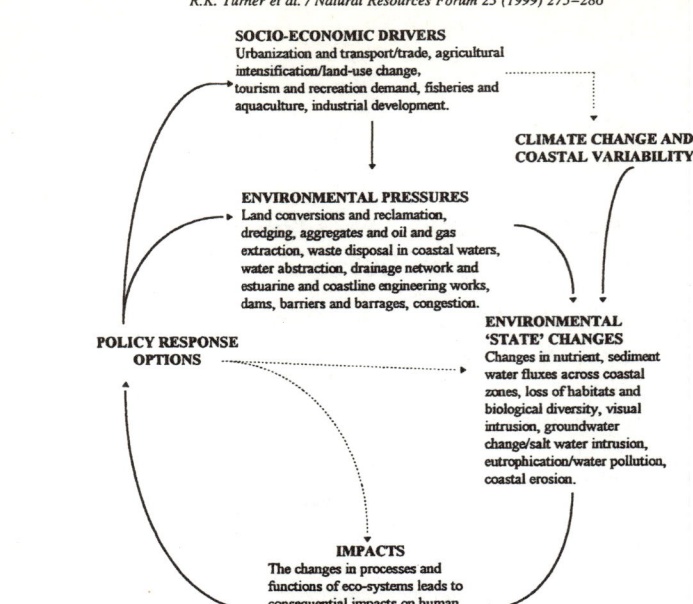

Fig. 1. Pressure-state-impact-response (PSIR) framework: continuous feedback process in coastal areas. The figure shows a simplified organisational and auditing framework illustrating the PSIR approach for a coastal zone and linked drainage basin. Source: Turner et al., 1998.

and management of coastal areas, a simplified organisational and auditing framework is adopted in this article: the pressure-state-impact-response (PSIR) approach (see Fig. 1) for a coastal zone and linked drainage basin (see Turner et al., 1998). Although simple, it is flexible enough to be conceptually valid across a range of spatial scales. It also highlights the dynamic characteristics of ecosystem and socio-economic system changes, involving multiple feedback within a co-evolutionary process (Turner et al., 1997). As environmental pressure builds up through various interrelated socio-economic driving forces—demographic, economic, institutional and technological—changes occur in the ecological system 'states'. These changes include increased nutrient, sediment and water fluxes through drainage basins and into the marine environment; land conversion loss, fragmentation and degradation of habitats; pollution of soils, water and atmosphere; and climate alteration.

The processing and functioning capabilities of ecosystems is affected, which in turn results in impacts on human welfare through changes in values such as productivity, health, amenity and other aspects. These impacts impose social welfare gains and losses across a spectrum

of different stakeholders, depending on the spatial, socio-economic, political and cultural setting. Policy response mechanisms will then be triggered within this continuous feedback process.

At the core of this interdisciplinary analytical framework is a conceptual model, based on the concept of functional diversity, which links ecosystem processes and functions with outputs of goods and services, which can then be assigned monetary economic and other values (see Fig. 2).

Functional diversity can be defined as the variety of different responses to environmental change, in particular the variety of spatial and temporal scales with which organisms react to each other and to the environment (Steele, 1991). Marine and terrestrial ecosystems differ significantly in their functional responses to environmental change and this will have practical implications for management strategies. Thus, although marine systems may be much more sensitive to changes in their environments, they may also be much more resilient (i.e. more adaptable in terms of their recovery responses to stress and shock). The functional diversity concept encourages analysts to take a wider perspective and examine change in large-scale ecological processes, together with the relevant socio-economic driv-

R.K. Turner et al. / Natural Resources Forum 23 (1999) 275–286

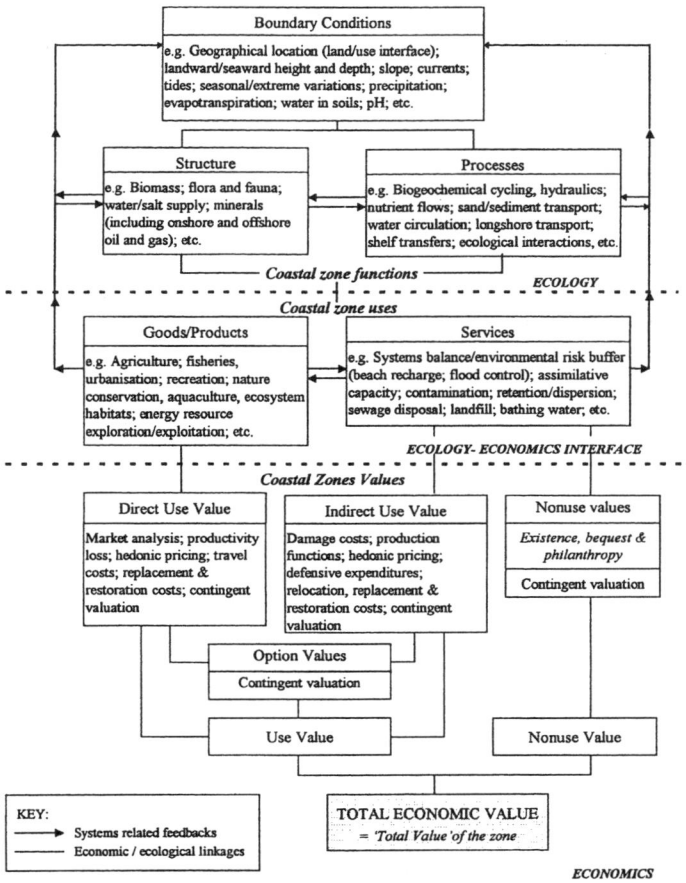

Fig. 2. Coastal zones, environmental functions and associated values. The figure shows a conceptual model, based on functional diversity, which links ecosystem processes and functions with outputs of goods and services. These outputs can be assigned monetary economic and/or other values. Source: Adapted from Turner et al. (1998).

ing forces causing diversity loss. The focus is then on the ability of interdependent ecological–economic systems to maintain functionality under a range of stress and shock conditions.

Economic development is thought, these days, to be constrained by the goal of sustainability. In the context of coastal management, sustainability can be defined as the preservation, through proper use and care, of the coastal environment, 'borrowed from' future generations. But, although an economic use of the environment can be both efficient and sustainable, economic efficiency does not in itself guarantee sustainability (Pearce and Turner, 1990).

The scale of socio-economic activity must be kept within biophysical limits and carrying capacity, a concept related to resilience. To keep the scale of socio-economic activities within the resilience limits of their underlying resource base, locally, regionally and globally, is the main challenge for sustainable development policy. For example: biodiversity loss may be related to a loss of system resilience, but the limits are not static; appropriate behaviour, management systems and institutions can serve to delay or postpone their onset. The principle of sustainable utilisation of resources should be a key component of any future coastal management strategy. The real challenge will be to demon-

278   *R.K. Turner et al. / Natural Resources Forum 23 (1999) 275–286*

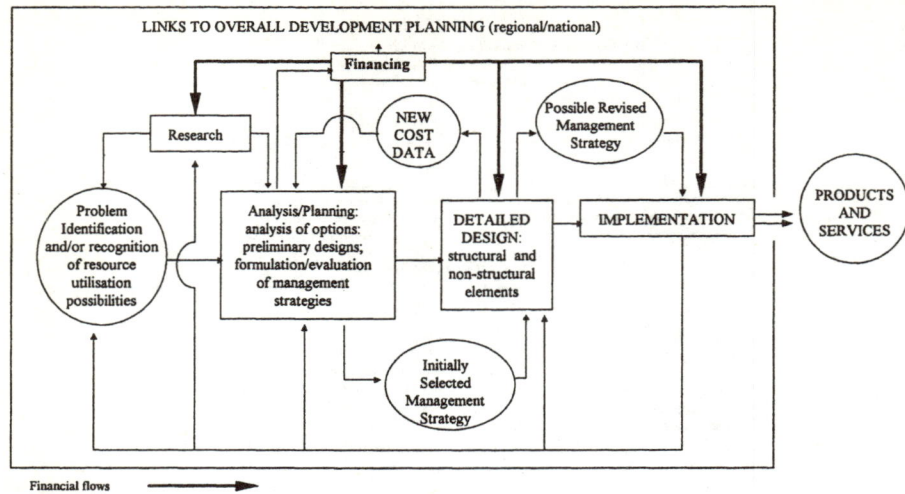

Fig. 3. Simple schematic of the elements of ICZM. Source: Bower and Turner, 1998.

strate in practical ways both the economic value of coastal resources (use and non-use values) and to identify mechanisms by which local people can participate in an equitable sharing of income and assets devolving from coastal zones under development pressure. The rate of social return derived from conservation-motivated activities needs to be compared with the rate of return available from alternative options (the opportunity cost of conservation), allowing for the prior correction of any existing market and institutional failures (Turner et al., 1996).

No management prescriptions can possibly anticipate all the vagaries of the environmental change process. Flexibility and adaptability will be necessary characteristics of any future successful management strategy. Nevertheless, some fundamental uncertainty will remain over climate variability and change, and the effects of change in coastal processes and consequent losses. There is generally considerable uncertainty about the threshold values of either population of organisms or biogeochemical cycles for many of the most important ecosystems. Further, the implications of breaching such a threshold are completely unknown (see Perrings and Pearce, 1994). Some judgement will be required about the socially acceptable margin of safety (the 'precautionary principle') in the exploitation and protection of coastal resources. This is essentially an ethical judgement and there are a number of ways in which such an ethical dimension can be reflected in the decision-making process. It may be possible to discern a hierarchy of economic and ethical considerations, from immediate self-interest and efficient resource use, through possible self-

interest (risk avoidance), to responsibility for and obligation towards future generations. An important task is to clarify where conventional economic values are sufficient for sustainability decisions and where broader human values—including non-monetary values—and criteria are more appropriate (Bingham et al., 1995).

3. Integration in coastal zone management

In principle, integrated coastal management should include:

- integration of programmes and plans for regional economic development, environmental quality management and coastal management;
- integration of coastal management with sectoral plans for fisheries, energy, transport, water resources, waste disposal and natural hazards management;
- integration of responsibilities for coastal management across different levels of government —local, state/provincial, regional, national, international—and between public and private sectors;
- integration of all elements of management, from planning and design, through implementation (see Fig. 3);
- integration among disciplines, e.g. ecology, geomorphology, marine biology, economics, engineering, political science, law; and
- integration of the institutional capacities available, i.e. the management resources of the agencies and entities involved (Bower and Turner, 1998).

R.K. Turner et al. / Natural Resources Forum 23 (1999) 275–286

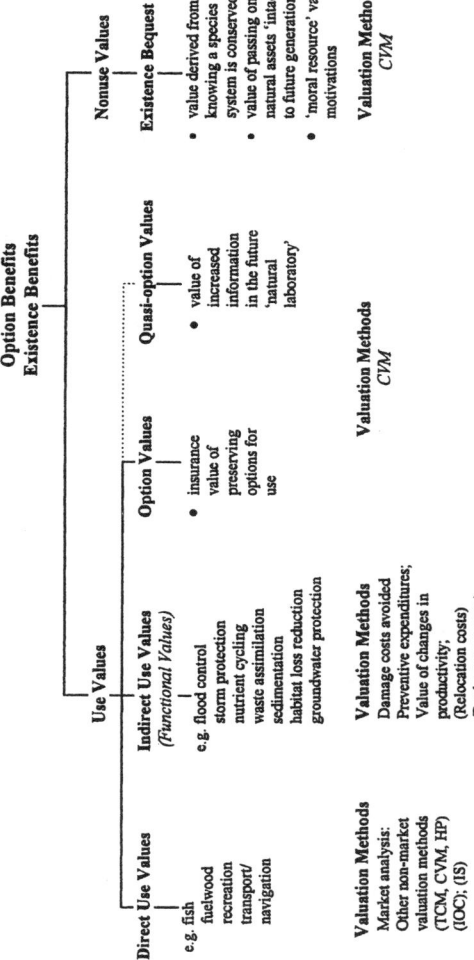

Fig. 4. Methods for valuing coastal zone benefits. Source: Adapted from Turner (1988) and Barbier (1989).

Notes: **Market Analysis**: based on market prices; **HP** = hedonic pricing, based on land/property value data; **CVM** = contingent valuation method based on social surveys designed to elicit willingness to pay values; **TCM** = travel cost method, based on recreationalist expenditure data; **IOC** = indirect opportunity cost approach, based on options foregone; **IS** = indirect substitute approach.
The benefits categories illustrated do not include the 'indirect' or 'secondary benefits' provided by the coastal zone to the regional economy, i.e. the regional income multiplier effects.

280 *R.K. Turner et al. / Natural Resources Forum 23 (1999) 275–286*

Coastal management will need to encompass multiple foci—the politically designated management area, ecological areas (covering several ecosystems or catchment areas), and demand areas (sources of resource and services demand). Given multiple problems and virtually always limited resources with which to 'tackle' them, a major task for integrated coastal management is to establish priorities. A difficult sustainability balance will need to be struck depending on the real and full economic value of the various resource management options, the extent to which sustainable human livelihoods can be fostered with alternative income sources substituting for unsustainable current usage, and the actual resilience of various natural systems and processes.

Coastal management is not therefore a unified concept that varies only according to its degree of integration. The scope of integration (top–down versus bottom–up) can also vary. The collective approach and maximising stakeholder co-operation are fundamental requirements of more sustainable future strategies for coastal resources management. To sum up, integration is required across broad policy objectives and plans, with different sectoral plans and management, with different levels of government and with the public and private sectors. Integrated coastal management itself must encompass all the elements of management from planning and design through financing and implementation. It also requires an interdisciplinary analytical and operational approach. There is a need to combine some centralised institutional capacity (to ensure co-ordination and cost effectiveness) with a more process-oriented and participatory network which extends down to the local level (to address equity and other stakeholder concerns).

4. Barriers to integrated coastal management

Obstacles to the fulfilment of the goal of sustainable coastal resources development include a lack of political will to invest in the forms of governance necessary to deal with the complex relationships found in coastal zones. Some existing institutions also serve to inhibit adaptive responses to ecosystem changes, and conjointly create a gridlock situation as well as confusion in environmental management (Pritchard et al., 1998). The slow rate of progress of integration measures reflects the strength of the vested interests that may exist, but may also be due to a failure to demonstrate the net social benefits of a more integrated strategy for coastal planning and management and the incomplete scientific understanding of the functioning of coastal processes and ecosystems (see Fig. 4). The benefits of integrated management can be demonstrated by comparing a coastal resource management strategy 'without integration' versus one 'with integration'. The net benefits are then represented by the difference between the two 'states of the world' in given coastal areas. The integrated assessment framework must include coupled or integrated models (biogeochemical

and socio-economic), but is not limited to just this. It is also a participatory process of combining, interpreting and communicating knowledge from diverse scientific disciplines to achieve a better understanding of complex phenomena (Bower and Turner, 1998).

It is important to take a historical perspective on the interaction of socio-economic-natural systems, in order to advance the analysis and debate on coastal management. As coastal zone contexts feature irreversible effects, surprise outcomes and unpredictable changes, the appropriate policy response should be a flexible one. Policy should be conditioned by the precautionary principle and notions such as safe minimum standards, with due regard to the cost effectiveness of options and choices, taking social opportunity costs into account. Coastal management to date, by contrast, has more often than not been dominated by a more closed attitude which has sought to buffer socio-economic activities and assets from natural hazards and risks via hard engineering protection. In the face of an increasing degree of environmental risk, uncertainty and ignorance, the distributional incidence of the risk, costs and benefits becomes a key issue. Stakeholder consultation and consequent acceptance of the inevitable trade-offs involved will be an important requirement in any legitimisation process.

In future, a more active and conciliatory approach to consultation across and among stakeholders in any given resource allocation and environmental decision-making situation will be required. It can be argued that there are strong social, political and economic arguments for widening the consultative arrangements and ensuring a more face to face participatory role for representative interests. O'Riordan and Ward (1997), for example, have explored the theory of legitimisation and legitimacy in participatory negotiative processes in the context of shoreline management planning in the UK. They have sought to justify the empowerment of stakeholders through respect, authenticity and trust in the conduct of mediating exercises.

Sustainable use of coastal resources cannot, therefore, be divorced from the economic, political and legal framework within which management takes place. This first part of the article has highlighted integration as a necessary condition for managing coastal resources in a sustainable manner and has also argued that an appropriate and flexible institutional infrastructure is an important component of sustainability in this context. Two case studies of coastal management, in the UK and Vietnam, will now be examined.

5. Institutional change and integration: UK

The coastal lowlands of the UK have been subject to ad hoc embankment and land-use modification since the Romano–British Period some 2000 years ago. The present coastal zone configuration reflects this unregulated development with, in England alone, over 860 km of soft cliffs, susceptible to erosion and requiring protection (23% of the coastline), and in excess of 1259 km of sea-defences protecting 2347 km[2] of embanked lowlands from flooding (Barne et al., 1996). On these coastal flood plains, over 5% of the population live (more than 2 million people) and 50% of the highest grade agricultural land is found. The remaining coastal natural resources in the UK are suffering from a continuous net decline, in part related to coastal squeeze[1] of intertidal habitats.

The need for a more integrated approach to coastal zone management (CZM) was highlighted by a parliamentary committee, convened in 1991, to address the growing environmental and development pressures (HOC, 1992). Although the study is limited to England and neighbouring estuaries, the recommendations of the committee are applicable throughout the UK.

The committee recommended that the coastal zone be treated as one integrated unit, to be subject to Europe-wide policy initiatives and legislation. A national strategic overview of policy and enabling framework for management was thought to be lacking, and an obvious first step in any integration process should be the combining of the currently separate responsibilities for coastal protection (largely erosion countermeasures) and sea defence (flooding and inundation countermeasures). More funding should be provided for regional bodies and groups, together with better databases and monitoring of coastal change. The concept of spatially defined areas, coastal 'cells', on which more participatory planning and management could be based was also advocated. A number of measures for strengthening environmental protection were put forward to buttress the provisions of the EU Habitats Directive, the establishment of marine areas of conservation, and the creation of new coastal habitats as part of managed retreat and other soft engineering[2] approaches (HOC, 1992).

Some of these suggestions have already been implemented or are currently being incorporated into local policies. Numerous statutory and non-statutory baseline management plans have been created to identify pressures within the coastal zone; nature conservancy agencies have been provided with funds to catalogue the state of the UK's natural resources; the UK has adopted European legislation, specifically the Habitats Directive, which provides a greater level of protection to nominated habitats within designated areas; operational planning authorities are gaining greater awareness of the need to incorporate environmental and risk considerations in the planning process of coastal areas; and research into habitat restoration by managed retreat is currently in progress. A significant step forward has been the creation of a single regulator to oversee environmental

quality in England and Wales. The remit of the Environment Agency (EA)[3] is now to police environmental quality in coastal areas, but the Agency has no power for direct management of coastal zones or for policy-making beyond an advisory capacity. Any modifications to the existing structures and existing policy implementation, suggested by the EA, have to be undertaken through the institutional channels currently in force.

Nevertheless, integration of CZM is by no means complete and, as yet, the coastal zone is not being treated as one integrated unit. Several barriers to integrated CZM remains to be overcome in the UK. A nationally fragmented institutional structure still hinders a holistic and co-ordinated approach to CZM. The distribution of responsibility for management of the UK coastal zone is partitioned across a tiered system of local, regional and national governmental levels. In England, environmental policy is implemented by the Department of the Environment, Transport and the Regions. The Department of Trade and Industry deals with mineral/energy extraction, waste and pollution control and areas related to navigation (such as shipping, communications), regulation of oil and gas exploration and exploitation, mainly offshore (HOC, 1992; DTI, 1998). The Home Office is responsible for bylaws in England and Wales. Likewise, flood and coastal defence and fisheries are the responsibilities of the Ministry of Agriculture, Fisheries and Food (MAFF). At sub-regional level local authorities are empowered to undertake flood defence works (funded by local taxes and grant aid support by MAFF) and are responsible for the zoning of land-use. All these exemplify the fractionation of responsibilities and are a good indication of the plethora of existing management reports for coastal zones produced by the many separate departments and organisations involved.

The range of management plans being produced have been valuable in highlighting conflicts of interest and the state of the coastal zone. The county of Northumberland, for example, currently possesses 21 different management plans encompassing various geographical scales and sectoral interests for its coastline. However, local managers and decision-makers are becoming fatigued by the process of producing plans which have yet to identify an integrated and holistic strategy for managing their coast. Managers in many areas are now requesting guidance on linking these plans together (including many non-statutory plans, such as Biodiversity Action Plans), rather than asking for further new plans.

In terms of legislative interdependence between the UK and the European Union, the implementation of the Habitats Directive illustrates the difficulties encountered in balancing maintenance/enhancement of biodiversity and flood defence considerations. To date, 272 Special Areas of Conservation

---

[1] Coastal squeeze is the loss of intertidal resources through narrowing of the intertidal zone between the low water mark and sea-embankments as sea level rises.

[2] Soft engineering refers to the integration of natural coastal systems such as a saltmarsh, dune system or beach to provide a more resilient form of sea-defence rather than falling back on less resilient hard, engineering structures.

---

[3] Created in April 1996 from the amalgamation of the National Rivers Authority, Her Majesty's Pollution Inspectorate and Waste Regulation Authorities.

282   *R.K. Turner et al. / Natural Resources Forum 23 (1999) 275–286*

Table 1
MAFF Priorities for Grant Aid Scheme (PGAS)

Elements of the scoring system	Description	Score
Priority	Flood warning	10
	Urban coastal/tidal defences	8
	Urban flood defence; environmental assets of international importance	6
	Rural coastal/tidal defences; existing rural flood defences and drainage works; environmental assets of national significance	4
	New rural flood defence works; environmental assets of local significance	2
Urgency (based on expected residual life)[a]	Failure...	
	Already occurred	10
	Expected within 2 years	8
	Expected within 5 years	6
	Not expected within 5 years	0
Alternatively, when information on urgency not easy to assess, equivalence can be based on following scheme types:	Flood warning scheme	New: 10
		Replacement: 9
	Research for a shoreline management plan (SMP)	New: 9
		Updates: 8
	Implementation of water level management plans (WLMPs)	8
	Strategy studies leading to capital defence works	7
Economic based on benefit:cost ratios of...		
	Over 5	10
	Between 3 and 5	8
	Between 2 and 3	6
	Between 1.5 and 2	4
	Between 1 and 1.5	2
	Studies (e.g. SMP), where b:c not known	8

[a] Scores for urgency can also be derived from a shortfall in standard of protection, based on a comparison between current and proposed standards of defence.
Source: Adapted from Purnell and Richardson, 1997.

(SACs) have been identified in the UK (English Nature, 1998) and submitted to the European Commission to be considered for inclusion with similarly identified areas in other Member States as part of the future Natura 2000 network. Despite these efforts, however, conservation policy in the UK, whether influenced by legislation at the European level or not, protects natural habitats to some extent from development within designated areas, but does not provide a mechanism to maintain habitat levels and diversity in the face of global changes (e.g. sea-level rise and coastal squeeze). This situation hinders the maintenance of a resilient coastline and the implementation of integrated CZM.

One of the areas of conflict still to be resolved is the merging of governance as enacted through a 'top–down' national policy approach with local sustainable management based on a 'bottom–up' approach involving the grassroots levels of society. The problem of integrating the maintenance of biodiversity and flood defence is a typical example. Until recently, the main funding mechanism to recreate habitat in the coastal zone consisted in government grants supporting schemes to maintain flood defence requirements through the landward realignment of sea-defences. These schemes were updated in Autumn 1997, when MAFF introduced a new pilot scheme (Priorities for

Grant Aid Scheme, PGAS) to structure the allocation of national funding for coastal flood and sea defence works (Table 1).

The 'points system' currently being used is based on a 'top–down' approach: coastal erosion and flood defence schemes are assessed individually according to three criteria, namely priority, urgency and economics and then ranked depending on points attained. Those schemes exceeding a given points threshold (the level of which is set by available funding that year: for the year 1998, it was 23 points) are provided with grant aid support. Point allotment reflects: principal land use of the area; how quickly work should be undertaken; any impacts resulting from this work; the residual life of sea defences or the decrease of protection provided by these structures; and the benefits related to any costs of the scheme (cost benefit analysis, expressed as a ratio).

From the MAFF perspective, operating with limited funding, the scheme is designed to channel funds to meet most critical needs: protecting human lives and the higher economic value areas, such as urban districts as against rural areas. Implementation of this scheme will cause the level of protection to rural areas to steadily fall, placing natural assets behind flood defences under increasing pressure. However, because PGAS considers projects on an individual basis and as funding is geared towards urgency, long-term pressure on overall coastal resilience are not taken into account. Although there is scope within the scheme to include intertidal habitat recreation as a benefit, both for environmental and 'soft-engineering' flood defence scheme considerations, large-scale geographical concerns are not taken into account.

At present, the strategic recreation or restoration of intertidal habitat beyond that created as a secondary consideration from flood defence requirements is poorly supported politically, institutionally and financially. Biodiversity Action Plans, produced by conservation organisations, have outlined requirements to be met if UK coastal regions are to maintain natural resource assets. However as these plans are non-statutory, demands for action cannot be legally enforced. The lack of financial means and instruments hampers agencies wishing to restore coastal habitat through coastal realignment. There are cases where agencies having raised funds to acquire land for habitat re-establishment were prevented from purchasing such land by pressure from landowners, who considered coastal realignment in their region as a threat rather than a benefit. Losses of natural resources in the UK coastal zone are therefore, in some measure, a consequence of the lack of commitment to ICZM.

Given these circumstances, the low-lying coastal regions of eastern England are particularly at risk from a combination of natural variability and the lack of an integrated approach to CZM. Much of the coastal natural resources in this region are protected under the Habitats Directive according to which losses through development pressures should be mitigated by compensatory provision of natural habitat. The scale of many of the developments, however, is small, and as such is not deemed to have a "significant effect" on the "integrity of the (natural) site". In the long-term, these losses may accumulate to degrade the site completely (known as 'death by a thousand cuts'). Natural resources are also being lost through the pressures of environmental change, including rising sea-levels. Not until these losses are addressed will coastal resources and their associated functions be maintained for the future.

From the point of view of local sustainability, the short-comings of the present coastline management are exemplified by the case of the villages of Cley and Kelling in north Norfolk, UK. The coastline in this area consists of a shingle barrier ridge, which protects the two villages from flooding as well as an embanked freshwater grassing marsh and reed-beds, with a high environmental status recognised internationally (designated as Ramsar, SPA, candidate SAC as well as being one of Britain's flagship County Nature Reserves). Under conditions of rising sea-levels, the barrier has been slowly migrating landward, and to prevent this, steps have been taken over the past 20 years to bulldoze the barrier. This action has reduced the integrity and resilience of the barrier, which has been breached several times in recent years, leading to saline flooding of the freshwater grazing marsh, with adverse consequences for the wildlife.

A consultation exercise over flood defence and land use issues at Cley and Kelling was undertaken in 1996 by university-based researchers, with support from the EA. A bottom–up approach was facilitated to reach a consensus on a strategy for the defence of the coast between Cley and Kelling (O'Riordan and Ward, 1997). All parties with an interest in any outcome of the decision-making process and the future of the area were involved in the discussions. The result of the consultation process was positive. Local stakeholders, governmental and non-government bodies (such as the EA and the Royal Society for the Protection of Birds) identified a sustainable strategy for the future of the area: a secondary line of defence was to be created some distance back and the shingle barrier was to be allowed to regain a natural form and continue its landward migration; habitat lost was to be recreated elsewhere though at the time no specific geographical location had been identified (O'Riordan and Ward, 1997). Although funding for the project had been earmarked, this issue was still to be resolved when the MAFF new PGAS was introduced; funding for the scheme agreed by stakeholders was postponed, partly due to the status of the area being considered 'rural' and thus not top priority.

To summarise, although CZM in the UK has come a long way in the last few years, it is as yet far from integrated. The institutional structure dealing with coastal issues remains fragmented. Local sustainable management issues are too often ignored and the dominance of 'top–down' policy and approaches based on flood protection schemes, has lead to a lack of commitment to the funding of more integrated schemes, i.e. those involving a combination of managed retreat and soft engineering measures. The net impact of climate change-induced sea-level rise on intertidal and other relevant habitats at the landscape level is also not adequately incorporated into the official strategy which only prioritises 'urban' defences as nodal points.

6. Institutional change, risk and sustainability: Vietnam

Vietnam has had a sophisticated and effective system of coastal management and policy response to environmental change. The following case study provides evidence that the level of institutional infrastructure is not determined simply by the level of income: developing countries with low income do not necessarily have less effective institutions for coastal management. In fact the necessity to substitute risk-minimising collective action for physical infrastructure can provide the incentives for appropriate institutional arrangements for sustainable coastal resource management. These lessons are drawn from a case study of risk management in coastal Vietnam by taking a historical perspective on the evolution of the institutions in that centrally planned economy which, in the past decade, has moved rapidly towards a decentralised and market-oriented development trajectory. The case illustrates the impacts-and-response section of the PSIR framework (see Fig. 1) outlined above, and demonstrates that sustainable coastal resource management requires effective and legitimate institutions which evolve and adapt to changing pressures and impacts.

The management of risk is a critical element of much coastal planning with respect to coastal flooding and longer term environmental change. Both physical infrastructure and the natural ecosystems of coastal Vietnam fulfil an important storm protection function, sheltering agriculture, aquaculture and urban settlements along the country's 3000 km coastline. These coastal areas have high economic value. The global assessment by Costanza et al. (1997) identifies coastal ecosystems as possessing the highest economic value of any major ecosystem, whether terrestrial or marine, and includes a value for coastal protection. An analysis of a part of the economic value of coastal protection for Vietnam has been outlined by Tri et al. (1998). Tri and co-authors show that coastal protection is a significant element of the overall value of coastal zones, but not as important as local benefits derived from maintenance of mangrove ecosystems. Coastal protection must be handled in an integrated manner, as it involves multiple aspects, such as ecosystem management; the development of infrastructure; the perception of risk; and impacts on settlements and economic behaviour.

The recent history of Xuan Thuy District in Nam Dinh Province, northern Vietnam, demonstrates how the restructuring of ownership and control of coastal resources throughout Vietnam's coastal provinces and the reduction of collective action in these areas has resulted in an increasingly hazardous environment (see Adger, 1998, 1999). Locally organised collective action for coastal defence and water management has been undermined by decollectivisation and the decline in importance of agricultural co-operatives. Offsetting these trends, informal collective action manifested by civil society has acted to counter the overall increase in vulnerability to external environmental change.

Features of the recent historical evolution of collective action to mitigate hazards in Nam Dinh Province include the hierarchical operation of local and regional central planning under collectivised agriculture in the communist era; the inertia of this system in the light of both liberalisation and of changing environmental pressures; and concurrent institutional adaptation to cultural and political–economic factors within the district. The local level formal government institutions have, over the past three decades, acted as facilitators for collective action to ameliorate the impacts of climate extremes and hazards. In the most recent 5 years under economic reforms, known as *Doi Moi* reforms, significant retrenchment of government institutions has occurred, which essentially has reduced the importance of collective action and therefore enhanced vulnerability. The major reason for this is the concentration of resources and power in the coastal communes as a result of decentralisation.

Collective vulnerability to extreme climate events is tempered through an elaborate system of social institutions from the formal commune and district level government to an informal moral economy of reciprocal arrangements and networks. The district and commune authorities operate a tax-funded system of activities to mitigate the impacts of storms. These activities include work brigades to repair and maintain the dike system and the mobilisation of province level labour and resources when floods or widespread damage occur.

The district level institutions in northern Vietnam are responsible for protection from coastal storms, primarily through a series of sea and river dikes along the southern and eastern boundaries of the district under their control. However, the threat of coastal hazards and related costs are both differentiated and diverse. Firstly, the communes of Xuan Thuy face different threats. The whole district is impacted by severe coastal storms and typhoons. The extensive coastline of Vietnam experiences a mean rate of landfall typhoons of approximately five per year over the last century, though it is projected that the typhoon regime will change in both intensity, frequency and seasonality with global climatic change (Kelly, 1996). In addition, the central nature of government planning in Vietnam has led to a government decree, enforced for over two decades, that all adults in coastal districts in the country must allocate 10 days of their labour annually to dike protection. This may well not be an efficient resource allocation, considering the range of typhoon risk and threat, as well as the varying erosion and sedimentation circumstances of the coast. Thus, while not all centralised planning can be seen to be contributing to sustainability from a perspective of economic efficiency, simple decentralisation is not the panacea either.

Given these threats and the uncertainty over future impacts of typhoons, the Xuan Thuy District Irrigation Committee establishes the need for dike repairs and maintenance throughout the district and decides on priorities for expenditure. This system embodies the conflict between economic liberalisation and collective security. Communes employ a variety of strategies to utilise the revenue they raise for storm protection. The difference in use depends primarily on geographic location: coastal communes have the sea dikes within their jurisdiction. The communes have rather different profiles of impacts of storms and have different opportunities to use surplus labour for dike protection.

The village council raises the following 'hypothecated' resources within the commune for dike protection with two major features. Firstly, taxes are raised, in an amount equivalent to 40 kg of rice (VND 70,000)[4] or the equivalent of 10 days of labour, based on the number of eligible workers in each household. All male workers between 18 and 45 years old and all females 18–25 years old are liable for this tax. The tax is district wide and supports both temporary work brigades and hired groups of workers drawn from the

---

[4] Rate of exchange for Vietnamese dong: 13,903 VND = US$1 (official rate of exchange used by the United Nations, 1 August 1999).

R.K. Turner et al. / Natural Resources Forum 23 (1999) 275–286 285

district who take on this labouring task as seasonal employment. Secondly, a labour force is constituted for maintenance during the months of March to May after the first rice crop has been transplanted. This workforce is a subset of the total eligible adults and complements the paid work brigades. In reality, not all eligible adults are called in any year, with the workforce determined by the extent of storm damage. In storm years all labour aged people are expected to carry out emergency repairs. In 1986, these lasted for over one month for a large section of the available population.

The tax and the contribution of labour are alternative means of payment for most people in the district. However, in practice only coastal dwellers actually contribute their labour. The 'tax' system represents the latest evolution of the shift to household responsibility as a necessary outcome of the privatisation process and has regressive distributional impacts. Households perceive all charges levied by the commune or the co-operative as the overall 'tax', while in effect some are user fees and some property and land taxes, some are general taxes and some are earmarked for specific purposes. The dike maintenance tax is an earmarked head tax, but is simply perceived as part of the general tax burden by Xuan Thuy householders.

Allocating their own labour to dike repair ensures that the coastal communes limit their input into the system, particularly in years where few repairs are required and avoid paying monetary resources to the district government. In the dike protection season following a year with little storm damage, such as 1995, those communes who do not directly allocate labour (the inland communes) still have to pay full tax rates, while the coastal communes, by contrast, can simply undertake their own repairs.

Further, the 'hypothecated' tax collected by the district government is not spent annually solely on coastal protection: the tax actually collected is at least four times that spent on coastal protection in years such as 1996. Indeed, the rich coastal communes receive a 'double dividend' of paying lower effective tax rate than inland communes, along with receiving disproportionate investment in other infrastructure projects. In effect, a 'core' of powerful coastal communes are being created which capture resources at the district level, to the detriment of 'peripheral' inland communes.

This type of action, along with the lack of consultation with the district irrigation committee, constitutes a form of non-decision-making, which keeps the political playing field tilted in favour of the coastal communes. Furthermore, some commune level officials appeal to the locally held perception that storm impact is a major constraint on economic performance. Hence, they argue that it is legitimate, as a reflection of local knowledge, to maintain tax collection for sea dike maintenance. However, coastal communes officials are equally ready to downplay potential storm impacts and other environmental change when convenient. Thus, when officials wish to promote economic growth to the exclusion of other policy objectives, they trivialise hazard mitigation.

The empirical observations from Xuan Thuy District demonstrate that institutional inertia and the reinforcement of strictly hierarchical structures exacerbate vulnerability in the short run at the district and commune level, while many other adaptations within formal as well as non-formal institutions offset this trend and enhance security. Institutional change can be observed both as material outcomes such as formal institutional action and inaction, and also by perceptions of being able to cope with and respond to the hazardous environment.

The observed institutional changes are almost exclusively stimulated by the processes of economic liberalisation and transition to a market economy. The rapidity of this liberalisation makes it relatively easier to observe the impacts of institutional change; and also puts in sharper relief the offsetting nature of most of the recent changes in terms of coastal zone vulnerability. Thus the reinforcing of commune influence over resource allocation in the present era of the 'rolling back' of state influence in Vietnam, increases collective vulnerability at the district level with respect to sea dike maintenance.

The dike maintenance aspect of the commune's responsibility for hazard mitigation has become increasingly monetised and professionalised. The redirection of resources towards the core of politically influential communes in coastal Xuan Thuy is the principal negative outcome of autarchic decision-making, in terms of increased vulnerability. At a higher level, district and province level institutions side-step the issue of a potentially more risky coastal environment, and provide a minimum of strategic planning. The Vietnamese experience of 'decentralisation' of the economy provides evidence that the political re-casting of some of the organs of administrative power does not necessarily lead to greater local participation or collective empowerment (Slater, 1989).

A key issue is whether presently observed institutional change in Vietnam is contributing effectively to collective security or if it is, on the contrary, increasing the vulnerability of society to climate extremes. Local autonomy from central state policy, and the enhanced role of provinces, districts and communes, have, in the past, resulted in dissipation of infrastructure investments in Vietnam. At the same time, it has not necessarily led to greater local participation in decision-making (see discussion of decentralisation above and Slater, 1989), but the system of local autonomy would appear to have a potential cost.

Local autonomy and privatisation have contributed to the loss of mangrove areas in coastal Xuan Thuy, thus weakening coastal protection and enhancing the damage potential of storms. In addition, decentralisation has created a vacuum at the level of strategic management.

In summary, the economic reforms and institutional changes associated with the dike protection system of Xuan Thuy provides insights into hazard mitigation and institutional adaptation to social and environmental change relevant for sustainable coastal management. The

286
R.K. Turner et al. / Natural Resources Forum 23 (1999) 275–286

communes essentially use the sea dike resource allocation system to maximise their own budgets, often by unaccountable actions through which the collective vulnerability to the impacts of storms may even be increased. However, in general, the enhancement of the institutions of civil society within the communes documented in this section strengthen security and potentially reduce the recovery time after the impact of a major storm. The need for co-ordinated action and the delineation of property rights in coastal resources necessary for sustainable coastal management, are critical in Vietnam, as this case study has shown. But the decentralisation and reduced effectiveness in coastal management at the level of government institutions in Vietnam observed over the past decade seem, in fact, to be increasing coastal vulnerability, and redirecting the integration of coastal management, planning and policy.

7. Conclusions

At the interface between ocean and terrestrial resources, coastal ecosystems undergo stress from competing multi-usage demands, while having to retain their functional diversity and resilience in the face of global environmental change. A more integrated management of coastal zones must therefore take into account the multiple resource demand and variety of stakeholders, as well as natural variability. This article has introduced a PSIR approach to scope this complex issue, together with a conceptual model based on ecosystem functionality. Case studies have illustrated this framework while highlighting the importance of the institutional, cultural and historical contexts. In the UK, the fragmented institutional structure, and the dominance of a top–down policy approach have lead to a lack of commitment to integrated coastal management, and inadequate consideration of potential impacts of environmental change in decision-making. In Vietnam, privatisation and decentralisation have created a vacuum of strategic management and increased the vulnerability to storm impacts. Both studies demonstrate the need for integration at the government level, while including stakeholder consultation in the decision-making process.

This article also highlights the need for further research on evaluation of coastal zone ecosystem functions, and their integration at the landscape level. Legitimisation and increased effectiveness of CZM policies require that more local participation be included in the evaluation procedures and that new, more flexible institutional arrangements be developed.

References

Adger, W.N., 1998. Observing institutional adaptation to global environmental change in coastal Vietnam. Paper presented at 'Crossing Boundaries', Seventh Conference of the International Association for the Study of Common Property, Vancouver, June, pp. 10–14.

Adger, W.N., 1999. Social vulnerability to climate change and extremes in Vietnam. World Development 27, 249–269.

Barne, J.H., Robson, C.F., Kaznowska, S.S., Doody, J.P., Davidson, N.C., Buck, A.L. (Eds.). 1996, Coasts and seas of the United Kingdom. Region 11. The Western Approaches: Falmouth Bay to Kenfig, Joint Nature Conservation Committee (JNCC), Peterborough.

Bingham, G., et al., 1995. Issues in ecosystem valuation: Improving information for decision-making. Ecological Economics 14, 73–90.

Bower, B.T., Turner, R.K., 1998. Characterising and analysing benefits from integrated coastal management (ICM). Ocean and Coastal Management 38, 41–66.

Costanza, R., d'Arge, R., de Groot, R., Farber, S., Grasso, M., Hannon, B., Limburg, K., Naeem, S., O'Neill, R.V., Paruelo, J., Raskin, R.G., Sutton, P., van den Belt, M., 1997. The value of the world's ecosystem services and natural capital. Nature 387, 253–260.

Department of Trade and Industry (DTI), 1998. The Energy Report, Vol. 2: Oil and Gas Resources of the United Kingdom, The Stationery Office, Norwich.

English Nature, 1998. SAC list swells, English Nature Magazine 37(3).

HOC, 1992. Coastal zone protection and planning, Second Report, Vol. 1: Report, together with the Proceedings of the Committee relating to the Report. (Vol. 2: Minutes of evidence and appendices.) Ordered by the House of Commons to be printed 12 March 1992. HMSO, London.

Kelly, P.M., 1996. Tropical cyclones on the coast of Vietnam: Prospects for the future. Climatic Research Unit and Centre for Social and Economic Research on the Global Environment, University of East Anglia, Norwich, UK.

O'Riordan, T., Ward, R., 1997. Building trust in shoreline management: Creating participatory consultation in shoreline management plans. Land Use Policy 14 (4), 257–276.

Pearce, D.W., Turner, R.K., 1990. Economics of Natural Resources and the Environment. Harvester Wheatsheaf, London.

Perrings, C., Pearce, D., 1994. Threshold effects and incentives for the conservation of biodiversity. Environmental and Resource Economics 4, 13–28.

Pritchard, L., Colding, J., Berkes, F., Svedin, U., Folke, C., 1998. The problem of fit between ecosystems and institutions, IHDP Working Paper No 2, Bonn, Germany.

Purnell, R.G., Richardson, B.D., 1997. Session G—Policy forum/workshop. In: Proceedings of 32nd MAFF Conference of River and Coastal Engineers, Keele University, 2–4 July.

Slater, D., 1989. Territorial power and the peripheral state: The issue of decentralisation. Development and Change 20, 501–531.

Steele, J.H., 1991. Marine function diversity. Bioscience 41, 470–474.

Tri, N.H., Adger, W.N., Kelly, P.M., 1998. Natural resource management in mitigating climate impacts: Mangrove restoration in Vietnam. Global Environmental Change 8 (1), 49–61.

Turner, R.K., Lorenzoni, I., Beaumont, N., Bateman, I.J., Langford, I.H., McDonald, A.L., 1998. Coastal management for sustainable development: Analysing environmental and socio-economic changes on the UK coast. Geographical Journal 164 (3), 269–281.

Turner, R.K., Perrings, C., Folke, C., 1997. Ecological economics: paradigm or perspective. In: van den Bergh, J., van der Straaten, J. (Eds.). Economy and Ecosystems in Change: Analytical and Historical Approaches, Edward Elgar, Aldershot, pp. 25–49.

Turner, R.K., Subak, S., Adger, N., 1996. Pressures, trends, and impacts in coastal zones: Interactions between socio-economic and natural systems. Environmental Management 20, 159–173.

[16]

ELSEVIER   Ocean & Coastal Management 38 (1998) 41–66

Ocean &
Coastal
Management

Characterising and analysing benefits from integrated coastal management (ICM)

Blair T. Bower[a], R. Kerry Turner[b,c,*]

[a] *Consultant, Office of Ocean Resources, Conservation and Assessment, NOAA, Silver Spring, MF 20910, USA*
[b] *The Centre for Social and Economic Research on the Global Environment, University of East Anglia, Norwich, Norfolk NR4 7TJ, UK*
[c] *University College London, London WC1E 6BT, UK*

Abstract

This paper outlines a methodology for the assessment of the net social benefits to be gained from the implementation of integrated coastal management (ICM). The methodological approach that is recommended is based on economic efficiency principles, and involves cost-benefit/cost-effectiveness analytical methods combined with scenario analysis. The economic efficiency rule is modified where necessary to account for other objectives by imposing constraints on the net benefit criterion. The basic approach has to be modified to fit the particular conditions, e.g. multiple resource outputs and competing resource demands, usually found in coastal zones. The approach can be illustrated by describing how it would be applied in the analysis of a prototype coastal system. Such a system has characteristics and modelled performance relationships based on real world contexts and data. © 1998 Elsevier Science Ltd. All rights reserved.

1. Introduction

ICM is a continuous, adaptive, day-to-day process which consists of a set of tasks, typically carried out by several or many public and private entities. The tasks together produce a mix of products and services from the available coastal resources. ICM involves continuous interaction between human systems and natural systems, among

*Correspondence address: The Centre for Social and Economic Research on the Global Environment, University of East Anglia, Norwich, Norfolk NR4 7TJ, UK. Tel.: 01603 593176; fax: 01603 250588.

0964-5691/98/$19.00 © 1998 Elsevier Science Ltd. All rights reserved.
PII: S 0 9 6 4 - 5 6 9 1 (9 8) 0 0 0 0 3 - 9

42   *B.T. Bower, R.K. Turner / Ocean & Coastal Management 38 (1998) 41–66*

human systems, and among natural systems, as these systems "coevolve" over time [1]. The management process must therefore be dynamic and adaptive in order to cope with changing circumstances, changing social tastes, increased knowledge of the behaviour of coastal processes and of human behaviour and "value" of coastal ecosystems, as well as, changing technology, changing factor prices and changing governmental policies [2–7]. The coastal system to be managed is shown conceptually in Fig. 1.

Because the resources of a coastal zone can generate a range of different products and services, not all of which are naturally compatible, conflicts are likely and trade-offs are necessary. This situation is exacerbated by the variety of different stakeholders that are usually present in any given coastal zone. Moreover, the coastal zone resource base is now under severe pressure from the sheer "scale" of the resource demands that are being generated. Parts of the coastal zone have become increasingly susceptible (more vulnerable) to stress and shock and consequent environmental/economic damage by climate and other natural geophysical factors, together with population, urbanisation, waste generation and disposal pressures. This is exemplified by projections for Chesapeake Bay (US), indicating that the substantial expenditures over the last 10–15 yr to improve water quality will, in a relatively short time, be negated by uncontrolled growth in population and economic activity [8]. These pressures, the consequent changes in the "state" of the coastal environment and human response options need to be evaluated in a consistent and systematic fashion [9].

One way of analysing coastal resource degradation and losses is to identify a set of interrelated "failures" phenomena which underlie the process. Two main related "failures" can be distinguished, market failure and governmental intervention failure. Both these problem factors are often exacerbated by scientific and social uncertainties.

Table 1 presents a typology of market and intervention failures which have been identified in coastal zones. They range from pollution problems caused by the lack of market prices for the waste assimilation/disposal services provided by coastal waters, through over use of some coastal resources because of open access, to resource damage caused by inappropriate incentive structures and poor/uncoordinated policy measures. A marked feature of the pollution and resource over-exploitation problems in coastal zones is the significance of "out of zone" activities and their effects. Most of the damage occurring in the coastal zone is related to activities that have taken place within the wider drainage basin areas and beyond. Thus around 12% of the nitrogen and ammonium loads entering the Baltic Sea, for example, are due to atmospheric deposition linked to emissions from as far away as Belgium, Netherlands, Norway, France and the UK. Other examples include:

- Chesapeake Bay: About 5 years ago, it was estimated that around 30% of the nitrogen inputs into the bay were from sources outside the bay via atmospheric transport. A recent study [10], yielded an estimate of about 75%.
- Lake Superior: Analysis by Cohen et al. [11] showed that about 90% of several pollutants of concern, are being brought into the lake from outside the drainage basin by atmospheric transport.

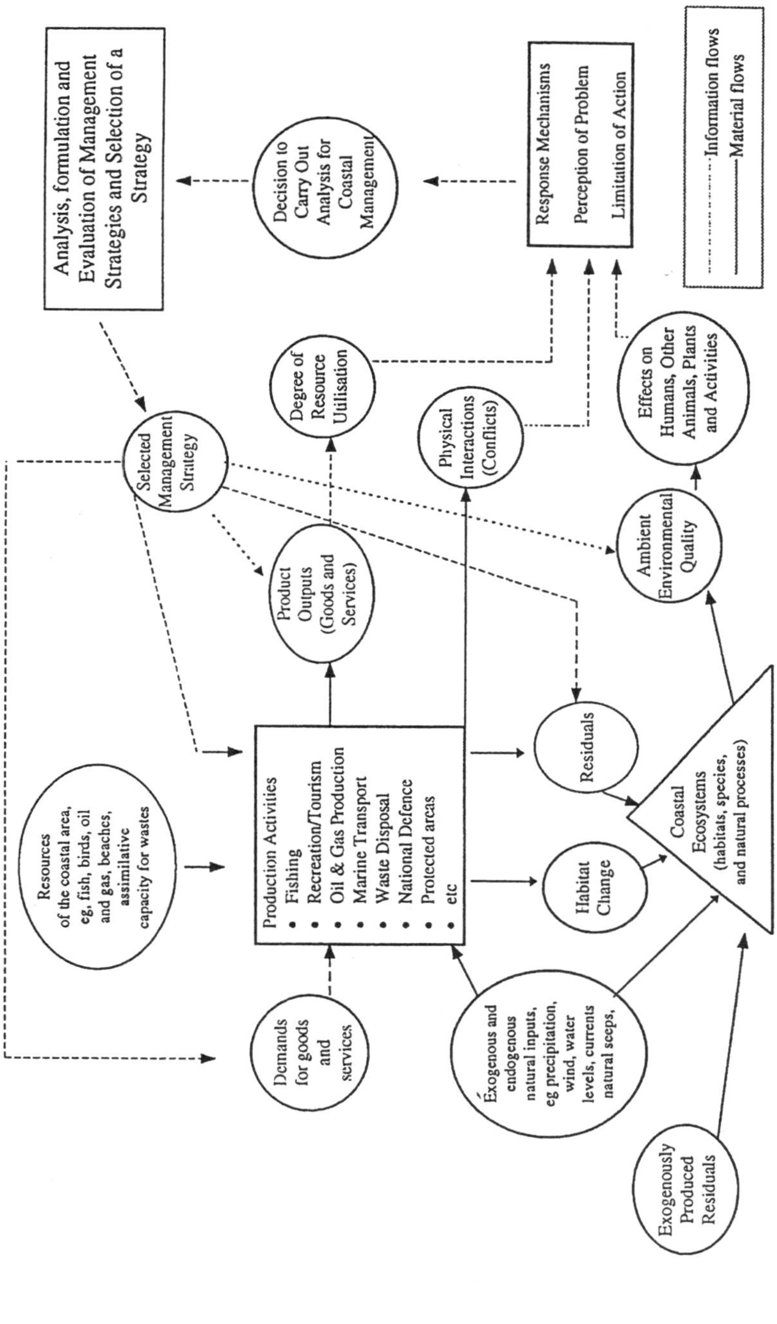

B.T. Bower, R.K. Turner / Ocean & Coastal Management 38 (1998) 41–66   43

Fig. 1. Coastal management system.

44   *B.T. Bower, R.K. Turner / Ocean & Coastal Management 38 (1998) 41–66*

Table 1
Market and governmental intervention failures in coastal zones

Market failures

1. *Pollution externalities:*

(a)	Air pollution, outside catchment sources	Excess levels of nitrogen, phosphorus and ammonia contributing to eutrophication of water bodies
(b)	Water pollution, land-based within catchment sources	Excess nitrogen and phosphorus from sewage and agricultural sources; industrial wastewater and toxic effluent pollution particularly from the pulp-and-paper and chemical industrial operations, nuclear power plants and research installations, military installations
(c)	Water pollution, coastal and marine sources	Excess nitrogen and phosphorus from coastal sewage outfalls; oil spills and contaminated bilge water from ships; outside catchment pollutants transported via longshore currents

2. *Public goods-type problems:*

(a)	Ground-water depletion/surface-water supply diminution	Over exploitation on-and-off-site of water supply
(b)	Congestion costs, on-site	Recreation pressure on beaches, wetlands and other sensitive ecosystem areas
(c)	Fisheries yield reduction	Over exploitation due to open access, loss of habitat, water pollution

Intervention failures (including lack of intervention)

3. *Intersectoral policy inconsistency:*

(a)	Competing sector output prices	Agricultural price fixing and associated land requirements
(b)	Competing sector input prices	Tax breaks or outmoded tax categories on agricultural land; or tax breaks for non-agricultural land development, including forestry; land conversion subsidies; state farming subsidies (historical) and other waste, wastewater and energy subsidies
(c)	Land-use policy	Zoning; regional development policy; direct conversion or fragmentation of wetlands; waste disposal policy and regulation (uncontrolled waste disposal dumping)

4. *Counterproductive factors*

(a)	Inefficient policy	E.g. strategies that lack a long-term structure; wastewater and industrial effluent combined treatment practices; general lack of enforcement of existing policy rules and regulations
(b)	Institutional failure, due to uncoordinated action by different government ministries, or the lack of an appropriate ministry or agency with a wide enough remit to deal with coastal issues.	Non-integrative agencies structure, non-existent agencies; lack of monitoring survey and enforcement because of capacity inadequate resources such as trained personnel, equipment and operating funds; lack of information dissemination; lack of public awareness and participation

B.T. Bower, R.K. Turner / Ocean & Coastal Management 38 (1998) 41–66   45

Table 1 (continued)

5. Conflicting perceptions
Different perceptions of a problem among stake-holders and management officials, e.g. local inhabitants and local governments do not consider hurricane or coastal storm flooding a major hazard, in contrast to specialists and state/federal/national government officials; or fishermen who do not think overfishing is a problem when state/federal/national agencies do.

- Bay of Sengal, Jakarta Bay: Both water bodies are stressed significantly by pollutant materials from agricultural and forest areas upstream [12, 13].
- Perhaps the epitome of effects on a coastal zone from activities outside the coastal zone is represented by the Gulf of California. Since about 1950, essentially no water from the Colorado River (mean annual yield of about 13.5 million acre-feet at Lees Ferry, Arizona), has reached the Gulf of California. What little water which is in the river from return irrigation flows after the last diversion in Mexico, simply evaporates in this desert country [14]. Fig. 2 illustrates the multiple demands on the river upstream from the coastal zone.

The objective of coastal zone management is to produce over time a "socially desirable" mix of coastal zone products and services. This mix is likely to change over time with changing demands, changing knowledge and changing pressures. Fulfilment of this objective will require the mitigation of current market and intervention failures; continuous monitoring of performance and of coastal zone conditions; and application of research findings over time. The desired social mix can be most effectively and efficiently provided by an integrated approach to coastal zone management. The analysis required for ICM performs an overall resource management auditing role; it forces decision makers to ask relevant questions relating to coastal ecosystems and processes, coastal zone pressure trends, trade offs, environmental impacts and various aspects of response strategies, including sources of finance and human/institutional capital resource potential. In macro-policy terms, ICM can be seen as a component of sustainable resource management within a regional/national economic development strategy [15, 16].

2. The elements of integrated coastal management (ICM)

Fig. 3 depicts the elements of ICM. Analysis generates the information which is subsequently utilised in planning decisions, which are made at various points in time in any given management context/area. A planning decision determines the distribution, timing and location of the coastal goods and services produced, the methods of delivery, and the receivers and payees.

This analysis/planning process will therefore be a significant component of ICM. It is underpinned by biophysical research and data relating to various processes,

46   *B.T. Bower, R.K. Turner/Ocean & Coastal Management 38 (1998) 41–66*

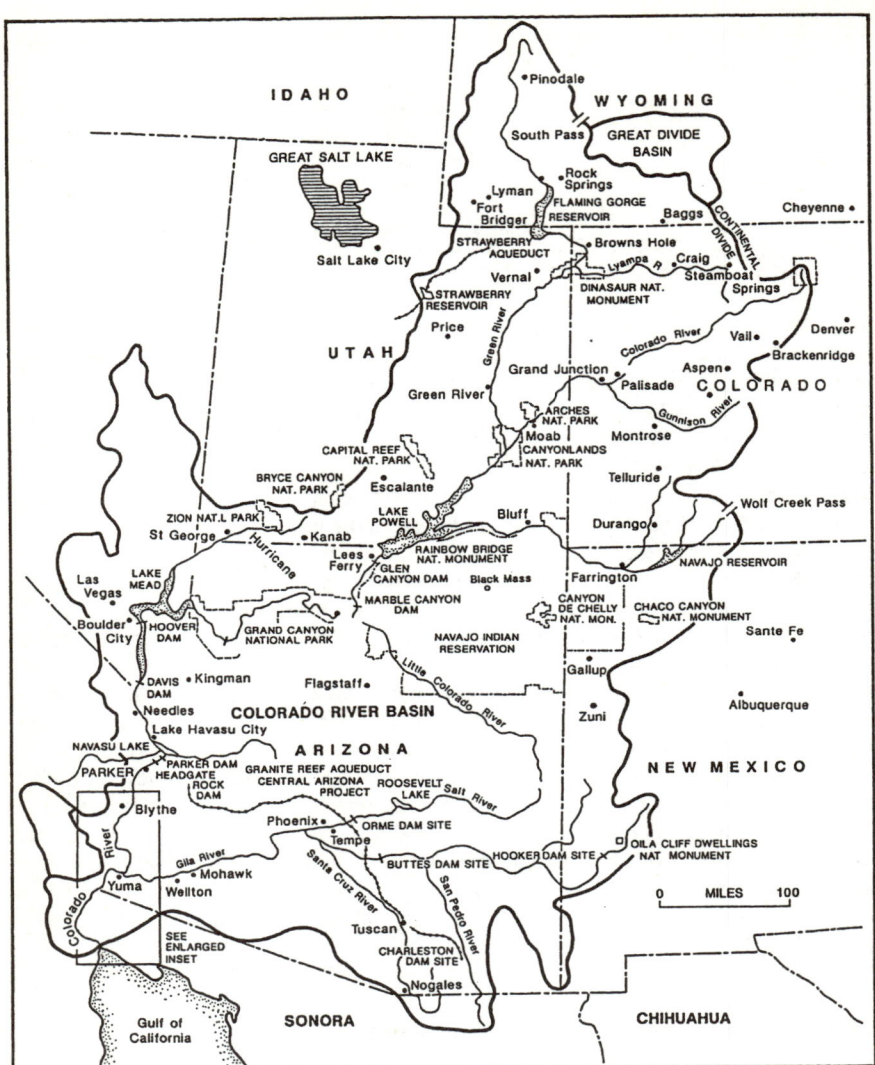

Fig. 2. The Colorado River and the Gulf of California.

structures, stocks, flows, and dose response relationships (quantified relationships
between pollutant discharges and related environmental state changes). From an
economic perspective it should be based on the cost-benefit/economic efficiency
criterion and evaluation method tempered by any relevant equity considerations,

B.T. Bower, R.K. Turner / Ocean & Coastal Management 38 (1998) 41–66   47

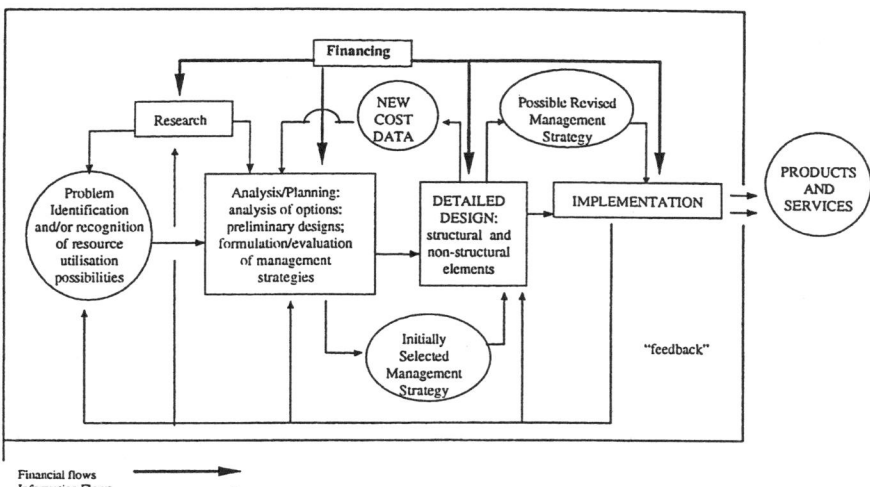

Fig. 3. Simple schematic of the elements of ICM.

other precautionary environmental (e.g. ambient quality) standards, and regional economic constraints. In the standard cost-benefit method the traditional decision criterion is, maximise net economic benefits. This criterion is too narrow, however, in situations when not all resource values can be translated into monetary terms and when criteria other than economic efficiency are deemed important by the relevant decision makers. Operational trade off relationships can nevertheless be developed by imposing constraints (e.g. ambient environmental quality, regional employment/income targets, conservation of designated nature reserves, etc.) on net economic benefit estimation.

In many coastal areas, maintenance or expansion of a regional economy is a major, often a primary, objective. Adverse effects on coastal economies, e.g. tax revenues, tourist expenditures, employment, can occur as a result of degraded water and/or beach quality or loss of or damage to unique natural features, such as a coral reef. Thus, beach replenishment programmes are typically justified on the basis of the need to maintain local economies dependent on tourism. In the context of regional or area economic development, the objective of the ICM can be expressed as follows:

Maximise the present value of:

$$\text{GRP} - C_{\text{p}} - C_{\text{cm}} - D + B - C_{\text{a}},$$

where
GRP = gross regional product;
C_{p} = normal production costs;
C_{cm} = net coastal management costs, e.g. discharge reduction costs/beach replenishment costs/coastal protection costs;

D = remaining damages;

B = benefits from improved environmental quality; and

C_a = administrative costs of ICM.

The "design" segment of ICM will encompass both preliminary designs of physical/biophysical facilities such as effluent treatment facilities, hard and soft sea defence/coastal protection engineered structures, construction of artificial marsh areas, and subsequent more detailed designs. But the design activity is also not restricted to structures and will include specifying the components and procedures of non-structural measures such as inspection and monitoring systems, charging systems and land use planning and implementation provisions.

To be able to manage requires proper resource assessment, involving the evaluation of (including costs of, and, wherever feasible, monetary evaluation of the benefits) multiple resources exploitation in the coastal zone and the interactions between and among the competing resource uses. The *first* management problem is that of deciding among the possible sets of outputs of goods and services which can be produced. Various combinations of outputs are possible, involving marine transport, waste disposal, fisheries yield, recreation, national defence, amenity, and preservation of unique coastal ecosystems. The different combinations will reflect different tradeoffs among the feasible outputs.

Because of the dynamic and "open system" nature of coastal zones, analysis for planning and management must consider at least three areas (multiple foci for ICM). These are:

- *The politically designated management area*: The political process in any given country, or in an international setting, will designate the boundaries of the management area, and will assign the management responsibilities to one or more public and private agencies.
- *Ecological areas*: a designated coastal management area may be within the boundaries of an identified ecosystem. More likely, the area will encompass, or be encompassed by, several ecosystems or catchment areas.
- *Demand areas*: Demand areas are those from which demands are exerted on the resources of the designated coastal area. These demands comprise: demands from within the designated management area; demands from outside the designated management area but within the catchment area; demands from outside the catchment area, with respect to, e.g. waste disposal of pollutants transported into the area via atmospheric transport, demands for coastal recreation, including visits to unique marine resource areas; and internationally determined demands, such as for global shipment of crude oil and oil products.

Thus, ICM involves *multiple* regions, the boundaries of which rarely – if ever – coincide. For example, governmental boundaries for counties, states and provinces are rarely contiguous with watershed or ecosystem boundaries. In analysis, explicit consideration must be given to cross-boundary flows in and out, upwind and downwind. However, the management structure established for a designated coastal area is not likely to have jurisdiction over activities beyond its area.

B.T. Bower, R.K. Turner/Ocean & Coastal Management 38 (1998) 41–66 49

3. Establishing priorities for management actions

Given multiple problems and virtually always limited resources with which to "address" them, a major problem in ICM is to establish priorities, i.e. what to do first. Criteria for establishing priorities include:

- Benefits in relation to costs, cost effectiveness
  [*Note*: Net benefits from addressing a particular problem may be the highest among a set of possibilities, but requires costs outlays – capital and/or operation and maintenance – which are greater than the available resources.]
- Distribution of benefits and costs, i.e. who gains and who pays;
- Political concern for some segment of the population, e.g. artisanal fishers;
- Physical, chemical, biological effects on the environment;
- Effects on institutional/administrative structure;
- Relative importance of problems, regardless of how "relative importance" is measured.
  [Relative importance will vary from area to area, and over time in a given area.]
- Feasibility of financing;
- Time to first returns;
- Accuracy of estimates of benefits and costs, i.e. how likely are they to be achieved.

The significance of each individual criterion and the relative importance of the criteria vary from area to area and over time.

Because the coastal zone is the most biodiverse zone, it may be prudent to impose a "zero net loss" principle or constraint on resource utilisation (affecting habitats, biodiversity and the operation of natural processes) in the zone, at least at the start of the analysis. Such a set of constraints will probably conflict with some other human needs. A typical multiple objective and value conflict problem is posed, for example, by: the need of local fisheries to increase fisheries yield in a given zone; the increasing use of the same or nearby waters for waste disposal; improving conditions for marine transport; an increasing need to raise the quality of bathing waters/beaches: and a political desire to increase the stringency of pollution controls on dischargers to coastal waters. A difficult sustainability balance will need to be struck depending on the real and full economic value of the various resource management options, the extent to which sustainable human livelihoods can be fostered with alternative income sources substituting for unsustainable current usages, and the actual resilience of various natural systems and processes.

It is important to note, with respect to ICM, that none of the desired outputs can be achieved without financing. This involves not only the financing of analysis and planning, but also the financing of capital investments, operation/maintenance/replacement (OMR) costs (including monitoring and enforcement) and an allocation for contingencies. With regard to capital investments, it is essential to investigate:

(a) how governmental agencies involved will raise the annual "cash flow" required for repayment of loans for capital investment, as well as covering OMR costs and a contingencies fund; and

(b) whether or not the activities allocated to private entities, can be financed by those entities.

4. Characterising the benefits of ICM

The benefits of ICM can be most readily discerned if they are related to baseline conditions in the given coastal zone i.e. coastal area A at time T_0. The condition of the coastal resources at T_0 reflects the effects of various human activities and of natural events over past time to T_0. Examples include:

- damage that has been incurred from coastal storms, hurricanes, eroding shorelines;
- loss of habitat, e.g. wetlands that have been filled and mangroves removed;
- water quality that has been degraded by discharges of various materials, e.g. heavy metals, synthetic organics, suspended solids, fecal matter;
- sedimentation that has occurred in navigation channels and in harbour areas, and on spawning beds and coral reefs;
- offshore sand and gravel mining that has affected fisheries and habitats;
- salinity intrusion that has occurred as a result of excessive withdrawals from ground water aquifers;
- finfish and shellfish yield declines;
- residential and commercial facilities that have continued to be constructed in hazardous areas in relation to coastal storms, hurricanes, shoreline erosion;
- conflicts among different types of recreation users of beach areas that have become more prevalent;
- conflicts between commercial and recreational fishers that have escalated;
- exotic species that have inadvertently been introduced into the coastal waters, resulting in damages to physical structures, e.g. water intakes, damage to indigenous species and ecosystem modification;

ICM benefits are achieved by: reducing damages; mitigating pollution and resource overexploitation problems; enhancing coastal zone outputs; and preserving unique coastal ecosystems [12, 17]. It is important to recognise that benefits comprise the net effects on coastal zone resources – processes, functions and outputs – linked to a management measure or set of measures. Any given measure often generates multiple effects not all of which will be positive. Thus a measure which reduces "pollutant" discharges into a coastal water body can reduce damages to recreation at the same time improving the environment for finfish, but also increasing borer populations which result in increased damage/maintenance costs of wooden structures such as piers. Benefits can be classified and defined as follows:

Mitigation benefits: These benefits are comprised of damage reduction and restoration benefits. Examples include reducing damages from: coastal storms, hurricanes, shoreline erosion, non-coastal soil erosion, salinity intrusion, excessive withdrawals from ground water aquifers, sedimentation in navigation channels and harbours, breeding areas for vector-borne diseases, over exploitation of fish species, water fowl and water animals. Examples of restoration benefits include actions which return the coastal system toward original ecosystem productivity, such as: replanting wetlands, constructing "artificial" wetlands, removing exotic species, restocking with native species, improving water quality by various means to reduce water intake treatment costs, increase water species productivity, increase availability of swimming and decrease fish advisories [18, 19].

B.T. Bower, R.K. Turner / Ocean & Coastal Management 38 (1998) 41–66 51

Enhancement benefits: These benefits are achieved by *increasing* the outputs from a coastal water body from current levels, as long as the costs involved are less than the potential gains. They are comprised of two subcategories:

(1) *increasing* the output of some product or service by, e.g. constructing artificial reefs to provide habitat for fisheries thereby increasing fish output, constructing recreation facilities such as ramadas, parking areas, outhouses and beaches, installing shark protection devices in coastal swimming areas, increasing the depth of navigation channels and harbours;

(2) *reducing conflicts* among or between various users of the coastal resources, such as competing uses among beach recreationalists, e.g. dunebuggies, surf fishers, swimmers, competing use of navigation channels by commercial and recreational vessels. Conflict reduction can be accomplished by, for example, some combination of pricing schemes and time and spatial scheduling of activities.

Preservation benefits: These benefits stem from setting aside and managing particular areas in order to preserve the natural ecosystem e.g. marine sanctuaries or preserves in order to counteract or preclude increasing "consumptive use" pressure on such resources. Two types of benefits are involved: (1) use benefits; and (2) non-use benefits. The former involves activities which produce benefits as a result of actual visits to the preserved area for observing natural ecosystems, for scuba diving, for taking photographs. In contrast, the latter subcategory of preservation benefits does not involve actual visits to an area. Rather, the benefits are estimated as a function of "option demand" and "existence value".

Indirect economic benefits: This category is comprised of benefits stemming from "second round" effects of measures applied to produce benefits in the first two categories and in the "use" category of preservation benefits. The context for the analysis and estimation is the regional economy (and/or the national economy), as the *direct* economic benefits result in additional economic activities in the region (which may or may not be *net* to the nation). Discussion of methods for estimating these benefits, and the problems in doing so, fill multiple pages in the literature on analysis for planning water resources developments in the period circa 1950 to ca. 1980, primarily under the rubric "secondary benefits". Fig. 4 illustrates the sequence of analyses involved in estimating the indirect benefits in relation to restoration of fish yield by reducing suspended sediment concentration in waters of a given coastal area.

"Option" benefits: These benefits refer to the potential gains from an ecosystem conservation policy which seeks to retain as extensive a set of future coastal resource use options as is practicable. Separating out such option values from the other components of total economic value is, however, a debateable issue [20]. *"Existence" benefits* refer to the non-use values that coastal ecosystems might possess. Some individuals would be willing to spend money just to be assured that the resource is preserved even though they have no intention of visiting or using the resource.

The estimation of benefits takes place as part of the analysis for planning for ICM in the context of continuous integrated coastal zone management. This means that the analysis is done at some "point" in time, or over some finite time period. This in turn means that the analysis must be based on some "baseline conditions". Those conditions, e.g., contamination in sediments, land use in the coastal area, represent

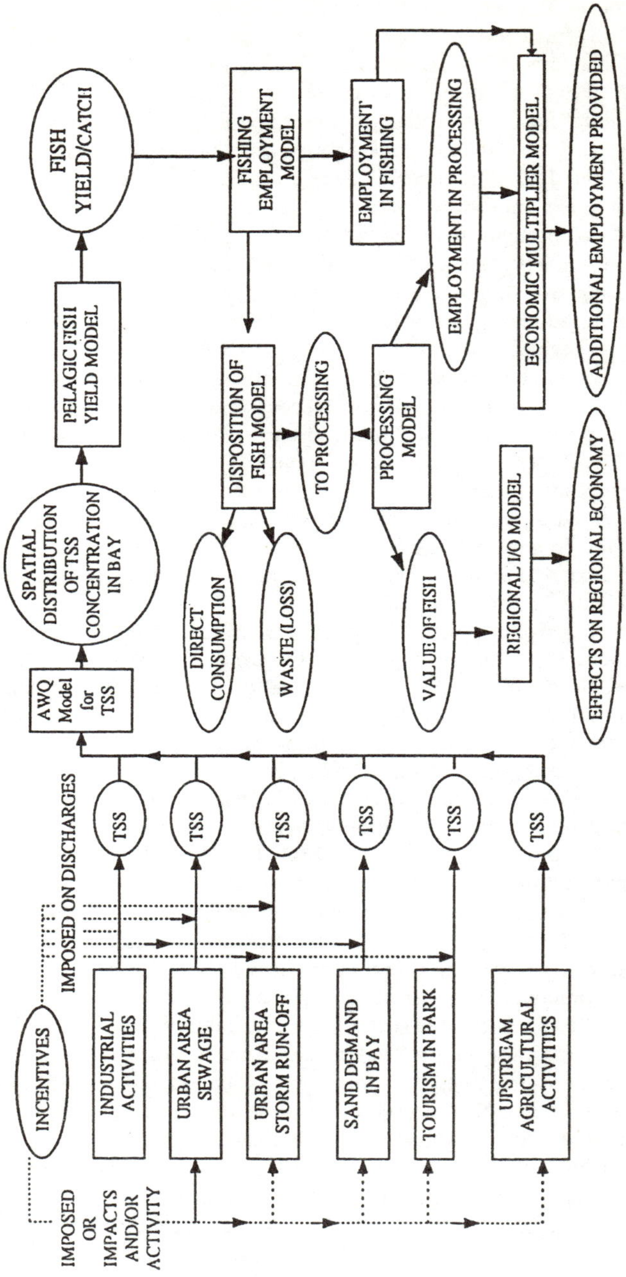

52   *B.T. Bower, R.K. Turner / Ocean & Coastal Management 38 (1998) 41–66*

Fig. 4. Sequence of analyses to evaluate direct and indirect monetary effects: relationship between fish yield and distribution of total suspended solids concentration.

(a) present conditions (hopefully some relatively recent year for which data are available), and (b) the factors which have operated over time to result in those conditions. For example, PCBs in the Lower Hudson River and in the Lower Fox River (Wisconsin) are a function of discharges which occurred some time ago. In both cases, no discharges of PCBs have occurred in at least 10–15 yr. But concentrations are still significant and result in current damages. Baseline conditions for estimating damages from hurricanes, for example, must include the changes that were made, e.g. in building construction, as a result of the most recent hurricane. In addition, the "baseline conditions" will change over time, as a result of increased information about coastal ecosystems, increased accuracy of forecasting hurricane landfalls and changing governmental policies.

Thus, a critical problem in estimating benefits from ICM is defining what would happen in the absence of ICM, i.e., what is the "baseline scenario".

ICM benefits over time are also a function of the research and monitoring components of ICM. These components provide information on which to base benefits estimates. For example, having monitored the Florida shoreline in the Pensacola area before Hurricane Opal enabled determining the effect of the hurricane on shoreline erosion by remeasuring after the hurricane [21]. Investigations to determine reasons for damages to residential structures in Florida by Hurricane Andrew lead to changed building codes and construction methods, thereby reducing damages from subsequent storms. Studies of recreation behaviour in coastal areas provide data for developing estimates of demands, and hence of benefits associated with coastal recreation. But the provision of continuous research and monitoring activities requires an institutional arrangement (a) to organise and finance the research and data collection; and (b) to make use of the results of those activities.

5. Demonstrating the benefits of ICM

From a pragmatic decision-making perspective, there needs to be a method for demonstrating to decision makers and other stakeholders what the benefits of ICM could be. However, there also must be an understanding that achievement of benefits requires implementation of a coastal resources management strategy. The components of a strategy are shown in Fig. 5. These components are related to the set of tasks of coastal management mentioned earlier.

The analytical approach suggested for demonstrating the benefits of ICM is based on a "without ICM" versus "with ICM" comparison. The *net benefits*, i.e. benefits minus costs, associated with (attributable to) ICM are then represented by the difference between the two "states of the world" in a given coastal area. The benefits and costs of ICM will be determined by the range of processes, functions, products and services found and produced in the coastal area and their interrelationships with factors external to the coastal area. The "without–with" comparison is combined with the application of "scenarios" of conditions extended to whatever points in the future are considered to be of interest, i.e. "alternative futures". The structural scenario analysis is indicated in Fig. 6.

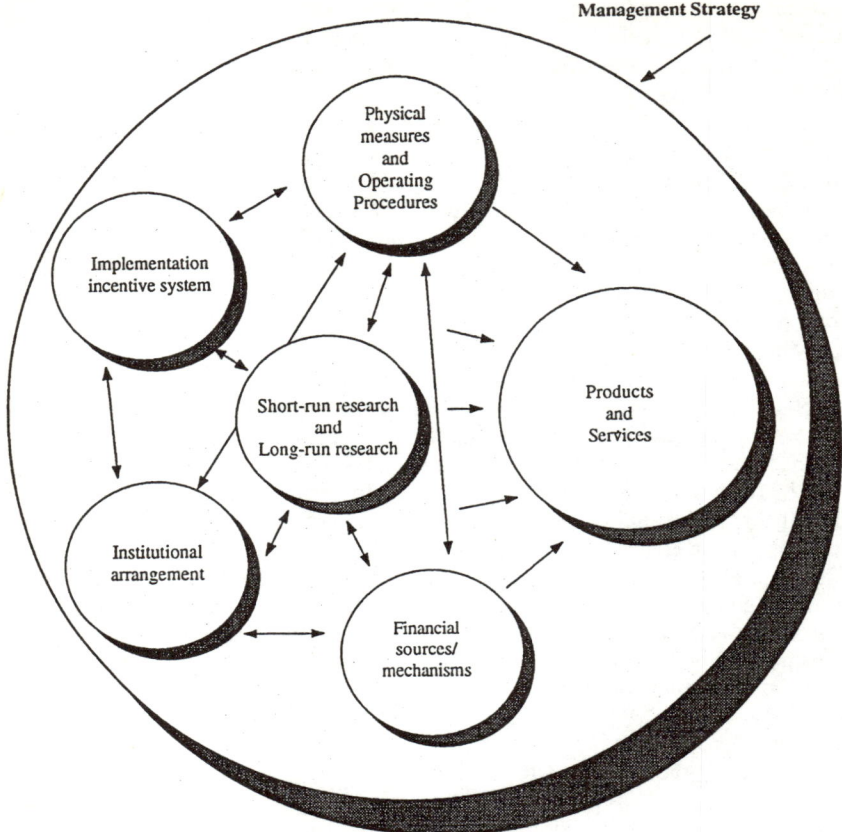

Source: Bower and Turner (1997)

Fig. 5. Components of a management strategy.

A scenario comprises some combination of values of three sets of linked variables:
(1) economic and demographic conditions over the time horizon of interest;
(2) environmental conditions; and
(3) non-coastal governmental policies and programmes, technological changes, and factor prices.

Clearly the third "variable" could be divided into three separate variables. However, each additional variable added multiplies the possible number of permutations and combinations of the variables.

B.T. Bower, R.K. Turner / Ocean & Coastal Management 38 (1998) 41–66 55

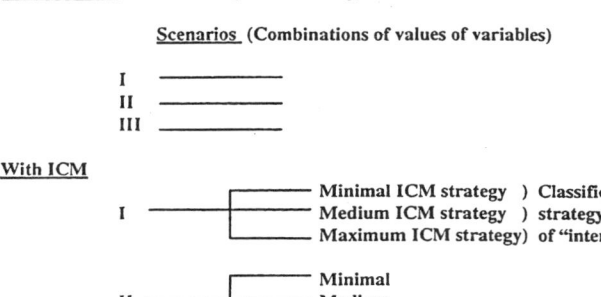

Fig. 6. Role of scenarios in estimating benefits of ICM.

5.1. Without ICM

For any given scenario, i.e., a combination of (a), (b), and (c), what is termed the "base case" involves the estimation of net benefits for the given scenario with the "trend" management strategy, termed the "without ICM" situation.

The zone's aggregate future production (i.e. feasible array of products and services) is evaluated both in terms of input requirements and residuals generation and discharge impacts, and the effects on the regional economy. The benefits and costs will be determined by the "trend" management strategy, which in turn is a function of the institutional structure. All these socio-economic activities and systems will of course be underpinned by the coastal ecosystems and their interrelationships with the abiotic environment.

Characterising the "without" situation is not straightforward, as a "business as usual" management strategy is not realistic because the biophysical and socio-economic systems in the zone are not likely to be static. Rather such systems will be subject to an almost continuous process of adaptative change as socio-economic systems learn from past experiences. These adaptations may then generate feed back effects in bio-physical processes and functions and so on. Changes, for example, were made in response to the last three hurricanes in Florida, USA. Building codes were tightened, particularly for mobile homes, and insurance companies became more particular about what they insured in relation to storms and hurricanes. When salinity intrusion was recognised in Southern California 50–60 yr ago, a recharge scheme was established and water withdrawal charges imposed to prevent any further intrusion. Regulations have been established, and in some places actually enforced and fines levied, preventing the filling of wetlands. "Artificial" wetlands have also been constructed in various areas, to compensate for previously filled wetlands and consequent loss of habitat.

The rate and scale of the adaptive behaviour will be conditioned by, among other things, human perceptions of environmental change. Perception of the nature, extent and severity of a problem affects the response of users to incentives imposed to change behaviour. Net benefits are a function of responses to incentives; responses to incentives are a function of perceptions. There may well be significant differences between perception of a problem by federal and state officials and local government officials and land users, as in the case of wildfires in urban fringe areas. There typically is a difference in perception with respect to the implications of the decline in yield of a particular fish species, e.g., over the last 15–20 yr among fishers, regulatory authorities and scientists. There is a difference between fishers and responsible governmental agencies in the perception of the efficacy of a particular method for allocating fishing "rights". There is also often a difference between the view of federal and state officials of the nature of the hazard of locating in the coastal zone and of the individuals who have located in the coastal zone, even those who have been through floods and hurricanes.

Thus, the perception by different "parties at interest" of the danger/risk related to a problem is a key variable in implementation of ICM, and hence in the achievement of net benefits of ICM.

The environmental impacts generated by human activities combine with "natural" variability in coastal zone biophysical processes to produce an array of effects, only some of which can be valued in monetary (economic value) terms. Some combinations of environmental conditions and impacts are shown in Table 2.

5.2. With ICM

For the *same* scenario analysed for the without ICM case, one or several ICM strategies are posited. Each strategy is a combination of physical measures and nonstructural measures. Examples of measures related to coastal hazard management which can be elements of an ICM strategy include:
- land use controls, i.e., precluding certain types of land use in specified areas of the coastal zone;
- land use controls in terms of setbacks;
- beach replenishment;
- extensive evacuation, temporary sheltering, resettlement procedures;
- intensive storm/hurricane forecasting procedures, coupled with extensive system of conveying warnings to coastal occupants;
- differential property tax rates, with higher taxes imposed on higher hazard areas;
- construction specifications in building codes, e.g., roof tie-downs to frames, shingle tie-downs; constructing facilities on "stilts";
- refusal by insurance companies to write damage insurance, except at very high rates.

B.T. Bower, R.K. Turner / Ocean & Coastal Management 38 (1998) 41–66   57

Table 2
Environmental conditions and impact categories

Impact categories	Climate-related events and human activities						
	Erosion	Flooding/ inundation	Saltwater intrusion	Sedi-mentation	Degraded water quality e.g. eutro-phication, red tides	Storms, hurricanes, typhoons	Upwellings e.g. El Nino
Tourism	$				$	$	
Fresh water supplies		$		$	$		$
Fishing/ agriculture	'$'	'$'		$	$	$	$
Coastal residences	$	$	$		$	$	
Commercial/ buildings	$	$		$		$	
Wetlands	$,nm	$,nm	$,nm	$,nm	$,nm	$,nm	
Agriculture & drylands	$	$	$	$	$	$	
Human health		$,nm			$,nm	$,nm	$,nm
Culture & heritage sites	nm	nm			nm		

Notes: nm = non-market impacts; $ = market priced major impacts; '$' = minor impacts.
Source: Adapted from [9].

Net benefits are estimated for the specified ICM strategy for the given scenario. These net benefits are compared with the net benefits for the scenario without ICM. The difference represents the net benefits attributable to ICM.

In principle on the benefit side of the net benefit estimations it is possible to conceive of a total economic value (TEV) of the coastal zone's output of products and services. TEV will consist of both use values (direct and indirect) and so-called non-use values, as shown in Fig. 7. A range of valuation methods is available, but it will not be possible to place meaningful monetary values on all the benefits (and some of the costs) of outputs from the coastal zone.

The benefits of ICM are linked to four environmental impacts/effects categories:
• direct and indirect productivity effects;
• human health effects;
• amenity effects; and
• existence effects such as loss of biodiversity and/or cultural assets.

Different valuation techniques are appropriate for each of the four broad effects categories. Choice of technique will depend on the magnitude and significance of the effects, on the availability of data and on the analytical/institutional capability in any

58   *B.T. Bower, R.K. Turner / Ocean & Coastal Management 38 (1998) 41–66*

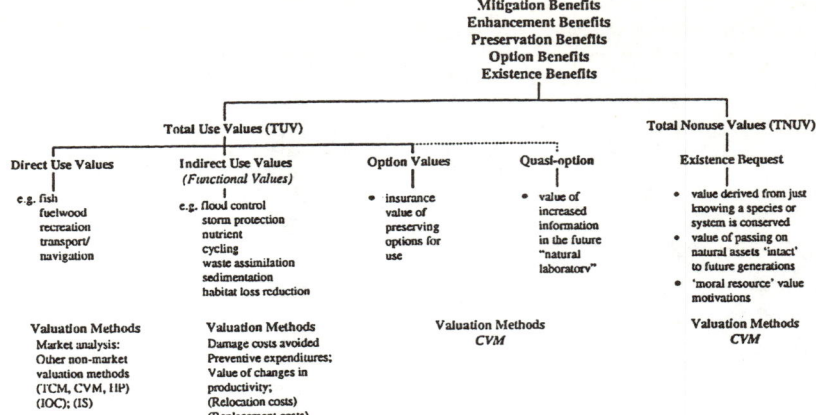

Fig. 7. Methods for valuing coastal zone benefits.

given context. Table 3 summarises both the characteristics of the valuation methods and the valuation options available in each environmental effects category.

6. Steps in estimating ICM net benefits

Operationally, the following are the basic steps in estimating the net ICM benefit:
1. Define the problems in quantitative terms. This typically involves developing various physical–chemical–biological relationships, e.g. pollutant discharge–water quality relationships, dose–response functions, environmental conditions–species response relationships, damage vs. intensity of storm relationships, level of water quality vs. water intake treatment costs.
2. Select a scenario and tabulate the related spatial pattern and levels of population and economic activities, for the time horizon specified in the scenario.
3. Based on (a) the relationships defined in #1; (b) the demographic conditions specified in #2; (c) the sequence of meteorologic/hydrologic events selected in the scenario; and (d) changes, if any, in factor prices and technology specified in the scenario, estimate the extent of various problems and the damages associated therewith for the "trend" management strategy, and the associated benefits and costs (where costs represent the incremental costs in addition to existing costs at time T_0). This yields the net benefits for the "without ICM".

Table 3
Coastal Zone Impacts and Valuation Methods
Valuation methods

- *Market orientated benefit valuation* – benefit valuation using actual market prices of productive goods and services based on changes in the value of output, or loss of earnings. Examples include loss of fisheries output due to pollution, or value of productive services or recreational benefits loss, through increased illness caused by coastal waters pollution.
- *Surrogate markets benefit valuation* – environmental surrogates may include marketed goods, property values, other land values, travel costs of recreation, wage differentials, compensation payments. Examples of such proxies are entrance fees to national parks as a proxy for value of visits to protected areas, changes in commercial property values as a result of water pollution, or compensation for damage to fisheries.
- *Cost valuation using actual market prices of environmental protection inputs* – preventive expenditures, replacement costs, shadow projects, cost-effectiveness analysis. For example the cost of environmental safeguards in project design, cost of replacing resource damage by pollution or conversion, cost of supplying alternative recreational facilities destroyed by development activities, or cost of alternative means of sewage sludge disposal in marine waters can be used as cost indicators.
- *Survey orientated (hypothetical valuation)* – contingent valuation or contingent ranking questionnaire-based surveys of individuals to elicit willingness-to-pay or to-be-compensated valuations.

Effects categories	*Valuation method options*
Productivity:	Market valuation via prices or surrogates
E.g. Fisheries, agriculture, tourism, water resources, industrial production, marine transport, storm buffering and coastal protection.	Preventive expenditure
	Replacement cost/shadow projects/cost-effectiveness analysis
	Defensive expenditure
Health	Human capital or cost of illness
	Contingent valuation
	Preventive expenditure (avertive behaviour)
	Defensive expenditure
Amenity	Contingent valuation/ranking
Coastal ecosystems, wetlands, dunes, beaches, etc., and some landscapes, including cultural assets and structures.	Travel cost
	Hedonic property "value" method
Existence values	
Ecosystems; cultural assets	Contingent valuation

4. Define an ICM strategy, which involves – as noted above – some combination of physical (structural) measures and non-structural (charges, land use regulations) measures, and the capital, OMR, and administrative costs associated therewith.
5. For the *same* scenario as selected in #2, estimate the benefits from applying the ICM strategy defined in #4, remembering that explicit consideration must be given to the extent to which the coastal occupants actually behave as "assumed" in response to the various incentives imposed.
6. Calculate the net benefits associated with the ICM strategy defined in #4.
7. Compare the net benefits estimated in #6 with the net benefits estimated in #3. The result represents the net benefits from ICM.

The method for estimating benefits from ICM can be illustrated by applying the method to a prototype coastal system. The system has characteristics analogous to real coastal systems. Data on costs, various relationships involved in the performance

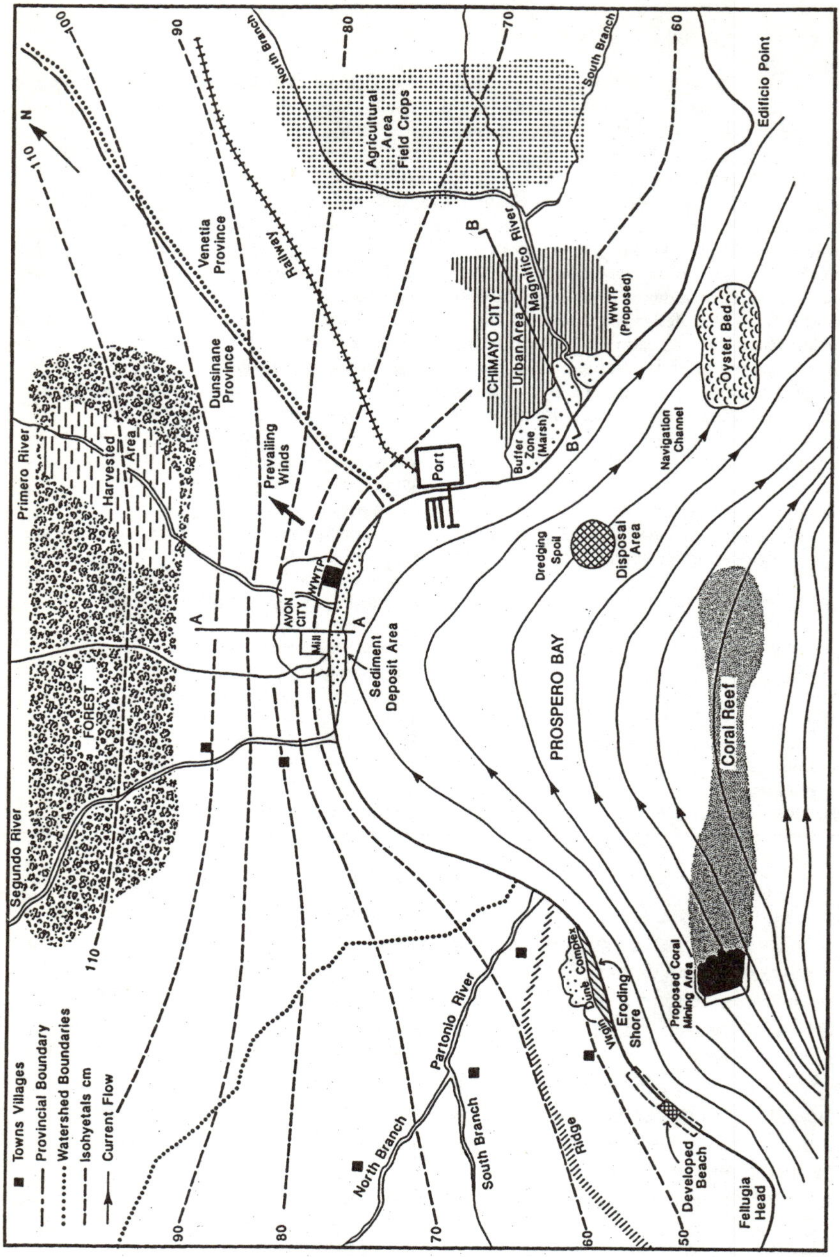

Fig. 8. Prospero bay prototype.

B.T. Bower, R.K. Turner / Ocean & Coastal Management 38 (1998) 41–66 61

of natural systems, values of unit discharges, and so on, are based on data from the real world". The result of this construction, the "Prospero Bay Prototype", is shown in Fig. 8. The prototype contains a variety of typical uses of coastal resources and the competitions between and among those uses [24].

7. Illustrative prototype analysis

The "Prospero Bay region" is a prototype with characteristics – hydraulic, meteorological and socio-economic – similar to those in "real word" coastal areas. It is designed to provide a realistic context for illustrating an application of a proposed framework for the management of land-based sources of marine degradation in the context of regional planning and economic development. The prototype contains a description of both the political context and selective physical characteristics of the region. It is assumed that although rigorous data have not been gathered, sufficient evidence is available to have produced the perception in the region that there are significant problems of marine degradation from land-based sources in the coastal zone.

Rough estimates suggest that the use of the beach area near Fellugia Head and at Avon City for water contact activities and other beach recreation has been decreasing. This reduction is attributed to the increasing mean turbidity of the water, resulting from higher concentrations of suspended solids; and to a lack of variety in the immediate beach vicinity. The virgin dune complex further along the eroding shore from Fellugia Head, seems to offer an extra amenity if it were to be encompassed by new tourist-related development. Any attempts to stabilise the dunes would, however, have an impact on the extent of beach frontage at Avon City. The relationship is not, however, a simple one because of the existence of a sand mining operation in the coastal waters between the dune site and Avon City. The dragline sand extraction operation releases about 15% of the sand that is disturbed in the longshore current, in which the material is transported down the coast to Avon beach. This beach is not only a recreational amenity but a component of a sea defence system (hard sea wall defence) for the city.

Avon City is close to the mouths of two rivers, the Primero and the Segundo, which have long been know for significant runs of anadromous fish. Current production in the spawning area of the Primero River averages about 30,000 fish, net of natural mortality. The runs represent a significant and locally marketed source of protein for local people. However, the size of runs is in decline. Suspended sediment from the forest area that has been logged and form the area currently being logged, has resulted in covering some spawning areas in the Primero River. In addition, increasing turbidity conditions in the bay in the vicinity of the mouth of the Primero River have resulted in loss of fish.

Further along the Bay, annual dredging to maintain adequate depths in the port and in the navigation channel has increased, from about two months per year to three months per year, as a result of increased sediment deposition in both port area and in the channel.

A visiting coral reef expert has concluded that, based on his inspection of the current condition of the reef e.g. abundance and distribution of species, the health of parts of the reef have deteriorated over an unknown period of time. Reef damage may be due to tourist (diving) activity and decreased water quality.

Water quality in Prospero Bay is affected by the lack of a modern wastewater treatment plant serving Chimayo City (Avon City has such facilities). Some data has been collected, sufficient to allow for the quantification of a rough estimate of total suspended solid (TSS) discharged into the Bay. Agricultural activities, the operation of an integrated paper mill and upstream forestry operations all contribute to the TSS loading. The discharges also include other substances such as cadmium from electroplating activities in Chimayo City. This city has a growing residential housing requirement which is being satisfied at the expense of the coastal marsh frontage. By reducing the extent of marshland buffering, expansion of the city increases the risks of flooding throughout this relatively flat city. Because the beach at Avon City is fronted by industry and commerce, residents of both cities have increasingly used Chimayo Marshes as a recreational area. Hence the loss of marshland represents not only a loss of flood protection for Chimayo City, but also a loss of recreation benefits for both cities. Both housing and tourism pressure on the marshes need to be reduced.

Given this basic description and supporting data, the prototype can be manipulated to simulate the multiple use conflict recognition and trade-off issues faced by resource managers and policymakers and to illustrate the quantification and valuation of the costs and benefits involved. Scenario analysis provides a way of comparing trend outcomes with a number of possible planned futures for the coastal region.

Take just one illustrative aspect only of the Prospero Bay management problem, the implementation of a strategy to reduce TSS discharges into the Bay. This will result in a number of economic gains in areas affected by TSS pollution under present conditions. These impacted areas – the zone around the mouth of the Primero River (fish assembly and exit area), the recreation beach at Avon City and parts of the coral reel – all receive a quantified TSS concentration, assuming some transfer crefficients linked to TSS generating activities/processes and discharge locations.

Reductions in TSS discharges can be obtained by the following measures – shoreline stabilisation via rock walls, construction of a sewage treatment plant in Chimayo and more careful operation of the Avon City plant, banning of dredging spoil disposal in the Bay, modified agricultural practices, better treatment systems at the paper mill, changes in forestry practices and the switching of sand mining activities to an inland site. Table 4 shows the estimated amounts of reduction in TSS discharges and the associated costs, under present conditions. Table 5 shows the unit costs of reduction by activities at the effects areas, for present conditions.

Because some measures result in physical and economic changes simultaneously at several impact areas in the bay, the analysis proceeds to determine the economic costs and benefits of total reductions by various measures at various locations. What is being sought is the combination of measures that achieves the largest total benefits per unit of cost, with whatever constraints may be imposed, or the combination that achieves the specified level of physical output(s) at the least cost.

Table 4
Estimated possible reductions in TSS discharges at effect areas and associated costs, present conditions, Prospero Bay

Activity	Total costs in year, (10³ 1992 $)	Reductions in TSS discharges (metric tons per day)		
		At effect area 1 (fish)	At effect area 2 (beach)	At effect area 3 (coral reef)
Eroding shoreline	250	6.6	5.2	2.6
Dredging	79	3.5	8.5	13.7
Agricultural operations	118	27.7	18.4	9.3
Urban sewage	2700	9.9	6.8	3.3
Logging	489	104	62	36.3
Paper Mill, TSS	21.2	1.8	0.9	0.5
Paper Mill, particulates	56.8	0	0.9	0.5
Sand mining	379	0	9.8	4.9
Totals (rounded)		153	113	73

Table 5
Estimated unit cost of reducing TSS discharge from each activity, at each effect area, present conditions, Prospero Bay

Activity	Cost of reducing TSS discharge ($/mt)		
	At effect area 1	At effect area 2	At effect area 3
Eroding Shoreline	104	132	265
Dredging	65.8	26.3	16.4
Agricultural operations	11.4	17.6	34.6
Urban sewage	750	1080	2250
Logging	12.9	21.4	43.8
Paper Mill, TSS	33.6	50.5	106
Paper Mill, particulates	NA	177	355
Sand mining	NA	105.2	210.6

mt = metric ton.

The economic benefits associated with a reduction in TSS would include beach recreation benefits (more recreation days valued at say 20 US 1992 dollars via a travel cost or contingent valuation method, the better the water quality). Fish yields would also be increased as sediment discharges from the forestry operations were reduced (the value of these benefits would be determined by dockside market prices, say 6 US 1992 dollars). Visits to the coral reef could be maintained (or perhaps increased) if TSS and cadmium discharges to the Bay were reduced (reef visits have been valued at a willingness to pay figure of 30 US 1993 dollars, again via a method such as a contingent valuation survey).

Table 6
Results with no Action Program and With an Action Program for management of land-based sources of marine degradation, present conditions, Unity Bay

	No Action Program	With Action Program	Gain from Action Program
Outputs	Unmodified discharges	Modified discharges	(10^3 1992 $)
Recreation-days (10^3)	18.8	21.8	3.0
Value of recreation-days (10^3 1992 $)	376	436	60
Andromous fish yield (10^3 km)	34.2	47.2	13.0
Value of anadromous fish (10^3 1992 $)	205	283	78
Coral reef health (rating)	7.0	9.2	2.2
Coral reef attractiveness (rating)	0.85	0.98	0.13
Coral reef visitor-days (number)	990	1000	10
Coral reef visitation value (10^3 1992 $)	29.7	30.0	0.3
Totals		1.6[a]	0.14[a]

[a] Total costs of action program are 1.6 million 1992 $; gain from program is 0.14 million 1992 $.

Each impact area benefit estimation requires prior information relating TSS/cadmium load reductions to increased recreational activity and fish yields i.e. dose-response relationships. Thus in the case of the coral reef data is required on the relationship between TSS and cadmium loadings at the reef and the relative health of the reef. The relative health of the reef measure then has to be related to the relative attractiveness of the reef to visitors. Finally, the estimated relationship between the relative attractiveness of the reef and the percentage reduction in visits to the reef is required, before an economic valuation can be calculated.

Table 6 simulates illustrative results under present conditions in the Prospero Bay area, if there is no pollution reduction program and if a least-cost discharges reduction strategy were adopted. In this particular simulation the mean annual economic gain is about 0.14 million US 1992 dollars, achieved at a mean annual cost of about 1.6 million US 1992 dollars.

But management of any coastal area cannot be based only on present conditions. Scenario analysis then has to be deployed in order to recognise the dynamic nature of coastal areas and the changes in the economic costs and benefits associated with different economic growth, population growth, recreation activity rates and pollution abatement practices, etc.

8. Conclusions

Coastal zones are under increasing pressure from the scale of the multiple resource demands that are impinging on them. Multiple use of the goods and services provided by coastal systems reflects a variety of stakeholder interests and perceptions. A number of these uses and interests and perceptions conflict with each other and are themselves changing over time. The management process deployed in order to maintain an acceptable mix of outputs from the coastal zone into the future must therefore be both a dynamic and an adaptable process. Past and current market and government intervention failures will need to be corrected if pollution and resources loss trends are not be exacerbated. An integrated approach to coastal zone planning and management (ICM) is necessary in order to produce (effectively and efficiently), and/or enhance the mix of goods and services that society wants, and to ensure an equitable distribution of the benefits of ICM.

Integration in ICM includes its integration across broad policy objectives and plans, with different sectoral plans and management, with different levels of government and with the public and private sectors. ICM itself must encompass all the elements of management from planning and design through financing and implementation. It also requires an interdisciplinary analytical and operational approach. Because of the dynamic and "open system" nature of coastal zones' analysis for planning and management has to encompass multiple regions i.e. politically designated management area, ecological areas and socio-economic demand areas, the boundaries of which rarely, if ever, coincide.

The benefits of ICM, which take a variety of forms, can be demonstrated by comparing a coastal resources management strategy "without ICM" versus a management strategy "with ICM". The net benefits of ICM are then represented by the difference between the two "states of the world" in a given coastal area. Human perception of the nature, extent and severity of a problem or risk of a problem, conditions responses and adaptive behaviour. It is therefore a key variable in implementation of ICM and hence in the achievement of net benefits of ICM.

Finally, analysis of coastal management, without and with ICM, using simulation of a prototype coastal system, has proved to be a good practical means of demonstrating, to a range of audiences, the real benefits which can be, or could be, achieved by ICM.

References

[1] Turner RK, Perrings C, Folke C. Ecological economics: paradigm or perspective. In: van den Bergh J, Van der Straaten J, editors. Economy and Ecosystems in Change: Analytical and Historical Approaches. Aldershot: Edward Elgar, 1997, ps 25–49.

[2] Ehler CN, Bower BT. Toward a common framework for integrated coastal zone management. coastal zone 95 "Spotlight on Solutions" Conference, Tampa, FL, July, 1995.

[3] Turner RK, Bower BT. The benefits of integrated coastal zone management. Coastal zone 95 "Spotlight on Solutions" Conference, Tampa, FL, July, 1995.

[4] Cicin-Sain B, editor. Integrated Coastal Management [special issue]. Ocean and Coastal Management 1993;21:1–3.

[5] OECD. Coastal Zone Management: Integrated Policies, Paris: OECD, 1993a.

[6] OECD. Coastal Zone Management: Selected Case Studies. Paris: OECD, 1993b.

[7] World Bank. Africa: a Framework for ICZM, Land, Water and Natural Habitats Division, The World Bank, Washington DC, 1995.

[8] Anonymous. Bay cleanup threatened by area growth. Washington Post 1996;19 June pE2.

[9] Turner RK, Subak S, Adger N. Pressures, trends, and impacts in coastal zones: interactions between socio-economic and natural systems. Environmental Management, 1996;20:159–173.

[10] Blankenship K. Detroit and Toronto meet the Bay. Bay Journal, 1995;5(1):1,4–7 (Alliance for the Chesapeake Bay).

[11] Cohen M. et al. Quantitative estimation of the entry of dioxins, furans and hexachlorobenzene into the Great Lakes from airborne and waterborne sources. Center for the Biology of Natural Systems, Queens College, CUNY, Flushing, NY, 1995.

[12] Lundin CG, Linden O. Coastal ecosystem: attempt to manage a threatened resource. Ambio 1993;22:468–473.

[13] Dennis KC, Niang-Diop I, Nicholls RJ. Sea-level rise and Senegal: potential impacts and consequences. Journal of Coastal Research 1995;14:243–261.

[14] Fradkin PL. A River No More. New York: Alfred A. Knopf, 1981.

[15] Cicin-Sain B, Knecht RW. Earth Summit Implementation: Progress Achieved on Oceans and Coastal [special issue]. Ocean and Coastal Management 1996;29:1-3.

[16] Chua, TE, editor. Lessons Learned in ICM [special issue]. Ocean and Coastal Management, in press.

[17] Emeis K-C, Larsen B, Serbold E. What is the environment capacity of enclosed marginal seas? Approaches to the problem in the Baltic, North and Mediterranean Seas. In: Hsu KJ, Thiede J, editors. The Use and Misuse of the Seafloor. 1992; ch. 9, pp. 181–211.

[18] Kent DM, editor. Applied Wetlands Science and Technology. Boca Raton, FL: Lewis, 1994.

[19] Clark E. The United States, ch. 2, 39-72. In: Turner RK, Jones T, editors. Wetlands: Market and Intervention Failures. London: Eathscan, 1991.

[20] Freeman M. The Measurement of Environmental and Resource Volves, Resources for the Future. Washington DC, 1993.

[21] Stone GW, Grymes JM III, Armbruster JP, Xu and Huh OK. Researchers study impact of hurricane Opal on Florida Coast. EOS 1996;77(19):181.

[22] Turner RK. Wetlands conservation: economics and ethics. In: Collard D et al., editors. Economics, Growth and Sustainable Environments. London: Macmillan, 1988.

[23] Barbier E. The economic value of ecosystems: tropical wetlands. LEEC Gatekeeper 89-02, London Environmental Economic Centre, London, 1989.

[24] Maass A. et al. A simplified River basin system for testing methods and techniques of analysis. Cambridge, MA: Harvard University Press, 1962.

[25] Bower BT, Ehler C, Basta D. A framework for planning for integrated coastal zone management. Office of Ocean Resources Conservation and Assessment, National Ocean Service, National Oceanic and Atmospheric Administration, Silver Spring, MD, 1994.

[17]

Carl Gustaf Lundin and Olof Lindén

Coastal Ecosystems: Attempts to Manage a Threatened Resource

Tropical coastal zones are productive ecosystems that currently face severe environmental threats, particularly from organic pollution. The role of the coastal ecosystems is analyzed and the relationship between coastal ecosystem health and fisheries productivity is explained. Ecological disturbances from organic sources like sewage and siltation is highlighted. The issues of integrated coastal zone management (ICZM) are discussed, particularly in the context of conserving natural ecosystems or transforming them to managed systems. Issues of population density, management capacity, and socioeconomic conditions are discussed. The possibilities for closing carbon cycles currently leaking organic materials to the coastal waters are pursued. Finally, examples of ICZM initiatives in the ASEAN countries and East Africa are presented.

THE ROLE OF OCEANS AND COASTAL AREAS

Since the dawn of civilization the oceans and coastal areas have played a key role in the development of human societies and communities. Whereas forests, deserts and mountains on land masses have separated people and delayed or even prevented communication between cultures, the sea and particularly the coastal areas, have enabled contact even between very remote civilizations. Oceans and coastal areas are of no less importance today. The countries around the "Pacific Rim" exhibit some of the highest rates of economic growth in the world, and the fact that these countries share a common ocean is undoubtedly an important contributing factor. Most of the people on this planet inhabit the coastal areas of the continents. The figures given today show two thirds of the entire world population within 60 km of the coasts of the continents (1). Furthermore, population growth in coastal areas is generally higher than inland, mainly due to migration from inland areas and, consequently, there is a greater increase in the numbers of coastal inhabitants than in other areas. As a consequence, population density in many coastal areas, particularly in the developing world, is now very high. In many rural provinces in countries in South and Southeast Asia, the Pacific, Central America and the Caribbean, population pressure is now extremely high, sometimes exceeding 1000 and even 2000 individuals per km². Obviously, such large numbers of people exert enormous pressure on the coastal ecosystems, and in many cases the level of acceptable exploitation of fisheries as well as harvest of mangroves and other types of vegetation has long ago been exceeded.

The seas and the coastal areas play an even more important role today, since they provide protein from fish and other seafood. For the developing world, where large numbers of poor people depend exclusively on this fish for their supply of protein, this is extremely important. In certain countries around the tropical oceans, between 40 and 95% of the protein in the food consumed comes from seafoods. For the Southeast Asian region 60% has been estimated as an average (2). In the Pacific and Caribbean regions, the figures are probably close to 90%. It

is protein, usually in the form of dried fish, consumed in a diet otherwise rich only in carbohydrates, that keeps millions of people in the Third World in reasonably good health, and allows normal physiological and mental development in children. Furthermore, it is the poorest in these societies that depend on fish, since alternative sources of animal protein are usually beyond their economical means. In Asia, it is estimated that over 1 billion people depend exclusively on fish for their protein requirements (3). The current problems of environmental destruction in tropical coastal seas, and the effects on the productivity of fish and other seafood from these areas are, therefore, of primary importance. Continued destruction of estuaries and lagoons, mangrove forests, seagrass beds and coral reefs in the tropical Third World will mean the difference between life and death for millions of poor people and for many others, the difference between a life in reasonable health and one of malnutrition, disease and starvation. In addition, it is likely that parts of the human populations will migrate to nearby or distant areas in their search for better conditions.

THE HEALTH OF THE COASTAL ECOSYSTEMS AND FISHERIES PRODUCTIVITY IN THE TROPICS

There is now considerable evidence that, for marine areas in the tropics, there is a clear correlation between the productivity of coastal ecosystems (particularly in mangrove forests, seagrass beds and coral reefs), and the productivity of the fisheries (4, 5). What has become increasingly clear is the relative importance of nearshore—mainly benthic—productivity of plants and algae in tropical areas, compared with the situation in temperate seas and seas in Arctic areas. In the latter, the productivity of the fisheries is dependent on food chains in which microscopic plankton are the primary producers. These microscopic plankton appear in abundance both in offshore and nearshore areas during the spring, and this "bloom" is the result of massive upwelling of nutrient-rich deep water over extensive areas due to the cooling of the surface water in the Arctic and temperate winter. This cooling eliminates the temperature differences between surface and deep waters and hence mixing of the water masses can take place. This is the classic textbook description of marine productivity and it means that the productivity of the fisheries can take place over huge offshore areas of the ocean (6–8). This can be illustrated by the fishery statistics from areas such as the North Atlantic, where typical offshore species such as cod, haddock, pollack, whiting, redfish, capeline, herring, sprat and mackerel dominate the catches. These are species which spend their entire life in offshore areas, and most populations of these species are not dependent on the coastal ecosystems for their survival.

In the tropics on the other hand, with the exception of a few well-known areas such as these off Western Sahara-Mauritania, off Chile-Peru, and off Angola-Namibia, there is over most of the oceans and coastal areas a much more pronounced and stable temperature stratification throughout the year. As a result most of the nutrients that are released during degradation of organic matter in deeper waters will not be available for the primary producers in the surface waters, and therefore the levels of

Small-scale traditional fisheries in Puttalam Lagoon, North-West Sri Lanka. As in most developing tropical coastal countries, the fishery has traditionally been carried out in shallow waters using primitive crafts and gear. Due to the low intensity, the fishing has been sustainable for hundreds of years. Photo: O. Lindén.

The fishing village in Kandakuliya, North-West Sri Lanka. The small-scale coastal fisheries in large parts of the tropical developing countries has "developed" in the last decades into an intensive fishing using modern fiberglass boats, OB-engines, nylon nets, etc. The development has, over large areas, resulted in overfishing and decreasing catches per effort. The improved gear in combination with decreasing catches has resulted in fishing over coral reefs which are rapidly destroyed. As a result the erosion has become an increasing problem. In Kandakuliya, the erosion has removed 10 to 20 meters of beach per year during the last decades resulting in destruction of houses, palm trees, etc. Photo: O. Lindén.

nutrients and rates of primary production in the tropical ocean, offshore, are very low throughout the year. The primary production in the surface water of the tropical offshore ocean is instead determined by the quantities of nutrients which are transported from land, and to the extent to which these are supplemented by the release from degradation processes above the thermocline. As a result, productivity is generally very low when compared with the Arctic or temperate ocean areas. In the tropics, the productivity in nearshore shallow areas is much more important for the production of fish than what is the case in more temperate regions of the world. In the former, the primary producers are, to a large extent, benthic vegetation, seagrasses, (macroscopic) algae, mangrove vegetation and coral reefs. Thus, there is a greater dependence of tropical commercial fish populations on shallow nearshore ecosystems than for similar fish populations from Arctic and Antarctic areas. For example, the dominating species in the catches in Tanzania and other parts of tropical East Africa are snappers, groupers, sweetlips, emperors, sardines, croakers, grunts, catfish, jacks and trevallys (9). Most of these species spend some or all of their entire life cycle in the shallow nearshore areas, either as larvae feeding and hiding in the mangroves, as juveniles in the seagrass beds, or as adults feeding in the coral reefs. In the areas surrounding the Bay of Bengal, 90% of the commercially most important species are directly or indirectly dependent on the areas of mangrove forest (10). In the Caribbean, coral reef fishes play a dominating role in the catches, constituting 50 to 65% of the catches (11). The contribution of coral-reef fish in the total fishery catches landed in tropical developing countries can be estimated at between 20 and 25%, and over 90% of the fish catch caught in the tropical seas is taken in shallow coastal waters (12). The importance of preserving the nearshore habitats is therefore much greater in the tropics than in other climatic regions. The implications of, for example decreasing areas of mangrove forest along the coast of a country, or a degradation of the coral reefs, will inevitably result in a corresponding decrease in the fish stocks and, sooner or later, a drop in the catches. This is also what we are currently witnessing throughout the tropics; a rapid degradation of the coastal ecosystems in countries such as Sri Lanka, Indonesia, Bangladesh, Philippines, Thailand, Ecuador and a large number of other tropical countries, and a simultaneous reduction in the catches of fish and shrimp (10, 11, 13). In many countries in

Southeast Asia, fish catches have been dropping steadily over the last decade, with figures ranging from 1 to 5% yearly. Catches in the Gulf of Thailand are near collapse with catches only some 20% of what they were a few decades ago (13, 14).

THE DEGRADATION OF COASTAL HABITATS AND ECOSYSTEMS

Coastal habitats and ecosystems are degraded and destroyed as a result of human interference in carbon flow, for example, by the release of sewage, organic matter, nutrients and sediment into coastal waters. In reality, it may often be difficult to assess which factor is the primary cause of the degradation that is taking place in coastal waters. Most studies suggest a combination of factors.

Sewage and Siltation in Estuaries and Coastal Waters

The discharge of untreated or inadequately treated sewage, pollutes the coastal environment near most cities, towns and villages throughout the tropics. Furthermore, there are many signs that the rate of siltation in coastal waters has increased considerably during the last century over large parts of the world. Several expert groups maintain that sewage from human communities, organic matter, nutrients and particles (causing turbidity and siltation) from agriculture, forestry and construction in coastal areas and the drainage areas of rivers have been ranked as the most severe and widespread problems causing degradation of productive coastal habitats in coastal and marginal seas (15, 16). This has severe consequences including increasing turbidity, eutrophication and viral and bacterial pollution. Algal blooms in tropical coastal waters are now much more common than some decades ago, for example in the South China Sea (13) and throughout the Pacific Ocean (17). Sewage and/or siltation is regarded as one of the most widespread problem(s) resulting in the degradation of the coral reefs and other coastal environments in the Philippines (13, 18), Singapore (19), Malaysia (20), Indonesia (18), Sri Lanka (21), in Pacific Islands (17, 22), Hawaii (23), in the Persian Gulf (24), in the Caribbean (25), along the Colombian and Costa Rican Caribbean coasts (26, 27), and in Cuba (28). Much of the soil eroded from the hillsides of the Himalayas of the Indo-Pakistani Subcontinent has entered the coastal zone of the Bay of Bengal and Indian and Bangladeshi coastal waters. Indian scientists have calculated that the transport of sediment by rivers discharging into the Bay of Bengal has increased one 100-fold during the past 100 years (15).

The contamination of coastal waters by pathogenic bacteria and viruses are causing more extensive and frequent outbreaks of viral hepatitis and typhoid in many parts of the tropics including Malaysia, Indonesia, the Philippines, Vietnam and countries in the Pacific Ocean (13, 17). In addition, the incidences of blooms of toxic microorganisms, dinoflagellates and cyano-bacteria, have also increased substantially in the last 20 years (29). Paralytic Shellfish Poisoning (PSP) caused by outbreaks of dinoflagellates such as *Gonyaulax* and *Pyrodinium*, so called "red tides", cause increasing human illness and many casualties each year. In the Pacific Ocean region alone, as many as 50 000 to 100 000 people are hospitalized each year as a result of ciguatera poisoning and related symptoms (30). The mean annual incidence of cases of ciguatera poisoning is now 1-4 per thousand of the population throughout the Pacific Ocean region. As most cases are never reported, the actual levels are much higher. It should be noted that these problems are of recent origin. In Malaysia, for example, hardly any cases had been reported before 1970. Microbial contamination of seafood such as oysters and mussels results in multi-million dollar losses for the aquaculture industry in Thailand and other countries of the region. In fact, such problems are considered the greatest constraint towards the development and expansion of aquaculture in many countries in Southeast Asia.

Coral Reefs

Coral reefs are disappearing at an alarming rate throughout the tropics. In Southeast Asia where approximately 30% of the world's coral reefs are found, about 60% of the reefs are already destroyed or on the verge of destruction (18). The remaining reefs are also threatened and it can be predicted that most of the reefs in the region will be eradicated within the next 40 years (Table 1; 18). The major causes for the destruction of the reefs in Southeast Asia are directly or indirectly induced by humans; organic and inorganic pollution, sedimentation and overexploitation. The overexploitation is exacerbated because dynamite and muro-ami fishing are used frequently to capture the few remaining fishes. There are few regional inventories available from other areas of the tropics. A review of the status of reefs in the Caribbean showed that the coral cover had recently declined at 8 out of 14 sites in as many countries (25). While much of the loss was attributed to natural events such as damage from hurricanes, organic pollution, sedimentation and effects of destructive fishing techniques were reported from most areas. It should be noted that 9 of the sites were in marine parks, reserves or other areas of restricted access.

Inventories from several countries, confirm this picture of a negative development. In Sri Lanka, the major portion of the reefs along the west and south coast have been seriously degraded, and it has been estimated that about 10% of the coral reefs are being destroyed each year (21, 31). A whole range of problems cause this damage including siltation, the release of sewage, impact of tourism, damage due to fishing on the reefs, collection of reef organisms for the aquarium trade, and the mining of coral for lime production. In the Gulf of Thailand, the coral reefs have suffered extensive damage due to the release of contaminated waters via rivers, notably the Pattaya (32). However, in other areas destructive fishing techniques (particularly dynamiting), siltation and pollution from prawn cultures that cause release of organically enriched waters are major problems. Recent reports have shown that the reef cover is declining by 20% annually due to pressures resulting from tourism, particularly siltation from construction in the coastal areas and the release of sewage (32, 33). In the Philippines, studies show that almost 70% of 735 studied reefs in the country are seriously damaged (coral cover ranging from "fair" to "poor"). The main reasons for this damage were destructive fishing, particularly

using explosives or muro-ami fishing, and siltation from extensive clearing of forests. The situation is similar in Indonesia with about 80% or more of the reefs in certain areas in the Eastern Indonesian Archipelago being damaged by dynamite fishing.

HALTING THE NEGATIVE TREND

The above description of coastal degradation caused by biological emissions and other forms of human impact is increasingly recognized as a threat to most tropical coastal countries. The question is what can be done to halt the negative trend? There are a number of potential answers to this question, but the first question that must be answered is whether protection and conservation of existing natural ecosystems are the prime objectives when managing natural systems or whether the transformation of them through new technologies into more hardy/managed/eutrophic systems is preferable?

The conservation approach might be possible and desirable in situations where population pressure is limited. The focus should be on areas of high biological diversity and of special interest to humanity and science. The International Union for the Conservation of Nature (IUCN) and United Nations Environmental Program (UNEP) have set priorities for such areas (34) as have the countries themselves. What is a reasonable level of conservation, 50% or 10%? This will be an *ad hoc* solution depending on the factors presented below. Conservation as a concept is reasonably straight forward and will not be discussed in this article.

Ecologically modified systems are desirable where large populations are dependent on coastal goods and services that go beyond the (almost undisturbed ecological) production capacity of the natural ecosystems. This is typically the case in large urban areas on the coast or along major rivers or in regions with enclosed bays and lagoons where circulation is poor. Modifications can either build on existing natural ecosystems and maintain as much as possible of the resources or create solutions involving major infrastructure construction and intense management. The determining factors will generally be population density, management capacity and socioeconomic conditions.

Population Density

There will be no solution to sustainable coastal use if the issue of population density is not addressed. The current trend has led to overpopulation of coastal areas, to the extent where there are significant health impacts and enormous pressure on the existing natural ecosystems. This trend needs to be halted through family planning; education, particularly of women; and economic development to permit other solutions than large families (35). An alternative solution could be relocation from sensitive areas, through economic incentives, of human activities that are not coast dependent.

Management Capacity

Management capacity is a serious constraint to development of technical systems that deal with human-generated carbon flows. In situations where the capacity of existing trained personnel is low, a more foolproof technical and maintenance solution must be used even when this means a greater destruction of the natural environment. Examples of this situation are common in many African countries where sewage works are constructed but are utilized below design capacity. In Accra in Ghana, the sewage system is currently running on a very low level of utilization mainly because the technical solution chosen was too costly (36). In the Seychelles capital city, Victoria, a sewage network was constructed in the middle of the 1980s by the African Development Bank, but it has not yet been put into operation because there is no sewage-treatment plant.

470

Socioeconomic Conditions

Socioeconomic conditions play a very large role in determining the range of technical solutions that can be considered. One determining factor is capital availability and a sense of economic priority. There is general agreement on the need for drinking-water projects and much solid data has been produced to show the profitability of such investments. For sewage and sanitation there is much less economic data that show the benefits of not polluting coastal and marine environments in the tropics. There is a lack of knowledge about the health costs when seafood is contaminated and when human recreation is affected. This is a main reason why capital is no longer forthcoming. As mentioned above large population groups depend on fisheries resources. The fact that the quality of marine products is uneven makes export and local markets value considerably lower.

Another social and cultural factor is whether the society has a "sea culture" or not. In East Africa there is very little tradition of utilizing coastal and marine resources and the sea has represented many threats, e.g. slave trade. As a consequence, there is a wide ignorance of available resources and their management. In many societies, up until recently, there has been little attention paid to the sea and marine life. This social alienation from the sea makes human concern for degrading the environment less pronounced than in the case of land degradation. The possibilities of letting the sea bring people together rather than distancing them from each other should be stressed in international work.

Closing the Carbon Flows

Closing the flow of organic wastes and nutrients will be an essential, but very difficult, component of any attempt to halt coastal-environment degradation. Many technological options are, however, available in agriculture, through soil conservation, capturing of effluents in drainage basins and applying appropriate levels of fertilizers. In sewage disposal, primary and secondary treatment should be carried out to ensure that maximum levels of nutrients are recycled for productive purposes. In freshwater, this can be duckweed or other fast-growing species with good nutrient-absorbing capacity and high protein content. In the coastal zone, algal production may be efficient in capturing nutrients and producing marketable products. Linking aquaculture of fish or shrimp to the treatment, particularly if done in a cascade of ponds, has proven to be an efficient way of closing the carbon cycles in certain situations. However, in many cases the existing circumstances may prevent effective use of such methods. The problems become particularly great when large volumes of sewage must be taken care of. The areas needed for ponds will be substantial and the problems related to the mixing of sewage with toxic/hazardous effluents from industries and workshops will present large problems that may be impossible to solve. However, in small communities where enough space is available, the use of such ponds for polyculture systems may prove to be a viable alternative. It must however be emphasized that considerable additional research and development are needed before such techniques can be applied over large areas and it is probably more realistic to attempt a range of alternative technical solutions.

Important scientific issues that must be determined for each coastal area where Integrated Coastal Zone Management (ICZM) is attempted are what nutrient levels are acceptable to the existing systems and where collapse occurs. How much needs to be absorbed by mangroves, seagrass beds and coral reefs? One important aspect of this is the competition between ecosystems and disruption of existing ecosystem functions; an example being where corals are replaced by algae when several environmental factors interact to change the existing environment.

Corals compete with algae. When nutrient levels increase or algae grazers like fish and sea urchins decline, corals lose their comparative advantage and are replaced by algae. Data from Jamaica's north coast indicate that the collapse of the coral reefs was due to a combination of factors including: overfishing, reducing algae grazing reef fish; deficiency in Caribbean-wide loss of sea urchins (*Diadema*) and increasing levels of nutrients from agricultural runoff and human wastes. The combination of factors favoring algae over coral leaves the reefs in a state where an occasional storm can wreck the entire system. The decline of coral reefs opens the beaches and coasts to erosion and leads to less biodiversity and further degradation of the ecosystems. The comparison can be made with Belize where there are no sea urchins and comparable nutrient levels exist. The situation is much more positive, in terms of the system's ability to sustain the coral cover, because a healthier reef-fish population is present. This example indicates that a combination of factors will disrupt the balance in the reef ecosystem and in the long term lead to collapse.

INTEGRATED COASTAL ZONE MANAGEMENT

The use of Integrated Coastal Zone Management (ICZM) as a toolbox to develop coastal resources in a sustainable manner and to mitigate conflicts between users has proven to be a possible solution in many countries. The inherent problem with sectoral policies is that they pay limited attention to issues outside of their sector and become incapable of finding solutions outside sector boundaries. This is addressed through inter-agency collaboration. The origin of coastal problems is located outside the coastal zone and associated with sectors that are unaware of the consequences of their actions. ICZM is a method that can lead to sustainable

The fishing for small fish and prawns in lagoons in Sri Lanka is often done as side activity by children. The catch is used directly in the cooking or, in the case of prawns and small groupers, sold alive for use in aquaculture. Photo: O. Lindén.

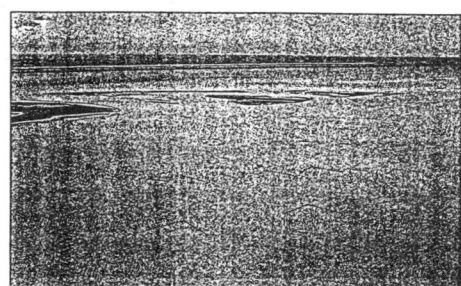

The coastal areas including the shallow lagoons in the tropical countries are highly productive. We find here the highest productivity of any natural ecosystem on this planet, measured as grams of carbon per area per year. The photo shows Laguna Madre in Mexico, where the seabed is covered by luxuriant seagrass beds supporting large stocks of fish and shellfish. The angular structures are fish traps. Photo: O. Lindén.

development because it has the advantage of securing government participation as well as stakeholder involvement (37).

There are currently over 100 ICZM programs operating around the world and some of these programs have been running for over 15 years (38). In the UNCED Agenda 21, Chapter 17, there are several recommendations on what should be included in a ICZM program. However, many complex issues, e.g. population growth in coastal areas were not addressed by UNCED (39).

There is no absolute methodology that can be prescribed to accomplish ICZM, but there are several good examples that have been conducted around the world that can inspire to new programs. Several of these programs are, however, developed in industrialized temperate and subtropical countries and have limited relevance for developing countries in the tropics. The ASEAN programs, Great Barrier Reef Marine Park (Authority) (40), Caribbean Regional Seas Program and others are of particular interest.

The ASEAN Experience

During the last several years the ASEAN countries (Brunei Darussalam, Indonesia, Malaysia, Philippines, Singapore and Thailand) have received financial assistance from USAID and technical assistance from the International Center for Living Aquatic Resources Management (ICLARM) to develop integrated ICZM pilot projects in each country within the framework of the ASEAN-US Coastal Resource Management Project (37). In the *Baguio Resolution on Coastal Resources Management*, March 7, 1990, there was a clear statement of the need for action, but also information on what was being done under the ASEAN/US Coastal Resources Management Project:

- Analysis, documentation and dissemination of information on how coastal resources are being developed for economic purposes.
- Strengthening of existing management capabilities of local and national institutions in the region.
- Provision of technical solutions to resolve conflicts arising from competing uses of coastal resources.
- Assistance to various organizations and agencies in the development of coastal-area management plans.

The ASEAN countries have implemented ICZM plans in various part of these countries more or less successfully. The smaller and richer countries have been most successful, whereas there have been greater problems in the countries like the Phil-

ippines where large groups of poor fishermen need to leave the fishing sector to find alternative employment. The overall impression is, however, positive and indicates that much can be accomplished when the will to change is present both at the top and among the stakeholders.

Arusha Resolution

At the *Workshop and Policy Conference on Integrated Coastal Zone Management in Eastern Africa including Island States*, held in Arusha, Tanzania in April 1993, ministers from six East African nations, Madagascar, Mauritius, Mozambique, Seychelles and Tanzania and subsequently Kenya agreed to make Integrated Coastal Zone Management a prominent part of government policy (41). ICZM will be used to assist government planning and direct sectorial policies so as to avoid resource destruction in areas outside the dominant sectors. This work started at Arusha, but will be implemented in the years to come (42).

Investing in Coastal Management

There are a variety of opportunities for improved management of coastal resources but they frequently require stronger political commitment and additional financial resources. In many cases,

> **Main Areas Addressed at Arusha**
>
> Degradation of marine and coastal areas is perceived as a great loss for the countries and will be addressed with policy changes and physical investments.
>
> Marine pollution plays an important part in coastal degradation in East Africa and will be counteracted within the framework of Integrated Coastal Zone Management (ICZM).
>
> Property rights, user rights and land tenure systems shall be reviewed to address common property issues.
>
> Local participation in the decision-making process shall be achieved through the use of local authorities and village councils.
>
> Capacity building in the field of marine science and coastal management will be facilitated through joint investments by the donor community.
>
> Development of National action plans are important steps to implement ICZM in the East Africa region.

insufficient work is devoted to building the financial mechanisms that will permit development of investment programs and major improvements of the environment through treatment and use of effluents from agriculture and urban populations. Investment programs can be developed through *ad hoc* bodies like the *Helsinki Commission*, the *Environment Program for the Mediterranean*, and the *Black Sea Program*, and both capital and technical assistance can be transfered between the countries in the region. It is important to ensure that the objectivity of the preparatory work identifies the most pressing environmental concerns. Scientific and technical data gathering can, in many cases, be sufficient to provide information for the prioritization and prefeasibility studies. Once it is time for feasibility studies of different investment alternatives, specific scientific investigations can be undertaken to cover outstanding issues. In most parts of the world, the available information is often far richer than expected, but in most cases widely dispersed.

CONCLUSIONS

There is rapid ongoing destruction of many of the coastal and marine resources essential to human beings throughout the Third World. Siltation and nutrient-rich emissions from agriculture and from urban conurbations are among the most important causes of coastal-resource degradation. The major underlying factor is the rapid population growth that is taking place in most tropical countries. The coasts are particularly vulnerable and often experience the highest growthrate, often more than 5% per year.

Coastal degradation cannot be solved within the traditional sectors like fisheries and shipping. What is required are Integrated Coastal Zone Management Programs, and projects, to address all the factors that have impacts on coastal zones. Major steps have been taken in several countries to halt negative trends and we will probably see more well-implemented ICZM programs that will address the coastal-resource-user conflicts.

One of the key issues to determine within the ICZM programs is whether conservation and sustainable use of the resources should be attempted in a particular area or if the focus instead should be on managing resources through engineering solutions. This should be determined on the basis of sound scientific evaluations of the existing resources and the carrying capacity of the ecosystem, in relation to the needs of the particular area. In areas with rapid population growth, it may be necessary to invest in major carbon and nutrient-absorbing systems to avoid further degradation and to preserve the biological productivity of existing systems.

References and Notes

1. United Nations. 1985. *Estimates and Projections of Urban, Rural and City Populations, 1950–2025: the 1982 Assessment.* United Nations, New York.
2. Young, S.K.T. 1989. Coastal resources management in the ASEAN region: Problems and directions. In: *Coastal Area Management in Southeast Asia: Policies, Management Strategies and Case Studies.* Chua, T.-E. and Pauly, D. (eds). ICLARM Contribution No. 543 xi.
3. FAO. 1985. *The State of Food and Agriculture 1984.* Agriculture Series No. 18. FAO, Rome, 185 p.
4. Macnae, W. 1974. *Mangrove Forests and Fishers.* Indian Ocean Fisheries Commission IOFC/DEV/74.34. FAO, Rome, 35 p.
5. Martsubroto, P. and Naamin, N. 1977. Relationship between tidal forests and commercial shrimp production in Indonesia. *Mar. Res. Indones. 8,* 81–86.
6. Ryther, J.H. 1969. Photosynthesis and fish production in the sea. *Science 166,* 72–76.
7. Steeman Nielsen, E. 1975. Marine photosynthesis with special emphasis on the ecological aspects. *Elsevier Oceanography Series 13.* Amsterdam.
8. Strickland, J.D.H. 1965. Production of organic matter in the primary stages of the oceanic food chains. In: *Chemical Oceanography.* Riley, J.B. and Skirrow, G. (eds). Academic, p. 478–610.
9. Tanzania Fisheries Department. YEAR??? *Fishery Statistics for 1990.*
10. Sen Gupta, R., Ali, M., Bhuiyan, A.L., Sivalingam, P.M., Subrasinghe, S. and Tirmizi, N. 1990. State of the marine environment in the South Asian Seas Region. *UNEP Regional Seas Reports and Studies, No. 123.*
11. FAO. 1993. *Fisheries Statistics 1992.*
12. McManus, J. 1988. Coral reefs of the ASEAN Region: status and management. *Ambio 17,* 189–193.
13. Gomez, E., Mungeugs, M., Hothy, A., Kuan, K.J., Wu, R., Soegiarto, A. and Deocadiz, E. 1989. State of the marine environment in the East Asian Seas Region. *UNEP Regional Seas Reports and Studies, No. 126.*
14. World Resources Institute. 1992. *World Resources 1992-93.* The World Resources Institute, the United Nations Environment Programme and the United Nations Development Programme. Oxford University Press, New York.
15. UNEP. 1988. Background document presented at GESAMP meeting of the Task Force to prepare regional overviews on the state of the marine environment in regional seas. Geneva.
16. International experts opinion regarding the condition of world's coral reefs, from "Global Assessment of Coral Reefs". Colloquium at University of Miami, 7–10 June, 1993.
17. Brodie, J., Arnould, L., Eldredge, L., Hammond, L., Holthus, P. Mombray, D. and Tortell, P. 1990. State of the marine environment in the South Pacific Region. *UNEP Regional Seas Reports and Studies, No. 127.*
18. Wilkinson, C.R., Chou, L.M., Gomez, E., Ridzwan, A.R., Soekarno, S. and Sudra, S. 1993. Status of coral reefs in Southeast Asia: Threats and responses. In: *Global Aspects of Corals; Health, Hazard and History.* University of Miami.
19. Chou, L.M. Community structure of sediment stressed reefs in Singapore. *Galaxea 7,* 101–111.
20. Mohamed, M.I.H., Aziz, H.A. and Ahmad, S.B. 1992. The impact of resort development on the coral reefs of Pulau Redang Marine Park. In: Chou, L.M. and Wilkinson, C.R. (Eds). *Third ASEAN Science and Technology Week Conference Proceedings, vol. 6.* Dept. of Zoology, National University of Singapore and National Science and Technology Board, Singapore.
21. Rajasuriya, A. 1993. Present status of coral reefs in Sri Lanka. In: *Global Aspects of Corals; Health, Hazards and History.* University of Miami.
22. Richmond, R.H. 1993. Effects of coastal runoff on coral reproduction. In: *Global Aspects of Corals; Health, Hazards and History.* University of Miami.
23. Hunter, C.L. and Evans, C.W. 1993. Reefs in Kaneohe Bay, Hawaii: Two centuries of western influence and two decades of data. In: *Global Aspects of Corals; Health, Hazards and History.* University of Miami.
24. Lindén, O., Abdulraheem, M.Y., Gerges, M.A., Alam, I., Behbehani, M., Borhan, M.A. and Al-Kassab, L.F. 1990. The state of the marine environment in the ROPME Sea Area. *UNEP Regional Seas Reports and Studies No. 112,* Rev. 1. UNEP.
25. Smith, S.R. and Ogden, J.C. (eds). 1993. Status and recent history of coral reefs at the CARICOMP network of Caribbean marine laboratories. In: *Global Aspects of Corals; Health, Hazards and History.* University of Miami.
26. Garzòn-Ferreira, J. and Kielman, M. 1993. Extensive mortality of corals in the Colombian Caribbean during the last two decades. In: *Global Aspects of Corals; Health, Hazards and History.* University of Miami.
27. Cortes, J. 1993. A reef under siltation stress: a decade of degradation. In: *Global Aspects of Corals; Health, Hazards and History.* University of Miami.
28. Alcolado, P.M., Herrera-Moreno, A. and Martinez-Estalella, N. 1993. Sessile communities as environmental bio-monitors in Cuban coral reefs. In: *Global Aspects of Corals; Health, Hazards and History.* University of Miami.
29. Maclean, J. 1989. Indo-Pacific red tides, 1985–1988. *Mar. Pollut. Bull. 20,* 304–309.
30. Anderson, D.M. and Lobel, P.S. 1987. The continuing enigma of ciguetera. *Biol. Bull. 172,* 89–107.
31. Statistics provided by Sri Lanka Coast Conservation Division.
32. Sudara, S. and Nateekarnchanalop, S. 1990. The status of tourism development on the reefs in Thailand. *Proc. 6th Int. Coral Reef Symp. 2,* 273–278.
33. ASEAN-Australia Marine Science Project. 1992. The status of living coastal resources in ASEAN countries. In: *ASEAN Marine Science Project: Living Coastal Resources.* English, S. (ed.). Australian Institute of Marine Science, Townsville, p. 6–17.
34. Kelleher, G. and Bleakley, C. 1993. *The Conservation of Global Marine Biodiversity, a Global Representative System of Marine Protected Areas.* A report prepared for World Bank Environment Department. (In print).
35. Birdsall, N. 1992. *Another Look at Population and Global Warming.* World Bank Policy Research Working Papers.
36. World Bank. 1977. *Project Performance Audit Report for Ghana: Accra/Tema Water Supply and Sewerage Project.*
37. Chua, T.-E. and Fallon Scura, L. (eds). 1992. Integrative framework and methods for coastal area management. *ICLARM Conf. Proc. 37.*
38. Sorensen, J. 1991. Integrated coastal area management programs, piliot programs and feasibility studies. *Intercoast Network International Newsletter of Coastal Management,* Issue 16, 8–12.
39. *Agenda 21.* 1992. Chapter 17. Protection of the oceans, all kinds of seas, including enclosed and semi-enclosed seas, and coastal areas and the protection, rational use and development of their living resources. UNCED, United Nations Conference on Environment and Development.
40. Kelleher, G. 1987. Management of the Great Barrier Reef Marine Park. *Austr. Parks Recreation 23,* 27–33.
41. Governments of Madagascar, Mauritius, Mozambique, Seychelles and Tanzania. 1993. *Resolution of the Policy Conference on Integrated Coastal Zone Management in Eastern Africa including the Island States.*
42. Lindén, O. 1993. Resolution on integrated coastal zone Management in Eastern Africa signed in Arusha, Tanzania. *Ambio 22,* 408–409.

Carl Gustaf Lundin is an environmental specialist in the Environment Department of The World Bank. He is PhD student at the Natural Resources Management Institute at Stockholm University. His research is focused on implementation issues in natural resources management in tropical coastal zones. He is involved in several coastal zone management projects, particularly in the Western Indian Ocean. He is co-author of the World Bank Discussion paper 210, Marine Biotechnology and Developing Countries. His address: Land Water and Natural Habitats, Environment Department, 1818 H Street N.W., Washington, DC 20433, USA. **Olof Lindén** is adjunct professor at the Department of Zoology, Stockholm University. His background is in marine ecotoxicology and the present research deals mainly with coastal area management in developing tropical countries. He is coordinator of SAREC-funded marine science programs in East Africa, Sri Lanka, Vietnam and Costa Rica. His address: SAREC Marine Science Program, Department of Zoology, Stockholm University, S-106 91 Stockholm, Sweden.

[18]

Journal of Environmental Management (1998) 52, 379–387

Article No. ev980186

Sustaining co-operation for coastal sustainability

C. A. Davos

Coastal-zone sustainability policies are socially constructed. It follows that their effective implementation depends on the sustainable voluntary co-operation of stakeholders with competing interests and priorities. No form of integrated coastal-zone management can nurture such co-operation as long as the objective is to determine 'best' policies, derived by expert-based rational analysis, instead of seeking to identify 'correct' policies, ones that can draw the maximum possible stakeholder support. The latter task requires a co-operative coastal-zone management that incorporates the relevant public discourse into the policy formation process in a direct, proactive and conflict minimizing manner. Towards this end, four major challenges are examined for maximizing the stakeholders' motivation for voluntary co-operation: (1) optimism about the level of optimism; (2) agenda setting; (3) value discourse; and (4) information and empowerment.
© 1998 Academic Press Limited

Keywords: conflict management, co-operation, co-operative coastal-zone management, integrated coastal-zone management, sustainability, value discourse.

Introduction

The sustainability of coastal zones is a growing concern worldwide. Calls abound for actions including: the definition of policy objectives which are specific to coasts and their resources; the integration and harmonization of sectoral policies; the collection of relevant information and development of coastal environmental indices; public education and participation in decision making at an early stage of policy formation; and the indentification and testing of the relative effectiveness of different policy instruments and institutional arrangements [Organization for Economic Co-operation and Development (OECD), 1993]. These calls find their summary expression in the proposition by the World Coast Conference (1993) of integrated coastal-zone management (CZM), which is defined as a continuous and evolutionary process for achieving sustainable development, involving 'the comprehensive assessment, setting of objectives, planning and management of coastal systems and resources, taking into account traditional, cultural and historical perspectives and conflicting interests and uses'.

Undoubtedly, such calls point to necessary functions of CZM. Their effectiveness is undermined, however, by their insistence on reiterating 'ends' while failing to suggest 'means' and argue the underlying principles for choosing them. The following questions regarding the above prescriptions of integrated coastal-zone management illuminate their weaknesses: who is supposed to define the specific objectives for the coasts and their resources? how is the integration and harmonization of sectoral policies to be achieved and by whom? whose capacity to access, assimilate and evaluate information will dictate the design and availability of the collected information and coastal environmental indices? will the objective of public education and citizen participation go beyond that of 'assessing the political feasibility of certain alternatives', and educating the 'uninformed' public about why the experts' decision or proposed action is the best one? and finally, who will decide which policy instrument and institutional regimes will be tested for their effectiveness, how will this be done and by whom? Any answers to these questions are bound to generate such conflicts that, unless they can be managed with the direct involvement of the stake-

Department of
Environmental Health
Sciences, School of
Public Health, 10833 le
Conte Avenue, Los
Angeles, California,
90024-1774, USA

Received 4 November
1997; accepted 9 March
1998

0301–4797/98/040379 + 09 $25.00/0

holders, will seriously undermine the latter's willingness to co-operate and, thus, the effectiveness of CZM decisions.

What is even more perilous for the effectiveness of coastal management than the failure to address the unavoidable conflicts associated with the above questions, however, is that this failure has a rational basis. More often than not, it reflects subscription to the positivist principle that expert-based rational analysis (often utilizing theoretical constructs of reality and normative criteria for determining 'ideal' states of the world) suffices to determine the 'best' policy and management decisions, and whoever disagrees with them should be viewed as 'irrational' (Portney, 1991). Hence, from this point of view the above questions need not be raised at all.

Worldwide experience establishes, however, that the advantages of adopting CZM solutions considered as 'best' because they are recommended by expert-based analysis can be dissipated during an arduous implementation process, marred by rancor and litigation (as partially evidenced by the emergence of environmental disputes as a whole new field of inquiry) (Harter, 1982; Bingham, 1986; Crowfoot and Wondolleck, 1990). More to the point, the fact that the aforementioned calls for integrated CZM continue to be made and that the concern over the sustainability of coastal zones is escalating, despite a prolonged effort to comply with positivist prescriptions, emphasizes that such prescriptions are less than effective.

Two key-factors contribute to the failure of decisions derived from expert-based rational analysis. First, as Lindblom (1980) pointed out, analytical policy making is inevitably limited because, for any given problem, all analysts should be expected to reach the same conclusion—which they cannot, unless they can manage to completely avoid any mistake of either fact or logic. The inevitable contradictions in the experts' recommendations undermine their effectiveness as arbiters of conflicts among stakeholders who themselves have differing interpretations of what a CZM problem may entail and what may be the most appropriate solution. They fuel, instead of resolving, such conflicts and thus

undermine stakeholder willingness to co-operate rather than engendering it. Secondly, paternalistic expert decisions ignore the conflicts associated with the above posed questions (regardless of how they are answered in practice and even with politics intervening to resolve conflicts among the experts) because, as was previously implied, they are foreign to the fundamental logic of these decisions. Hence, their capacity to engender voluntary co-operation in their implementation and thus assure the achievement of their goals is further compromised.

With these two key-factors in mind, the following argument merits support. No process of CZM can produce legitimate answers to the previously posed questions with regards to integrated CZM and, thus, effective solutions, unless it incorporates the public will in a proactive, participatory and conflict minimizing manner. (Where the characterization 'effective solutions' applies to those capable of achieving anticipated outcomes within anticipated time horizons.) Stated differently, acceptance of the premise that CZM solutions depend on the voluntary co-operation of stakeholders for their effective implementation, raises doubts about the value of positivistic or normative prescriptions of integrated CZM. The lack of answers to questions regarding either their implementation or the resolution of the conflicts they may generate undermines such co-operation.

The alternative is to pursue a new approach that of co-operative CZM. Its defining property must be its reliance on the social discourse and on a framework for guiding this discourse through the integration of diverse and conflicting individual interests into 'co-operative' collective decisions—ones that can: (1) draw maximum support; and (2) enhance the stakeholders' willingness to voluntarily co-operate in their implementation by inviting respect for the whole process of their selection and implementation. [For a discussion of certain fundamental principles for developing such a framework, methodologies and application see Davos (1986, 1987) and Davos et al., (1993)].

It is true that the pragmatists among those subscribing to the positivist approach do not ignore entirely the value of public input (after all it is now required by most environmental laws in the USA and called for by both the above cited OECD and World Coast Conference reports). However, as Portney (1991) remarks, more often than not, by seeking public participation these pragmatist positivists attempt either to 'manage' public involvement, assess the 'political feasibility' of specific alternatives or limit and control public participation in order to minimize its effect on the decision under consideration. Alternatively, they engage in symbolic exercises, channelled towards deflating rather than incorporating public views into the decision process or towards educating the uninformed public why the expert's decision or proposed decision is the best one. One should recognize of course that such behaviour might be dictated by the institutional constaints that statutory bodies with the power to carry out CZM have to satisfy. Nonetheless, the argument that sustainable voluntary co-operation might be the better long-term strategy, even for the institutional interests of these bodies, since it promises a more secure progression towards coastal-zone sustainability, merits support.

In this paper, four challenges have been discussed that co-operative CZM faces as a process for the maximization and sustainability of stakeholder willingness to co-operate. An on-going research project is currently testing the validity of some of the hypotheses underlying this discussion (as acknowledged at the end of the paper). Firstly, a formal statement of the fundamental challange faced by a co-operative CZM process if it is going to achieve sustainable stakeholder co-operation is outlined. This challenge is referred as one of maintaining a high level of 'optimism about the level of optimism' after Seabright (1993). Then three additional challenges are examined relating to: (1) the process of agenda setting; (2) the value discourse; and (3) the challenges of information and empowerment.

First however, it is important to emphasize that the focus of this discussion is on the process of managing a wide range of coastal-zone resources with an interplay between decision-making agencies and a large number of stakeholders, rather than as in 'local commons' where: (1) the number of stakeholders is small enough so that their knowledge of each other and the potential of observing each other's actions can serve as an incentive to behave in

certain co-operative ways; and (2) their participatory management can be performed, at least theoretically, without the intervention by a state that is more powerful than any of the stakeholders (Seabright, 1993). Certain inshore fisheries, which can differ from open-sea fishery because they are not often as broadly open to access by outsiders, is a typical example of such 'local commons' resources. The participatory management of other environmental resources can also be addressed along the lines of this paper.

The challenge of creating optimism about the level of optimism

It is reasonable to postulate that a determinant of stakeholder willingness to co-operate is the expected level of co-operativeness of all other stakeholders. For when the stakeholders are optimistic that the others will co-operate, they will be more willing to co-operate themselves. This expectation depends on an historical evaluation of the capacity of CZM process to inspire such optimism. Thus, the first challenge of co-operative CZM is for it to be carried out in such a way as to maximize over time the stakeholders' optimism about the general level of optimism regarding co-operation.

By involving all the stakeholders and integrating their input, co-operative CZM lays the foundation for stakeholders to build a habit of co-operation, a reputation of conforming with collective agreements, a trust that others will co-operate and an appreciation of the greater promise of co-operation over defection (pursuit of self interest) for coastal-zone sustainability. However, the transition from principle to practice is not automatic, even for co-operative CZM. The literature gives three reasons why this might be so: (1) the threat of the 'tragedy of the commons' (Hardin, 1968); (2) the paradox of the 'prisoner's dilemma' (Dawes, 1973); and (3) the logic of collective action (Olson, 1965). However, an expanding literature also indicates that 'free riding' is not always the choice of stakeholders when managing commons (see Crance and Draper, 1996; Ostrom,

1996). Ways for inducing co-operative behaviour in the management of common resources such as coastal resources have also been suggested. They are normative in nature and, thus, conflict inviting but they lay the foundations for addressing the growing recognition of co-operation as a condition for effective CZM. For example, Seabright (1993) proposes such formal mechanisms as privatization of property rights, decentralization of incentives within common ownership and control and delegation of management responsibility to an agent so that participants are limited to a monitoring role. Seabright also suggests such informal mechanisms as making the future matter, by threatening, for example, the violators of present co-operative agreements with exclusion from the commons resource, instituting credible retaliatory strategies and making co-operation history dependent by establishing memory preserving mechanisms (Seabright, 1993). Finally, Crance and Draper (1996) suggest such strategies for inducing co-operation as: scope-reduction (i.e. focusing on a distinct part of a larger problem); phased segmentation strategy (i.e. emphasizing that each individual's behaviour will determine whether or not the management goal is reached); and education on social values and responsibility in order to enhance the value that an individual places on collective welfare above self-interest.

The challenge of agenda setting

Another challenge of sustaining the stakeholder's willingness to co-operate relates to the process by which their CZM concerns and conflicts gain prominence and exposure and thus become legitimate concerns meriting the attention of the entire polity, i.e. it relates to the agenda setting process as defined by Cobb and Elder (1983). Equally significant is the challenge of managing the competition for policy response among issues and their proponents (agenda control), imposed by the reality that for any given Governmental structure, the number of legitimate issues exceeds the capabilities of

decision-making institutions to address them all.

In light of the discussion in the introduction of the shortcomings of positivism, a 'pluralistic' agenda setting process that provides all the stakeholders with an equal opportunity to develop the agenda of coastal-zone sustainability should be expected to be more conducive to co-operation than an 'elitist' process that only allows for major initiatives to come from such power centers as Government officials and policy-expert communities (for more on these two types of processes, see Studler and Layton-Henry, 1990). Moreover, a 'systematic' or 'public' agenda for action that includes the full range of sustainability issues salient to the entire community of a coastal zone should be expected to nurture a broader co-operation than a 'formal' or 'institutional' agenda consisting of issues that concern explicitly authoritative decision makers (for more on the distinction between these two types of agenda controls see Cobb and Elder, 1983).

To highlight the implications of agenda setting, consider for example the multifarious effects of urbanization and tourism development, the pollution from aquaculture development, the pollution of estuarine and coastal waters from other sources and the irreversible loss of natural areas. All coastal countries face these issues and the generic stakeholder interests are the same, e.g. those of decision-making agencies, scientists, economic interest groups, special interest groups and the media. Yet according to the OECD report (1993), the coastal-zone sustainability priorities and actions differ among countries because of their different agenda setting and control processes.

In addition to directly influencing the stakeholders' willingness to co-operate, the process of agenda setting and control is bound to also affect their optimism about the level of optimism. Consider the current situation in the USA where a continuous public concern over environmental protection (public agenda) clashes with calls for repealing all major environmental legislation emanating from changes in party power in the Congress and the accompanied changes in views regarding the Government's role (institutional agenda). The consequence is a return to contentious rhetoric and a harshening of negotiation positions that threatens whatever progress has been achieved towards co-operation among all stakeholders including environmental-protection agencies.

In summary, a major challenge to maximizing stakeholder co-operation as a foundation of effective CZM is to establish 'windows of opportunity' where policy, politics and participants can operate together to develop the coastal-zone sustainability agenda and act upon it with maximum co-operation (for more on the concept of 'windows of opportunity' see Kingdom, 1995).

The challenge of value discourse

The public debate of coastal-zone sustainability policies focuses primarily on their impacts without paying proper attention to the underlying issues of: (1) what should be considered a positive and a negative impact (benefit and cost) and who should make this determination (the origin and meaning of values); and (2) how the benefits and costs of alternative decision options should be evaluated and integrated within the framework of a rule for making the final choice (the application of values for selecting among alternative policy options, instruments of implementation, etc.). Both of these issues are critical for the sustainability of stakeholder co-operation, however. As shown below, they generate significant conflicts that can undermine the effectiveness of any decision if they are not managed in advance of deciding.

On the origin of values

The current public-value discourse has elevated anthropocentricity as the absolute foundation of affording all things meaning and value. Regardless of whether the 'process of exchange' (the Classical paradigm) or 'labour and its productive organization' (the Ricardo-Marx pradigm) is relied upon, humans are expected to be the ones who decide what must be valued and what its value means and must be. However, the environmental debate has led to the re-examination of this anthropocentric thesis by forcing the question of what must be valued (e.g. whether the interest of inaminate nature and future generations have value) and how its value must be determined (Tribe *et al.*, 1976; Elliot and Gare, 1983; Scherer and Attig, 1983; Bedau 1991). The logical next step is to question whether humans must and can continue to act as the sole arbiters of value. Notice that what is actually challenged is not the unavoidable role of humans, as the only value holders and decision makers, but their value foundations.

It is this challenge that presents the highest risks for sustaining co-operation in the implementation of coastal-zone sustainability policies. This assertion may appear unjustified to those who cannot see how anthropocentricity can be divorced from the traditional collective value calculus, including those who subscribe to the aforementioned calls for integrated CZM. However, the same is not true for those concerned with sustainability and those who cannot see how the conventional value discourse can be continued in light of a technological progress that destroys the Galbraithian 'system' that engineers, plans, steers and administers human lives by: (1) making it possible and profitable to promote 'difference' rather than 'sameness'; (2) transforming the human identity problem, from that of lack of options, to that of too many options; (3) facilitating multicultural congregation and exchange; (4) confusing human ideas and ideologies; and (5) multiplying human environmental concerns.

Indeed the environmental discourse can no longer avoid the dilemma of how can a long-term co-operation in the implementation of coastal-zone sustainability policies be induced when individual interests are becoming volatile and ephemeral; when the culture promotes speed and heterogeneity, when 'everything goes'. How can the collective will for coastal-zone sustainability to determine when society becomes a vast argumentative texture through which individuals construct their own independent reality (to paraphase Laclau, 1988).

It is indeed consistent with the orderly world of positivists to keep developing such value typologies as 'utilitarian' (production/commercial/market values), 'user' (based on *in situ* uses) and 'intrinsic' (existence, option, bequest values) but very difficult to relate them to the every-day value praxis that defies positivism [Tunstall and Coker (1992) offer a good review of value typologies]. It is also logical, according to conventional discourse, to demand the definition of new objectives, the development of new policy instruments and the creation of new institutions for integrated CZM, but arduous to actually achieve these ends with broad agreement and co-operation within a multicultural, relativistic, postmodern society.

Hence, it merits support to argue that only a proactive, well informed, constructive public discourse could highlight all value differences and invite respect for their conflicting ramifications as regards coastal-zone sustainability. Only through such discourse may the self interest be guided and reconciled with its collective counterpart and diversity channelled towards enriching instead of confusing the coastal-zone sustainability management process. Co-operative CZM with its emphasis on negotiation, compromise and conflict management can provide a promising framework for such a public discourse to be a valuable precursor to coastal-zone sustainability by maximizing the stakeholders' willingness to co-operate.

Application of values

The value discourse also becomes a source of conflict when stakeholder concerns (values) are translated into criteria for evaluating the comparative advantage of alternative policy choices. The current practice favours the application of a narrowly defined benefit-cost criterion (sanctified in the USA by law) that excludes from consideration whatever impacts of policy choices cannot be quantified in monetary terms. The anthropocentric, utilitarian foundation of this criterion relates directly to the previously discussed issues of the origin and meaning of values. It is true that more often than not, benefit-cost analyses qualify their results by acknowledging all pertinent, but not quantifiable in monetary terms, impacts. However as Socolow (1976) so elquently points out, such qualifications rarely enter the debate while

the quantifiable monetary results become 'golden numbers'.

Another source of value conflicts is the choice of instruments for policy implementation, e.g. command-and-control, economic instruments such as charges, taxes, eligibility fees or emission credits and allocation of property rights. The current debate on these instruments, too, follows the conventional practice of comparing the monetary cost of their application to their anticipated policy enforcement effectiveness. Limited attention is paid to the fact that these insutruments promote different value-systems and thus debating them strictly on quantitative monetary cost-effectivenesss terms obfuscates significant value conflicts.

Consider, for example, the choice between market-based instruments and the alternative of command-and-control. The foundation of the former on neoclassical economic theory promotes self interest as the dominant motivational force for inducing acceptable behaviour on the part of resource users and as the best arbiter of their competing interests. On the other hand, the command-and-control alternative subscribes to the principle that normative criteria can be applied to the determination of standards with which resource users must conform. Therefore, even the choice of policy instruments generates important value related issues such as whether self interest should be trusted to guide the collective will or whether normative constraints to individual behaviour should be relied upon. In broader terms, the issue is whether in formulating coastal-zone sustainability policies a disjunctive or a conjunctive relationshp should be accepted between freedom and obligation, i.e. freedom vs. obligation or freedom from within obligation.

In summary, sustaining stakeholder co-operation in the implementation of coastal-zone sustainability policies requires more than reliance on the proactive incorporation of the related public discourse into the policy formation process. It also depends on how comprehensively this discourse will cover all value-related issues, such as those associated with the origin and meaning of values as well as with the application of values for making such seemingly operational decisions as those for the appropriate evaluation approach and policy instrument.

The challenge of information and empowerment

Finally, another significant challenge to the maximization and sustainability of stakeholder co-operation is that of information and empowerment, i.e. whether knowledge neutralizes all exercises of power or whether the many forms of power pervade, invade, traverse and ultimately constitute knowledge itself [as Amariglio (1988) poses when discussing Foucault's *Discipline and Punish*]. More specifically, the proactive inclusion of stakeholders in the decision-making process necessitates a debate on the modalities of the relationship between power and knowledge and the management of the conflicts they generate.

For example, there must be a debate on the constraints imposed by modern economics on what can be said regarding coastal-zone sustainability values and policies, being the foundation of the positivist approach to CZM. Similarly, it must be debated whether true and false discourse (what is said and what could or should be said) can and must be distinguished, as well as whether the stakeholders can and must be empowered to face power—which according to Said (1983), in order to work it 'must be able to manage, control and even create detail: the more detail, the more real power, management breeding manageable units, which in turn breed a more detailed, a more finely controlling knowledge'.

Interwoven with the conflicts created by the relationship between knowledge and power are those associated with information generation, control and dissemination. Currently there is a proliferation of environmental databases but they are designed to be accessed and understood only by small groups of highly specialized experts in accordance with the pervasive positivist attitude. They fail to consider the perceptions, concerns and capacity to access and assimilate information of diverse groups of on-line searchers for information, thus failing to inspire trust in their capacity to assist the management of conflicts among stakeholders and enhance the stakeholders' motivation to co-operate. (For more on the subject see Ercegovac, 1993.) In light of this trend, current calls for more environmental indices and the collection of more 'facts' and 'hard' data (themselves manifestations of the positivist approach) can only increase this marginalization of information as a potent means of forging a broader support for final CZM decisions.

Epilogue

If research efforts must be expanded in order to provide increased knowledge of coastal systems and the requirements of their sustainability, should not research also be carried out into ways of formulating and implementing related policies more effectively and in a sustainable fashion guaranteed only by the sustainable co-operation among stakeholders? An affirmative response raises challenges to the current coastal-management praxis reflected in the calls for integrating the management methods mentioned at the beginning of this paper. The preferred alternative should be co-operative coastal-management.

What has been presented in this paper is a synthesis of thoughts advanced by different disciplines with a focus on human behaviour as opposed to that of nature. Stated differently, the interdisciplinary nature of coastal-zone sustainability has been highlighted from the point of view of social sciences, not as is customary from that of natural sciences. The purpose was not, however, only to fill a perceived gap in the development of an interdisciplinary perspective on coastal-zone sustainability. The intent was also to address the critical need of moving on from the obsolete instrumentalist epistemology of conventional environmental economics, management and engineering. As argued here, efforts should be redirected towards a critical exploration of the ways in which coastal-zone sustainability policies: (1) are socially constructed (to partially borrow from Redclift, 1993); and (2) can be implemented with the sustainable co-operation of stakeholders with conflicting preferences and priorities. Unless this is a success, sustainability policies can sustain the illusory power of rational analyses as the sole infallible arbiter of what is the 'best' future of coastal zones.

It is submitted that this is a valid assertion regardless of regional scale and Governmental level (local, regional, national, and international) or culture of public policy formation (centralized, participatory). Public-policy decisions, such as those for coastal management, are negotiated, not made. Decision-making authorities negotiate among themselves because of their fragmented responsibilities or the diversity of their constituencies' interests as well as with power centres because of their dependence on them for their effectiveness. Scientists negotiate among themselves as well as with decision makers because the limitations of their theoretical models of reality restrict them to indeterminate conclusions and advice. Power elites negotiate constantly their conflicting interests. All these negotiations can be contentious or co-operative. Only co-operative decisions have the inherent power to lead to sustainable outcomes, however. How such co-operative coastal-management decisions can be engendered was the main focus of this paper.

Acknowledgements

This paper is part of a project entitled 'The role of value conflict assessment techniques in the formulation of implementable and effective coastal zone management policies' which is funded by the EC R&D Programme in the Field of the Environment (Second Phase) of the European Commission, Contract No. EV5V-CT940392. The author gratefully acknowledges the valuable comments of anonymous referees. The expressed views are solely those of the author, however.

References

Amariglio, J. L. (1988). The body, economic discourse, and power: an economist's introduction to Foucault. *History of Political Economy* **20**, 583–613.

Bedau, H. A. (1991). Ethical aspects of environmental decision making. In *Environmental Decision Making: A Multidisciplinary Perspective* (R. A. Chechile and S. Carlisle, eds), pp. 176–194. New York: Van Nostrand Reinhold.

Bingham, G. (1986). *Resolving Environmental Disputes: A Decade of Experience.* Washington, D.C.: The Conservation Foundation.

Cobb, R. G. and Elder, C. D. (1983). *Participation in American Politics: The Dynamics of Agenda-Building.* 2nd Edition. Baltimore: John Hopkins.

Crance, C. and Draper, D. (1996). Socially co-operative choices: an approach to achieving resource sustainability in the coastal zone. *Environmental Management* **20**, 175–184.

Crowfoot, J. E. and Wondolleck, J. M. (1990). *Environmental Disputes.* Washington, D.C.: Island Press.

Davos, C. A. (1986). Group environmental preference aggregation: the principle of environmental justice. *Journal of Environmental Management* **22**, 55–65.

Davos, C. A. (1987). Group environmental valuation: suitability of single interest approaches. *Journal of Environmental Management* **25**, 97–111.

Davos, C. A. Thistlewaite, W. and Paik, E. (1993). Air quality management: participatory ranking of control measures and conflict analysis. *Journal of Environmental Management* **37**, 301–311.

Dawes, R. M. (1973). The commons dilemma game: an N-person mixed-motive game with a dominating strategy for defection. *ORI Research Bulletin* **13**, 1–12.

Elliot, R. and Gare, A. (1983). *Environmental Philosophy.* University Park: Pennsylvania State University Press.

Ercegovac, Z. (1993). Principle of cultural diversity in the design of 'right-to-know' databases. *Proceedings of the 14th National Online Meeting*, New York, May 4–6, (M. E. Williams ed.), pp. 119–128. Medford, NJ: Learned Information.

Hardin, G. (1968). The tragedy of the commons. *Science* **162(3859)**, 1243–1248.

Harter, P. J. (1982). Negotiating regulations: a cure for the malaise. *Georgetown Law Journal*, **71**, 1–118.

Kingdon, J. W. (1995). *Agenda, Alternatives, and Public Policies.* 2nd Edition. New York: Harper Collins College Publishers.

Laclau, E. (1988). Politics and the limits of modernity. In *Universal Abandon* (A. Ross ed.), pp. 63–82. Minneapolis: University of Minnesota Press.

Lindblom, C. E. (1980). *The Policy Making Process* (2nd edn.). Englewood Cliffs, NJ: Prentice-Hall.

Olson, M. (1965). *The Logic of Collective Action: Public Goods and the Theory of Groups.* Cambridge, MA: Harvard University Press.

Organization for Economic Co-operation and Development (OECD) (1993). *Coastal Zone Management: Integrated Policies.* Paris: OECD.

Ostrom, E. (1996). Governing the Commons: the Institutions for Collective Action. Cambridge: Cambridge University Press.

Portney, K. E. (1991). Public environmental policy decision making: citizen roles. In *Environmental Decision Making: A Multidisciplinary*

Perspective (R. A. Chechile and S. Carlisle, eds), pp. 195–216. New York: Van Nostrand Reinhold.

Redclift, M. (1993). Environmental economics, policy consensus, and political empowerment. In *Sustainable Environmental Economics and Management* (R. K. Turner ed.), pp. 106–119. London and Florida: Belhaven.

Said, E. (1983). *The World, the Text, and the Critic*. Cambridge, MA.

Seabright, P. (1993). Managing local commons: theoretical issues in incentive design. *The Journal of Economic Perspectives* **7**, 113–134.

Scherer, D. and Attig, T. (1983). *Ethics and the Environment*. Englewood Cliffs: Prentice-Hall.

Socolow, R. H. (1976). Failures of discourse. In *When Values Conflict.* (L. H. Tribe et al., eds), Cambridge, MA: Ballinger.

Studler, D. T. and Layton-Henry, Z. (1990). Non-white minority access to the political agenda in Britain. *Policy Studies Review* **9**, 273–292.

Tribe, L. H., Schelling, C. S. and Voss, J. (eds) (1976). *When Values Conflict*. Cambridge, MA: Ballinger.

Tunstall, S. M. and Coker, A. (1992). Survey-based valuation methods. In *Valuing the Environment: Economic Approaches to Environmental Evaluation* (A. Coker and C. Richards, eds), pp. 104–126. London and Florida: Belhaven.

World Coast Conference (1993). *Preparing to Meet the Coastal Challenges of the 21 stakeholder Century* (Conference Report). Ministry of Transport, Public Works and Water Management, National Institute for Coastal and Marine Management/RIKZ, Coastal Zone Management Centre, The Netherlands.

[19]

Ocean & Coastal Management, Vol. 37, No. 2, pp. 155–174, 1997
© 1998 Elsevier Science Limited. All rights reserved
Printed in Northern Ireland
0964-5691/97 $17.00 + 0.00

PII: S0964-5691(97)00105-X

A common framework for learning from ICM experience

Stephen Olsen, James Tobey & Meg Kerr

Coastal Resources Center, University of Rhode Island, Coastal Resources Center, Narragansett, RI 092882-1197, USA

1. INTRODUCTION

There are a growing number of integrated coastal management (ICM) initiatives worldwide—some 140 ICM efforts in 56 coastal nations can be identified[1]—but at present the lessons learned from these initiatives are generally undocumented and the efficiency and effectiveness of learning from ICM is being compromised. We have very little information that demonstrates the success of ICM efforts and how the process of ICM has influenced outcomes. Many descriptions of ICM experience are anecdotal and, to date, no hypotheses about ICM design and practice have been systematically tested across the diverse spectrum of coastal nations.

At its 1996 meeting, the international Group of Experts on the Scientific Aspects of Marine Environmental Protection (GESAMP) identified this as a priority 'emerging issue':

> there is an urgent need for an accepted evaluation methodology for assessing the changes identified and implemented. When an evaluative framework is in place it will be possible to document trends, identify their likely causes and objectively estimate the relative contributions of ICM programs to observed social and environmental change.

The challenge is to develop and standardize methodologies and indicators by which the impacts of the rapidly expanding number of integrated coastal management initiatives can be analyzed, and by which the collective learning process can be improved. An activity at the global level that clearly measures progress (or lack thereof) towards ICM goals, and disseminates the results widely offers great opportunity for increasing the efficiency of the collective learning process for how to make ICM an effective response to the challenges of sustainable coastal development. Such an activity could stimulate national actions and provide guidance to donors.

Why is this an urgent need? First, because the transformation of coastal regions is of vital importance to our species. Approximately half of humanity is already concentrated in a narrow ribbon of land around the planet's oceans, seas and great lakes. These coastal regions encompass less than 10% of the inhabited land space.[2] The proportion of the world's population that is coastal will increase as the population swells during the next century: two-thirds of the human population is projected to be concentrated in coastal regions by 2025.[3] With this population comes at least half of the infrastructure for the manufacturing, transportation, energy processing, and consumption that these populations require, as well as more than half of the waste products and tourism.

A second reason to systematically evaluate ICM initiatives is that global trends show a decline in the qualities of coastal regions that support sustainable human societies. The pressures produced by a growing population will increase from a global perspective, and the expressions of overuse and misuse of the coastal life support systems are mounting. ICM practitioners are all familiar with the symptoms of declining water quality, degradation or destruction of critical habitats, decline and collapse of fisheries, and losses in biodiversity. We are also aware that these problems bring mounting user conflicts and that governments are often unable to avoid these adverse effects even when they are predictable and lead us away from sustainable forms of development.

A third reason to systematically evaluate and learn from ICM initiatives is that the existing successes are as yet puny compared with the forces worldwide causing coastal degradation. If integrated coastal management is to have a significant global impact on the condition of coastal ecosystems we must quickly scale up endeavors that are now largely conceived and implemented as a scattering of pilot projects.

These realities are important because selecting indicators, monitoring, self-assessment, and evaluation are all activities that have seldom invoked enthusiasm among either the coastal managers attempting to move programs forward, or among the politicians and bureaucrats who fund such initiatives. Monitoring consumes resources and seldom produces a quick return on investment. It requires repetitive, painstaking work, and the analysis of the data generated is time consuming, technically challenging, and often yields controversial conclusions. Periodic, internal self-assessment, and external evaluation, requires careful preparation and, if they are to be meaningful, will sometimes require painful internal adjustments to an on-going program's objectives and design. The response to these realities can be seen in projects that protect themselves from potentially damaging evaluations by:

- adopting vague goals;
- selecting objectives that defy measurement;
- selecting indicators that measure effort rather than results;
- adhering blindly to the project's original objectives and strategies while refusing to adapt to changing conditions;
- skipping formal evaluation entirely or leaving them to the end of the project when they will have no impact on the design or operation of the original project and only marginally influence future projects.

There are, of course some notable exceptions. Furthermore, it is now widely recognized that more effective monitoring and evaluation must be built into all programs and projects funded by development banks and international donors. In the United States, performance evaluation is now mandated by the 1993 Government Performance and Results Act. This recent interest in evaluation offers the opportunity to rise to the challenge and to develop consistent evaluation methodologies and indicators that can be applied to all ICM initiatives.

2. WHAT IS ICM?

At the root of any discussion of learning methodology is the definition of ICM. There has been considerable progress in recent years on defining the major characteristics of ICM.[4-8] There appears to be growing consensus on the outlines of a general model. All definitions stress the dynamic nature of the ICM process and its emphasis upon sectoral integration. A recent report[9] defines ICM as:

a continuous and dynamic process that unites government and the community, science and management, sectoral and public interests in preparing and implementing an integrated plan for the protection and development of coastal ecosystems and resources.

So defined, ICM belongs to the family of initiatives that are working towards a better balance between human societies and the ecosystems of which they are but one element. ICM, like some forms of watershed planning and development, holds the promise of being a vehicle for progressing towards sustainable development. Achieving ICM is especially complex because of the superposition of many human activities along coastlines, and the many dimensions of integration that need to be addressed. In this paper we propose that the governance process itself must be a central focus of learning and evaluation. The term governance, as used here, refers to the method of coastal management, and includes the laws, institutions, policies and process that affect how coastal resources are utilized and allocated. The governance process is the means by which we test and improve ICM strategies in order to move away from socially and environmentally unsustainable forms of development and toward more sustainable forms of development. Some of the strategies and principles that support the ICM process include:

- work at both the national and local levels with strong linkages between levels (the 'two-track' approach)
- build programs around issues that have been identified through a participatory process
- build constituencies and political support for resource management through public education programs
- develop mechanisms for sustained learning on how to improve the efficiency and effectiveness of integrated coastal management
- develop an open, participatory and democratic process, involving all stakeholders in planning and implementation
- build capacity at the national, regional, and local levels to practice integrated, community-based management of coastal resources through training, learning-by-doing and cultivating host country colleagues who can forge long-term partnerships based on shared values
- complete the loop between planning and implementation as quickly and frequently as possible, using small projects that demonstrate the effectiveness of innovative policies
- adopt policies that lead to economically and ecologically sustainable and equitable resource management
- strengthen or introduce mechanisms for cross-sectoral action
- adopt an incremental, adaptive, and long-term approach to integrated coastal management, recognizing that programs undergo cycles of development, implementation and refinement, building on prior success and adapting and expanding to address new or more complex issues

These principles are frequently identified as important in the literature on policy implementation and ICM program assessment.[10–14]

In ICM, the governance process is continuous and dynamic, and is therefore predicated upon learning and adaptation. The assumption is that ICM does not offer a blueprint that merely needs to be applied and will then produce known results. If this were the case, monitoring and evaluation would be superfluous. Adaptive management calls for learning by doing. The experiential learning cycle[15] is well known to those who study organizations and management. In its simplest form it is expressed as continuing cycles of action and reflection (Fig. 1).

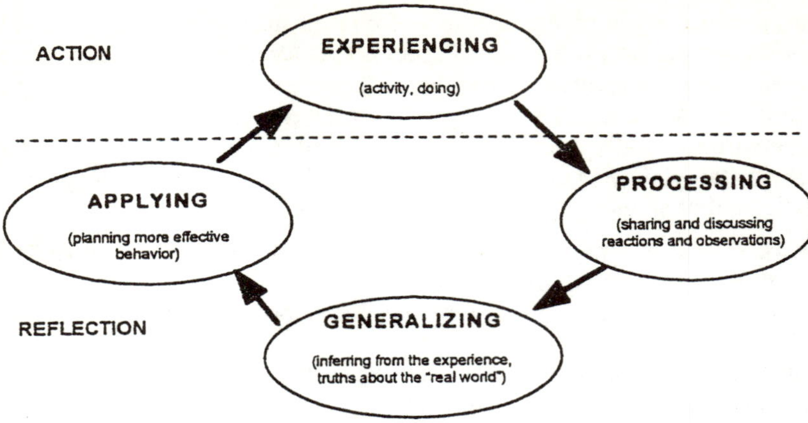

Fig. 1. The learning cycle.

The most critical step in the learning process is reflection on concrete experience and the formulation of new concepts. This is where we draw conclusions and may reconfigure our understanding of ICM issues. A commitment to a thoughtful and objectively rigorous evaluative process therefore promises to increase the efficiency of learning to make ICM programs more effective.

3. THE GOAL OF ICM

There have been many attempts to state simply and forcefully the goal of ICM. One recent version[16] states:

> The goal of ICM is to improve the quality of life of human communities who depend on coastal resources while maintaining the biological diversity and productivity of coastal ecosystems.

This statement captures the central idea that ICM is first and foremost about people and attempting to define a dynamic balance between people and the qualities of our coastal environments.

From the point of view of learning, however, the lofty nature of the goal, and the great scope of the endeavor poses formidable challenges. How do we design a methodology and select indicators for such a complex and protracted effort that are simple enough to be cost effective but sufficiently rigorous to produce useful results? Knecht *et al.*[17] address the issue of measuring ICM program impacts. In the absence of substantive indicators of program performance, the authors rely on the indirect approach of soliciting perceptions of performance from samples of individuals knowledgeable of the programs being studied. Unlike business ventures, the 'bottom line' of human and capital investment in ICM is not easily quantifiable. The challenge for evaluating ICM is inherently complex since we must make judgments on a 'process' that is designed to avoid conflicts and ecosystem degradation by identifying problems and opportunities proactively and acting upon them.

Medium-term outcomes, both material (e.g. mangrove planting, building a dock, installing mooring buoys) and non-material (e.g. training, institution building), are more tangible and easier to track. The timeframe for achieving the ultimate goals of ICM is long, it is therefore essential that any methodology for learning from experience address the governance process itself and progress towards medium-term objectives.

4. THE ICM POLICY CYCLE

There are many descriptions of the process by which ICM programs evolve.[18-23] In its most essential and stripped-down form, however, most would agree that the process can be described as a cycle with the same features of other institutional endeavors. The process begins (step 1) by identifying and analyzing the issues in the stretch of coast in question, and then proceeds to set objectives and prepares a plan of policies and actions (step 2). Next comes step 3 of formalization through a law, decree or interagency agreement and the securing of funds for implementation of some selected set of actions. Policy implementation (step 4) is the step in which procedures and actions planned in the policy formulation stage are made operational. Mechanisms may include public meetings, conflict resolution, and enforcement procedures, while actions span the building of physical infrastructure, the strengthening of institutions and dissemination of appropriate forms of resource use. Step 5, too often ignored or poorly executed, is formal evaluation. In this step, the results of the policy-making process are compared with the desired outcome(s).

This process has been recently described in a report[24] that details the contributions of the social and natural sciences to each of the five steps. Table 1 identifies priority actions associated with each step. When ICM programs build constituencies and earn support they combine a concern for a sustained and responsive governance process with tangible achievements. Thus successful programs negotiate early on (step 1) an agreement among stakeholders both in and out of government on the major issues that require improved management and the specific objectives of the program—in itself often a major accomplishment—and then test new management techniques and procedures during the planning stage (step 2). Such pilot scale actions can bring considerable attention and credibility to a program when they demonstrate that meaningful action is indeed possible. The formalized acceptance of a plan and/or law and/or funding for a stage of full scale implementation (step 3) can attract attention at a larger scale and constitute a major achievement that brings significant change to the people and the resources affected. Such tangible achievements are essential to sustained progress.

Coastal management programs in a range of developed and developing nations suggest that completion of an initial cycle typically requires 8–15 years. Each cycle may be termed a 'generation' of an ICM program (Fig. 2).

5. INTERMEDIATE AND FINAL OUTCOMES OF THE ICM POLICY CYCLE

In designing a framework for learning from ICM experience it is essential to recognize the time that it takes to complete a sequence of policy cycles

TABLE 1

Essential actions associated with each step of the ICM policy cycle

Step	Priority actions
Stage 1: Issue identification and assessment	• Rapidly assess existing conditions • Consult key stakeholders and identify priority issues
Step 2: Program preparation	• Select issues to be addressed and geographic focus • Conduct sustained public education program • Define boundaries of management area • Define management objectives, strategies, and actions • Carry out early implementation actions
Step 3: Formal adoption and funding	• Adopt formal management plan and governance process • Secure adequate funding for implementation
Step 4: Implementation	• Construction/operation of infrastructure • Promote compliance to regulations and agreements • Implementation of sustainable development practices
Step 5: Evaluation	• Evaluation of governance process and outcomes • Reassess issues and strategies • Select adjustments to plan and governance process

162   *S. Olsen* et al.

to achieve the ultimate goals of (1) sustainable quality of life in coastal communities, and (2) sustainable well-being of coastal ecosystems. Mature ICM programs make it very clear that it takes a sustained effort measured in decades and spanning several generations of a given program, to achieve tangible expression of the end goal at a significant scale. This time scale is beyond the duration of the vast majority of projects currently funded by development banks or international donors. This means that such projects typically will not encompass a single full generation of an ICM program, and highlights the importance of identifying a sequence of intermediate outcomes. The sequence may be visualized as first-, second- and third-order intermediate outcomes as shown in Fig. 3.

The 'generations' of an ICM policy cycle follow this sequence of intermediate and end outcomes at different scales. If a program is strategic, it will define in general terms an end goal and then carefully and pragmatically define its intermediate objectives (quantified and time

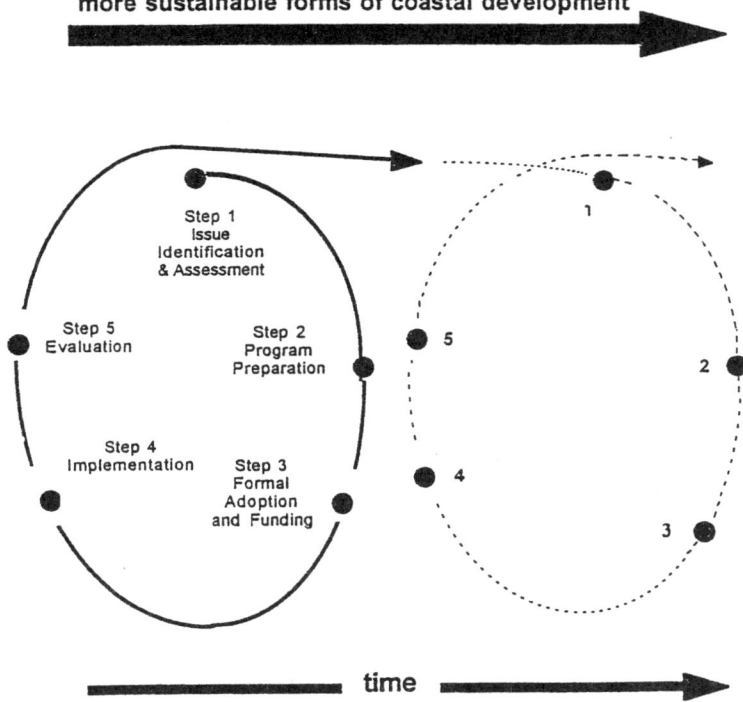

Fig. 2. The steps of the ICM policy cycle. The dynamic nature of ICM requires feedback among the steps and may alter the sequence, or require repetition of some steps (from Ref. 9).

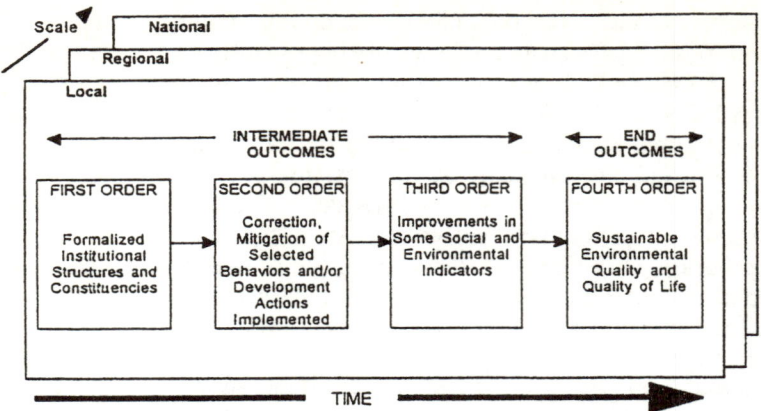

Fig. 3. Ordering coastal governance outcomes (adapted from USEPA, 1994)[33].

bounded) for a given generation of the ICM policy cycle. The importance of clear, specific, objectives that are amenable to objective analysis cannot be overstated. Underlying project objectives are invariably a set of ideas, beliefs, or assumptions about what constitutes integrated coastal management and what constitutes effective strategies for coastal management. These should be stated explicitly; only by being explicit and adopting an essentially scientific and objective approach to articulating hypotheses will our collective learning process be improved.

In developing nations, a first generation ICM program will typically focus its objectives on one or more pilot sites and on a limited set of issues. It is far better to do a few things well than many things poorly. A pilot project may achieve improvements in reef fisheries and the qualities of life of a small community at a pilot site within a single generation, but may require several generations to achieve similar results for an entire region or nation. At the same time, building capacity and linkages at the national level should be a key element of the pilot site initiative

5.1. First order: Formalized institutional structures and constituencies for ICM

For many programs, the first priority is to create a program that has the mandate, the human and financial resources and the political backing to begin practicing integrated resource management. Where institutional capacity is lacking and inter-agency conflicts dominate, this is in itself a major undertaking. We have learned that this cannot be achieved only by 'coordinating' or reallocating power and responsibility among the many agencies of government with significant roles in the management of coastal ecosystems.

To make the description of intermediate outcomes more concrete we refer to the experience in Sri Lanka. Tangible first-order outcomes in Sri Lanka include establishment of a national Coast Conservation Division in 1978 that was converted in 1983 to Department status.[25] The Coast Conservation Act, which regulates all development activity within a narrowly defined coastal zone, was passed in 1981, and a coastal management plan was formally adopted in 1990. Following passage of the Coast Conservation Act, permits were granted, training was undertaken to build a cadre of ICM professionals, and professional relationships were developed with coordinating agencies and stakeholders at sites along the coast where conflicts were most acute.

From the outset, an ICM program must also be doing things that are tangible, that build credibility and that attract constituencies among the people affected as well as the elements of government concerned. However, such actions in first generation programs usually only set the stage and begin to build the body of experience for making a sustained and effective thrust towards the end goal. If an ICM initiative is narrowly defined to focus on a small group of communities, or limited portion of a coast as is appropriate for pilot scale projects, the chances for short-term (5- to 10-year) measurable impacts may be good. Usually, however, such efforts are undertaken, and justified, as 'demonstrations' that are designed to subsequently instigate ICM on a larger scale (a region, watershed, province, or nation). In such cases judgments must be made as to whether, and how, such demonstrations lead to progress on a larger scale. Thus, clearly differentiating between the ultimate goal and project objectives becomes critical.

Tangible actions in Sri Lanka have from the beginning focused on shoreline protection structures to address severe problems of shore erosion.

5.2. Second order: Correction, mitigation of selected behaviors and/or development actions implemented

Once the ICM program is in place and capable of functioning, it can expect to produce measurable impacts on the human behaviors selected as the focus for that generation. Here again, scale is of critical strategic importance. An ICM program must walk before it can run. The most successful and sustainable programs make good judgments of what they can reasonably hope to accomplish in any particular generation. Usually the limiting factor is institutional capacity.

Reduction in coral mining is an important second-order outcome in Sri Lanka. With multiple strategies, there has been a noticeable reduction in coral mining, even in a context of strong demand for coral lime, used in construction, and high economic returns to those engaged in this activity. Pilot-scale projects designed as demonstrations for community-based coastal management were initiated in Hikkaduwa and Rekawa in 1991. In these demonstration sites, coral mining has been reduced by about 95%; and in Rekawa, over 75 illegal coral lime kilns have been voluntarily demolished.[26]

5.3. Third order: Specific improvements in quality of life and the condition of target environmental qualities

There is usually a lag between modifying a behavior and the effect on society and the ecosystem. The achievement of measurable improvements in selected indicators of quality of life and the environment, such as fish stocks, water quality, and income, are major accomplishments that bring credit to ICM programs and justify the process by which they were achieved. In Sri Lanka, measurable improvements in shrimp harvests in the Rekawa lagoon as a result of reseeding and improvements in water circulation is a tangible third-order outcome of that pilot-scale coastal management project.

Once a body of experience and capacity is in place, replication and advances towards the goals of ICM are likely to occur with increasing frequency and often independently. Thus, in Sri Lanka, major advances towards effective lagoon management in Negumbo and Puttalam lagoons have recently been made through programs loosely linked to the Coast Conservation Department.

Problems that were important a decade ago may be irrelevant today. Both problems and opportunities both evolve and change. The objectives and priorities of a vigorous ICM program therefore shift over time as they adapt to changing circumstances. The end goal of sustainable coastal development, however, remains constant.

5.4. Fourth order: Sustainable environmental quality and quality of life

Pragmatically, it is unlikely that we will see, in our lifetimes, the achievement of sustainable forms of coastal development at significant scales. What matters to us now, and matters urgently, is rather the direction of the development trajectory. Are we, as human societies, moving towards sustainable forms of coastal development, or are the actions of the societies of which we today are a part, compromising the ability of our children and their children to meet their needs? ICM programs must pose these questions in honest and realistic terms and attempt to answer them. ICM offers a framework for addressing such questions in the context of a holistic, long-term and scientifically rigorous approach to the challenges of development and the environment. This is why the endorsement of ICM was one of the features of the 1992 United Nations UNCED Conference in Rio.

5.5. Correctly matching the ICM policy cycle with intermediate objectives

It is very important to define the relationship between a given generation of an ICM program and the sequence of intermediate outcomes that must be achieved before the ultimate goal. For example, Sri Lanka's first generation program focused on the management of shoreline erosion along the southwest coast, and postponed other intermediate outcomes to its second generation. In Ecuador, an achievable first generation objective in Ecuador was to formally create and obtain funding to implement a national ICM program while building the institutional capability and the constituencies for improved management in that country.[27] A major issue in Ecuador is the accelerating destruction of mangrove wetlands. Mangrove cutting has slowed, and some replanting has occurred in pilot sites. Halting all mangrove destruction along a 1600 km coast was not judged to be a realistic objective for a first generation ICM program that had to be built from the ground up. Practical exercises conducted during the planning stage, which experimented with the feasibility of a broad range of ideas for actions that promoted more sustainable forms of development, were the basis for all the activities now funded for implementation by a loan from the Inter-American Bank. These activities will continue to occur at a pilot project scale in this first generation program.

Matching the scale and the objectives of a given ICM generation to the capacity of the institutions involved and the strength of the constituencies affected is crucial but often misjudged. For example, in the 1970s the coastal management program for the State of Rhode Island in the US initially failed to scale its legislative objectives in a manner that was politically acceptable. Subsequently, the first plan drafted after that legislation proved to be not implementable—in large part because its scope and complexity surpassed the capacity of the implementing agency. A second generation effort, after a difficult internal evaluation, scaled back and simplified the scope of the regulatory program and has now been successfully implemented for 10 years.

An evaluative process that is successful in promoting learning within and across ICM programs must be designed to address the success with which each step of the policy cycle is undertaken. It should also be able to evaluate the impact of changing the order of the steps and the hypotheses and strategies—too often unstated—that drive how each generation is designed and executed.

6. THE PRESSURES OF COASTAL CHANGE

A major challenge for learning from ICM experience lies in confidently identifying cause and effect relationships—the attribution problem. The causal relationship between the efforts of an ICM program and the impacts of the program on quality of life and the condition of coastal ecosystems are often tenuous. An ICM program is usually, and at larger scales always, one among many forces acting upon society and the environment. The pressures that influence, and sometimes drive, both the intermediate and final outcomes that an ICM program is striving to achieve are numerous and complex. They include:

- demographic pressures; in many tropical nations the numbers of people in coastal regions double every 30 to 40 years and in some coastal cities every 10 to 15 years;
- economic pressures; these can shift rapidly and often are more powerful than population pressures in driving coastal change;
- institutional and political pressures; these lie at the heart of how a governance process plays out in a given setting;
- social pressures (including conflicts among ethnic and stakeholder groups); these are powerful and volatile forces of coastal ecosystem change;
- external pressures of ecosystem change (such as global warming, acid precipitation, and ozone depletion); these are an increasing global concern.

These pressures must be assessed and monitored, not only to understand the sources of coastal change in a given place, but to enable reasonable comparisons among ICM initiatives in different settings. Progress in one setting that can be achieved quite readily may be a major accomplishment elsewhere. For example, where robust and capable institutions are in place, it may be possible to adopt and enforce a new control over a class of industrial pollutants discharges quite quickly. In other settings, this same measure may require major institution building and training, which addresses deep-rooted corruption and inefficiency. In some settings, such a technically simple and even economically cost-effective measure may be beyond the capacity of a first-generation ICM program.

Such external pressures invariably overshadow the efforts of a program and require rethinking of the fundamental strategies, and even the objectives for a generation of a program. A cholera outbreak, civil war, a major flood or cyclone, or a sudden change in the price of an important commodity can radically change the short-term prospects for the success of an ICM program. Some of these changes offer opportunities that an intelligent and agile ICM program should grasp. It is always essential that such external forces are understood and are factored into any ICM evaluative process.

7. THE PRESSURE–STATE–RESPONSE FRAMEWORK

When we assemble the three major components of an ICM learning methodology we see the central features of the Pressure–State–Response (PSR) framework that has been widely applied to a variety of environmental quality issues.[28] Human activities create the pressures that affect the ability of ICM programs to achieve intermediate objectives, and change the quality of the environment and human life. These qualities or 'state' conditions can in turn be influenced by the governance process throughout the ICM program cycle. The PSR framework unites the three elements in a cycle of causality whereby the responses of an ICM program form a feedback loop to the pressures created by human activities. These basic relationships are shown in Fig. 4. The large arrow illustrates that through ICM governance (represented by the 'response' box) progress towards the goals of ICM (the 'state' box) may be achieved. Other arrows illustrate the influence of outside influences (represented by the 'pressures' box) on the governance process and the achievement of ICM goals.

8. INDICATORS OF ICM GOVERNANCE

There is an extensive and rapidly growing number of efforts aimed at developing environmental indicators, including coastal environmental indicators.[29] Unfortunately, the great majority of monitoring and research efforts on environmental state variables do not attempt to link change in ecosystem qualities with societal behaviors and management. An exception is the ReefBase Project led by the International Center for Living Aquatic Resource Management (ICLARM) that develops a cohesive methodology for assessing and monitoring the condition of coral reefs worldwide. ReefBase is unusual in that it includes social and governance indicators as well as environmental indicators.[30] The Global

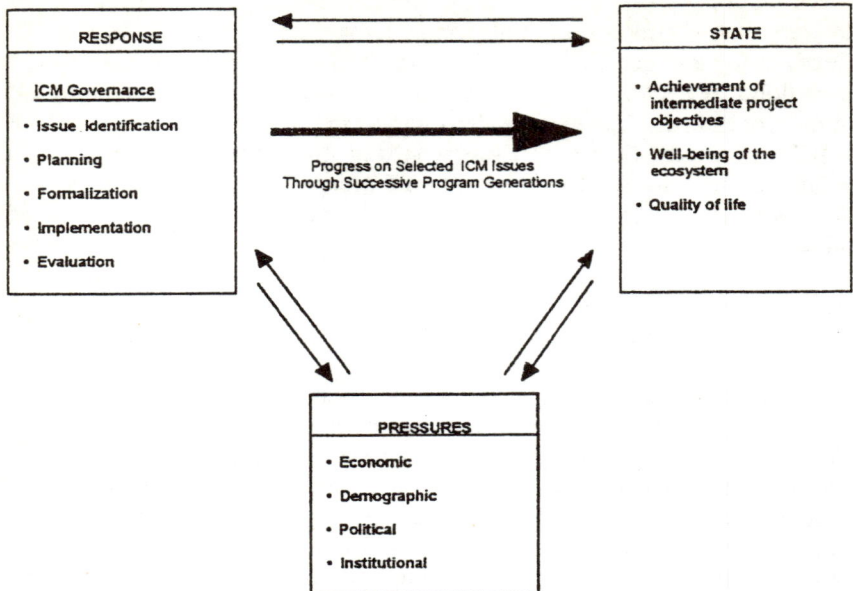

Fig. 4. PSR Framework applied to coastal management.

Coral Reef Monitoring Network (GCRMN) is further developing the social and governance aspects of monitoring coral reef management projects.

The response indicators, or governance component of the PSR framework as it applies to ICM, is the least developed leg of the PSR tripod. Yet, thorough understanding of the elements of governance and the public and private institutions that implement or manage coastal management programs in a specific country is essential. These institutions are more likely to be the source of program debate and conflict than are purely technical matters. Governance capacity is vital to sustained action on coastal resource issues.

The University of Rhode Island's Coastal Resources Center has been developing instruments that gauge the maturity of an ICM program through sets of indicators for each of the major program components.[31,32] The instruments rank the degree to which the stages of the policy cycle have been achieved. A number of indicators for each of the five components of the ICM policy cycle described earlier have been developed. Each indicator can be ranked from 0, representing no effort, to 3, representing a high level of program effort. The methodology is designed to be as simple as possible so that it might be administered by a professional ICM manager or social scientist, with interviews with knowledgeable in-country sources. To illustrate how the ordinal ranking system is applied, Table 2 is an example of three indicators to measure ICM implementation.

A composite measure of governance capacity could be formulated by summing the values across the five steps. Alternatively, some steps might be considered more crucial than others, which could be addressed with a system of weighting. What may have been a decisive factor in one country or site, may have played a less significant role in another. As more data are accumulated, we will be better able to determine the relative weight of the indicators on an international basis. This instrument is currently a first version of a tool that must be revised and improved.

It is important to recognize that this instrument does not assess the impact of the program. It is a tool for assessing the condition, or maturity of the ICM governance institutions and process. By combining instruments such as this that measure ICM program activities, with information on state and pressure variables, we can assemble an objective basis to:

1. assess trends and progress in specific ICM programs over time;
2. compare the status and trends in different countries over time; and
3. assist in the design of a balanced set of ICM activities in a given setting, and in identifying the levels of financial commitment in ICM that may be appropriate in a given country at a given time.

9. INCREASING THE EFFICIENCY OF LEARNING FROM ICM INITIATIVES

This paper addresses the need for analytical approaches to improve our learning capacity from the experience of ICM initiatives worldwide. One of the keys to success will be to conduct the requisite learning in the right areas so as to anticipate emerging management needs. It is essential that any methodology for learning should address (1) the governance process itself; (2) progress towards intermediate project objectives and the specific social and environmental qualities ICM programs are attempting to attain; and (3) the pressures that are affecting those qualities.

To advance comparative analysis of ICM experience at the international level it is essential that we progress toward a common methodology for learning from ICM experience, and that we assess the impacts of coastal management upon the quality of life in coastal communities, and the condition of the natural environment. We must identify specific indicators and information that should be monitored across ICM projects worldwide.

TABLE 2

Assessing ICM governance: Implementation

Component	Description	Rank: 0	1	2	3
Public investment	Degree to which there is direct public investment in essential physical facilities (water, sewerage, artificial reefs, etc.)	No investment	Limited investment and construction; no cost recovery	Significant construction; major problems with operation, maintenance and cost recovery	Significant continuing construction, operation and maintenance; established cost-recovery mechanisms
Issuing of fines and permits	Extent to which fines and permits are issued for an illegal activity	No laws and/or permits and fines are never or rarely issued	Many fines and permits issued	Declining numbers of permits and fines	Clear evidence that the activity has ceased
Conflict resolution	Extent to which there are mechanisms for successful conflict resolution at the local and national level	No mechanisms established	Attempts being made to establish mechanisms	Mechanisms established; often they succeed	Mechanisms established; usually they succeed

172   S. *Olsen* et al.

Once a common methodology and indicators are applied across a large number of diverse settings, it will be possible to systematically test hypotheses about what works, doesn't work, and why. Such a common methodology and indicators are urgently needed to increase the efficiency by which the widening community of ICM initiatives learn from one another and make ICM an effective response to the challenges of sustainable coastal development.

REFERENCES

1. Sorensen, J. The international proliferation of integrated coastal zone management efforts. *Ocean and Coastal Management* 1993, **21,** 1-3 45–80.
2. Pernetta, J. and Elder, D., *Cross-Sectoral, Integrated Coastal Area Planning (CICAP): Guidelines and Principles for Coastal Area Development,* IUCN, Gland, Switzerland, 1993.
3. UNCED, Protection of the oceans, all kinds of seas, including enclosed and semi-enclosed seas, and coastal areas and the protection, rational use and development of their living resources, Ch. 17, Agenda 21, United Nations Conference on Environment and Development, 1992.
4. Chua, T. Essential elements of integrated coastal management. *Ocean and Coastal Management* 1993, **21,** 81–108.
5. Clark, J. *The Coastal Zone Management Handbook.* Lewis Publishers, 1996.
6. OECD, *Coastal Zone Management: Integrated Policies.* Organization for Economic Co-operation and Development, Paris, 1993.
7. Post, J. and Lundin, C. (eds), *Guidelines for Integrated Coastal Zone Management,* Environmentally Sustainable Development Studies and Monographs Series No. 9. The World Bank, Washington, DC, 1996.
8. Pernetta and Elder, *op. cit.* reference 2.
9. GESAMP (IMO/FAO/UNESCO-IOC/WMO/WHO/IAEA/UN/UNEP Joint Group of Experts on the Scientific Aspects of Marine Environmental Protection), The Contributions of Science to Integrated Coastal Management. GESAMP Reports and Studies No. 61, 1996.
10. Hennessey, T. Governance and adaptive management for estuarine ecosystems: The case of Chesapeake Bay. *Coastal Management* 1994, **22,** 119–145.
11. Olsen, S., Tobey, J., Robadue, D. and Ochoa, E., Coastal management in Latin America and the Caribbean. Draft report submitted to the Inter-American Development Bank, Washington, DC, December 1996.
12. Pomeroy, R. Community-based and co-management institutions for sustainable coastal fisheries management in Southeast Asia. *Ocean and Coastal Management* 1995, **27,** 3 143–162.
13. Sabatier, P. and Mazmanian, D. The conditions of effective implementation: A guide to accomplishing policy objectives. *Policy Analysis* 1979, **5,** Fall 481–504.
14. White, A., Hale, L., Renard, Y. and Cortesi, L., *Collaborative and Community-Based Management of Coral Reefs: Lessons from Experience.* Kumarian Press, Connecticut, 1995.
15. Kolb, D., Rubin, I. and McIntyre, J., *Organizational Psychology,* 2nd edn. Prentice-Hall, Englewood Cliffs, NJ, 1974, pp. 27–42.

16. GESAMP, *op. cit.* reference 9.
17. Knecht, R., Cicin-Sain, B. and Fisk, G. Perceptions of the performance of state coastal zone management programs in the United States. *Coastal Management* 1996, **24,** 141–163.
18. Chua, T. and Scura, L. (eds), Integrative framework and methods for coastal area management. In *ICLARM Conference Proceedings*, 37, 1992.
19. Intergovernmental Panel on Climate Change (IPCC), *Preparing to Meet the Coastal Challenges of the 21st Century*. Conference Report, World Coast Conference, 1993. The Hague, Netherlands, National Institute for Coastal and Marine Management (RIKZ), Coastal Zone Management Centre, 1994.
20. Knecht, R., On the role of sciences in the implementation of national coastal management programs. In *Improving Interactions Between Coastal Science and Policy: Proceedings of the Gulf of Maine Symposium*. National Academy Press, Washington, DC, 1995.
21. Pernetta and Elder, *op. cit.* reference 2.
22. UNEP, *Guidelines for Integrated Management of Coastal and Marine Areas with Special Reference to the Mediterranean Basin*, United Nations Environment Program, UNEP Reg. Seas Rep. Stud., 161, 1995.
23. World Bank, The Noordwijk guidelines for integrated coastal zone management. Paper presented by the Environment Department, World Bank, at the World Coast Conference, 1–5 November 1993, Noordwijk, the Netherlands, 1993.
24. GESAMP, *op. cit.* reference 9.
25. Olsen, S., Sadacharan, D., Samarakoon, J., White, A., Wickremeratne, H. and Wijeratne, M., *Coastal 2000: Recommendations for a Resource Management Strategy for Sri Lanka's Coastal Region*, Vol. I. The Coastal Resources Management Projects, Sri Lanka, and the Coastal Resources Center, The University of Rhode Island, 1992.
26. Hale, L., Halting coral mining in Sri Lanka...a hard-won success story. *People and the Planet*, Special International Year of the Reef Issue 1997, **16**.
27. Robadue, D. (ed.), *Eight Years in Ecuador: The Road to Integrated Coastal Management*. Technical Report No. 2088, Coastal Resources Center, University of Rhode Island, September 1995.
28. OECD *Environmental Indicators: OECD Core Set*. Organization for Economic Cooperation and Development, Paris, 1994.
29. Lourens, J. and Cardoso da Silva, M., Indicators for coastal zone management and characterization. Report prepared for the European Environment Agency, Topic Centre on Marine and Coastal Environment, November 1996.
30. Pollnac, R., Rapid assessment of management parameters for coral reefs. RAMP Final Report, Coastal Resources Center, University of Rhode Island, 1996.
31. Cobb, L. and Olsen, S. *International Coastal Resource Management Program Effort: Tools for Assessment, Planning, Monitoring and Evaluation*. Manuscript, Coastal Resources Center, University of Rhode Island, Kingston, RI, June 1994.
32. Olsen, S. and Tobey, J., Final evaluation: Patagonian coastal zone management plan. Draft report submitted by the Coastal Resources Center, the University of Rhode Island, to the United Nations Development Programme, January, 1997.

174 *S. Olsen* et al.

33. USEPA, 1994. USEPA, *Measuring Progress of Estuary Programs.* US Environmental Protection Agency, Office of Water, Doc 842-B-94-008, Washington, DC, 1994, 267 pp.

[20]

Pergamon

PII: S0264-8377(97)00024-0

Land Use Policy, Vol. 14, No. 4, pp. 257–276, 1997
© 1997 Elsevier Science Ltd. All rights reserved
Printed in Great Britain
0264-8377/97 $17.00 + 0.00

Building trust in shoreline management: creating participatory consultation in shoreline management plans

Timothy O'Riordan and Rosie Ward

This article canvasses the arguments in favour of a more active and conciliatory approach to consultation than is generally the case for resource allocation and environmental decision-making in the UK. It looks at the experience of consultation in shoreline management and flood defence decision-making in East Anglia, England. It concludes that there are many different ways of conducting and justifying mediation, but that there are strong social, political and economic arguments for widening the consultative arrangements and ensuring a more face to face participatory role for representative interests. © 1997 Elsevier Science Ltd

Keywords: shoreline management, legitimacy, participatory consultation, consensus building, environmental planning, coastal protection.

The authors are from the Centre for Social and Economic Research on the Global Environment at the University of East Anglia, Norwich (co-located in University College, London). E-mail: t.oriordan @uea.ac.uk; rosie.ward@btinternet.com

Consultation is common practice in British resource management experience. Consultation is usually required as a legal duty for certain categories of consultees, but is widened to include many other interested parties as a matter of custom. There are obvious reasons for this arrangement.

Political pragmatism. Interested parties expect to be involved, so to ignore them would be counterproductive to a workable outcome.

Information. Interested parties usually know much about the fine points of resource use and the acceptability of any changing practice. Their knowledge is vital for a well supported decision.

Conflict resolution. Interested parties may only reach agreement when everyone knows what each wants and what each will settle for.

Trust. The agencies responsible for resource management require the good faith and support of those with whom they have to consult or cooperate. Good faith is built on trust, and trust requires mutual respect. Participants need to feel wanted, and the implementing agency can only function effectively with their goodwill.

Solidarity. Cost–benefit justification is partly based on economic analysis. But where there are non-priced amenities involved, project justification is made much easier if there is wide consensus expressed as group solidarity. The shading of valuation studies becomes much darker when local support is demonstrable. Hence economic valuation merges into political justification.

In practice, consultation is imperfect. This is because it takes time and costs money. There is a tendency to use representative spokespeople rather than inclusive interests. Commonly, too, professional consultants are employed to run the process, partly to ensure that the sponsoring agency does not appear to influence the views of those canvassed. It is not always the case that such professional consultants are experienced in

257

negotiative or participatory styles of consultation, as this is still very rare practice in the UK.

The purpose of this particular paper is to assess past, current, and possible future approaches to consultation over shoreline management and flood alleviation with a view to assessing the scope for a more participatory and mediated approach to consultation and decision taking. This approach includes an analysis of the experience so far of consultative procedures in the Ouse Washes and the North Norfolk coast shoreline management planning process.

As these examples are at a relatively early stage in their evolution the more specific aim of this analysis is to examine the various ways in which a more participatory form of consultation can take place so as to generate a sense of trust, facilitate a workable consensus amongst interested parties, and create an outcome that will be continually reassessed and recalculated, yet supported. Trust, input of knowledge, and output of accommodated positions in a manner that informs, educates and sensitises participants to the common purpose of effective, environmentally sustainable and economically optimal shoreline management are fundamental ingredients for successful implementation.

In the text that follows, there will first be an analysis on the significance of shoreline management planning in the face of sea level rise. Second there is a full assessment of the genesis and purpose of shoreline management plans. Third the criteria for establishing legitimacy in negotiated consultation processes are explored in relation to discourse theory and principles of empowerment. Then there is an analysis of existing coastal management and flood defence scheme consultation procedures. Finally the lessons learnt for the future of this style of consultation in relation shoreline management planning will be given a full airing.

Sea level rise and shoreline management plans

Adjusting to sea level rise

According to the 1995 Report of the Intergovernmental Panel of Climate Change (IPCC, 1996, p. 35) sea levels around the world have risen by between 10 and 25 cm over the past 100 years. The IPCC believes that between 2 and 7 cm of this measured rise is due to thermal expansion of the oceans linked to increased global temperatures. Some 2 to 5 cm may be caused by ice melt, though the precise role of the major ice sheets in Greenland and the Antarctic remains a mystery. The IPCC also report circumstantial evidence of more extreme weather patterns with a tendency to an increase in coastal storm surge events. Coming on top of modest but measurable sea level rise, associated with already established climate change, this combination suggests that shorelines generally will be subject to greater threat of flooding, erosion and accretion of sediment than heretofore.

The Climate Change Impacts Review Group (1996, p. 211) estimate that, for the UK, sea levels may rise by 19 cm over the average for 1961–90 by the 2020s, by 37 cm by the 2050s, and by 42–63 cm by the 2090s. These estimates are likely to be increased further by the slow but steady post-ice age sinking of land in the southern and eastern UK. East Anglia may experience a combined effect rise of 50 cm by the 2050s. Though the Group admit that it is not possible to quantify the risk, they nevertheless conclude:

Coastal lowlands around the wash, sections of the Norfolk and Suffolk coasts, and, to a lesser extent, those on Teeside and in south east Lancashire seem particularly vulnerable. (Climate Change Impacts Review Group, 1996, p. 211)

The IPCC (1992) recommended three complementary approaches to shoreline policy

(1) *Managed retreat* by progressively abandoning land and protective structures, and by creating or restoring appropriate alternative habitats or landscapes to replace especially prized areas that will be lost as a consequence.

(2) *Accommodate* by continuing occupancy and by applying a selection of adaptive measures; including adjusting to periodic flooding by raising the level of buildings and access routes, or accepting temporary inundation.

(3) *Protect, or hold the line*, by providing robust and reliable defences.

We shall show that these concepts, though clear in written English, are by no means understood in the terms of a 'political culture' of acceptance and programmed support. What may be regarded as a retreat option in 10–20 years time, for instance, is a real blight on property value. It can severely reduce the value of the property asset for investment and insurance cover in the immediate future. Similarly 'to accommodate' can mean a host of ambiguous responses or inactions that have to be clarified if a property owner or a local flood defence committee is to be persuaded of its longer term inevitability.

Organisational responsibilities for shoreline management in England

The coast has always proved an awkward zone to administer. There are many overlapping statutory and political responsibilities to accommodate that cannot be wrenched clear for any special-purpose body to take over. The offshore zone is technically the responsibility of the Crown Estates Commissioners and the marine and fishing interests, headed up by the Ministry of Agriculture, Fisheries and Food (MAFF) for fishing and pollution control, and the Department of Trade and Industry (DTI) for shipping and the Department of Transport (DoT) for offshore transportation generally. There are no formal planning controls over activities in the offshore zone, though planning tools such as environmental impact assessment apply for activities covering dredging and oil/gas installations.

The coast itself is the responsibility of both the Department of the Environment (DoE) as the policy ministry and the local authorities as coastal defence organisations. But while the local authorities handle the coast itself, the flood or coastal protection element is the responsibility of the Environment Agency operating through MAFF cost–benefit guidelines and grant aid. To bring all this together, local and regional flood defence committees, composed of local authority members and MAFF appointees, are responsible for determining both strategy and budgets. At the administrative level, Holgate-Pollard (1996, pp. 29–34) notes that a coastal policy branch has been created within the DoE. This seeks to co-ordinate coastal management across all relevant government departments. Holgate-Pollard believes that the prospect of sea level rise and greater pressures for land use change and offshore dredging licences have led to a 'new recognition' that coastal issues must be looked at in the round. In response

Building trust in shoreline management: T O'Riordan and R Ward

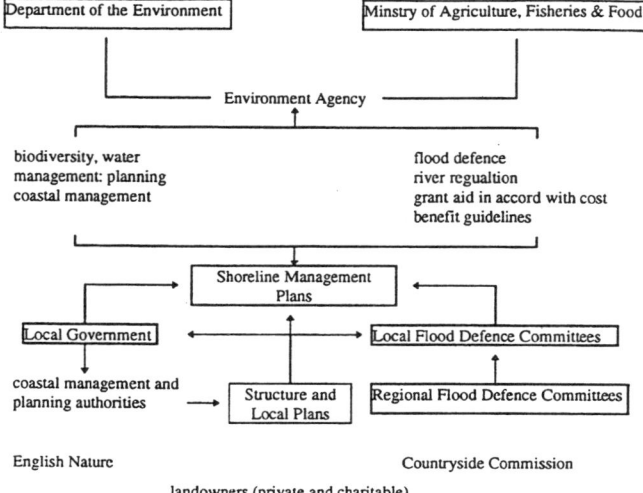

Figure 1. Responsibility for coastal management in England.

to the call for much impaired and more effective co-ordination, the DoE have reissued Planning Policy Guidance 20 (DoE, 1992) in the form of policy guidelines (DoE, 1995). This provides a more coherent framework for intergovernmental discussions but it does not automatically create an atmosphere ripe for fresh approaches to negotiated consent.

By way of a summary, the following list of interested parties in shoreline management is offered. Figure 1 describes the relationships between these parties in diagrammatic form.

The Environment Agency (MAFF-linked for coastal management at flood protection, but DoE-linked for conservation and amenity considerations) as the primary executive agency, and ultimately responsible for the shoreline management plans (SMPs).

The Ministry of Agriculture, Fisheries and Food for giving guidance on SMPs, for determining project justification, and for granting aid for implementation.

The Department of the Environment for providing planning guidance for local authorities for managing the coast, and for being responsible for wildlife and amenity at a policy level, including biodiversity.

The county councils for determining structure plans and for co-financing recreation and access schemes on coastal zones.

The local district councils for protecting the coast from erosion, and for controlling development in inundation-threatened zones, grant aided by the Department of the Environment.

The Flood Defence Committees, primarily the creatures of the Environment Agency, and consisting of county council member appointees plus MAFF selected non-political representatives. The Local Flood Defence

Committees have executive responsibility for programming flood defence works and for supplying the county council levies to match grant aid. The Regional Flood Defence Committees have no executive responsibilities, but play an important strategic and policy guidance role.

English Nature and the Countryside Commission for England as executive agencies for managing key wildlife sites and for providing grant aid for conservation and access.

Landowners, and especially the main voluntary conservation bodies (National Trust, Royal Society for the Protection of Birds, Country Wildlife Trusts) who own and manage 'critical' conservation habitats.

This group of six layers of responsibility are the 'top tier' of 'statutory' consultees in any coastal management issue. We return to the significance of this group later in the paper.

The shoreline management plan

Shoreline management plans were introduced in 1995 to provide a more comprehensive and integrated approach to managing the coast (MAFF, 1995). The development of these plans came at a time when MAFF was extending its criteria for assessing the costs and benefits of management options to include environmental and other non-market considerations (MAFF, 1993a). At the same time the proposed Environment Agency, connected to the DoE, was being given ministerial guidance from both MAFF and DoE on how to incorporate sustainability into its procedures (DoE, 1995). Both the cost effectiveness and the environmental sustainability remits involve judgements that require a sophisticated approach to coastal planning. This means including such approaches as organised retreat, carefully orchestrated erosion and accretion, protection of dunes, saltmarshes, tidal flats and other natural features, and 'holding the line' defences, all set in a framework of best practice and precautionary management (MAFF, 1993b).

An important reason for developing SMPs was that the coastal defence operations on the landward side were originally keyed to local authority and other administrative boundaries rather that the natural 'cells' of largely self-contained sediment movement (Purnell, 1995).

The main purposes of SMPs, as outlined by MAFF, are as follows:

- to improve understanding of the coastal processes operating within a sediment cell;
- to predict the likely future evolution of the coasts;
- to identify all human-made and natural features likely to be affected by coastal change;
- to pinpoint zones for special investigation;
- to facilitate consultation amongst those bodies with an interest in a shoreline.

After a plan is completed, it should

- help the assessment of strategic defence options;
- clarify the responsibility for monitoring change;
- inform the planning authorities of any areas where development would be undesirable;
- identify opportunities for maintaining or enhancing the natural coastal environment;
- ensure continuing arrangements for consultation amongst interested parties;

Reg Purnell (1995) the MAFF Chief Engineer and architect of the SMP, recognises that the strategic, cross-administrative responsibility of the SMP will require that very special attention be given to the consultation processes. He believes that the coastal management groups established by the coastal district councils should be particularly concerned to ensure that consensus is reached early and regularly thereafter.

In interview for this project a senior official linked to the SMP process noted:

> There is a difficulty in integrating the SMP into the planning process, and I think it is a failing within the operating authorities themselves, that they, or the engineering department of some authorities don't actually liaise with these planners. This is quite sad when you consider that they are in the same building.... The learning process is probably going to take a bit longer than we envisaged.

The precise role of the SMP vis-à-vis the strategy's statutory functions of coastal management remains to be tested in practice. The aim is to provide a framework for co-ordination and to ensure that effort and resources are programmed according to agreed priorities. But the planning function is a statutory matter subject to the demonstrable accountability of local authorities and environment ministries. Shoreline management plans have no such statutory power. Their significance depends upon informed consent and widespread support amongst all interested parties. This is why legitimacy in the consultative process is such a vital matter. In principle, there need be no difficulty in implementation, should that degree of legitimacy be attained, even though there is a degree of ambiguity as to the precise statutory role of the SMP. It is in this context that this paper is presented.

Ambiguities in compensatory arrangements for shoreline planning

In shoreline management planning there is an ambiguity over the degree to which 'losers' are identified and compensation offered. Because of the biases inherent in any policy context, it is possible for a policy framework to be structured in such a way as not to recognise a 'loser' as legitimate, and hence not to regard the 'loser' as an object for compensation or mitigation. For instance, current MAFF policy states that where managed retreat involves the necessary loss of property in the interest of 'best practice' shoreline management, there is no automatic provision for monetary compensation. There is a grey area between 'natural' loss and 'planned' loss for which compensation is sometimes payable and sometimes not.

This ambiguity is contentious enough in the case of land actually altered by erosion or accretion, but where that alteration is deliberately allowed within the paradigm of good shoreline management practice, there are deeper problems. However, property blight, namely the loss of potential property value, because of an extended period of delay before retreat actually takes place, raises a fundamentally problematic issue both for landowners and for coastal defence funding bodies. The presence and continuance of such an apparently unfair arrangement makes it very difficult for a genuinely negotiated consultation process to work.

MAFF ministers have long been lobbied to clarify the situation. At an informal meeting of the Anglian Region Flood Defence Committee, the then Minister Tim Boswell went further than MAFF had gone before

> Managed setback...involves a deliberate decision to realign the existing line of defences in order to achieve environmental and engineering benefits. If land fronting the new defence contributes to the flood protection of the area, then there may be a basis for consequent depreciation to be reimbursed as part of legitimate project expenditure. Boswell, 1996, p. 4.

This is a significant policy shift. It indicates that ministers accept that the loss of a particular land use or habitat for a continuation of a shoreline protection process, in the name of an agreed SMP, should be paid for from the flood defence budget with the usual 75% grant aid and reimbursement of most of the remainder one year later. So far this policy shift has not been tested, but it will be in the Cley–Kelling stretch of North Norfolk (see Figure 3, in a later section of the paper).

This policy shift is particularly relevant where 'critical' zones of Special Areas of Conservation (SACs) or Special Protected Areas (SPAs) are involved. For a good summary of the policy and practical significance of the Natura 2000 approach to coastal zone management, see Healey and Doody, 1996. At present all SACs and SPAs, are in the process of being nationally designated before being taken to the European Commission for Europe-wide approach. That process will take until June 2004 (Julien, 1995). Meanwhile, every SAC and SPA will be subject to a formal management plan. This should form the basis of their future protection (Huggett, 1995).

Because of UK commitments to biodiversity and to the Habitats Directive of the European Union, special consideration has to be given to areas of internationally recognised wildlife interest, whose functions are vital for shoreline stability or managed change. Such zones are regarded as 'critical'. This concept is still a little vague, but includes the notion of essential habitat, irreversible loss, and international wildlife significance. Under the guidelines for the EC Habitats Directive (92/43/EEC), and legally enshrined in the Conservation (Natural Habitat) Regulations SI 2716, any unavoidable loss of such areas in the interests of best practice shoreline management should be replaced by appropriate investment and management in alternative sites over a long period of time. This would meet the principle of 'maintaining overall conservation status'. Whether such a procedure is the subject of UK or EC/UK funding remains unclear at present. But the obligation is there, and someone will have to foot the bill for researching and creating alternative conservation sites, where either an SAC or an SPA is threatened.

At present, the only mechanisms for providing a compensatory alternative wildlife site lie in the agri-environment package on offer from MAFF. The obvious candidates are a special 'wetland tier' of an Environmentally Sensitive Area (where an ESA is already in existence on the site, as in the case of the Broads), or a similar wetland/saltmarsh habitat for the Countryside Stewardship Scheme, or via the Habitat Improvement Scheme, on offer for long-term set aside. The problem with all of these is that the landowner may not agree to what is on offer, and the financial roles of management agreements are not too flexible. Sooner rather than later a more tailored financial package will have to be made available.

Unless such compensatory rules are agreed, it is unlikely that any reasonable consensus will be reached. Losing parties will have every incentive to lobby hard for the best deal, and to raise their valuation assessment of the property which they are in danger of losing. MAFF cost–benefit guidelines will subsequently be strained, if an acceptable outcome based on alternative conservation status is not found. The pressure to raise the environmental benefits of 'critical' site protection will be enormous. Where the rules of valuation are not fixed, a more political process is brought into play. This is as good a reason as any for extending valuation through widening the basis of empowerment.

Extending consultation in shoreline management planning

There is therefore a pragmatic political imperative for widening the basis of consultation in shoreline management planning. First of all, the guidance to the Environment Agency states that the Agency should 'strive to develop close and responsive relationships with the public, local authorities and other representatives of local communities and regulatory organisations, and to work in partnership with all such groups' (DoE, 1995). This calls for a more evolving and partnership approach that is 'close and responsive'. This line of reasoning is also evident in DoE guidelines to Local Agenda 21 initiatives. The idea is to create fundamentally different ways of working in which partnerships are created, conflict is avoided, and public participation is treated as a core part of decision-making processes. In addition the recently published mission statements of the Environment Agency, cite, as one of its visions, 'to operate openly and consult widely', and one of its aims is 'to develop a better informed public through open debate, the provision of soundly based information and rigorous research' (Environment Agency, 1996, p. 3).

There are also efficiency gains to be made. A genuinely participatory approach to consultation will be demanding in terms of time and personnel. Though the investment of staff time may be high, the benefit gained by building up trust and co-operative relationships at an early stage should result in cost savings by getting things done in the longer run, due to the reduced risk of time consuming and politically contentious opposition.

Though less easy to quantify, there is also a potential cost saving in designing a strategy over, say, 10–20 years so that interested parties are aware of possible outcomes and can plan both for compensation and for more adaptive responses to coastal change than would otherwise have been the case. An open and agreed programme of shoreline management should reduce the need for expensive emergency repairs and inappropriate short-term investment following upon special interest pleading. That, of course, is an ideal. In practice there is never the money for comprehensive proactive management.

All this, arguably, should help to build trust, and through trust should come a more negotiated process of fair play in giving and taking to reach compromise by consensus. Because any outcome of shoreline management will involve new possible winners and new possible losers, it is especially important that participants feel confident and trusting of the process itself. In order to achieve this, the process must be seen to be open and accessible, as well as honest about what has to be done and why, in the name of good shoreline management. In addition what cannot be done, because of lack of resources or difficulties in implementation due to policy deficiencies, is best explained and understood through a trustworthy process. This is a critical point. In practice no SMP can meet all expectations, and it will never be possible to compromise or compensate in the face of unavoidable losses for all complainants' interests. But to explain and to show how a difficult outcome can be taken forward, via a mechanism that creates trust and forgiveness, is a very important part of the SMP consultation procedure. Such an arrangement should give greater legitimacy to ultimate decisions taken by politically representative flood defence committees.

Legitimacy and negotiated consent

According to Beetham (1995, p. 3) only when power is acquired and exercised to justifiable rules, and with evidence of consent, can it be called rightful or legitimate. For such an arrangement to occur, power must

- conform to established and agreed-upon rules;
- obey rules supported by beliefs shared by both the dominant and to the subordinate parties;
- operate through consent by those subordinate to the particular power relationship.

Legitimacy is not an absolute state: there will always be the dissent and dissatisfaction over the abuse of power and of information manipulation. What matters it seems, is that the process of reaching a decision is seen to be fairly conducted, respectful of every legitimate view, accommodating where adjustments can reasonably be made, and based on the trust that is recorded by continued accountability.

In a detailed study of eight participatory techniques, each based on principles of fairness and compensation, the German sociologist Ortwin Renn and his colleagues (Renn *et al.*, 1995) pointed out that traditional 'top-down' consultative styles are proving unsatisfactory to many of the parties involved, and counterproductive for any final management solution. This is because they often do not lead to outcomes around which affected interests agree. They are counterproductive because they tend to generate rather than alleviate conflict.

> Often heavily shaped by scientific analysis and judgement, these kinds of decisions are vulnerable to two major critiques. First, because they de-emphasise the consideration of affected interests in favour of 'objective analyses', they suffer from a lack of popular acceptance. Second, because they rely almost exclusively on systematic observations and general theories, they slight the local and anecdotal knowledge of the people most familiar with the problem, and risk producing outcomes that are incompetent, irrelevant or simply unworkable. (Renn *et al.*, 1995, p. 1)

Webler and Renn (1995) summarise the principles of participation as being based on popular sovereignty and political equality. The public, they argue, ought to engage in political affairs in order to legitimate, that is create support for, any final decision taken by the state. People 'learn democracy' through participation. For this to be reasonably assured, they claim, people need to participate in a manner that is *authentic*, namely through clear articulation of their knowledge and values, and *transparent*, that is, through being intelligible to all other interested parties. The aim is to devise consultative procedures which permit all relevant interests to contribute in a manner that allows them to be respected, and responsive to equally respected differences of view or alternative interpretations of gainers and losers.

Procedural fairness, therefore, relies both on legitimacy (which is founded on respect, authenticity and transparency) and on efficiency. The dilemma here is devise consultative arrangements that do not unduly prolong an outcome because of their volume of evidence and numbers of participants. This is possibly best achieved by creating, or encouraging the creation of, informal networks of interested parties on a 'come and go' basis. Such networks could be seen as gatekeeper coalitions that permit the entry and exit of informed and legitimate interests to be involved through representative arrangements. It may be possible to establish such networks on a local and informal basis.

The analysis so far consists of both procedural requirements for a more extended and co-ordinated participatory consultation, as laid down in official guidance from MAFF and the Environment Agency, and a more normative assessment of the advantages of fresh approaches from the sociological political science and environmental economics literature.

Sabine O'Hara (1996) offers a more cross-disciplinary justification for the need to extend the basis of valuation and negotiation in complex environmental management decisions. She uses as her medium the approach of discoursive ethics, which she describes as 'a process of uncovered and undistorted communicative interaction between individuals in open discourse' (O'Hara, 1996, p. 96). The basis for involvement is none other than 'mutual recognition and acceptance of others as "response-able" subjects' The function is to open up people's 'life world', or zone of consciousness, to a host of ethical, ecological, economic and sociological experiences that shape the evolution of all environmental–societal inter-relationships. To do this, she argues, it is essential to create an arrangement of a shared commitment as to why a given outcome is desirable. In this manner, the technological and political restraints that almost universally shape the institutionalised frame of environmental management processes can constantly be exposed and re-examined in a more holistic framework.

Existing socio-economic and political structures determine existing ecological valuation structures and resulting policy decisions. Who comes to the table and whose voices are expressed in the discourse are therefore essential to the ethical character of the discourse process (O'Hara, 1996, p. 97).

O'Hara justifies this approach of negotiative participation on this basis that it

- *integrates ecosystem valuation* by seeking to overcome the disciplinary blindspots, assumptions and value judgements exposed in the ecological and economic approaches to ecological valuation;
- *copes with uncertainty* by combining scientific rationalities to stakeholder biases and negotiated judgements by extending the basis of peer group authority. Human beings are moral agents influenced by social and ecological contexts that frame their interpretation of fair treatment;
- *reconstructs life-worlds* by breaking down existing patterns of power and privilege that shape the norms and values of others, so as to re-establish an authentic and trustworthy (legitimising) context for debate and resolution.

This combination is likely to be messy. It is open ended, exploratory, laden with differential expectations, and power expanding. These are unfamiliar conditions for traditional consultation procedures. But arguably they are necessary in a world of systemic thinking, long time horizons, and a genuine concern for the rights of those affected by any transition towards a more sustainable world.

Empowering constructive informed opinion and local knowledge

This theoretical backdrop sets out the stall for empowerment in negotiative consultative procedures. It is rarely the case that local opinion meets the criteria of authenticity and transparency. This may be because local opinion is neither coherent nor consistent. One way of responding to this is to test for *empowerment* of local opinion. Empowerment is a buzz word that means much more than granting or extending power to others (Schwerin, 1995). Empowerment means giving, or enabling, self-esteem, self-reliance, personal competence, coping skills and community building. As the American sociologist Rappoport put it

> Empowerment expresses itself at the level of feelings, at the level of ideas about self worth, as the level of being able to make a difference in the world around us, and even at the level of something more akin to the spiritual. (Rappoport, 1985, p. 17)

Local opinion frequently believes it can contribute constructively to an outcome. And usually this is so. Local opinion that believes it is self-important is not empowered in the meaning of the Rappoport definition. Hence the need for dialogue based on mutual respect and understanding. Hence, too, the advantages of informal networks of groups and individuals who may facilitate the process of empowerment with care and persuasion.

Empowerment fits closely to the theoretical framework of legitimacy advanced by O'Hara. To be empowered is to be joyful in the process of reaching genuine and lasting consensus. That in turn means ensuring that those who gain and those who lose as an outcome of a particular set of actions are fully aware of each others' position, and that appropriate forms of compensation are available to those whose losses cannot be avoided. This is the basis of economic optimisation, namely a position where nobody is worse off from where they were before. In reality gainers do not compensate losers, which is the basis of neo-classical economic theory, nor do losers necessarily get what they legitimately deserve even when the state or some other intermediary acts on behalf of the gainers. Yet such an outcome is the essence of distributional equity, or the justice of ensuring that property and personal rights are properly respected.

For perfectly good reasons, none of these objectives of legitimacy (built on trust, authenticity, empowerment and decisive ethics) are currently built into flood defence and shoreline management consultative procedures in the UK. What follows are comments on three different consultative exercises. One involved the Ouse Washes, a flood protection scheme with a high degree of environmental safeguard. The second was the Cley–Kelling area of the North Norfolk Coast SMP where some of the procedures raised above were actually set in motion. The research involved a series of structural interviews with key participants on the basis of the theory of legitimacy, empowerment and trust as discussed above.

The Ouse Washes

The Ouse Washes is a Special Protection Area as well as other designated lands substantially owned and managed by the Royal Society for the

Protection of Birds (RSPB) and English Nature (see Figure 2). There are also graziers and wildfowlers with an interest in flood protection and river regulation. During 1994–5 the National Rivers Authority (NRA, the precursor of the Environment Agency) instigated a participatory consultative programme involving statutory and voluntary conservation bodies, farmers, recreationists and wildfowlers.

The programme created an informal network of interested parties who began to know and trust one another through informal contact beyond the formal meetings. This led to a better and more broadly based under-

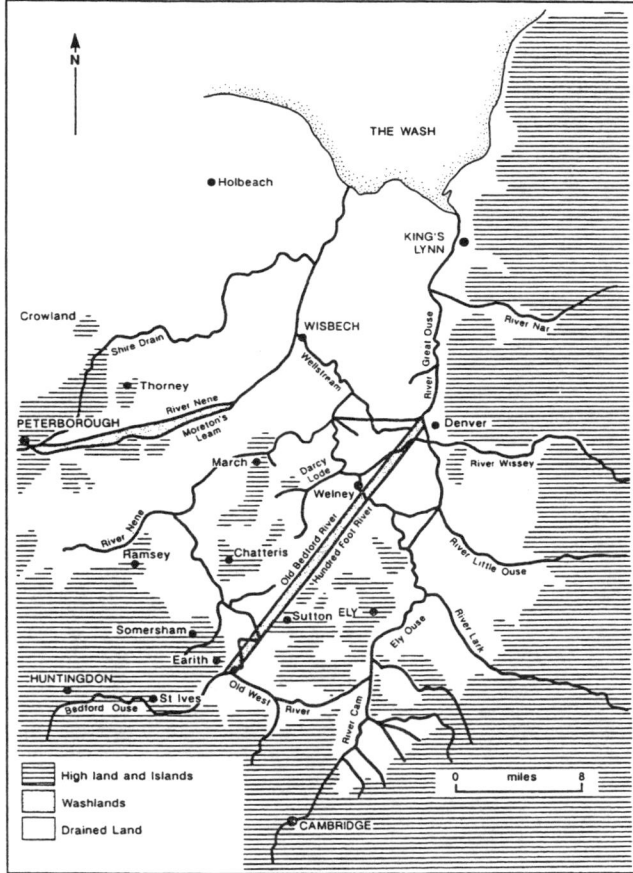

Figure 2. The drainage geography of the Ouse Washes. The problem was to manage water levels and flood protection to meet the landowner interests of the RSPB, wildfowlers and graziers, bearing in mind the international wildlife status of the site.

standing of each others' positions at an early stage, as well as a strong sense of trust in the process. The 'formal' meetings were conducted in a 'round table' style so that throughout the process, all participants worked together to consider the issues being raised.

A key factor was the early accommodation by the RSPB to give way on its demands for drier summer conditions to assist the breeding success of godwit. All parties agreed that summer grazing should be maintained to ensure both habitat and landscape features. The consultation exercise solicited an atmosphere of open discussion and give and take. Participants stressed that they found the process useful because

- they could field their views in at various critical times, not just on a once-only basis, so they were able to assess how views were shifting (*accommodation*);
- they could keep abreast of the changing thinking amongst the various parties concerned (*transparency*);
- they could get their key messages across in ways they felt were properly understood (*authenticity*).

Other aspects of interest in this case were the successful servicing of the process by the consulting engineers Binnie and Partners, who also had experience on the Broads flood alleviation strategy, and the representativeness of the individuals involved. This meant that the negotiating group was small, well established and adaptable in its negotiative style.

As an example of the sense of trust that was developed, here is the reaction of a grazier

> The process was very honest and open apart from a suspicion at first that the RSPB had the ear of the NRA and consultants. I think that had the consultations gone the way the RSPB wanted, this would have formed the basis of a grievance. As it was the graziers and wildfowlers were happier with the outcome than the RSPB. This was obviously reassuring.

The RSPB officer was less enthusiastic

> The consultation procedures were good, but after the final meeting there was a long pause before the report was published, for certain elements were included that had not been discussed or agreed. I had the impression that the NRA insiders had had a go at it and that some of the consultations had been disregarded.

The project officer for the Ouse Washes pointed out the advantages of lengthening the consultation.

> We did so to allow all concerned to really understand the process and to feel that they had had their say. In the end we did not feel we needed a formal launch after the scheme was finalised, because everyone had been involved at a much earlier stage.

The MAFF evaluation officer also approved

> The decision that was finally made [in the Ouse Washes] did not suit everybody. But it was a pragmatic decision that was made within the constraints which we were working under. It made everybody take responsibility for the aspects they were responsible for.

Nevertheless, the final report was not discussed in the same open manner as was the dialogue leading to its preparation. This led to a suspicion amongst this small but influential group of participants, who by this

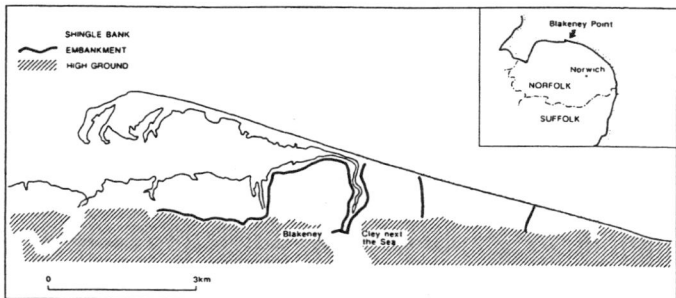

Figure 3. The Cley–Kelling stretch of the North Norfolk coast. The shingle bar is a natural geological feature that is slowly migrating landward. It is regularly reprofiled by the Environment Agency to protect the internationally recognised Cley Nature Reserve, owned by the Norfolk Wildlife Trust.

time had formed a coalition of collective interest, that the NRA remained keener to conclude their own engineering solutions based on MAFF cost-benefit principles than genuinely to listen to local (and possibly more expensive) demands.

There is a lesson here: participation has to go all the way to the final report. If cost-benefit guidelines interfere, their role should be explained clearly before consensus is reached. In any case, as indicated at the outset, it is possible for the costs and benefits to be adjusted should real consensus be reached. The Ouse Washes consultation process was not pursued to the final agreement, so this 'consensus bias' effect or the cost-benefit analysis was not tested. This point has been taken up by the Environment Agency in its Cley–Kelling study (see below).

In terms of the theory presented earlier, the Ouse Washes scheme created a participatory forum at the outset. But it did not fully establish trust. The ultimate test of a participatory exercise is the willingness to share an outcome—in effect, to create a more decentralised and pluralistic basis for distributing power. That did not happen entirely on the Ouse Washes. This is an early experiment, and it is by no means too late to continue the process for future management discussions. The role of the project officer is proving critical in helping to maintain trust through dialogue.

The North Norfolk Coast Shoreline Management Plan

The North Norfolk Coast Shoreline Management Plan is still at an early stage. Shoreline Management plans are evolving documents with broad outlines of possible options. However, in the Cley–Kelling stretch of the plan, depicted in Figure 3, the increasing storminess of the North Sea has precipitated matters. At present the two villages of Cley and Salthouse are protected by a shingle bank, bulldozed into a loosely consolidated structure, that fronts a freshwater marsh. This is partly a grazing and shooting area, and partly an international wetland bird reserve, owned by the Norfolk Wildlife Trust.

The original proposal for the plan was to hold the shingle bank for ten years whilst deciding how to cope with the relocation of the habitat and the localised flood protection for the villages. However, serious breaches of the shingle bank in February and August 1996 convinced the Environment Agency that a more rapid decision would have to be made.

The difficulty the Agency faced was that the SMP consultation process had followed an earlier model of 'once-only' input to parties who were now aware of who was being contacted or what they were saying. There was certainly no sense of give and take. If local people were regarded primarily as an information source, and not an intimate part of the strategic vision, then both the opportunities for mutual education and the broad basis of understanding of what is or is not possible never got aired. In any process of building trust, such a limitation on one crucial group of participants proved counterproductive to the objectives of consultation and the delivery of a publicly supported Shoreline Management Plan.

A widespread view amongst local people was

We have no say whatever: local people are just not consulted. We want to have a dialogue with EA. I feel the issues are very important and I would be prepared to put a great deal of effort into a working group if one was set up and locally based.

Another local spokesperson spoke for many:

I feel we had no real say. I wanted to be more involved than I felt able to be. I also resent the lack of feedback. We have heard nothing. It is like feeding information into a black hole.

Yet there is a real dilemma here regarding the formal statutory and landowning roles of the conservation agencies and the NGOs, on the one hand, and the authentic involvement of local people, on the other. The conservation bodies are aware of this issue. Here is a comment from a senior officer of the Norfolk Wildlife Trust.

Hopefully local people should be involved as much as everyone else. My gut reaction is that there is a danger of losing the wider perspective. But it is important because you are dealing with local's livelihoods. I must admit that we do tend to think in organisations. Perhaps we do overlook the individuals. They are in a difficult position and it is difficult for them to have a voice.

What emerged from the analysis of the North Norfolk participants was evidence of an 'inside track' consisting of the English Nature, local government and the Environment Agency. Key landowners (including conservation bodies) were not part of this steering group. So English Nature and the local district council felt comfortable and involved, even though they had reached no consensus as to how to move forward. The landowners were suspicious but willing to co-operate. The 'outside track' consisted of local residents, wildfowlers, fishermen and local users. This group, with no statutory connections, regarded themselves as fodder for the legitimation of a process that, to them, appeared largely deaf to these interests. In practice nobody really knew how to involve them. In this sense, the very issues of trust, transparency and legitimacy simply could not be put in place. And it showed in the kinds of responses already quoted. The sense of lack of empowerment was palpable.

There was no obvious purpose to the consultative exercise. It was not confidence inspiring, there was no build up of trust, and no clear idea of where the process was going. We were just not welcomed. (Sailing interest representative)

If only the locals had been consulted on the future of the shingle ridge, they could have told the Environment Agency that the ridge was unstable and there is insufficient material offshore to maintain it. The Agency does not need expensive consulting engineers to tell them that. Local people lost heart because their valuable knowledge was not being tapped. (Cley Parish Councillor)

These reactions were widely typical of the initial phase of the SMP consultation process, when the 'old-fashioned' approach to consultation was tried. However, a form of breakthrough came when all of the senior conservation/landowner officers tried a more co-operative approach.

Subsequently, the Environment Agency decided to try out this negotiative style in a slightly more structured manner. It commissioned the two authors of this article to undertake a participatory consultative exercise. This consisted of two phases.

Phase 1:
All assessment of likely responses to a more mediation-focused approach based on the principles outlined in the introductory sections to this paper.

Phase 2:
A series of one to one, and mini-roundtable discussions leading to a coastal forum round table before, during and after various options for the coastline were canvassed, with a view to reaching consensus.

Phase 1 is now complete. It revealed a considerable reservoir of goodwill towards the Environment Agency, and a sincere willingness on the part of all 17 representative interests involved to reach an acceptable set of solutions that are sound in engineering terms, cost effective in the broad sense of economic analysis, and socially acceptable both locally and nationally. However, the study also showed that the universal goodwill is enormously dependent on the successful promotion of both the atmosphere of mutual respect, and a continuing process of creative involvement. Any reversal to the previous style of consultation would prove counterproductive to the success of the outcome.

Following the first meeting of the informal 'coastal forum' round table the following comment was widely representative:

It was very interesting to see how well behaved everyone was when we met. Whilst I think it would be naive to think that this will always be the case, I am sure that this type of forum produces more positive, less aggressive behaviour than most others.

This is one of the most pleasing meetings I have ever been to. The ground rules need setting, so the participatory method is an admirable approach to adopt.

Phase 2 is now in full swing but will take six months to complete. The key conclusion is that Phase 1 set the conditions for a process that has created stable and meaningful networks through which inclusive representative interests have been identified. These individuals become members of the coastal forum, and through their authentic and trustworthy communication, 'non-coercive communicative discourse' to use O'Hara's phraseology, has been established.

Already the vision of a solution is emerging. There is widespread acceptance that the existing shoreline shingle ridge is unstable and cannot continue to act as a reliable defence. The quality of the local shingle is

Building trust in shoreline management: T O'Riordan and R Ward

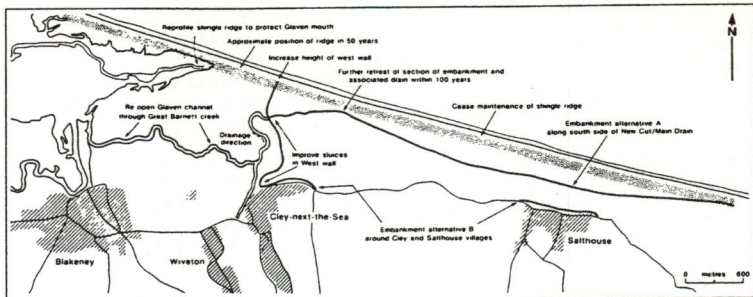

Figure 4. A possible solution for the Cley–Kelling stretch of the North Norfolk coast. This diagram is based on one appearing in a report commissioned for the Environment Agency by Professor John Pethick (see Pethick, 1997, p. 69). The line of retreat might be a clay wall with a low level of management of the shingle bank, primarily to protect the Glaven Estuary. This would protect two freshwater wetlands, but the Norfolk Wildlife Trust site at Cley would be reduced in size. (Alternative A in the diagram.)

deteriorating as the bank is being reprofiled. Smaller sized material and too much clay and sand are now being incorporated in the dredged material.

The 'do nothing' option is not possible. Because the shingle bank is unstable and is naturally retreating, some sort of continued management is necessary to maintain shingle movement along the coast. This is also essential to retain the current mouth of the Glaven River estuary to protect the landward freshwater marshes. A couple of small 'bunds' protecting the two villages, costing only £300 000 would be unacceptable to local opinion. This leaves the location of a retreat clay wall, along the 'new cut' drain that divides the wildlife wetland approximately in half. Such a location would mean that the seaward portion of the existing nature reserve would be lost but the landward portion would be better protected (see Figure 4). The main problem is the cost–benefit equation. This wall would probably cost around £1.5 million, and the issue of replacing the lost wildlife habitat has not yet been resolved. Nor has anyone determined how this relocation, or re-investment in wetland habitat elsewhere, should be paid for. And then there is the awkward issue of how these costs can be justified. This is why participatory negotiation is such a critical issue, since local consensus may result in an effectively higher benefit justification. We shall see.

Perspective

The general conclusion from these case studies is that in both instances consultation was tried on a more adventurous basis than was traditionally the case. Nevertheless, only in the Ouse Washes case was there closure and consensus on the outcome, and even then a more extensive consultation process would have led to a more flexible interpretation of the official cost–benefit analysis. In the North Norfolk case, the Environment Agency changed tack to establish a process based on the theory outlined earlier. All the evidence to date suggests that this shift has proved most effective in building trust in shoreline management.

Table 1. Summary of consultees impressions of consultation processes.

North Norfolk Shoreline Management Plan

	Very good	Satisfactory	Fair	Poor	Don't know
Trust		a*aa		aaa bbbbb cc	
View heard	a*	aa	aa	a bbb cc	bb
Feedback		a*	a b	aaa bbbb	a cc

Consultees approach to process:

open/flexible	pursuing own interests
aaa bb cc b	bb a* aa

Ouse Washes Summer Flood Control

	Very good	Satisfactory	Fair	Poor	Don't know
Trust	bbb a*		a		
View heard	bbb a*		a		
Feedback	bbb a*	a			

Consultees approach to process:

open/flexible	pursuing own interests	
bb a*	b	a

a* = statutory, a = 'top tier', b = local representatives, c = not consulted

Table 1 shows how the interviewees summarised their expressions of the consultation process. The table is structured to reflect the distinction between those who were statutory, those who were in the 'inner circle' (NGOs and landowners), and those who were 'peripheral' to the process (as seen by the Environment Agency or the consulting engineers). The results are reasonably clear cut. If one is 'in' the process is fine: if one is 'out' it is not. Yet the evidence from the theory and the quotes is that this arrangement is unstable and counterproductive.

The Environment Agency has now commissioned a much more open participatory process along the lines outlined by O'Hara, and by Renn and his colleagues. Though this applies only to the Cley–Kelling Marshes, the early indications are that trust has rapidly been restored, local people can act as conduits to their constituencies, and that a consensus can be reached, over the difficult issues of compensation and cost–benefit analysis.

Mediation by participatory roundtables, or such like, will only work if participants are representative, treated with respect, assisted to organise and articulate information, are enabled to express their fears and feelings in as authentic a manner as possible, and appreciate the concerns and legitimate expectations of others. That requires a clarification of the role of SMPs, better understanding of the policy context regarding cost–benefit and compensation, as well as a fuller appreciation of the link to the strategic and to the statutory consultative procedures. Above all, it requires early involvement in the process, so that the role of consultation in all its forms is clearly understood at the very beginning. This in turn means the use of explanatory leaflets, newsletters, and informal discussion groups.

Local networks play an essential role. In many cases incipient networks based on personally co-operative individuals exist and can be mobilised and extended. Such networks are vital for sharing information, formulating positions, and for mutual education. This is the basis of empowerment. These networks may be facilitated by parish councils, or local activists, or through a variety of social connections. It is important that there is a measure of self-generation and spontaneity about their formation.

The statutory consultees can provide more direct involvement in the SMP process through participation in a steering group linked to project management. This body would meet in public, with published proceedings, and attended by local network observers. They would also participate in the local roundtables, though this could be done through representatives where individuals' time commitments are stretched.

The policy impediments need not be brushed aside. Rather they should form an effective framework for debate and reasoned analysis. While many of the imperfections require national resolution, developing a shared appreciation of the limitations and the opportunities for policy change and for greater resonance with the coastal management process generally would help to frame the nature of expectations and demands. Coastal management involves both manipulation of natural processes and socially engineered discussions on what should provide a marvellous way to reinforce trust and empowerment. Participants feel all the more that they 'are in it together', as companions on a voyage towards better shoreline management planning.

Acknowledgements

The financial assistance of the Environment Agency is much appreciated, though the Agency bears no responsibility for the analysis presented in this paper.

References

Beetham, D. (1995) *The Legitimation of Power*. Macmillan: London.
Boswell, Tim (1996) Speech to Anglian Regional Flood Defence Committee, Blakeney, Norfolk.
Climate Change Impacts Review Group (1996) *Review of Potential Effects of Climate Change in the United Kingdom*. Department of the Environment, London.
DoE (1992) *Planning Policy Guidance: Coastal Planning*. PPG 20. Department of the Environment, London.
DoE (1995) *Policy Guidelines for the Coast*. Department of the Environment, HMSO, London.
Environment Agency (1996) *A New Agency—A New Approach: Contributing to Sustainable Development*. Environment Agency, Bristol.
Healey, M. G. and Doody, J. P. (eds.) (1996) *Directions in European Coastal Management*. Samara Publishing Ltd, Cardiff.
Holgate-Pollard, D. (1996) Coastal zone management in England: The policy context. In *Studies in European Coastal Management* ed. A.S. Jones, M.G. Healey and A.T. Williams pp. 29–34. Samara Publishing, Cardiff.
Huggett, D. (1995) The role of the Birds Directive and the Habitats and Species Directive in delivering integrated coastal zone planning and management. In *Directions in European Coastal Management* ed. M. G. Healey and J. P. Doody pp. 9–19. Samara Publishing, Cardiff.
IPCC (1992) *Global Climate Change and the Rising Challenge of the Sea*. Intergovernmental Panel on Climate Change, Geneva.
IPCC (1996) *Climate Change 1995: The Science of Climate Change a Summary for Policymakers*. Intergovernmental Panel of Climate Change, Geneva.
Julien, B. (1995) Integrated management of the European Coastal zone. In *Studies in European Coastal Management*, ed. P. S. Jones, M. G. Healey and A. T. Williams, pp. 5–21. Samara Publishing, Cardiff.
MAFF (1993a) *MAFF Project Appraisal Guidance Notes for Flood and Coastal Defence*. Ministry of Agriculture, Fisheries and Food, London.

MAFF (1993b) *Coastal Defence and the Environment: A Guide to Good Practice.* Ministry of Agriculture, Fisheries and Food, London.

MAFF (1995) *Shoreline Management Plans: A Guide for Operating Authorities.* Ministry of Agriculture, Fisheries and Food, London.

O'Hara, S. (1996) Discursive ethics in ecosystems valuation and environmental policy. *Ecological Economics* **16**(Suppl. 2), 95–107.

Pethick, J. (1997) *North Norfolk Sea Defences: Cley to Kelling Environmental Investigation.* Coastal Research Unit, Cambridge.

Purnell, R. S. (1995) Shoreline management plans: national objectives and implementation. In *Coastal Management 95: Putting Policy into Practice.* Institution of Civil Engineers, Bournemouth, 1.1.2–1.2.37.

Rappoport, J. (1985) The power of empowerment language. *Social Policy* **16**, 17.

Renn, O., Webler, T. O. and Wiedermann, P. (eds.) (1995) *Fairness and Competence in Citizen Participation: Evaluating Models of Environmental Discourse.* Kluwer, Dordrecht.

Schwerin, E. W. (1995) *Mediation, Citizen Empowerment and Transformational Politics.* Praeger Publishers, New York, pp. 55–57

Webler, T. and Renn, O. (1995) A brief primer on participation: theory and practice. In *Fairness and Competence in Citizen Participation: Evaluating Models for Environmental Discourse* ed. O. Renn, T. Webber and P. Wiedermann pp. 17–34. Kluwer, Dordrecht.

Part IV
Valuation of Coastal Resources

[21]

RESEARCH
Management and Valuation of an Environmentally Sensitive Area: Norfolk Broadland, England, Case Study

R. KERRY TURNER
JAN BROOKE
School of Environmental Sciences
University of East Anglia
Norwich, Norfolk NR4 7TJ
England

ABSTRACT / Wetlands, like any other environmentally sensitive resource, require very careful evaluation. While it is accepted that all wetlands may be equally valuable in terms of maintaining global life-support systems, individual areas may be ranked according to their uniqueness or the irreplaceability of the resource should the wetland be developed. The various techniques available for evaluating the wetland resource in the development versus conservation conflict situation are critically assessed. Indirect appraisal via the opportunity cost method can generate valuable data which have contributed to the mitigation of such conflict situations.

The Broadland, in Norfolk, England, recently designated an environmentally sensitive area (ESA), provides a case study example of wetland management. The search for an "acceptable" flood alleviation strategy for the ESA is examined in detail. The economic and environmental asset structure of the study area is examined at two levels. A basic "screening" system is applied to each of the identified flood protection planning units to enable the rank ordering of the units. A more detailed appraisal is then made of the value of selected units so that the cost-effectiveness of any planned expenditure on flood protection works can be assessed. Specific management issues and their likely effect on the environment, in terms of land use for example, are also addressed. The 1986 Agriculture Act marks a potential watershed in British conservation policy. The ESA policy encompasses a dual management strategy that attempts to stimulate compatible agricultural and conservation practices and activities. Other countries that still retain significant unspoiled wetland resources may find that preemptive regulatory government intervention in favor of conservation would help to avoid the worst aspects of the British experience.

Background

The Broadland complex wetland in Norfolk, UK (see Figure 1 for location) provides for multiple resource uses and encompasses multiple land-use systems. The origins of the Broads lie in the flooding of medieval peat diggings, the connection of these shallow lakes to the main watercourses, and the creation of a marsh-based economy. The area today supports a variety of intensive and extensive agricultural land uses on the pump-drained marshland. The management of Broadland has proved to be a complex task, as various factors have served to intensify land-use conflict over recent years. Figure 2 summarizes the factors leading to ecosystem change in the area and the resulting socioeconomic implications and inherent value conflicts.

Increasing demands on the productive use of the wetland's resources in terms of more agricultural output, more use of the rivers for recreation (boating and fishing), and more use of water for domestic and industrial consumption and effluent assimilation have created undesirable side effects. There has been an accelerated enrichment of the watercourses by nutrients (eutrophication) which leads to a widespread deterioration in the quality and quantity of aquatic life. Eutrophication also contributes indirectly to the erosion of the floodwalls which protect the low-lying marshlands. Enrichment can lead to the depletion of the natural reed beds which act as a buffer protecting the floodwalls themselves. Financial pressures have persuaded some farmers to underdrain the pumped marshlands and convert them from low-intensity grazing regimes to intensive arable production. As a result, the characteristic grazing-marsh landscape is being lost, together with the associated ecological interest in the previously high-water-level drainage dykes (ditches). There is also a simultaneous loss of ornithological and invertebrate interest.

The threat to Broadland is, therefore, twofold: the loss of the traditional landscape together with its wild-

KEY WORDS: Wetlands; Environmentally sensitive area; Opportunity cost; Cost-effectiveness; Flood alleviation

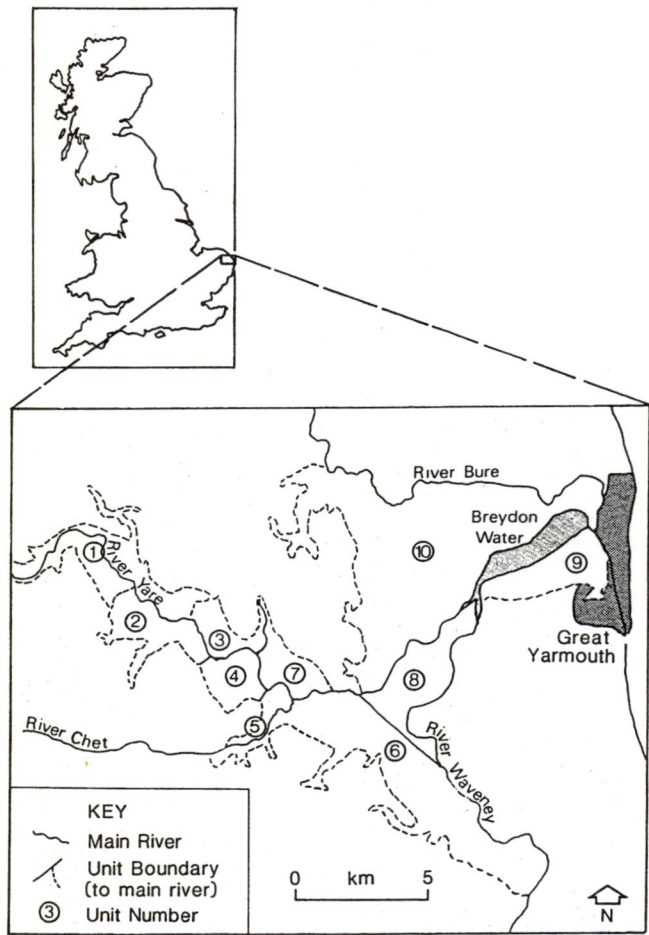

Figure 1. Location of the Norfolk Broads study units.

life interest, and the potential loss of significant areas to the sea as floodwalls are weakened and eroded.

In August 1986, the UK Government, under the 1986 Agricultural Act, announced that Broadland was to be one of six sites in England and Wales designated as an environmentally sensitive area (ESA). The philosophy underlying this dual management scheme is one of promoting traditional farming methods where these would help to maintain characteristic landscape, rare wildlife, and local historic values. Annual payments will be made to individual farmers, via the Ministry of Agriculture, Fisheries and Food (MAFF), on

the provision that they adhere to certain management guidelines. In the case of the Broads, the grant is designed to boost incomes from traditional grazing practices and to mitigate against arable conversion. Thus, protection of one of the few remaining lowland river-valley grassland sites, together with its ecologically unique wetland system of rivers and broads, reed beds, fen, and carr woodland, should be enhanced. Nationally the stock of unspoiled wetlands is so depleted that government intervention to conserve such environmentally sensitive areas in Britain was overdue.

Economic management principles are required for

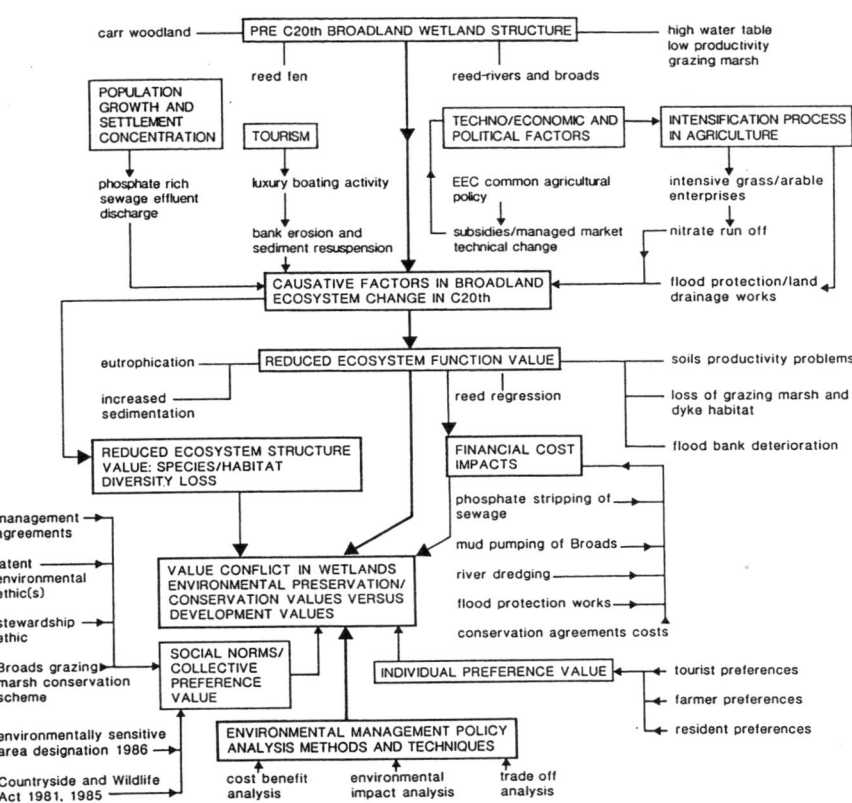

Figure 2. Ecosystem change and environmental management in Broadland.

both higher-order, unique and irreplaceable wetlands, and lower-order, locally or regionally important areas. Wetland development schemes can, in principle, be directly appraised via extended cost–benefit analysis, or indirectly appraised via the opportunity cost method. The latter method has been extensively used in the UK and USA to provide data relevant to the mitigation of wetland conflict situations. Using the Broadland wetland as a case study, typical environmental management issues are examined and evaluated. The search for an environmentally acceptable flood alleviation strategy for the area provides the main focus for this analysis. Other countries which still retain a significant fraction of their wetlands heritage may be able to draw a number of lessons from the UK experience. In particular, despite the prevailing political predisposition toward public expenditure constraint and deregu-

lation, wetlands conservation policy must be anchored to a strong regulatory framework.

Global Wetland Resources

Assessment of wetlands such as Broadland should not be viewed solely in local or regional terms. Wetlands have been under development pressures on a global scale over a long period of time. There is also growing evidence to suggest that their aggregate value may be significant in terms of biospherical life-support systems (Armentano and Menges 1986). Although wetlands yield a range of important functional and structural values, not all individual wetlands appear to be of equal value at a subglobal level. The analysis in this article does not directly address the global significance of individual wetlands. In principle, these particular values could be translated into monetary terms, although it is difficult to do so in practice. Instead, the majority of studies reported in the literature have made use of a range of valuation methods in order to estimate the "local" value of individual natural wetlands. Such methods include market prices, travel costs, land prices, opportunity costs, hedonic prices and property values, and contingent valuations derived via surveys and bidding procedures and unit-day values (Walsh and others 1984).

The economic benefits of environmental conservation include both use and nonuse values, so aggregate preservation value can be interpreted as an amalgam of use value, option value, bequest value, and existence value. Economists have sought to identify an individual's option price for conservation. This is defined as that payment which an individual is prepared to make in order to keep his/her expected utility (satisfaction) constant for differing levels of nature conservation. The difference between option price and actual current use value is the option value. Use value can be expressed in terms of expected consumer surplus (that is, the marginal value rather than the lower market price value) from the actual recreational use of a wetland, for example. The desire to keep one's options open on possible future use of the environment will be partly individual, but may also involve bequest motivations toward one's children or future generations. Individuals may also place existence value on environmental resources. The knowledge that such assets will continue to be conserved, even though no individual current or future use is anticipated, may itself generate utility.

Economic Valuation of Wetlands

A number of studies have computed approximate values for some intermediate wetland services—such as fish harvests—via market prices. Such estimates do not, however, represent theoretically valid economic values, that is, combined consumer surplus and rent values. Gupta and Foster (1975) estimated the visual/cultural benefits of wetlands using land prices (in particular, prices paid by public agencies in order to purchase wetlands on conservation grounds) as indicators of the opportunity cost of wetland preservation. The analysis, however, neglected consumer surplus value. Both Gupta and Foster (1975) and Thibodeau and Ostro (1981) estimated the value of wetlands water supply and flood control functions via the alternative or replacement cost approach. The validity of this approach depends on the existence of practicable substitution possibilities and on the fact that the alternative

chosen is really the least-cost alternative. Thus, for example, the valuation of predicted wetland biospherical life-support services (for instance, possible links between wetland ecosystem storage/emission of gaseous by products and global atmospheric stability) on the basis of primary productivity and energy flow and their replacement costs has proved to be particularly controversial (Farnworth and others 1983).

So-called participation models utilizing unit-day values of recreational activities (Thibodeau and Ostro 1981) have been criticized because of the lack of meaningful coefficients linking water quality and recreational activity. It is also doubtful that the "recreation day" value represents an adequate estimation of the value of recreation sites to the average user (Turner 1978, Freeman 1979). None of the methods surveyed so far represent adequate solutions to the user valuation conundrum. Further, none of them encompass "nonuse" value, which is now thought to be a significant component of environmental resource value.

Early research using the hedonic pricing (HP) approach centered around the statistical estimation of the relationship between environmental amenities and land prices. The assumption was that individuals exhibit preferences for environmental quality through their choice of residential location. Brown and Pollakowski (1977) estimated the first-stage marginal implicit price function for proximity to water and water-related open space. Shabman and Bertelson (1979) constructed an index of attribute factors to value coastal waterfront land. Thibodeau and Ostro (1981), utilizing a similar approach, found that an average size home in the USA close to a wetland was 1.5% more valuable because of its location. Debate continues over the question of whether or not such HP models have successfully made the jump from an estimated property value/environmental quality gradient to willingness-to-pay functions. In any case, the general applicability of hedonic pricing to a wider range of environmental amenities remains uncertain given the reliance of most studies on property values. Total wetland value cannot be quantified via hedonic pricing as nonuse values are not being captured by this method. Milon and others (1984) have also recently noted that, although reported coefficients for water amenity-related variables in published HP studies generally had the expected sign, a number of different function forms were used to describe the amenity relationship. They suggested that there is reason for concern about the choice of functional form.

Travel cost (TC) methods, wherein travel costs are taken as surrogates for visit market prices, have been extensively applied in cases where natural wetlands provide recreation services. They work best when visitors travel from a wide range of distances to a site, and only visit that one site. A number of TC model limitations have been expressed over the years, including assumptions about the value of time and time costs, the existence of site congestion and quality deterioration, and the "weak complementarity" assumption which precludes nonuse value. If best results are obtained in a site-specific context and limited to recreational value, then the TC method is clearly not adequate as an estimate of total wetland conservation value. Wetlands are not homogeneous resources. Structural and functional characteristics vary to some extent from site to site, and certainly from type to type (that is, saltmarsh, wet meadows, peat bogs, freshwater broads, and so on). In the Broadland ESA, Halvergate marshes have become an environmental cause célèbre as they represent a strategic part of the last remaining extensive stretch of open grazing marsh in eastern England (Lowe and others 1986). The conservation-versus-development conflict that raged over these marshes during 1980–1985 would not, however, have been reduced by the introduction of TC type analysis and valuation into the decision process. For nine months of the year, the Halvergate marshes are practically inaccessible and, even in summer, visitation rates are insignificant. Nonuse values (option, existence, and bequest values) are presumably very important in this case as well as political factors and the "test-case" nature of the dispute.

Contingent valuation (CV) studies, defined as "any approach to valuation of a commodity which relies upon individual responses to contingent circumstances posited in an artificially structured market" (Seller and others 1985), have proved popular in recent years. The hypothetical market arrangements are described in detail for the interviewees, who are then asked to respond to bid prices (open- and closed-ended bids) for the environmental amenity in question. In principle, nonuse values can be encompassed by the CV method. Critics of the CV approach have concentrated on the potential biases claimed to be inherent in the technique (Bishop and Herberlein 1979, Bishop and others 1983, Boyle and others 1985, Rowe and Chestnut 1983). Hypothetical (interviewee perception), strategic information (gamesmanship), vehicle (the question format) and starting-point biases could influence respondents' final bids and potentially adversely affect the CV analysis. Not enough empirical evidence currently exists to enable a conclusive answer to the question of how serious, in practice, these bias problems are. Studies by Rowe and Chestnut (1983)

and Boyle and others (1985) have found evidence of starting-point, instrument, and hypothetical bias in published CV studies.

Overall, too little attention has so far been paid to the question of the comparative validity of estimates derived from alternative valuation methods under similar conditions or problem settings. The apparent divergence between willingness-to-pay (WTP) and willingness-to-be-compensated (WTBC) measures of environmental value also requires further investigation (Meyer 1979). The general theoretical proof that justifies the conventional stress on WTP measures (that is, that over a broad range of practical applications, WTP and WTBC measures will not diverge significantly) needs to be tested empirically (Willig 1976).

Economic Management Principles

Wetlands can be and are classified on a hierarchical basis at the local, regional, or national level. High-rank-order wetlands represent unique and irreplaceable assets. Lower-rank-order wetlands, perhaps already degraded, can be substituted for and/or partially restored. Extended cost–benefit analysis (CBA) methodology can offer some resource management principles for decision making in the wetlands conflict situation. Economic analysis would suggest that, because of a lack of substitution possibilities, and given the constraints of scientific and other forms of uncertainty, the development option for high-order wetlands represents an irreversible policy decision and should more often than not be postponed indefinitely. The correct economic measure of such wetlands' value is the foregone economic surplus value (including option, existence, and bequest values) of the functions and/or services lost when development takes place. On the other hand, the development of lower-order wetlands is to some extent a reversible decision and therefore is not in principle so likely to be precluded. The correct economic measure of this wetland value is the lesser of (a) the cost of substitutes or (b) the direct measure of a natural wetland function/service value.

Opportunity cost models represent a pragmatic solution to the valuation dilemma. Approximate, proxy value measures for the uncertain wetland functions/services can be calculated by costing foregone development values and appropriate least-cost substitutes. The evidence that the total global complex of all wetlands may contain important biospherical stabilization properties should be explicitly incorporated in any policy analysis. Organic soil wetlands under natural conditions, for example, are net carbon sinks. They are important links in the global cycling of carbon dioxide and other atmospheric gases. Drainage activities and, to a lesser extent, peat exploitation have caused shifts in the balance of carbon movement between the wetlands and the atmosphere over the last two centuries. Many wetlands are now net carbon sources (Armentano and Menges 1986), leading to increased CO_2 releases to the atmosphere and potential climatic warming. Further development of either higher- or lower-order wetlands may, therefore, lead to a breaching of the safe minimum standard (SMS) in terms of the total global wetland resources required for the atmospheric cycling of gases. The SMS of conservation is an operational version of the minimax principle derived from game theory, that is, in the face of uncertainty, choose a strategy that minimizes the maximum possible loss. A "safe" minimum total wetland stock should be conserved unless the social costs are unacceptably high.

Because of the various uncertainties involved, estimating the monetary value of wetland functions and services has proved to be a formidable task, but there is still an important and constructive role for an incomplete (economic efficiency) analysis in this wetlands exploitation context. An important element in the SMS conservation decision process would be the likely economic opportunity costs of wetland conservation. The opportunity cost of unpriced wetland functions and services can be estimated from the foregone income of potential development uses. In the case of Broadland marshes, more intensive agricultural uses are the only feasible development options. Where the development option, for example, involves an agricultural conversion and drainage scheme, the present value of the net benefits of such a scheme could be computed via the following equation (Turner 1985):

$$PVNB = \sum_{t=0}^{t=n} \frac{GMI_t - C_t - Ff_{t-1} - F_cI_t}{(1 + r)^t}$$

where PVNB = present value of net benefits
   GMI_t = increase in gross margins
   C_t = capital costs of drainage (arterial drains, pumping facilities, and so on)
   Ff_{t-1} = field drainage cost
   F_cI_t = incremental fixed cost increase
   r = discount rate

Studies undertaken in the UK (see Bowers 1983, Turner and others 1983) suggest that such agricultural schemes have often produced negative net benefits and, therefore, that the social opportunity cost of conservation, that is, the costs of retaining the area as wetland, is unlikely to be very high, and may be zero.

Thus, for example, the simulated cost–benefit

analysis illustrated in Tables 1 and 2 is reasonably representative of land drainage improvement schemes (actual and proposed) in Broadland. The implicit assumption made in the engineering feasibility study (Dossor 1984), from which the basic data in Tables 1 and 2 have been drawn, was that the improvement scheme was necessary and sufficient to generate a conversion of existing poor grassland into higher-productivity grass and arable land. Critics of the UK land drainage investment program have argued that such arable conversion of wetlands would have continued even without many of the subsidized capital works schemes, because of the financial returns available to arable farmers under the EEC's Common Agricultural Policy.

The correct scheme benefits in such cases would therefore be limited due to the benefits derived from an accelerated takeup by farmers and consequent improved crop yields. It is likely that these benefits would be relatively small and that negative economic returns would be generated by such schemes.

The simulated drainage scheme analyzed in Tables 1 and 2 is assumed to be necessary before conversion can proceed. However, despite the inclusion of assumptions such as an optimistically rapid takeup rate and yield improvements above national average levels, the simulation results fail to produce positive economic returns. The negative results would be reinforced if the analysis took account of the loss of environmental assets inherent in the conversion process and of the fact that a residual flooding risk is still present in the area. Average expected scheme benefits would have to be reduced according to the relevant flooding probability in the scheme area.

In the USA, the opportunity costs of preserving coastal wetland sites have been quantified (Shabman and Bertelsen 1979, Batie and Mabbs-Zeno 1985). These analyses indicated that, where alternatives to wetland development exist (on so-called "fastland" sites), economic development values may not be very significant. Thus, wetland conservation opportunity costs appear to be very low, at least for certain types of wetland, both in the UK and the USA. US researchers have noticed, however, that provision of water access to a large group of homeowners via the development of small areas of wetland could yield significant social value, especially in locations where area-wide water access is limited.

The UK situation, which provides a rather different conflict situation, will be analyzed in detail in succeeding sections via a case study of the Norfolk Broads. Environmental protection measures such as the introduction of the Broads Grazing Marshes Con-

Table 1. Forecast scheme uptake and gross margin increases (benefits).

				Years after completion of arterial drainage investment								
	Original situation			5 Years			10 Years			15 Years		
Crop	ha	Gross margin (£1983)	Economic value[c] (£)	ha	Gross margin (£)	Economic value (£)	ha	Gross margin (£)	Economic value (£)	ha	Gross margin (£)	Economic value (£)
Unimproved grass	348	34,800	(27,840–17,400)	—			—			—		
Improved grass	—			218	45,780	(36,624–22,890)	218	45,780	(36,624–22,890)	218	45,780	(36,624–22,890)
Cereals[a], below average yield	70	25,340	(20,272–12,670)	—			—			—		
Cereals, average yield	50	25,700	(20,560–12,850)	70	35,980	(28,784–17,990)	—			—		
Cereals, average-to-good yield	30	16,560	(13,248–8,280)	—			50	27,600	(22,080–13,800)	50	27,600	(22,080–13,800)
Cereals, improved yield	—			180	114,840	(91,872–57,420)	200	127,600	(102,080–63,800)	200	127,600	(102,080–63,800)
Roots[b], below average yield	25	13,875	(11,100–6,937)	—			—			—		
Roots, average-to-good yield	15	11,400	(9,120–5,700)	40	30,400	(24,320–15,200)	40	30,400	(24,320–15,200)	10	7,600	(6,080–3,800)
Roots, improved yield	—			30	16,560	(13,248–8,280)	30	16,560	(13,248–8,280)	60	39,360	(31,488–19,680)
Total gross margin		127,675	(102,140–63,837)		243,560	(194,848–121,780)		247,940	(198,352–123,970)		247,940	(198,352–123,970)
Deduct exist GM				127,675			127,675			127,675		
Increase in GM				115,885	(92,708–57,943)		120,265	(96,212–60,133)		120,265	(96,212–60,133)	

[a] Winter wheat 75% and barley 25%.

[b] Potatoes 25% and sugar beet 75%, but roots are quota controlled so the additional acreage, above the existing 40 ha, is grown at the expense of acreage released elsewhere. It is assumed this land grows average yielding wheat.

[c] The official feasibility report gross margins have been used. They are optimistic, in the sense that they assume above national average yields in nearly all cases. Producer subsidy equivalent (PSE) measures are required in order to convert these financial gross margins into economic values. Estimation of such values is possible (see Black and Bowers 1984), but there is a large variation from year to year. The UK MAFF position is that PSE measures must themselves be adjusted as part of the subsidy element is paid for by other EEC member countries, because of the way each individual member country's budgetary contribution is calculated. The MAFF PSE calculation of 20% (1985) reduction on gross margin value has been used as a lower bound and a 50% reduction figure as an upper bound; for comparison, Bowers and Black (1983) estimate the following PSE values (1983): wheat 48%, barley 50%, sugarbeet 60%, and potatoes 20%.

servation Scheme (BGMCS), and more recently the designation of the area as "environmentally sensitive," are contributing to the protection of its unique landscape and wildlife characteristics. Flood alleviation in the area provides the second focus for the case study. In Great Britain, the relevant public agencies (Regional Water Authorities) give priority, in their investment programs, to the protection of people and property from flooding. Improvements in the level of protection for a valuable environmental resource have, therefore, to be assessed with this background in mind. The ESA designation for Broadland should serve to increase the priority ranking of flood protection schemes which protect environmental resources, and assist the Water Authority, the Broadland management agency (Broads Authority), and other interested parties in reducing land-use conflict.

Flood Alleviation Policy in Broadland

The Yare Barrier Controversy in the Mid-1970s

The Broadland ecosystem encompasses the North Sea, the rivers and Broads, and the undrained and partly drained marshes. The area is subject to the risk

Table 2. Simulated drainage scheme costs–benefit analysis results.

Scheme costs (£1983)	Scheme benefits (£1983)		
		After 5 years	After 10 years
Capital costs = C_0 = 414,000			
Field drainage costs plus	Increased gross margin = GMI_t		
Grassland reseeding costs = Fd_{t-1} = 269,000[a]	Economic value[c] =	92,708–57,943	96,212–60,133
Incremental fixed costs = FCI_t = 573,593[b]			
(in present value terms)	*Benefit–cost ratio[d]* = 0.9 to 0.6		

[a] 350 ha underdrained at £700/ha and 200 ha of land reseeded at £120/ha; assumed to be evenly incurred during first five years of scheme.

[b] Fixed costs were not properly accounted for in the official feasibility study; a sum of £85,000 for new buildings and machinery was estimated and spread evenly over the first five years of the scheme; in addition, extra fuel and machinery depreciation costs were approximately estimated at a present value total (assuming a 5% discount rate) of £500,000, using data on tractor and machinery usage rates plus fuel and depreciation costs itemised by Nix (1983).

[c] Assuming PSE deduction range by 20%–50%.

[d] At 5% discount rate over 25 years and assuming linear growth in benefits between the benchmark years.

of saltwater flooding, notably when surge conditions in the North Sea coincide with high tides. Such a combination of surge tide and storm conditions caused a major flooding event in the North Sea Basin in 1953. During the 1970s, the regional water authority proposed the construction of a flood barrier across the river Yare to deal specifically with such storm/surge flood events, which at the time were estimated to have a return period of between 1 in 100 and 1 in 175 years. A number of official economic cost–benefit studies of the barrier project were undertaken on behalf of the water authority. The results of these official studies caused a storm of controversy and they were challenged by a variety of interest groups and agencies. The scheme opponents were able to demonstrate that, on narrow economic efficiency grounds, the net present value of the barrier project was negative and, on a wider basis, that significant negative environmental impacts would be caused. Analysts critical of the official (Water Authority) appraisal studies demonstrated that a significant percentage of the scheme benefits (increased levels of arable crop output and improved grazing regime values because of new drainage and land conversion activities) yielded negative economic value. Environmental assessments demonstrated the extent of the ecological (drainage dyke flora and fauna) and landscape asset losses that the barrier would have caused. They also highlighted the extent of an acid–sulfate soils problem in parts of the area which mitigated against arable cultivation. Table 3 summarizes the enterprise gross margins received by a sample of farmers in Broadland, all growing a winter-wheat crop. The presence of an acid layer in the soil served to reduce the crop yield and necessitated remedial soil treatments. The overall result was an increase in variable costs and reduced gross margins. More recently, another soils problem has been identified in the region (Turner and others

Table 3. Winter-wheat gross margins on different acid-affected soils in Broadland.

Crop	Soil type	Gross margins (£/ha) 1979/1980 costs and prices
Winter wheat	Calcareous clay	£325–£465[a]
Winter wheat	Clay, acid within 1 m	£325–£465
Winter wheat	Clay, acid within 0.6 m	£217[b]
Winter wheat	Peat, acid within 1 m	£125–£275[c]
Winter wheat	Peat, acid within 0.6 m, but subject to ameliorative lime treatment	£174–£342[d]

[a] Based on yields of 5 to 6.4 tonnes/ha.

[b] Based on yields of 3.92 tonnes/ha.

[c] Based on yields of 3.5–5.0 tonnes/ha.

[d] Based on yields of 4.0–5.5 tonnes/ha.

1986). Limited areas of marsh are suffering from saline deflocculation, first identified in Britain on the North Kent marshes (Hazelden and others 1986). Ameliorative treatments, for example, the application of large amounts of ground gypsum and the jetting of the underdrainage network, are very expensive. Additional costs of between £10/ha (where the soils are very carefully managed and preventative measures are taken) and £150/ha (where the problem is severe and established) would significantly reduce the potential net returns from arable crops on these soils.

The barrier scheme and its uniform area-wide level of flood protection were finally shelved by the water authority in the early 1980s. Undoubtedly the use of CBA by both sides in the conflict did lead to a clarification of many of the issues involved and also, over time, reduced factual misunderstandings. The BGMCS, operated jointly by the Ministry of Agriculture, Fisheries and Food and the Countryside Commission (MAFF/Countryside Commission 1986), by which farmers are compensated financially for maintaining traditional grazing regimes and their associated ecological interest, was one compromise arrangement that subsequently proved to benefit both farmers and conservationists. The reduction of subsidies for land drainage investment and changes in the levels of price support for livestock and cereals have also reduced wetland development pressures. But it may well have been that participation in the public debate and improved economic valuation contributed to the construction of new or modified value preferences. Indeed, the BGMCS and the ESA scheme in Norfolk's Broadland are both conservation programs into which individual farmers may opt on a voluntary basis. In the limited area covered by the former, over 90% of the eligible farmers have agreed to abide by conservation-enhancing management guidelines in return for the £50/acre (£125/ha) grant. A similarly high level of voluntary takeup is becoming apparent following the introduction of the area-wide ESA management conditions earlier this year (1987).

The ESA Concept

A brief consideration of the overall environmental implications of land use is necessary as the coastal marsh environment is a particularly sensitive one. In Broadland, there are three factors which are of particular interest to the conservationist—high water levels and good quality water in the dykes (Driscoll 1983), the variety of birdlife associated with the wetland marshes (Round 1979), and the amenity value of the region in terms of visual quality (Broads Authority 1982). The BGMCS provides a financial incentive to dissuade the farmers from ploughing up marshland and therefore safeguards these environmental assets. The Nature Conservancy Council and other bodies operate a variety of similar management agreements, and the process has culminated in the area-wide protection provided by the designation of the ESA.

Within the Broads ESA, the grazing marsh subject to protection has been divided into two categories: tiers 1 and 2. Tier 1 is designed primarily to protect the landscape characteristics, and farmers receive £50/acre (£125/ha) at this level. Tier 2 is designed to protect landscape and improve the ecological/wildlife potential in the area, and as such the guidelines which must be followed are far more stringent. Tier 2 compensation payments are £80/acre (£200/ha). Farmers entering the scheme must agree to maintain permanent grassland, initially for five years, and to abide by regulations governing the number of type of stock kept, and the number of cuts of hay or silage taken, the amount of nitrogen applied, and the water levels maintained in the drainage dykes.

Flood Risk and Protection Policy in the Mid-1980s: Toward a Compromise Set of Objectives

By the mid-1980s, various pieces of scientific evidence had become available, all of which indicated that the risk of flooding in the Broadland area was probably increasing over time. It is predicted that the warming effect caused by the buildup of CO_2 and other trace gases in the troposphere (the "greenhouse effect") may lead to a rise in global sea levels of between 0.2 m and 1.4 m by the middle of the 21st century. This rise would be due to a rise in surface water temperature and water density changes, and excludes the possibility of ice-cap melting (World Meteorological Organization 1986). Paradoxically for Broadland, a significant factor in the CO_2 buildup which is now contributing to the increased flood risk is the past worldwide development of wetlands for agricultural and other purposes (Armentano and Menges 1986).

Recent analysis of climatic records also suggests that the UK is in a section of long-term cyclical climatic pattern in which the preponderance of northeasterly as opposed to westerly wind directions is likely. It is this pattern that is most often associated with storm and surge conditions in the North Sea and adjacent coastal areas. Finally, geological evidence suggests that since the Ice Age there has been a marked isostatic recovery, for example, in western Scotland. So while the northwest of Britain is still rising, it seems probable that the southeastern corner is experiencing a very gradual submergence. Added to this trend is the impact of a slow tectonic subsidence of the North Sea Basin. Current average sea level increase has been estimated at 3 mm per annum. Over the long term, all the available evidence points to the risk of serious coastal/tidal flooding increasing through time (see Figure 3). Saline inundation would inflict damage on valuable freshwater ecosystems as well as on other assets in the area.

The flood defense system currently in place in Broadland was upgraded in the years immediately following the 1953 event, but has since suffered a gradual lowering of standards due to lack of maintenance and replacement investment. Successive public expenditure constraints by central government have added to the flood protection underinvestment problem. Today (1986/1987) the system on average probably protects the assets behind the flood banks up to a one-in-five-year flood event.

On the assumption that some overall level of flood protection for Broadland is essential, cost-effectiveness

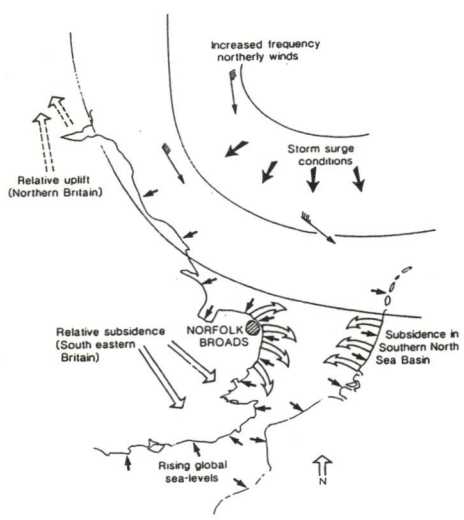

Figure 3. Physical factors contributing to a potential increased flood risk.

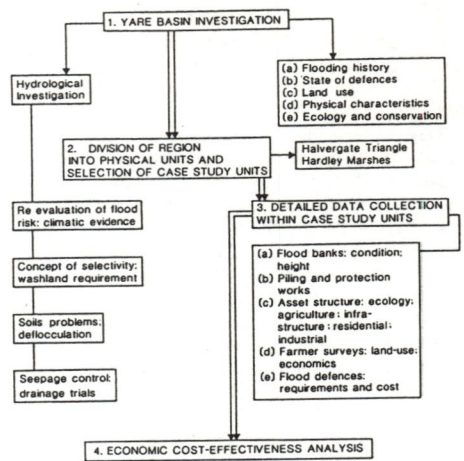

Figure 4. Research methodology outline.

analysis, rather than benefit–cost analysis, is now being used to plan a selective flood protection strategy. The strategy has multiple goals and encompasses an economic analysis which is being run in parallel with environmental impact investigations in order to determine protection priority areas and potential washland sites. Selectivity is, therefore, interpreted in a number of dimensions—economic, environmental, and aesthetic/ethical—and it is assumed that several levels of rationality exist in any social decision-making process. Within the overall strategy, the costs of various alternative means of achieving given levels of flood protection are being compared with damage-cost-avoided measures of benefits for agricultural, residential/industrial, and environmental assets (Turner and others 1986).

Initially, broad-scale investigations were undertaken, covering the lower Yare basin from Great Yarmouth to Norwich. Background data were collected (see Figure 4 for methodology) concerning the flooding history of the region, the state of the existing defenses, urban and agricultural land use, physical characteristics (soil types, tidal regimes, and so forth), and the environmental and ecological assets. A full hydrological analysis was also carried out in order to assess the expected water levels associated with events of different return periods. The region was then divided into physical units (each one capable of containing a

certain level of floodwater) surrounded by upland, landspring dykes/highland carriers, earth banks, and the main river (see Figure 1). A standarized score, based on a scale of 1–5, was assigned to each unit for each of the factors studied, according to its asset structure (that is, the value of the various types of land use contained within the unit). The Water Authority's own flood protection priorities were also considered. A "high" score of 5 was allocated where, for example, a unit contained more than 100 properties. A "low" score of 1 was assigned where a unit contained fewer than five properties. Scores of 2–4 were given to units with 5–20, 21–50, and 51–100 properties, respectively. Other factors dealt with in a similar way include population, the communications network (roads and railways), and the proportion of agricultural land with a high flood loss potential. The conservation "value" of a unit was estimated using the gradings assigned to ecological, ornithological, and environmental assets by the Broads Authority in the Broads Plan (1987). Much of the ecological interest in the Broads area is associated with the freshwater flora and fauna of the marsh dykes (ditches). A saltwater flood would cause a varying degree of damage to these ecosystems. Two points were thus allocated to each unit containing grade 1 dyke flora, and one point to a unit containing grade 2 dyke flora but no grade 1. The impact of flooding on the unique Broadland landscape would be less severe (unless any inundation was unusually prolonged) because livestock grazing would soon be resumed. Grade 1 landscape in a unit was therefore allo-

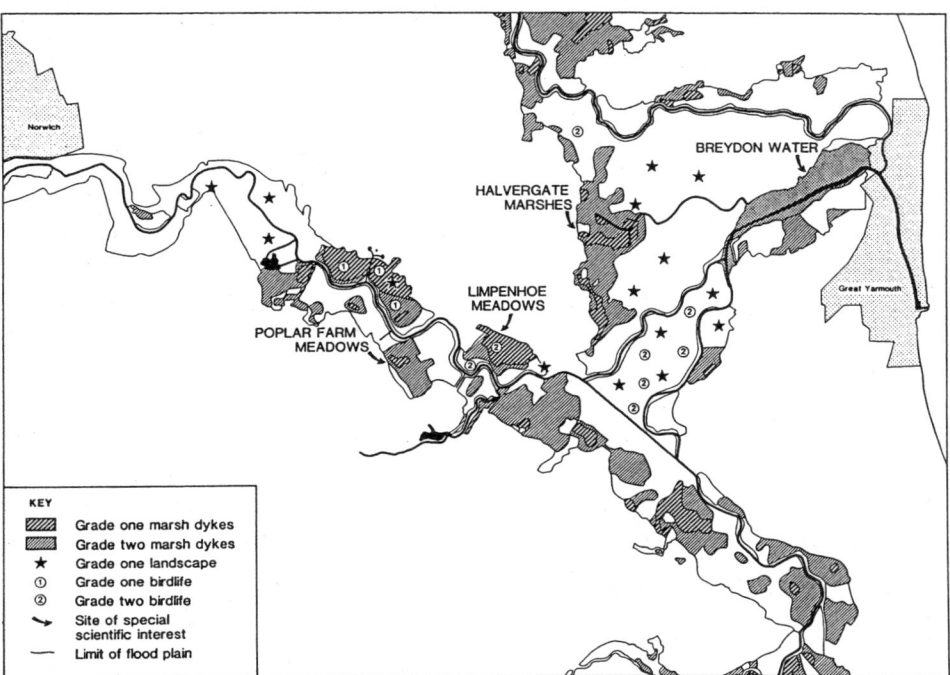

Figure 5. Conservation interest.

cated one point. Short-term ornithological interest may actually be enhanced rather than damaged by flooding and, as a result, no score was attributed to this asset. Sites of special scientific interest (SSSI), as designated by the Nature Conservancy Council, were also considered. National or regional importance is implicit in the designation of an SSSI, and a score of 2 was allocated where the total SSSI area exceeded 10% of the total unit area. Where the total SSSI area was below this percentage, one point was assigned. Figure 5 shows some of the environmental assets in the study area. In addition, the total acreage within a unit receiving compensation under the ESA scheme is being considered.

Within this scoring system, two other factors were also examined. The risk of any flooding being saline in nature was assessed, and an estimate was obtained of the ratio "area protected per kilometer of bank," the latter giving a crude estimate of "value for money." These scores were then utilized as part of the rank-ordering process, and the units were listed in terms of

their priority for improved flood protection (that is, the highest priority at the top of the list).

Two case study units were subsequently selected for the more detailed cost-effectiveness analyses. For these case study units, the Halvergate Triangle (unit 10 on Figure 1) and that part of Hardley marshes adjacent to the river Chet (unit 5), a thorough data search was undertaken. Land-use surveys were carried out, local farmers interviewed, and the asset structure better defined. Calculations were made of agricultural (arable and grazing) gross and net margins using collected local data. Flooding damage costs on arable land were assumed to be equivalent to complete crop failure. Historical evidence indicates that it is unlikely that a farmer could salvage any of a crop subjected to serious saltwater flooding. High residual salinity also prevents the cultivation of a second (replacement) crop, and may affect the type of crop grown and the expected yields for several years. In addition, damage cost calculations included the potential costs of ameliorative treatments for existing soils problems (acidity and sa-

Table 4. Cost-effectiveness analysis of flood protection in Halvergate 1986–2000.

	£m (1985/86 costs and prices)
Capital costs of scheme[a]	£2,046,800
Status quo flood protection benefits[b]	
Agricultural[c]	£391,612
Urban[d]	£1,694,656
Environmental[e]	£309,493
TOTAL BENEFITS	£2,395,769
PRESENT VALUE OF ANNUAL expected benefits[f]	£3,763,059
SCHEME NET PRESENT VALUE	£1,802,122

[a] Based on a feasible water authority works program, phased over a four-year time period with a consequent improvement in flood protection from around a 1-in-5-year return to 1-in-25-year return.

[b] Damage costs avoided, assuming an extensive 0.3-m flood occurrence, but with prior warning, and inplace protection up to 1-in-5-year flood return.

[c] Benefits based on net economic margins for arable and grazing enterprises calculated on the basis of a sample survey of 15 farms carried out in 1986. A producer subsidy equivalent value deduction on gross margins was employed at a rate of 20% for cereals and beef, as recommended by the Ministry of Agriculture, Fisheries and Food. Grazing regimes are locally specific and an extended farm survey is underway (1987) in order to improve the precision of grazing net margin calculations and to quantify actual flood damage costs on a farm-specific basis.

[d] Based on standard depth-damage fraction data plus saline water damage estimates.

[e] A proxy minimum environmental asset value was assumed to be equivalent to Broads Grazing Marsh scheme payments to farmers, which is not a profit forgone calculation.

[f] Total incremental damage costs avoided over assumed project lifetime of 15 years and at a discount rate of 5%.

From Turner and others (1986).

line deflocculation). As an interim position, a complete loss of enterprise output was assumed for grazing regimes. It is now appreciated, however, that total enterprise output will not be lost in all cases and a continuing farm survey is attempting to clarify grazier responses to flooding and the associated increased enterprise costs, that is, movement of stock, supplementary feed, and additional land rental. Such additional costs will vary on a farm-by-farm basis. For the residential area, depth-damage assessments were made using updated data and the "Blue Manual" (the standard methodology in the UK) (Penning-Rowsell and Parker 1977).

The complexities involved in the monetary valuation of environmental assets were surveyed earlier in this article. In Norfolk's Broadland, however, the British Government has provided at least a partial and pragmatic answer to the valuation dilemma. The pay-

ment made to farmers in each of the ESAs is an incentive payment; it is not an economic assessment of profit foregone (that is, the difference in profit between a high-intensity regime and low-intensity grazing). In the case of Broadland, therefore, the £50/acre or £80/acre can be interpreted as an approximate proxy measure of the minimum value to society of the landscape and wildlife assets. This "value to society" would be diminished if the unit were subject to salt-water inundation. Such a flood would affect land use and consequently both landscape and ecological values.

Total incremental damage costs avoided were calculated for the case study units. This exercise was repeated for each of three flood depth scenarios. The outcome of one scenario is outlined in Table 4. With the cooperation of water authority engineers, it was possible to assess the engineering works required to raise the level of protection to, say, the one-in-25-year standard. The environmental acceptability of the proposed engineering works was also checked (that is, the use of geotextiles in preference to steel piling wherever possible, in order to minimize visual intrusion). Projected scheme costs were calculated, and allocated over a feasible water authority work schedule. These costs are also shown in Table 4.

No major changes are expected in the system, but the risk of flooding, as already discussed, may be increasing. There is general acknowledgment in the area that some form of improved flood protection is required, at least to protect the more valuable assets. Given this assumption, an economic cost-effectiveness analysis of proposed flood protection expenditure for selected "valuable" areas was undertaken. The methodology adopted for this economic cost-effectiveness assessment is outlined in Figure 6. It was demonstrated for the Halvergate case study unit that expenditure on improved flood protection could be justified on these terms. Table 4 shows the costs and benefits of protection measures sufficient to safeguard the Halvergate area against a flood event with a return period of up to one in 25 years. The benefits are expressed in terms of damage costs avoided for all the urban, agricultural, and environmental assets in the planning unit. For the other case study unit, part of Hardley marshes, the analysis demonstrated that improved flood protection would be difficult to justify.

Further units from a wider area are to be selected for future investigations. Those units which score well in the preliminary assessment will be analyzed to determine the level of justification for improved flood protection. Those which score badly, however, will be examined in order to assess their agricultural and en-

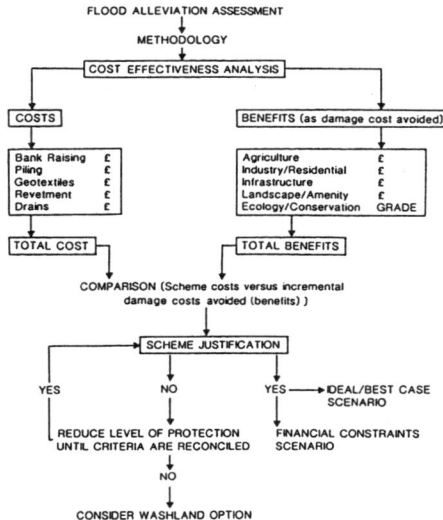

Figure 6. Scheme assessment methodology.

vironmental suitability for use as potential washland sites, (to ensure that such sites are managed, as far as possible, in accordance with traditional agricultural and/or conservation aims). With a selective flood alleviation strategy, storage provision for floodwater is recommended if unnecessary damage/repair expenditure is to be avoided.

Summary and Conclusions

Conservation of the Norfolk Broads has long been a controversial issue, characterized by inherent value conflict (see Figure 2). The formulation of an "acceptable" managment plan and operational guidelines sufficient to moderate and direct the process of change in this complex wetland ecosystem has turned out to be a formidable task. Wetlands are now recognized as important sources of both functional and structural value, although the evaluation of such assets has proved problematic. From a pragmatic viewpoint, accepting that ongoing development pressures are severe, opportunity cost models have provided useful information which can aid the process of improving the rationality of decision making. At the national level, the designation of Broadland as an environmentally sensitive area may finally help to ameliorate the conflict situation.

Agricultural conversion has been one of the major issues attracting attention. The introduction of the

BGMCS in 1985 provided an incentive for farmers in selected areas to retain traditional grazing practices. The compensation payments to be provided under the ESA scheme should help to ensure that a much larger area of the characteristic Broadland landscape is protected from the plough. In parallel, the extent and severity of the soils problems (acidity and deflocculation) must also be monitored if serious, possibly irreversible, deterioration in soil structure is to be avoided.

Agricultural intensification is, however, only one of a number of issues affecting this national wetland resource. The flood risk problem and consequent protection strategy have yet to be resolved. The capital intensive barrier solution is likely to be politically unacceptable for a number of years, although it may be the only long-term solution. The impact on the area of the predicted rise in global sea levels requires further examination and monitoring. Meanwhile, because of current public expenditure constraints, a *selective* flood protection policy is required in order to protect parts of this "protected landscape" from the threat of saltwater flooding. It has been argued that such a selective protection option should encompass multiple goals—economic cost-effectiveness, ecological sustainability, and amenity conservation. Based on the methodology outlined in Figures 4 and 6, decisions can be made concerning which parts of the ESA are to be protected against saltwater flooding, and the level of protection to be provided in the light of a given budgetary constraint. The environmental acceptability of any engineering works undertaken is also arguably more important now than at any time in the past.

The comparatively greater depletion of British wetlands suggests that the "balance" struck between conservation and agricultural interests in the UK should serve as a signal to other countries whose stock of wetlands is still relatively large. It has been argued that past experience in the UK may, in some respects, serve as a prologue to the future of US interior wetlands unless a balance is struck in the immediate future which is more favorable to conservation (Nelson 1986). Our survey and analysis have presented a rather more optimistic current (1987) UK picture. Both the level of price support for UK agricultural production and other subsidies favorable to wetland conversion have recently been reduced and in some cases eliminated. Although these changes have not been in force for very long, it does appear that they are more than mere palliative measures toward deescalating competition between arable farming and conservation programs.

The battle in the UK among conservation, agriculture, and various other interests is ongoing. But it is to be hoped that greater government intervention, such as the designation of the Norfolk Broads as an ESA, will buttress the impact of the fiscal changes. ESA designation and its dual management potential should also help to stimulate the formulation of sustainable usage management guidelines, within which each interest must operate. A combination of subtle pressures for change and some measure of economic incentives is encapsulated within the ESA policy.

The proposed selective flood protection strategy for Broadland also indicates that conservation interests can be accommodated, and to some extent protected, within public agency policy, despite overall public expenditure constraints. The central message that a more coherent, consistent, and more interventionist US policy on interior freshwater wetlands is urgently required is underscored by the UK experience (Nelson 1986). Once the remaining stock of unspoiled wetlands reaches as low a level as it has in the UK, the striking of an equitable balance between the interests of agriculture and conservation inevitably becomes a difficult and politically protracted affair. Neither outright public purchases nor voluntary management agreements are likely to provide sufficiently extensive protection, and in any case would be prohibitively expensive. Early government intervention via a package of conservation incentive instruments centered around a core of regulatory change is likely to be the most cost-effective strategy. Both the principle of dual management of environmentally important areas and a continued recognition of the importance of ecological and general "amenity" value in the public mind have been encompassed within the 1986 UK ESA policy initiative.

Note

The "screening" system outlined in this article has subsequently been modified. Most recent analysis incorporates two parallel techniques: the allocation of a score as a proportion of the maximum and the use of standard score (Z). Overall results have not changed dramatically.

Acknowledgment

We gratefully acknowledge the comments of three references on an earlier draft of this article; nevertheless, all responsibility for remaining errors is ours.

Literature Cited

Armentano, T. V., and E. S. Menges. 1986. Patterns of change in the carbon balance of organic soil wetlands of the temperate zone. *Journal of Ecology* 74:755–774.

Batie, S. S., and C. C. Mabbs-Zeno. 1985. Opportunity costs of preserving coastal wetlands: a case study of a recreational housing development. *Land Economics* 61:1–9.

Bishop, R. C., and T. A. Heberlein. 1979. Measuring values of extra market goods: are indirect measures biased? *American Journal of Agricultural Economics* 61:926–930.

Bishop, R. C., T. A. Heberlein, and M. J. Kealy. 1983. Contingent valuation of environmental assets: comparisons with a simulated market. *Natural Resources Journal* 23:619–633.

Bowers, J. K. 1983. Cost–benefit analysis of wetland drainage. *Environment and Planning A*, 15:227–235.

Bowers, J. K., and C. J. Black. 1983. Investment appraisal and the environment. School of Economic Studies, discussion paper 129, University of Leeds, Leeds, 20 pp.

Bowers, J. K., and C. J. Black. 1984. The level of protection of UK agriculture. *Oxford Bulletin of Economics and Statistics*, 4:291–311.

Boyle, K. J., R. C. Bishop, and M. P. Welsh. 1985. Starting point bias in contingent valuation bidding games. *Land Economics* 61:188–194.

Broads Authority. 1982. Landscape group report: towards a landscape strategy for the broads. BA SMP 6, Colegate, Norwich, UK.

Broads Authority, 1987. Broads plan. Broads Authority, Colegate, Norwich, UK, 128 pp.

Brown, G. M., and H. D. Pollakowski. 1977. Economic valuation of shoreline. *Review of economics and Statistics* 59:272–278.

Dossor, J., and Partners. 1984. Land drainage improvement scheme for five-mile level: St. Stephens Street, Norwich, January, 15 pp.

Driscoll, R. J. 1983. Land-use surveys in Broadland: an inventory of surveys carried out between 1931–34 and 1982. Nature Conservancy Council, Bracondale, Norwich, UK.

Farnworth, E. G., T. H. Tidrick, W. M. Smathers, and C. F. Jordan. 1983. A synthesis of ecological and economic theory toward more complete valuation of tropical moist forests. *International Journal of Environmental Studies* 21:11–28.

Freeman, A. M. 1979. The benefits of environmental improvement. John Hopkins University Press, Baltimore, 272 pp.

Gupta, T. A., and J. H. Foster. 1975. Economic criteria for freshwater wetland policy in Massachussetts. *American Journal of Agricultural Economics* 57:40–45.

Hazelden, J., P. J. Loveland, R. G. Sturdy, 1986. Saline solis in North Kent. Soil Survey of England and Wales, Harpenden, 60 pp.

Lowe, P., G. Cox, M. MacEwan, T. O'Riordan, and M. Winter. 1986. Countryside conflicts. Gower, Aldershot, 370 pp.

Meyer, P. A. 1979. Publicity vested values for fish and wildlife: criteria in economic welfare and interface with the law. *Land Economics* 55:223–235.

Milon, J. W., J. Gressel, and D. Mulkey. 1984. Hedonic amenity valuation and functional form specification. *Land Economics* 60:378–387.

Ministry of Agriculture, Fisheries and Food (MAFF) Countryside Commission. 1986. Broads grazing marshes conservation scheme: Broads Unit first annual report. MAFF, Norwich, 14 pp.

Nelson, R. W. 1986. Wetlands policy crisis: United States and United Kingdom. *Agriculture Ecosystems and Environment* 18:95–121.

Nix, J. 1983. Farm management pocketbook. Wye College, University of London, 186 pp.

Penning-Rowsell, E. C., and D. J. Parker. 1977. The benefits of flood alleviation. Saxon House, Farnborough, 297 pp.

Round, P. D. 1979. An ornithological survey of the Yare Basin, Spring and Summer 1979. Royal Society for the Protection of Birds, Sandy, Bedfordshire, 46 pp.

Rowe, R. D., and L. G. Chestnut. 1983. Valuing environmental commodities: revisited. *Land Economics* 59:404–410.

Seller, C., J. R. Stoll, and J. Charvas. 1985. Validation of empirical measures of welfare change: a comparison of non-market techniques. *Land Economics* 61:156–175.

Shabman, L., and M. K. Bertelson. 1979. The use of development value estimates for coastal wetland permit decisions. *Land Economics* 55:213–222.

Thibodeau, F. R., and B. D. Ostro. 1981. An economic analysis of wetland protection. *Journal of Environmental Management* 12:19–30.

Turner, R. K. 1978. Water pollution. Pages 97–119 *in* D. W. Pearce (ed.), The valuation of social cost. Allen and Unwin, London.

Turner, R. K. 1985. Land evaluation: financial, economic and ecological approaches. *Soil Survey and Land Evaluation* 5:21–33.

Turner, R. K., D. Dent, and R. D. Hey. 1983. Valuation of the environmental impact of wetland flood protection and drainage schemes. *Environmental and Planning A* 15:871–888.

Turner, R. K., J. Brooke, and R. D. Hey. 1986. A flood alleviation strategy for Broadland: interim report to Anglian Water, School of Environmental Sciences. UEA, Norwich, 136 pp.

Walsh, R. G., J. B. Loomis, and R. A. Gillman. 1984. Valuing option, existence and bequest demands for wilderness. *Land Economics* 60:14–29.

Willig, R. D. 1976. Consumer's surplus without apology. *American Economic Review* 66:589–597.

World Meteorological Organization. 1986. Report of the international conference on the assessment of the role of Carbon dioxide and on other greenhouse gases in climate variations and associated impacts. World Climate Programme. WMO. 661, 90 pp.

[22]

Environmental and Resource Economics **12**: 151–166, 1998.

© 1998 *Kluwer Academic Publishers. Printed in the Netherlands.*

Valuing Mangrove-Fishery Linkages

A Case Study of Campeche, Mexico

EDWARD B. BARBIER[1] and IVAR STRAND[2]

[1]*Environment Department, University of York, York YO1 5DD, UK;* [2]*Department of Agricultural and Resource Economics, University of Maryland, College Park, ML 20742, USA*

Accepted 21 November 1997

Abstract. This paper explores the value of mangrove systems as a breeding and nursery habitat for off-shore fisheries, focusing on mangrove-shrimp production linkages in Campeche State, Mexico. We develop an open access fishery model to account explicitly for the effect of mangrove area on carrying capacity and thus production. From the long-run equilibrium conditions of the model we are able to establish the key parameters determining the comparative static effects of a change in mangrove area on this equilibrium. We then estimate empirically the effects of changes in mangrove area in the Laguna de Terminos on the production and value of shrimp harvests in Campeche over 1980–90. Our findings suggest that mangroves are an important and essential input into the Campeche shrimp fishery, but that the low levels of deforestation between 1980 and 1990 mean that the resulting losses to the shrimp fishery are still comparatively small. Over-exploitation of the fishery due to open access conditions remains the more pervasive threat, and without better management any long-run benefits of protecting mangrove habitat are likely to be dissipated.

Key words: bioeconomic, Campeche, deforestation, ecological, economic, fishery, habitat, harvest, Laguna de Terminos, mangrove, Mexico, open access, shrimp

JEL classification: Q2, Q12

1. Introduction

Shocked by ecologists' attempts to value wetland services (Gosselink et al. 1974), economists have spent over twenty-five years refining methods and estimated values (e.g. Barbier 1994; Batie and Wilson 1978; Bell 1989; Hammack and Brown 1974; Lynne et al. 1981; Shabman and Bertelson 1979; Turner 1991). There have even been joint efforts by the disciplines to determine wetland and estuarine system values (Farber and Costanza 1987; Kahn and Kemp 1985). In the process, the exchange of economic and ecological concepts has likely improved our knowledge of the contributions that wetlands make to our well-being.

One component of wetland valuation methodology, valuing the effects of wetlands on the flow of output from commercial fisheries, may have overlooked a fundamental characteristic of many fisheries: open access. Although most studies appreciate the necessity of controlling for human effort while assessing the marginal productivity of wetlands, the fishery is generally not modeled in a framework that reflects the characteristics of open access.[1] In this paper, we intend to present a model which allows the vagaries of open access fishing to be considered, albeit crudely. In addition, we establish a value for one of the non-market functions of mangroves by exploring the relationship between mangroves and shrimp production in the State of Campeche, Mexico. Other fish species are also dependent on the mangroves, but we focus on shrimp because it is critical to the region, its production is likely to be separable from the rest of the area's fishing, and its data are available. However, by focusing solely on shrimp, we obtain only a partial accounting of the entire indirect use value of the mangroves.

Campeche has been chosen because it contains one of the largest and most productive mangrove areas on the Gulf of Mexico. Its Laguna de Terminos is thought to support one of the largest shrimp fisheries on the Gulf (Yañez-Arancibia and Aguirre-Léon 1988). Here, shrimp are produced from both an industrial and artisanal fleet. The industrial fleet is comprised of large vessels, each having a crew of around six. The artisanal fleet has vessels with outboard motors or without power entirely. The crew per vessel is usually no more than two. The artisanal fleet numbers over 5000 boats (all under 10 gross tons) and employs about 13% of the entire Campeche labor force. The industrial shrimp fleet has over 350 vessels, which are all over 10 gross (metric) tons but averaging about 50 gross tons (Ramos-Miranda et al. 1991).

If the mangrove area in the Laguna de Terminos can be linked empirically to the production of shrimp, then there is evidence that mangrove depletion can have a deleterious effect on the shrimp industry and the entire Campeche economy. This evidence is especially important because the low price of mangrove areas is leading to their conversion to other than natural uses. In particular, the expansion of the city of Carmen, adjacent to Laguna de Terminos, has depleted the acreage of mangroves by more than 2% in the last decade. There is also concern that lucrative shrimp aquaculture activities will begin encroaching on the mangrove areas.

In addition, by attempting to value the contribution of the Laguna de Terminos mangrove system to the commercial shrimp fisheries of Campeche, this paper aims to develop a general methodology for valuing mangrove-fishery linkages that can be applied to mangrove and coastal wetland systems elsewhere. This approach is also consistent with other attempts to assess the economic value of coastal wetland habitats in support of marine fisheries and other ecological functions, such as determining the value of marshlands as habitat for Gulf Coast fisheries in the southern United States (Bell 1989, 1997; Ellis and Fisher 1987; Farber and Costanza 1987; Freeman 1991; Lynne et al. 1981), analyzing the competition between mangroves and shrimp aquaculture in Ecuador (Parks and Bonifaz 1994), determining the value of a multiple-use mangrove system under different management options in Bintuni Bay, Irian Jaya, Indonesia (Ruitenbeek 1994), and examining general coastal system trade-offs, such as the effects of development on habitat-fishery linkages (Kahn and Kemp 1985; Swallow 1990, 1994; Strand and Bockstael 1990; Suthawan 1997). In contrast to many of these approaches, however, we attempt to value wetland-fishery linkages through a straightforward adaptation of the standard open access bioeconomic model to incorporate changes in habitat area, in this case mangroves.

The outline of the paper is as follows. The next section provides additional background on the mangrove-shrimp fisheries linkage in Campeche State. Subsequent sections develop the theoretical and empirical methodology for investigating this linkage. We assume throughout that shrimp harvesting occurs through open access management that yields production which is exported internationally, and we modify a standard open access fishery model to account explicitly for the effect of the mangrove area on carrying capacity and thus production. We derive the conditions determining the long-run equilibrium of the model, including the comparative static effects of a change in mangrove area, on this equilibrium. Through regressing a relationship between shrimp harvest, effort and mangrove area over time, we estimate parameters based on the combinations of the bioeconomic parameters of the model determining the comparative statics. By incorporating additional economic data, we are able to simulate an estimate of the effect of changes in mangrove area in Laguna de Terminos on the production and value of shrimp harvests in Campeche state. We conclude by discussing the policy implications of our findings, which we believe to have general relevance for the economic analysis of similar ecological linkages between fisheries and mangroves or other coastal wetland habitats elsewhere.

2. Background

The Gulf of Mexico is the source of nearly half of all fisheries production in Mexico and nearly one third of the total production in shrimp (Yañez-Arancibia and Aguirre-Léon 1988). Landings in Campeche account for one third of all Gulf finfish production and one half of all shrimp landings. Although several coastal ecological factors determine the biological productivity of the Gulf fisheries, the most important production mechanisms underlying theses fisheries is thought to be the combination of estuaries and lagoons with coastal vegetation (mangroves), which provide the ideal habitat as breeding grounds and nurseries (Soberón-Chavez and Yañez-Arancibia 1985; Yañez-Arancibia and Day 1988). The five Gulf states of Mexico – Campeche, Tabasco, Tamaulipas, Veracruz and Yucatan – all have important mangrove-lagoon systems, but by far the largest and most important of these systems is the Laguna de Terminos in Campeche.

The two fleets that exploit the shrimp fishery of Campeche have been in transition since about 1980, with the artisanal fleet increasing in size and the industrial fleet decreasing. The artisanal fishery has gone from less than 800 boats in 1980 to over 5000 by the early 1990s. In the same period, the industrial fleet has decreased by about half, from around 700 to 380 boats. A cooperative of industrial vessels was established when the Federal government helped the cooperative purchase vessels from private firms. It is alleged that the new vessel owners have had difficulty in maintaining the large vessels and have sold them. With the proceeds of the sale, they have invested in boats for the artisanal fishery. It is estimated that the artisanal fishery accounts for approximately 13% of the economically active population in Campeche state (Ramos-Miranda et al. 1991).

During this transition in the commercial and artisanal fisheries, shrimp production stagnated and then collapsed. From 1980–1987, production fluctuated steadily between 7 and 8 thousand metric tons (KMT) but by 1990 production had fallen to 4.6 KMT. The average real price of the shrimp catch from Campeche also increased steadily through the 1980s until 1987 and then halved by 1989–90 and remained at this level. Revenues in the Campeche shrimp industry have followed the same pattern.

Less is known about changes in the mangroves of Laguna de Terminos. The Mangrove area was estimated to be around 860 km^2 in 1980 (SARH 1980). The area of mangroves was estimated to decline to about 835 km^2 in 1991, a loss of around 2 km^2 per annum. The primary reason for the loss is the encroachment of the population from Carmen, the large city adjacent to Laguna de Terminos (Yañez-Arancibia and Benitez-Torres 1991; Benitez-Torres et al. 1992). Future threats are expected to come from expansion of shrimp aquaculture through conversion of coastal mangroves.

Although several commercial fish species are thought to be dependent on the mangrove habitat of Laguna de Terminos as a breeding and nursery ground, shrimp is considered to be the most economically important species. For example, in 1990 shrimp catches accounted for over 55% of the total production tonnage of the mangrove dependent fisheries in Campeche, and shrimp is by far the most commercially valuable of these species.[3]

3. A Model of Mangrove-Shrimp Fishery Linkages

Given the evidence suggesting that shrimp production in Campeche is dependent upon the mangroves contained in the Laguna de Terminos as a habitat and nursery, in this section we modify a standard open access fishery model to account explicitly for the effect of the mangrove area on carrying capacity and thus production. For analytical convenience, we choose to employ a discrete time model of the open access fishery.

Defining X_t as the stock of shrimp in the fishery measured in biomass, change in this stock over time can be represented as

$$X_{t+1} - X_t = F(X_t, M_t) - h(X_t, E_t), F_X > 0, F_M > 0. \tag{1}$$

Thus net expansion in the shrimp stock occurs as a result of biological growth in the current period, $F(X_t, M_t)$, net of any harvesting, $h(X_t, E_t)$. We make the standard assumption that harvesting is a function of the stock as well as fishing effort, E_t; however, we modify the biological growth function to allow for the influence of mangrove area, M_t, as a breeding ground and nursery, and we assume that this influence on growth is positive, i.e. $\partial F / \partial M_t = F_m > 0$.

Although Equation (1) and its underlying growth and harvesting functions can take several forms, we follow the convention of many analytical fishery models developed for empirical purposes and assume a simple Schaefer-Gordon model (Clark 1976; Conrad 1995). Several other studies have also employed the Schaefer-Gordon model to estimate the impacts of habitat influences on fisheries (Lynne et al. 1981; Bell 1989).

Thus we assume a basic Schaefer production process for harvesting, h_t

$$h_t = qX_tE_t, \tag{2}$$

with q_t as the 'catchability' coefficient. Representing Equation (1) as a logistic growth function and substituting in Equation (2) yields

$$X_{t+1} - X_t = [r(K(M_t) - X_t) - qE_t]X_t, \tag{3}$$

where r is the intrinsic growth of shrimp each period, K is the environmental carrying capacity of the system and mangrove area, M_t, has a positive impact on carrying capacity, i.e. $K_M > 0$.

Finally, as stated in the Introduction, the management of the Campeche shrimp fishery has the characteristics of an open access fishery. Following standard analysis, this suggests that fishing effort next period will adjust in response to the real profits made in the current period (Clark 1976; Conrad 1995). Letting p represent constant shrimp prices per unit harvested, c the real unit cost of effort and $\phi > 0$ the adjustment coefficient, then the fishing effort adjustment equation is

$$E_{t+1} - E_t = \phi[ph(X_t, E_t) - cE_t]. \tag{4}$$

The above system of equations constitutes our basic model for analyzing fishery-mangrove linkages in Campeche. Next, we use this model to derive an open access equilibrium, and to determine the comparative static effects on this equilibrium of a change in the mangrove area.

4. The Open Access Equilibrium

Our analysis of fishery-mangrove linkages is conducted by examining the effects of a change in mangrove area on the long-run open access equilibrium of the Campeche shrimp fishery.[4] In equilibrium, both the shrimp stock and the level of fishing effort are assumed to be constant over time, i.e. $X_{t+1} = X_t = X$ and $E_{t+1} = E_t = E$. In addition, we assume initially that the mangrove area is in equilibrium, i.e. $M_t = M_{t+1} = M$. Equations (3) and (4) can therefore be solved for steady-state levels of shrimp stock, X, and effort, E

$$X = \frac{c}{pq}, \quad \text{for } E_{t+1} = E_t = E, \tag{5}$$

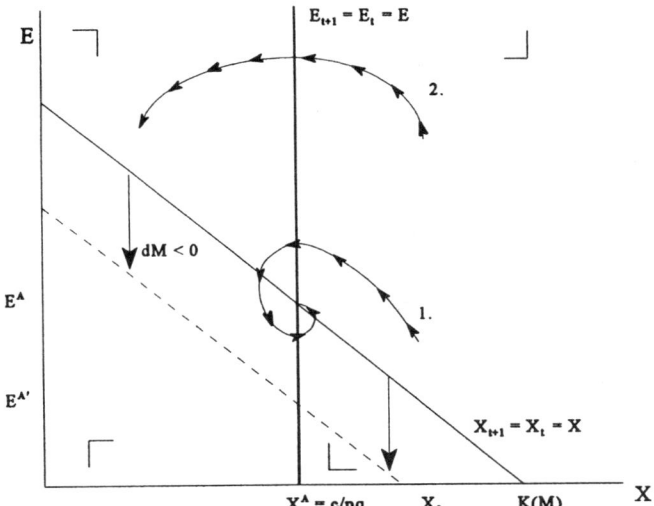

Figure 1.

$$E = \frac{r(K(M) - X)}{q}, \quad \text{for } X_{t+1} = X_t = X. \tag{6}$$

Equation (5) is the standard open access condition that assumes any profits in the fishery will be competed away in the long run. Equation (6) indicates the combinations of fishing effort and shrimp stock size (and thus also mangrove area) that will lead to a constant level of shrimp stock in the fishery in the long run. Figure 1 depicts the equilibrium conditions in (X, E) space. The $E_{t+1} = E_t = E$ curve is of course vertical, given Equation (5). From Equation (6), the slope of the $X_{t+1} = X_t = X$ curve in (X, E) space is downward sloping, i.e. $dE/dX = -r/q < 0$.

Two possible trajectories for fishing effort and shrimp stocks are depicted in Figure 1, assuming an initial level of stock X_0. Trajectory 1 is essentially a stable spiral that leads to an open access equilibrium denoted by (X^A, E^A). Trajectory 2 leads to rapid decline of the shrimp fishery from the outset to near extinction or 'collapse' levels, which, as depicted, could occur if the initial level of effort is too high, given X_0.[5]

As shown in Figure 1, because the mangrove area affects the carrying capacity of the shrimp fishery, the impact of mangrove deforestation on the system represented by Equations (5) and (6) is to shift down the $X_{t+1} = X_t = X$ curve. This results in a lower steady-state level of effort, $E^{A'}$, although equilibrium stock is unchanged. Assuming that a trajectory to this new equilibrium is still feasible given the initial stock X_0, then the impacts of mangrove loss on carrying capacity will also mean a lower level of initial fishing effort. Thus in our model of mangrove-fishery linkages, the open access fishery adjusts to the impacts of mangrove loss by reducing both initial and equilibrium levels of effort. If the fishery fails to do this, then it may instead find itself on a different exploitation path, such as trajectory 2, that leads to 'near collapse' of shrimp stocks.[6]

In our analysis, we estimate the impacts of mangrove deforestation on the stable open access equilibrium, assuming that fishery effort adjusts instantaneously to allow a new equilibrium to be attained. We do not consider the case where the effect of deforestation is to make a steady-state equilibrium infeasible, thus causing the shrimp fishery to switch to a different exploitation path, such as that represented by trajectory 2 in Figure 1. As discussed above, there is evidence that mangrove deforestation in Laguna de Terminos is affecting the Campeche shrimp fishery, but it is unlikely that the fishery is currently in danger of the 'near collapse' scenario depicted by trajectory 2. Thus in the following analysis of the effect of a change in mangrove area, we assume that the fishery has attained a stable open access equilibrium, such as the steady-state (X^A, E^A) depicted in Figure 1, and we therefore analyze the impacts of mangrove deforestation on the fishery in terms of changes in the equilibrium steady-state conditions.

5. The Comparative Static Effects of a Change in Mangrove Area

For simplicity, we assume a proportional relationship between mangrove area and carrying capacity, i.e. let $K(M) = \alpha M$, $\alpha > 0$. Thus, from Equation (6), the comparative static effect of a change in the mangrove area on the equilibrium level of fishing effort, E^A is

$$r[\alpha \mathrm{d}M - \mathrm{d}X^A] - q\mathrm{d}E^A =$$

$$\text{or } \frac{\mathrm{d}E^A}{\mathrm{d}M} = \frac{\alpha r}{q} > 0. \tag{7}$$

This confirms that loss of the mangrove area results in a lower level of equilibrium fishing effort. From Equation (2), it is clear that there will be a loss of harvest as well. Utilizing Equations (7) and (5), the impact on the equilibrium harvest level can be solved for explicitly

$$\mathrm{d}h^A = qX^A\mathrm{d}E^A = \alpha r X^A\mathrm{d}M = \frac{\alpha r c}{pq}\mathrm{d}M > 0. \tag{8}$$

The resulting change in gross revenue of the fishery is then

$$p\mathrm{d}h^A = \frac{\alpha r c}{q}\mathrm{d}M > 0. \tag{9}$$

A fall in the mangrove area will therefore result in a decline in both steady-state shrimp harvest and the gross revenue of the fishery. Moreover, given that these impacts are based on the bioeconomic parameters of our model (α, r and q) combined with prices and costs for the fishery (p and c), it is possibly to estimate these comparative static effects explicitly.

6. Empirical Estimation of Mangrove-Fishery Linkages

By employing one of the above equilibrium conditions of our model, it is possible to establish a relationship between shrimp production, effort and mangrove area that can be estimated over time. We run this regression utilizing the Campeche shrimp fishery and mangrove area data over the 1980–90 period. The resulting parameter estimates allow us to determine the appropriate combinations of α, r, and q that underly the comparative static results of our model. By also incorporating values for the economic parameters p and c, we are therefore able to simulate the comparative static effects of a change in mangrove area on equilibrium harvest and gross revenue of the shrimp fishery over the 1980–90 period of analysis.

If we assume in our analysis that the shrimp stock is constant, i.e. $X_t = X_{t+1} = X$, then we can use steady-state condition (6) to derive a relationship between shrimp production, mangrove area and effort. By substituting Equation (2) into Equation (6) and re-arranging, we obtain

$$h = qEK(M) - \frac{q^2}{r}E^2 = q\alpha EM - \frac{q^2}{r}E^2. \tag{10}$$

Equation (10) can be estimated by employing the available time series data over 1980–90 on shrimp harvests, effort and mangrove area for Campeche, Mexico. By assuming that it would take 5.5 artisanal boats on average to harvest the same amount as an industrial boat, we can combine the data on the number of artisanal and industrial boats in the Campeche fishing industry to form a composite effort variable E.[7] In Equation (10) h is represented as annual harvest in kilograms (kg) of shrimp by both artisanal and industrial boats, M is the annual mangrove area (sq km) in Laguna de Terminos and E is the composite, or aggregate, harvesting effort each year. We estimate this equaion over the period 1980–90, the years for which data are available, and assume that any resulting error term is identically indepedently distributed over time. A summary of the results is presented in Table I. Except for not being able to reject autocorrelation (test is inconclusive), the regression results and the coefficient estimates are quite good and were fairly robust for slightly different specifications of the model.[8]

The marginal productivity and output elasticity estimates corresponding to the regression of Equation (10) are also shown in Table I. Calculated using the mean level of effort (5556 combined vessels), the marginal productivity of mangrove area is 24.7 metric tons per km^2.[9] This is because MP_M is proportional to the level of effort, and over the 1980–90 period the number of vessels in the Campeche shrimp fishery increased substantially, from around 4500 combined vessels in 1980 to over 7200 in 1990.[10] In contrast, the marginal productivity of fishing effort (calculated

Table I. Regression results for time-series analysis of the relationship between Campeche shrimp harvest, effort and mangrove area

Dependent variable:	Annual shrimp harvest (kilograms)			
	(mean: 7 506 636 kg)			
Parameter estimates				

Variable	Parameter estimate	Standard error	T-statistic	prob > (T)
Mangrove area (M) × Effort (E)	4.4491	0.2991	14.877	0.00001
Effort squared (E²)	−0.4297	−0.0439	−9.797	0.00001

Adjusted R^2: 0.745		F Statistic:	30.216
S.E. of regression: 639745.7		No of observations:	11
Durbin-Watson: 1.461		1st Order Autocorrelation:	0.254

Marginal productivity estimates (at means)

$$MP_M = \frac{\partial h}{\partial M} = 24\,719 \quad MP_E = \frac{\partial h}{\partial E} = -997$$

Output elasticity estimates

$$\epsilon_{hM} = 2.80 \quad \epsilon_{hE} = -0.74$$

Notes: The marginal productivity and output elasticity estimates are evaluated at the mean M and E levels. ϵ_{hM} and ϵ_{hE} are the output elasticities for mangrove area and effort respectively.

at mean effort and mean mangrove area, 849 km²) is actually negative in the regression, at an average of −0.997 tons per vessel. MP_E was slightly negative in 1980, at −26 kg per vessel, but fell to −2.48 tons per vessel by 1990. Part of this decline in the marginal productivity of effort is due to the loss in mangrove area in the Laguna de Terminos from 1980 to 1990, but by far the more important influence appears again to be the increase in the number of small fishing vessels over this period.

The output elasticity for the mangrove area (2.80) indicates that a decline in the Laguna de Terminos mangroves has a more than proportionate impact on output in the Campeche shrimp fishery. However, the overall impact of mangrove loss during the 1980–90 period on the fishery was still relatively small. Only 2.3% of the mangrove area was deforested between 1980 and 1990, which would suggest that the corresponding loss in fishery output was around 6.5%. The negative output elasticity for effort (−0.74) indicates that the increase in fishing effort over the 1980–90 period had a significant negative impact on shrimp production. As the number of combined vessels increased by 61.5% over this period, the corresponding loss in shrimp harvest was 45.5%.

As it yields only two coefficient estimates, the regression of Equation (10) is insufficient to determine explicitly the three bioeconomic parameters (α, r, and q). However, it is not necessary to solve for the values of these three parameters in

order to estimate the comparative static relationships Equations (8) and (9). For example, denoting the estimated coefficients of Equation (10) as $b_1 = \alpha q$ and $b_2 = -q^2/r$, it follows that Equations (8) and (9) can be rewritten as

$$\mathrm{d}h^A = \frac{\alpha rc}{pq}\mathrm{d}M = -\frac{cb_1}{pb_2}\mathrm{d}M, \tag{8'}$$

$$p\mathrm{d}h^A = \frac{\alpha rc}{q}\mathrm{d}M = -\frac{cb_1}{b_2}\mathrm{d}M. \tag{9'}$$

Thus to determine the comparative static effects (8') and (9') requires imputing values for prices and costs p and c and combining these values with the parameter estimates b_1 and b_2 from our regression of Equation (10). We impute values to p and c through the following approach.

As noted above, the assumption underlying the estimation of Equation (10) is that the 1980–90 shrimp harvest and effort data for the Campeche fishery satisfy one of the open access equilibrium conditions of our model, i.e. Equation (6). However, the latter is only one of the conditions necessary for an open access equilibrium. The other is Equation (5), which assumes zero profits in the long run. Values of h and E that satisfy both Equation (10) and the long-run zero profit condition will therefore also satisfy the open access equilibrium.

For the purposes of illustrating the comparative static effects of a change in the mangrove area on equilibrium harvest and revenues, it would be convenient to assume that the 1980–90 harvest and effort levels satisfy simultaneously both open access conditions, Equations (5) and (6). This will be the case if the price and cost parameters in each of the years of data attain levels that equate with zero profits, i.e. $ph = cE$. Price data in terms of US\$/kg of imported fresh/frozen shrimp from Mexico are available for the 1980–90 period. Assuming a 100% mark-up from the ex-vessel prices, we use these data to generate a price series for the Campeche shrimp fishery.[11] Consequently, based on the price, effort and harvest data, we calculate for each year of the analysis the corresponding 'equilibrium' cost per unit effort levels that yield zero profits, and denote these cost values as c^A. Thus by employing our estimated parameters, b_1 and b_2, and the economic parameters, p and c^A, we are able to calculate the comparative static effects of the impacts of mangrove loss over 1980–90 on the open access Campeche shrimp fishery, as indicated by conditions (8') and (9'). The results of the simulation are shown in Table II.

On average over the 1980–90 period, a marginal (in km^2) decline in mangrove area produces a loss of 14.39 metric tons of shrimp harvest and US\$139 352 in revenues from the Campeche fishery each year. This is equivalent to a reduction of 0.19% in the annual harvest and revenues of the fishery.[12] Since over the simulation period mangrove deforestation occurred at the rate of around 2 km^2 annually, the resulting loss each year amounts to about 28.8 metric tons, or US\$278 704. As our theoretical analysis would suggest, mangrove conversion in the Laguna de

Table II. Simulation results for the effects of mangrove loss on the open access equilibrium of the campeche shrimp fishery, 1980–90

Parameter estimates:

$$b_1 = 4.4491$$
$$b_2 = -0.4297$$

Simulation estimates of a marginal change in the mangrove area (dM)

Year	Price (p) US\$/kg[a]	Cost (c^A) US\$/vessel[b]	Change in equilibrium harvest (dh^A), metric tons	Change in equilibrium revenues (pdh^A) US\$	Change (%)
1980	7.10	13 984	20.40	144 808	0.23
1981	9.68	15 628	16.72	161 826	0.20
1982	10.57	13 816	13.53	143 060	0.18
1983	9.80	13 636	14.41	141 197	0.18
1984	9.83	14 096	14.85	145 963	0.19
1985	9.80	16 687	17.63	172 798	0.20
1986	10.00	15 013	15.55	155 460	0.19
1987	10.22	14 363	14.55	148 731	0.20
1988	10.56	14 132	13.86	146 334	0.20
1989	10.21	10 000	10.14	103 547	0.17
1990	10.40	6 677	6.65	69 143	0.14
Mean	9.83	13 457	14.39	139 352	0.19

[a] US\$/kg, in real (1982) prices.
[b] c^A is the 'equilibrium' (real) cost per unit effort, defined as the cost level necessary to attain zero profit in the fishery, i.e. $c^A = ph^A/E^A$.

Terminos clearly has a negative impact on the Campeche shrimp fishery. However, given the relatively small rate of annual mangrove deforestation in the Laguna de Terminos over the 1980–90 period, the resulting loss in shrimp harvest and revenues does not appear to have been substantial. If, as expected, mangrove deforestation accelerates in the region, perhaps as a result of urban expansion and the conversion of mangrove swamps to mariculture ponds, then it is likely that more substantial losses to the off-shore shrimp fishery will occur.

An interesting feature of our simulation is that it indicates how the economic losses associated with mangrove deforestation are affected by long-run management of the open access fishery. As noted previously, the early years of the period (e.g. 1980–81) were characterized by much lower levels of fishing effort and higher harvests (e.g. on average around 4800 combined vessels extracting about 8.5 KMT annually). Table II shows that, if this earlier period represented the open access equilibrium of the fishery, the economic impacts of a marginal (km²) decline in the mangrove area would be a reduction in annual shrimp harvests of around 18.6

tons, or a loss of about US$153 300 per year. In contrast, the last two years of analysis (e.g. 1989–90) saw much higher levels of effort and lower harvests in the fishery (e.g. around 6700 combined vessels extracting 5.3 KMT annually). As a consequence, if this latter period represents the open access equilibrium, then a marginal decline in mangrove area would result in annual losses in shrimp harvests of 8.4 tons, or US$86 345 each year.

Thus, the value of the Laguna de Terminos mangrove habitat in supporting the Campeche shrimp fishery appears to be affected by the level of exploitation. This suggests that, if an open access fishery is more heavily exploited in the long run, the subsequent welfare losses associated with the destruction of natural habitat supporting this fishery are likely to be lower. Intuitively, this makes sense. The economic value of an over-exploited fishery will be lower than if it were less heavily depleted in the long run. The share of this value that is attributable to the ecological support function of natural habitat will therefore also be smaller.

7. Conclusion

Our model of mangrove-shrimp fishery linkages demonstrates that it is possible to modify a standard bioeconomic fishery model to account explicitly for the effect of a change in mangrove habitat area on carrying capacity and thus production . By employing a dynamic discrete-time model and assuming open access conditions, we are able to solve for the impacts on steady-state levels of shrimp harvest and revenues that result from a change in mangrove area. In our model, mangrove habitat loss leads unambiguously to a decline in both steady-state shrimp harvest and revenue. More importantly, we are able to calculate these effects explicitly from our model, given regression estimates of two coefficients based on the bioeconomic parameters of our model (α, r, and q) and imputed economic data on prices and costs (p and c) for the Campeche shrimp fishery. The empirical application of our model to the relationship between mangroves in Laguna de Terminos and shrimp production in Campeche has led to two principal findings.

First, the high marginal productivity and output elasticity of the mangrove habitat in terms of shrimp production tend to support the claims of ecologists that the mangroves of the Laguna de Terminos are both an important and essential input into the Campeche shrimp fishery (Soberón-Chavez and Yañez-Arancibia 1985; Yañez-Arancibia and Day 1988). However, given that only 2.3% of the mangrove area was deforested between 1980 and 1990, the actual losses that have occurred in the shrimp fishery as a result of mangrove deforestation are still comparatively small. We calculate that over 1980–90 with an average annual rate of mangrove deforestation of 2 km^2 the resulting loss in shrimp harvest each year amounted to about 28.8 metric tons, or a loss of US$278 704 in revenue. This is equivalent to a reduction of only 0.4% in the annual harvest and revenues of the fishery.[13] In the future, mangrove deforestation in the Laguna de Terminos is expected to increase substantially in the region, most likely from urban expansion and the conversion of mangrove swamps to mariculture ponds.[14] If this is the case, then it is likely that more substantial losses to the off-shore shrimp fishery of Campeche will occur.

Our empirical analysis also shows that, although mangrove loss has an important economic influence on the Campeche shrimp fishery, a more pervasive problem over the 1980–90 period of analysis has been over-exploitation. Our estimate of the marginal productivity of fishing effort is actually negative over the entire period of analysis, reflecting the rapid expansion of fishing vessels over this period, from around 4500 combined industrial and artisanal vessels in 1980 to over 7200 in 1900. In common with the analysis of wetland-fishery linkages by Freeman (1991), we also find that open access conditions and the level of exploitation affect the economic value attributed to the role of the mangrove habitat in supporting the shrimp fishery. As equilibrium levels of fishing effort rise and harvests subsequently fall, the economic losses associated with mangrove deforestation become smaller. Over-fishing appears to lower the economic value of natural habitat.

The resulting management implications are clear. In the case of an open access fishery such as the Campeche shrimp industry, protection of the nursery and breeding habitat function of the Laguna de Terminos mangroves may be important for reducing losses in the fishery, but control of over-fishing is more critical. As long as effort levels continue to rise, harvests will fall, even if mangrove areas are fully protected. Moreover, any increase in harvest and revenues from an expansion in the mangrove area is likely to be short-lived, as it would simply draw more effort into the fishery. Better management of the Campeche shrimp fishery to control over-exploitation may be the only short-term policy to bring production back to respectable levels, as well as being essential for realizing the more long-term economic benefits of protecting mangrove habitat. We expect that this conclusion holds also for other open access fisheries elsewhere, which are supported by mangroves or other estuarine and coastal wetlands that provide breeding habitats and nurseries for the fisheries.

Acknowledgements

This study was based on data collected and preliminary work conducted at the Diploma short course 'Ecological Economics in Tropical Coastal Ecosystems' held at the Programa de Ecologia Y Oceanografia de Golfo de México (EPOMEX), Universidad Autónoma de Campeche, Mexico, 10–28 February, 1992. We are grateful to the assistance provided by participants in this course, Andrés Gomez-Lobo, Julia Ramos-Miranda, Evelia Rivera-Arriaga, Maria Consuelo Sánchez-González and David Zárate-Lomeli, and to the Director of EPOMEX, Alejandro Yañez-Arancibia. Additional support for completion of this paper was provided by the Economy and Environment Program South East Asia (EEPSEA) for Ivar Strand, and the Department of Agricultural and Resource Economics, University of Maryland for Ed Barbier, who was a Visiting Professor at the Department in Fall 1996. A previous version of this paper was presented at the Association of Environmental and Resource Economics (AERE) 1997 Workshop 'The Economic Analysis of Ecosystems', 1–3 June 1997, Annapolis, Maryland, USA. We are grateful to participants at the AERE workshop, and particularly Jim Wilen, for helpful comments. We were also helped by the comments provided by two anonymous referees for this journal. However, the usual caveat applies.

Notes

1. A notable exception is Freeman (1991), who extends the optimal management model of Ellis and Fisher (1987) based on the original analysis of the marginal productivity of marsh land acreage in terms of the Gulf of Mexico blue crab fishery in Florida developed by Lynne et al. (1981). However, Freeman's valuation of the effects of changes in wetland area on an open access fishery is based on a static market model. Here, we attempt to use the equilibrium conditions of a dynamic mangrove-fishery model to conduct such a valuation. An interesting example of investigating open access-wetland interactions on the input side of fisheries is explored by Parks and Bonifaz (1994), who develop a conceptual model to analyze open access collection of post-larval shrimp inputs and mangrove deforestation as two potential causes for the scarcity of these inputs to shrimp mariculture in Ecuador.

2. Information on the extent of mangrove loss since 1980 has been provided by Programa de Ecologia Y Oceanografia del Golfo de México (EPOMEX), Universidad Autónoma de Campeche, Mexico, who have estimated a total loss of around 2500 ha in mangroves over the decade since 1980 (see Yañez-Arancibia and Benitez-Torres 1991; Benitez-Torres et al. 1992).

3. The other commercially important mangrove dependent fish species in Campeche State are clam, sea trout, crab, oyster, snapper and snook. In 1990, the real price of shrimp was nearly 60% higher than the next highest valued species (snook).

4. In analyzing the effects of a change in mangrove area we do not model explicitly the possible socio-economic factors affecting mangrove deforestation. We also implicitly assume that the decision to clear mangroves for, say, fuelwood, agriculture or housing, is independent of the decision to allocate effort to shrimp fishing. However, some researchers have indicated that in coastal zones the effort devoted to fishing and mangrove clearing could be linked through local labour supply decisions (Mäler et al, 1994; Ruitenbeek 1994; Suthawan 1997). For example, as mangroves are depleted, local labour is switched to greater fishing effort as marginal returns from clearing mangroves decline; alternatively, as fishing profitability declines, effort is switched to greater mangrove clearing. It is also possible that increased fishing profitability may in fact be reinvested in greater effort in mangrove rehabilitation. To explore such linkages would require a more complex household production function model along the lines suggested by Mäler et al. (1994), which includes local labour allocation decisions across a variety of mangrove-dependent economic activities, including collection of fuelwood, agricultural production and fishing.

5. Although not shown in Figure 1, trajectory 2 is a spiral that is asymptotically bound (at low values of X) by the vertical and horizontal axes. the implications are that the shrimp fishery approaches what we term 'near collapse' but not complete exhaustion. As X declines towards zero, the slope of trajectory 2 tends to negative infinity, and thus the stock is never fully depleted. We are grateful to Jim Wilen for pointing this out to us.

6. However, as indicated in the previous note and from Figure 1, trajectory 2 is actually a spiral, so after the near collapse of stocks there should be a period of recovery, with the pattern repeating itself.

7. 5.5 is very close to the ratio of average annual catch per industrial boat to the average annual catch per artisanal boat.

8. Correction of the autocorrelation in the regression reduces the already limited degrees of freedom but does not change the parameter estimates substantially.

9. To place this estimate of the marginal productivity of mangrove area in perspective, Parks and Bonifaz (1994) report that the average productivity of shrimp ponds in Mexico was 96 metric tons/km^2, and generally around 40–133 tons/km^2 across a range of developing countries. This

suggests that mangrove systems are highly extensive shrimp production systems, compared to shrimp ponds. See also notes 10 and 14, below.

10. Although different to our Equation (10), the production relationship between harvest, effort and marshland area estimated by Lynne et al. (1981) for the blue crab fishery of the Florida Gulf Coast also has an interactive term between fishing effort and habitat area (i.e. marshland). Their resulting estimate of the marginal productivity of salt marsh is therefore also influenced by the level of effort. Lynne et al. estimated this marginal productivity to be 2.3 pounds/acre (0.26 metric tons/km^2). Bell (1989) applies the Lynne et al. production relationship to a number of Florida Gulf Coast fisheries, including shrimp. Although the marginal productivity of salt marsh area estimated by Bell is highest for the shrimp fishery, at around 4.6–5.9 pounds/acre (0.52–0.66 tons/km^2), it is still substantially lower than our estimate for the marginal productivity of mangrove area in the Laguna de Terminos.

11. This was in part verified by obtaining several years of Mexican ex-vessel prices and converting them into US $ using published exchange rate figures. We chose to present values in US$ and use the import prices because the US figures would be more meaningful to most readers and we would not have to convert using a exchange rate that varied wildly in the latter part of the decade.

12. As shown in Table II, the percentage change in harvest and revenues are the same, as $pdh/ph = dh/h$.

13. Another way to put this figure into perspective is to note that the average per capita income in Mexico in 1985 was US $2600.

14. Although mariculture development is not yet a major source of mangrove deforestation in Laguna de Terminos, shrimp ponds in Mexico and elsewhere in the tropics are increasingly being established through mangrove conversion, one of the attractions being that shrimp ponds offer a highly intensive shrimp production system, at least initially (see Parks and Bonifaz (1994) and note 8, above). However, Parks and Bonifaz also demonstrate that the mangrove deforestation associated with the establishment of shrimp ponds results in loss of post-larval shrimp inputs for the ponds, indicating a trade-off between short-term profits and long-term productivity. Our analysis additionally shows that the loss of mangrove area can have a detrimental impact on off-shore shrimp fisheries.

References

Barbier, E. B. (1994), 'Valuing Environmental Functions: Tropical Wetlands', *Land Economics* **70**(2), 155–173.

Batie, S. S. and J. R. Wilson (1978), 'Economic Values Attributable to Virginia's Coastal Wetlands as Inputs into Oyster Production', *Southern Journal of Agricultural Economics* **10**, 111–117.

Bell, F. W. (1989), *Application of Wetland Valuation Theory to Florida Fisheries*, Report No. 95, Florida Sea Grant Program, Florida State University, Tallahassee.

Bell, F. W. (1997), 'The Economic Value of Saltwater Marsh Supporting Marine Recreational Fishing in the Southeastern United States', *Ecological Economics*, **21**(3), 243–254.

Benitez-Torres, J., D. Zarate, J. L. Rojas-Galaviz and A. Yañez-Arancibia (1992), 'Expansión Urbana y Deterioro Ambiental en la Región de la Laguna de Terminos, Campeche', Report presented at the Seminar on Population and the Environment, Sociedad Mexicana de Demografía, The Population Council. Colegio de México, April.

Clark, C. (1976) *Mathematical Bioeconomics*, New York: John Wiley and Sons.

Conrad, J. M. (1995), 'Bioeconomic Models of the Fishery', in D. Bromley, ed., *Handbook of Environmental Economics*, Oxford: Basil Blackwell.

Ellis, G. M. and A. C. Fisher (1987), 'Valuing the Environment as Input', *Journal of Environmental Management* **25**, 149–156.

Faber, S. and R. Costanza (1987), 'The Economic Value of Wetlands Systems', *Journal of Environmental Management* **24**, 41–51.

Freeman, A. M. (1991), 'Valuing Environmental Resources Under Alternative Management Regimes', *Ecological Economics* **3**, 247–256.

Gosselink, J. G., E. P. Odum and R. M. Pope (1974), *The Value of the Tidal Marsh*, Publication No. LSU-SG-74–03, Center for Wetland Resources, Louisiana State University, Baton Rouge.

Hammack, J. and G. M. Brown, Jr. (1974), *Waterfowl and Wetlands: Towards Bioeconomic Analysis*, Washington DC: Resources for the Future.

Kahn, J. R. and W. M. Kemp (1985), 'Economic Losses Associated with the Degradation of an Ecosystem: The Case of Submerged Aquatic Vegetation in Chesapeake Bay', *Journal of Environmental Economics and Management* **12**, 246–263.

Lynne, G. D., P. Conroy and F. J. Prochaska (1984), 'Economic Value of Marsh Areas for Marine Production Processes', *Journal of Environmental Economics and Management* **8**, 175–186.

Mäler, K-G., I-M. Gren and C. Folke (1994), 'Multiple Use of Environmental Resources: A Household Production Function Approach to Valuing Natural Capital', in A. M Jansson, M. Hammer, C. Folke and R. Costanza, eds., *Investing in Natural Capital: The Ecological Economics Approach to Sustainability*, Washington DC: Island Press.

Parks, P. and M. Bonifaz (1997), 'Nonsustainable Use of Renewable Resources: Mangrove Deforestation and Mariculture in Ecuador', *Marine Resource Economics* **9**, 1–18.

Ramos-Miranda, J., D. Flores Hernandez and P. Sanchez-Gil (1991), 'Pesca Artesanal: Panorama Actual en El Estado de Campeche', *Jaina* **2**(2), 20–21.

Ruitenbeek, H. J. (1994), 'Modelling Economy-Ecology Linkages in Mangroves: Economic Evidence for Promoting Conservation in Bintuni Bay, Indonesia', *Ecological Economics* **10**, 233–247.

SARH, (1980), *Atlas del Uso del Suelo en la República Méxicana*, Dirección General de Recursos Hidráulicos, México.

Shabman, L. and M. K. Bertelson (1979), 'The Use of Development Value Estimates for Coastal Wetland Permit Decisions', *Land Economics* **55**, 213–222.

Soberón-Chavez, G. and A. Yañez-Arancibia (1985), 'Control Ecológico de los Pesces Demersales: Variabilidad Ambiental de la Zona Costeara y Su Influencia en la Producción Natural de los Recursons Pesqueros', in A. Yañez-Arancibia, ed., *Recursos Pesqueros Potenciales de México*, Mexico: UNAM.

Strand, I. E. (1990), 'Interaction Between Agriculture and Fisheries: Empirical Evidence and Policy Implication', in R. E. Just and N. E. Bockstael, eds., *Commodity and Resource Policies in Agricultural Systems*, New York: Springer-Verlag.

Suthawan, S. (1997), *Economic Valuation of Mangroves and the Roles of Local Communities in the Conservation of the Resources: Case Study of Surat Thani, South of Thailand*. Final report submitted to the Economy and Environment Program for Southeast Asia (EEPSEA), Singapore: EEPSEA.

Swallow, S. K. (1990), 'Depletion of the Environmental Basis for Renewable Resources: The Economics of Interdependent Renewable and Nonrenewable Resources', *Journal of Environmental Economics and Management* **19**, 281–296.

Swallow, S. K. (1994), 'Renewable and Nonrenewable Resource Theory Applied to Coastal Agriculture, Forest, Wetland and Fishery Linkages', *Marine Resource Economics* **9**, 291–310.

Turner, R. K. (1991), 'Economics and Wetland Management', *Ambio* **20**(2), 59–63.

Yañez-Arancibia, A. and A. Aguirre-Léon (1988), 'Fisheries of the Lagoon Terminos Region', in A. Yañez-Arancibia and J. W. Day, Jr., eds., *Ecology of Coastal Ecosystems in the Southern Gulf of Mexico: the Terminos Lagoon Region*, Mexico: UNAM Press.

Yañez-Arancibia, A. and J. A. Benitez-Torres (1991), 'Expansion Urbana Ecología Litoral en Isla del Carmen, Campeche, México', *Jaina* **2**(2), 12.

Yañez-Arancibia, A. and J. W. Day, Jr., eds. (1988), *Ecology of Coastal Ecosystems in the Southern Gulf of Mexico: The Terminos Lagoon Region*, Mexico: UNAM Press.

[23]

Risk Decision and Policy (2000), *vol. 5, pp.* 49–68.   Printed in the United Kingdom
© 2000 Cambridge University Press

Coastal bathing water health risks: developing means of assessing the adequacy of proposals to amend the 1976 EC directive

STAVROS GEORGIOU[‡], IAN J. BATEMAN[‡§], IAN H. LANGFORD[*‡] and
ROSEMARY J. DAY[‡]

[‡]*Centre for Social and Economic Research on the Global Environment, School of Environmental Science, University of East Anglia, Norwich, Norfolk, NR4 7TJ, UK,* [‡]*University College London, London, WC1E 6BT, UK, and* [§]*School of Environmental Sciences, University of East Anglia, Norwich, Norfolk NR4 7TJ, UK*
E-mail: I.Langford@uea.ac.uk

Abstract

Everyone likes clean seawater to bathe in and standards for acceptable seawater quality are set by the European Commission (CEC, 1976). In 1994, proposals to revise these standards were announced. These proposals were the subject of a House of Lords Select Committee Inquiry (HMSO, 1994, 1995), which deplored the fact that a soundly based cost–benefit analysis of the proposed revision had not been produced. This paper considers the question of developing means to assess the adequacy of the proposed revision from a social/public perception standpoint, using a mixed methodology of quantitative survey and qualitative focus groups. The aim of using such an approach is to provide a more in-depth and informative input into the decision-making process for policy makers. Our results show that mean willingness-to-pay amounts, representing the economic benefits of the revision to the 1976 EC bathing water standard, are roughly of the same order of magnitude as the estimated potential cost increases in average annual household water bills necessary to implement the revision. This result is qualified by analysis of how preferences are constructed in terms of socioeconomic variables, perceptions and attitudes towards risk, and issues such as trust, blame and accountability of institutions and regulatory processes involved in setting standards for bathing water quality.

1. Introduction

In recent years, both the general public and policy makers have become increasingly concerned about sewage discharges to coastal bathing waters in the United Kingdom and the consequent risk to public health (HMSO, 1994, 1995). The European Commission (EC) Bathing Water Quality Directive of 1976 (CEC, 1976) set out standards for designated bathing waters which should be complied with by all member states. Although remedial action to improve the quality of coastal waters was initially taken rather slowly, the UK government eventually sanctioned a programme of improvements in 1987 to achieve compliance of all UK beaches

---

*Author to whom correspondence should be addressed.

with the 1976 Directive. In 1991, the EC introduced the Urban Waste Water Treatment Directive, which required the cessation of dumping sewage sludge at sea by 1998. For the 1997 bathing season 397 of the designated bathing waters (88.8 per cent) in the UK complied with the bathing water standards for the coliform parameters[1] at Mandatory level (1976 EC Bathing Water Directive), whereas only 166 bathing waters (37.1 per cent) complied with the more stringent Guideline[2] microbiological standards. A major impediment to rapid improvement has been the estimated cost of cleaning up. It has been estimated (HMSO, 1995) that the capital costs of achieving compliance with the two existing directives will exceed £9 billion (1993/94 £).

Further legislation has been proposed by the EC in the form of amendments (see section 3) to the 1976 Bathing Water Directive and these have been the subject of recent House of Lords Select Committee Inquiries (HMSO, 1994, 1995). In the Committee Reports of these inquiries it is estimated that the additional capital costs of achieving compliance with the amendment to the Bathing Water Directive are between £1.6–4.2 billion (HMSO, 1995). There is major concern amongst policy makers about how worthwhile such an investment is in terms of gains in public health and amenity, and what the public expects from new legislation aimed at reducing pollution at such high cost.[3] This debate also has to be placed in the context of the recent privatisation of the water companies responsible for sewage and water treatment and supply in the UK. Controversy has arisen over the potential conflict between shareholders dividends and clients water charges, and around the issue of government accountability.

This contribution reports from the results of a study conducted in 1997–98, which focused on the estimation of the benefits of health risk reductions afforded by the proposed amendment to the 1976 EC Bathing Water Directive. The principle aims of that study were as follows (see Turner *et al.*, 1998 for further details):

1. To provide a better-informed and more robust measure of the willingness-to-pay[4] using a mixed methodology which combined contingent valuation[5] surveys with focus groups.
2. To improve the understanding of the established link between willingness-to-pay amounts and individuals' assessments of risk.

---

[1]Coliforms are bacteria which are generally accepted as an indicator for sewage pollution though they do not themselves cause health problems. Compliance with the coliform standards indicates good control of any nearby sewage discharges.

[2]Guideline values are more stringent than the mandatory standards and are set as targets to be achieved but are not compulsory.

[3]The House of Lords Select Committee Inquiry on Bathing Water (HMSO, 1994) deplored the fact that a soundly based cost–benefit analysis of the proposed amendment had not been produced.

[4]Willingness-to-pay is defined as the economic valuation placed by an individual on a good or service in terms of money (Pearce, 1986).

[5]The contingent valuation method is a survey technique that attempts to elicit information about individuals' (or households') preferences for a good or service. Respondents in the survey are asked a question or a series of questions about how much they value the good or service. The technique is termed 'contingent' because the good is not, in fact, necessarily going to be provided by the enumerator: the situation the respondent is asked to value is hypothetical.

3. To investigate perceptions of cost–benefit and critical load approaches to acceptable health risk assessment.

This paper considers the question of acceptable health risk with respect to sea bathing water quality, and develops means to assess the adequacy of the proposed revision to the 1976 Bathing Water Directive using elements taken from the overall study.

2. Acceptable health risk assessment

Environmental degradation will often involve risks to human beings in terms of health damage inflicted directly, as well as indirectly through impairing the natural functions of ecological systems. Because resources are scarce, there will be a trade-off between reducing health risks and other competing wants in society, such that the problem of choosing between alternatives arises. This means that we have to decide on an acceptable health risk level and/or pollution control level. How then are we to decide how safe is safe enough?

An important step in any such decision consists of quantifying the likelihood (or probability) of occurrence of the deleterious event and the scale of its effect. This process is known as risk assessment. Risk management is the set of procedures, or process, by which risk is reduced (US EPA, 1990). The probability of an event multiplied by the scale of likely effect gives the overall risk or 'expected damage' arising from an action. Expected damages are measured, for example, by numbers of premature fatalities, or expected incidence of morbidity. One specific form of risk assessment is 'comparative risk assessment' whereby two or more risks are compared and then set against the expenditures made to reduce them.

There often will be a multiplicity of heterogeneous and non-commensurable effects from risky actions such that this type of comparative risk assessment is difficult. We can look to economic theory for a possible solution. Economic theory assumes that the objective of society is to maximise human welfare or utility, as represented by individuals' preferences (the axiom of utility maximisation). Looking at acceptable health risk/pollution control policy from an economic perspective therefore requires that rational policy decisions regarding resource allocation be based on an informed assessment of the utility (or benefits) of controlling health risks/pollution.

In estimating such benefits, the focus of economic analysis of individual behaviour is with the trade-offs used to infer the values for health risk reductions. Economic assigned values are expressed in terms of prices, where, in turn, the prices reflect individual willingness-to-pay for risk reduction. The concept of willingness-to-pay is most familiar in market contexts, such as the prices in a supermarket. In this way, we can conduct a benefit–cost analysis whereby the risks are 'monetised' (Pearce, 1986). Avoided risks are then benefits and these can be compared to the cost of controlling the risk.

Yet another alternative in deciding upon acceptable health risk levels considers the ability of the environment (its assimilative capacity) to accept a deterioration in ambient quality without adverse biological effects occurring. The concentration level at which damage begins to occur in the environment is known as the Critical Load concentration. The Critical Load approach (Ramchandani and Pearce, 1992) to acceptable risk utilises the assimilative capacity of the environment whilst restricting discharges such that the environmental concentration meets the Critical Load concentration.

In the following sections, we mainly investigate the links between public perceptions of risk management and the economic approach to the problem of coastal bathing water health risks.

3. Recreational bathing water pollution, public health risks and the EC Bathing Water Directives

The public health risks of sewage discharged into coastal marine waters are derived from human population infections. The sewage contains various micro-organisms that have been shown to be pathogenic and the causative agents of several human diseases. Any illness spread either anally or orally, whose aetiological agents are found in the faecal wastes of animals and man, can theoretically be transmitted through actual contact with polluted seawater, and ingestion of seawater contaminated with pathogens. The main risk faced by people bathing in sewage contaminated water is in increases to minor morbidity, such as gastrointestinal and upper respiratory tract ailments.

The sources of causative agents include untreated or poorly treated municipal and industrial wastewater and sewage sludge discharges, sanitary wastes from coastal residences and pleasure boats, storm-water runoff, faecal wastes from animals and man, sanitary landfill drainage and the marine environment.

As a result of the public health risks from sewage contaminated coastal waters, the quality of bathing waters in England and Wales is monitored against standards laid down in the bathing water regulations (SI 1991/1597). These give effect to the EC Bathing Water Direction (76/160/EEC) covering the sanitary quality of waters used for recreational purposes. Numerical quality standards are specified at either mandatory or guideline values for a range of physicochemical, bacteriological and aesthetic criteria.

With respect to the bacteriological criteria, the mandatory standards, which should not be exceeded, are 10,000 *total coliforms* per 100 millilitres (ml) of water and 2,000 *faecal coliforms* per 100 ml of water (the latter being indicative of the presence of traces of human sewage). In order for a bathing water to comply with the EC Directive, 95 per cent of the samples (i.e., 19 out of the 20 taken in any bathing season[6]) must meet these standards.

The origin of these parameter values is the subject of debate. According to Moore (1977), the EC committed themselves to producing standards without any discussion at the expert level, and without there being any published scientific evidence in Europe of a significant health risk. Fleisher, in evidence to the House of Lords Select Committee on the European Communities Inquiry into Bathing Water (HMSO, 1995), was of the opinion that there was little scientific basis for the bacteriological criteria contained in the Directive. In fact, the Commission did not adopt any risk assessment procedure in establishing the Directive and no benefit–cost risk assessment

---

[6]The bathing season is from 15 May to 30 September.

was ever required since the Directive predates the provisions of the Single European Act and Maastricht Treaty.

At the beginning of 1994, the European Commission announced proposals to revise the 1976 Bathing Water Directive. The Commission's explanatory memorandum (CEC, 1994) claimed that the proposal was intended to simplify, consolidate and bring up to date the original proposal and to adapt the directive to scientific and technical progress. The proposed amendment would make the following changes to the existing parameters: introduce an imperative standard for *faecal streptococci* of 400 per 100 ml; make the standard for enteroviruses more stringent, and seek to replace it with a bacteriophage parameter; test for *E. Coli* rather than *faecal coliforms*; withdraw testing for *total coliforms* and for *Salmonellae*; add a requirement for the absence of sewage solids.

The public health basis and justification of the proposed amendment has again been the subject of considerable debate – see for example HMSO (1994, 1995). The House of Lords Select Committee Inquiry (HMSO, 1995) found that there was no firm public health justification for the proposed imperative levels of the parameters included in the directive amendment in terms of quantifying the reductions in incidence of illness. Again no risk assessment procedures have been used in drawing up the amendment, though the Commission appears to have been aware of the potential criticism that no benefit–cost assessment had been undertaken, arguing that as a result of the amendment, the financial burden on member states would 'on balance' be reduced. We now look at the benefits from a revised Directive and develop means to assess its adequacy from a social/public perception perspective of safety.

4. Methodology

In order to develop means to assess the adequacy of the proposed amendment to the EC Bathing Water Directive, we decided to employ a mixed methodological approach to investigate the themes of interest, combining a questionnaire survey of the general public with smaller focus group meetings.

The combination of quantitative (individual response) and qualitative (group discussion) elements used in this mixed methodological approach has allowed us to investigate the range of public perceptions, attitudes and beliefs, and to relate these to the needs of policy makers responsible for implementing new legislation.

4.1. Preliminary focus group work

A series of preliminary focus groups were undertaken in order to discuss some of the issues raised in the pilot study (Georgiou *et al.*, 1998) but mainly to aid the development of the contingent valuation survey questionnaire. The focus group participants were 26 adults who participated in one of three focus group interviews. A stratified sampling approach was used to recruit the groups (see Turner *et al.*, 1998, for more details).

4.2. The questionnaire survey

Design of the questionnaire survey was influenced both by elements from economic, psychological and sociological models (see below), and by results obtained in an initial pilot study conducted in the summer of 1995 (Georgiou *et al.*, 1998) and the preliminary focus groups undertaken. It was decided to survey user and non-user members of the public at one urban (Norwich) and two coastal (Great Yarmouth and Lowestoft) locations in East Anglia (for reasons of proximity), in order to investigate differences in physical characteristics, locational context, visitor type profiles, use profiles and/or alternative populations (e.g. those who have chosen not to visit the seaside versus those who have).

Those interviewed at the coastal locations were chosen randomly from people in the vicinity of the beach, whilst the Norwich sample was chosen using a stratified sampling approach. Each interview was completed in about 25 minutes, face to face with a trained interviewer. Interviews were carried out during the hours of daylight over the summer and autumn period, 1997.

The surveys aimed to cover a number of approaches to risk perception, preference deconstruction and the valuation of health risks from bathing water. These approaches included the application of cultural theory (Douglas, 1982; Douglas and Wildavsky, 1982; Schwarz and Thompson, 1990; Thompson *et al.*, 1990; Rayner, 1992) and a modified social learning theory approach (Bandura, 1977; Wallston, 1992). As already mentioned a Contingent Valuation exercise was also performed. A psychosocial model was thus developed which linked together respondents world-views with intentions to behave, such as willingness-to-pay (see Turner *et al.*, 1998 for details).

The survey questionnaire contained questions representing key variables from eight categories defined by reference to the different theoretical approaches mentioned above (see Turner *et al.*, 1998 for further information and details of the questionnaire). The eight categories are as follows:

1. Values – These questions related to the importance to the respondent of the proposed amendment to the European bathing water standard (hereafter referred to as the 'new EC standard'), both personally as well as to the nation, and whether the trustworthiness of government in implementing EC directives was an important issue to the individual.

2. Behavioural expectations – Here respondents were asked about the possible outcomes of new legislation, and whether their participation in the survey would have an input into decision making. In particular, we asked about what decrease in health risks respondents would expect from the new EC standard, whether the government was trusted to implement the new EC standard, if the application of the new EC standard was realistic in practice or whether its success or failure would be largely a matter of chance.

3. Self efficacy – We asked whether respondents felt they had enough information to make a decision about the new EC standard, and if they felt personally capable of making a decision. We also asked if the decision should be left to experts and whether public consultation should be courted.

4. Behaviour and intention to behave – These questions provided information on what uses were made by respondents of the sea, such as paddling, swimming and water sports, both now and potentially in the future (assuming the new EC standard was introduced). In order to estimate the economic benefits of any health risk reductions afforded by the proposed amendment to the EC

Bathing Water Directive, this category of questions included a Contingent Valuation (Mitchell and Carson, 1989) exercise to elicit individuals' willingress-to-pay.

5. Knowledge – This set of questions enquired about respondents' awareness of risks to health from polluted bathing waters, and whether they had heard of the current EC standard. Respondents in the coastal location survey were also asked about how they rated the quality of the sea water at the location beach, as well as being asked to say if the beach had in fact passed the current EC standard.

6. Perception of risks – We asked respondents some questions about how they perceived the riskiness of a number of environmental risks to health (faced by society in general), such as air pollution, drinking water, household garbage, hazardous wastes, pesticides in food and pollution in coastal bathing waters, in the manner of the psychometric paradigm of analysis (Slovic, 1992; Marris *et al.*, 1997).

7. World views – These questions attempted to ascertain respondents' underlying beliefs about the environment, and their world views in general. We elicited respondents' views of nature, or 'myths of nature' as proposed by cultural theorists (Schwarz and Thompson, 1990; Thompson *et al.*, 1990), and used a reduced version of the cultural type questionnaire developed by Dake (1992) and used by Marris *et al.* (1998) to determine scores for each individual on the four cultural theory scales, namely hierarchy, individualism, egalitarianism and fatalism (see section 4.3). Details of this procedure are given in Langford *et al.* (1998 in preparation), Marris *et al.* (1998) and Turner *et al.* (1998).

8. Personal characteristics – Each individual was asked a set of questions about their sex, age, income, employment, education, whether they had children or not, or were members of various environmental and other groups.

4.3. Post survey focus groups

Following on from the questionnaire survey an additional set of focus groups were undertaken in order to provide further insight and interpretation of survey findings. The focus groups were chosen from the sample of Norwich-based questionnaire surveys, on the basis of respondent's scores on one of four cultural theory scales. These scales are based on two dimensions, 'group' and 'grid' (Thompson *et al.*, 1990). Thompson *et al.* interpreted these dimensions as follows: '*Group* refers to the extent to which an individual is incorporated into bounded units. The greater the incorporation, the more individual choice is subject to group determination. *Grid* denotes the degree to which an individual's life is circumscribed by externally imposed prescriptions. The more binding and extensive the scope of the prescriptions, the less of life that is open to individual negotiation' (Thompson *et al.*, 1990, p. 5). Representing these variables as a pair of orthogonal axes produces four quadrants, each one corresponding to a cultural archetype defined by the strength of its grid and group characteristics. The four cultures are described as follows:

Hierarchists – characterised by strong group boundaries and binding prescriptions (high group, high grid). Hierarchists will tend to see an individual's place in the world as defined by a set of institutional classifications, e.g. based on age or gender. These demarcations are not questioned, but justified on the grounds that they enable the society to run smoothly. Control is vested in formal, hierarchical systems of authority.

Egalitarians – characterised by strong group boundaries but little belief in prescribed roles (high group, low grid). Egalitarians have a strong sense of society but expect each individual to negotiate their relationship with others and no individual is granted authority by virtue of their position.

Fatalists – characterised by low group association but a strong sense of social distinctions (low group, high grid). Like hierarchists, fatalists' autonomy is controlled by systems and institutions, but they see themselves as excluded from these institutions and therefore 'outsiders' with little or no power.

Individualists – characterised by no group incorporation and no prescribed roles (low group, low grid). Individualists feel no responsibility towards other members of society and allocation of power and resources is by competition, not by position or status (Douglas, 1982; Douglas and Wildavsky, 1982; Thompson *et al.*, 1990; Jenkins-Smith, 1994).

The motivation for undertaking the focus groups in this way was to examine whether cultural theory helped explain the preference responses given in the survey. Full details of the relationship between this and the qualitative analysis are given in Langford *et al.* (1998), where information on the selection procedure and logistics of each of the four group meetings is also given.

The group discussion protocol focused on: public perceptions of bathing water health risk information; possible solutions to coastal bathing water pollution problems; the extent to which willingness-to-pay reflects individual preferences; the appropriateness of cost–benefit and critical load criteria for the setting of standards; trust and accountability in the agencies and groups concerned with bathing water issues, and how this influences/affects willingness-to-pay; the level and type of consultation over bathing water issues desired by the public; and the public's informational requirements regarding bathing water quality and health risks.

5. Results[7]

The results section is divided into five subsections. First, we discuss some general summary results from the survey, after which we look at risk perception and world views, and investigate the extent to which the divergence between scientific and public perception of risk is potentially influenced by factors such as trust, blame and accountability. Next, we look at the valuation exercise to assess health risk reductions from the new EC standard. Finally we look at cost–benefit and critical load criteria and their role in standard setting, as well as looking at some additional results from the study arising from an additional request by the Environment Agency to investigate issues of information and public participation in decision making. Interpretation of information provided by the focus group discussions along with the quantitative information provided by the questionnaire survey have been combined in order to arrive at the study's final results.

[7]See Turner *et al.* (1998), for a detailed presentation of results.

5.1. General summary results from the survey

Total sample size was 626, of which 237 interviews were carried out at Great Yarmouth, 192 at Lowestoft and 197 at Norwich.

The socioeconomic composition of the three location subsamples and visitor type composition of the beach-based questionnaire samples (Great Yarmouth and Lowestoft) were examined. Socioeconomic composition was similar, except for statistically significant differences between samples in the mean number of household residents, the percentage who were members of an environmental organisation, and the percentage of Anglian Water ratepayers. Norwich had a significantly lower mean number of household residents, whilst environmental organisation membership and number of Anglian Water ratepayers were significantly higher in Norwich than in the other two sample sites. The composition of visitor types was also quite different (statistically significantly so – $\alpha = 5\%$), with the main group of respondents in Great Yarmouth being holidaymakers, whereas in Lowestoft the composition is more evenly distributed between holidaymakers, day-trippers and local residents. This difference in composition has implications for some of the results presented subsequently.

5.2. Risk perception and world views

Perceptions of respondents' personal risks to their health from swimming at beaches in East Anglia were found to be similar, across sites and between respondent types. In addition, respondents rated (on a Likert scale) societal bathing risks about as serious as pesticides in food (risks to society generally). More serious were considered to be the risks from household garbage.

In looking at how the public's perceptions are formed, we looked at the role of previous experiences in the form of family illness history, and knowledge about historical and present water quality. We found that respondents' perceptions were that water quality at Lowestoft was better than at Yarmouth. Also, there was a much greater awareness of whether the beach passes the EC bathing water standard or not amongst the respondents who visit Lowestoft. Whilst this was the general case overall, this was not so for holidaymakers visiting Great Yarmouth. They gave the water quality a significantly higher subjective water quality rating over the other respondent types (not the case for holidaymakers at Lowestoft), and were the least aware of whether the beach passed the standard or not. We hypothesise that holidaymakers who have previous experience of illness learn from the experience. They try to gain more knowledge about water quality at the sites. Given this increased knowledge, they prefer to go to Lowestoft (26 per cent of holidaymakers at Lowestoft had previous experience of illness compared with 11 per cent at Yarmouth) where the objective measures of water quality were better up until 1996 (when Yarmouth passed for the first time). However, since 1996 actual objective measures of water quality at the two sites indicate that water quality at Yarmouth is better.[8] This seems to be filtering through to respondents as the percentage who correctly stated that Great Yarmouth passed the EC standard is much

---

[8] Analysis on *total coliform, faecal coliform* and *faecal streptococci* indicator bacteria carried out by the authors from Environment Agency data set on National Bathing Water Quality. See Georgiou (in preparation).

higher now (37 per cent) than a year before in the pilot study (12 per cent) (Georgiou *et al.*, 1998).

In the group discussions which focused on risk perceptions and world views we found that there were interesting differences between the cultural theory defined categories. The hierarchists in particular were sensitised to the risks of both minor and major illnesses from bathing in polluted water, and yet were also concerned that not enough media coverage was given to the potential risks. According to cultural theory, hierarchists have faith in the *status quo*, until that is seen to be operating inadequately, then they become very concerned that solutions are found to outstanding problems. In this respect, more information and public education were seen to be solutions so that exposure to risks could then be voluntary. In addition, they were concerned that sources of pollution should be correctly identified, so that blame could be apportioned appropriately and action taken. Individualism was not associated with risks from pollution, which were seen as minor and overstated. They perceived personal lack of resistance to infection as one of the main problems in contracting minor illnesses, and therefore this was not an issue for great public concern.

Individualism was also associated with a lack of interest in the issue, in contrast to egalitarians who expressed both interest and concern. However, these differences were really based in the political arena, with egalitarians feeling that *on principle*, the public should not be exposed to health risks, and that the current EC standard was inadequate to protect the public. Individualists questioned the statistics on risk of illness as being unreliable, and only a potential problem for regular bathers who took a known risk associated with a pleasurable activity, as in many other areas of life. Hierarchists were concerned that the figures were so high, again challenging their belief in the current standard, but were concerned that comparisons were made with other locations in Europe so that a fair new standard could be reached based on reliable evidence. Fatalists readily accepted the figures given, but were shocked at how high they were, and had little faith in the standard. However, they did not immediately propose any solutions to the problems, only commenting negatively on the current situation.

The issue of trust in the government was very important to the questionnaire respondents, though a minority felt that the government could be trusted to implement the new standard. There were differences to these general findings amongst the cultural groupings however.

Individualists and hierarchists were least concerned about trust and governance, believing that regulatory change, technological innovation and market mechanisms could be combined to provide workable solutions. Hierarchists were generally satisfied with the agencies involved in regulation and sought action by parliament to ensure proper regulation of the Water Companies. Individualists focused on the perceived inefficiency and insensitivity of the EU, commenting negatively on their inflexibility and financial profligacy. Egalitarians, in contrast, were concerned about the amount of bureaucracy and lack of democracy in the EU, and fatalists expressed both ignorance and mistrust of the operation of the EU. However, both egalitarians and fatalists had strong opinions about Anglian Water. Whilst both groups were against perceived profiteering, egalitarians favoured renationalisation, but fatalists did not believe that the government would manage things any better.

5.3. The valuation exercise

The purpose of the valuation exercise was to estimate the economic benefits of health risk reductions obtained from the proposed amendment to the EC bathing water standard.

Survey respondents were given information on the causes and effects of sewage contamination of bathing water, the current status of the Anglian Water region beaches, as well as the most recent epidemiological evidence on incidence of illness for beaches which satisfied the current EC standard (Fleisher *et al.*, 1998). Respondents were then told that the EC was considering the introduction of a new standard (the proposed amendment to the 1976 EC standard) which should result in further reductions in risks to health at those beaches which satisfy the new standard, although this may require extra expenditure.[9]

A question was asked about what respondents expected from the new EC standard in terms of proportional risk reductions from the existing EC standard (for which they were given the appropriate overall average risk). Nearly 25 per cent of all respondents expected nobody to get ill under the new EC standard, whilst the majority expected some incidence of illness to remain.

Following a reminder to respondents to consider the limitations and other constraints upon their available income, a 'Payment Principle' question was then presented. This asked respondents if they would, in principle, be in favour of all beaches in the Anglian Water region having to comply with the new EC standard even if it cost their household some money in extra water rates (though they were asked to set aside for the moment the issue of how much this increase might be). Table 1 summarises the responses given to this question disaggregated by site as well as by respondent type for the beach-based surveys.

Considering the full sample responses at each site we can see in each case a majority of positive payment principle responses. However, χ^2 testing confirmed that between site differences are statistically significant ($\alpha = 1\%$) with the Norwich sample providing a larger 'yes' vote than that at Lowestoft, which in turn exceed that at Great Yarmouth.

Analysis of the two coastal surveys reveals an interesting and consistent pattern of support across the three respondent types identified here. In both cases local residents were less willing to accept the payment principle than were day-trip or holiday respondents. This may reflect the fact that the latter groups both derived more direct utility from improved bathing water quality and were less likely to have to pay for implementation of improvements in the Anglian region than were local residents.

We analysed why people voted against the payment principle. The main reasons were to do with not being able to afford to pay, not living in the region, and having problems with the payment vehicle – people felt that they paid enough taxes already

---

[9]The exact terms of the new EC standard were not stipulated since it was felt that the criteria for conforming to the amendment, on indicator bacteria counts (see note 1) would be meaningless to lay members of the public. It was also not possible to give precise reductions in illness incidence afforded by the new mandatory levels of the EC standard since there has been no attempt by the relevant authorities at quantifying these reductions.

Table 1. The payment principle.

| Payment principle | Norwich | Great Yarmouth | | | | Lowestoft | | | |
	Full sample	Full sample	Local resident	Day tripper	Holiday maker	Full sample	Local resident	Day tripper	Holiday maker
% Yes	83	64	49	63	69	71	58	79	74
% No	16	33	51	33	27	27	40	20	24
% d.k[a]	1	3	0	4	4	2	2	1	2
Sample number	196	234	41	55	133	191	57	72	57

[a]d/k = don't know responses; includes those who stated they either did not know if they would pay or that they were indifferent.

and objected to 'profiteering' by the privatised water utilities, i.e., their perception of too high company profits and directors' pay.

Respondents who answered positively to the payment principle were then asked an open ended (OE) willingness-to-pay question as follows: 'What is the maximum amount in extra water rates which your household would be willing to pay every year in order for all beaches in the Anglian Water region to pass the new EC standard?'.[10] Various descriptive statistics quantifying responses to this question are detailed in table 2. The figures are for those respondents who were in favour of the payment principle, were able to state a willingness-to-pay amount, and exclude protest bidders (classed as respondents who stated a payment principle refusal reason other than: 'benefits received not worth rates increase'; 'cannot afford payment'; 'current standard is good enough'; 'don't use beaches').[11] The mean willingness-to-pay at the two beach sites is broadly similar, and indeed the difference in means and medians between these sites is not statistically significant ($\alpha = 5\%$). However, the Norwich sample has a mean willingness-to-pay of about £10 more than the two beach sites combined (full samples), a difference which is statistically significant ($\alpha = 5\%$).

In 1995, the House of Lords Select Committee on the European Communities considered the EC's proposal for the amendment to the 1976 EC Bathing Water Directive (HMSO, 1995). As part of this inquiry a cost compliance assessment was carried out in relation to designated bathing waters, based on the provision of suitable engineering and sewage treatment facilities to meet the limit values of the indicative parameters set by the amendment. Table 3 sets out the estimates of the likely impact on water charges of implementing various directive amendment scenarios as given in the Committee report. As can be seen annual average potential cost increases for implementing the amendment (scenario A) were estimated at £10–28 per household.[12] Considering the mean household willingness-to-pay sums reported in table 2, the proposed amendment appears to pass a rough cost–benefit test.[13]

---

[10] Regarding the particular valuation payment vehicle, we acknowledge that, whilst water rates may not be the most appropriate vehicle, especially since visitors in the beach-based survey may be from outside the charging area, we nevertheless believe that the financing of any improvements to bathing water would in all probability be financed in this way, i.e., at the regional water authority/company jurisdiction. This also meant that it was necessary to take a regional perspective as far as provision of the good/service of interest was concerned. However, improvements to Anglian Water region beaches were considered, rather than a single beach or all beaches in the UK. The elicitation method employed was open-ended (OE), and whilst having been criticised as liable to result in understatement of true willingness-to-pay (Hoehn and Randall, 1987), alternatives such as dichotomous choice are liable to result in anchoring (Bateman *et al.*, 1995) and yield welfare measures which are highly sensitive to statistical issues such as functional form (Kerr, 1996). Given this it was felt that OE methods provided more robust and conservative welfare measures.

[11] See Georgiou (in preparation), for analysis on this, as well as on outliers with respect to willingness-to-pay.

[12] This cost figure relates to the eight affected water utility companies (excluding Northern Ireland and Scotland), and it is thought that the impact in some water company areas might be twice the national average (HMSO, 1995).

[13] The willingness-to-pay figures corresponding to expectations from the proposed amendment would furthermore seem to suggest that reductions in incidence of illness of around 25–50 per cent from current EC standard levels are expected given the costs of implementing the amendment (given upper cost estimate of £28 this is the corresponding expectation level which equates costs and willingness-to-pay – see figure 1).

Table 2. Willingness to pay (£) per annum for new (proposed amendment to the) EC Standard per household.

WTP (£)	Norwich	Great Yarmouth				Lowestoft			
	Full sample	Full sample	Local resident	Day tripper	Holiday maker	Full sample	Local resident	Day tripper	Holiday maker
Mean	42.29	32.22	20.17	28.30	37.71	31.84	37.41	27.55	32.84
95% Confidence, interval	32.86– 51.71	24.97– 39.47	7.80– 32.53	17.89– 38.71	26.30– 49.13	24.44– 39.25	15.05– 59.78	17.85– 37.26	23.39– 42.29
Median	25	20	10	20	20	20	20	20	25
Sample number	139	136	24	33	77	112	29	47	34

Table 3. The impact on water charges of implementing the EC bathing water amendment.

Scenario	Capital cost (£ million)	Operating cost (£ million/year)	Potential increase in prices in real terms (%)	Potential increase in average annual household bills (£)
A	1520–3940	50–140	4–12	10–28
B	1050–2240	40–90	3–8	7–17
C	400–1010	20–40	1–3	3–7
D	20–40	0	0–0.1	0.1–0.02

Description of scenarios:
A – The Commission's proposal including the introduction of a mandatory standard for *faecal streptococci*, set at 400/100 ml, and the *enterovirus* standard made more stringent.
B – The existing directive made more stringent by making mandatory the standards which are presently the optional *Guideline* standards.
C – The Commission's proposal except for the omission of the more stringent enterovirus requirement.
D – The existing directive plus a new mandatory standard of 1000/100 ml for *faecal streptococci*.

Figure 1 shows how willingness-to-pay varies with respondents' expectations from the proposed amendment. Willingness-to-pay broadly increases as respondents' expectations from the amendment increase. This is as standard economic theory would lead us to expect. Those who state they do not know what to expect from the amendment have the lowest willingness-to-pay. One-way analysis of variance again highlighted statistically significant ($\alpha = 5\%$) relationships between the amounts and respondents' expectations.

Multivariate statistical analysis[14] of the payment principle responses and willingness-to-pay amounts established that there were notable differences in variables explaining the willingness-to-pay and payment principle responses, both between the three locations and across the different user groups, namely holidaymakers, day-trippers and locals. While standard neoclassical economic factors (as signified by 'personal characteristics' such as income, education, etc.) are significant in explaining stated willingness-to-pay and payment principle responses, other factors derived from

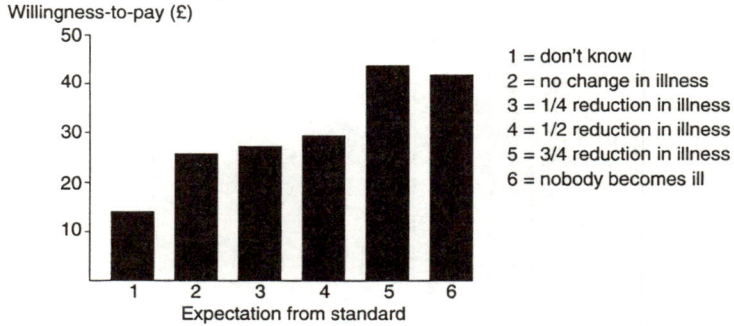

Figure 1 Willingness-to-pay by expectation from standard.

[14]The regression models estimated in this multivariate statistical analysis are reported in detail in Georgiou (in preparation).

social learning theory and cultural theory must also be taken into account. Such factors extend the range of variables considered to be important in determining stated preferences in contingent valuation (see Turner *et al.*, 1998 for further details).

5.4. Cost–benefit and critical load criteria and their role in standard setting

In the case under investigation, the cost–benefit approach dictates that a comparison of all the costs and benefits associated with a change in bathing water quality provides the optimal guide for decision makers. The scientifically defined critical load in the present case restricts the discharge of sewage waste to *zero*, a situation that will substantially elevate costs, necessitating alternative waste stream technologies.

We investigated the issue of cost–benefit and critical load criteria in both the survey and focus group analyses. In the survey-based analysis we looked at the issue simply in terms of respondents' expectations about the new EC standard. We found that nearly 25 per cent of all respondents expected nobody to get ill under the new EC standard (Critical Load Solution), whilst the majority expected some incidence of illness to remain. Figure 1 shows mean willingness-to-pay corresponding to each of the expectation levels. Overall mean willingness-to-pay for the sample was £35.73 which comes somewhere between the willingness-to-pay for those expecting a $\frac{3}{4}$ to $\frac{1}{2}$ reduction in incidence of illness from current levels. The actual percentage of respondents who expected a reduction in incidence of illness of $\frac{3}{4}$ to $\frac{1}{2}$ was between 25 per cent and 30 per cent. This would seem to suggest that people on average are willing to allow some damage beyond the critical load position of nobody becoming ill, i.e., the cost–benefit solution.

We also questioned respondents on more broader issues concerning the success or failure of the new EC standard, whether reducing health risks by the introduction of a new standard was realistic, and whether experts should be left to determine whether the new standard was worthwhile. One-third of the sample had a rather fatalistic attitude to the success or failure of the new EC standard, though over two-thirds did think that reducing health risks by the introduction of a new EC standard was realistic. There was a roughly equal split between those who thought experts should be left to determine the new EC standard and those who did not.

In the focus groups, we also found that the great majority of participants accepted the cost–benefit solution at least to a degree, but for different reasons depending on cultural standpoint. Hierarchists and egalitarians were initially unsure about the economic approach, but were worried about the high costs associated with a fixed standard solution. They also had some concerns, as did fatalists, about the success of a new standard, suggesting that they lost faith in the current regulatory procedures for controlling sewage input into the sea. However, both hierarchists and egalitarians saw the critical load approach and consequent standard as something good which should be aimed for over time, and egalitarians broadened the issue into one of issues about polluted seawater in general, refusing to confine the debate solely to health risks.

Fatalists found the idea of a critical load solution unfeasible, but were also unhappy with the economic solution, as they perceived that private companies would manipulate the situation for their own benefit. They also decided that they had very little choice in what would actually happen, and that price increases were inevitable whatever the solution proposed. In contrast, individualists immediately took to the economic approach, and questioned the use of so-called 'expert' opinion about what standard should be set. A price rise in water rates was seen by individualists as acceptable, and they felt that willingness-to-pay should be accounted for in any cost–benefit analysis. It thus appears that the individualists seem to exhibit a rationality perfectly in line with economic theory, but that others (such as the fatalists) are less perfectly in line.

5.5. Information and public participation in decision making

A request was made and additional funding was obtained from the environment agency to carry out further work on the issue of information and public participation in decision making. We asked questionnaire respondents if they were aware of health risks from bathing in the sea, and if they had heard of the EC standards. Whilst there was a high overall level of awareness about health risks and the EC standards found in the survey (75 per cent and 85 per cent respectively), there was nevertheless a large majority of respondents in favour of the provision of more information and public participation in the decision-making process. Furthermore, a majority of respondents felt that their decision on whether they were willing to pay for the new EC standard was an important input into the decision-making process.

In the focus group sessions, we asked participants about the information they wanted regarding the risks from bathing. We found that hierarchists were ready to accept broad categorisation of passing and failing a particular standard, and individualists were prepared to rely on local knowledge and common sense backed up by compliance to the EC standard, although they were not well informed about what the 'flag' standard actually meant. Egalitarians were more in favour of a 'pollen style' count being provided at beaches, and wanted more frequent measurements to be taken of water quality, but expressed the belief that if standards were made rigorous enough, then continuous monitoring may not be necessary. Fatalists were also in favour of more rigorous monitoring, and wanted easily understandable information to be provided at beaches.

While there appeared to be a large overall desire for public consultation, we found in the focus groups a distinction between egalitarianism and individualism. Whilst egalitarianism was associated with the need for public consultation and the belief that experts and policy makers should not be left to take decisions, individualism on the other hand was associated with the opposite opinion. Hierarchists were against large-scale public consultation believing that such decisions should be taken through the proper channels, i.e., Members of Parliament, and that voting in elections and lobbying MPs was sufficient means of having an input into the decision-making process. Fatalists doubted the motivation of most people to make the effort to participate in any consultation.

6. Discussion

The overall aim of our study was to provide a more in-depth and informative input into the decision-making process for policy makers. Rather than simply providing a summary statistic, such as mean or median willingness-to-pay, our intention has been to qualify this figure with an understanding of how preferences are constructed in terms of socioeconomic variables, perceptions and attitudes towards risk, and issues such as trust, blame and accountability of the institutions and regulatory processes involved in setting standards for bathing water quality.

Briefly, some of the main recommendations we would make from our survey are:

• Traditional socioeconomic parameters, such as income, education, etc. explain only a small amount of the variation in willingness-to-pay for reduction in health risks. Inclusion of items measuring factors such as importance value, self efficacy, expectations, knowledge and overall world views greatly increase the explanatory power of our models.

• The factors mentioned above also affect to a varying degree the public's faith in any proposed new standard, and their expectations of the health benefits that the standard will deliver. Overall, we can conclude that individuals across all four identifiable cultural categories (egalitarian, fatalist, hierarchist and individualist) are concerned either with the current level of risk, the current practices for implementing the standard, and with issues of efficiency and accountability of the water companies. Any attempt at redrafting and successfully implementing a new standard must take account of these concerns.

• There was a large majority of respondents in favour of the provision of more information, and greater public participation in the decision-making process. However, there was no overall consensus on what was the best implementation process, including, for example, on the efficacy of continuous monitoring, or the best method of assessing public opinion. We conclude that more detailed research is needed on these issues, but it is unlikely that any one approach will satisfy a majority of the public. Hence, a package of measures including public education to increase knowledge of the risks, surveys of stakeholders and public opinion, and public meetings may be best employed to involve the public more in decision making.

• The great majority of respondents accepted an economic approach to the mitigation of the problem. However, individuals with different viewpoints arrived at this decision for different reasons. Willingness-to-pay was generally seen as a reasonable way of assessing public commitment to reducing risks to health. However this needs qualification with respect to proper apportionment of blame and responsibility to those who pollute the sea, the distribution of impacts across different sectors of society, and the setting of the issue of health risks in the context of wider environmental issues.

• Given these qualifications, the mean willingness-to-pay amounts representing the economic benefits of the proposed amendment to the 1976 EC standard, expressed by the Great Yarmouth, Lowestoft and Norwich sample populations (£35.73 per household), were roughly of the same order of magnitude as the estimated potential cost increases in average annual household water bills necessary to implement the proposed amendment (£10–28 per household). The willingness-to-pay figures corresponding to respondents' expectations from the proposed amendment would furthermore seem to suggest that reductions in incidence of illness of roughly 50 per cent from current EC standard levels are expected given the upper cost estimate of implementing the amendment (refer to figure 1).

Acknowledgements

The Centre for Social and Economic Research on the Global Environment is a designated research centre of the UK Economic and Social Research Council. This Research was funded by ESRC Grant Number L320253244. The authors are grateful to two anonymous referees for their comments on earlier drafts of this paper. Any errors remain the responsibility of the authors.

References

Bandura, A. (1977) *Social Learning Theory*. Englewood Cliffs: Prentice Hall.

Bateman, I.J., Langford, I.H., Turner, R.K., Willis, K.G. and Garrod, G.D. (1995) Elicitation effects in contingent valuation studies, *Ecological Economics*, **12**, 161–79.

CEC (1976) EC Council Directive 76/160/EEC, 8 December 1975. Concerning the quality of bathing water.

CEC (1994) Proposal for a Council Directive concerning the quality of bathing water, 6177/94; COM(94) 36 final.

Dake, K. (1992) Myths of nature: culture and the social construction of risk, *Journal of Social Issues*, **48**(4), 21–37.

Douglas, M. (1982) *Cultural Bias*, Occasional Paper, 35, Royal Anthropological Institute (Republished in: *In the Active Voice*, 1982, pp. 183–254. London: Routledge and Kegan Paul.

Douglas, M. and Wildavsky, A. (1982) *Risk and Culture: An Essay on the Selection of Technological and Environmental Dangers*. Berkeley: University of California Press.

Fleisher, J.M., Kay, D. and Wyer, M. (1998) Estimates of the severity of illnesses associated with bathing in marine waters contaminated with domestic sewage, *International Journal of Epidemiology*, **27**(4), 722–26.

Georgiou, S. (in preparation) Coastal bathing water health risks: public perceptions and determination of preferences to assess the adequacy of EC Standards, PhD Thesis, University of East Anglia.

Georgiou, S., Langford, I.H., Bateman, I.J. and Turner, R.K. (1998) Determinants of individuals' willingness to pay for perceived reductions in environmental health risks: a case study of bathing quality, *Environment and Planning A*, **30**, 577–94.

Hoehn, J.P. and Randall, A. (1987) A satisfactory benefit cost indicator from contingent valuation, *Journal of Environmental Economics and Management*, **14**(3), 226–47.

HMSO (1994) The 1st Report (Session 1994–1995) from the House of Lords Select Committee on the E.C. (1994), Bathing water (HL Paper 6-I). London: HMSO.

HMSO (1995) The 7th Report (Session 1994–1995) from the House of Lords Select Committee on the E.C. (1995), Bathing water revisited (HL Paper 41). London: HMSO.

Jenkins-Smith, H.C. (1994) Stigma models: testing hypotheses of how images of Nevada are acquired and values attached to them. Unpublished Manuscript, University of New Mexico, Albuquerque.

Kerr, G.N. (1996) Probability distributions for dichotomous choice contingent valuation, CSERGE Working Paper GEC 96-08, Centre for Social and Economic Research on the Global Environment, University of East Anglia and University College London.

Langford, I.H., Day, R.J., Georgiou, S. and Bateman, I.J. (1998) Testing a social learning theory model of risk perception, Centre for Social and Economic Research on the Global Environment, University of East Anglia and University College London (in preparation).

Marris, C., Langford, I.H., Saunderson, T. and O'Riordan, T. (1997) Exploring the

68   *Georgiou* et al.

'psychometric paradigm': comparisons between aggregate and individual analyses, *Risk Analysis*, **17**(3), 303–12.

Marris, C., Langford, I.H. and O'Riordan, T. (1998) A quantitative test of the cultural theory of risk perceptions: comparison with the psychometric paradigm, *Risk Analysis*, **18**(5), 635–47.

Mitchell, R.C. and Carson, R.T. (1989) *User Surveys to Value Public Goods: The Contingent Valuation Method*. Washington DC, Resources for the Future.

Moore, B. (1977) The EEC Bathing Water Directive, *Marine Pollution Bulletin*, **8**, 269–72.

Pearce, D.W. (1986) *Cost–Benefit Analysis*. London: Macmillan.

Ramchandani, R. and Pearce, D.W. (1992) Alternative approaches to setting effluent quality standards: precautionary, critical load, and cost–benefit approaches, CSERGE Working Paper WM 92-04, Centre for Social and Economic Research on the Global Environment, University College London and University of East Anglia.

Rayner, S. (1992) Cultural theory and risk analysis, in S. Krimsky and D. Golding (eds) *Social Theories of Risk*. Westport: Prager.

Schwarz, M. and Thompson, M. (1990) *Divided We Stand: Redefining Politics, Technology and Social Choice*. New York: Harvester Wheatsheaf.

Slovic, P. (1992) Perceptions of risk: reflections on the psychometric paradigm, in S. Krimsky and D. Golding (eds) *Social Theories of Risk*. Westport: Praeger.

Thompson, M., Ellis, R. and Wildavsky, A. (1990) *Cultural Theory*. Boulder, Co: Westview Press.

Turner, R.K., Langford, I.H., Bateman, I.J. and Georgiou, S. (1998) Economic and epidemiological investigation of coastal bathing water risks, Final Report to Economic and Social Research Council, ESRC Award No: L320253244.

United States Environmental Protection Agency (1990) *Reducing Risk: Setting Priorities and Strategies for Environmental Protection*, 4 volumes. Washington DC, US EPA.

Wallston, K.A. (1992) Hocus-pocus, the focus isn't strictly on locus: Rotter's social learning theory modified for health, *Cognitive Therapy and Research*, **16**, 183–89.

[24]

Marine Resource Economics, Volume 9, pp. 275–286
Printed in the USA. All rights reserved.

Total Economic Values of Increasing Gray Whale Populations: Results from a Contingent Valuation Survey of Visitors and Households

JOHN B. LOOMIS

Department of Agricultural and Resource Economics
Colorado State University
Fort Collins, CO 80523

DOUGLAS M. LARSON

Department of Agricultural Economics
University of California-Davis
Davis, CA 95616

Abstract *The consistency of an individual's willingness to pay (WTP) responses for increases in the quantity of an environmental public good (whale populations) is tested along three lines. First, we test whether WTP for 50% and 100% increases in whale populations are statistically different from zero. Second, we ask whether the incremental WTP from a 50% increase to a 100% increase is statistically significant. Finally, we test whether there is diminishing marginal valuation of the second 50% increment in gray whale populations. The paired t-tests on open-ended WTP responses supported all three sets of hypotheses. Both visitors and households provided WTP responses that were statistically different from zero and increased (but in a diminishing fashion) for the second increment in WTP. In this survey both visitors and households provided estimates of total economic value (including non-use or existence values) for large changes in wildlife/fishery resources that were consistent with consumer theory.*

Keywords Existence value, contingent valuation, gray whale, California, willingness to pay.

Introduction

It is generally recognized that direct viewing of wildlife provides an economic value to the participant. However, preservation of the species for viewing is a joint product for the "users" and also may provide value to non-viewers, from the continued existence of the species for both themselves and (in many cases) for others in the future. That continued preservation of a species provides joint pro-

Funding for this research was provided by grants from Resources for the Future, the U.S. Environmental Protection Agency and Agricultural Experiment Station Regional Research Project W-133. We would like to thank Renatte Hagemann, San Diego State University, who supervised the collection of the San Diego portion of this data, and our visitor interceptors Paula Stevens and Dan Ritzman for their persistence in a variety of weather conditions. Ms. Stevens and Sunny Williams provided substantial effort in the data entry and follow up phone contacts of the household survey. Finally, the editor and two reviewers provided numerous suggestions for clarifying the paper. The usual disclaimer applies.

duction of viewing, existence and bequest values was first formalized by Randall and Stoll (1983) as "total economic value." Most economists recognize that existence of a species is a public good and, like any other public good, its total economic value should be included when calculating benefits and costs of proposed policies (Randall and Stoll, 1983; Kopp, 1992). Given this, two key questions arise. First, how is one to measure the total economic value? Second, is the marginal or incremental total economic value for changes in abundance of a particular species significantly different from zero? That is, once a minimum viable population is provided, does increased abundance beyond that add to society's well being?[1] In answering this latter question, we focus on whether the pattern of incremental valuation of gray whale populations is consistent with first principles of consumer demand theory. What we learn about the *patterns of an individual's valuation* of gray whales is useful for understanding how valuations of other marine mammals such as elephant seals, sea otters and dolphins would also change with their population levels. Such information may be helpful in dealing with a wide variety of policy decisions such as what magnitude of incidental take of marine mammals to allow in commercial fishing operations.

With respect to the first question, at present, the Contingent Valuation Method (CVM) appears to be the only way to measure total economic value, although Larson (1993) has proposed a revealed preference approach. While the reliability of CVM to measure existence values has been subject to much debate over the years, the journal literature provides several examples of the reliability (Loomis, 1989, 1990; Reiling, *et al.*, Kealy, *et al.*,) and validity (Brookshire, *et al.*, 1982; Welsh, 1986) of CVM derived benefit estimates. Nonetheless, the high stakes in natural resource damage assessment has brought renewed interest in CVM by a broad spectrum of economists. Concerns about inconsistency of CVM with economic theory appears to have been one of the reasons why the National Oceanic and Atmospheric Administration (NOAA) commissioned a "blue ribbon" panel to advise NOAA on the use of CVM for measuring non-use values (Augustyniak, Jones and Meade, 1992). After reviewing the available research, "The Panel concludes that CV studies can produce estimates reliable enough to be the starting point for a judicial or administrative determination of natural resource damages—including lost passive-use value." (Arrow, *et al.*, 1993), as long as certain sampling and survey design guidelines are adhered to. Since these guidelines were released about a year *after* our survey administration, we will only make reference to areas of consistency and inconsistency with these new guidelines.

This paper contributes to improving our understanding of the properties of individual's CVM responses as measures of total economic value, including non-use value. Specifically, we ask whether CVM-derived estimates of an individual's incremental total economic value for increases in wildlife populations are significantly different from zero, and whether they are consistent with expectations based on economic theory. These questions are also of practical importance to policy analyses aimed at increasing wildlife populations beyond their legally required minimum viable population. While numerous previous studies have found statistically significant marginal *use* values for consumptive users (Brookshire,

---

[1] This simplistic discussion assumes that once society is above the minimum viable population that risk of extinction is zero. In addition, a decidedly anthropocentric viewpoint is represented here.

Randall and Stoll 1980; Loomis and Cooper, 1990), and for bird viewing (Cooper and Loomis, 1991), few studies have attempted to measure incremental values for *total economic values* of wildlife.

Hypotheses About Marginal Values of Increases in Public Goods

Utility and consumer theory is primarily a theory about individual behavior. First principles of consumer theory require only that increases in quantities of economic goods provide non-negative values to individuals. It seems plausible that increases in populations of many native fish and wildlife species could have positive values given their currently low populations. For a given individual, the principle of diminishing marginal returns suggests that if the first increment has a positive value, the value of the second increment would normally have a lower marginal value than the first, *i.e.*, one would expect diminishing marginal value for greater and greater additions to the stock of the public good.

The first hypothesis to be tested is whether stated WTP for a given increase of size n in a population of a particular species is statistically different from zero. Specifically the null hypothesis is:

$$H_o: WTP_n \le 0 \tag{1}$$

against the alternative that $WTP_n > 0$. If the dollar amount of stated WTP is obtained from responses to open-ended CVM questions, then WTP is measured as a ratio level variable. Therefore (1) can be tested using parametric tests such as a one-tailed Student's t-test.

A second and related hypothesis test is whether WTP for a larger increase in the public good (WTP_{n+m}), where $m > 0$, is statistically different from WTP for a smaller increase (WPT_n). Formally,

$$H_o: WTP_{n+m} - WTP_n \le 0 \tag{2}$$

against the alternative that $WTP_{n+m} - WTP_n > 0$. This is a one-tailed *paired* t-test since it is a test of the consistency of an individual's behavior, a more direct test of consumer theory then comparing consistency of aggregate behavior.

If we obtain positive incremental values for $WTP_{n+m} - WTP_n$, and accept the alternative to the second hypothesis, then one would expect a diminishing incremental valuation. Specifically if both increments are of the same size $n = m$, then a diminishing marginal value implies that

$$H_o: WTP_{2n} - WTP_n < WTP_n \tag{3}$$

This will also be tested as a paired t-test with the null hypothesis that

$$H_o: WTP_{2n} - WTP_n \ge WTP_n \tag{4}$$

Case Study

The gray whales that migrate from their summer home in Alaska down the west coast past California to Baja were until recently a threatened species under the

J. B. Loomis and D. M. Larson

Endangered Species Act. Whales are also a species with documented nonconsumptive use and strong evidence of existence/bequest values. Many conservation organizations such as Greenpeace and the Oceanic Society are maintained by contributions from people, many of whom rarely, if ever, see whales.

Our sample frame is comprised of two populations: (1) persons who went to the coast during the gray whale migration usually, although not exclusively, to view whales; and (2) California households. We include both current users and the general population of California households (some of whom may be past users) for two reasons. First, current viewers will have more first-hand experience "trading money for opportunities to view whales" and better information on whale abundance. These individuals generally meet the "Reference Operating Conditions" for CVM suggested by Cummings, *et al.* (1986). Hence CVM estimates of WTP should be more reliable for this group. In addition, we would expect that if any group would have significant total economic values, it should be this group. This user group contrasts with general households who may have never gone whale watching, possibly do not care about marine mammals, and have less experience with trading money for whales. However, it is this group for which even small estimates of total economic value per household often expand into large aggregate estimates, due to the sheer size of this group. Thus we are also interested to see whether this less-experienced and less-interested group also expresses statistically significant marginal values for successive increases in whale populations.

There have been two CVM studies to date attempting to measure the total economic value of whales. Hagemann (1985) used a payment card to elicit WTP for gray whales and blue whales as well as three other marine mammals. Hagemann's mail survey was of general households in California. Samples and Hollyer (1990) have applied CVM to value humpback whales in Hawaii. They asked individuals their WTP to provide one-time emergency assistance to protect whales in Hawaii. A dichotomous choice CVM format was used with in-person interviews of the general population.

While these studies both represent pioneering efforts in valuation of whales and presented interesting methodological comparisons, they generally involved either small sample sizes or low response rates. We informally compare our dollar values with these two studies in the results section. Formal comparisons are not possible since our survey evaluates large increases in number of whales while the two past studies involved WTP to avoid large reductions in whale populations.

Format of Whale Questionnaire

A major focus of the survey was questions regarding the total value for two increases in gray whale populations. First the respondents were told the current population of gray whales was 20,000. Then respondents were informed that the population could be increased by reducing coastal pollution and restricting commercial and industrial activities along the coast, particularly in calving habitat. The payment vehicle was payment into a Gray Whale Protection Fund. The survey stated that *"Legally the money could only be used to clean up coastal waters of pollution and drift nets, purchase additional calving habitat areas, etc."* Respondents were asked to state their WTP for a 50% and 100% increase in gray whale populations and corresponding increase in sightings of gray whales along the California coast. Specifically the question read *"What is the maximum you would pay each year into the Gray Whale Protection Fund to increase gray whale populations and your sightings by 50%?''* Respondents were asked to assume that a 50% increase in whale populations would translate into equal percentage increase in whale sightings. This WTP question format was open-ended and was followed by questions to help determine whether persons stating zero WTP were protesting some feature of the CVM market. On the adjacent page, the 100% increase in whale program was presented as follows: *"Program Two: This alternative program is a more comprehensive gray whale protection that would increase by 100% or double gray whale populations and your sightings along the California coast. This second program is illustrated in the graph below.''* Respondents were then asked to state their WTP for Program #2 if it were the only program offered.

Thus, individuals were first asked to pay for a 50% increase, and then asked to consider an alternative second WTP question for a 100% increase. Reliance on a series of WTP questions has both methodological advantages and disadvantages. First, having the same individual give their WTP for both population levels provides a more efficient statistical control for estimating the value of the population change. In this setting, the difference in WTP by the same individual is related only to the change in whale population. The alternative questioning approach, of having different individuals answer one question for one of the two population levels would be statistically inefficient since the research would have to be able to completely control for differences among individuals before they could conclude if the differences in valuation related to differences in whale populations or differences in individual respondents. If separate samples were each asked one of the two questions, the required sample would need to be far greater than simply twice the current sample. Asking more than one quality level of the same respondent has been a standard CVM procedure for deriving Bradford bid curves (Bradford, 1970) for decades (Daubert and Young, 1981; Walsh, *et al.*, 1984) and continues to be used today (Hoehn, 1990; Hoehn and Loomis, 1993). Loosely speaking, this question format is like a panel design and has the same statistical advantages.[2] One methodological concern relates to possible order effects. Since everyone answered the 50% increase before the 100% increase this may create a conditional path in the WTP values, even though individuals were told that the 100% increase program was an alternative program. However, *if* there are sequencing effects such that the program valued second is worth less than the one valued first, this will tend to bias our hypothesis test #2 ($WTP_{50\%} < WTP_{100\%}$) against us because it would raise the WTP for the 50% increase and lower the WTP for 100% increase, thereby shrinking the differences.

A draft of the survey instrument was pretested using individuals who had gone whale watching in the previous year. A survey was given or mailed to them and they were asked to fill it out and make additional comments on it. The pretest indicated respondents had difficulty understanding what a 50% and 100% increase in whale populations represented. Therefore, two sets of bar charts were added to illustrate what a 50% and 100% increase in gray whale populations represented relative to the current situation (which was the no action or default level if they did not pay). The first bar chart showed the 50% increase next to the current popu-

---

[2] As a reviewer points out such a practice may not be an acceptable way to test for the presence of embedding. See Arrow, *et al.*, 1993 on this point.

lation, while the second chart showed the 100% increase next to the current population. We believe these bar charts significantly aided respondent understanding of the scope of the programs. However, we did not perform formal one-on-one debriefing or use verbal protocols on the revised survey instrument to verify this. Lastly, demographic information including attitudes toward the marine environment, work status, wage rates and household income was asked. The overall survey was a booklet.

Data Sources

Visitor Sample Frame

The visitor sample frame is defined by both time and space. We sampled over the months of the gray whale migration along the California coast. Whales start arriving in northern California in late December and migrate south to San Diego and Baja California during January. In February whales begin to migrate back north, passing through northern California by late March. Many people taking trips to the coast during these months are out to view whales rather than for traditional summer ocean activities. For cost effectiveness in sampling, we sampled weekends and holidays at the locations described below. The choice of whether to sample on Saturday or Sunday was random, but did allow for a balance of Saturdays and Sundays over the season. The interviewers were present from 10am until dark.

The visitor survey took place at four locations along the California coast: San Diego (Point Loma National Seashore), Monterey, Half Moon Bay (south of San Francisco) and Point Reyes National Seashore (north of San Francisco). At both Point Loma and Point Reyes people view whales from the shore. At both Monterey and Half Moon Bay, people took short boat trips being run at that time specifically for gray whale viewing by commercial operators in the area.

Visitor Data Collection Procedures

The survey was administered as follows. Every nth adult (age 16 or older) visitor (where n = 5 for boat trips and n = 10 for onshore viewers) was contacted by a trained interviewer as they returned from the viewing area or boat. The interviewer asked each sampled visitor to take a survey packet home and then recorded the visitor's name and address in case follow-up mailings were required. The interviewers explained that the survey was being done by the University of California, which is viewed as a non-advocacy, research-oriented organization by most Californians. This message was reinforced by official jackets and name tags worn by the interviewers. The survey packet include a cover letter, the questionnaire and a postage-paid return envelope.

We chose this survey approach because it combined the best of personal interview and mail questionnaires. The personal contact allowed the interviewer to stress the importance of the respondent to the study and obtain a "good faith" commitment to return the survey (Mitchell and Carson, 1989:110). The interviewer was able to answer any concerns the visitor had about the survey or how they were selected. However, we did not want to conduct the interview at the site itself, for several reasons. First, weather at the coast in the winter is not always conducive to lengthy outdoor interviews and weatherproof shelter was very limited at most of these interview sites. Second, some passengers coming off the boats (particularly at Half Moon Bay) were not in physical or mental condition to answer detailed questions about their trip and thought-provoking CVM questions. By giving the visitor the survey to take home, they could devote adequate thought to answering the CVM questions at their own pace (Bailey, 1987:148).

In total, 1,402 surveys were handed out at the four intercept locations. We obtained 1,003 back for an overall response rate of 71.3%. The response rate was reasonably similar across the four locations, varying from a low of 65.2% at Point Loma in San Diego to a high of 80.3% at Half Moon Bay. On-site refusals were not a problem. For example, at Point Reyes only 10 people out of roughly 600 refused to take a survey (about 1.5%).

Household Survey Design and Data Collection Procedures

The CVM questions in the household survey were identical to those asked of visitors. However, less information was requested on whale viewing trips.

A stratified sample of California households was purchased from Survey Sampling Inc. The sample was stratified between persons living in counties adjacent to the California coast and those living in inland counties. Given that the California population distribution is heavily centered along the coast, a simple random sample would not have given us an adequate representation of non-coastal residents who are more likely to be nonusers. The foundation for the sample drawn by Survey Sampling Inc. is telephone listings. To partially overcome the omission of unlisted numbers, Survey Sampling Inc. supplements the initial list with drivers license and utility records. Since our purpose here is to compare households to whale viewers rather than generalize our WTP results to give aggregate values, we feel this is an adequate sampling scheme to obtain non-users.

A total of 2,000 names and addresses were obtained. Following Dillman's repeat contact approach, a personally-addressed cover letter (with original signature) and postage-paid return envelope accompanied the questionnaire. We also attached a dollar bill to the survey as a token of appreciation. This first mailing was followed up with a reminder postcard one week later. Then a second mailing was carried out (without the dollar bill). Finally, a random sample of non-respondents was phoned. The purpose of the phone call was to encourage non-respondents to complete the survey or to find out the reasons why they would not complete a survey. We also noted any English language difficulty (which to our surprise was not a problem). We asked persons refusing to accept another survey whether they had ever been whale viewing and their education level for comparison purposes. As expected, non-responding households were less likely to have been whale watching and had slightly lower education levels.

Of the 2,000 household surveys, 301 were undeliverables, 41 were deceased and 16 turned out to be businesses or government agencies. Of the eligible sample of 1642, 890 questionnaires were returned after all three contacts. The overall response rate was 54% of deliverable questionnaires, slightly above the average for other mail CVM surveys in California (Loomis, 1987). In the household sample, 56% of the respondents had *never* gone whale watching, 35% had gone at some time in the past and only 9% had gone whale watching this year. This compares to the visitor survey where 100% had gone whale watching this year.

Table 1
Visitor and Household WTP for Increases in Whale Populations

	Visitor Sample		Household Sample	
	Mean	Std Error	Mean	Std Error
50% Increase in Population	$25.00	1.16	$16.18	1.07
100% Increase in Population	$29.73	1.39	$18.14	1.16
n =		672		519

Thus the household sample is dominated by pure non-users or those with only past whale viewing activity.

Survey Results

WTP Estimates

Before calculation of WTP two steps are typically performed in most CVM studies. First, individuals stating a zero WTP and indicating it was a "protest" response (*e.g.*, it is unfair to ask people to pay for greater protection of gray whales, *etc.*) are dropped. This was only about 10% of the visitor sample and about 20% of the household sample. We offer two reasons why visitors were more willing to accept the trust fund scenario. First, visitors had experience trading money for whale viewing opportunities, so thinking about whales in dollars was not alien to them. Second, nearly half of the visitors were members of environmental organizations (more than double the rate in the household sample). Hence visitors were most used to seeing real appeals for money to protect particular species or natural areas.

Once protests are accounted for, the next step, which is especially important with an open-ended response format, is to check for outliers. There was only one bid in the visitor sample that would be considered an outlier: an annual WTP of $500 per year, where the next highest amount was $300.[3] In the household survey the highest WTP was $180. Both of the remaining upper dollar amounts are well within the household's payment capacity, as the sample average *household* income was $50,000.[4]

Table 1 presents the mean WTP and standard error of the mean for the visitor and household samples. Each sample was asked their WTP for both a 50% and a 100% increase in whale populations. As would be expected, the average WTP values for general households are lower than for current visitors. In both cases the WTP estimates are measured fairly precisely, with the coefficients of variation for mean WTP (standard error/mean) being 6 or 7% of the means in the household survey and about 4% in the visitor survey. These standard errors are generally

[3] The $500 WTP response was dropped from the empirical analysis that follows. However, none of the results of the hypothesis tests reported below for visitors were changed by inclusion of this $500 WTP amount.

[4] While by U.S. standards this average household income may seem high, it is fairly close to the average California household income of $48,100 in 1991 (based on updating the 1990 Census estimate for California).

smaller than what others (Walsh *et al.*, 1984; Desvousges, *et al.*, 1992; Loomis, *et al.*, 1993) have found in other open-ended CVM surveys. Table 2 presents the WTP distribution of visitors and households for the two increases in whale populations.

Table 3 presents the results of the hypothesis tests concerning mean WTP for increments n = 50% and 2n = 100% above current gray whale populations. With respect to whether the incremental valuations are statistically different from zero, we can reject the hypothesis of equality with zero for both the 50% and 100% changes in gray whale populations for both visitors and general households. The t statistics are all significant at well below the 1% level.

Since each individual surveyed provided WTP for the 50% and 100% increment, the second hypothesis test involves a paired t-test. Thus we test whether their *difference* in expressed WTP for the 50% and 100% increments is different from zero. That is, by forming the WTP difference $D = WTP_{100} - WTP_{50}$ we test whether $D \leq 0$. The t statistics reported in Table 3 are significant at the 1% level, indicating the incremental valuation from a 50% to a 100% increase is statistically different from zero.

While the incremental valuations are statistically different from zero, the third consistency test requires that the second increment (D) to be of less value than the first increment (WTP_{50}). Rearranging and collecting like terms allows us to use a paired t-test of $2(WTP_{50}) - WTP_{100} > 0$. The test statistics reported in Table 3 are statistically significant, indicating the marginal value of the public good in our study is decreasing. That is, we reject the null hypothesis that the valuation for the second 50% increase is greater than or equal to the value of the first 50% increase for both the visitor and household samples.

Thus results from our case study suggest that for substantial increases in wildlife populations that are above current (and minimum viable) levels, there appears to be a statistically significant positive but diminishing marginal valuation for individuals. This pattern of results was found for both current gray whale

Table 2
Distribution of WTP of Visitors and Households

WTP	Number of Visitors		Number of Households	
	50% change	100% change	50% change	100% change
$0	107	121	195	216
$10	201	145	145	115
$20	127	121	79	63
$30	119	117	60	65
$40	15	38	15	24
$50	71	74	29	34
$60	6	10	2	2
$80	10	10	3	5
$100	39	48	18	25
$150	7	15	3	4
$200	1	2	1	1
$250	0	1	0	0
$300	1	2	0	0

Table 3
Hypothesis Tests Concerning WTP for Increases in Whale Populations

Null Hypothesis	Visitor Sample		Household Sample	
50% Increase				
$WTP_{50} = 0$	Reject	t = 21.45	Reject	t = 15.15
100% Increase				
$WTP_{100} = 0$	Reject	t = 21.31	Reject	t = 15.65
$WTP_{50} = WTP_{100}$	Reject	t = 9.14	Reject	t = 4.60
WTP 1st 50% gain =				
WTP 2nd 50% gain	Reject	t = 17.78	Reject	t = 12.47

viewers and for a sample of households, the majority of which were non-users and is consistent with expectations from basic principles of consumer demand theory.

There are of course several qualifiers to our results. First, we tested our hypotheses by comparing responses by the same individuals to the two increases in whale populations. Had we asked one sample their WTP for 50% increase and another sample their WTP for a 100% increase, the results of the hypothesis tests might have been different. In addition, our WTP values for the 100% increase in population may be conditioned on the responses given to the 50% increase. Since only one survey treatment for each population was performed, we do not have the ability to test for the significance of order effects. These are important methodological issues that should be evaluated in future research.

Comparison of Results to Previous Studies

Samples and Hoyller (1990) used in-person interviews of Hawaii residents and a dichotomous choice format. A one-time payment to prevent a loss of Humpback whales in Hawaii was elicited from two relatively small (n = 65 to 72) samples. They obtained values for general households ranging from $125 to $142, depending on the survey treatment. Annualized at 10% and updated for inflation, this value is somewhat higher than our annual estimates. In part this may be due to being asked study about their WTP to avoid a complete loss of humpback whales (Samples and Hoyller, 1990:183). Hagemann's mail survey of California households has some similarity to ours in terms of sample frame and evaluation of large changes in whale populations. Using a payment card, Hagemann (1985:72) obtained an annual value of $27 for gray whales in 1984 dollars or $36 in 1992 dollars to avoid a reduction in whales from their 1984 population of 16,000 to 1,300. However, there are some differences that may partially explain why she obtained higher values for the general population that we did. First, she asked WTP to avoid a 90% reduction in gray whale populations, rather than an increase in populations as we did. She also had a relatively low response rate (21%) which may imply only very interested respondents with correspondingly higher values. Compared to both studies, our value estimates for general households are relatively low, but similar in magnitude, and the differences may well be ascribed to the relative magnitudes and direction of changes in whale populations between the three studies.

Our annual values for doubling whale populations are lower than what Olsen, *et al.* found for doubling anadromous fish runs in the Columbia River basin. His non-users had values averaging $27 while users had values of $74. Of course his users, were consumptive users (*e.g.,* anglers) while our users where viewers. This comparison suggests that CVM studies of different marine resources do obtain values that seem to vary with the intensity of interest the public and users have in the resource.

Conclusions

At the level of the individual respondent, this research found that non-marginal increases in whale populations above current and minimum viable levels resulted in a statistically significant positive and diminishing marginal valuation. This pattern was found in samples of both current users (whale viewers during winter 1991-92) and of general households in California, which consists mainly of nonusers. The results suggest that carefully-performed CVM studies can obtain results consistent with principles of consumer demand theory for reasonably large changes in the quantity of a public good. Thus this study provides evidence that CVM can provide useful information on societal benefits from maintaining and increasing wildlife populations. The robustness of our conclusions should be tested in future research for order effects and use of independent samples for each increment, however.

In addition, the results suggest that citizens do derive a benefit from having a larger stock of whales, but from the standpoint of marine policy the costs (especially opportunity costs) of increasing whale populations would need to be considered before concluding that expanding populations beyond current levels is warranted on the grounds of economic efficiency.

References

Arrow, Kenneth, Robert Solow, Paul Portney, Edward Leamer, Roy Radner, and Howard Schuman. 1993. Report of the NOAA Panel on Contingent Valuation. Federal Register 58(10):4601–4614.

Augustyniak, Christine, Carol Jones, and Normal Meade. 1992. Nobel Laureat Panel on Contingent Valuation. Damage Assessment Center, National Oceanic Atmospheric Administration, Rockville, MD.

Bailey, Kenneth. 1987. Methods of Social Research. Free Press, New York, NY.

Bradford, David. 1970. Benefit-Cost Analysis and Demand Curves for Public Goods. Kyklos 23:775–791.

Brookshire, David, William Schulze, Mark Thayer, and Ralph d'Arge. 1982. Valuing Public Goods: A Comparison of Survey and Hedonic Approaches. American Economic Review 72:165–177.

Brookshire, David, Alan Randall, and John Stoll. 1980. Valuing Increments and Decrements in Natural Resource Service Flows. American Journal of Agricultural Economics 62:478–488.

Cooper, Joseph, and John Loomis. 1991. Economic Value of Wildlife Resources in the San Joaquin Valley: Hunting and Viewing Values. in Economics and Management of Water and Drainage, edited by A. Dinar and D. Zilberman. Kluwer Academic Publishers, Boston, MA.

Cummings, Ronald, David Brookshire, and William Schulze. 1986. Valuing Environmental

Goods: An Assessment of the Contingent Valuation Method. Rowman & Allenheld. Totowa, NJ.

Daubert, John, and Robert Young. 1981. Recreational Demands for Maintaining Instream Flows: A Contingent Valuation Approach, American Journal of Agricultural Economics 63(4):666–676.

Desvousges, William, F. Reed Johnson, Richard Dunford, Kevin Boyle, Sara Hudson, and K. Nicole Wilson. 1992. Measuring Nonuse Damages Using Contingent Valuation. Research Triangle Monograph 92-1. Research Triangle Park, NC.

Hagemann, Rhonda. 1985. Valuing Marine Mammal Populations: Benefit Valuations in a Multi-Species Ecosystem. Administrative Report LJ85-22. Southwest Fisheries Center, National Marine Fisheries Service, La Jolla, CA.

Hoehn, John. 1990. Valuing Multidimensional Impacts of Environmental Policy: Theory and Methods. American Journal of Agricultural Economics 73(2):544–551.

Hoehn, John, and John Loomis. 1993. Substitution Effects in the Valuation of Multiple Environmental Programs, Journal of Environmental Economics and Management 25(1):56–75.

Kealy, Mary Jo, John Dovidio, and Mark Rockel. 1988. Accuracy in Valuation is a Matter of Degree. Land Economics 64:158–170.

Kopp, Ray. 1992. Why Existence Values Should Be Used in Benefit-Cost Analysis. Journal of Policy Analysis and Management 11(1):123–130.

Larson, Douglas. 1993. On Measuring Existence Value. Land Economics 69(4):377–388.

Loomis, John. 1987. Expanding Contingent Value Sample Estimates to Aggregate Benefit Estimates. Land Economics 63(4):396–402.

Loomis, John. 1989. Test-Retest Reliability of Contingent Valuation: A Comparison of General Population and Visitor Responses. American Journal of Agricultural Economics 71(1):76–84.

Loomis, John. 1990. Comparative Reliability of the Dichotomous Choice and Open-Ended Contingent Valuation Techniques. Journal of Environmental Economics and Management 18:78–85.

Loomis, John, and Joseph Cooper. 1990. Comparison of Environmental Quality-Induced Demand Shifts Using Time-Series and Cross-Section Data. Western Journal of Agricultural Economics 15(1):83–90.

Loomis, John, Michael Lockwood, and Terry DeLacy. 1993. Some Empirical Evidence on Embedding Effects in Contingent Valuation of Forest Protection. Journal of Environmental Economics and Management 24:45–55.

Olsen, Darryll, Jack Richards, and R. Douglas Scott. 1991. Existence and Sport Values for Doubling the Size of Columbia River Basin Salmon and Steelhead Runs. Rivers 2(1): 44–56.

Randall, Alan, and John Stoll. 1983. Existence Value in a Total Valuation Framework. in R. Rowe and L. Chestnut, eds. Managing Air quality and Science Resources at National Parks and Wilderness Areas. Westview Press, Boulder, CO.

Reiling, Stephen, Kevin Boyle, Marcia Phillips, and Mark Anderson. 1990. Temporal Reliability of Contingent Valuation. Land Economics 66:128–134.

Samples, Karl, and James Hollyer. 1990. Contingent Valuation of Wildlife Resources in the Presence of Substitutes and Complements. in R. Johnson and G. Johnson, editors. Economic Valuation of Natural Resources. Westview Press, Boulder, CO.

Walsh, Richard, John Loomis, and Richard Gillman. 1984. Valuing Option, Existence and Bequest Demands for Wilderness. Land Economics 60(1):14–29.

Welsh, Michael. 1986. Exploring the Accuracy of the Contingent Valuation Method: Comparisons with Simulated Markets. Department of Agricultural Economics, University of Wisconsin, Madison.

[25]

Marine Resource Economics, Volume 8, pp. 119–132
Printed in the USA. All rights reserved.

Total Economic Values for Coastal and Marine Wildlife: Specification, Validity, and Valuation Issues

JOHN C. WHITEHEAD

Department of Economics
East Carolina University
Greenville, North Carolina 27858

Abstract *Benefit-cost analysis of coastal and marine wildlife management programs requires that economic benefits be monetized for comparison with the costs of preservation. Without explicit measurement and consideration of nonuse values, benefits may be underestimated and resources devoted to wildlife programs may be underallocated. Using data from a contingent valuation survey of nongame wildlife programs in coastal North Carolina, this paper provides additional evidence that total economic values under uncertainty for wildlife are theoretically valid. Specification error is found for valuation models which do not include measures of uncertainty. Specification error can lead to errors in benefit estimation.*

Keywords contingent valuation, option price, specification, validity.

Introduction

Benefit-cost analysis can be used to identify whether preservation of threatened or endangered wildlife resources is efficient. Identification of efficient policy requires that all benefits and costs of wildlife preservation, including the benefits to users and nonusers of nongame wildlife, be monetized and incorporated in the policy decision. The contingent valuation method (CVM) allows empirical measurement of total economic value, including both use and nonuse values, for wildlife resources. Nonconsumptive use and nonuse values measured by contingent valuation are significantly greater than zero, indicating that nongame wildlife resources are scarce economic goods.[1]

The appropriateness of benefit-cost analysis for wildlife resources has been

This research project was supported under Contract Number 90SG03 with funds from the North Carolina Nongame and Endangered Wildlife Fund. The analysis and evaluation offered in this paper is that of the author. They should not be interpreted as reflecting policy of the North Carolina Nongame Wildlife Program. Personnel of East Carolina University's Survey Research Laboratory designed the sampling and implemented the mail survey. This paper has benefitted from the constructive comments and insights of Richard Ready, two anonymous referees, and the editor.

[1] Benefit estimates range from about $1 to $75 per household/individual depending on the wildlife species, characteristics of the preservation policy, and type of survey. See, for example, Brookshire, *et al.* (1983), Boyle and Bishop (1987), Bowker and Stoll (1988), Samples and Hollyer (1989), Stevens, Glass *et al.* (1991), and Stevens, Echeverria, *et al.* (1991). Loomis and Walsh (1986) and Gregory, *et al.* (1989) provide thorough reviews of this literature.

debated. Kellert (1984) argues that economic methods are not sufficient to measure the total economic value of wildlife. Loomis and Walsh (1986) counter that developments in nonmarket valuation methods, namely the CVM, allow total economic values to be estimated. Recently, the use of benefit-cost analysis for wildlife programs has been questioned again, this time on the grounds that measures of nonuse value lack validity (Stevens, Echeverria, *et al.* 1991; Stevens, Glass, *et al.* 1991).

Contingent valuation studies of threatened and endangered wildlife have proceeded assuming conditions of certainty in demand for and supply of wildlife (Boyle and Bishop, 1987; Bowker and Stoll, 1988; Samples and Hollyer, 1989; Stevens, Echeverria, *et al.* 1991; and Stevens, Glass, *et al.* 1991). Uncertain wildlife populations, however, make consumer choice and valuation uncertain. Attempts to estimate the determinants of total economic values for wildlife resources without controlling for uncertainty may result in specification error (Whitehead 1992). Specification error from omitted uncertainty variables may bias validity tests and valuations.

The purpose of this paper is to explore the theoretical validity and empirical specification of contingent valuation models and value estimates for wildlife resources under conditions of supply and demand uncertainty. This extends work by Brookshire, *et al.* (1983) and Samples, *et al.* (1986). Theoretical validity assesses whether a measure of a theoretical construct, such as total economic value under uncertainty, behaves according to theoretical predictions. Formal validity tests require at least two steps (Carmines and Zeller 1979). First, the theoretical relationship between total economic value and variables which influence value were set forth. Neoclassical theory of consumer behavior is relied upon to specify theoretical relationships and empirical specifications. Secondly, empirical testing of these relationships were conducted using contingent valuation data from a survey assessing nongame wildlife programs in coastal North Carolina. Total economic values were then estimated to determine if they are consistent with theory.

Total Economic Value

Total economic value is the sum of use and nonuse values. Economic values which arise from recreation activities in which the wildlife resource is not harvested, such as wildlife observation or photography, are called nonconsumptive use values. Nonuse values may accrue to a potentially larger portion of the population who feel better off from knowing that nongame wildlife is preserved, even if they never travel to the wildlife habitat area to pursue nonconsumptive recreation.

With no uncertainty, the total economic value of a wildlife resource can be measured by the amount of money that an individual would be willing to pay to maintain wildlife populations. Total economic value can be modelled by a comparison of utility functions

$$v(G,y\text{-WTP}) = v(0,y) \tag{1}$$

where $v(\cdot)$ is the indirect utility function, G is the wildlife resource population, y is income, and WTP is willingness to pay which is a measure of the total economic value. Extinction of the wildlife resource is indicated by $G = 0$.

Economic values associated with threatened or endangered wildlife are likely to contain much uncertainty. For instance, threatened or endangered wildlife listings provides information that the risk of species extinction is greater than zero. Supply uncertainty occurs when it is indeterminate whether the wildlife resource will continue to exist so that it can be enjoyed by recreational users and nonusers. For wildlife users, demand uncertainty occurs when it is indeterminate whether recreational use of the wildlife resource will be pursued because of uncertain travel costs, income, or tastes. For nonusers, demand uncertainty depends on uncertain tastes.

The theoretical construct of total economic value for threatened or endangered wildlife should contain measures of uncertainty.[2] Without a wildlife protection program q_2 is the probability of supply and $q_1 = 1 - q_2$. With a protection program r_2 is the probability of supply, $r_1 = 1 - r_2$ and $r_2 > q_2$.

Expected indirect utility without the protection program is

$$E(v_q) = q_1 v(0,y) + q_2 v(G,y) \tag{2}$$

Expected indirect utility with the protection program is

$$E(v_r) = r_1 v(0,y) + r_2 v(G,y) \tag{3}$$

With the simplifying assumption that preferences are state-dependent and prices and income are known with certainty, the source of demand uncertainty is uncertain tastes. The probability of demand is p_2 where $p_1 = 1 - p_2$. With no protection program, expected indirect utility is

$$E(v_q) = p_1[q_1 v(0,y) + q_2 v(G,y)] + p_2[q_1 v(0,y) + q_2 v(G,y)] \tag{4}$$

Following Freeman (1985), when the wildlife resource is not demanded $v(0,y) = v(G,y)$ and

$$E(v_q) = p_1 v(G,y) + p_2[q_1 v(0,y) + q_2 v(G,y)] \tag{5}$$

The total economic value of the protection program under supply and demand uncertainty is the option price (OP), $E(v_r, OP) = E(v_q)$

$$p_1 v(G, y - OP) + p_2[r_1 v(0, y - OP) + r_2 v(G, y - OP)]$$
$$= p_1 v(G,y) + p_2[q_1 v(0,y) + q_2 v(G,y)] \tag{6}$$

Option price is the *ex-ante* maximum amount of money an individual would be willing to pay to increase the supply probability under conditions of supply and demand uncertainty.

---

[2] Inclusion of supply and demand uncertainty in the theoretical model follows Freeman (1985).

The Contingent Valuation Survey

The CVM requires that a contingent market be presented to survey respondents using mail, in-person, or telephone survey instruments. In order to generate reliable and valid measures of economic value a contingent market for wildlife preservation with uncertainty must contain (1) information about uncertainty, (2) the policy scenario (the baseline preservation level of wildlife and proposed increments in preservation), (3) market institutions, such as the payment rule and policy implementation rule, and (4) a value elicitation question (Mitchell and Carson 1989).[3]

Information about Demand and Supply Uncertainty

The survey obtains information about demand uncertainty with the question: "What are the chances that you will visit coastal North Carolina in the future with a purpose of observing or photographing threatened or endangered wildlife species?" Answers range from "definitely will not" ($p_2 = 0.00$, where p_2 is the demand probability) to "definitely will" ($p_2 = 1.00$) with intermediate answers coded with 0.25, 0.50, and 0.75 demand probabilities.

Following the description of the sea turtle program (see below) questions were presented concerning species extinction. The respondents' subjective supply probability was gauged with answers to the question: "*In your opinion,* what do you think the chances are that the loggerhead sea turtle will become extinct within the next 25 years?" Answers range from "definitely will become extinct" (coded $q_1 = 1.00$, where q_1 is the extinction risk) to "definitely will not become extinct" (coded $q_1 = 0.00$), with intermediate answers coded with 0.75, 0.50, and 0.25 risks. The answer to this question serves as the pre-program supply probability ($q_2 = 1 - q_1$).[4]

Descriptions of the Policy Scenarios

Loggerhead Sea Turtle

Each survey respondent was presented with two contingent markets in a mailed questionnaire. The first market presents a preservation program for the loggerhead sea turtle (*Caretta caretta*). Respondents were informed about the current status of and threats to loggerhead sea turtle nesting habitat in North Carolina (see Thompson 1988). Respondents were then told about a "Loggerhead Sea Turtle Preservation Fund" with money used to manage loggerhead sea turtle nesting habitat. One-half of the respondents were asked to assume that with the manage-

---

[3] A complete version of the survey instrument is available upon request from the author. An attempt to avoid compliance bias (Mitchell and Carson, 1989) was made by stating that the survey was being conducted by a university on the cover of the survey booklet.

[4] Respondents will state option price based on demand and supply uncertainty. Due to survey space limitations, elicitation of both subjective demand and supply information for both contingent markets was precluded. The demand uncertainty variable for coastal nongame wildlife is used as a proxy variable in the loggerhead sea turtle model and supply uncertainty variable for the loggerhead sea turtle market is used as a proxy variable in the coastal nongame wildlife models.

ment program the loggerhead sea turtle will definitely not become extinct within the next 25 years ($r_2 = 1.00$, where r_2 is the supply probability after the program). The other half were asked to assume that with the management program the loggerhead sea turtle will probably not become extinct within the next 25 years ($r_2 = 0.75$).

Coastal Nongame Wildlife

The second contingent market presents a preservation program for all threatened or endangered species in coastal North Carolina, including the loggerhead sea turtle. Survey respondents were informed that wetlands, forests, and beaches are wildlife habitat areas that support populations of species listed by the North Carolina Nongame Wildlife Program (1990) as state threatened or endangered species.[5]

After several existing threats to coastal nongame wildlife were mentioned, a "Coastal Nongame Wildlife Preservation Fund" and program was described. Money from the Fund would be used to manage nongame wildlife habitat along the North Carolina coast. Again, one-half of the respondents were asked to assume that with the management program coastal nongame wildlife will definitely not become extinct within the next 25 years ($r_2 = 1.00$). The other respondents were asked to assume that with the management program coastal nongame wildlife will probably not become extinct within the next 25 years ($r_2 = 0.75$).

Valuation Questions

Following description of the loggerhead sea turtle preservation policy, respondents were asked if they would be willing to donate a dollar amount ($\$A$ = either 1, 5, 10, 25, 50, or 100) to preserve loggerhead sea turtles:

"Suppose that a $\$A$ contribution from each North Carolina household each year would be needed to support and fund the loggerhead sea turtle program. Would you be willing to contribute $\$A$ each year to the 'Loggerhead Sea Turtle Preservation Trust Fund' in order to support the loggerhead sea turtle program?"

Respondents answer "yes" or "no" to the dichotomous choice question. A follow up question contained categories of reasons for the response to the contingent market.

Another dichotomous choice valuation question follows description of the coastal nongame wildlife protection program:

"Suppose that a $\$A$ contribution from each North Carolina household each year would be needed to support and fund the nongame wildlife management program. Would you be willing to contribute $\$A$ each year to the 'Coastal

---

[5] The species mentioned are the Dismal Swamp Southeastern shrew (*Sorex Longirostric fisheri*), red-cockaded woodpecker (*Picoides borealis*), Southeastern bald eagle (*Haliaeetus leucocephalus*), American alligator (*Alligator misissippiensis*), Carolina salt marsh snake (*Nerodia sipedon williamengelsi*), Cape Fear shiner (*Notropis mekistocholas*), Outer Banks kingsnake (*Lampropettis getutus sticticeps*), the piping plover (*Charadrius meldus*), and the loggerhead sea turtle. The contingent market makes clear that the representative species mentioned for protection are only a few of those found in coastal North Carolina that would be affected by the management program.

Nongame Wildlife Preservation Trust Fund' in order to support the nongame wildlife management program?''

Respondents may again answer "yes" or "no." The dollar amount variable ($A) was the same in both contingent markets.[6]

Market Institutions

The contribution payment rule was chosen since it is similar to the North Carolina Nongame and Endangered Wildlife Fund and would be familiar to survey respondents. Survey respondents are familiar with this Fund because information about the Nongame and Endangered Wildlife Fund is presented in the state tax instruction booklet each year. The implicit policy decision rule is that if a sufficient amount of contributions are received the management programs will be implemented. The payment and implementation rules suggest a payment obligation and provide incentives for truth telling which minimizes strategic behavior (Mitchell and Carson, 1989, Hoehn and Randall, 1987).[7]

Specification of Validity Tests

Theoretical validity tests based on the yes or no responses can be specified with supply and demand uncertainty. A dichotomous choice CV question presents the problem: "would you pay $A for q_2 to r_2?" which creates the choice problem $E(v_r,A) >/< E(v_q)$. Substitution of A for OP in equation (6) yields

$$p_1v(G,y - A) + p_2[r_1v(0,y - A) + r_2v(G,y - A)] \gtrless p_1v(G,y)$$

$$+ p_2[q_1v(0,y) + q_2v(G,y)] \tag{7}$$

If $A > (<)$ OP then the respondent will answer "no" ("yes"). Let $\Delta v = E(v_r,A) - E(v_q)$. If $\Delta v < (>)$ 0 the respondent will answer "no" ("yes"). Subtraction yields

$$\Delta v = p_1v(G,y - A) + p_2[r_1v(0,y - A) + r_2v(G,y - A)] - \{p_1v(G,y)$$

---

[6] The same dollar amount was used in both markets in order to enhance the direct comparability of the total economic values revealed by the two markets. This sequential contingent market design presented two programs with increasing benefits at the same cost. This design is similar to markets with iterative bidding which present one program with constant benefits with increasing (or decreasing) costs. The two designs should be equally plausible. In discussions with noneconomist reviewers of the sequential markets it was apparent that the two scenarios in conjunction with each other were plausible. Dollar amounts were chosen after reviewing the CVM wildlife literature to determine the probable range of willingness to pay. Dollar amounts were randomly assigned to the surveyed households.

[7] Mitchell and Carson (p. 221) and an anonymous referee successfully argue, however, that a special fund is less familiar than payment vehicles such as higher prices and taxes in association with the referendum voting implementation rule. Incorporation of these recommended features of contingent market design would have increased the reliability of the results of this study by reducing the hypothetical nature of the contingent market.

Total Economic Values for Wildlife   *125*

$$+ p_2[q_1 v(0,y) + q_2 v(G,y)]\}$$ (8)

and since $p_1 = 1 - p_2$, $r_1 = 1 - r_2$, and $q_1 = 1 - q_2$

$$\Delta v = v(G,y - A) + v(G,y) + p_2\{(1 - r_2)[v(0,y - A) - v(G,y - A)]$$

$$+ (1 - q_2)[v(G,y) - v(0,y)]\}.$$ (8')

It can be shown that

$$\frac{\partial \Delta v}{\partial p_2} > 0, \frac{\partial \Delta v}{\partial r_2} > 0, \frac{\partial \Delta v}{\partial q_2} < 0.$$

The change in utility increases with the probability of demand and the probability of supply with the protection program. The change in utility decreases with increases in the probability of supply without the protection program.[8]

Empirical Methods

Logistic regression is used to estimate the probability of a yes response in both contingent markets (Amemiya 1981). Following Hanemann (1984) and Bowker and Stoll (1988), the probability of a yes response is

$$\pi(\text{yes}) = F[\Delta v > 0]$$ (9)

where $F(\cdot)$ is the probability function for the mean zero random error in Δv. Adopting the logistic regression technique the probability function

$$\pi(\text{yes}) = [1 + \exp(-\Delta v)]^{-1}$$ (10)

is employed.

The linear functional form for the indirect utility function where the public good and income are additively separable is often assumed

$$v(G,y) = \alpha_G + \beta y$$ (11)

where $\alpha > 0$ and $\beta > 0$. The contingent market provides the choice of two utility levels

$$v(G,y - A) = \alpha_G + \beta(y - A)$$ (12A)

$$v(0,y) = \alpha_o + \beta y$$ (12B)

The change in indirect utility with no uncertainty is

$$\Delta v = v(G,y - A) - v(0,y) = \alpha - \beta A$$ (13)

---

[8] The mathematical results are available upon request from the author.

where $a = \alpha_G - \alpha_o$ (see Hanemann 1984). Estimates of total economic value can be found by solving for the dollar amount that would make respondents indifferent to utility in (12A) and (12B) and $\Delta v = 0$ (Hanemann, 1984, 1989). Setting the probability of a yes response equal to .50 yields the willingness to pay estimate

$$WTP = a/\beta \tag{14}$$

A positive willingness to pay estimate requires the theoretically consistent results that $a > 0$ and $\beta > 0$.

From equation (8') with supply and demand uncertainty and a linear form for indirect utility[9]

$$\Delta v = a[p_2(r_2 - q_2)] - \beta A \tag{15}$$

where

$$\frac{\partial \Delta v}{\partial [p_2(r_2 - q_2)]} = a > 0, \frac{\partial \Delta v}{\partial A} = -\beta < 0.$$

This functional form, with the constant suppressed, is similar to that used by Edwards (1987). Setting the probability of a yes response equal to .50 yields the option price estimate

$$OP = a[p_2(r_2 - q_2)]/\beta \tag{16}$$

where

$$\frac{\partial OP}{\partial [p_2(r_2 - q_2)]} = \frac{a}{\beta} > 0.$$

It can also be shown that

$$\frac{\partial OP}{\partial p_2} > 0, \frac{\partial OP}{\partial r_2} > 0, \frac{\partial OP}{\partial q_2} < 0.$$

The option price for the protection program increases with the probability of demand and the probability of supply with the protection program. The option price decreases with increases in the probability of supply without the protection program. These results are available upon request.[10]

---

[9] Derivation of this result can be obtained from the author.
[10] For the linear form of Δv, the median option price is equal to the mean option price. Other functional forms of Δv do not have this characteristic. The log functional form of utility suggested by Hanemann (1984, 1989) and used by Edwards (1987) was also attempted. However, unexpected signs of coefficients resulted and further analysis of this form was dropped. The functional form derived from either Hanemann specification has the unwanted characteristic that the option price estimate is equal to zero if the demand probability is equal to zero. Option prices for certain nonusers are, therefore, underesti-

The Survey Data

The sampling frame is telephone directories of North Carolina households in both cities and rural areas. The sample of 600 households is weighted toward the North Carolina coastal plain in an attempt to increase the possibility of nonconsumptive use recreation by respondents. The mail survey was conducted following procedures described in Dillman (1978) during the Winter of 1991. After a postcard follow-up reminder and a follow-up mail survey, a response rate of 35% was achieved.[11] Table 1 presents the variables used in the empirical analysis along with variable means and standard deviations.[12] The sample is fairly representative of the North Carolina population (see Whitehead, 1991).

Empirical Results

Logistic regression equations with and without uncertainty and from both contingent markets are statistically significant according to the model chi-square statistic at the 95% confidence level (Table 2).[13] Coefficient estimates for the dollar amount, $A, variable are negative and significant at the 95% confidence level. As the cost of the preservation program increases the probability of a yes

---

mated. A log-linear first-order approximation of Δv, which included income, was also attempted. Option price estimates for certain nonusers are positive but less than $1. Since this functional form has the unwanted characteristic that the option price estimate is positive even when the change in supply probability is zero, these results are not presented.

[11] Contingent valuation mail surveys assessing wildlife resources tend to achieve lower response rates than the usual 40–60% cited by Loomis (1987). In addition to this study, see Brookshire, *et al*. (1983), Bowker and Stoll (1988), and Stevens, Echeverria, *et al*. (1991). The exception is Boyle and Bishop (1987) who achieve a substantially higher response rate. Because of the low response rate, nonresponse bias is a real concern. The most serious consequence of nonresponse bias is distortion of the aggregate benefit estimates (Dalecki, *et al*. forthcoming). When nonresponse bias is a suspected problem aggregate benefit estimates should be found using a weighting approach (Dalecki, *et al*.) or a conservative approach where nonrespondents are assigned a weight of zero (Whitehead 1991).

[12] Following the sea turtle valuation question, a follow up question was presented. The answers to this question legitimate 91% of the 222 total yes/no responses. The most important reason given (83%) for the yes responses was "survival of endangered species is important for the environment." Of those who responded yes, few value nongame wildlife for their outdoor recreation experiences (9%). One responsent answered "yes" but felt he/she "should not have to contribute to the Trust Fund." One respondent needed more information and three respondents provided other reasons in written form which did not contradict their response. The most important reason (40%) for a no response to the sea turtle question is "I can't afford to contribute to the Trust Fund." Twenty-six percent of no respondents needed more information. The 20 no respondents who felt they "should not have to contribute to the Trust Fund" were considered "protest no" responses. It is standard practice to delete protests from contingent valuation data analysis because they are not believed to be true no responses (Mitchell and Carson 1989). Thirty-four respondents provided other reasons for their no response in written form. The written reasons did not contradict their no response.

[13] The chi-squared test statistic is for the null hypothesis that all regression coefficients are equal to zero. It is calculated as $X^2 = -2[LL(0) - LL(B)]$ where $LL(0)$ is the value of the beginning log-likelihood function and $LL(B)$ is the value of the ending log-likelihood function.

Table 1
Description, Mean, and Standard Deviation of Variables Used in
Logistic Regressions

Variable	Description	Mean	Standard Deviation
$A	The requested payment	$32.34	34.42
r_2	Supply probability with the management program	0.90	0.12
q_2	Supply probability without the management program	0.43	0.23
p_2	Demand probability	0.51	0.29
$p_2(r_2 - q_2)$	Supply probability change weighted by the demand probability	0.25	0.23

response decreases. This result partially validates the contingent market response since survey respondents answered the valuation questions rationally.

Specification

Overall model specification can be explored using several statistical tests. In this study, the most appropriate overall goodness of fit comparison is with the percentage of correct predictions statistic.[14] The percentage of correct predictions slightly increases from the certainty model (Model 1) to the uncertainty model (Model 2) in both markets. This result is slightly misleading, however, since Model 1 predicts all no responses in the sea turtle market and 87% no responses in the nongame wildlife market. Over-prediction of no responses results from the low values of the intercepts in Model 1. The intercept is an estimate of a which should be positive to be consistent with theory. In both markets the intercept is insignificantly different from zero.

For the sea turtle market, the correctly predicted yes responses is 0% and the correctly predicted no responses is 100% using Model 1. For the nongame wildlife market, the correctly predicted yes responses is 24% and the correctly predicted no responses is 90% using Model 1. Using Model 2 the number of correctly predicted yes responses increases to 53% in the sea turtle market and to 52% in the nongame wildlife market. The number of correctly predicted no responses falls to 81% and 74% in the sea turtle and nongame wildlife markets, respectively. Since use of Model 1 would predict all or mostly no responses, Model 2 is a more appropriate specification.

Validity

Model 2 in both contingent markets includes measures of uncertainty. The coefficient on the uncertainty variable is of the expected sign and significantly differ-

---

[14] Results from both the model chi-square and McFadden's R^2 statistics, which both decrease as the uncertainty variables enter the model, are misleading. Suppression of the constant term in Model 2 causes the value of the beginning log-likelihood function, which is the log-likelihood with only the constant, to decrease in absolute value and make the test results ambiguous.

Table 2
Logistic Regression Results Parameter Estimate (Absolute Value t-Statistic)

Variable	Sea Turtle		Nongame Wildlife	
	Model 1	Model 2	Model 1	Model 2
Intercept	−0.205		0.055	
	(1.00)		(0.27)	
$A	−0.018*	−0.029*	−0.019*	−0.025*
	(3.25)	(5.05)	(3.52)	(4.88)
$p_2(r_2 - q_2)$		1.272*		1.447*
		(2.38)		(2.71)
Sample Size	202	202	202	202
% yes	33%	33%	38%	38%
% Correct				
Predictions	68%	72%	63%	65%
Beginning LL[a]	−151.87	−144.23	−149.95	−143.49
Ending LL	−121.02	−118.54	−126.66	−122.75
Total Economic				
Value	−$11.10	$10.98	$2.97	$14.74

* indicates significance at the 95% confidence level.
[a] LL = value of the log-likelihood function.

ent from zero at the 95% confidence level in both models. The larger the change in wildlife supply probability weighted by the demand probability the larger the potential effect of the proposed management program and the higher the probability of a yes response. The individual coefficient results partially validate the yes/no responses to the CV question on a theoretical level.

Valuation

Mean total economic values are calculated for each of the four models (Table 2). Without uncertainty in the valuation model, the willingness to pay estimate is −$11.10 for the loggerhead sea turtle program and $2.97 for the coastal nongame wildlife program. The negative willingness to pay result is consistent with the result of Bowker and Stoll (1988) when using the same functional form. Further, these estimates are not significantly different from zero since the numerator in the ratio that forms willingness to pay is insignificantly different from zero in both markets.

With measures of uncertainty in the valuation model the option price estimate is $10.98 for the loggerhead sea turtle program and $14.74 for the coastal nongame wildlife program.[15] These estimates are calculated at the mean of the change in

---

[15] Subjective probabilities were chosen for study since they are often found to be more relevant in explaining behavior, relative to objective probabilities. From prospective reference theory (Viscusi 1989, Smith 1992), a divergence between objective probabilities and perceived probabilities is often observed. Therefore, the choice of the uncertainty measure to use when calculating total economic value can have significant welfare implications (Smith 1992). In this study the subjective extinction probabilities are likely to be greater

wildlife supply probability weighted by the demand probability. The option price estimate becomes, more plausibly, positive as the uncertainty variables are included in the sea turtle model. In the nongame wildlife model, the option price estimate also increases. These results strongly suggest that failure to properly specify valuation models by including measures of uncertainty can lead to errors in estimates of total economic value for wildlife resources.[16]

Conclusions

Using data from a recent contingent valuation survey, this paper provides additional evidence that total economic values under uncertainty are theoretically valid. Validity tests are specified based on theoretical predictions from the theory of consumer choice under uncertainty. Empirical results from two contingent markets support the theoretical predictions. Specification error is found for valuation models which do not include measures of uncertainty. Estimates of total economic value suggest that specification error may also lead to errors in benefit estimation.

The results of this paper provide evidence that uncertainty must be considered in contingent valuation studies of threatened or endangered wildlife. Failure to include uncertainty may lead to erroneous conclusions about the validity of total economic values for wildlife resources. Exploration of the effects of demand and supply probabilities on validity, specification, and estimates of option prices may continue to support the feasibility of using total economic values, which include nonuse values, in benefit-cost analysis.

References

Amemiya, T. 1981. "Qualitative Response Models: a survey." *Journal of Economic Literature,* 19:1493–1536.

Bowker, J. M. and John R. Stoll. 1988. "Use of Dichotomous Choice Nonmarket Methods to Value the Whooping Crane Resource," *American Journal of Agricultural Economics,* 70:372–381.

Boyle, Kevin J. and Richard C. Bishop. 1987. "Valuing Wildlife in Benefit-Cost Analyses: a Case Study Involving Endangered Species." *Water Resources Research,* 23:943–950.

than objective measures of the risk of extinction (Viscusi, 1989) so that the option price estimates presented here will be greater than option price estimates calculated with objective probabilities.

[16] The option price for the coastal nongame wildlife program, which would include a sea turtle program, is larger than that for the sea turtle program. This increases confidence in the validity of the option price estimates since embedded values are not found. The embedding effect occurs if the option price for the sea turtle program is equal to the option price for the nongame wildlife program. This can be tested by comparing the regression models. Since the option price is an increasing function of the ratio of two regression coefficients, if the regression coefficients are significantly different in the expected direction, then the option prices are significantly different. The a coefficient from the nongame wildlife model is significantly larger than the a coefficient in the sea turtle model (t = 3.41). The absolute value of the β coefficient from the nongame wildlife model is significantly lower than the absolute value of the β coefficient in the sea turtle model (t = 7.50). These two results combine to show that the option price for the nongame wildlife program is significantly different from the option price for the sea turtle program.

Brookshire, David S., Larry S. Eubanks and Alan Randall. 1983. "Estimating Option Prices and Existence Values for Wildlife Resources," *Land Economics*, 59:1–15.

Carmines, Edward and Richard A. Zeller. 1979. *Reliability and Validity Assessment*, Sage University Paper Series on Quantitative Applications in the Social Sciences, Beverly Hills and London: Sage Publications.

Dalecki, Michael G., John C. Whitehead, and Glenn C. Blomquist. Forthcoming. "Sample Nonresponse Bias and Aggregate Benefits in Contingent Valuation: An Examination of Early, Late, and Non-Respondents," *Journal of Environmental Management*.

Dillman, Don A. 1978. *Mail and Telephone Surveys: the Total Design Method*, John Wiley and Sons: New York.

Edwards, Steven F. 1989. "Option Prices for Groundwater Protection," *Journal of Environmental Economics and Management*, 15:475–487.

Freeman, A. Myrick III. 1985. "Supply Uncertainty, Option Price, and Option Value," *Land Economics*, 61:176–181.

Gregory, Robin, Robert Mendelsohn, and Terry Moore. 1989. "Measuring the Benefits of Endangered Species Preservation: From Research to Policy," *Journal of Environmental Management*, 29:399–407.

Hanemann, Michael. 1984. "Welfare Evaluations in Contingent Valuation Experiments with Discrete Responses," *American Journal of Agricultural Economics*. 66:332–341.

Hanemann, Michael. 1989. "Welfare Evaluations in Contingent Valuation Experiments with Discrete Response Data: Reply," *American Journal of Agricultural Economics*. 71:1057–1061.

Hoehn, John P. and Alan Randall. 1987. "A Satisfactory Benefit-Cost Indicator from Contingent Valuation," *Journal of Environmental Economics and Management*, 14: 226–247.

Kellert, Steven. 1984. "Assessing Wildlife and Environmental Values in Cost-Benefit Analysis," *Journal of Environmental Management*, 18:355–363.

Loomis, John B. and Richard G. Walsh. 1986. "Assessing Wildlife and Environmental Values in Cost-Benefit Analysis: State of the Art. *Journal of Environmental Management*, 22:125–131.

Mitchell, Robert Cameron and Richard T. Carson. 1989. *Using Surveys to Value Public Goods*. Resources for the Future: Washington, D.C.

North Carolina Nongame Wildlife Program. 1990. Endangered Wildlife of North Carolina. mimeo, North Carolina Wildlife Resources Commission, Nongame and Endangered Wildlife Program, Raleigh, N.C., February 22.

Samples, Karl C., John A. Dixon, and Marcia M. Gowen. 1986. "Information Disclosure and Endangered Species Valuation," *Land Economics*, 62:306–312.

Samples, Karl C. and James R. Hollyer. 1989. "Contingent Valuation of Wildlife Resources in the Presence of Substitutes and Complements," in *Economic Valuation of Natural Resources: Issues, Theory and Application*, ed. by R. L. Johnson and G. V. Johnson, Boulder, CO: Westview Press.

Smith, V. Kerry. 1992. "Environmental Risk Perception and Valuation: Conventional Versus Prospective Reference Theory," in *The Social Response to Environmental Risk*, ed. by Daniel W. Bromley and Kathleen Segerson, Kluwer: Boston, MA.

Stevens, Thomas H., Jaime Echeverria, Ronald J. Glass, Tim Hager, and Thomas A. More. 1991. "Measuring the Existence Value of Wildlife: What do CVM Estimates Really Show?" *Land Economics*, 67:490–400.

Stevens, T. H., R. Glass, T. More, and J. Echeverria. 1991. "Wildlife Recovery: is Benefit-Cost Analysis Appropriate?" *Journal of Environmental Management*, 33:327–334.

Thompson, Nancy B. 1988. "The Status of Loggerhead. Caretta Caretta; Kemp's Ridley, Lepidochelys Kempi; and Green, Chelonia My   .S. Waters." *Marine Fisheries Review*, 50:16–23.

Viscusi, W. Kip. 1989. ''Prospective Reference Theory: Toward an Explanation of the Paradoxes,'' *Journal of Risk and Uncertainty,* 2:236–264.

Whitehead, John C. 1991. ''Measuring Public Benefits of Nongame Wildlife in Eastern North Carolina,'' Final Report submitted to North Carolina Wildlife Resources Commission, Nongame and Endangered Wildlife Program, Raleigh, NC.

Whitehead, John C. 1992. ''Ex Ante Willingness to Pay with Supply and Demand Uncertainty: Implications for Valuing a Sea Turtle Protection Program,'' *Applied Economics,* 24:981–988.

[26]

| Journal of Coastal Research | 12 | 1 | 171–178 | Fort Lauderdale, Florida | Winter 1996 |

Multicriteria Evaluation in Coastal Management

Amalia Moriki†, Harry Coccossis‡ and Michael Karydis‡

†Institute of Biology
National Center for Scientific
  Research "Demokritos"
153 10 Aghia Paraskevi
Athens, Greece

‡Department of Environmental
  Studies
University of the Aegean
Karadoni 17
81 100 Mytilini, Greece

ABSTRACT

MORIKI, A.; COCCOSSIS, H., and KARYDIS, M., 1996. Multicriteria evaluation in coastal management. *Journal of Coastal Research*, 12(1), 171–178. Fort Lauderdale (Florida), ISSN 0749-0208.

The driving forces of the development of the coastal area of the island of Rhodes (Greece) are examined in relation to the quality of the marine environment. The integrated variables of the system were analysed by four multidimensional techniques: Multicriteria Analysis, Cluster Analysis, Multidimensional Scaling and Principal Component Analysis. The analysis was performed in five zones of the island with different socio-economic structure and marine environmental quality. All these methods revealed the same trend in each zone suggesting the robustness of the multidimensional approach. The conflict between the quality of natural resources and the intensity of land use was expressed through the dissimilarity measures of these techniques, and therefore a quantification of the conflicts was obtained. In addition, a graphical illustration of the interrelated variables contributed to a better understanding of the system structure. The use of these algorithms is well established from applications in the fields of ecology and physical planning. They are therefore proposed as an effective analytical tool in problems where coastal development and marine quality are encountered. The combination of the multiple criteria, analytical techniques can contribute to the work of planners and decision-makers to identify the major trends of the system and quantify the interrelated socio-economic and environmental aspects.

ADDITIONAL INDEX WORDS: *Integrated analysis, multivariate methods, environmental management.*

INTRODUCTION

Coastal development in our century is accompanied by growing environmental problems and the need for evaluation and reorientation of the management strategies, taking into account environmental concerns.

Awareness of the conflict between coastal development and coastal environmental quality emerged as a priority issue in the 1970's (WCED, 1987) and led scientific research towards the study of environmental problems, the assessment of quality criteria and prediction of future trends. In planning and policy-making, the need for holistic approaches to coastal management is seriously considered. There still seems to be a gap between experimental work and decision-making, mainly because of the incompatibility of methods for the policy interpretation of scientific analysis. In this way, environmental-ecological aspects appear only theoretically in the actual decision-making process; huge data bases, originated from physical and chemical monitoring of the marine

93124 received 20 November 1993; accepted in revision 20 January 1995.

environment, remain unexplored, restricted to academic purposes only.

Environmental protection and sustainability require integration of environmental and socio-economic variables and the application of effective methodologies. However, complex problems involve multiple aspects and require analytical and systematic work to measure the degree of conflict and quantify the interrelations between them. Modelling has offered a tool for integrated management especially with scenario-predictive models and economic-ecological simulation. However, models require sophisticated techniques and need to be tested under real conditions. At the same time, coastal management requires effective decisions in a reasonable time-scale; therefore, holistic approaches should be based on realistic methods rather than complicated and time-consuming techniques. So far, multivariate statistical methods have been used extensively in the assessment of ecological impacts in natural sciences (WARWICK, 1986; CLARKE and GREEN, 1988), but without integrating socio-economic variables. Applications of multiple criteria methods, based on the assignment of scores in the multiple objectives

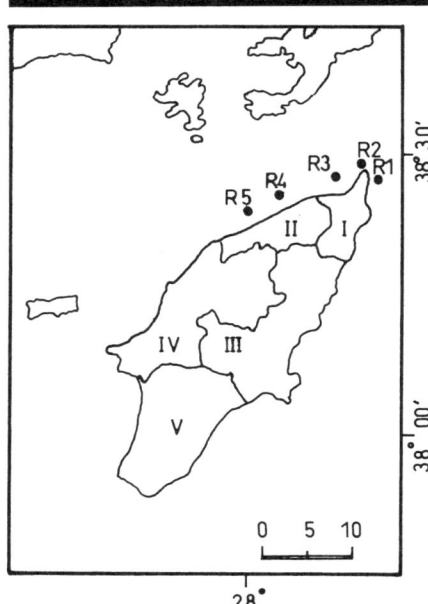

Figure 1. Map of the island of Rhodes, divided in five zones, and the sampling locations along the northwest coast.

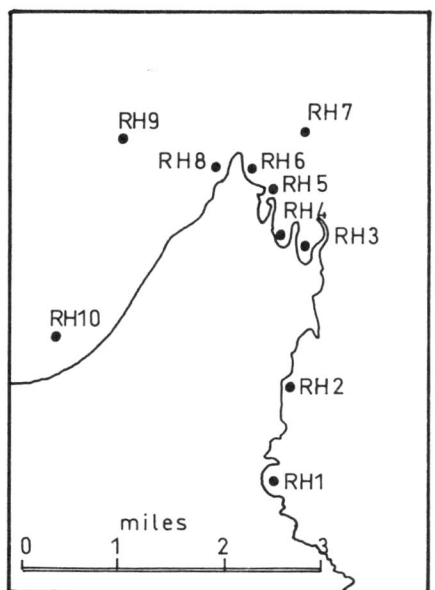

Figure 2. Map of the sampling locations along the coastal area of the city of Rhodes (KARYDIS and COCCOSSIS, 1990).

of a complex problem and the classification of the objectives in a ranking list, have been limited in the field of physical planning and policy-making (HARTOG *et al.*, 1989; NIJKAMP *et al.*, 1990).

In the present work, a case study of the island of Rhodes, Greece is presented, analysing simultaneously economic, social and ecological data of the coastal system. Four different multi-variable techniques were used: Multicriteria methods, the non-parametric method multidimensional scaling and two multivariate statistical methods: cluster analysis and principal component analysis. The aim of the study was to explore the potential of the island for development, maintaining the quality of the marine environment and to examine the effectiveness of the above methodologies in their application to holistic approaches in system analysis. The qualities and disadvantages of each technique are discussed as well as their use as a multi-method system for the support of environmental management.

METHODOLOGY

Description of the Study Area

The island of Rhodes (Greece) lies in the Eastern Mediterranean Sea and covers an area of 1,400 km² (Figure 1). The population of the island is approximately 110,000 persons (COCCOSSIS *et al.*, 1993), 50,000 of them inhabiting the city of Rhodes which is a major tourist resort. During the summer, Rhodes city and its surroundings receive more than a million visitors. Tourism is the main economic activity on the island; commerce, services and other tourist related activities are also growing.

The island is divided into five geographical zones by the Ministry of Planning and the Environment (COCCOSSIS *et al.*, 1993) named Zones 1–5 (Figure 1). Zone 1 is the northeastern area and includes the capital of the island, the city of Rhodes. This is the area where most of the tourist facilities (33,000 hotel beds of the 36,000 for the whole island), services and industrial-manufacture units are concentrated. Zone 2 is the northwest area. The airport is located in this zone, while many hotels and other tourist facilities are located along the beaches. Zone 3 is a rapidly growing tourist area which also has the largest concentration of manufacturing. The area represents a high percentage of agricultural land and shares with Zone 1 the largest part of cultural resources in the island. Zone 4 is the western mountainous part of the island, while Zone 5 presents the southern, relatively underdeveloped part of Rhodes. The last two areas represent a high priority of agricultural employment.

Collection of Data

Data on the socio-economic system have been obtained through the National Statistical Service of Greece (1986, 1991) and the Development Scenarios for Rhodes, Phase I, Appendix (COCCOSSIS *et al.*, 1991). Data on the quality of the marine environment concern nutrient values (nitrate and ammonia) as well as a biological parameter Chlα (VOUNATSOU and KARYDIS, 1991; KARYDIS and COCCOSSIS, 1990). Ten nearshore stations RH1– RH10 were spaced along the coastal area of the city of Rhodes (Figure 2) and five more stations in the north west coast of the island (Figure 1).

Manipulation of Data

There are two types of data: most of the variables are cardinal, but some of them such as the relief structure, the suprastructure and the cultural value, are measured in an ordinal scale, taking values from 3 to 0 (3 being the most favorable).

The values of nutrient and Chlα are numerical but have been given an opposite sign in order to conform to the idea "the greater the value, the more favored the criterion. In Zone 1 for example, the value for Chlα is 0.28 and appears in the table as -0.28.

All values have been standardised according to SNEATH and SOKAL (1973) prior to the analysis. Standardisation is a necessary procedure for both multicriteria analysis and multivariate procedures, so as to permit the comparison of variables with different dimensions.

Multicriteria Analysis: The Regime Method

The Regime Method is used for the classification of alternative sites; the five zones of Rhodes Island are classified according to the values of various criteria, which form the "Impact Matrix".

The criteria are the stated variables of the coastal system of Rhodes and are presented in the Impact Matrix of Table 1. After the pairwise comparison of the values of two alternatives for each criterion, a new matrix is formed by the signs of the output of the comparison. This output might take a positive value for some of the criteria and a negative for others. The relative sizes of the subsets of the criteria with negative and positive values are interpreted as the probabilities that Zone 1 dominates over Zone 2. The final ranking of the Zones is based on the calculation of the mean value of all probabilities for each Zone (HINLOOPEN and NIJKAMP, 1990). One of the advantages of the Regime Method is that it allows for different priorities among the criteria. This is very useful since some criteria are more important than others.

Classification

Numerical Classification (cluster analysis) was performed based on the euclidean distances among sampling locations:

$$ED_{jk} = \sum_{i=1} (X_{ij} - X_{ik})^2$$

The hierarchical sorting strategy for the development of dendrograms was the group average sorting. This method joins two groups of samples together at the average level of similarity between all members of one group and all members of the other (SNEATH and SOKAL, 1973). The dendrogram was obtained using Primer Programs (CLARKE, 1993).

Ordination

The non-metric multidimensional scaling (MDS) was applied on the similarity matrix developed for cluster analysis. MDS was originally developed in psychometrics but has found many applications in the social sciences as well as in ecological studies (FIELD *et al.*, 1982; WARWICK, 1986; KARYDIS, 1992). MDS uses proximities among samples to indicate how similar (or dissimilar) two samples are. The output is a graphical representation, consisting of a geometric configuration of points, making the comprehension of the structure of the data easier. Additional details and the software used have been given by CLARKE (1993).

Principal Component Analysis

Principal Component Analysis (PCA) is a well-known and popular statistical method (ANDERSON,

Table 1. *The values of the sixteen variables of the coastal system of Rhodes island, in the five zones.*

Variables	Zones				
	Zone 1	Zone 2	Zone 3	Zone 4	Zone 5
Social subsystem					
Population					
(thousands)	82	17	8	5	12
Services					
(thousands of employees)	10	0.80	1	0.40	1.10
Industry					
(thousands of employees)	1.90	0.19	0.31	0.04	0.30
Agriculture					
(thousands of employees)	0.38	0.13	0.27	0.72	0.79
Industrial annoyance	0	3	0	3	3
Tourism					
Hotels					
(thousand beds)	33	1.20	1.30	0.01	0.05
Distance	3	3	2	2	1
Marinas	3	0	0	0	0
Suprastructure	3	3	2	1	1
Cultural resources	3	1	3	1	0
Morphology					
Relief	3	3	2	0	2
Agricultural land					
(thousand stremmas)	49.80	43.50	66.90	45.60	67.60
Beaches					
(thousand stremmas)	0.90	6.40	1.80	0	2.70
Marine subsystem					
Chlorophyll a					
(μg-at/l)	−0.28	−0.10	−0.08	−0.08	−0.08
Nitrate					
(μg-at/l)	−2.25	−0.20	−0.15	−0.15	−0.15
Ammonia					
(μg-at/l)	−0.69	−0.54	−0.44	−0.44	−0.44

1984). It is applied in cases of large numbers of variables reducing their number by linear transformations. The Principal Components are the characteristic eigenvalues of the Covariance Matrix, of the original variables. PCA was performed with the statistical software STATGRAPHICS (1989).

RESULTS

Table 1 presents 16 criteria-variables of the five zones of the island of Rhodes (Figure 1). These variables concern the social subsystem, tourism-related activities, the geomorphology and the marine subsystem. The variables were selected to represent the major social, economic and ecological characteristics of the system of Rhodes island (Greece). Wherever possible, the values of each variable at each zone are quantitative. However, certain variables necessary for the analysis could not be expressed in quantitative terms and ordinal scaling has been used from 0 to 3, situation

0 representing the least favorable condition. The majority of population inhabit Zone 1; therefore, the population variable shows a great difference between Zone 1 and the remaining Zones. The percentage of employees in the tertiary sector and the number of hotels are about the same in each zone.

Table 2 presents the standardised values of Table 1 and forms the impact matrix for multicriteria analysis as well as the similarity matrix for cluster analysis and multidimensional scaling. Correlation Factors for the Principal Component Analysis were also calculated from the Matrix of Table 2.

Table 3 presents the results of the Regime Method. In Table 3(a), all criteria have been given equal priorities. Zone 1 is classified first in the ranking list, followed by Zone 5, Zone 3, Zone 2 and the mountainous Zone 4, in descending order. This analysis means that Zone 1, the most de-

Table 2. *The values of the variables of the coastal system of Rhodes after standardization.*

Number	Variable	Zone 1	Zone 2	Zone 3	Zone 4	Zone 5
(1)	Population	1.98	−0.27	−0.56	−0.69	−0.44
(2)	Services	1.99	−0.51	−0.45	−0.61	−0.42
(3)	Industry	1.99	−0.53	−0.35	−0.75	−0.37
(4)	Agriculture	−0.31	−1.27	−0.73	1.00	1.27
(5)	Industrial annoyance	−1.22	0.82	−1.22	0.82	0.82
(6)	Hotels	1.99	−0.46	−0.45	−0.55	−0.54
(7)	Distance	1.07	1.07	−0.26	−0.26	−1.60
(8)	Marinas	2.00	−0.50	−0.50	−0.50	−0.50
(9)	Suprastructure	1.18	1.18	0.00	−1.11	−1.11
(10)	Cultural resources	1.17	−0.50	1.17	−0.50	−1.33
(11)	Relief	0.39	0.39	−0.10	−0.59	−0.10
(12)	Agricultural land	−0.47	−1.07	1.17	−0.87	1.24
(13)	Beaches	−0.67	1.83	−0.25	−1.07	0.15
(14)	Chlorophyll	−1.99	0.31	0.56	0.56	0.56
(15)	Nitrate	−1.99	0.45	0.51	0.51	0.51
(16)	Ammonia	−1.83	−0.30	0.71	0.71	0.71

veloped at the present time, is the most favorable area from a development point of view, in cases where all variables of the socio-economic and ecological subsystems are of equal importance. In Table 3(b), the criteria that concern the primary sector have been given priority to reveal the po-

tential of each zone for agricultural development. Zones 3 and 5 are the most favorable for primary sector, followed by Zone 2. Zones 4 and 1 are the least important for agricultural development. In Table 3(c), priority has been given in the quality of the marine environment. In this case, Zone 5

Table 3. *Multicriteria analysis results, Regime Method (HINLOOPEN and NIJKAMP, 1990).*

Zones								Scores									Result

(a) Equal priorities

Criteria number

	16	15	14	13	12	11	10	9	8	7	6	5	4	3	2	1	
Z1	2	2	2	−0	−1	2	1	2	1	1	0	−0	−0	−2	−2	−0	0.926
Z5	−0	−0	−0	1	1	−1	−2	−1	−1	−1	−0	1	1	1	1	0	0.674
Z3	−1	−0	−0	−1	−1	−0	−0	−1	0	1	−0	1	0	1	1	0	0.583
Z2	−0	−1	−1	−1	1	−0	1	−1	1	−1	0	−1	0	0	0	−0	0.297
Z4	−1	−1	−1	1	1	−1	−0	−1	−1	−1	−1	−1	−2	1	1	0	0.020

(b) Priority to primer sector

Criteria number

	11	4	12	1	7	9	2	3	5	6	8	10	13	14	15	16	
Z3	−0	−1	1	−1	−0	0	−0	−0	−1	−0	−1	1	0	1	1	0	0.836
Z5	−0	1	1	−0	−2	−1	−0	−0	1	−1	−1	−1	1	1	1	0	0.836
Z2	0	−1	−1	−0	1	1	−1	−1	1	−0	−1	−1	0	0	0	0	0.506
Z4	−1	1	−1	−1	−0	−1	−1	−1	1	−1	−1	−1	−2	1	1	0	0.249
Z1	2	−0	−0	2	1	1	2	2	−1	2	2	1	−0	−2	−2	0	0.073

(c) Priority to the quality of the marine environment

Criteria number

	16	15	14	5	1	2	3	4	6	7	8	9	10	11	12	13	
Z5	0	1	1	1	−0	−0	−0	1	−1	−2	−1	−1	−1	−0	1	1	0.771
Z3	0	1	1	−1	−1	−0	−0	−1	−0	−0	−1	0	1	−0	1	0	0.747
Z2	−0	0	0	−1	−0	−1	−1	−1	−0	1	−1	1	1	0	−1	0	0.498
Z4	0	1	1	1	−1	−1	−1	1	−1	−0	−1	−1	−1	−1	−1	−2	0.484
Z1	−0	−2	−2	−1	2	2	2	−0	2	1	2	1	1	0	−0	−0	0.000

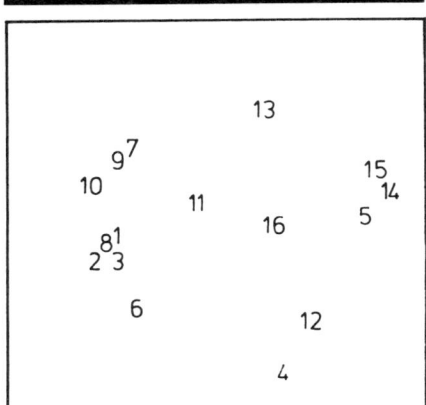

Figure 3. MDS plot for the 16 variables in the five zones of the island of Rhodes. The names of the variables are given in Table 2.

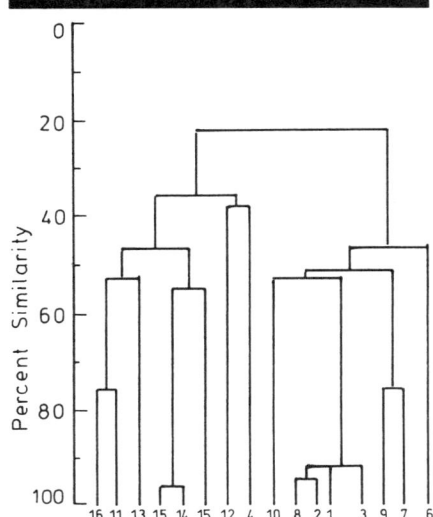

Figure 4. The classification pattern based on the euclidean distances among the 16 variables in the five zones of the island of Rhodes. The names of the variables are given in Table 2.

is the first in the ranking list, followed by Zone 3; Zone 2 is again of medium quality and Zone 1 takes the last position. An overview of Table 3 reveals that the most favourable Zone for a balanced development of primary sector and tourist activities with the protection of the marine environment is Zone 5, in the southern part of the island. Zone 3 can be classified as the second best; Zone 1 gains in order when criteria are equal because most of services and tourist activities take place in this area, but loses in primer sector and in ecological value. Zone 2 presents a medium profile, while Zone 4 is almost unfavored in all three cases.

Figure 3 presents the MDS plot for the 16 criteria in the five zones. Two aggregates can be observed, one to the right with criteria 15, 14 and 5 and one to the left with criteria 1, 2, 3, 8, 10, 9 and 7. The aggregate to the right concerns environmental quality while the aggregate to the left represents the intensity in the land use and the coexistence of services (2), population (1), industry (3) infrastructure and tourist activities (7, 8), suprastructure (9) and cultural resources (10). Criteria 4 (employees in primary sector) and 12 (agricultural land) seem to separate from the others in a third group.

Figure 4 presents the classification pattern based

on the euklidean distances among the 16 variables in the five zones of Rhodes island. Two clusters can be observed in 50% similarity in the dendrogram. As in the ordination pattern, one of them, the cluster to the left, concerns environmental quality, while services, population, suprastructure and touristic activities form the cluster to the right. A third cluster can be observed in the medium of the dendrogram consisting of criteria 12 and 4, thus the primary sector.

Figure 5 presents the Principal Components, where the variables appear in four aggregates. At the top and the right of the diagram, variables that concern population, hotels, marinas, services and industry are aggregated together. The variables that characterise the quality of the marine environment are found in the left quarter to the bottom, while the variables concerning suprastructure (9) and distance (7) are found in the right quarter to the bottom. The variables that refer to the primary agriculture sector (4), agricultural land (12) and beaches (13) are aggregated to the top and left of the diagram. The two axes can be interpreted as the environmental quality (the vertical axis) and the intensity of land use (the horizontal axis).

DISCUSSION

Coastal areas face increasing pressures for development which threaten coastal environmental resources. Planners and policy-makers have become aware of the need to integrate environmental issues in policy making. Recently, ecologically sustainable economic development has become a key issue in coastal management. The complexity of coastal systems requires the adoption of integrated coastal managing strategies focusing on the resolution of conflicts among the various uses of the coastal environment.

The island of Rhodes in Greece is a typical example of excessive concentration of multiple and conflicting activities in a small area. Monitoring studies in the area have shown eutrophication trends in the coastal waters of the city of Rhodes (KARYDIS, 1992; IGNATIADES *et al.*, 1992). Hence, an integrated approach to the application of ecologically sustainable policies in the island requires consideration of all the different aspects of the socio-economic and the marine environment.

The present work had four main purposes: a) to analyze the conflicts in coastal uses; b) to explore the potential for development of the different parts of the island; c) to examine the effectiveness of multivariate procedures and multiple criteria analysis as integrated tools in coastal management; and d) to assist the development of guidelines for future planning and decision-making in the area.

The island has been divided in five zones and analysis concerns the major socio-economic characteristics in these five areas as well as chemical and biological parameters of the marine environment that indicate coastal water quality. Multiple Criteria analysis by the Regime Method (HINLOOPEN and NIJKAMP, 1990) promoted the most developed area (Zone 1) in the highest position, when there was no particular concern for protection of the marine environment or for change in the present economic activities in the island. However, the analysis when priority is given to environmental protection reduces the final score of Zone 1 placing it in the worse position. The development of the primer sector could be a future challenge for the island, making it self-sufficient in agricultural production and keeping a balance in employment and economic activities.

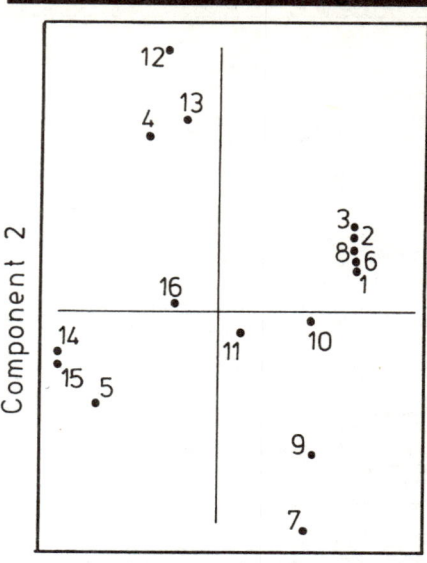

Figure 5. Representation of the variables of the system of Rhodes in two dimensions by Principal Component Analysis. The names of the variables are given in Table 2.

The first Zone is without doubt the least probable for agricultural development; tourism occupies the largest part of employment and agriculture is not important in this urban area. The results of the multicriteria analysis indicate that the fifth Zone, the southern area of the island, holds a high score in all three multicriteria trials with different priorities. Therefore, this area presents a land for future exploration in regard to tourist activities and agricultural development and allows for the application of an integrated policy. The third zone rates about the same; the fourth zone, the western and mountainous part of the island, does not present a development potential, *i.e.*, suprastructure, population and employment. The second Zone in the northwest, presents a moderate profile in the three multicriteria runs, and this can be interpreted as an indication that this area is going to develop the same way as the first Zone unless a different policy is applied.

Ordination and Classification techniques as well

as Principal Component Analysis have been used with the same data set as multicriteria analysis. The results of the three different methods show a clear distinction between intensity of land use and environmental quality on the island, probably as a result of the weak implementation or absence of land use controls. This evidence is supported by the findings of multicriteria analysis. In fact, the MDS plot, the dendrogram and the PCA plot resulted in two main aggregates, on the one hand the variables characterising the quality of natural resources and on the other the variables characterising the intensity of land use. Future policy for the development of the island should acknowledge the necessity to link these above two components in a coherent strategy of integrated management of coastal resources.

Multiple Criteria Analysis has proved to be a powerful tool for a systematic analysis of complex systems where quantitative and qualitative variables are involved. Multivariate techniques can assist in the illustration of existing patterns of interactions of the various components of an economic/environmental system and the identification of key factors in the development/environment interface. Therefore, both tools are valuable in policy-planning for multidimensional systems, for instance in coastal areas.

LITERATURE CITED

ANDERSON, J.W., 1984. *An Introduction to Multivariate Statistical Analysis.* New York: Wiley, 675p.

CLARKE, K.R., 1993. Non-parametric multivariate analyses of changes in community structure. *Australian Journal of Ecology,* 18, 117–143.

CLARKE, K.R. and GREEN, R.H., 1988. Statistical design and analysis for biological effects study. *Marine Ecology Progress Series,* 46, 213–226.

COCCOSSIS, H., KARYDIS, M., MARGARIS, N., PAPAGEORGIOU, G., PYRGIOTIS, Y., VERNICOS, N., and CHIOTIS, G., 1991. *Development Scenarios for Rhodes, Phase I Report.* Athens: University of the Aegean, Department of Environmental Studies. UNEP/MAP/BLUE PLAN.

COCCOSSIS, H., KARYDIS, M., MARGARIS, N., PAPAGEORGIOU, G., PYRGIOTIS, Y., VERNICOS, N., and CHIOTIS, G., 1993. *Development Scenarios for Rhodes, Phase III Report.* Athens: University of the Aegean, Department of Environmental Studies. UNEP/MAP/BLUE PLAN.

FIELD, J.G., CLARKE, K.R., and WARWICK, R.M., 1982. A practical strategy for analysing multispecies distribution patterns. *Marine Ecology Progress Series,* 8, 37–52.

HARTOG, G., HINLOOPEN, E., and NIJKAMP, P., 1989. A sensitivity analysis of multicriteria choice-methods. *Energy Economics,* October, 293–300.

HINLOOPEN, E. and NIJKAMP, P., 1990. Qualitative multiple criteria choice analysis, the dominant regime method. *Quality and Quantity,* 24, 37–56.

IGNATIADES, L., KARYDIS, M., and VOUNATSOU, P., 1992. A possible method for evaluating oligotrophy and eutrophication based on nutrient concentration scales. *Marine Pollution Bulletin,* 24(5), 238–243.

KARYDIS, M. and COCCOSSIS, H., 1990. Use of multiple criteria for eutrophication assessment of coastal waters. *Environmental Monitoring and Assessment,* 14, 89–100.

KARYDIS, M., 1992. Scaling methods in assessing environmental quality: A methodological approach to eutrophication. *Environmental Monitoring and Assessment,* 22, 123–136.

NATIONAL STATISTICAL SERVICE OF GREECE, 1986. *Distribution of the Country's Area by Basic Categories of Land Use.* Hellenic Republic, Athens.

NIJKAMP, P., RIETVELD, P., and VOOGD, H., 1990. *Multicriteria Evaluation in Physical Planning.* Amsterdam: North-Holland, 219p.

OFFICE NATIONALE DE STATISTIQUE DE GRECE, 1991. *Resultats du Recensement de la population et des Habitations, effectue le 5 Avril, 1981.* Republique Hellenique, Athens.

SNEATH, P.H.A. and SOKAL, R.R., 1973. *Numerical Taxonomy.* San Francisco: Freeman, 573p.

STATGRAPHICS (STATISTICAL GRAPHICS SYSTEM), 1989. *User's Manual.* Rockville, Maryland Statistical Graphics Corporation, 546p.

VOUNATSOU, P. and KARYDIS, M., 1991. Environmental characteristics in oligotrophic waters: Data evaluation and statistical limitations in water quality studies. *Environmental Monitoring and Assessment,* 18, 211–220.

WARWICK, R.M., 1986. A new method for detecting pollution effects on marine macrobenthic communities. *Marine Biology,* 92, 557–562.

WCED (WORLD COMMISSION ON ENVIRONMENT AND DEVELOPMENT), 1987. *Our Common Future.* Oxford: Oxford University Press.

Part V
Regional Seas

[27]

AN ACTION Plan
TO Clean
UP THE Baltic

By Janusz Kindler and Stephen F. Lintner

The largest body of brackish water in the world, the Baltic Sea, has severe problems of water pollution and environmental degradation.[1] As late as 1950, the Baltic Sea was still regarded as environmentally "healthy." Since that time, however, the situation has changed dramatically. In spite of numerous protective efforts, the Baltic Sea's rich biodiversity is threatened by environmental pollution that could cause irreversible damage to a sea that is an important source of economic activity and recreation for more than 80 million people who live along its coast and within its catchment area.

The Baltic Sea catchment area, or drainage basin, is home to 14 nations (see Figure 1 on page 9). Nine of them share the Baltic's coastline: Sweden and Finland to the north; Russia, Estonia, Latvia, and Lithuania to the east; Poland to the south; and Germany and Denmark to the west. The catchment area also includes parts of Belarus, Norway, Ukraine, the Slovak Republic, and the Czech Republic because some

JANUSZ KINDLER served as vice chairman of the Helsinki Commission's ad hoc high-level task force and is a professor of water and environmental systems at the Warsaw Technical University in Poland. He is currently a water resources planning specialist at the World Bank in Washington, D.C. STEPHEN F. LINTNER served as the World Bank's representative to the Helsinki Commission's ad hoc high-level task force and is currently its representative to the Programme Implementation Task Force. He is also a senior environmental specialist at the World Bank. The views presented in the article are solely those of the authors and should not be attributed to the World Bank or the Helsinki Commission.

of the rivers flowing into the Baltic find their sources there. Many of these countries are undergoing dramatic political, economic, and social changes. As the division of Europe into Western and Eastern blocs vanishes and the countries in the eastern and southern Baltic region lose their peripheral position, the region has a chance of regaining its historical position as one of the major crossroads of economic exchange, as in the time of the Hanseatic League—a mercantile organization of towns along the Baltic Sea and the North Sea from the 13th through 18th centuries that allowed trade to flow across the Baltic Sea without political obstacles. Certainly, such present cooperation as the recent establishment of the Union of the Baltic Cities and the Baltic Council of Ministers of Foreign Affairs confirms this possibility (see the box on pages 12 and 13).

The Baltic Sea is important in several ways. Its coast is a popular vacation area; therefore, the demand for tourist and recreation services is very high. The coastal areas also serve as spawning, nursery, and feeding grounds for several species of marine and freshwater fish. Marine species such as herring, sprat, and cod dominate in open waters, and both marine and freshwater species inhabit coastal areas. The value of the catches, which amounts today to about 540 million European currency units (ECU) per year, is an indication of the considerable economic importance of the Baltic fishery. But above all, the Baltic Sea has an environmental significance of its own: It is a shared heritage to be preserved and managed as a common resource.

Causes of Degradation

The Baltic Sea has a total surface area of 415 square kilometers and is naturally vulnerable to pollution because of its semi-closed character and particular hydrography. The link with the North Sea is very narrow, with the shallowest sill being only 18 meters deep. Because of this narrow

Forty percent of the arable land in the Baltic's catchment is in Poland, where small family farms like this one predominate.

link, the Baltic Sea is dominated by freshwater inputs. Vertical variations in salinity cause permanent stratification, hampering the exchange of oxygen in the deeper parts of the sea. In some years, as much as 100,000 square kilometers, close to one-fourth of the whole sea, approaches "dead bottom" conditions.

The natural vulnerability of the Baltic Sea has been seriously aggravated by the destruction of its wetlands, particularly in the western parts of the catchment area. During the 19th century, coastal wetlands were ditched and drained to meet the demands of expanding agriculture. Wetlands have also been dredged or filled to make room for urban and industrial developments, including harbors. The consequences of these activities were not readily visible until the environmental situation in the Baltic Sea had further deteriorated.

Pollution reaches the Baltic Sea by a wide diversity of pathways. Diffuse "nonpoint" sources include airborne emissions and agricultural runoff, while point-source pollution from urban areas and industry—including untreated sewage from 30 million people,

wastewater from pulp-and-paper and other industries, toxic substances, and heavy metals—is either directly discharged into the sea or carried into it by rivers. This devastating influx can be attributed to a host of different factors, including economic inefficiency, poor legislation, lack of enforcement, and institutional weaknesses. In general, however, Denmark, Finland, Norway, Sweden, and the western federal states of Germany present a different set of problems than do the once centrally planned economies. In the former, municipal and industrial pollution loads have been significantly reduced over the last few decades, and product control measures have reduced the use of hazardous compounds. There is, however, still scope for improved control in some industries, such as pulp-and-paper. The total farming area in these countries is much smaller than in other parts of the basin, but intensive agricultural practices involve high use of chemicals, and better policies are needed to control nutrient releases.

In the countries that used to have centrally planned economies, pollution loads from various sources are still heavy although declining somewhat because of new control measures and the overall slowdown of economic activities in the past few years. In the past, the economic structure of these countries became skewed toward heavy industry dominated by large, state-owned monopolies. These industries used as much as two to three times more water and energy per unit of output than did similar industries in other countries of the basin. The input intensity and high concentration of industries (especially in such large industrial basins as Upper Silesia in Poland and Ostrava in the Czech Republic), coupled with inadequate wastewater treatment and limited emission control, resulted in very high levels of water and air pollution as well as large quantities of solid wastes.

The principal pollution sources for the Baltic Sea are municipalities, industries, and agriculture that are located not only on the coast but deep in the interior as well.

Municipalities

In the formerly centrally planned economies, most of the water bodies, such as rivers, natural and man-made lakes, and groundwater aquifers, receive excessive amounts of untreated or insufficiently treated municipal wastewater. Existing facilities, including sewer systems and treatment plants, are generally inadequate, overloaded, and poorly maintained and operated. Moreover, large quantities of highly toxic industrial wastewater and chemical liquid wastes are discharged into the municipal sewage systems. For example, in the late 1980s, the city of St. Petersburg was responsible for about 90 percent of all copper and chromium discharged into the Gulf of Finland. Pretreatment of industrial effluent is generally insufficient for biological treatment processes to operate efficiently. Without either pretreatment or the separation of noxious industrial wastes from household sewage, serious problems with sludge disposal also arise. Another important source of water pollution, and

one which is attracting increasing attention, is storm water runoff from urban areas.

In the Nordic countries and in the western states of Germany, there have been substantial reductions in the pollution load discharged by municipalities during the last 20 years. The load of organic substances has been reduced from 50 to 5 percent and that of phosphorous from 25 to 10 percent of the amount generated by municipalities. For example, in Sweden and Finland, the average reduction efficiency for biochemical oxygen demand (BOD) and phosphorus in municipal sewage treatment plants exceeds 90 percent.

FIGURE 1. The Baltic Sea catchment.

——— International border
− − − Catchment area

SOURCE: Helsinki Commission, The Baltic Sea Joint Environmental Action Programme, Helsinki, 1993.

Industries

Industrial restructuring is taking place in the formerly centrally planned economies, but it will take many years for the environmental effects of this process to become more visible. For the time being, industry continues to play a significant role in the discharge of polluting substances directly into the sea and inland waters. Metallurgical and chemical industries discharge a variety of inorganic wastes that affect water in diverse ways, including rendering it completely toxic. Synthetic chemical production creates new and more exotic types of waste. Many technologies are obsolete, the most dramatic examples being the paper mills in Kaliningrad, Russia, that were built at the end of the 19th century. Wastewater treatment installations typically suffer from the same problems as sewage treatment plants—insufficiency, overloading, and poor operation and maintenance.

The pulp-and-paper industry, in particular, plays a significant role in the discharge of oxygen-consuming, nutrient-rich, and slowly degradable substances into the sea and river waters in the Baltic region. Most of the old mills are located in the Karelian, St. Petersburg, and Kaliningrad regions of Russia, Estonia, Latvia, Lithuania, and Poland. In Sweden and Finland, pulp-and-paper mills are located primarily in the catchments of the Bothnian Sea, Bothnian Bay, and the Gulf of Finland. Some of the mills already have closed circulation of process water in chemical bleaching; intensive research and development seeks to find further substitutes for chlorine- and sulfur-containing chemicals.

In the Baltic's catchment area, about 400 million tons of solid waste of industrial and municipal origin are generated annually. In Denmark, Finland, Sweden, and the western states of Germany, hazardous wastes are handled separately from both household and regular industrial waste and are properly treated. Unfortunately, this is

Catches of the Baltic's marine and freshwater fish are important to the economy of Estonia and its neighbors.

not the case in the remaining countries of the region, where thousands of poorly maintained or uncontrolled dumping sites contribute significantly to contamination of local aquifers, lakes, rivers, and ultimately the Baltic Sea. The lack of separation of various kinds of wastes is a major problem. Furthermore, the effects of past practices are monumental. For example, landfill and dumping sites around St. Petersburg contain about 200 million tons of mineral waste. In Poland, accumulated solid waste is estimated at 1,500 million tons.

Agriculture

Important agricultural areas are located in Russia, Estonia, Latvia, Lithuania, and Poland, with the latter accounting for about 40 percent, or 19.5 million hectares, of the arable land in the catchment. Except in Poland, large, formerly state-owned farms dominate; their average size is more than 5,000 hectares. In Poland, the ownership structure is different; about 80 percent

of agricultural land is in small family farms with an average size of about 5 hectares (although the number of large private farms is growing).

Agriculture in Denmark and southern Sweden relies quite heavily on fertilizers. Farming operations in the Danish part of the Belt Sea catchment, which has a total area of about 12,400 square kilometers, annually discharge about 30,000 tons of nitrogen into the sea, while operations in the much larger Vistula River basin, whose 166,000 square kilometers cover two-thirds of Poland and small portions of Belarus, the Slovak Republic, and Ukraine, discharge some 50,000 tons. As agriculture intensifies in Poland and other formerly communist countries, the overuse of chemical fertilizers and pesticides must be avoided and alternatives developed. Otherwise, nutrient loading of the Baltic Sea will increase dramatically.

At present, inputs of nitrogen and phosphorus from agriculture contribute significantly to the overall nutrient load of the Baltic Sea. They originate from ammonia volatilization, nitrogen leaching (nitrate and organic nitrogen), phosphorus leaching, soil erosion, and the discharge of farm wastes, such as effluents from animal houses, manure storage, and silage heaps. Animal manures produced not only in the Baltic catchment but also in the North Sea catchment in Germany, Belgium, the Netherlands, Luxembourg, and France release ammonia into the atmosphere that is eventually deposited in the Baltic Sea. Nitrate leaching is mainly the result of inadequate handling and inefficient use of commercial fertilizers. Although livestock density is generally low, large animal farms—for example, hog farms with 100,000 to 150,000 animals—cause the most serious and difficult problems in handling animal manure.

A Framework for Regional Action

The Baltic Sea's environmental problems have been the concern of the surrounding countries for several decades.

In 1974, inspired by the 1972 UN Conference on the Human Environment, the governments of the region signed the Baltic Marine Environment Protection Convention, known as the Helsinki convention. Its implementation began immediately on a provisional basis, and the convention formally entered into force in May 1980. The Helsinki Commission (HELCOM) is the steering agency for the convention. Unfortunately, until the relatively recent political changes in the Baltic region, the activities of HELCOM concentrated on the open sea. As long as the coastline was shared by countries belonging to different military blocs, the coastal waters were largely inaccessible, and only extremely limited data on land-based pollution sources were available.

About 15 years after signing the initial Helsinki convention and as soon as the political situation in the region had changed, the prime ministers of Poland and Sweden invited their colleagues from the other countries in the Baltic catchment to a conference held in Ronneby, Sweden, in September 1990. The participants at the Ronneby conference adopted the first Baltic Sea Declaration, which sets out a number of principles and priority actions necessary to enhance the Baltic environment. Most importantly, the conference signaled the launching of a concrete effort to "assure the ecological restoration of the Baltic Sea, ensuring the possibility of self-restoration of the marine environment and preservation of its ecological balance."[2] That effort is the Baltic Sea Joint Comprehensive Environmental Action Programme. The program was developed from 1990 to 1992 by a HELCOM ad hoc high-level task force (HLTF), which was a joint endeavor of the Baltic Sea countries (Denmark, Estonia, Finland, Germany, Latvia, Lithuania, Poland, Russia, and Sweden), the countries of the catchment area (Belarus, the Czech Republic, the Slovak Republic, Norway, and Ukraine), the Commission of the European Communities, and four international financial institutions (European Investment Bank, European Bank for Reconstruction and Develop-

ment, Nordic Investment Bank, and World Bank). Also invited to attend HLTF meetings were several organizations that acted as observers, including the International Baltic Sea Fisheries Commission, the Coalition Clean Baltic, Greenpeace, and the World Wide Fund for Nature. The task force used a variety of studies and sources of information to develop the program, including national plans prepared by all Baltic countries and eight major prefeasibility studies covering the eastern, southern, and southwestern parts of the catchment.[3] The prefeasibility studies and special topical studies[4] were financed by grants totaling 5 million ECU from the Commission of the European Communities, the governments of Denmark, Finland, Germany, Norway, and Sweden, the Nordic Project Export Fund, and Sweden's World Wide Fund for Nature. In April 1992 in Helsinki, the Diplomatic Conference of Ministers of the Environment revised the 1974 Helsinki convention and approved the Baltic Sea Joint Comprehensive Environmental Action Programme.

The Joint Comprehensive Programme

Nearly 20 years of scientific work by HELCOM committees and studies conducted for the HELCOM task force have followed ecological developments in the Baltic Sea catchment. This preparatory work for the program concluded that preventive and curative actions are necessary in all of the countries in the catchment to reduce the pollution load reaching the sea. Some of these actions are already under way: Some state-owned enterprises with polluting production activities have shut down; wastewater treatment plants are planned or partly constructed; new protected areas have been created; and environmental controls have been strengthened. These ongoing activities require support; in particular, new environmental policies and pollution control programs should be formulated and adopted in the formerly Eastern-bloc economies.

The obsolete technologies used in many industries in the formerly communist countries, such as this sulfur mine in Poland, have helped pollute the Baltic.

STEPHEN F. LINTNER

The Strategy

The underlying strategy of the Baltic Sea Joint Comprehensive Environmental Action Programme consists of actions by each concerned government to carry out needed policy and regulatory reforms; to build capacity; and to invest in controlling pollution from point and nonpoint sources, safely disposing of or reducing the generation of waste, and conserving ecologically sensitive and economically valuable areas. To complement these activities, the program also includes elements to support applied research, environmental awareness, and environmental education. These actions will be phased in to keep pace with the gradually increasing capacity to raise funds and pay for the recurrent costs of environmental management in the transforming economies. For the first few years, emphasis will be placed on creating the enabling policymaking environment and institutional arrangements, on making limited investments in the highest priority projects, including pilot and demonstration projects, and on promoting private investment and initiative through concessions and incentives. Environmental investment programs in the northern and western parts of the region are also an integral part of the joint comprehensive program and are expected to be financed from local resources.

Program Components

The program consists of six components that comprise broad and distinct areas of action. The first component is a set of policy, legal, and regulatory reforms that establish a long-term environmental management framework in each country, including macroeconomic policies and incentives; financial facilities, policies, and controls; environmental standards and laws; and the appropriate systems for monitoring and enforcing these laws. To accomplish these goals, the program will encourage participating governments to undertake studies of new legal regulatory arrangements

and drafting of governmental and parliamentary decision documents; policy studies of options, costs, and benefits; investment in new monitoring equipment, upgrading of laboratory equipment and procedures, and upgrading of data processing and analytical capacity; and development of new organizational structures and arrangements to carry out management functions. The program will also promote the use of a range of economic measures for the management of environmental quality, such as adoption of the "polluter pays principle," adoption of realistic user charges, and introduction of cost-sharing principles in the case of transboundary pollution. In addition, the program will seek to promote the adoption of the best environmental practices and the use of the best available tech-

nologies to prevent pollution at the source, consistent with the 1992 Helsinki convention (an update of the 1974 convention that was signed at the Diplomatic Conference) and HELCOM recommendations. Actions that facilitate the transfer of technology between and within cooperating countries will also be encouraged.

The second component of the program is institutional strengthening and human resources development: building the organizational and human capacity to enforce regulations; planning, designing, and implementing environmental management systems, including infrastructure; and managing natural resources efficiently. The program focuses on training people to use new concepts of management and new technology and developing the organizational and ad-

---

UNION OF THE BALTIC CITIES
By Piotr Krzyzanowski

For hundreds of years, people living by the Baltic Sea were unified by their common work—fishing. Although the importance of this industry has diminished in the region, the spirit of the oldest traditions of the Baltic region—families and local associations working together to fish, rescue the drowning, and build houses, roads, and harbors—has survived. In September 1991, despite years of divisions and political barriers, representatives of 32 towns and cities on the Baltic Coast met in Gdansk, Poland, to form the Union of the Baltic Cities. The union was established to encourage and contribute to positive democratic and economic development in the region.

The goals of the Union of the Baltic Cities are far-reaching and include cultural awareness, improved transportation and telecommunication, and environmental protection. Member cities are encouraged to enter into twin-city relationships with each other so that cities with extensive

needs can benefit from the knowledge, experience, and willing partnership of highly developed cities. The union also acts as a clearing house to match the needs of such member cities with the resources of others.

Environmental protection of the Baltic is a priority for the union. Much of the region's environment is undergoing increasing degradation because the speed of economic development has severely strained the ability of nature to accommodate it. The countries surrounding the sea have been abusing its resources for too long; in essence, the Baltic Sea has been treated as a waste collector of unlimited capacity.

Although much of the policymaking necessary to restore the Baltics is the responsibility of officials at the national and international levels, such matters as improved sewage treatment, district heating systems, and handling of refuse are usually dealt with by individual cities. Thus, the union could be instrumental ministrative framework for people to work effectively and efficiently. Priority will be given to support for environmental planning, river basin management, coastal zone management, and protected areas management. Consistent with the objective of decentralizing environmental management, the program will support the development of effective local authorities that are responsible for water supply, wastewater, and disposal of solid wastes. The program will also support the development of local capacity to carry out environmental audits by training personnel from local industries, consulting firms, and regulatory organizations.

The third component is a program for infrastructure investment in specific measures to control point and nonpoint sources of pollution and to minimize and dispose of wastes.

---

in arranging the transfer of environmental protection technologies between member cities. For example, the Danish Association has developed the Baltic Sustainable Cities Programme on behalf of the member city of Aarhus, Denmark. The program has instituted partnerships between eastern cities, such as Elblag, Kaliningrad, Klaipeda, Riga, and Tallinn, and western cities to provide assistance and expertise in planning and implementing environmental projects.

The activities of the Union of the Baltic Cities are numerous and varied so that it can fulfill its goal of helping the Baltic region to enter the 21st century as a new civilization of democracy, respect for rights and duties, and balanced economic development.

PIOTR KRZYZANOWSKI is the Polish representative to HELCOM and the deputy chairman of the Union of Baltic Cities.

These measures include the rehabilitation and modernization of existing infrastructure; the development of new infrastructure; and the conservation of environmentally sensitive areas and resources. This program includes investments in three areas:

• Investments in municipal environmental management will emphasize the collection, treatment, and disposal of wastewater. Existing municipal water systems in the region commonly include all or most of the industrial wastewater within the service area of the sewer network, which means that they are functioning as combined municipal and industrial wastewater systems. The program will support national and local activities to rationalize municipal water consumption and to reduce leakages and unaccounted-for water. Support will also be provided for the completion of facilities already under construction, subject to design modification (taking a least-cost approach) and replacement of any substandard civil works. There are a number of such projects under way in the region's "hot spots," including Kaliningrad, Klaipeda, Liepaja, Tallinn, and Vilnius. Key operational problems, such as inadequate plant instrumentation and energy-inefficient equipment, will also be addressed.

• Industrial waste management at specific sites and complexes will also be financed. These projects are primarily for wastewater management, but they also address waste minimization, safe disposal of solid wastes, treatment and safe disposal of various forms of hazardous and toxic wastes, and environmental restoration from past degradation. Particular attention will be given to the pulp-and-paper industry, which has special significance in the region. Although this industry is perhaps most modern in the northern and western parts of the region, it is arguably one of the most outdated and inefficient in the eastern and southern portions. In both cases, the pulp-and-paper industry produces serious pollution prob-

lems, particularly persistent organic chlorinated substances. In addition, priority will be given to addressing wastewater treatment in the chemical and metal production and processing industries.

• Controlling polluted water runoff and discharges from agricultural lands, large livestock operations, and rural settlements is also an investment priority. Agriculture and livestock are major contributors of nutrient loads to the sea. Control measures will involve a wide range of changes in farming practices, including modifications in the timing of manure application and tillage; improvements in farm infrastructure, such as larger capacity and more secure storage for fertilizers, manure, and farm wastes; changes in water use; and new equipment, such as manure spreading and waste handling equipment. The program will also support investment activities to reduce pollution problems resulting from fish farming.

The fourth component of the program is to aid in the management of coastal lagoons and wetlands. These environmentally sensitive and economically valuable areas serve as important buffers of pollution before it reaches the sea and provide critical habitat for diverse flora and fauna, including commercially important fisheries. The management systems will include land-use controls and limited infrastructure and, in some cases, will be integrated with compatible ecotourism and recreation developments, possibly through joint public and private ventures or private investment. The program will support the development of a series of demonstration activities concerning the use of natural and constructed wetlands for wastewater treatment and storm water retention and as traps for nutrients and other pollutants.

The program's fifth part is supporting applied research to build the knowledge base needed to develop solutions, transfer technology, and broaden understanding of critical problems. Specific priority topics include

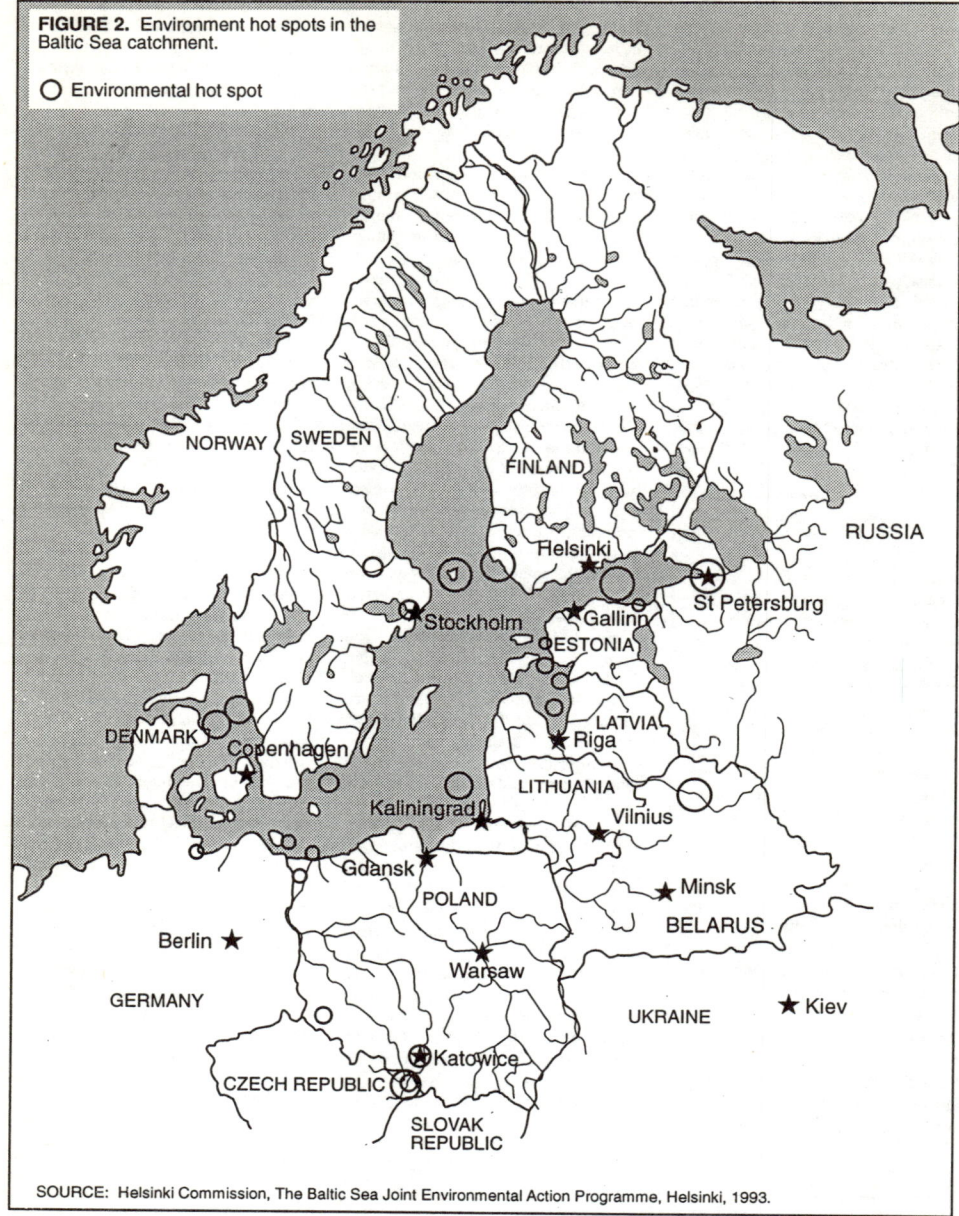

FIGURE 2. Environment hot spots in the Baltic Sea catchment.

○ Environmental hot spot

SOURCE: Helsinki Commission, The Baltic Sea Joint Environmental Action Programme, Helsinki, 1993.

environmental trends; systems ecology; development and application of the "critical load" concept for different pollutants in the Baltic Sea and its subregions; assessments of risks to human health; agricultural development patterns in the Baltic Sea basin; future trends in transportation and its environmental management; and management of critical ecosystems, such as coastal lagoons and wetlands.

Finally, the program will encourage public awareness and environmental education to develop a broad and sustainable base of support for the implementation of the other components. The participation of nongovernmental organizations reaching the grassroots level and the development of effective environmental education programs are essential to promoting public awareness and political commitment. Efforts to promote and expand environmental education, particularly in the context of local environmental cleanup activities, will be given priority.

The 1992 Baltic Sea Declaration that was signed at the April 1992 Diplomatic Conference mandated that the key elements of the Joint Comprehensive Environmental Action Programme be initiated in 1993. Although the main responsibility for implementing the program will rest with the governments concerned, coordination of many program activities is needed. Such activities include monitoring pollution and the impact of the program, maintaining databases, reforming policies and regulations, providing technical assistance, planning research programs, and exchanging information, as well as periodically updating the program. To this end, the Programme Implementation Task Force has been set up within the framework of HELCOM.

The Costs and Financing of the Program

The total cost of the 20-year program for all countries in the Baltic Sea catchment is estimated to be about 18 billion ECU (about $25.6 billion),[5] and

Although the Baltic Coast, including this spot in Lithuania, has long been a popular vacation area, pollution has closed several beaches and reduced the tourist trade.

phase 1, from 1993 to 1997, is projected to cost about 5 billion ECU (about $6 billion). (See Table 1 on page 28.)

The program will focus on remediating 132 "hot spots" (see the map on page 14), 98 of them located in Russia, Estonia, Latvia, Lithuania, Belarus, Ukraine, Poland, the Czech Republic, and the Slovak Republic. These 98 actions will cost about 8.5 billion ECU, out of which 6.5 billion ECU will go toward 47 "priority hot spots," including 26 municipal spots, 9 industrial spots, and 12 others. The remaining 34 hot spots, located in Denmark, Finland, Germany, and Sweden, have been selected by these countries themselves, and their remediation will cost approximately 1.5 billion ECU. In Poland, which has the largest land area and population in the Baltic Sea catchment, the estimated cost for remediating the 40 recommended hot spots and priority hot spots exceeds 4.0 billion ECU (see "Report from Poland: Politics in the Midst of Environmental Disaster" in

the March 1991 *Environment*).

In phase 1 of the program's implementation, the needs of 29 priority hot spots will be addressed, 19 in the Vistula and Oder river basins and 10 in Russia, Estonia, Latvia, and Lithuania. In addition, major reductions in discharges of adsorbable organic halogens are planned for the pulp-and-paper industry in Finland and Sweden.

The immediate costs for the program total about 30 million ECU for the feasibility studies of the highest priority projects. Financial arrangements for the implementation of these projects cannot be completed without the results of these studies. In addition, the governments in the eastern and southern parts of the catchment will need to arrange for local and external support to carry out the required policy and regulatory reforms. Timely implementation of the program critically depends on those and other noninvestment activities. To complicate matters, nearly two-
(continued on page 28)

Cleaning Up the Baltic
(continued from page 15)

thirds of the hot spots are located in countries whose current economic situation makes ordinary loans and commercial financing difficult. Therefore, some combination of normal and concessional development lending, supplemented by outright grants, will be needed from bilateral and multilateral financing agencies. Varied approaches to project financing are also needed, including enhanced local sources of revenue from taxes and user charges, user taxes on potentially polluting substances, and private investment that supplements traditional public-sector borrowing and budgetary resources.

In March 1993, at the High Level Conference on Resource Mobilization for the Baltic Sea Joint Comprehensive Environmental Action Programme in Gdansk, Poland, the participants stressed that availability of local or foreign funds is not the only constraint to program implementation; several mutually related factors also limit the mobilization and flow of, and the capacity to use effectively, the funds that are available. Several important observations were made at the Gdansk conference.[6] First, the greatest proportion of project financing must come from local resources; however, despite the efforts on the part of governments of the formerly communist countries, local resources will remain limited in the short term. Slow progress in this regard has been

largely due to the precarious position of the economies in transition. Nevertheless, according to those at the conference, "early action should be taken to support the establishment of financing mechanisms and incentives, to develop the institutional capacity to implement them, and to create greater public awareness of their importance."[7] The principal local sources of funds available to support activities under the program include user charges, pollution fees, budgetary allocations and nonbudgetary incentives, domestic loans, and, potentially, private investments. Domestic loans may not be a major factor in the near term because local financial institutions, such as capital markets and banks, have not been sufficiently developed to support environmental improvements.

Under the current economic conditions, however, the concerned countries, municipalities, and enterprises have sought increased support for program financing from external sources. Although some of the potential donor countries are suffering from a protracted recession, "there will be a critical need, at least for the short term, for continued and better coordinated support from bilateral donor organizations to implement the Programme, especially to support policy, institutional development and investment project preparation activities."[8] Grant and concessional funding from the Commission of the European Communities and bilateral donors, from both within and outside the region, has been an important source of support for program activities in recent years.

In addition, the conference participants anticipated that "the international financing institutions will continue to provide loans and implement projects for selected priority actions under the Programme consistent with the requests of the borrowers and within lending limits established by their creditworthiness."[9] The level of investment by international financing institutions is determined by the priorities of the borrowing country, the

TABLE 1 ▉▉▉▉
ESTIMATED COSTS OF THE BALTIC SEA JOINT COMPREHENSIVE ENVIRONMENTAL ACTION PROGRAMME

Program element	Phase 1 1993–1997 (millions ECU[a])	Phase 2 1998–2012 (millions ECU)	Total 1993–2012 (millions ECU)
Policies, laws, and regulations	5	5	10
Institutional strengthening and human-resource development	70	140	210
Investment activities			
Point-source pollution			
Immediate support and warning systems	50		50
Municipal wastewater treatment	1,000	2,000	3,000
Combined municipal and industrial wastewater treatment	1,600	4,000	5,600
Pulp-and-paper industry wastewater treatment	400	1,000	1,400
Environmental control at other industries	300	1,000	1,300
Solid and hazardous waste management	200	800	1,000
Air quality management	460	1,200	1,660
Nonpoint-source pollution[b]	800	2,700	3,500
Management programs for coastal lagoons and wetlands	100	120	220
Applied research	10	20	30
Public awareness and environmental education	5	15	20
Total	**5,000**	**13,000**	**18,000**

[a]European currency units
[b]Nonpoint-source pollution includes runoff from agricultural and livestock operations and rural settlements.
SOURCE: Helsinki Commission, The Baltic Sea Joint Comprehensive Environmental Action Programme, Helsinki, 1993.

levels of borrowing and indebtedness that present and anticipated economic conditions are able to support, the balance of investment activities between priority sectors, the quality of the proposed investment, and the related characteristics and conditions of the borrower.

During phase 1 of the program's implementation, special emphasis "must be put on comprehensive feasibility studies that yield economically and financially feasible and affordable investment projects, and that devote more attention to the assessment of the current institutional framework and the formulation of measures to strengthen that framework to support project implementation and operation."[10] Several actions will require special attention in phase 1, including local fundraising, support for institutional reforms, support for industries in transition, and the phasing in of investments requiring long development periods.

The conference also acknowledged that, in the formerly communist countries, pollution control "at the source" will have to be incorporated into the ongoing privatization, restructuring, and modernization of industries. In the municipal sector, innovative approaches—such as joint ventures with foreign capital and long-term concessions—will be needed to attract capital financing, especially from private and commercial sources. Investments in municipal wastewater treatment plants must be seen within the broader framework of developing autonomous self-financing municipal and regional water and wastewater utilities, whether public or private.[11]

The environment ministers and other representatives of the governments of the Baltic Sea countries and the representative of the Commission of the European Communities assembled at the Gdansk conference jointly decided to make all efforts to mobilize local, national, bilateral, and multilateral financial and other resources for the implementation of the program.[12]

Many of the wetlands around the Baltic have been destroyed. The program hopes to manage the region's remaining wetlands, such as this one in Estonia, more wisely.

Progress in Implementation

Phase 1 investment activities will concentrate on improving combined municipal and industrial wastewater treatment to reduce organic pollution loads. High-priority projects are the completion of unfinished or inoperable treatment facilities, the installation or improvement of pretreatment of industrial wastewater discharges into the municipal sewage systems, and the expansion of safe disposal of sludges. In the formerly centrally planned economies, nutrient-removal facilities (tertiary treatment) at municipal wastewater treatment plants will be deferred until more substantial progress is achieved in nutrient reductions in the agricultural sector.

As discussed at the Gdansk conference,[13] initial program activities have emphasized the establishment of priorities at the national level and detailed preparation of projects, including feasibility studies. Out of the 29 priority hot spots to be addressed in phase 1, detailed feasibility studies are under way or are planned at 17 of the 26 priority municipal hot spots; and environmental audits have been carried out at 6 of the 9 priority industrial hot spots. In addition, an economic and environmental assessment of the region's pulp-and-paper industry was completed, which reviewed market prospects and cost competitiveness at the region's 34 pulp-and-paper mills and focused on 12 mills that are candidates for feasi-

Two primary contributors of pollution in the Baltic are pulp-and-paper mills and agriculture. Animal manures produced on large livestock farms, like this one in Estonia (below), emit ammonia that is eventually deposited in the sea. Pulp-and-paper mills, both modern ones—such as this Finnish mill on the Saimaa canal (left)—and older ones in the formerly communist countries, discharge varying amounts of chlorine- and sulfur-containing chemicals.

bility studies and possible investment support.

The development of control programs for nonpoint source pollution in the agricultural and livestock sectors requires a different approach. The emphasis in phase 1 is on establishing an effective institutional framework and on developing and implementing well-designed pilot and demonstration activities. Seven pilot studies and programs have been established for the control of agricultural and livestock runoff in hot spot areas of Russia (at a major hog farm in the St. Petersburg region), Estonia (protection of groundwater from agricultural pollution), and Latvia, Lithuania, and Poland (testing ecotechnologies in small catchment areas). Thirteen surveys, studies, and pilot programs have been established or are planned in the most important wetland areas and in the key coastal lagoons, which are mostly in Estonia, Latvia, Lithuania, and Poland. In addition, the World Wide Fund for Nature, in cooperation with the countries of the region, is developing a regional program for managing coastal lagoons and wetlands, focusing on central Matsalu Bay in Estonia, the

north coast of Latvia, and Putsk Bay, which is a subsection of Gdansk Bay in Poland.

Although several factors have combined to slow the rate at which resources are committed to and used for various program activities, the first year of implementation has demonstrated that, despite all difficulties, a great deal can be accomplished. The foundations established during this early phase will ensure the long-term success of the Baltic Sea Joint Comprehensive Environmental Action Programme.

Expected Environmental Benefits

The program is expected eventually to have a major impact on the rivers in the Baltic Sea catchment, which are the principal source of water supply for industry, agriculture, and about 80 million people. Reducing pollution loads and restoring aquatic ecosystems will increase the quality and reliability of water resources and benefit the health and well-being of the local population.

The coastal waters are expected to improve most rapidly, which would allow several contaminated beaches to open again and contribute to re-establishing favorable conditions for tourism and local recreation. The overall reduction in the load of nutrients should reduce algal blooms, assist in lowering eutrophication levels, and improve oxygen conditions. These improvements should have a major positive impact on fishery resources, including an increase in the number and diversity of commercially valuable fish species, such as whitefish, cod, and plaice. The quality of the open sea will be restored more slowly, however, because of the difficulty in controlling long-range atmospheric transportation of various pollutants.

Another benefit of the program to Baltic countries is the strengthening of local capacities to plan, finance, and manage environmental measures. The strategic gains from capacity building for environmental cleanup will be of particular importance in the formerly centrally planned econo-

mies, which are undergoing profound institutional as well as economic changes.

Given the current state of the Baltic Sea environment, however, the program cannot be expected to make any major impacts for about 20 years. But gradual and visible improvements, both environmental and economic, can be realized in the relatively near future.

NOTES

1. The information contained in this article is based principally on the findings and recommendations of the Helsinki Commission (HELCOM) ad hoc high-level task force and the HELCOM Programme Implementation Task Force.

2. Ronneby Diplomatic Conference, Baltic Sea Declaration, 1990.

3. The national plans were worked out in the first six months of the operation of the task force, on the basis of nationally available data and information. They included information on the 1985 pollution loads discharged into the Baltic Sea, environmental policies and regulations currently in force, and major pollution control programs and initiatives. The eight prefeasibility studies carried out in the Baltic Sea catch-

ment covered St. Petersburg, the St. Petersburg region, Karelia, and Estonia; the western coast of Estonia; the Gulf of Riga and the Daugava River basin; the Lithuanian coast and the Nemunas River basin; the Kaliningrad region and the Pregel River basin; the Vistula River basin and the Baltic coast of Poland; the Oder River basin; and the north German Baltic coast.

4. Topical studies included Atmospheric Deposition of Pollutants; Agricultural Runoff; Wetlands and Coastal Lagoons; and several special studies concerning selected environmental problems in Estonia, Latvia, Lithuania, Poland, and the St. Petersburg region of Russia.

5. The exchange rate used is 1 European currency unit (ECU) equals U.S. $1.20.

6. The Baltic Sea Joint Environmental Action Programme, conference document no. 2 (Background paper for the High Level Conference on Resource Mobilization held in Gdansk, Poland, 24–25 March 1993).

7. Ibid.

8. Ibid.

9. Ibid.

10. Ibid.

11. W. Stottman, "World Bank Activities in the Municipal Water/Wastewater Sector" (Background paper presented at the Baltic Sea Joint Environmental Action Programme, High Level Conference on Resource Mobilization, Gdansk, Poland, 24–25 March 1993).

12. The Baltic Sea Joint Environmental Action Programme, "Declaration on Resource Mobilization for the Baltic Sea Joint Comprehensive Environmental Action Programme, 1993" (Gdansk Declaration).

13. The Baltic Sea Joint Environmental Action Programme, note 6 above.

[28]

ELSEVIER

Ecological Economics 27 (1998) 13–28

ECOLOGICAL
ECONOMICS

ANALYSIS

An approach to Baltic Sea sustainability

Jörg Köhn *

Institute for Economics, University of Rostock, Parkstraße 6, D-18051 Rostock, Germany

Received 25 November 1996; received in revised form 28 May 1997; accepted 8 June 1997

Abstract

One may understand the Joint Comprehensive Programme to restore the ecological balance of the Baltic Sea, which was initiated and has been signed by the Baltic Littoral States and the European Community, as confirming the target of environmental sustainability. The paper seeks to answer three questions and proposes a program for participatory action. The first questions it broaches are: (a) is the target of environmental sustainability socially feasible; (b) how can ecological economics support decision making processes in complex systems such as the Baltic Sea and its watershed area, where uncertainty is an important process variable; and (c) how can a program be designed so as to combine environmentally and socially based targets. In other words, how should social processes be designed to make 'sustainability' into a regulatory principle for policy making. This paper proposes one conceivable way of organizing processes in which the social players can participate in the decision making process. It also supports the idea of designing input-output scenarios for economic sectors and subregions for supporting target setting on the local, subregional and watershed scales and to monitoring such processes. Finally, the paper suggests a system for applying the polluter-pay-principle in the Baltic Sea region. © 1998 Published by Elsevier Science B.V. All rights reserved.

Keywords: Baltic Sea region; Sustainability; Systems complexity; Uncertainty; Institutional economics; Funding

1. Introduction

In 1992 the Diplomatic Conference on the Protection of the Marine Environment of the Baltic

* Tel.: + 49 381 4982912; fax: + 49 381 4982911; e-mail: koehn@wiwi.uni-rostock.de

Sea Area established The Baltic Sea Joint Comprehensive Environmental Action Programme (JCP) for the renovation of the Baltic Sea within the framework of the (second) Convention on the Protection of the Marine Environment of the Baltic Sea Area, 1992. This program aimed at reducing nutrient loads in the period 1993–2012 to "assure the ecological restoration of the Baltic

Sea, ensuring the possibility of self-regeneration of the marine environment and preservation of its ecological balance" (HELCOM, 1992a). It noted the need for pollutant-specific reductions of at least 65% for nitrogen and 80% for phosphorous on the basis of current nutrient loads and a total ban on some substances (Tables 1 and 3). This target may be sufficient for environmental sustainability, but clearly neglects other aspects of sustainability.

Sustainability is a concept with different meanings for natural scientists, social scientists and policy makers. Summarizing the discussion, Goodland (1995) distinguishes between economic, environmental and social sustainability, and also between weak, strong and absurdly strong sustainability concepts. Evidently, in view of this diversity, the set targets and possible outcomes must always be taken into account when considering sustainability, and it must always be remembered that the way sustainability is defined may be strongly influenced by the goals set by the social players. *Economic sustainability* is based mainly on a capital substitutability theory (Solow, 1974 see Common and Perrings, 1992; p. 5), but social and environmental sustainability concepts take social and environmental equity and cultural and biological diversity into account as well as ethical considerations.

The sustainable development declared as the political goal of the Baltic Sea Convention (1992) bases on and has to maintain "… the indispensible values of the marine environment of the Baltic Sea Area, its hydrographic and ecological characteristics and the sensitivity of its living resources to changes in the environment… " and "… the historical and present economic, social and cultural values of the Baltic Sea Area for the well-being and development of the people of that region… ". The first of these goals refers to the biophysical properties of the ecosystem, whereas the second reflects values based on the socioethical complex (Daly, 1977). Hence, bearing in mind the two aspects of the political target, the convention focuses on a comprehensive sustainability concept rather than simply restoration of the Baltic Sea Area. However, if we add the rights of future generations, the goals include all components of the WCED (1987) definition of sustainable development.

Since the economic, cultural and political development of the Baltic Sea Region is also crucial to sustaining the economic and social performance of the region and restoring the ecological balance of the Baltic Sea ecosystem, the development of national economies and local policies are obviously of paramount importance. Therefore, the paper describes the economic and ecological background that must be taken into account before discussing the mitigation targets needed to restore the ecological balance in the context of an economic indicator system. Although aware that the complexity and inherent uncertainties of the systems involved can scarcely be reflected in a model, a simple input-output model is formulated that allows the Baltic Sea sustainability target to be broken down for sectoral, local, subregional and watershed scales. This paper proposes a joint program for providing a financial infrastructure for the Baltic Sea restoration process. Finally, this paper focuses on a new institutional framework to overcome the obstacles of the present situation and to open the floor for social player network participation.

2. Sustainability — carrying capacity, resilience, and uncertainty

Sustainability as a concept was first mentioned in the writings of Malthus (1798) and Mill (1848) (Goodland, 1995; p. 7). Faustmann used the concept in 1849 to calculate forest rotation periods needed to maximize returns (Ludwig, 1993). Later, economists elaborated the *Maximum Sustainable Yield* concept (MSY) for managing the use of renewable resources. The concept in general is strongly reductionist in its treatment of environmental variability. Moreover, it can only be applied in a one-resource-as-commodity approach.

Like the MSY, the *carrying capacity* is also part of the sustainability concept. Carrying capacity describes the maximum stock of a certain population that can be supported by an ecosystem: a given stock (i.e. a certain bacterial, plant, animal or human population) can exploit the structural and functional constraints (space, food, predation, etc.) of its ecosystem until the system cannot supply additional members of the population with food, space or other needs without losing some of its properties or stability or 'flipping' to another resilient state of the system concerned. The carrying capacity concept reflects at least one component of the sustainability concept: the non-overexploitation rule.

The structural and functional constraints of systems, both natural and social, are known to change over space and time. Such changes may be cyclic (Holling, 1986), but they may also be unpredictable if system properties change completely (Lovelock, 1995). The latter 'wave like' changes are completely unpredictable because the system assets, i.e. information stocks and appearance, also change. They therefore exhibit 'true uncertainty' (Ayres, 1988; Köhn, 1997). However, both types of change may be affected by *resilience* (Holling, 1986), which is a broader concept than that of the carrying capacity concept since it describes the bounds of a system's stability. Hence, "[r]esilience is the ability of a system to absorb perturbations." (Berkes and Folke, 1994; p. 4), i.e. resilience explains why a system may maintain in or return to a state more or less identical to the one that existed before the perturbation.

But substituting the *resilience* concept for *carrying capacity* leads to a crucial implication: sustainability does not imply perpetuation (Costanza and Patten, 1995; p. 196). System sustainability as a function ($F_{(t)}$) therefore depends on the structure of all realized organic variables in a particular system, their interactions, changes (adaptation, co-evolution, evolution), etc. over time (t). Let X_{min} represent the species reservoir of the biogeographic region in which the particular ecosystem is embedded and a set of human values in a particular region, and let X_{max} be the conservation of the biological or cultural diversity and system function maximum. In addition, we assume that all populations (like multi-species resources, $p_{o \cdots k}$), abiotic and deduced biological structures as well as social institutions ($s_{o \cdots l}$) and their interrelations ($f_{o \cdots m}$) as well as economic uses ($e_{o \cdots n}$) and uncertainty (u) are vectorial process variables Eq. (1).

$$F_{(t)} = f(p_{o \cdots k}, s_{o \cdots l}, f_{o \cdots m}, e_{o \cdots n}, u, \ldots) \qquad (1)$$

This approach was chosen to show that: first, a shift from a one-resource to a multi-species approach has taken place; second, humanity is not the sole aspect of the sustainability concept; third, the concept implies biophysical and socioethical limits; fourth, that all components are time dependent vectors; and finally, that the system contains inherent uncertainties.

Since declared political goals (e.g. JCP) favour a return to the predisturbance conditions of the 1950s, the resilience concept might be applied here. Although aware that a return to the former ecological conditions is scarcely possible on account of drifting or hardly changed information stocks (e.g. in the natural system by extinction of species and invading of neozoic species, and in the social system by the process uncertainty inherent in the major transformations taking place in the Eastern Baltic states), this simplified concept might be applicable here as political thinking on the 'ecological restoration processes' follows this assumption. Decision analysis states that two or more opportunities for action exist. Any subsequent state is difficult to predict. However, if a return to the former state is possible, decision makers will, if complete knowledge about this (target) state is available and supposing that it is environmentally and socially more desirable than the present one, naturally opt for such decisions because they bear 'no' risk. Uncertainty then might be reduced to finding appropriate technologies and fund raising. However, in reality this kind of 'risk management' may boil down to a matter of 'financial options'. In contrast, the processes involved in restoring the Baltic Sea environment and assuring and sustaining social development in the Baltic Sea region are uncertain, not just risky.

3. The ecological background

The Baltic Sea is a semi-enclosed brackish water sea with a low, stable salinity over most of its area (surface salinity about 7–9‰). It covers an area of 415000 km^2 and has a volume of about

22000 km³. Like other regions, the Baltic started its self-generated system after the last glaciation about 12000 years ago, and has since undergone many physical changes. The Baltic Sea is the largest truly brackish water system of the world that is connected to an ocean. It became stable from the biological point of view during the *Littorina*-Transgression in about 5000 years BC. Consequently, from the evolutionary standpoint, the Baltic is a young ecosystem. The sea is non-tidal, and the water has a correspondingly long dwell time (25–40 years). Therefore, the Baltic has more of a stagnant than a through-flow character. The Baltic itself consists of several basins that differ in their hydrographic features. The narrow entrances to the Baltic Sea region and the distinctive sills separating the basins from each other impede inflows of oxygenated salt water from the North Sea. Precipitation and a large fresh-water inflow from about 200 rivers (about 400–500 km³ fresh water input a year) lead to a stratified water column. Aperiodic inflows of salt water from the North Sea renew the salt water in the deep basins, but generate a strong halocline. This stratification hinders deep oxygenation, and some elements of the sea have been damaged owing to major economic impacts. The Baltic Sea resembles a storage basin for the discharges of civilization in the Baltic Sea drainage basin.

4. The economic background

Since the Baltic Sea is part of a larger Baltic environment occupying a drainage area of about 1615000 km², its health is affected by environmental impacts from at least nine littoral political states. In addition, parts of five other countries are also located in the Baltic Sea drainage area. The histories of these states, their social systems and their perception of environmental problems have diverged drastically at least since World War II. The Scandinavian countries (Sweden, Finland, and Denmark) and Germany developed into industrial nations. The East European countries (Poland, Lithuania, Latvia, Estonia and Russia), also 'industrial' countries to some extent, are currently experiencing a complex process of political,

economical and cultural change started by the decline of the former socialist bloc in the late 1980s. Nevertheless, a historical Baltic Identity exists in cultural and economic life, based on tradition, trade, transportation, industries, agriculture, science, arts, etc. (Serafin and Zaleski, 1988).

The differences between the economies of the littoral countries are immense. While the Scandinavian countries and Germany have among the highest standards of living, productivity and wage levels in the world, these indicators have very low values in the eastern countries (HELCOM, 1992b, see also Table 2). In terms of gross net product (GNP), the economic distance between these two groups is considerable (7.5:1). Moreover, the more or less identical growth rates of the GNP per capita indicator shows that long-term convergence cannot be supposed. Although the Baltic region was originally a resource supplier, a self-generating Baltic Economy has recently emerged as a strong factor in the world economy.

4.1. Assessing the national impacts

The relative national environmental impacts were stated in a special report of the Helsinki Commission in 1990. The negotiation process by which the littoral states adopted the Convention on the Protection of the Marine Environment of the Baltic Sea Area, 1992, was based on this report. In addition, National Reports to the JCP give more detailed information on point-source

Table 1
Species and sources of pollution and mitigation efforts

Species of pollutants	Source of pollution	Share of pollution
Nitrogen	Airborne	51%, About 60% from outside the water drainage basin
	Waterborne	49%
Phosphorus	Deposition	11%
	Waterborne	89%
Chemical compounds	(chemical industry)	100%

Table 2
Indicators for economic performance, land use intensities in the Baltic Littoral States and inequity relations of the eastern to the western Baltic Littoral States

Indicator	Eastern Baltic States, in physical terms [potentially]	Western Baltic States, in physical terms [potentially]	Ratio of eastern to western Baltic States (west = 1)
Land area (000 ha)	66 418	80 077	0.82
Population (000 inhabitants)	55.85	23.55	2.37
Population density (1993, per 1000 ha)	840	294	2.86
Urban population as % of total, 1995	(70.5)	79.2	0.89
Total labor force 1989–1991 (000)	17 292	11 296	1.53
Annual economic performance (in million $US, as productivity of gainful employment in 1994)	354 840 [2 129 040]	903 680	0.39
In agriculture (% of 1989–1991 labor force)	27	5	5.4
In industry (% of 1989–1991 labor force)	37	31	1.2
In services (% of 1989–91 labor force)	36	64	0.56
GNP per capita ($US/annum)	3087	23 085	0.13
Annual growth rate of GNP per capita	2.6	2.4	1.08
Industrial environmental impacts (as equivalent of CO_2-emissions, million metric tons)	798 164	>250 000	3.19
Industrial CO_2-emissions per capita (metric tons/annum)	11.44	11.33	1.00
Agriculture			
Cropland (000 ha, 1989–1991)	>20 000	9181	2.17
Cropland (ha per capita)	0.38–0.94	0.15–0.51 (0.37)	(2)
Average annual fertilizer use (kg per ha of cropland)	52–151 (113)	113–520 (262)	0.43
Average fertilizer use (metric tons)	(3 000 000)	2 058 468	—
Annual pesticide consumption (metric tons)	(>20 568)	(>12 000)	—
Average yields of cereals (kg per ha, 1990–1992)	2253	4708	0.48
Average yields of roots and tubers (kg per ha, 1990–1992)	12 988	28 993	0.44
Livestock population in human NO_x-equivalents (000)	>80 000	>47 000	1.7
Cattle (000, 1990–1992)	>15 500	>8300	1.9
Sheep and goat (000, 1990–1992)	>5480	>900	6.0

Table 2 (continued)

Indicator	Eastern Baltic States, in physical terms [potentially]	Western Baltic States, in physical terms [potentially]	Ratio of eastern to western Baltic States (west = 1)
Pigs	> 30 000	> 15 400	1.9
Equines	> 1000	> 180	5.6
Chicken	> 100 000	> 45 000	2.2
Grain fed (% of total grain consumption, 1990–1992)	(47)	(70)	0.67
Transportation (persons per vehicle 1991)	8	2	—
Environmental impacts in NO_x-equivalents (% of total NO_x-emissions)	35	65	—
Tourism (summer cottages per 000 inhabitants)	—	37.4–76.4 (64.4)	—

and total pollution at national, subregional, and in some cases, local levels. Although these reports are partly accurate, estimation of the total sewage discharge, for example, is impossible for at least two reasons. In the first place, most of the reports do not distinguish between industrial and municipal discharges and their compositions, and in the second, no total values are given for, say, biological oxygen demand (BOD). This complicates the technological and cost planning processes besides making introduction of the *polluter pays principle* (PPP) difficult. In addition, the transition process in the former socialist countries is strongly affecting these figures. Some pollution is not produced continuously, so that it is impossible to design suitable technological remedies. Other potential pollution sources still exist, but the companies operating them have vanished or ceased trading. This situation poses a risk for investments.

Industrial growth rates were remarkably high in the former socialist countries because they started from a very low level of development. Investments in the heavy, textile and chemical industries led to high growth rates in the 1960s and 1970s. Therefore, these industries were technologically rather backward, and standard technologies to mitigate their environmental impacts were lacking. The rapid industrial development induced people to leave the countryside. Cities grew very quickly. Although, demand for water now exceeds 200 l per person per day, most sewage is still simply

collected in sewerage systems. There was little or no environmental inducement to treat it. The result was a fairly strong environmental impact from both municipal and industrial effluents (HELCOM, 1992b).

Naturally, the northern and western countries also had a strong environmental impact on the Baltic. However, environmental awareness, early agreements (e.g. Helsinki Convention, 1974, activities of the Nordic Council, National Environmental Programs) and the consequent investments mitigated the impacts to some extent once the programs were in place in the 1970s. The impacts from the northern and western industries stemmed mainly from the pulp and paper industries and from various metallurgical and chemical plants. However, these became less severe after legislation in the various states had forced industries to modify their processes and adopt environmentally sounder technologies in the 1970s and 1980s. In addition, in the 1970s, the northern and western economies were already developing into service-orientated economies. Nevertheless, demand for energy and water for household purposes (laundry, domestic appliances, bathing etc.) increased in these countries as standards of living improved. Since sewage treatment has become widespread, environmental impacts have decreased drastically. On the whole, the northern countries and West Germany were able to greatly reduce environmental loads from their municipali-

J. Köhn / Ecological Economics 27 (1998) 13–28　　　　　　　19

Table 3
Impacts of economic sectors and mitigation efforts

Economic sector	Share of pollution	Mitigation efforts
Municipal sewage	About 40% of waterborne nutrient inputs in human equivalents	80% In phosphorous 65% in nitrogen
Agriculture	About 60% of waterborne nutrient inputs in human equivalents	80% In phosphorous 65% in nitrogen
Agriculture	About 11% of phosphorous inputs (deposition) about 15–20% of airborne nitrogen inputs	80% In phosphorous 65% in nitrogen
Industry	Causal for all inputs of pestizides, organochlorines, etc.	90% Reduction
Industry	15–20% Of airborne nitrogen inputs	65% Reduction
Transportation	35–75% Of airborne nitrogen	65% Reduction
Tourism	?	Related to municipal impacts

ties and industries during the 1980s and 1990s. However, this process was accompanied by the appearance of new sources and increased emissions from some sectors generating environmental loads. The increasing nitrogen inputs to the sea, for instance, stem mainly from electricity generation, transportation and run-off from agricultural land.

4.2. Assessing the sectoral impacts

The growth of the *chemical and petrol industries* since World War II has affected the Baltic environment through at least three pathways. First, the discharge of industrial sewage has had a strong toxic impact on the rivers in the drainage basin, and consequently on the Baltic Sea. Second, the agricultural use of pesticides has introduced these substances to the food chain. And finally, atmospheric fall-out has led to persistent organic contamination. However, the environmental impact of most of these organic contaminants is still unknown. Pulp bleaching, metallurgical and incineration processes also produce stable organic compounds as byproducts. Despite some uncertainty regarding their chemical transformation within the ecosystem, the environmental risks of these contaminants are becoming clear, e.g. infertility among seals and organochlorine concentrations in fish from the Baltic (HELCOM and Stockholm Environmental Institute, 1990). No exact data are available for the impacts of specific substances. Moreover, the long dwell

times of some substances makes estimation of a separate damage function impossible.

Transportation accounts for 76% of NO_x emissions in Norway, 65% in Germany, 68% in Finland, and 35% in the Eastern countries (HELCOM, 1992b; Institut der Deutschen Wirtschaft, 1995). Since about 60% of the total nitrogen load stems from airborne pollution, transportation is one, but not the only, source of diffuse inputs (Table 3). Carbon dioxide output from transportation amounts to 28.5% of total carbon dioxide discharge in these countries annually. Despite its negative impact, shipping is among the beneficiaries of the Baltic Sea. About 350 000 000 tons of goods and 40 000 000 passengers are transported on the Baltic each year (Böhme, 1988).

A major contribution to the reduction of environmental loads in the Baltic Sea can be expected through changes in *agriculture*. Agricultural airborne pollution accounts for about 30–35% of the airborne nitrogen (approximately 200 000 ton/ year) and 10% of the airborne phosphorous (approximately 8000 ton/year) input. Agricultural run-off varies within the drainage basin and is influenced quantitatively mainly by the kind of land use. It depends on farming intensity and environmental conditions in the littoral countries. For example, average run-off is 24.22 kg/ha∗year in Denmark, but is much lower in Poland (3.00 kg/ha∗year HELCOM, 1992b). In general, it can be assumed that fertilizer use in the Baltic region is about 5 million ton/annum. Since productivity

20   *J. Köhn / Ecological Economics 27 (1998) 13–28*

in the Eastern countries is only half as high as in Western countries for cereals and 40% for root crops and tubers because only about 45% as much fertilizer is used per hectare, future environmental impacts from agriculture in the Eastern countries can be expected to increase. Moreover, in view of the growing livestock populations, impacts can be expected to increase through direct emission of nitrogen compounds, agricultural run-off from forage production and the discharge of manure (Table 2).

Since *industrial impacts* stem from an unknown number of compounds, heat discharge, noise and industrial sewage, calculation of a total industrial impact is difficult. Since industrial processes are based on energy use, economic analyses use carbon dioxide output to gauge industrial production (Table 2).

Impacts from *radioactivity and energy generation* have not been assessed in a specific way. Reports on existing depositories and landfills for radioactive substances draw attention to an additional risk potential that has yet to be taken into account. Six nuclear power plants and three nuclear research reactors are sited directly beside the Baltic Sea (Conference of Ministers for Spatial Planning and Development, 1994; p.39). There is also no doubt that, say, the discharge of cooling water from power plants affects river ecosystems in the drainage area as well as the Baltic Sea ecosystem itself.

Fisheries and tourism are the main beneficiaries of a healthy Baltic Sea. While the fishing industry itself has a relatively small impact on the environment[1], tourism can affect ecosystems through discharges of sewage and waste in the same way as municipalities. Moreover, tourism leads to increasing use of private means of transportation.

5. The endangered ecosystem and its value

The ecosystem of the Baltic Sea is endangered by a huge variety of inputs. However, it still provides economic benefits in the form of seaborne transport, supplies of cooling water for

---

[1] excluding fish processing

energy production, fishing, recreation, etc. In addition, further options for economic use and non-use values, for instance marine resource use and biodiversity (option and expectation values), may enhance the future utility of the whole Baltic region. One wellknown purpose of an option value about 100 years ago was the current economic use-value of tourism in this particular region. Currently, tourism accounts for about 20% of the economic benefits in the coastal region (Köhn, 1995).

It can therefore be assumed that the littoral countries share a common interest in reducing impacts on the Baltic Sea environment. However, there are still several structural, functional and long-term uncertainties in both the ecological and the social system. Estimations of the costs vary widely. While the High Level Task Force of the Helsinki Commission estimated that ECU 18 billion will be needed within a 10-year schedule (i.e. ECU 1.8 billion a year); the Stockholm Environmental Institute calculated that ECU 7 billion will be needed annually for a 60-year schedule. The calculated potential losses will, according to current estimates, be about ECU 35 billion a year for the coastal region only (Köhn, 1995). In fact, these overall losses will not materialize in all economic sectors simultaneously. However, potential restructuring processes in the industrial sector, changes in agricultural practice, potential dematerialization strategies and changes in consumer behaviour all represent uncertainties that must be taken into account. There is little doubt, that "… a demand for higher standard of living similar to that in the western economies may result in an increase in pollution load to the Baltic Sea, due to higher energy and food production, etc." (Wulff and Niemi, 1992).

If the former socialist countries cannot adopt the most efficient technologies when restructuring their economies, environmental impacts can be expected to increase by at least about 40%. Consequently, the convergence process must be based on a new political, cultural/ethical, economic/technological and scientific way of thinking in order to avoid this increase and to organize the societies around the Baltic towards a less resource-consuming lifestyle. Thus, the process of technological and economic/political transition will affect both the wealthy nations in the north and west and the eastern countries. Ultimately, the transition resulting from ecological necessity needs a cooperative system that can cope with process uncertainties. Furthermore, a successful economical and political transition process in the former socialist countries may also influence economic competition within the Baltic region. Finally, the transition process in the Baltic region will be based on and constantly produce new structural, functional and long-term uncertainties in both the environment restoration process and social progress. Therefore, the rules of the program itself are in a state of flux.

6. The relationships between ecology, economics and policy

The Baltic ecosystem consists of subsystems that differ in their ecological (e.g. hydrography, climate, biodiversity), socio-economic, cultural and political features. This is why the subsystems react differently to environmental impacts and knowledge concerning one regional subsystem cannot be applied in a linear manner to another. In view of this, it seems quite impossible to simulate a complex system like the Baltic Sea and its drainage area in a model. However, it might be possible to develop an input-output model for the Baltic Sea, its parts and the watershed regions in

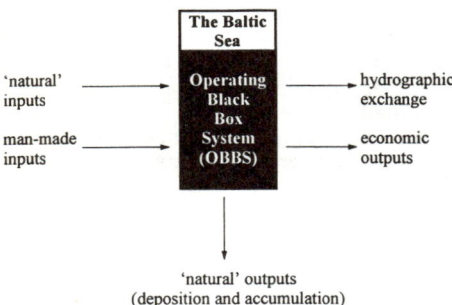

Fig. 1. The Baltic Sea as an input — 'operating black box' — output system

its drainage basin. With this approach, each specific part of the system is regarded as an Operating Black Box System (OBBS, Fig. 1) of which the structural changes in stocks and flows, system functions and time-dependent changes can be studied. The approach can also be extended to the economic sector. However, as Von Bertalanffy (1932) (p. 85) pointed out, a system is more than the sum of its parts, so the puzzle must be assembled with great caution.

Such an OBBS model for the Baltic Sea system as a whole shows, for instance, that it is a basin collecting the natural and man-made emissions from the landscapes and societies located in its drainage basin. The estimated current annual total load includes about 980 000 tons of nitrogen, 50 000 tons of phosphorous besides an unknown total load in terms of BOD, heavy metals and persistent organic contaminants. The inputs from non-point sources, and even from some point sources, are largely unknown. Some of them ultimately return to the Baltic states through fishing. Moreover, accumulation in the food chain and deposition within the ecosystem affect both the structure and function of the ecosystem. Energy and/or material balance approaches may apply. However, knowledge concerning details of these processes is lacking (Elmgren, 1989; SWEPA, 1990). Wulff and Stigebrandt (1989) reported a total sink of nutrients within the Baltic and an export of about 10% to external sources.

Since ecological processes depend on time scale patterns and the stability of the system as a whole (Arrow et al., 1995), we assume further that two states A (the state up to 1950) and B (the present state) can be described with certainty. Although the Baltic Sea ecosystem also received major inputs in previous centuries, it remained in a state of equilibrium (state A). This state differs slightly from the 'natural' state, but we have no, or only unsubstantiated, information about what the 'natural' state was like. Moreover, since the Baltic Sea ecosystem is comparatively young, it is hard to say which effects are evolutionary and which are anthropogenic. We may define state A as a 'near natural' state in which nutrient inputs stem mainly from weathering in the watershed region. These inputs are defined as 'natural fertilizers'.

22 *J. Köhn / Ecological Economics 27 (1998) 13–28*

It was only after environmental impacts had increased and various kinds of man-made inputs began to appear from the 1950s/1960s onwards that the ecosystem responded, after a considerable time lag, in the mid-1970s. After the annual loads (artificial fertilizers) had stopped increasing in the 1980s, the ecosystem became stable again (adapted state B). This stability was achieved through higher turn-over rates in the ecosystem, a higher rate of primary production, a higher rate of deposition, an increase in biomass and, consequently, a higher demand for dissolved oxygen to supply particular elements of the ecosystem. Thus, the oxygen supply governs the carrying capacity of at least part of the ecosystem.

When oxygen is lacking, the system temporarily becomes unstable again. This temporary state C, however, has been seen to occur only in some parts of the Baltic Sea ecosystem, notably the deep basins. State C may regarded as a microbial ecosystem affecting the entire system, for instance, by making it impossible for cod to spawn.

Reports concerning the Baltic ecosystem suggest that the critical loads for a sustainable Baltic Sea environment are those reported in the 1930s and 1940s (HELCOM and Stockholm Environmental Institute, 1990, state A). Although this 'normal' state of the Baltic Sea ecosystem may have been stable for a few thousand years, the shift towards states B/C indicates that the Baltic Sea ecosystem is intrinsically resilent and may be sustainable at various levels of, say, productivity. In view of this, we can assume that the task facing the economic policy complex is to stabilize state B. This in turn implies that self-regeneration (the recovery potential) must be stimulated. Self-regeneration and self-(re)organization will depend on, first, the rate at which inputs are reduced and, second, the recovery capacity of the Baltic Sea itself. Even if the recovery potential can be sufficiently increased through these two factors to equalize the former buffer capacity of the system, the Baltic Sea ecosystem still cannot return to the previous 'natural' state A that policy assumes to be possible. The presence or self-stimulation of the recovery potential (similar to trade-off effects in economic theory) may be used as part of the trade-off effects to attain a state B' — which

might be somewhat better than B from an environmental point of view. This aspect should be taken into account during monitoring to formulate advice on future investments.

However, the very unpredictable nature of these processes may also lead to an outcome that is neither state B or C, but a completely unknown state D. Such 'true uncertainty' is beyond the scope of policy, which is better able to handle risk management and, to some extent, 'process uncertainty' (by encouraging appropriate institutional frames; Köhn, 1997). Naturally, the assumption that chaotic changes may lead to an unpredictable state (D) might affect policy, but since this does not provide any basis for decision making in the sense of giving priority to environmental policy, it should be excluded from our further analysis.

Accepting that economic developments in the Eastern Baltic Littoral States will lead to convergence with the economic condition currently enjoyed by the western countries, environmental impacts must be expected to increase (Table 2). Therefore, stabilization of the present state should possibly be given long-term priority and might even be realistic in view of the potential changes.

The environmental impact assessments of HELCOM and Stockholm Environmental Institute, 1990, the results of the local impact estimations by JCP, 1992, the sectoral impact assessments and the economic potential of the Baltic Sea region permit estimation of the economic value and potential damage. The target 'stabilization of state B' also allows calculation of the potential annual total damage and this can be used to allocate the total annual costs caused by environmental damage or needed to prevent it. This approach seems to be very rigid, however, it may provide an operational basis for introducing the *polluter pays principle*.

7. Risks resulting from the decision making process

Since states A and B are both 'stable' states of the ecosystem, the political intention "to assure the ecological restoration..., and preservation of its ecological balance" (HELCOM, 1992a) is am-

J. Köhn / Ecological Economics 27 (1998) 13–28   23

biguous insofar as more than one stable state of a system can exist. Following the basic report (HELCOM and Stockholm Environmental Institute, 1990), policy seems to prefer state A, that is, the program refers to 'active restoration'. In other words, the system is to be restored to the 'natural' state seen in the 1950s by reducing at least nitrogen and phosphorus loads from point sources by 65 or 80%. Although nobody knows if this would be enough, this alone implies investment in, for instance, sewage treatment facilities corresponding to the Best Available Technology (BAT) with a very high purification rate. Such a requirement would increase investment and operating costs by factors of 7 and 3, respectively, due to the marginal cost theory. Even then, we cannot be sure that it will stimulate the self-regeneration of Baltic Sea subsystems. If investments are channeled into single point-source management, local effects may be too small to influence the whole subsystem or even neutralized by nonpoint-source pollution such as transportation or agricultural emissions. There is therefore a need for integrated investment management. From the economic standpoint, the orientation of all investments towards achieving the technologically feasible best mitigation target raises the risk of each investment. The risk may be reduced by specifying targets (policy approach), waiting (industrial sector approach) or institutionalization (e.g. shared responsibility approach).

8. Economic assessments for the Baltic Sea — Consumer approach

Various techniques have been developed during the past 30 years to estimate values and demand functions for natural resources and, more recently, ecosystems. The contingent valuation method could provide a basis for including these non-use economic values in the current decision making process (Bergstrom and Loomis, 1995) and, economically speaking, to include them in cost-benefit-analyses to make sure that options and expectations (future values) are suitably weighted when a decision is made. These techniques need and are based on a comprehensive

system of information about the commodity or system under valuation and about the population asked for its preferences concerning a distinct subject. Although the technique involves some risks, the contingent valuation method might yield some insights enabling, for example, estimation of the (accepted) annual losses (*willingness to accept*) or the annual expected (individual) benefits (*willingness to pay*, cf. Freeman, 1985). The former leads to the *victim pays principle* (VPP) and the latter could facilitate introduction of a *polluter pays principle* (PPP). However, since the Baltic region is split into two distinct economic regions comprising the northern/western and eastern states, respectively, direct measurement of both the *willingness to pay* and the *willingness to accept* is impossible. Moreover, the perception of ecological damage within the Baltic Sea region and the ecological and social restoration process bears many uncertainties so that contingent valuation approaches seem inappropriate.

9. Estimating the economic value of the Baltic Sea Area — Productivity approach

These analyses are based on estimations of demographic data for the Baltic Sea region (The World Resources Institute, 1994). In addition, German labour productivity data have been taken into account in assessing economic performance potential (Institut der Deutschen Wirtschaft, 1995). However, the total labour force in the Baltic littoral countries is about 28.5 million, while its current economic performance based on individual labour performance in the Baltic region is at least 1100 billion ECU annually. This is about 40% of the potential total economic performance if the Eastern Baltic Littoral States were as efficient as the Western ones.

The realization of economic benefits in the coastal region depends on both distinct sectoral efforts to achieve a healthy environment and the economic performance potential of each sector. Therefore, sectoral utility was analyzed before calculating the potential losses. Benefits can be expected to accrue at least from the use of the Baltic Sea as a resource supplier for waterborne transport, energy production, fisheries and fish processing, tourism, environmental industries, extraction of nonrenewable resources, climate and nature conservation. Although the data base is very limited, total annual benefits surely exceed ECU 35 billion. However, benefits from investments in environmental industries, for example, are difficult to assess owing to their ambivalent character and, in addition, transport does not always depend on a healthy environment. Since the above value is probably an underestimate of the annual benefits, it can at least serve to calculate the estimated annual loss risk. Moreover, it does not include the benefits estimated to accrue by value enhancement in cities located on or near the coast (Köhn, 1995).

10. Efforts to restore Baltic Sea — sources and sectoral impacts

The political target of JCP includes mitigation efforts (Table 1). These should be translated into sectoral mitigation strategies (Table 3). Although, these figures show the divergence between political goals and reality, they do not include the compensatory effects of the growing economies in the Eastern Baltic States (Table 2). The tables include sources of nonpoint pollution. These rates of pollution are not easy to reduce. In addition, huge uncertainties are encountered as regards causes, accumulation and release effects as well as economic management when dealing with these specific kinds of human impact on the environment. Since these are crucial components of the restoration program, the need for economic or political regulation is obvious. This, however, implies new institutional frameworks as well as the application of appropriate economic instruments.

11. Responsibilities for a(n) (un)sustainable development in the Baltic Sea region

This section will deal with economic performance in the Eastern and Western Baltic Littoral States. National economic performance will be described by means of a set of available economic indicators. Per capita availability of economic supplies will be estimated in a second step. Then, the relative per capita equipment will be used to show the per capita impacts and the potential economic growth rates assuming equal rights to consumption in East and West. Without dematerialization of production and consumption, if standards of living converge, environmental impacts can be expected to increase at the rate shown by the ratios (Table 2, column 4). Therefore, assuming that production and consumption will converge and that equity for the present generation is crucial to sustainability, it is possible, to estimate first, the threat of unsustainable development in the Baltic Sea region, second, the past and present responsibility for damage to the Baltic Sea environment in a national perspective, and third, the need to reduce production and consumption and rate of technological progress needed to do so (dematerialization or sufficiency revolution). This, finally, will enable the establishment of an institutional framework to overcome the obstacles of the unsustainable growth mania currently pervading the economy.

12. Funding Baltic Sea region sustainability

The JCP target of restoring the ecological balance of the Baltic Sea environment by reducing environmental impacts to levels similar to those from 1930–1940 appears unrealistic in view of the economic indicators and estimates of growth potentials in the Eastern Baltic Littoral States, at least in the period (1993–2012) envisaged by the program. However, impact mitigation is doubtless necessary to sustain the Baltic Sea environment. In addition, simple models suggest that at least some recovery of the Baltic Sea ecosystem might be expected. Therefore, it may be implied that the first step towards sustaining the Baltic Sea ecosystem involves the subtarget of stabilizing the present state. Economically speaking, this means avoiding further damage by implementing suitable technologies, both economic and political institutions and encouraging social participation.

Avoiding further damage corresponds at least in theory to implementing mitigation strategies on the same scale. In financial terms, this involves reinvesting income from the economic process in mitigation strategies. Although, this will lead only to financial equilibrium it may help to build up institutions that will reduce the uncertainties when setting intermediate targets and in process monitoring. Since one may estimate annual potential losses of ECU > 35 billion, the *willingness to pay* matrix and existing charges may yield about 34 billion ECU annually, financial equilibrium may be achieved by institutionalization of the *polluter pays principle*.

On the contrary, the *victim pays principle* is excluded as unrealistic. About 60% of airborne pollution does not stem from beneficiaries of the Baltic restoration process. But as at least four EU states (Denmark, Finland, Germany and Sweden) will benefit from the Baltic Sea restoration process, other pollutant source countries in the EU will ultimately also be involved in Baltic Sea restoration through EU regulations. It seems reasonable to assume that those Western Baltic Littoral States that acquired economic benefits from the past environmental degradation of the Baltic Sea will at least to some extent be willing to subscribe to a Baltic Sea Restoration Fund rather than pay for mitigation strategies in comparatively wealthy nations further west. In addition, the program may be broadened from a consumer program to one involving joint participation, including economic sectors such as industry and financial institutions.

Considering the economic indicators the following assumptions for the PPP may hold for the institutional structure. The ratio between per capita GNP in the Eastern and Western Baltic Littoral States shows that although over 60% of the population lives in the Eastern States, 75% of the GNP is created in the Western States. The ratios for the impacts of the various economic sectors are (in total amounts, West:East) in agriculture 2:1; transportation 2:1; sewage discharge 1:2; industry 1:1; tourism 3:1. Western impacts account for about 2/3 of the total. This means that about 2/3 of the PPP payments are due to the Western Baltic Littoral States. These figures also show the following distribution in economic sectors (production and consumption, Fig. 2). In addition, a

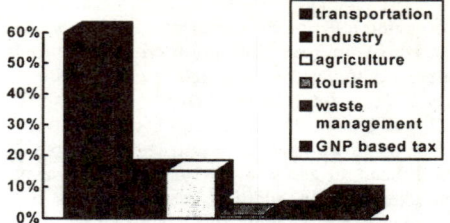

Fig. 2. Polluters pay in western Baltic Littoral States (ECU 22 billion annually)

GNP based tax is proposed. This would reflect direct interest in the form of an incentive to mitigate impacts on a shorter time scale. This portion of the tax would decrease as the environment improves in absolute terms and as the relative shares of the Eastern and Western Baltic Littoral States converge. Consequently the share of the Eastern Baltic Littoral States would be about 1/3. However, the sectoral shares are different, as is the participation of financial institutions. Moreover, the lower the per capita GNP, the smaller the share of the GNP-based tax (Fig. 3).

13. Institutional structure for Baltic Region sustainability policies

Since environmental policy has often led to inefficient ecological outcomes, this paper focuses on a strong economic approach: the greatest possible deregulation and an invitation to all social players to take part in the Baltic Sea restoration process. However, uncertainties appear in both

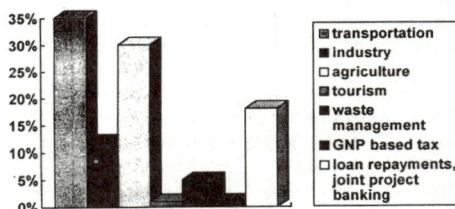

Fig. 3. Polluters pay in Eastern Baltic Littoral States (ECU 12 billion annually)

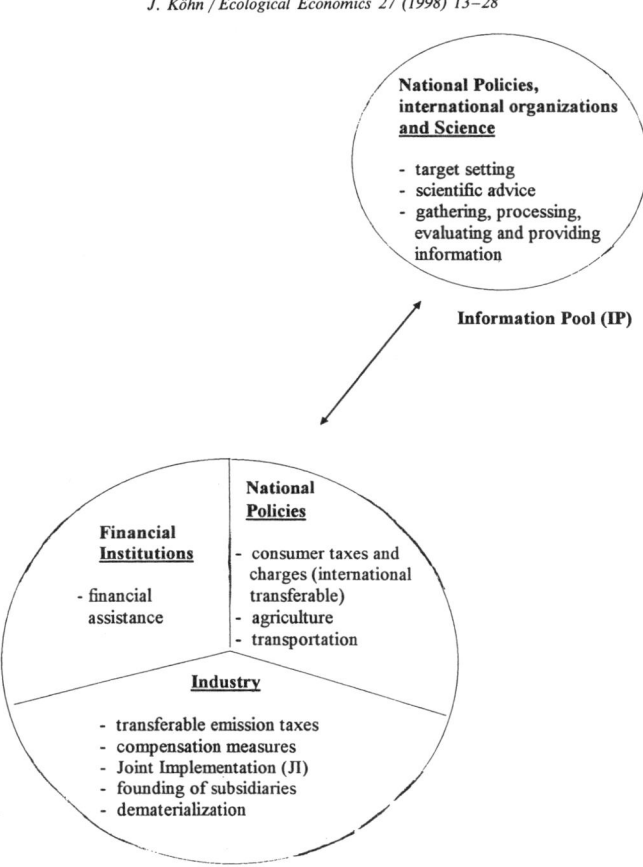

Fig. 4. An opportunity for an institutional frame to govern the Baltic Sea region sustainability process

the regulatory and the deregulatory approach. But uncertainties may give rise to additional costs for transactions in dealing with resources and socioeconomic activities. Least cost planning in the case of Baltic Sea restoration, therefore, is a challenge to policy and economic theory.

Keohane and Ostrom (1995) show in their analyses concerning common property resources that institutions may govern the commons through cooperation of the players. The Baltic Sea sustainability approach is a multi-species, multi-resource one with heterogeneous participants and a long time horizon. This implies that outcomes are unpredictable and often uncertain. It also implies that priority must be given to erecting new institutions to govern the common property resource 'Baltic Sea' and to cope with the inherent uncertainties of its systems and processes. The main structure of the Baltic Sea negotiation program may as shown in Fig. 4.

Players in the Baltic Sea sustainability program are beneficiaries of the sustaining process, those who will benefit directly from environmental improvements and those who, for these reasons and to save costs, are willing to pay and participate. The program structure may be based on an ongoing negotiation process, should be hierarchically

● supernegotiation: Finance Pool — Information Pool
● negotiations within the Information Pool: National Policies — National Policies; National Policies — Science
● negotiations within the Finance Pool: National Policies — (National) Economic Sectors and Consumers; and the
● subnegotiations: National Policies — Agriculture — Industry — Transportation — Tourism — Waste Management — Consumer Taxation — Financial Institutions; Industry — Financial Institutions — Municipalities — Industries — Agriculture; Municipalities — Financial Institutions, etc.

The negotiation structure may be organised on local, subregional, watershed, economic sector, industrial network or other scales. The approach conforms to Köhn's proposals for user-complexes for certain resources within the Baltic Sea region (Köhn, 1990) and Lundgren and Mattsson's proposals for industrial networks to support industrial change in the transition processes taking place in the Eastern economies (Lundgren and Mattson, 1995).

The *Information Pool* may supervise the sustainable development process in the Baltic Sea region by setting targets, providing scientific advice in both the natural and social sciences, monitoring the processes and gathering, processing, analyzing and disseminating information to the *Finance Pool*. The players in the *Finance Pool* will develop implementation strategies and act accordingly. Their negotiations will focus on cooperative funding and creating associations of stake holders at the local, regional, national and Baltic Sea region levels. National policies will be responsible for consumer integration, agriculture and transportation, while industry will receive an international focus. Industrial interests will be invited to set up subsidiaries in the Eastern Baltic Littoral

States, to provide funds for compensation where further investments in their own production facilities are of marginal utility in relation to the compensation needed elsewhere and to pay taxes to the *Finance Pool* for environmental damage, to implement environmentally sound technologies (joint implementation, partly repaid from the tax pool to which they contribute), and so on. Financial institutions will be invited to offer financial assistance.

14. Concluding remarks

This paper raises questions concerning an approach for achieving sustainability policies of the Baltic Sea. Since the Baltic Sea is a multi-species, heterogeneous common property resource, the environmental sustainability concept conflicts with social goals such as convergence between the social and economic systems in the Eastern and Western Baltic Littoral States. The transition process and system-inherent changes in time involve uncertainties and economic risks for investors, but management of the common resource by a participatory network will reduce the uncertainties and encourage investment in the Baltic sustainability process. This process will be long, however.

This paper uses various approaches offered by economics theory. It purposely avoids detail, as its principal aim is rather to give a general analysis of possible ways to make regional sustainability feasible. As the studies presented here are based on an on going research project, the results must be regarded as provisional.

Acknowledgements

The author expresses his thanks to Caroline Carp, Faye Duchin, Richard Bishop, Daniel Bromley, John Gowdy, Xiannuan Lin, Talbot Page, Bernard C. Patten, Wilfried Siebe, Peter Söderbaum for helpful discussions. Thanks are also due to Brian Patchett (Intertext Rostock) for editing the English text. Remaining mistakes are my responsibility.

References

Arrow, K., Bohlin, B., Costanza, R., Dasgupta, P., Folke, C., Holling, C.S., Jansson, B.-O., Levin, S., Mäler, K.-G., Perrings, C., Pimentel, D., 1995. Economic growth, carrying capacity, and the environment. Ecol. Econ. 15, 91–95.

Ayres, R.U., 1988. Self-organization in biology and economics. International Institute for Applied Systems Analysis, Laxenburg: p. 34.

Bergstrom, J.C., Loomis, J.B., 1995. Economic dimensions of ecosystem management. Department of Agricultural and Applied Economics, University of Georgia, Faculty Series 95–5, p. 22.

Berkes, F., Folke, C., 1994. Linking social and ecological systems for resilience and sustainability. Beijer Discussion Paper Series 52, 15.

Böhme, H., 1988. Der Wettbewerb im Ostseeverkehr. Internationales Verkehrswesen 40, 383–387.

Common, M., Perrings, C., 1992. Towards an ecological economics of sustainability. Beijer Discussion Paper Series 2, 30.

Conference of Ministers for Spatial Planning and Development, 1994. Vision and strategies around the Baltic Sea 2010. Towards a framework for spatial planning in the Baltic Sea region. Report Third Conf., p. 96.

Convention on the Protection of the Marine Environment of the Baltic Sea Area, 1992. p. 41.

Costanza, R., Patten, B.C., 1995. Defining and predicting sustainability. Ecol. Econ. 15, 193–196.

Daly, H.E., 1977. Steady-state Economics. The Economics of biophysical Equilibrium and Moral Growth. W.H. Freeman, San Francisco, CA.

Elmgren, R., 1989. The eutrophication status of the Baltic Sea: input of nitrogen and phosphorous, their availability for plant production, and some management implications. Baltic Sea Environ. Proc. 30, 12–31.

Freeman, A.M., III, 1985. Methods for assessing the benefits of environmental programs. In: Kneese, A.V., Sweeny, J.L. (Eds.), Handbook of Natural Resource and Energy Economics, North-Holland, Amsterdam, pp. 223–270.

Goodland, R., 1995. The concept of environmental sustainability. Ann. Rev. Ecol. Syst. 26, 1–24.

HELCOM and Stockholm Environmental Institute, 1990. Current state of the Baltic Sea. Ambio, special report 7.

HELCOM, 1992a. The Baltic Sea Joint Comprehensive Environmental Action Programme. Preliminary version. Agenda item 5, Ch. 1–7.

HELCOM, 1992b. Background document for the Baltic Sea environmental declaration, 1992. Agenda item 5/2., p. 15.

Holling, C.S., 1986. The resilience of terrestrial ecosystems: Local surprise and global change. In: Clark, W.C., Munro,

R.E. (Eds.), Sustainable Development of the Biosphere, Cambridge University Press, Cambridge.

Institut der Deutschen Wirtschaft, 1995. Zahlen zur wirtschaftlichen Entwicklung der Bundesrepublik Deutschland, Köln.

Keohane, R.O., Ostrom, E. (Eds.), 1995. Local Commons and Global Interdependence. Sage Publications, London.

Köhn, J., 1990. Restoring the ecological balance: a common prospect for the future. Framtider Int. 1, 24–25.

Köhn, J., 1995. Wirtschaftsraum Ostsee. Ringvorlesung. Die Ostsee unser Lebensraum. Institut für Ostseeforschung, pp. 58–67.

Köhn, J., 1997. Complexity and uncertainty in natural and social systems. Postdoctoral thesis, forthcoming.

Lovelock, J., 1995. The Ages of Gaia. W.W. Norton, New York.

Ludwig, D., 1993. Uncertainty, resource exploitation and conservation: lessons from history. Science 260, 17–53.

Lundgren, A., Mattsson, L.-G., 1995. Industrial change and ecological consequences in the transition process of the Eastern economies — the dynamics of industrial network. In: Köhn, J., Schiewer, U. (Eds.), The Future of the Baltic Sea. Metropolis, Marburg, pp. 131–153.

Malthus, T.R., 1798/1997. An essay on the principle of population. In: Neliesseu, N., van der Straaten, J., Klinkers, L. (Eds.), Classics in environmental studies. An overview of classic texts in environmental studies. International Books, Utrecht, 29–38.

Mill, J.S., 1848/1965. Principles of political economy. University of Toronto Press, Toronto.

Serafin, R., Zaleski, J., 1988. Baltic Europe, Great Lakes America and ecosystem redevelopment. Ambio 17, 99–105.

Solow, R., 1974. The economics of resources or the resources of economics. Am. Econ. Rev. 15, 1–14.

Swedish Environmental Protection Agency, 1990. Large scale environmental effects and ecological processes in the Baltic Sea. Research programme for the period 1990–1995 and background documents.

The World Resources Institute, 1994. World resources 1994–1995. People and the environment. Oxford University Press, Oxford.

Von Bertalanffy, L., 1932. Theoretische Biologie, Verlag von Gebrüder Borntraeger, Berlin.

WCED (World Conference on Environment and Development), 1987. Our Common Future. Oxford University Press, Oxford.

Wulff, F., Niemi, A., 1992. Priorities for the restoration of the Baltic Sea — A scientific perspective. Ambio 21, 193–195.

Wulff, F., Stigebrandt, A., 1989. A time-dependent budget model for nutrients in the Baltic Sea. Global Biochemical Cycles 3, 63–78.

[29]

ELSEVIER

Ecological Economics 30 (1999) 333–352

ECOLOGICAL ECONOMICS

www.elsevier.com/locate/ecolecon

ANALYSIS

Managing nutrient fluxes and pollution in the Baltic: an interdisciplinary simulation study

R. Kerry Turner [a,*], Stavros Georgiou [a], Ing-Marie Gren [b], Fredric Wulff [c], Scott Barrett [d], Tore Söderqvist [b], Ian J. Bateman [a], Carl Folke [b,c], Sindre Langaas [e], Tomasz Żylicz [f], Karl-Göran Mäler [b], Agnieszka Markowska [f]

[a] *Centre for Social and Economic Research on the Global Environment, University of East Anglia and University College London, Norwich, NR4 7TJ, UK*
[b] *Beijer International Institute of Ecological Economics, The Royal Swedish Academy of Sciences, Box 50005, S-10405 Stockholm, Sweden*
[c] *Department of Systems Ecology, University of Stockholm, 106 91 Stockholm, Sweden*
[d] *London Business School, London, NW1 4SA, UK*
[e] *UNEP/GRID-Arendal, P.O. Box 1602, Myrene, N-4801 Arendal, Norway*
[f] *Warsaw Ecological Economics Centre, Economics Department, Warsaw University, 00-41 Warsaw, Poland*

Received 21 July 1997; received in revised form 6 April 1999; accepted 15 April 1999

Abstract

This interdisciplinary paper reports the results of a study into the costs and benefits of eutrophication reduction in the Baltic Sea. A large multidisciplinary team of natural and social scientists estimated nutrient loadings and pathways within the entire Baltic drainage basin, together with the costs of a range of abatement options and strategies. The abatement cost results were compared with clean-up benefits on a basin-wide scale, in order to explore the potential for international agreements among the countries which border the Baltic. Most countries would seem to gain net economic benefits from the simulated 50% nitrogen and phosphorus reduction policy. © 1999 Elsevier Science B.V. All rights reserved.

Keywords: Nutrient pollution; Eutrophication; Cost-effective abatement costs; Environmental benefits valuation; Contingent valuation

1. Introduction

All countries with a coastline have an interest in the sustainable management of the coastal re-

* Corresponding author. Tel.: + 44-1603-593176; fax: + 44-1603-593739.
E-mail address: r.k.turner@uea.ac.uk (R.K. Turner)

0921-8009/99/$ - see front matter © 1999 Elsevier Science B.V. All rights reserved.
PII: S0921-8009(99)00046-4

source systems. The task of sustainable management, i.e. sustainable utilisation of the multiple goods and services provided by coastal resources (processes, functions and their interrelationships), is likely to be made more difficult because of the consequences of global environmental change (GEC). Understanding the interactions between the coastal zone and global changes cannot be achieved by observational studies alone. Modeling of key environmental processes also has an important role to play. In particular, modelling work on the dynamics of carbon (C), nitrogen (N) and phosphorus (P) in the coastal ocean needs to be combined with socio-economic analysis of the drivers of C, N and P fluxes and the human welfare consequences of changes in these fluxes across the coastal zone over time.

A particular characteristic of GEC (encompassing population growth and density increases, urbanisation and the intensification of agriculture, etc.) is that it has led to, among other things, the progressive opening of biogenic nutrient cycles, e.g. much increased mobility of nitrogen and phosphorus. This increased mobility of nutrients has meant increased exchanges between land and surface water and consequent impacts on the ecological functioning of aquatic systems. Other process changes have also added to the cumulative changes experienced in coastal systems.

The major flux of nutrients from land to sea occurs through river transport via the drainage basins network. The network contains various 'filters' such as wetlands which retain or assimilate nutrients during their downstream passage to the sea. The effectiveness and selectivity of these filters depend on the strong biogeochemical coupling existing between carbon, nitrogen, phosphorus and silica circulation. They are also affected by hydrology and land use/cover (Howarth et al., 1996).

This paper reports the overall results of an interdisciplinary study which focused on N and P fluxes on a drainage basins-wide scale in the Baltic Region (Turner et al., 1995). The Baltic Sea region catchment area covers around 1 670 000 km² and contains a population of about 85 million people in 14 countries. A significant proportion of the world's industrial production comes from this

area, but up until around 40 years ago there was little recorded environmental damage in the Baltic Sea. However, since 1960 the environmental condition of the Baltic Sea has increasingly become a cause of public concern and is currently perceived to be in an unacceptably polluted state. Symbolically, eutrophication is a major problem facing policymakers and the public. The aims of the study were decomposed into a number of interrelated intermediate goals including: (1) to provide a comprehensive and rigorous picture of the land use and ecological carrying capacity of the region, and to relate this resource inventory to the patterns of human activity in the region; a 'pressure-state-impacts-response (P-S-I-R) framework (Fig. 1) was adopted to facilitate the analysis; (2) to develop a model looking at different nutrient loading scenarios and their consequences on the ecological state of the Baltic Sea and its sub-systems; (3) to estimate the costs of various strategies designed to reduce the nutrient loading of the Baltic Sea, and the identification of the most cost-effective nutrient abatement options; (4) to estimate the economic valuation of eutrophication damage to the Baltic Sea; (5) to increase our understanding of the institutional issues involved in the management of the Baltic Sea.

The GEC process is a complex flux of factors, the impacts of which can manifest themselves at a number of different spatial and temporal scales. It is, however, possible to identify a group of interrelated socio-economic trends and pressures which both contribute significantly to the Baltic's environmental change impacts, as well as to an increasing degree of environmental risk to the marine ecosystem and the surrounding drainage basins' biophysical and socio-economic systems. This paper therefore seeks to analyse the problems of the Baltic region in terms of a 'pressure-state-impacts-response' (P-S-I-R) framework (Fig. 1).

2. Pressure-state-impact-response framework

Within the Baltic Sea area the northern sub-basins (Bothnian Bay, Bothnian Sea) have a low population concentration, extensive forests, wet-

R.K. Turner et al. / Ecological Economics 30 (1999) 333–352   335

lands, lakes and a mountainous terrain. The southern sub-basins (Baltic Proper) contain 55 million of the 85 million population and have significant agricultural areas. The Baltic Sea itself has a total surface area of 415 000 km² and because of its semi-enclosed character has a very slow water exchange, the mean residence time for the entire water mass being of the order of 25–30 years (Folke et al., 1991). As we shall see this combination of biophysical and socio-economic characteristics has important implications for the environmental vulnerability of the Baltic Sea and its resource system.

An increasing degree of environmental pressure has been felt in the Baltic region as a result of a range of socio-economic drivers. The outcome has been that the Baltic Sea and coastal zone resources (including the waste assimilative capacity) have been subject to a range of, often competing, usage demands. In this paper we pay particular attention to nutrient (nitrogen and phosphorus) pollution of the Baltic Sea and its consequences. Evaluating the importance (in human welfare terms) of the various environmental impacts requires that their effects be measured in biophysical and then in monetary terms. This gives us some measure of the state of the Baltic environment and the importance of the environmental degradation that has taken place. In addition, we seek to identify the causes of the problem, priorities for action and cost-effective policy instruments.

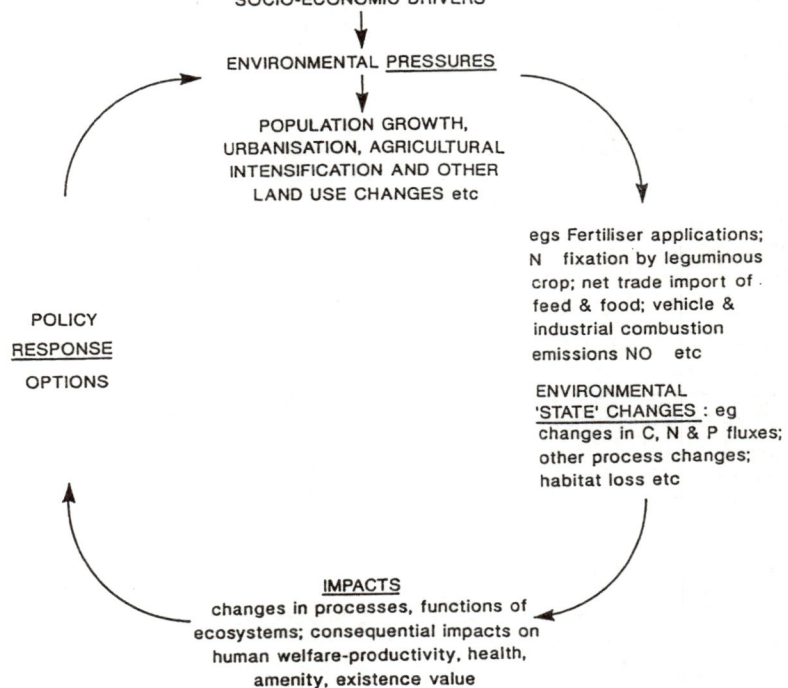

Fig. 1. P-S-I-R cycle, continuous feedback process.

The Baltic Sea is the largest brackish body of water in the world and is a naturally very sensitive area. It depends on short- and long-term variations in climate and has several times in its history transformed from lake to sea, from freshwater to saline water. Vertical variations in salinity cause permanent stratification, hampering the exchange of oxygen in the deeper parts of the sea. In some years as much as 100 000 square kilometres (nearly 25% of the total area) approach 'dead bottom' conditions.

The natural vulnerability of the area has been magnified by the magnitude and extent of socio-economic activities, impacts and interventions that have become commonplace since the 1950s. The economic and the environmental systems are now sufficiently interrelated as to be jointly determined. They are now in a process of coevolution. Because of the sheer scale of economic activity the pollution generated is a pervasive problem across the drainage basin and beyond. Localised solutions, for example at the municipal level, are no longer sufficient; international co-operative agreements and actions are required.

The principal pollution sources for the Baltic Sea are municipalities, industries and agriculture, located both in the coastal zone and also beyond in the drainage basin. A range of pollution pathways can also be identified, including diffuse 'non-point' sources such as airborne emissions and agricultural run-off, and 'point' source pollution from urban areas and industry.

Inadequate or absent municipal sewage treatment in the eastern, southern and south-western sections of the drainage area pose problems which are exacerbated by the synergistic effects of untreated industrial effluent wastestreams passing through the same facilities and into the rivers and the sea. Agricultural practices, including intensive livestock husbandry, are also a major contributor to the high nutrient load and consequent eutrophication problem. Current loads of nitrogen and phosphorous entering the Baltic sea are three times those of the 1950s (Nehring et al., 1990). The resulting excessive biomass growth causes oxygen depletion when it decays and threatens marine life. Eutrophication is now pronounced in the Gulf of Finland, Gulf of Riga and in limited coastal areas in the eastern, southern and south-western Baltic Sea area. Many toxic algae outbreaks have also occurred.

It is important therefore to know what the principal causes of the resource degradation and pollution problems in the Baltic Sea are. One way of analysing these problems is to identify a set of interrelated 'failures' phenomena which seem to underlie the degradation and quality decline trends. There are two main related 'failures' which can be distinguished—market failure and policy intervention failure–which when combined with scientific and social uncertainties (information failure) can account for the environmental damage process.

3. Market and policy intervention 'failures'

Table 1 presents a typology of market and intervention failures which is relevant to the Baltic context. The most widespread type of market failure is that of pollution externalities. The external costs result from waste generators (municipalities, industry and farms) who over-utilise the waste assimilative capacity of the ambient environment, e.g. rivers and the Baltic Sea, because this environmental function is perceived to be virtually free of charge (absence of market prices). Some waste generators also have had over time close to open access to the marine waste repository.

Government interventions have also been partly responsible for the environmental degradation process in the Baltic. The effectiveness of the regulation of sewage treatment facilities and practices, for example, varies dramatically from country to country around the Baltic. There is a general absence of properly integrated coastal resource management policies and water catchment management and planning. This has resulted in intersectoral policy inconsistencies and resource depletion and degradation, with the loss of wetland ecosystems being an important damage impact.

Although these 'failures' phenomena are or were pervasive across the entire drainage basin, they tend to be focused in greater numbers and

R.K. Turner et al. / Ecological Economics 30 (1999) 333–352

Table 1
Market and policy intervention failures in the Baltic region

Market failures		
1. Pollution externalities		
(a)	Air pollution, outside catchment sources, e.g. North Sea area	Excess levels of nitrogen and ammonia contributing to eutrophication of water bodies
(b)	Water pollution, land-based within catchment sources	Excess nitrogen and phosphorus from sewage and agricultural sources; industrial wastewater and toxic effluent pollution particularly from the pulp-and-paper industry
(c)	Water pollution, coastal and marine sources	Excess nitrogen and phosphorus from coastal sewage outfalls; oil spills and contaminated bilge water from ships
2. Public goods-type problems		
(a)	Ground-water depletion/surface-water supply diminution	Over exploitation on- and off-site of wetlands' water supply
(b)	Congestion costs, on-site	Recreation pressure on beaches, wetlands and other sensitive ecosystem areas
(c)	Fisheries yield reduction	Over exploitation due to badly defined property rights
Intervention failures		
3. Intersectoral policy inconsistency		
(a)	Competing sector output prices	Agricultural price fixing and associated land requirements
(b)	Competing sector input prices	Tax breaks or outmoded tax categories on agricultural land; or tax breaks for non-agricultural land development, including forestry; land conversion subsidies; state farming subsidies (historical)
(c)	Land-use policy	Zoning; regional development policy; direct conversion of wetlands policy; waste disposal policy and regulation (uncontrolled waste disposal dumping)
4. Counterproductive policy		
(a)	Inefficient policy	E.g. policies that lack a long-term structure; wastewater and industrial effluent combined treatment practices; general lack of enforcement of existing policy rules and regulations
(b)	Institutional failure	Non-integrative agencies structure, non-existent agencies; lack of monitoring and survey capacity; lack of information dissemination; lack of public awareness and participation

with greater severity in Poland, Russia, the Baltic Republics and the Slovak and Czech Republics, partly as a result of the historical legacy left by a central planning system based on input intensive, inefficient heavy industry complexes.

In the Nordic countries and the western parts of Germany municipal and industrial pollution loads have been significantly reduced over the last few decades. Nevertheless, the agricultural sector poses problems due to the intensive nature of the farming regimes.

The dire message for Baltic policymakers in the future is clear — if the agricultural sectors in Poland, etc., develop intensive methods similar to those fertiliser/pesticide-dominated regimes commonplace in Denmark and Sweden, the outlook for the reduction of eutrophication pollution is poor.

4. Land use, nutrient loads and damage in the Baltic Sea

The current status of the Baltic Sea is determined by the set of activities present in the entire drainage basin. The load of nutrients to the various sub-drainage basins is determined by several factors such as land use, population density, climate, hydrology, and air transportation of nitrogen oxides and ammonium. A set of geographic information system (GIS) map layers were created and used to generate information on the current landscape characteristics and population distribution patterns in the drainage basin. The map layers included: land cover, drainage basis, administrative units, population distribution, arable lands, pasture lands, wetlands, and the terrestrial net primary production of the Baltic drainage basin. A description of the technical procedures and the primary data sources used to create each layer, as well as an assessment of data quality, is presented in Sweitzer and Langaas (1994), Folke and Langaas (1995), and Sweitzer et al. (1996).

The map layers in this database were combined to generate new results. The maps were used to generate basic statistics on land use and population in the drainage basin, which we briefly report on below. We also present results showing the distribution of land cover and population as a function of distance from the coast, since this information is directly relevant to the eutrophication problem in the Baltic Sea.

It was found that forests dominate the landscape in the drainage basin (48% coverage), followed by arable land (20%), and non-productive open lands (17%). Wetlands cover roughly 8% of the drainage basin and are most prominent in the northern regions. Turning to population, about 85 million people live in the Baltic Sea drainage basin. Of these, the vast majority (64%) live in the Baltic Proper drainage areas. Of the total drainage basin population 26% (22 065 000 people) live within metropolitan areas (population > 250 000). Of the total population, 54% live in towns or small cities (population between 200 and 250 000), and 29% are inhabitants of rural settlements (population < 200 000).

Using the land cover and population map layers and an adjusted map of the drainage basin we further determined the characteristics of the drainage basin as they relate to distance from the coast. The further away from the coast or from rivers that eutrophying substances are released, the more likely they are to be absorbed through ecosystem processes and prevented from entering the Baltic Sea. High population concentrations, agricultural land, and urbanised land are all important nutrient generation sources. Wetlands, forests and inland water bodies can act as natural filters/sinks for nutrients as well as other pollutants. Given this, it is important to determine the landscape characteristics of the drainage basin as a function of distance from the coastline and rivers.

At its furthest point, the drainage basin is nearly 650 km away from the Baltic Sea shoreline. Land cover and population of the drainage basin were assessed at 10-km intervals extending away from the coast. It was found that while most of the land use classes are distributed fairly evenly throughout the drainage basin, population—particularly urban population—is heavily concentrated toward the coast. Within a 10-km distance from the Baltic coast, for example, we find 27% of the populated area and 19% of the total drainage basin population—nearly 15 million people. Of these 90% are concentrated in urban areas. Also within this area we find 8% of the total arable land, 5% of the pasture land, 5% of the forests, and 2% of the inland water bodies.

Expanding to a 50-km distance from the coastline, population remains the dominant feature of the landscape. In this zone we find 43% of the total populated area and 31% of the total population (over 26 million people). Of this population, 83% are urban. Additionally, 23% of the total arable land, 17% of the pasture land, 20% of the forest, and 10% of the inland lakes are found within a 50-km distance from the coast. More detailed results are reported in Sweitzer et al. (1996).

The information on the location of various land uses and population within the drainage basin provides a useful basis for the estimation of nutrient load discharged directly into the Baltic Sea or transported by surface water to which we now turn.

In 1993, the total load of nitrogen and phosphorous to the Baltic Sea amounted to approximately 1022 000 t of N and 39 000 t of P. The largest basin of the Baltic Sea, the Baltic Proper, receives about 85% of the total load of both nitrogen and phosphorous (Table 2).

In principle, there are two major sources of waterborne nutrient loads, arable land and sewage treatment plants. In addition atmospheric transports of nitrogen are also deposited directly on the Baltic Sea. Air transports originate, not only from countries within the drainage basin, but from other external countries.

The agricultural sector, excluding the emissions of ammonium, accounts for one-fifth of the total load of nitrogen. Other water transports of nutrients include flows from sewage treatment plants and air emissions deposited on land within the drainage basin, accounting for nearly 50% of the total load. The direct discharges, mainly sewage treatment plants located at the coast, correspond to approximately 10% of the total load. Poland is the largest discharging country with respect to total nitrogen loading (28.5%), followed by Sweden (10.4%) and Germany (10.2%).

Poland is also the country providing the largest load of phosphorous to the Baltic Sea, approximately 50% of total load. The phosphorous load

from the agricultural sector accounts for about one-third of the total load and the direct discharges, mainly from sewage treatment plants correspond to one-quarter of the total load.

The increased nutrient flux entering the Baltic Sea implies a higher concentration of a given nutrient which in turn may lead to an overabundance of phytoplankton production. As a result of increases in phytoplankton production, oxygen deficits may occur which reduce the spatial extent of regions available for successful cod reproduction. On the other hand, more zooplankton increases the stocks of other fish species (Fig. 2).

Although the impacts of greater nutrient input are well documented, the quantitative relationships between variations in loads of nutrients and concentration are poorly understood. For instance, the eutrophicating processes will alter redox conditions and thus the biogeochemical pathways and efficiency of internal sinks of nitrogen and phosphorous through denitrification; phosphorous adsorption will also be altered. The different sub-basins vary in terms of water and nutrient residence times, load received, and internal biogeochemical processes. These differences will significantly alter the N/P ratios in each sub-basin. Fig. 3 illustrates that there are processes in the sea that change the concentrations of nutrients, with specific differences between basins as well as between nutrients species: in the Bothnian Bay, the phosphorus loss (P-sink) is much more efficient than the sink for nitrogen. In the Baltic Proper, it is the opposite—a more efficient N-sink than P-sink. There is a gradient from north to south where the production in the Bothnian Bay is P-limited and the Baltic Proper is N-limited. The N/P ratio of 18 in the Baltic Proper is for total-N and P that includes a lot of refractory organic matter: for the inorganic fractions it is about 4, far below the Redfield ratio (data are not available to show inorganic N/P ratios both in terms of load and concentration).

An empirical budget model of the Baltic Proper has been expanded to cover all three sub-basins. The model consists of the three coupled basins with advective water and nutrient transports between these and with Kattegat (Wulff, 1995). Empirical relationships between load, nutrient

Table 2
Nutrient loads to the Baltic Sea, 1993

	Nitrogen (t)	Phosphorous (t)
Bothnian Bay	60 787	3008
Bothnian Sea	100 699	3063
Baltic Proper	861 268	32 817
Total	1 022 754	38 888

R.K. Turner et al. / *Ecological Economics* 30 (1999) 333–352

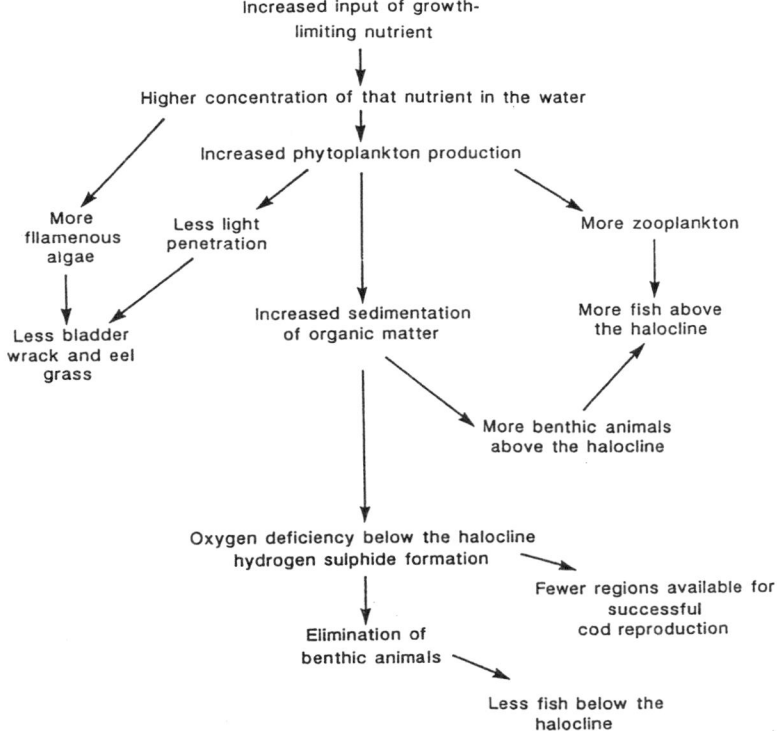

Fig. 2. Eutrophication impacts. Regarding eutrophication problems, the quantitative relationships between variations in loads of nutrients and concentration are poorly understood.

concentrations and advective transports derived from extensive regional data sets over the last few decades have been used to empirically calculate relationships between concentration and internal nutrient sink terms. These models rely on past relationships between inputs of water, nutrients and the observed trends in concentrations in the different basins of the Baltic. Several analyses have been undertaken to collate and verify these data sets.

The basic information on the flow of nutrients from the land to the Baltic Sea and their impact on the sea is required in order that abatement measures designed to improve the conditions in

the Baltic Sea can be selected on the basis of minimum cost. But before we can address the abatement options and their costs question we need to estimate the 'filter'/sink capacities of wetlands in the drainage basins.

The natural wetlands in the drainage basin account for about 8% of the total area. Their nitrogen retention/elimination capacity was estimated to be close to 65 000 t per year when only atmospheric downfall of nitrogen was taken into account. Adding direct emissions per capita in terms of excretory release, in relation to the location of the wetland to human population densities, we estimated the nitrogen retention/

elimination capacity to be about 100 000 t per year (Jansson et al., 1999).

The GIS database was used to assess the spatial relationship between nutrient sources and sinks. Maps were created to show the location of wetlands in relation to population centres in the Baltic drainage basin. The assumption is that wetlands will function more effectively as nutrient traps if they are in close proximity to nutrient sources. A visual assessment of the maps shows that areas with high concentrations of wetlands in the drainage basin are not near the densely populated regions. Areas with moderate or low concentrations of wetlands tend to have low or moderate population densities. These results suggest that development and restoration of wetlands in highly populated and also intensively cultivated areas could be an effective and practicable means to reduce nutrient flows into the Baltic Sea.

Therefore, we estimated the potential nitrogen retention/elimination capacity in a scenario where drained wetlands in the drainage basin would be restored. The capacity of wetlands to retain/eliminate nitrogen in such a scenario was estimated at about 180 000 t per year (Jansson et al., 1999). Additional analyses on the nitrogen filtering capacity will be reported on below in relation to the analyses of cost-efficient nitrogen abatement.

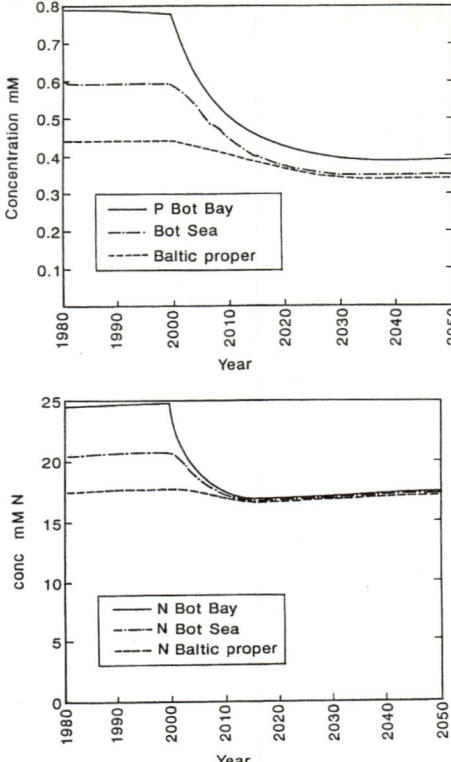

Fig. 4. Reduction from current levels of both N and P load with 50% to Baltic Proper.

5. Nutrient reduction simulations

Two nutrient reduction simulations were carried out.

In the first simulation, both N and P loads are reduced, but only to the Baltic Proper. Since the effect of eutrophication is most clearly seen in the Baltic Proper, this scenario would be the most obvious choice for a future abatement strategy. In this scenario it is assumed that the nutrient reduction occurs instantaneously in year 2000 and the changes in concentrations follow on until a new steady state occurs. As can be seen from Fig. 4

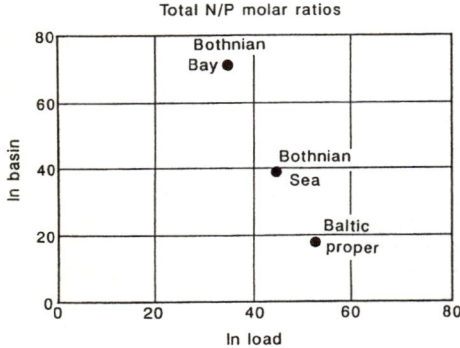

Fig. 3. Variations in N/P ratios.

the nitrogen concentrations reach this new steady state within 10 years while it takes about 25 years for phosphorous. This is due to the inherently different behaviour of these nutrients in the Baltic (and in most other marine systems). Denitrification represents an efficient internal nutrient sink for N while P reduction is less efficient in this brackish system. According to the results presented in Fig. 5, the final concentration of P and N are about 50 and 70% of the current levels in the Baltic Proper.

It is difficult to estimate the ecological consequences from the output of this model alone—the results have to be related to empirical knowledge of 'the state of the Baltic' with different concentrations of nutrients. The 'new' nutrient concentration corresponds roughly to levels found during the 1960s, before the drastic deterioration

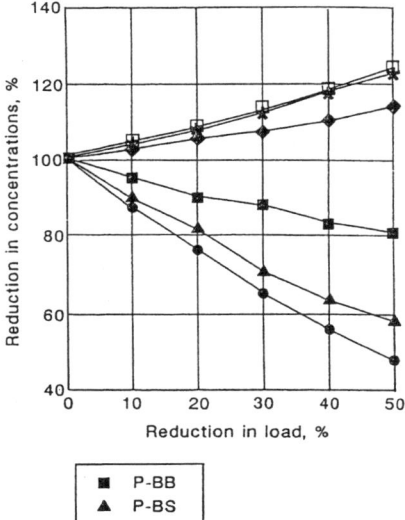

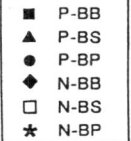

Fig. 6. Reduction of phosphorus loads to Baltic Proper.

of the Baltic environment occurred. We would expect less primary production of organic matter and thus less frequent periods of oxygen deficiency in the deep basins. It is also likely that the decrease in P concentrations will reduce the frequency of cyanobacterial bloom during the late summer. These are now favoured by the high P concentrations (and low N/P ratio) found during summer and cause accumulations of sometimes toxic algal mats on the surface of the Baltic Proper.

In the second simulation, only the P load is reduced to the Baltic Proper. This scenario was included since it is likely that the inputs of P are more easily reduced than those of N, since the sources are mostly municipal and agricultural (Wulff and Niemi, 1992). The model simulations show corresponding reductions of P concentrations on N (50% for the Baltic Proper and 80% for the Bothnian Bay) (Fig. 6). A decrease of

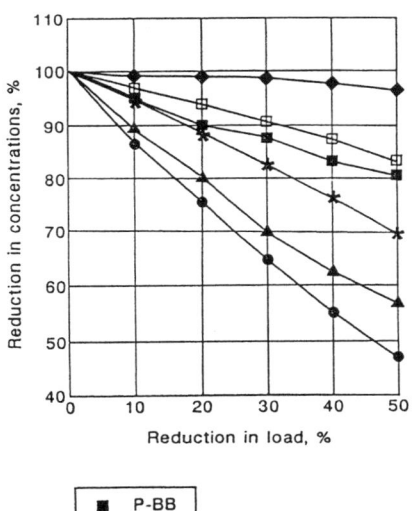

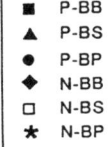

Fig. 5. Reduction from current levels of both nitrogen and phosphorus load but only to the Baltic Proper.

input and concentration of P means that less N will be utilised in the biogeochemical cycles, since these nutrients are utilised in fixed stochiometric Redfield ratios (16 mol of N for each mol P). Thus, less N will be incorporated into organic matter and subsequently mineralised and denitrified (lost) in this scenario. The Baltic Proper and Bothnian Sea will change from N-limited to P-limited systems in this scenario.

The model simulations presented above illustrate the consequences of nutrient reductions on a basin-wide scale. However, decisions about abatement policies are often made because of concerns at the local or regional level, rather than on the basis of large-scale environmental concerns. The resulting conflicts from a Baltic-wide perspective are discussed by Wulff and Niemi (1992), and are further explored below in a regional study of the Gulf of Riga. A nutrient budget exists that describes nutrient inputs, retention and exports of nitrogen and phosphorous from this highly eutrophicated bay adjacent to the Baltic Proper. This is a region of the Baltic where lack of sewage treatment contributes to very large inputs of P to the sea. A reduction of P inputs of more than 30% would occur if a modern sewage treatment plant was built for the city of Riga. A model was therefore built to explore the consequences of different N and P reduction schemes on the Gulf and the Baltic Sea. One model run is illustrated in Fig. 7. The net exports of nitrogen and phosphorous to the Baltic are shown in relation to different levels of P reduction in the inputs of the Gulf. Naturally, a P reduction in inputs will result in reduced exports of P. However, more nitrogen will be exported to the (N-limited) Baltic Proper.

These model simulations, although based on very simplistic assumptions, empirical relationships and basic physical and biogeochemical properties, show very clearly the basic features and interactions of hydrodyanamics and biogeochemistry of nitrogen and phosphorous in the Baltic Sea region. The overall model clearly demonstrates that it is reduction of inputs to the Baltic Proper that is most efficient in reducing concentrations in this basin. A strategy where all inputs are uniformly reduced is not optimal since the situation in the two northern basins is not critical in terms of eutrophication (only small amounts of nutrients are exported southwards).

The simulations also demonstrate that both nitrogen and phosphorous inputs have to be reduced. This is also emphasised in the regional Gulf of Riga study where it was shown that a P removal might actually increase the net export to the off-shore Baltic Proper. It has also been demonstrated that it will take several decades before the nutrient levels are returned to an acceptable level, particularly for phosphorous.

To understand the institutional implications of this result, consider a problem in which there are two basins, each controlled by a different country acting unilaterally. Basin A is phosphorus-limited (like the Gulf of Riga) while Basin B is nitrogen-limited (like the Baltic Proper). Country A controls discharges into Basin A and has preferences only over the quality of this Basin. Similarly, Country B controls discharges into Basin B and cares only about eutrophication in this Basin. As Basin A is phosphorus-limited, A can enhance its welfare by reducing its phosphorus discharges and improving the state of Basin A. However, the reduction of phosphorus in A will also release

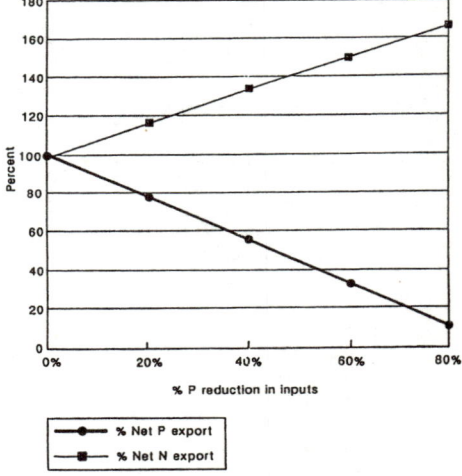

Fig. 7. % Change in nitrogen and phosphorus export from the Gulf of Riga at different levels of reduction in P load.

nitrogen, and this released nitrogen will be exported to Basin B. As B is nitrogen-limited (i.e. phosphorus-rich), eutrophication will increase in Basin B as a consequence of the actions undertaken by A. Similarly, if Country B reduces its nitrogen discharges in Basin B, phosphorus will flow into Basin A, exacerbating A's eutrophication problem.

As abatement of phosphorus by A increases eutrophication in B, Country B's best response is to reduce its nitrogen discharges further. But in doing so eutrophication is made worse in A, and A will therefore respond by reducing its phosphorus discharges even further. The process will continue until neither country can improve its welfare by abating discharges any further. This state defines the equilibrium in unilateral policies. As neither A nor B take into account the effect of their actions on the welfare of the other country, each is driven to abate its own Basin's limiting pollutant too much and the other Basin's limiting pollutant too little. As is typical of all equilibria in unilateral policies, pollution of both Basins is excessive compared with the full co-operative outcome. However, in contrast to every paper so far published in the literature, abatement (of each Basin's limiting pollutant) is also excessive in the equilibrium in unilateral policies. For more detail see Barrett (1995). The policy implication is that full co-operation is the optional strategy, but one in which abatement effort is redistributed rather than merely increased overall. Since marginal costs increase with abatement effort, this means that a small redistribution in abatement will lower total costs as well as total environmental damages.

6. Cost effective abatement strategies

Cost effectiveness is defined as achieving one or several environmental targets at minimum costs. A condition for cost-effectiveness is that the marginal costs of all possible measures are equal. Marginal cost is defined as the increase in costs when, in our case, nutrient load to the Baltic Sea is decreased by 1 kg N or P. As long as the marginal costs are not equal it is always possible to obtain the same level of nutrient reductions at a lower cost by reducing the load via measures with relatively low costs and increasing the load by the same amount via measures with relatively high costs. Thus, in order to calculate cost-effective nutrient reductions to the Baltic Sea we have to (i) identify all possible measures, (ii) quantify their impact on the Baltic Sea, and (iii) calculate marginal costs for all measures.

The environmental impact of a certain reduction of nutrient load at the source is, ceteris paribus, determined by the location of the source. If the source is located some distance away from the coastal waters of the Baltic Sea, only a fraction of any reduction at the source is finally felt at the coast. The share of the source reduction that reaches the coast depends on the retention of the nutrient that may occur at various points between the source and the coast. This implies that, for a given marginal cost at the source, the marginal cost of coastal load reduction is higher than for remote sources with low impact on the coast. In order to calculate impacts of source-related measures we require information on source location as well as on transportation of nitrogen and phosphorous. No water and soil transport models exist for the drainage basin and so we use very simplified retention numbers.

The abatement measures can be divided into three different classes:
1. reductions in the deposition of nutrients on Baltic Sea and on land within the drainage basin,
2. changed land uses reducing leaching of nutrients, and
3. creation of nutrient sinks which reduce the transports of nutrients to the Baltic Sea.

The first class of measures includes improvement in sewage treatment plants, reductions in air borne emissions from traffic and stationary combustion sources, and reductions in the agricultural application of fertilisers and manure on land. Reductions in air emissions are obtained by the installation of catalysts in cars and ships, reductions in the use of motor fuel and other petroleum products, and the installation of cleaning technologies in stationary combustion sources. Reductions in agricultural deposition of nutrients are

R.K. Turner et al. / Ecological Economics 30 (1999) 333–352 345

Table 3
Marginal costs of different measures reducing the nitrogen load to the coast, SEK/kg N reduction

Region	Agriculture	Sewage treatment plants	Atmospheric deposits	Wetlands
Sweden	20–242	24–72	135–9500	23
Finland	57–220	24–60	874–6187	66
Germany	20–122	24–60	210–3576	27
Denmark	23–200	24–60	544–3576	12
Poland	12–101	7–35	523–3412	10
Latvia	59–196	7–35	183–1195	20
Lithuania	72–208	7–35	254–1723	15
Estonia	55–192	7–35	153–1999	36
St Petersburg	43–236	7–35	353–1884	51
Kaliningrad	28–210	7–35	273–1593	43
Belgium			742–4184	
France			1507–9045	
Netherlands			562–7184	
Norway			475–3460	
UK			785–4855	

obtained by a reduction in the use of fertilisers and reductions in livestock. Another measure included is a change in spreading time of manure from autumn to spring. Decreases in leaching from arable land are obtained by increasing the area covered by catch crops, energy forests, and ley grass. Nutrient sinks are created by constructing wetlands downstream in the drainage basin close to the coastal water.

In principle, the cost of an abatement measure includes the cost at the emission source and the cost impacts on other sectors of the economy. In the following analysis we only include the abatement costs at the source, which are calculated by means of engineering methods and econometric techniques (see Gren et al., 1995, for details about the cost estimation work). The calculated marginal costs at the source for different abatement measures aimed at reducing nitrogen load in different regions (Table 3) indicate that increased nitrogen cleaning capacity at sewage treatment plants is a low cost measure in all countries. Further low cost measures include, in the agricultural sector, the reduction in use of nitrogen fertilisers and cultivation of cash crops. Another low cost option is the construction of wetlands, whereas measures reducing air emissions are relatively expensive in all countries.

The marginal cost of phosphorous reductions tends to be much higher than those for nitrogen (Table 4). Nevertheless, measures involving improvements in sewage treatment plants represent relatively low cost reduction options. Restoration of wetlands is relatively expensive. It can be seen that phosphorus reduction in Swedish wetlands is more than ten times more expensive than Finland, for example. This is because the cheaper Swedish measures with respect to wetlands have already been extensively deployed, whereas they have not in the other countries.

Table 4
Marginal costs of phosphorous reductions, SEK/kg P reduction

Region	Agriculture	Sewage treatment plants	Wetlands
Sweden	155–6604	41–52	18232
Finland	225–6080	41–52	1748
Denmark	144–2610	41–68	1202
Germany	188–2964	41–68	899
Poland	114–2033	20–100	611
Estonia	282–5622	20–100	6090
Latvia	234–5662	20–100	1234
Lithuania	186–6696	20–100	964
St Petersburg	230–4314	20–100	823
Kaliningrad	338–4290	20–100	545

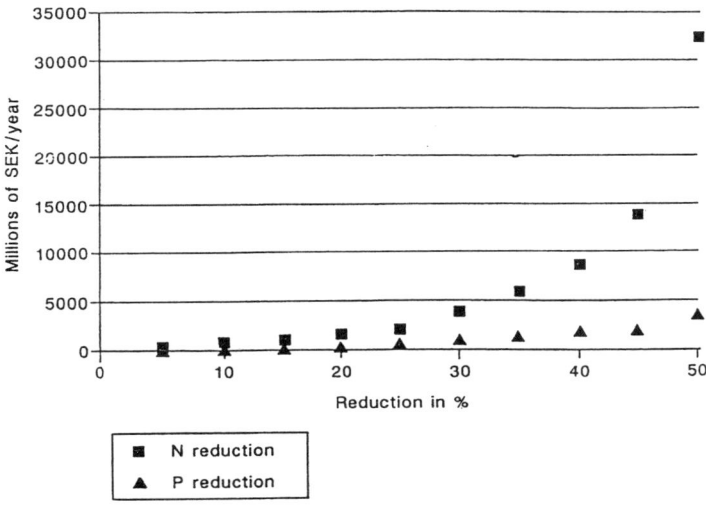

Fig. 8. Cost effective N and P reductions.

6.1. Minimum costs of nutrient reductions

Because it is reductions in nutrient loads to the Baltic Proper which have the main impact on the ratio N/P, we estimate minimum costs for load reductions to this basin only. The minimum costs for various reductions in either N or P are shown in Fig. 8.

We can see from the figure that the costs of reducing the load of nitrogen are much higher than the costs of corresponding decreases in phosphorous loads.

Several of the measures, such as livestock reductions, change of manure spreading time and wetlands restoration, imply reductions in both nitrogen and phosphorous. When one of these measures is implemented with the aim of reducing the load of one nutrient, reductions are obtained in the other nutrient load 'free of charge'. These joint impacts on several nutrients imply that abatement measures are relatively less costly if simultaneous reductions in N and P are undertaken. Total costs for various reduction levels are then lower for simultaneous decisions on N and P

than for separate decisions, especially for abatement levels in excess of 40% reductions (see Gren (1995) for more details). Note that P is more 'mobile' than N and therefore requires more abatement effort, thus P can be the 'keystone' pollutant, i.e. if P is managed then so is N but not usually the other way around.

In order to achieve a 50% reduction in nitrogen loading the most cost effective mix of measures would be one in which agriculture, wetlands and sewage treatment plant-related measures account for 35, 28, and 31%, respectively, of the total nitrogen reduction. Measures involving air emissions account for 6%. The single most important country source in a cost effective reduction strategy is Poland, which accounts for 40% of the total reduction (corresponding to about 2/3 of the Polish load of nitrogen). We also note that Poland, Russia and the Baltic states account for 72% of the total nitrogen reduction. The nitrogen reduction contribution of Swedish and Finnish regions amounts to only 8 and 7%, respectively.

For phosphorous load reductions wetland measures can only play a minor role in coastal waters,

R.K. Turner et al. / Ecological Economics 30 (1999) 333–352   347

and in fact it is only in Germany that such measures form part of a cost-effective abatement package. Instead, measures relating to sewage treatment plants are of major importance, accounting for 66% of the total reduction. This is a reflection of the relatively large load of phosphorous from households and industries and the availability of low cost abatement options. Again the single most important country source in a cost effective reduction strategy is Poland, which accounts for 67% of the total reduction. The Baltic states, Poland and Russia together account for approximately 90% of the total phosphorus reduction.

It is important to emphasise that the cost estimates are based on several assumptions of a biological, physical and economic character. The biological assumptions refer to retention of nutrients, the leaching impact of agricultural measures and the nitrogen removal capacity of wetlands. The physical assumptions concern the feasibility limits of different measures such as nutrient cleaning capacity of sewage treatment plants and area of land available for alternative land uses. The economic assumptions relate to the estimation of the costs of the various measures. According to the results of sensitivity analysis, both the costs of nitrogen and phosphorous reduction seem to be sensitive to assumptions of a biological character. Changes in the physical assumption about land available for agricultural measures often have a significant impact of the total costs. It should be noted, however, that the sensitivity analysis was carried out only for an overall reduction of 50% in the load of both nutrients. At other overall reduction levels, the costs may be sensitive to other types of assumptions.

7. Benefits valuation

The process of measuring the economic value of eutrophication damage in the Baltic involves three basic stages. Firstly, discharges of nutrients into the Baltic lead to eutrophication as outlined earlier and this leads to reductions in the various measures of environmental quality. Second, these changes in environmental quality lead to changes in the stream of services (use and non-use values) provided by the Baltic region. Third, the change in the stream of services will affect individuals' well-being and the economic proxy for well-being, money income, such that willingness to pay for the stream of services will change.

A concerted attempt was made to estimate the economic benefits of environmental improvements in the Baltic. A total of 14 empirical valuation studies in three countries—Poland, Sweden, and Lithuania—were carried out to look at benefit estimation issues. These included the total economic value of reducing the effects of eutrophication, as well as sub-components of this total value such as beach recreation benefits, existence and option values of preserving species and their habitats, and the benefits from preserving and restoring wetlands. Of the applied studies that have been done in the different countries, some of them have focused on similar valuation issues, thus enabling a comparative evaluation of the studies to be carried out between the differing economic, cultural and political systems. We also have considered the question of total basin-wide benefit estimates and benefits transfer. Whilst the studies outlined here provide a large amount of information about the value of the Baltic's resources, there are still gaps in our knowledge of total basin-wide benefit estimates. Nevertheless, the estimates that are available indicate the significant value of the limited number of resource types considered.

The full results of all the studies are presented in Georgiou et al. (1995). Here we present the results of two of the studies carried out in Poland and Sweden which looked at the use and non-use value of reducing eutrophication to a sustainable level. These two studies were used to estimate basin-wide benefits.

The first study was a contingent valuation study focusing on Baltic Sea use and non-use values in Sweden. This study was designed as a mail questionnaire survey. The questionnaire was sent to about 600 randomly selected adult Swedes. The response rate turned out to be about 60%, which is quite similar to other contingent valuation method (CVM) mail questionnaire surveys that have been undertaken in Sweden.

R.K. Turner et al. / Ecological Economics 30 (1999) 333–352

The eight-page questionnaire is presented in detail in Söderqvist (1995). It contained, inter alia, summary information on the causes and effects of eutrophication of the Baltic Sea. In the valuation scenario, the respondents were asked to assume that an action plan against eutrophication had been suggested, and that this action plan would imply that the eutrophication in 20 years would decrease to a level that the Baltic Sea can sustain. The types of action that this plan would involve were briefly described. It was also explained that the way to finance the actions would be to introduce an extra environmental tax in all countries around the Baltic Sea.

The respondents then met the following question: "If there were a referendum in Sweden about whether to launch the action plan or not, would you vote for or against the action plan if your environmental tax would amount to SEK X per year during 20 years?". Seven different amounts of money, X, were randomly used for the question. The answers to the question give an estimate of mean annual Willingness To Pay (WTP) of about 5900 SEK per person (or 3300 SEK if we assume non-respondents to the survey have a zero willingness to pay).

It is likely that the respondents considered use values as well as non-use values when they answered the WTP question. This means that the WTP reflects perceived total benefits. However, note that there may be important differences between perceived benefits and real benefits. One reason for this is that the information communicated to the respondents about the eutrophication and its effects was far from complete. Moreover, the results from this CVM study may be influenced by embedding phenomena, i.e. that the respondents have also considered their WTP for other environmental improvements, and not only for a reduction of eutrophication. Embedding is a recognised problem in CVM studies. Note also that it is not easy to relate the outcome in the valuation scenario—a reduction of the eutrophication to a level that the Baltic Sea can sustain—to a specific reduction of the nutrient load (though such an outcome is probably consistent with the 50% nutrient reduction target adopted by the Helsinki Commission). A time horizon of 20

years is reasonable in the sense that even if considerable action is taken today, it takes many years until any results will be evident. The description of the outcome as a 'sustainable' level reflects the fact emphasised by ecologists that actions against eutrophication will probably result in neither the complete disappearance of eutrophication, nor a return to the same ecological situation that characterised the Baltic Sea some decades ago, but rather to some new equilibrium.

The second study was almost identical to the first, except that it was carried out in Poland, thus providing a direct international comparison to be made between the benefit estimates found in both countries. Again a mail questionnaire was used and 600 questionnaires were sent out to a random sample of Polish adults. The response rate was just above 50% which was considered reasonable for this context and location. It was found that the level of support for the environmental tax was 54.9%. Mean annual willingness to pay per person for the action plan was 840 SEK (or 426 SEK if we assume non-respondents to the survey have a zero willingness to pay).

In order to calculate basin-wide benefit estimates we need to add up the values for the different activities carried out, taking care not to double count, and using the relevant correct populations. Since there are benefit estimates available for the same valuation scenario in only two of the 14 countries that are included in the Baltic drainage basin, any aggregation to the whole basin has to rely on strong assumptions. The aggregate benefit estimates to be presented below should thus not be taken too literally. However, they may give useful information regarding the order of magnitude of basin-wide benefit estimates.

Table 5 shows estimates of aggregate benefits for the total economic value of a Baltic Sea nutrient reduction strategy. Data from the Polish and Swedish mail surveys are used since they are both concerned with total economic value (use + non-use value), and they contain the same valuation scenario. Given an adjustment for the difference in GDP per capita levels between the countries, the Polish mean WTP estimate of 840 SEK (426 SEK) will be regarded as representative for the transition economies around the Baltic Sea, i.e. Estonia, Latvia, Lithuania, Poland, Russia; and the Swedish mean WTP estimate of 5900 SEK (3300 SEK) is taken as representative of the market economies of Finland, Germany, Norway and Sweden (Table 5). The possible WTP of the population in the other countries included in the Baltic drainage basin (Belarus, Czech Republic, Norway, Slovakia and Ukraine) will be ignored in this analysis.

In order to calculate national WTP estimates, the estimate per person was multiplied by the (adult) population in the Baltic drainage basin part of each country. According to Table 5, the basin-wide estimate for total economic value is SEK 69 310 million per year (SEK 37 892 million per year). This is a highly uncertain figure, but it indicates that the benefits from a Baltic Sea clean-up of eutrophication may be considerable.

Table 6 brings together both the costs of pollution abatement and related economic benefit estimates in a cost-benefit analysis framework. It is clear that there are considerable net benefits available to a number of Baltic countries, sufficient for them to pay their own clean-up costs and subsidise the Baltic republics' abatement programme, while still gaining increased economic welfare benefits. While the economic benefit calculations are not precise point estimates, they are indicative of the range or order of magnitude of clean-up benefits in the Baltic.

Poland faces the largest cost burden because of its relatively high pollution loading contribution and the modest levels of effluent treatment that it currently has in place.

The costs in Table 6 refer to the allocation of nitrogen reductions that minimises total costs. We note that the reductions, measured in percentages of original loads, vary between 39% (Germany)

Table 5
Basin-wide benefit estimates

Country	GDP per capita at PPP[a] (US$)	Annual WTP per person[b] (SEK)	National WTP, year 1[c] (MSEK)	National WTP, present value[d] (MSEK)	National WTP, present value per year (MSEK)
Transition economies					
Estonia	3823	700 (355)[e]	790 (401)	8369 (4248)	418 (212)
Latvia	3058	569 (284)	1100 (549)	11653 (5816)	583 (291)
Lithuania	3632	665 (337)	1743 (883)	18465 (9355)	923 (468)
Poland	4588	840 (426)	21958 (11136)	232623 (117974)	11631 (5899)
Russia	4970	909 (461)	6585 (3340)	69761 (35384)	3488 (1769)
Market economies					
Denmark	19306	6770 (3790)	23365 (13080)	247529 (138570)	12376 (6929)
Finland	15483	5430 (3040)	20387 (11414)	215980 (120920)	10799 (6046)
Germany	18541	6500 (3640)	15800 (8848)	167385 (93736)	8369 (4687)
Sweden	16821	5900 (3300)	39122 (21882)	414458 (231818)	20723 (11591)
Total			130850 (71533)	1386223 (757821)	69310 (37892)

[a] PPP = purchasing power parity.

[b] For the transition economies, the Polish mean WTP estimate of SEK 840 (SEK 426) was multiplied by the ratio between each country's GDP per capita (at purchasing power parity) and Poland's GDP per capita at PPP. For the market economies, the Swedish mean WTP estimate of SEK 5900 (SEK 3300) and Sweden's GDP per capita at PPP were used correspondingly. Source of GDP data: OECD.

[c] The annual mean WTP estimates per person multiplied by the (adult) population in the Baltic drainage basin part of the country.

[d] Time horizon: 20 years (specified in the CVM studies). Discount rate: 7% (this rate was also used in the estimation of nutrient reduction costs).

[e] Figures in brackets are for benefit figures which assume zero WTP of non-respondents.

R.K. Turner et al. / Ecological Economics 30 (1999) 333–352

Table 6
Costs and benefits from reducing the nutrient load to the Baltic Sea by 50%, millions of SEK/year[a]

Country	Reduction in %	Costs	Benefits	Net benefits
Sweden	42	5300	20 723 (11 591)	15 423 (6291)
Finland	52	2838	10 799 (6046)	7961 (3208)
Denmark	51	2962	12 376 (6929)	9414 (3967)
Germany	39	4010	8369 (4687)	4359 (677)
Poland	63	9600	11 631 (5899)	1761 (−3701)
Russia	44	586	3488 (1769)	2902 (1183)
Estonia	55	1529	418 (212)	−1111 (−1317)
Latvia	56	1799	583 (291)	−1216 (−1508)
Lithuania	55	2446	923 (468)	−1523 (−1978)
Total	50	31 070	69 310 (37 892)	38 240 (6822)

[a] Figures in brackets are for benefit figures which assume zero WTP of non-respondents.

and 63% (Poland). If the abatement cost strategy was based not on a cost-effectiveness criterion linked to an overall ambient quality target, but on some 'political' solution based, for example, on uniform national load reductions, then aggregate costs would be increased significantly (Table 7). This cost increase is due to the expensive measures that have to be implemented in Germany and Sweden. However, several countries with reduction levels exceeding 50% in Table 6 will gain from a country restriction as compared to a restriction of the total load of nitrogen.

The costs presented in Table 6 may also be overestimated because they do not include other environmental improvements associated with these nutrient reductions such as improved ground water quality and less acidification related to nitrogen oxides emissions. It is well known that several of the measures implying land use changes also yield other ecological services. For example, wetlands provide food, biodiversity and flood water buffering, and energy forestry on arable land provides fuel and may act as a carbon sink. If all these other positive aspects were included, some measures might imply internal net benefits instead of net costs.

The simulation results derived from our modelling of nutrient transports in the Baltic Sea provide a proxy for the missing dose-response scientific data. The model simulates the impacts of nutrient reduction on the concentration ratios of N and P but does not provide any detailed infor-

mation on the impacts on the biological conditions and production of ecological services. The available model does, however, predict that a 50% reduction in the loads of nitrogen and phosphorous to the Baltic Sea may correspond to the levels found during 1960s, i.e. before the major deterioration in the Baltic environment occurred. This scenario is likely to be consistent with the one used in the CVM studies. Therefore, a crucial assumption when comparing costs and benefits is that a 50% reduction in the loads of nitrogen and phosphorous imply that we reach ecological conditions which resemble those of the Baltic Sea prior to 1960s. Another important assumption

Table 7
Cost change of a move from a 50% reduction in total load to 50% reduction in the load of each country, in percent[a]

Region	Nitrogen reduction	Phosphorous reduction
Sweden	−57.8	361.8
Finland	−51.9	718.1
Denmark	−48.8	29.1
Germany	543.6	32.1
Poland	−80.4	−57.9
Latvia	18.6	195.0
Lithuania	−17.5	289.2
Estonia	−13.4	277.1
St. Petersburg	981.8	−80.0
Kaliningrad	779.4	−81.6

[a] A negative sign implies cost savings when country restrictions are imposed as compared to reduction by 50% in the total load of nutrient.

concerns the nutrient filtering capacity of different Baltic Sea coasts, which is likely to vary a lot. There is, however, no appropriate data on the coasts' filtering capacity. Therefore, no distinction has been made between different coastlines. Given all these qualifying assumptions, estimated costs and benefits of an overall reduction in the nutrient loads by 50% for different countries are as presented in Table 7. Note that Belarus, Czech Republic, Norway, Slovakia and Ukraine are excluded, since these countries were excluded from the cost estimation work.

8. Policy implications

There is considerable merit in the adoption of a basin-wide approach to pollution abatement policy in the Baltic and therefore in the implementation of an integrated coastal zone management strategy. It is clear that the ambient quality of the Baltic Sea is controlled by the coevolution of both biophysical and socio-economic systems throughout the macro-scale drainage basin.

Despite the pioneering nature (i.e. in the 'transition' economies) of some of the economic benefits research, there seems to be little doubt that a cost-effective pollution abatement strategy roughly equivalent to the 50% nutrients' reduction target adopted by the Helsinki Commission would generate significant positive net economic benefits (benefits minus costs). The research into the monetary valuation of environmental benefits also indicated that the public's and experts' perception of environmental quality and quality decline are not necessarily synonymous.

A policy of uniform pollution reduction targets is neither environmentally nor economically optimal. Rather, what is required is a differentiated approach with abatement measures being concentrated on nutrient loads entering the Baltic Proper from surrounding southern sub-drainage basins. The northern sub-drainage basins possess quite effective nutrient traps and contribute a much smaller proportionate impact on the Baltic's environmental quality state. The countries within whose national jurisdiction these southern sub-basins lie are also the biggest net economic gainers from the abatement strategy.

Although there are a range of feasible individual N-reduction and P-reduction measures available, our research indicates that the simultaneous reduction of both N and P loadings into the Baltic is more environmentally effective as well as cost effective. The increased deployment of N-reduction and P-reduction measures within existing sewage effluent treatment works, combined with coastal wetland creation/restoration schemes and changes in agricultural practice, would seem to be a particularly cost-effective option set.

The marginal costs of nutrient reduction measures increase sharply towards the full works treatment end of the spectrum. This finding suggests that the greatest environmental and economic net benefits are to be gained by an abatement policy that is targeted on areas which lack treatment works of an acceptable standard, rather than on making further improvements to treatment facilities that already provide a relatively high standard of effluent treatment. This finding, combined with our findings relating to the importance of the spatial location of nutrient loading, suggests that nutrient reduction measures in the Polish and Russian coastal zone areas would be disproportionately effective. The financing of such measures remains problematic if only 'local' sources of finance are to be deployed. Non-commercial funding from the European Commission and other international agencies, together with bilateral agreements, could play a vital role in the enabling process for an effective and economic Baltic clean-up programme.

Acknowledgements

This research was funded by the European Commission, Project Number: EV5V-CT-92-0183. The project participants include CSERGE, Norwich and London; the Beijer Institute, Stockholm; University of Siegen, Germany; University of Stockholm; GRID Arendal, Norway; and Stockholm School of Economics.

References

Barrett, S., 1995. Institutional analysis. In: Turner, R.K., Gren, I.-M., Wulff, F. (Eds.), The Baltic Drainage Basin Report: EV5V-CT-92-0183. European Commission, Brussels Ch. 8.

Folke, C., Langass, S., 1995. Land use, nutrient loads and damage in the Baltic Sea. In: Turner, R.K., Gren, I.-M., Wulff, F. (Eds.), The Baltic Drainage Basin Report: EV5V-CT-92-0183. European Commission, Brussels Ch. 2.

Folke, C., Hammer, M., Jansson, A.-M., 1991. Life-support value of ecosystems: a case study of the Baltic Sea region. Ecol. Econ. 3, 123–137.

Georgiou, S., Bateman, I.J., Söderqvist, T., Markowska, A., Zylicz, T., 1995. Benefits valuation. In: Turner, R.K., Gren, I.-M., Wulff, F. (Eds.), The Baltic Drainage Basin Report: EV5V-CT-92-0183. European Commission, Brussels Ch. 7.

Gren, I.-M., 1995. Cost effective nutrient reduction to the Baltic Sea. In: Turner, R.K., Gren, I.-M., Wulff, F. (Eds.), The Baltic Drainage Basin Report: EV5V-CT-92-0183. European Commission, Brussels Ch. 6.

Gren, I.-M., Elofsson, K., Jannke, P., 1995. Costs of nutrient reductions to the Baltic Sea. Beijer International Institute of Ecological Economics, Royal Swedish Academy of Sciences, Stockholm. Beijer Discussion Paper Series No. 70.

Howarth, R.W., Billen, G., Swaney, D., Townsend, A., Jaworski, N., Lajtha, K., Downing, J.A., Elmgren, R., Caraco, N., Jordan, T., Berendse, F., Freney, J., Kudeyarow, V., Murdoch, P., Zhu, Z.L., 1996. Regional nitrogen budgets and riverine N-and-P fluxes for the drainages to the north atlantic ocean — natural and human influences. Biogeochemistry 35, 75–139.

Jansson, Å., Folke, C., Langaas, S., 1999. Quantifying the nitrogen retention capacity of natural wetlands in the large scale drainage basin of the Baltic Sea. Landscape Ecol. (in press).

Nehring, ·D., Hansen, H.P., Hannus, M., Jörgensen, L.A., Körner, D., Mazmatchs, M., Perttilä, M., Wulff, F., Yurkovskis, A., Rybinski, J., 1990. Nutrients. Ambio 7, 5–7 Special Report, September.

Söderqvist, T., 1995. The benefits of reduced eutrophication of the Baltic Sea: a contingent valuation study. Stockholm School of Economics and Beijer International Institute of Ecological Economics, Stockholm (mimeo).

Sweitzer, J., Langaas, S., 1994. Modelling population density in the Baltic states using the digital chart of the world and other small data sets. In: Proceedings from EUCC/WWF Conference on Coastal Conservation and Management in the Baltic Region, May 2–8, Klaipeda, Lithuania. Department Systems Ecology, Stockholm University, Stockholm.

Sweitzer, J., Langaas, S., Folke, C., 1996. Land use and population density in the Baltic Sea drainage basin: a GIS database. Ambio 25, 191–198.

Turner, R.K., Gren, I.-M., Wulff, F., 1995. The Baltic Drainage Basin Report: EV5V-CT-92-0183. European Commission, Brussels.

Wulff, F., 1995. Natural systems state. In: Turner, R.K., Gren, I.-M., Wulff, F. (Eds.), The Baltic Drainage Basin Report: EV5V-CT-92-0183. European Commission, Brussels Ch. 3.

Wulff, F., Niemi, A., 1992. Priorities for the restoration of the Baltic Sea—a scientific perspective. Ambio 21 (2), 193–195.

[30]

The aral sea basin

A critical environmental zone

By V. M. Kotlyakov

A mong today's severely damaged ecological zones, the Aral Sea region is one of the most notorious.[1] The Aral, a large, desert-bound sea in south central Asia, was brought to life by the abun-

V. M. KOTLYAKOV is director of the Institute of Geography at the Soviet Academy of Sciences in Moscow.

dant Amu Darya and Syr Darya rivers, which drain into the Aral Sea basin from tributaries deep in Soviet central Asia and Kazakhstan (see Figure 1 on page 7). But the sea, once comparable in size to one of the larger Great Lakes in North America, has been shrinking at an alarming rate over the last 30 years.

The process taking place in the Aral region may be called anthropogenic

desertification because the sea is, in effect, drying up as a consequence of the development of irrigated agriculture in the basin. Its river deltas and other natural habitats, the local climate, and regional hydrology have all been similarly affected. Moreover, these negative ecological changes have been accompanied by grave socioeconomic costs: deteriorating human health, increasing unemployment as resource-

based economic activity declines, and decreasing production of cotton, rice, and other agricultural crops. The fate of the Aral basin bears watching by other regions in the world prone to desertification.

In the 1950s, the Aral Sea covered an area of 66,000 square kilometers, containing 1,064 cubic kilometers of water with a mean salinity of 1.0 or 1.1 percent. At the time, the mean depth of the

Aral was 16 meters, and 90 centimeters of water evaporated from its surface annually. The sea's water balance was maintained by an annual inflow of 56 cubic kilometers of water from the Amu Darya and Syr Darya rivers, supplemented by 5 cubic kilometers of atmospheric precipitation.

But early in the 1960s, rapid development of irrigated agriculture began, and water withdrawals from the Amu

Darya, Syr Darya, and their tributaries greatly increased. By 1965, 4.5 million hectares in Soviet central Asia were irrigated, consuming 50 to 55 cubic kilometers of water annually. In the next quarter century, another 2.6 million hectares were irrigated, using an additional 50 cubic kilometers of water annually.

Since the early 1960s, the area of irrigated lands has increased by 50 percent in the Uzbek and Tadzhik republics, by 70 percent in the Kazakh Republic, and by 140 percent in Turkmenia. Today, a total of 7 million hectares in the Aral region are irrigated, with annual withdrawals of 60 cubic kilometers of water from the Amu Darya and 45 cubic kilometers from the Syr Darya.

With the rapid expansion of irrigated lands, the total inflow of water to the Aral decreased sharply, and the sea's level dropped (see Table 1 on this page). Between 1961 and 1970, the mean decline was 0.21 meters per year. The annual decline was 0.58 meters during the 1970s and as much as 1.09 meters during the 1980s. By 1990, the sea level had fallen more than 14 meters from the 1960 level, the area of the sea had shrunk by more than 40 percent, and its volume had decreased by more than 60 percent.

The mean inflow of river water to the Aral Sea between 1971 and 1980 was only 16.7 cubic kilometers per year, and early in the 1980s, it practically ceased altogether. In 1987 and 1988, there was a relative abundance of water: the Syr Darya delivered 1.2 and 6.2 cubic kilometers of water to the Aral, and the Amu Darya delivered 8 and 16 cubic kilometers. In 1989, the inflow from the Syr Darya dropped to 4.2 cubic kilometers, and that of the Amu Darya fell to only 1.1 cubic kilometers.

By the end of 1989, the Aral Sea had receded into two separate parts. The level of the southern "Greater Sea" dropped to 38.6 meters, and that of the northern "Lesser Sea" dropped to 39.5 meters. Now the area of the Greater Sea is about 33,500 square kilometers with a volume of 310 cubic kilometers and mean salinity of 3.0 percent. The Lesser Sea covers approximately 3,000 square kilometers with a volume of 20 cubic kilometers. The salinity of its waters varies in different parts from 1.8 to 3.5 percent because of the freshening effect of water from the Syr Darya.

The trend is clear. Without drastic measures, the Aral Sea is destined to become a small brine lake of between 4,000 and 5,000 square kilometers.

Ecological and Health Impacts

As a consequence of the physical changes, the Aral Sea now has almost completely lost its productive fisheries. The Aral once contained more than 20 species of fish, including 12 game species. Now all but a few have died out as shallow spawning grounds have dried up and food reserves have disappeared. If the present trend in salinization continues, by the year 2000, the salinity of the sea water will reach 3.8 to 4.2 percent. This high concentration will exclude the possibility of colonization by new species of fish and the organisms they feed on that might have acclimatized to the saline conditions.

The exposed sea bottom of the Aral is becoming a salty (solonchak) badlands, a source of constant wind-blown sand. Dust storms that are so powerful that they can be observed from space have become common in spring. In addition to fine particles of soil, the dust contains salts—primarily sulfates and chlorides, which are poisonous to plants. The dust clouds, which stretch for 150 to 300 kilometers, are particularly harmful to desert oases and pastures during spring flowering.

During the last 25 years, the duration of dust storms in the coastal areas of the sea has increased 2 to 3 times.[2] Based on preliminary estimates, the Aral region is the source of from 15,000 to 75,000 tonnes of atmospheric dust annually. Each year, each hectare in the lower Aral basin receives 340 kilograms of dry salt and an additional 180 kilograms of salt in precipitation for a total of 520 kilograms of salt deposited per hectare. The mineral content of precipitation in the basin was 3 times higher in the 1980s than it was in the 1960s and 1970s.

Before its level dropped, the Aral Sea influenced the climate of the land within 100 to 200 kilometers of the sea. But as the sea and its river deltas have dried up, that influence has diminished, and the continentality of the basin's climate has increased—that is, the climate has become more like that of a continental interior, with decreased precipitation and greater swings in temperature. Early in the 1980s, the difference between the mean monthly temperatures of January and July increased by 1.5° to 2° C. Frosts are now more likely to occur later in spring and earlier in fall. The frost-free period in the Amu Darya delta, a cotton-growing region, has shortened to as few as 170 days, far fewer days than the minimum of 200 frost-free days required for growing cotton.

As spring floods on the Amu Darya and Syr Darya stopped and the deltas dried up, the natural landscapes of the region were destroyed. During the past 20 years, the 550,000 hectares of reeds on the floodplains of the Amu Darya shrunk to 20,000 hectares, and the productivity of pastures and haying grounds dropped by roughly 20 percent. More than 50 lakes in the Amu Darya delta—about 10 percent of the

TABLE 1 ▰▰▰▰▰
HYDROLOGIC PARAMETERS OF THE ARAL SEA FROM 1960 TO 1989

Year	Sea level (meters)	Sea area (thousand square kilometers)	Sea volume (cubic kilometers)	Mineral content (grams per liter)	Total river run-off into sea (cubic kilometers)
1960	53.3	67.9	1,090	10.0	40
1965	52.5	63.9	1,030	10.5	31
1970	51.6	60.4	970	11.1	33
1975	49.4	57.2	840	13.7	11
1980	48.2	52.4	670	16.5	0
1985	42.0	44.4	470	23.5	0
1989	39.0	37.0	340	28.0	5

SOURCE: D. B. Oreshkin, "Aral'skaya Katastrofa," (The Aral Catastrophe), *Nauka o Zemle*, no. 2 (1990):41.

region's lakes—have dried up, as have the floodplain forests known as *tugai*. As a result, the region's biodiversity has been greatly impoverished. Only 168 of 319 species of bird continue to nest in the delta, and only 30 of 70 mammalian species remain.

A considerable part of the mineralized and polluted drain water from the surrounding expanse of irrigated fields returns to the Amu Darya and Syr Darya. This run-off has critically lowered the water quality in the river deltas. The problem has been exacerbated by the salinization of croplands because water is used to wash away the salts. The mean consumption of water between 1978 and 1982 in Uzbekistan was about 17,500 cubic meters per hectare. In Karakalpak and Khorezm, it reached 36,000 cubic meters per hectare.

In the lower reaches of the Amu Darya, the mineral content of the river water from run-off of drain water and fertilizers is already about 1.5 grams per liter. In some years, the mineral content of the water in the lower stretches of the Syr Darya reaches 2.5 grams per liter. One liter of water from these rivers contains up to 6 grams of phosphates, 3 milligrams of ammonia, 2 milligrams of nitrites, and 6 milligrams of nitrates. The chlorinated hydrocarbon pesticides applied to agricultural fields in the basin can be detected in the water of both rivers.

Leaked and surplus irrigation water often ends up not back in the rivers but in local depressions and irrigation reservoirs at the periphery of irrigated zones, where the water evaporates. Such reservoirs have become particularly numerous along the lower reaches of the Amu Darya, where more than 40 large water basins annually lose from 6 to 7 cubic kilometers of water to evaporation. The largest reservoir of Sarykamysh receives 3 to 4 cubic kilometers of irrigation wastewater annually, increasing its total volume to 30 cubic kilometers.

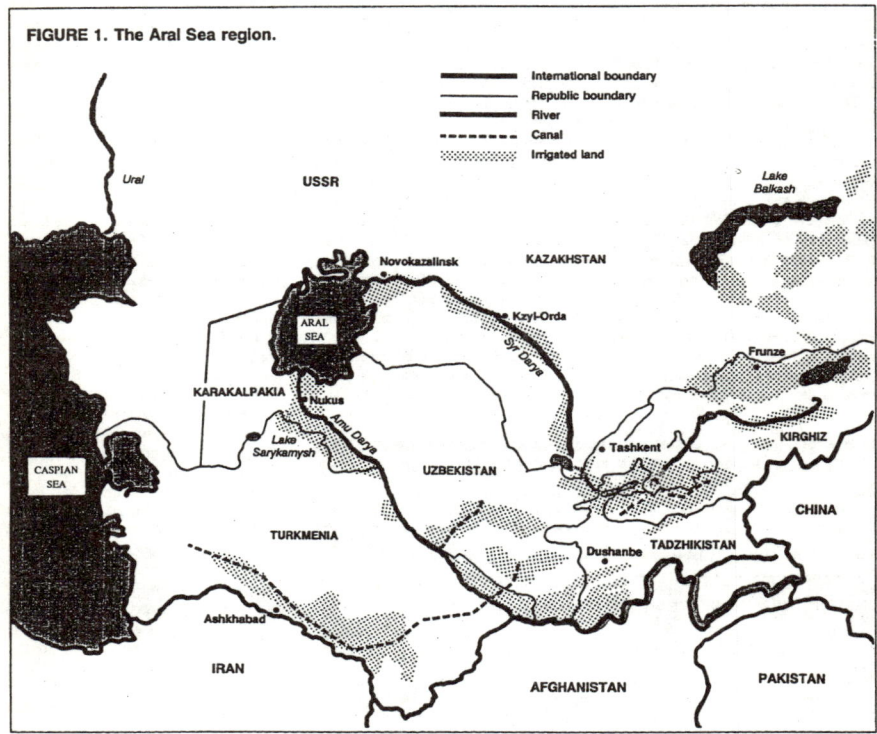

FIGURE 1. The Aral Sea region.

Because of these reservoirs, bogs and boggy patches not typically found in desert habitats are forming. The bogs displace valuable pastures, which are adapted to scant precipitation. The bog water itself is wasted through evaporation, thereby increasing the salinization of the land.

The anthropogenic desertification of the Aral and its region, initially an ecological problem, in recent times has turned into a human health problem. Much illness has resulted from the use of highly toxic chemical pesticides, which accumulate in drainage waters that flow into the rivers and contaminate local drinking-water supplies. The deterioration of the drinking water, furthered by inadequate purification plants, piping, and sewage systems, has greatly impaired the health and sanitary conditions of the people in the Aral region. In the last 15 years, the incidence of typhoid fever has increased almost 30 times and that of hepatitis, 7 times.[3] There are also considerable increases in the incidences of kidney disease, gallstones, and chronic gastritis. Many babies are born weak and ailing, and child mortality is more than 50 per 1,000 births. A recent appeal by scientists to improve the health of the region's children appears in the box on page 36.

The Path of Development

The desertification of the Aral region is a direct result of the socioeconomic development of Soviet central Asia and a considerable part of Kazakhstan. For example, the current direction of agricultural development in the region is toward water-intensive production. The agriculture is characterized by a predominance of cotton monoculture, an absence of crop rotation, and overly large plantations of water-intensive rice. Marginal lands are being developed, producing low harvests but abundant salt-laden runoff. As a result, despite the increase in irrigated areas, the gross yield of cotton in the central Asian republics and in Kazakhstan (excluding the Tadzhik Republic) has not increased. For example, Uzbekistan produced 3.9 million tonnes of raw cotton in 1970 and 5.3 million tonnes in 1975, but only 5.1 million tonnes in 1985.[4]

Despite the rapid salinization of soils because of excessive water application with insufficient drainage, attempts to economize water are inefficient. The accepted quotas of water consumption per irrigated hectare are excessive because specific soil and plant types and the actual consumption and mineralization of irrigation water are not fully taken into account. Often, the actual volume of water supplied to fields is even greater than the quotas; in different

central Asian republics, the excess varies from 20 to 100 percent. Yet as mineralization of the river water used for irrigation increases, irrigation quotas are raised even further, resulting in still greater water withdrawal. An increase in water mineralization by 0.1 grams per liter is offset by applying an additional 1,000 cubic meters of water per hectare.

If the anthropogenic desertification of the Aral region is not stopped, all environmental factors will be affected. The Aral Sea will be replaced by a 60,000-square-kilometer salt-and-sand desert generating dust that will harm the desert oases. Salt that formerly went to the Aral, which was the main salt repository for the whole region, will be deposited in the oases, which will shrink under the burden of salt and dust and, eventually, disappear.

The Search for Remedies

The ecological problems of the Aral Sea and its basin have worried scientists since the end of the 19th century. The Institute of Geography of the Soviet Academy of Sciences has participated in research on the Aral since 1976. At that time, many specialists thought that the ecological and socioeconomic consequences of the development of large-scale irrigation in the Aral Sea basin—primarily, the drop in the sea's level—would be insignificant and ignorable. However, geographers at the institute thought that the socioeconomic consequences could be serious and must be taken into account. Recent events have shown that position was correct.

It is now realized that the problems of the Aral basin affect the health and life of a vast territory. Their solution requires new approaches and decision-making based on a profound and integrated analysis of the numerous direct and feedback relationships between the sea and the land; the fate of one cannot be divided from the other.

The Aral region is in a state of ecological crisis in which the natural processes and ecological links have been disturbed so profoundly and the degradation has become so serious that the human population can no longer live and work as it once did. An ecological catastrophe is fast approaching wherein *(continued on page 36)*

Aral Sea Basin
(continued from page 9)

the degradation would be irreversible and life and work in the region impossible. The urgency of the situation must be realized. Only a short time remains to provide the population with clean drinking water, to ban the application of highly toxic pesticides, and to clean up the mineralized and polluted drainage and river waters.

Also, the current practice of adjusting water use to meet the demands of agricultural production development must change. On the contrary, the problems of water use should be the deciding factor in agricultural production plans. The role of cotton in the country's economy also should be reconsidered to increase the production of foodstuffs and to introduce more salt-resistant and less water-intensive crops.

The water resources of Soviet central Asia and Kazakhstan must be inventoried. Special attention should be given to increasing the efficiency of water-economizing systems, decreasing irrigation quotas by 15 to 20 percent, optimizing drainage systems to reduce water consumption by several thousand cubic meters per hectare, and improving regulation of run-off. The water freed in this way should be used primarily for resolving ecological imbalances to reinstate the normal conditions of life and health for local populations. If necessary, the water supply for agriculture could be increased by introducing water-saving technologies such as drip irrigation, by planting protective forest belts, by decreasing evapotranspiration, and by desalinizing water resources, primarily drainage waters. A combination of measures for increasing the productivity of irrigated and nonirrigated lands is needed to increase agricultural production without expanding irrigated areas or water consumption.

Saving the Aral region and resolving its tangle of ecological, hydrological, and socioeconomic problems will require a new attitude and an entirely new approach to economic activity through-

WOMEN SCIENTISTS' APPEAL FOR IMMEDIATE ACTION TO SAVE CHILDREN IN THE ARAL ECOLOGICAL CRISIS REGION[1]

We, the women participating in the first international symposium, "The Aral Crisis: Origins and Solution," held in Nukus, Krakalpak ASSR [Autonomous Soviet Socialist Republic], on 2–5 October 1990—specialists in the areas of ecology, medicine, geography, sociology, and demography—have concluded from analysis of data gathered from field observations that the Aral Sea region is one of ecological catastrophe that is especially hazardous to children.

The mortality rate of children in the Karakalpak ASSR is among the highest in the world and is growing each year: from 47.3 percent in 1978 to 59.8 percent in 1989. The maternal death rate in Karakalpakia has tripled in the last 5 years. More than 80 percent of women suffer from anemia, and every third [pregnant] woman gives birth prematurely. The dispensary system has revealed that in 1989–1990 almost 70 percent of the children in Karakalpakia fall into health categories II and III—i.e., [they are] practically sick. The number of children suffering from nervous and psychological disorders has tripled in the last 2 years.

The primary reasons for so sharp a decline in public health as to threaten the survival of the population are degradation of the environment, qualitative and quantitative depletion of the supply of potable water, microbial contamination of water, pesticide concentration, and protein and vitamin deficiency.

The situation is aggravated by the lack of necessary medical services and ecological education. We are extremely concerned by the slow pace, bordering on criminal, of [remedial] action. The knowledge already accumulated is sufficient to justify urgent action. We demand immediate action to save the population of the Aral region and other ecological disaster areas.

We call upon the government and the peoples of the USSR and the republics of central Asia and Kazakhstan, at all administrative levels of the country and the region; we call upon the United Nations, its specialized institutions—UNICEF [United Nations International Children's Emergency Fund], WHO [World Health Organization], and UNEP [United Nations Environment Programme] —and all organizations concerned with problems of public health and survival;

and we call upon all of the world's women to render emergency assistance to save the lives of children in the Aral region and to declare this region an ecological disaster zone. During 1990–1991,

• provide the population with clean drinking water and food as well as necessary medical assistance;
• hasten the development and implementation of a plan of action for solving the Aral problem;
• immediately shut down all hazardous production;
• place under tight control and decrease use of all pollutants, toxic chemicals, radioactive materials, and ozone-depleting substances;
• provide *glasnost* [complete reporting] of ecology-related information;
• spread energy-, water-, and resource-conserving technology; and
• prohibit use of child labor in cotton fields.

We send our appeal from Nukus, the burning center of ecological disaster, but we know that similar problems are emerging in an ever increasing number of areas encompassing the whole world.

Working women in all fields and at all levels—teachers, physicians, engineers, writers, and artists—we must step up our efforts to maintain normal environmental conditions to ensure the health of our children.

Here in Nukus, we are creating a committee of women scientists and other specialists, "Mothers, Save Your Children." The primary purpose of the committee is to gather and disseminate knowledge needed for emergency action to restore the environment in the Aral region and the other most degraded areas in the world.

We will not allow the killing of our children!

Please direct inquiries about membership in the committee or suggestions for enhancing the effectiveness of our work to Nina Maksimovna Novikova, Coordinator, "Mothers, Save Your Children" Committee, USSR Academy of Sciences Commission on Ecological Problems, Sadovaya-Chernogradskaya 13/3, Moscow 103064, USSR.

1. Before being edited, this appeal was translated by Katya Partan.

RESOLUTIONS OF THE INTERNATIONAL SYMPOSIUM
"THE ARAL CRISIS: ORIGINS AND SOLUTION" [1]

The international symposium "The Aral Crisis: Origins and Solution" was held on 2–5 October 1990 in Nukus, Karakalpuk Autonomous Soviet Socialist Republic. More than 200 leading specialists and scientists participated. Among these were 27 foreign scientists from the United States, England, Canada, Australia, Germany, and Spain, representing international research institutions specializing in hydrologic, health/sanitation, and biomedical problems in arid zones. Also in attendance were about 100 scientists from institutes of the USSR Academy of Sciences and the academies of sciences of Uzbekistan, Turkmenia, Tadzhikistan, and Kirgizia and from Moscow University and other leading institutions of higher learning in the USSR, and more than 50 representatives from the conservation committees of the central Asian republics and Kazakhstan, the USSR State Hydrometeorological Center, social and economic organizations, and foreign companies.

The symposium included presentations of more than 40 papers and reports, as well as a Nukus–Muynak–Aral field trip. The presentations focused on the deteriorating ecology of the region, the resultant declining health of its population, decreasing efficiency in economic management, and growing social tension. The medical/hygienic and socioeconomic aspects of the Aral problem were discussed in connection with environmental pollution, improper use of pesticides and fertilizers, insufficient medical facilities, and the lack of a modern water-supply system. Attention was also focused on imprudent water use, insufficient water resources, and conservation and restoration of the Aral Sea.

The scientists of the central Asian republics and Kazakhstan reached a unanimous conclusion as to the reasons for and ways to resolve the Aral crisis:

1. The scientists consider it necessary to request that the supreme soviets of the central Asian republics and Kazakhstan, as well as the Supreme Soviet of the USSR, declare the lower reaches of the rivers in the Aral basin to be an ecological disaster area and confer a corresponding status on the region.
2. The fundamental causes of the Aral crisis are erroneous choices of strategies for developing the region's productive forces; extensive development of water resources and agriculture; low standards for planning, construction, and operation of irrigation systems; and indiscriminate use of chemicals in agriculture.
3. Ecological restoration is impossible without stabilizing the level of the Aral Sea. To accomplish this, it will be necessary to perfect a complete water-management system.
4. A combination of urgent measures is necessary to restore the ecological conditions in the Aral Sea basin. The most urgent of these must be completed during the period 1993 to 1995, as delaying their implementation may have unforeseen and irreversible consequences. These measures are economic/organizational in nature and are aimed at restructuring industry and agriculture. Immediate measures must be taken to improve the health of the population and the quality and quantity of potable water and to impose strict controls on the use of pesticides and fertilizers.
5. To solve the problem, the following measures must be implemented in stages:
  • Strictly limit intake of water by the republics, with consideration for the need to introduce water-conserving technology in all areas of the economy.
  • Strictly prohibit expansion of the land area under irrigation and new siting of water-retaining industry, thus freeing river flow for preserving the Aral Sea.
  • Immediately revise the structure of cultivated land and crop composition to limit rice culture, remove low-yielding irrigated land from cultivation, and end cotton monoculture, thus freeing the land for development of orchards, viti-culture [vineyards], and seed alfalfa.
  • By the end of 1992, complete the removal of outflow from the region's [water] collection/drainage networks to the sea and reassess the continued use of lowland, valley, and container reservoirs, which lose 5 cubic kilometers of water annually.
6. Concomitantly, it is necessary to accelerate a comprehensive reconstruction of land-reclamation systems based on progressive water conservation and irrigation technology and coupled use of ground and surface water, which in many cases will provide waste-free use of irrigation waters.
7. Use of newly available water for unorganized flooding of high deltas and creation of polders are inadvisable. Feed crops should be grown only on land that has long been used for pasture or been under irrigation.
8. Construction of collectors to divert outflow to the Aral and Caspian seas from collection-drainage networks in the middle basins of the Syr Darya and Amu Darya is impermissible.
9. Current projects for redistributing flow between basins are scientifically unfounded. Fundamental research is needed to provide water to the Aral basin by alternate means.
10. Inasmuch as incomplete study of many important problems has made decisionmaking impossible, it is necessary to strengthen basic research and to create a specially funded comprehensive program to solve the Aral problem.
11. The leadership of the union's sovereign republics of central Asia and Kazakhstan is requested to create an interrepublic committee to coordinate joint activities, including scientific investigations, to solve the Aral Sea problem.
12. It is recommended that the republics of the union create scientific research subdivisions in their respective academies of sciences for ecological and water problems.
13. A program should be developed and implemented to increase ecological awareness in the population at large and to train personnel skilled in the area of ecology.
14. The conference considers it expedient to enlist the cooperation of international and foreign scientific organizations as well as individual foreign scientists in researching the Aral problem.
15. The participants in the international symposium support requesting the Union of Red Cross and Red Crescent Societies of the USSR to assist the population of the Aral region in improving [medical] service, the health of mothers and children, and supplies of good quality drinking water, and [they] appeal to governments, social organizations, and foundations to assist in these humanitarian activities.

---

1. Before being edited, these resolutions were translated by Katya Partan.

STEVE GUCHTY

Excessive water diversions into large, wasteful irrigation canals have greatly reduced the flow of the Amu Darya and Syr Darya rivers. Thus, the shipping and fishing industries have been hard hit.

out Soviet central Asia. Agriculture and other branches of the economy should be allowed to return to their traditional forms, which have been destroyed over the past decades, and attitudes toward water use should be changed. A number of new economic laws already have been adopted to change economic and social relationships in the USSR in principle; changes in practice will naturally take time.

Global Connections

The problems of the Aral Sea region are not unique on Earth. They are part of the universal problem of desertification that is occurring in many regions of the world, especially in Africa.[5] Hence, the search for a solution to the complex problems of the Aral region is of general and global significance.

Indeed, international organizations already have begun to work cooperatively on the problems of the Aral Sea

basin. In cooperation with Soviet specialists, the United Nations Environment Programme has set up a research project and formed a small working group on the problems of the Aral Sea. The group is expected to provide specific recommendations for action in two years. The executive committee of the International Geographical Union also has formed a research group that is working on the problem of critical ecological zones. The investigations of this group will embrace several critical zones, but one of the most important will be the Aral Sea region.

This past fall, an international symposium on the state of the Aral Sea basin was held in Nukus, the capital of Karakalpak. (The symposium resolutions may be found in the box on page 37.) The symposium, organized by the Aral Research and Coordination Center with the participation and assistance of the International Geographical Union Study Group on Critical En-

vironmental Zones and Global Change, was an important landmark in cooperation on the Aral. The symposium launched a serious effort to gather the data needed to conceptualize scientifically grounded socioeconomic development in Soviet central Asia. In the end, such thinking can be the only way to resolve the problems of the Aral Sea.

NOTES

1. P. P. Micklin, "Dessication of the Aral Sea: A Water Management Disaster in the Soviet Union," *Science* 241 (1988):1170-76.

2. T. I. Molosnova, O. I. Subbotina, and S. G. Chanysheva, *Klimaticheskiye Posledstviya Khozyaistvennol Deyatel Nosti v Zone Aral'skogo Morya* (Climatic Consequences of Economic Activity in the Zone of the Aral Sea) (Leningrad: Gidrometeoizdat, 1987).

3. D. B. Oreshkin, "Aral'skaya Katastrofa," (The Aral Catastrophe), *Nauka o Zemle*, no. 2 (1990):41.

4. See also, F. I. Khakimov, *Pochvenno-Meliorativnye Usloviya Opustynivaniya Del't* (Soil-Meliorative Conditions of Desertification of Deltas) (Pushchino, 1989).

5. H. E. Dregne, "Aridity and Land Degradation," *Environment*, October 1985, 16.

[31]

Article   Kerstin Lindahl Kiessling

Conference on the Aral Sea—Women, Children, Health and Environment

Women do not want to be mainstreamed into a polluted stream. We want to clean the stream and transform it into a fresh and flowing body. One that moves in a new direction – a world at peace, that respects human rights for all, renders economic justice and provides a sound and healthy environment.
Bella S. Abzug 1920–1998

The Stockholm 1998 Conference covered broad aspects of the Aral Sea crisis, in particular the health consequences for women and children. It was recognized that the situation is unique and severe in terms of environmental health, and that it is important to engage the whole international community to resolve the issues at stake. It was felt that one of the main objectives of the Conference, namely to raise public interest and awareness in Sweden and in the broader international community about the nature and consequences of the Aral Sea catastrophe, had been achieved. Furthermore, it was clear that efficient action, to improve the situation would involve all stakeholders in and outside of the region. An important element is to establish cooperative projects at the local level, designed to support major projects involving international donors.

In May 1991 Ambio published an article entitled Requiem for the Aral Sea (1). There was a sad and bitter conclusion to the article: a relentless drive for ever greater production of cotton in Soviet Central Asia under the banner "Millions of Tons of Cotton–At Any Cost," has resulted in considerable progress along a road ending in catastrophic costs and no production whatsoever.

Seven years later, on the 23–24 April 1998, after the break-up of the Soviet Union, a conference initiated by The Royal Swedish Academy of Sciences, together with the Swedish Save the Children and the Swedish Unifem Committee, once again approached the same complex of problems, under the more hopeful and enterprising title *Alleviating the Consequences of an Ecological Catastrophe: Women, Children, Health and Environment*. The major objectives of the conference were to increase awareness of the dimensions, the causes and the effects, of the Aral crisis, but first and foremost to focus on possibilities for action that will provide hope for the future.

The scientific program of the conference presented analyses of physical environmental issues, hydrology and ecology. Naturally, the conference has not revealed all the facts about the Aral Sea and its problems but the survey constituted an alarming and thought-provoking background for the exposition of the economic, social and health problems which followed. No change for the better is yet discernible in the Aral region, despite massive input of international assistance funds. A historic, ethnic and cultural background of the region is necessary for a proper understanding of the present situation and in any discussions of alternative conceptions of the developments. The importance of a strong civil society and community participation in present and future developmental decisions is strongly advocated. It has now been recognized that future efforts need to focus on the situation of women and children, especially on health and nutrition.

This paper is based on several reports from the conference, and I am especially indebted to Nick Aladin, Oral Ataniyazova, Maria Haralanova, Galiya Khasanova, Zaira Mazhitova, Anders Rapp, Merrick Tabor and Rolf Zetterström for their generous contributions.

ENVIRONMENTAL ISSUES

Thirty years ago, the Aral Sea, in the heart of Central Asia, was the world's fourth largest inland sea. Today, there are miles and miles of dry sand around the once so flourishing fishing villages. In some parts of the Aral Sea, the shoreline is now 120 km from where it used to be, and its water volume is now less than one third of that of 1961 (2). Where the shoreline has receded, salt deposits remain on the surface. These "land" areas are highly contaminated by agricultural and industrial chemical residues, which are carried by strong winds and deposited in areas distant from the source. The problems now faced by the inhabitants of the region are enormous, both with respect to their health and employment opportunities.

The "virgin lands" scheme, which aimed at growing wheat on the arid and semiarid steppes of Kazakhstan and the cultivation of cotton in Uzbekistan during the Soviet era, represents the elements of economic policy which are largely responsible for the current situation. Cotton monoculture in Uzbekistan and Turkmenistan led to a draining off of the water flowing to the Aral Sea, to provide for irrigation. In addition, the decreased water-flow was polluted as a result of the over-use of chemical fertilizers and pesticides (3).

This complex ecological disaster is mainly the result of misused irrigation technologies, inefficient bureaucracy, and lack of communication between politicians, engineers and the local people, in what was the periphery of the Soviet Empire. Before glasnost it was quite evident that Soviet scientists were instructed not to speak about degradation of national natural resources. At best they acknowledged a temporary decrease in the level of the Aral Sea, but were confident that their water engineers would soon compensate for that by construction of a large diversion canal, to turn one of the large Siberian rivers southward, to flow into the Aral Sea (4). But they were aware of the large dust plumes, seen on satellite images and experienced for many years, which resulted from the drying of the lake-bed. The so called Davidov Plan, to change the course of a Siberian river is a dream that was doomed from the start (5–7).

After glasnost and the collapse of the Soviet Union information is freely available, and thus some alleviation of the worst consequences of the catastrophe might be hoped for, although a reconstruction of the sea as it once was appears impossible. Between 1926 and 1960 about 55 km³ of water flowed from the Amu and Syr rivers into the Aral Sea each year; around 1990 that flow had more or less been eliminated. Salinity had risen to approximately the salinity of the open ocean. Wind erosion of the exposed dry seabed now removes about 43 mill. tonnes of salty silt each year, inhibiting plant growth in areas where it is deposited, often hundreds of kilometers from the source.

The diversion of the river water was intended to promote massive cotton farms, rice being the only other major crop. However, it has been calculated that at least 40% of the water diverted to the Kara Kum canal in Turkmenistan, irrigating about 1 mill. ha, disappears through leakage and evaporation losses. Moreover, waterlogging and salinization have affected much of the newly irrigated areas, making the proceeds of the whole enterprise highly questionable. This disaster is increasing year by year through wind erosion, salinization, pollution of air and drinking water, with growing health and nutrition problems as a consequence.

SUSTAINABLE DEVELOPMENT

Since the UN Conference in Rio 1992, Agenda 21 has become a guiding principle in judgments on environmental decision-making (8). The Aral catastrophe may have been averted had Agenda 21 rules for the protection of natural resources, and the precautionary principle served as guidelines for the decision makers. Agenda 21 is a formidable document showing the way to democracy and sustainability. It points to the priorities between environment and exploitation, householding and expenditure, people, machines and techniques, healthy and unhealthy human settlements, long-term and short-term, simplicity and complexity, and the perspectives of men and women. Agenda 21 contains a separate chapter on women and the roles they play in sustainable development (9). It states that the participation of young people in the process of decision-making and in the implementation of programs is crucial (10).

The governments of Central Asia were also active in the process that created Agenda 21. The fate of the Aral Sea will be a measure of the value of those commitments the world leaders undertook at UNCED in Rio in 1992 (11). It is incumbent on all environmentally active NGOs to ensure that the process to hold these leaders responsible starts now. "We have to take responsibility today for the tomorrow of our children" said Görel Thurdin, president of the International Save the Children Fund.

What then are the possibilities to remedy what has been damaged during decades of misuse?

Since 1989, Aral has ceased to exist as one lake. Because of the enormous water loss the Aral Sea was divided into two parts—the northern Small Aral and the southern Large Aral. In the Small Aral salinity is lower, and its level is a rather stable 2–3 m above the Large Aral. A temporary construction of a dam between the two lakes has provided an opportunity to remedy the Small Aral Sea. Theoretically, at least, it is assumed that if the water flow from the Syrdarya (5 to 7 km^3 yr^{-1}) is kept stable, complete desalinizaton of the Small Aral Sea will be possible within several decades (12). A similar favorable solution for the Large Aral Sea is not envisaged. Left to itself the Sea will rapidly be transformed into a hypersaline waterbody. The eastern part will dry up completely, and transformation of the ecosystem into diapause will occur (13).

These adverse conditions eventually resulted in the formation of a very original ecosystem, which is capable of transforming according to change in salinity. When the Aral Sea is deep and salinity is low freshwater and brackish-water fauna and flora dominate. At the same time, marine and hypersaline fauna and flora survive in the shallow salinized gulfs, waiting for favorable conditions to arise. Apparently, with complete drying of the Aral Sea part of its ecosystem is capable of running into diapause, as some hydrobionts form resting stages (spores, seeds, latent eggs, etc), that allow them to survive even a complete dry-up of the lake (14). This has to be taken into account when the future of the lake is considered.

It has been anticipated that the Aral Sea will have ceased to exist around the year 2010, and that it is already too late for restoration. However, there are alternatives. If the rate of dry-up could be slowed down somewhat, and if chemical pollution is halted, the chances for a rehabilitation are rather high if the needed amount of water could be discharged into the Sea (13).

But if special measures of conservation and rehabilitation are not carried out in the near future the chances for restoration could become very low indeed. In this event, will out-migration be the only solution for the majority of people in Karakalpakia (the semi-autonomous republic south of the Aral Sea)? Will massive programs for improved water management and clean drinking water for villages and hospitals in the region become a reality in time? Can anyone claim to be an innocent bystander in the face of such a tragedy?

THE SOCIAL ENVIRONMENT

Over the years, scientific and popular attention has focussed on analyses of the natural environment and the physical disaster. However, the social, cultural and health problems are disastrous and a rapid and sustainable change is needed, involving all members of society. Successful and sustainable change can only come about if the people themselves are mobilized to take active part in the process. How can greater community participation be supported and facilitated by organizations, and follow-up actions to conferences like that in Stockholm 1998?

The problems associated with the Aral Sea region are global both in terms of causes and in terms of consequences (15). Projects intended to alleviate the crises must consider these global factors and the prerequisites for long-term solutions. The potential for local action, in contributing to more immediate remedies, is essential for any project. It is important to gain an understanding of the historical, economic, political, social and cultural aspects of the problem, and the ways in which these aspects are related to each other and to environmental and medical factors. There are a number of such aspects which need to be dealt with.

Historical aspects are important for understanding the specific background and causes, however perhaps even more important than actual historical developments are perceptions of historical developments. The role of the Aral Sea and the two primary rivers feeding into it—the Amudarya and Syrdarya—have been central to the development of the region. The Aral Basin, comprised of a combination of high mountains and deep valleys, is seen as having provided the preconditions for a pattern first of nomadic pastoralism and later of agricultural micro-regions in the oases and valleys. This influenced the development of self-sufficient agricultural economies, based on specialization of production and commerce between regions (16). It also influenced the development of decentralized social and political structures, a high degree of regionalism and traditional political systems based on the development of local clan identities. The traditional role of women and consequences for children can also be found in this history. With independence came attempts at nation building and the creation of national identities based, primarily, on conceptions of the traditional cultures of the titular nationalities in the Central Asian states. In the specific case of women, there is a notion that the Soviet period involved a distortion of traditional social and cultural values which should be revived. A proper understanding of these historical conceptions is important to appreciate the present situation and to discuss alternative conceptions.

Many of the traditional systems have assumed a new relevance in the post-Soviet political systems. Regionalism and clan identity have become more salient, and local interests have come to be represented through both formal and informal political arrangements, and power relations tied to these structures. They have even become the basis of independence movements. The ways in which these local structures can be used to influence both national and regional policy can provide important solutions (17).

It is also important to gain a greater understanding of the economic constraints existing in the current situation. In the short-term, the problems involved in diversifying the economies of the

Central Asian states, in order to reduce water usage, are significant. Independence has increased the reliance on overexploitation. At the same time a water crisis imposes constraints on agricultural incomes, rural employment and agricultural export opportunities. Studying these problems at the local level, and once again the more specific consequences for women and children, is a key to seeking solutions (18–20).

The relations between the Central Asian states and prospects for cooperation are also of the utmost interest. The Amudarya and Syrdarya each flow through four of the five Central Asian states. Given these economic imperatives, the period directly following independence witnessed the proliferation of regional disputes over water, since the boundaries of the newly independent Central Asian states do not coincide with the division of water resources under the unified Soviet economy. The water crisis has, in addition, implications for the trade strategies of these states. The agricultural specialization of the economies of the Central Asian states makes them trade competitors. As a consequence, water disputes act as a constraint on the development strategies of these states (21–23). In the absence of economic asymmetries that make possible complementary trade exchanges, uncoordinated and conflicting development strategies can easily lead competing states into overt conflict with one another over transboundary resources. A study of these aspects is important in assessing the opportunities for diversification and alternatives on the local level and even possibilities for substitution for trade relations which have largely collapsed between the former republics. The Nukus Declaration of 1995 and the work of the IFAS (International Fund to Save the Aral Sea) are indications that the five states are aware of the need for cooperation. But the political and economic difficulties are formidable; the world community has a heavy responsibility to support cooperative efforts in the region.

HEALTH AND NUTRITION

The breakdown of the Soviet Union created immense new opportunities. Over the past seven years, a major revolution has taken place in many countries. By and large, this has been a quiet revolution, although it has drastically changed the lives of many people living in the newly independent states. The past few years have been a period of excitement, of establishing new identities, and of facing the challenges of socioeconomic and cultural reform. For many people, however, these changes have meant hardship, poverty, unemployment and ill health.

In the Aral Sea region the ecological catastrophe which led to severe environmental pollution has caused an alarming deterioration in human health. At the conference and elsewhere (see e.g. 3) detailed data have been given; many are presented in the following paragraphs. They are difficult to verify and interpret: indeed much further research is needed. But they are sufficient to realize that early and efficient action is needed.

The deterioration of human health with increasing infant mortality rate, declining life expectancy at birth and increasing prevalence of serious infectious diseases as well as of chronic diseases of the kidneys and gastrointestinal tract, malignancies, psychiatric disease and alcoholism, as seen in the Central Asian Republics, is thought to be due to a combination of several factors such as inadequate nutrition, poor sanitation, collapse of the health-care system and pollution from Soviet agriculture and industries.

The highest maternal and child mortality and morbidity in the European region are to be found in Kazakhstan, Kyrgyzstan, Tajikistan, Turkmenistan and Uzbekistan. Perinatal complications are responsible for 30% of infant deaths. Infants experience upper respiratory tract infections, diarrhoeal diseases, vaccine-preventable diseases such as tuberculosis and nutritional deficiencies. Exclusive breastfeeding is still a marginal practice.

In Kazakhstan, only 12% of infants aged 0–3 months are exclusively breastfed (24).

Among children under five, acute respiratory infections, including pneumonia, are responsible for between 1/3 and 1/2 of infant deaths. In Kazakhstan, diarrhoea, acute respiratory infections, including flu and pneumonia are responsible for over 40% of the mortality of children under five.

In the Kazakhstan Republic, according to official statistics from 1993, life expectancy for males was 63.2 while for females it was 72.7 years and the gap between men and women is widening. In the same year, the infant mortality rate was estimated to be about 70 per 1000; compared to around 20 per 1000 ten years previously (25).

Maternal mortality ranges between 63 and 135 per 1000 live births. Lack of contraceptives forces women to resort to abortion as a means of fertility control. Unsafe abortion is a common cause of mortality and morbidity in women, but is only one of the many basic medical interventions that still often lead to complications due to poor sanitary conditions and lack of antibiotics. Many pregnant women are malnourished. Anaemia is one of their most acute problems, and iodine deficiency and chronic malnutrition are very common in the region, while the lack of laboratory testing equipment makes full analysis of blood impossible. Chronic malnutrition in children is also reported from several countries (26–29).

It has repeatedly been stated that people living in the Aral Sea region have been subjected to severe environmental hazards due to the exposure to industrial pollutants such as PCB-compounds and heavy metals but also to pesticides, which have been deposited over large areas to support the cotton fields in the former desert land. The pollutants have not only accumulated locally in water, but also in soil and have been deposited over large areas by atmospheric transport and have entered the food chain leading to humans.

In a recent study, on a group of children referred to the National Children's Rehabilitation Centre in Almaty, elevated concentrations of PCB-compounds were found. In addition, the blood lipid concentration of the beta-isomers of hexachloro-cyclohexanes was extremely high and that of DDT-compounds was elevated up to 20 times. The concentration of lead in red blood cells was moderately elevated and that of cadmium slightly elevated, compared to findings in children from Stockholm, Sweden (30, 31). High concentrations of similar substances and, even more alarming, some extremely toxic and carcinogenic compounds have been demonstrated in the milk from mothers from some agricultural districts in Kazakhstan. Obviously, no general conclusions can be drawn from the limited studies done so far but the results need serious reflection (Ramel, C., pers. comm.).

Chemical pollution, while certainly a vital element in the whole picture, has perhaps been emphasized at the expense of other environmental health issues. In many parts of the region, more widespread adverse effects on health are caused by microbiological rather than chemical contaminants. The dramatic upsurge of, for example, cholera and tuberculosis in the eastern part of the Region is closely linked with the environmental health conditions (Michèl-Sellier, E., pers. comm.).

Currently more than 40% of the population (60% in rural areas) obtains water from decentralized (individual or communal) sources. These sources include canals, rivers and dug wells; however, no official information is available regarding the quantity of these water points or the quality of the water provided. Another problem is the irregularity of water supplies. Environmental pollution caused by pesticides and heavy metals has been widely attributed as the main cause of deteriorating health in the Aral Sea area. However, no actual environmental and drinking water quality data exist to confirm this view. Moreover, some studies suggest that environmental sources of toxicity are secondary to

© Royal Swedish Academy of Sciences 1998
http://www.ambio.kva.se

other factors such as mineralization. Local agencies have a limited capacity to measure the concentration of either pesticides or heavy metals (32).

Sanitation represents a major problem across the region, especially when combined with inadequate water supplies in certain areas. Poor sanitation facilities and practices in schools represent a major health threat. Sewerage systems exist only in some cities. The rural population uses primitive pit latrines. These are also used in the majority of the schools and health facilities in the region. Many latrines are situated in high watertable areas, thus contributing to groundwater pollution. In real terms, 90–95% of the population use primitive pit latrines (33).

A WHO conference in Tashkent in 1995 reported that due to the efforts of the five countries in Central Asia, to date, USD 2 billion have been ensured from external resources and invested to improve the quality of environment, health, and living conditions in the coastal area. However, despite these efforts and investments, the situation continues to deteriorate and the crisis is growing. WHO's greatest concern is that the population in the region, particularly children, do not have access to safe water, and that there is no safe sanitation provided. No city or village should be provided with a water supply without a sewage system. Adequate sanitation is the entry point to water management in the collaborative way.

Solving the abovementioned problems, reversing the negative trend, building up an effective system for the prevention and control of environmental health hazards, and improving the health of populations, particularly women and children, will require large-scale investments of existing knowledge and financial resources. In all cases, however, one particular challenge does not need more money—or technology or research—and yet it is crucial in all efforts; i.e. the need to create better cooperation and coordination of the work and potential of the major international organizations and governments willing to work to improve environment and health in Europe. We are, today, far from exploiting the full potential of cooperation in the environment and health fields.

However, this very dark picture of the living conditions in the region should not lead to dejection: *don't focus on the problems but on the potentials.* M. Haralanova of WHO has said: "During this time of crisis, we should not give up hope. On the contrary, we should strengthen our resources for a brighter future so that our children may live in the improved environment of a new Europe. We should have courage and determination to fight for action with regard to the environmental health challenges mentioned and to all other. And above all, we should recognize that the limited resources available to countries, international organizations and funding agencies must be organized in a new spirit and within a new framework for joint action." (24, 34).

GENDER PERSPECTIVE

In order to suppress further deterioration of the present natural resource base and to relieve human poverty and suffering, the competence and expertise within society need to be fully utilized. This requires the right for both men and women to participate, on equal terms and to an equal extent, in societal activities. It also needs acknowledgement and consideration of the unique interests and needs of all people, men, women, and children alike. Despite growing awareness worldwide, the process of attaining gender equality within all areas, i.e. what is now called mainstreaming of gender issues, is only in its infancy.

The empowerment of women, and the anticipated follow-up actions of the Stockholm Conference 1998 also aimed to improve the situation of women, including the welfare of their children in the Aral Region. No real and long-term change can be brought about without the participation of women, in all aspects of development; technical, cultural, economic, social or political.

In the follow-up process special attention must be given to the gender aspect: it is essential to avoid development policies and programs that can have negative impacts, or are not people/gender friendly. There is an indispensable need for multidisciplinary action, particularly the input of the social sciences, in order to avoid gap forming, which could prevent positive outcomes. A number of direct and indirect health issues for both women and men are strongly linked to the social aspect of the crisis; i.e. destruction of livelihoods, out-migration of men, increasing poverty, feminization of poverty, etc. Gender perspectives and gender analysis are suggested as useful tools to show a way to solutions. Strengthening health services alone could assist the "women-children" focus, but will not affect real gender or empowerment issues, or give access to resources beyond those of health care. The hope for children is, in the long term, to be found in women's empowerment.

An important lesson from the Cairo and Beijing UN conferences was that women's economic and political empowerment is strongly linked to traditions and customs. The real platform to aim for is for intersectoral action from commonly understood perspectives, which can be accepted by the stakeholders, (inside WHO referred to as "resetting the health agenda") (Sims, J., pers.comm.). This implies that the issues can be addressed in a broad integrated framework rather than through the pursuit of individual programs, projects, or sector goals.

Among the most important historical, ethnic and social patterns are those tied to gender relations. It is important to understand historical and existing gender relations in the region. The attempt to create new national identities has been based to a large extent on emphasizing older traditions. When specific gender relations are conceived in terms of the traditional relations, this tends to lead to the conservation of these relations.

Conventional social science approaches also tend to be strongly androcentric and, thus, conceal the true situation of women and the relations between men and women in society. Historical descriptions are based on developments which are seen as important turning points. However, these developments have often had different consequences for men and women. In order to gain a proper understanding of the historical background it is important to understand the specific histories of men and women in the region (16).

The economic constraints in the Aral Sea catastrophe and the dependence on specific forms of production must be seen in terms of the differentiaton of the labor force. Economic descriptions based on aggregated figures do not reveal ways in which women are involved in low-wage and even no-wage labor. This creates a dilemma between notions of equality and a potential competitive advantage obtained through this division of labor. Technological solutions which take traditional patterns as their point of departure propose solutions which serve to reinforce existing gender relations.

Even the methods which are employed in the study of these problems display gender bias. Specifically, historical, political and social phenomena are often studied on the basis of a model of science most relevant to the natural sciences. Simply stated; a hydrologist has no need to build up a personal relationship with the water he or she studies in order to understand how it behaves. This is not the case with human beings. Another type of methodology is necessary in order to understand the various problems and to judge the practicality of various solutions. This is related to the need for local grassroots approach.

BROAD OUTLINES FOR ACTION

During the Soviet era many basic public sector activities, such as health and education depended on massive external resource inputs. An estimated loss of 60% of the state income occurred in the newly formed states at the dissolution of the Soviet Union which made the financial, human and social costs of restructuring immense. Even if the Central Asian Republics have managed to avoid total collapse, the situation for vulnerable populations continues to deteriorate. Nowhere are these problems more clearly illustrated than in the Aral Sea area. Immediate foreign assistance and active help to self-help are indispensable. Simple tools and know-how are needed to help families and communities mitigate against problems which can still be controlled, and to prevent further expansion of these problems. Prevention of diarrhoeal diseases, acute respiratory infections, malaria, measles, malnutriton and its manifold attendant phenomena must be promoted. The first and essential prerequisite to achieve this aim is clean water for hospitals, schools and villages. There is an urgent need for basic equipment and supplies to provide standardized service procedures and treatment, especially for maternal and childhood illnesses, together with training and local capacity building to develop hygienic, nutritional and sanitary standards.

The Advisory Committee that was initiated for the Stockholm 1998 conference, can act as an instrument for constructive follow-up to the conference. The main task of this Committee will be to present activities to be undertaken, and to initiate discussions about priority areas within a wider international network. Creative thinking is crucial, money alone is not the solution. The work of the Advisory Committee, properly composed, will provide a foundation for the selection of future projects and proposals for Swedish and international funding agencies.

The Conference proposed that the following activities be pursued.

Background for Action

- Important efforts should be made to improve and enlarge the flow of information about the situation in the region.
- The need for more, and more reliable, statistical data was underlined, taking into account the absolute need for gender disaggregated statistics.
- Priority areas for further research should be defined in cooperation with all the stakeholders.
- Communication with the region has to be improved; information technology is a tool.

Joint Projects

NGOs should try to agree on possible joint projects relating to health issues, both physical and psychosocial, taking into account the gender aspects and the particular problems of children, in accordance with the *UN Convention on the Rights of the Child*. Outlines of such projects should be developed before 1 October 1998.

In particular, efforts should be made to develop projects at the local level, possibly in coneection with twinning between municipalities in the Aral region and Sweden.

The potential for micro-financing schemes should be fully explored.

Follow-up

The proceedings of the Conference will be published in an readily accessible form as soon as possible.

A report will be made in an appropriate form to the ongoing session of the Commission on Sustainable Development (CSD), which has freshwater as a special priority item.

The possibilities for follow-up seminars in Sweden on relevant themes should be explored. Such seminars could be held at the regional level and involve interested NGOs and/or other entities.

References and Notes

1. Precoda, N. 1991. Requiem for the Aral Sea. *Ambio 20*, 109–114.
2. UNEP Atlas. 1992. *World Atlas of Desertification*. Middleton, N. and Thomas, D. (eds). UNEP and E. Arnold Ltd, Kent, UK, 80 p. and UNEP Atlas. 1997. 2nd Edition, Arnold, London. Co-published by Wiley and Sons, UK.
3. Létolle, R. and Mainguet, M. 1993. *Aral*. Springer Verlag, France.
4. Libert, B. 1995. *The Environmental Heritage of Soviet Agriculture*. CAB International, Wallingford, Oxon, UK, Chapter 3, Irrigation: Friend or Foe? pp 41–65.
5. Orlovskij, N.S. 1962. *Some Data on Dust Storms in Turkmenistan*. Mimeo. Report of the Desert Institute, Ashkabad, pp. 17–42. (In Russian).
6. Grigoriev, A.A., Ivlev, L.S. and Lipatov, U.B. 1976. Analysis of Meteor-4 satellite TV picture of dust storms in the Precaspian Region. In: *Cospar Research XVI*. Rycroft, M.J. (ed.). Akad. Verlag, Berlin, Germany, pp. 29–34.
7. Rapp, A. The environmental disaster of the Aral Sea. The emergence of the crisis. Proceeding from the conference *Alleviating the Consequences of an Ecological Catastrophe*. Stockholm, April 23–24, 1998. Lindahl Kiessling, K. (ed.). (To be published).
8. *Agenda 21*. 1993. UN Publications; Sales No. E.93.1.11.
9. *Agenda 21, Chapter 24*. 1993. Global Action for Women. Towards sustainable and equitable development.
10. *Agenda 21, Chapter 25*. 1993. Children and youth in sustainable development. UN Publications; Sales No. E. 93.1.11.
11. *Rio Declaration on Environment and Development*. UN 1992.
12. Aladin, N.V., Plotnikov, I.S. and Potts, W.T.W. 1995. The Aral Sea desiccation and possible ways of rehabilitating and conserving its northern part. *Int. J. Environmetrics* 6, 17–29.
13. Aladin, N. Hydrology and ecology of the Aral Sea. Proceedings from the conference *Alleviating the Consequences of an Ecological Catastrophe*. Stockholm, April 23–24, 1998. Lindahl Kiessling, K. (ed.).
14. Aladin, N. 1997. Resting eggs of *Artemia salina* and *Moina mongolica* from the bottom sediment of the Aral Sea. Diapause in Crustacea. *Book of Abstract*. Ghent, August 24–29, p. 2.
15. Proceedings from the conference *Alleviating the Consequences of an Ecological Catastrophe*. Stockholm, April 23–24, 1998. Lindahl Kiessling, K. (ed.). (To be published).
16. Tabor, M. What were the grounds for the catastrophe? Historic, ethnic and cultural development in the Aral Sea region. Proceedings from the conference *Alleviating the Consequences of an Ecological Catastrophe*. Stockholm, April 23–24, 1998. Lindahl Kiessling, K. (ed.). (To be published).
17. Anderson, J. 1997. *The International Politics of Central Asia*. Manchester University Press, Manchester and New York, pp. 1–225.
18. Khasanova, G. Women's empowerment in the rebuilding process. Proceedings from the conference *Alleviating the Consequences of an Ecological Catstrophe*. Stockholm, April 23–24, 1998. Lindahl Kiessling, K. (ed.). (To be published).
19. Gradin A. Community participation. The importance of civil society. Proceedings from the conference *Alleviating the Consequences of an Ecological Catstrophe*. Stockholm, April 23–24, 1998. Lindahl Kiessling, K. (ed.). (To be published).
20. Kanaev, A. and Fägerlind, I. 1996. *Citizenship Education in Central Asia. Status and Possibilities for Cooperation*. Final Report of a UNESCO sub-regional workshop in Ashgaot, Turkmenistan, Report No. 103, IIE.
21. Libert, B. 1995. *The Environmental Heritage of Soviet Agriculture*. CAB Internationl, Wallingford, Oxon, UK, Chapter 3, Irrigation: Friend or foe? pp 61 and following.
22. Falkenmark, M. 1991. In: *Water: The International Crisis*. Clarke, R. (ed.). Earthscan Publications Ltd, London, UK.
23. Klotzli, S. 1995. *The Water and Soil Crisis in Central Asia. A Source for Future Conflict*. UNDP, Occasional Paper.
24. Haralanova, M. Health problems of women and children and the Aral Sea cricis. Proceedings from the conference *Alleviating the Consequences of an Ecological Catastrophe*. Stockholm, April 23–24, 1998. Lindahl Kiessling, K. (ed.). (To be published).
25. Zetterström, R. Consequences on human health from environmental pollution and poverty in the Aral Sea Region. Proceedings from the conference *Alleviating the Consequences of an Ecological Catastrophe*. Stockholm, April 23–24, 1998. Lindahl Kiessling, K. (ed.). (To be published).
26. Ataniyazova, O. The Aral Sea crisis: Health consequences for women. Proceedings from the conference *Alleviating the Consequences of an Ecological Catastrophe*. Stockholm, April 23–24. Lindahl Kiessling, K. (ed.). (To be published).
27. Ataniyazova, O. 1998. Some aspects of anemia in women from the Aral Sea ecological crisis region. Vestnik. Karakalpak. Otc. Academii Nauk Uzbek N1, pp. 7–10. (In Russian).
28. Morse, C. and Holmes, F. 1994. The prevalence of anemia in the Aral Sea region. The meaning of anemia. *Common Health Newsletter 2*.
29. Falkingham, J. 1997. *Household Welfare in Central Asia*. Macmillan, Basingstoke, UK.
30. Jensen, S, Mazhitova, Z. and Zetterström, R. 1997. Environmental pollution and child health in the Aral Sea region in Kazakhstan. *Sci. Tot. Environ.* 206, 187–193.
31. Mazhitova, Z., Jensen, S., Ritzén, M. and Zetterström, R. 1998. Chlorinated contaminants, growth and thyroid function in schoolchildren from the Aral Sea region in Kazakhstan. *Acta Paediatr.* 87, 991–995.
32. Reimer, M. (ed.). 1995. *The Aral in Crisis*. UNDP report, Tashkent 1995, p. 4.
33. ASPERA. 1996. Protecting children and women in the Aral Sea Disaster Zone. Project summary and outline of fund raising needs for the Aral Sea Project for Environmental and Regional Assistance, p. 36. UNICEF Area Office for the Central Asian Republics and Kazakhstan, pp. 1–50.
34. Thurdin, G. The UN Convention on the Rights of the Child and Agenda 21. Proceedings from the conference *Alleviating the Consequences of an Ecological Catastrophe*. Stockholm, April 23–24, 1998. Lindahl Kiessling, K. (ed.). (To be published).

Kerstin Lindahl Kiessling is a biologist. She has been professor of zoophysiology at Uppsala University and Vice President of the Royal Swedish Academy of Sciences, in charge of the Academy's Committee for Science in Society. She is a member of the Board of the Swedish UNIFEM Committee. Her address: Ripvägen 14, S-756 53, Uppsala, Sweden.
e-mail: Kerstin.Lindahl-Kiessling@zoofys.uu.se

[32]

Quality Status, Appropriate Monitoring and Legislation of the North Sea in Relation to its Assimilative Capacity

A.R.D. Stebbing · R.I. Willows

8.1
Introduction

The North Sea is one of the most heavily polluted marginal seas on earth (Degens 1988), and used to be one of the most biologically productive. The problems have been slowly recognized and the eight North Sea states are committed to "… the principle of safeguarding the marine ecosystem of the North Sea" (Ministerial Declaration 1987), and regular Ministerial Declarations set out agreed measures to achieve these ends. The Quality Status Reports provide the scientific basis for the Ministerial Declarations, and with associated research programmes (see Quality Status Reports for the North Sea 1987, 1990, 1993), demonstrate that the North Sea is also one of the most intensely studied and monitored in the world. The example must therefore provide an instructive case study for those enclosed seas where the need to achieve a sustainable environment may be in conflict with the economic growth of bordering states.

Our aim is to consider the North Sea from the perspective of its assimilative capacity for wastes and effluents. Assimilative capacity is a cross disciplinary concept that describes a resource of considerable economic value which, if it is exceeded, represents a limitation on the sustainable use of the marine environment. An integral part of the argument for accepting assimilative capacity as a resource is that contamination is not a threat to sustaining North Sea ecosystems unless it has a biological impact. Thus we advocate the use of biological techniques to monitor environmental quality, and will demonstrate that a predominately chemical approach is no longer appropriate.

Currently, monitoring effort and resources are directed to establish compliance with environmental legislation and regulations, which are expressed mainly in chemical terms. For example, the effective application of Environmental Quality Standards (EQS) depends on chemical monitoring of individual contaminants. However, the assumptions that are made in the setting of EQSs raises questions as to their relevance. We therefore consider North Sea pollution and its control in a cross-disciplinary manner, centred on the observation that it is the assimilative capacity of the North Sea that we utilize as an environmental service. It is the extent to which that capacity is exceeded that represents the threat to the sustainability of ecosystems. Until recently there has been little reason to consider the assimilative capacity of the North Sea as an entity, since there has not been much evidence that its overall capacity to dilute, to disperse and detoxify chemical contamination might be exceeded by contaminant inputs, and the accumulation of those that persist. Here we consider evidence, using a biological approach to monitoring, that has developed over the last decade and more, which provides more relevant data, quantifying pollution in terms of its biological impact on water quality.

The approach uses measures of the biological impact of contaminant loading, rather than attempting to assess the health of the system by monitoring each chemical or class of chemical contaminant. We advocate monitoring in terms that relate directly to the criteria used to gauge environmental health. We will consider the capacity of the North Sea as a whole to assimilate wastes and effluents, and in the light of recent data (post 1990) which demonstrate that it is being exceeded. Since the work of the regulatory authorities is to monitor the conformation of contaminant concentrations to pollution legislation, we ask whether the present statutory controls and approach to monitoring are appropriate for their purpose.

8.2
The North Sea

8.2.1
Inputs and Outputs of Contaminants

The North Sea is a centre of intense human activity for eight west European countries, for it is their only access to the sea for shipping, oil and mineral wealth, and fishing. The North Sea provides an environmental service by accepting the industrial, agricultural and societal wastes and effluents from the 164 million people that live on its shores and in the catchment of the rivers that flow into it. Such inputs are likely to challenge the capacity of the environment to sustain the indigenous biota by their volume, if not their toxicity.

The catchment area of the rivers flowing into the North Sea is 850 000 km², compared with its surface area of 575 000 km². It has a volume of 93 830 km³ and receives a variable run-off from these rivers of 296–354 km³ a⁻¹. The catchments of the major rivers are densely populated and heavily industrialized, providing the main anthropogenic inputs (contaminants and nutrients) to the North Sea. The river basins of the Thames, Humber, Elbe, Weser, Rhine, Scheldt and Seine are the most densely populated (statistics are drawn from ICES (1983) and Quality Status Report for the North Sea 1993). The North Sea (Fig. 8.1a) is bounded on three sides by land with its principle exchange with the Atlantic to the North and a minor route for exchange via the Dover Straits to the south. It is a shallow sea (30–200 m), less so in the south, such that the volume of water that can dilute estuarine inputs is greater for those rivers that flow into the northern North Sea (rivers Forth and Tyne) than in the south (rivers Thames and Rhine).

While the North Sea can be considered as a semi-enclosed system flushed by relatively clean waters from the North Atlantic, its northern open boundary is also a route by which contaminants may be imported and exported. Input via this route is well illustrated by radioisotopes originating from Sellafield and Dounreay (Kautsky 1988). The Norwegian coastal current provides the major outflow of all water from the North Sea and the Baltic, originating in the Skagerrak and hugging the coast of Norway as it moves north.

The stronger tidal flows (Fig. 8.1a) are coastal, with weak and variable currents in the central North Sea. The dominant residual flows are from north to south along the UK East Coast with a reciprocal flow northward along the European coast towards the Skagerrak, supplemented by an input via the Dover Straits. Current speeds fall as the northward flowing currents slow at the entrance to the Skagerrak and Kattegat, before returning to the North Atlantic.

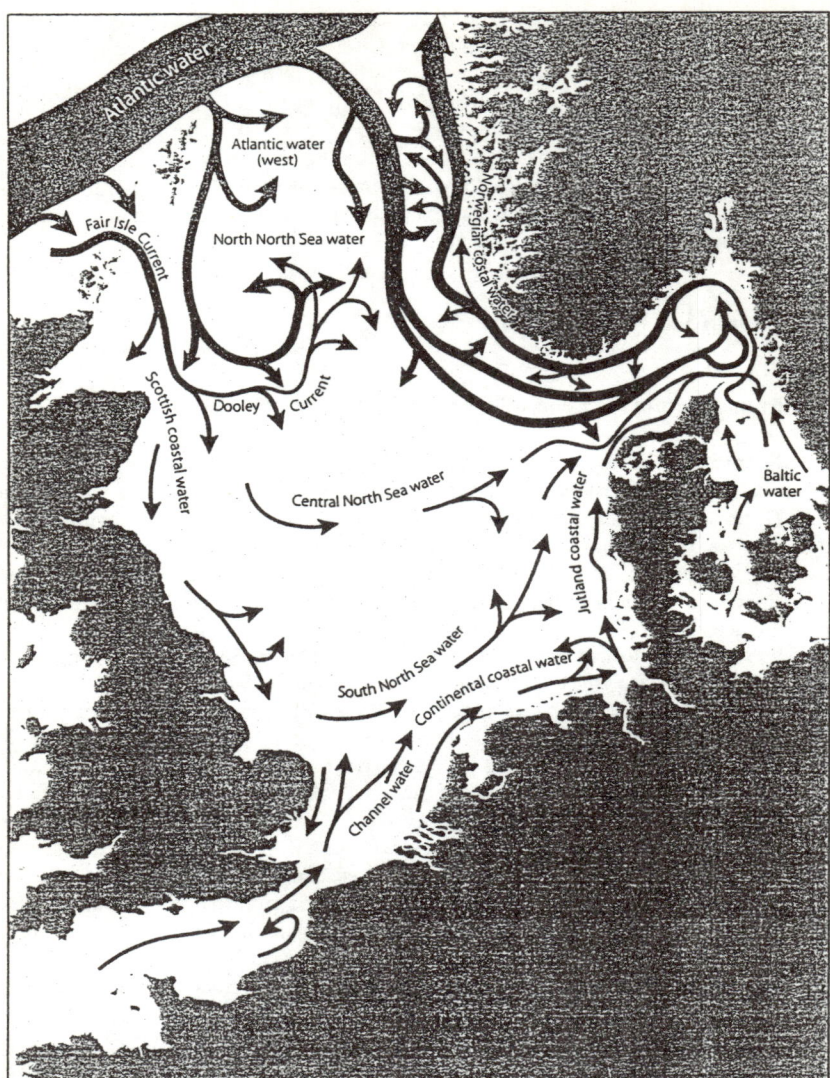

Fig. 8.1 a. Map of the North Sea indicating residual currents. The width of arrows indicates the magnitude of volume transport (From Quality Status Report for the North Sea (1993), after Turrell et al. 1992)

On average the water in the North Sea is exchanged with the North Atlantic every 1–2 years, but the renewal time varies between 0.5 and 3 years for different areas. For example, a renewal time of less than 6 months is expected for Norwegian coastal wa-

1. Gluss Voe
2. Voxter Voe
3. Orkney
4. Ythan
5. Montrose
6. Lucky Beacon
7. Musselburgh
8. Berwick
9. Holy Island
10. Coquet Estuary
11. Cresswell
12. Blyth
13. Trow Point
14. Teesmouth
15. Whitby
16. Filey Brigg
17. Humber Bull Fort
18. Cleethorpes
19. *Dowsing*
20. *Dudgeon*
21. Gat Sand
22. Hunstanton
23. *Newarp*
24. *Smiths Knoll*
25. Walberswick
26. *Shipwash*
27. Crabknowe Spit
28. *Sunk*
29. Creeksea
30. Southend
31. Swale
32. *Falls*
33. *East Goodwin*
34. *Sandettie*
35. *South Goodwin*
36. Whitstable

37. Plymouth
38. Whitsand

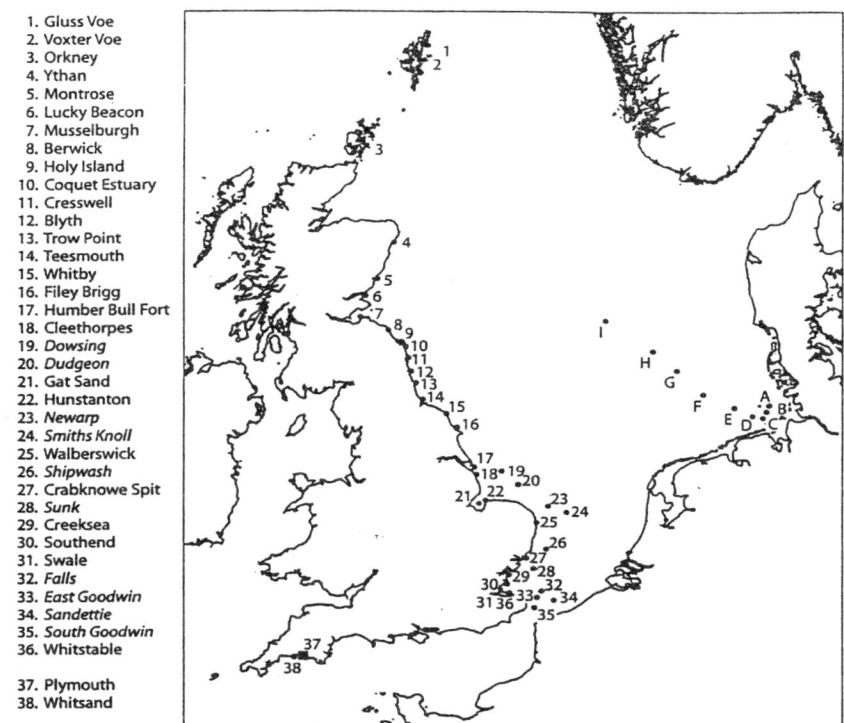

Fig. 8.1 b. Map indicating biological monitoring sites in the North Sea. Stations numbered *1–38* refer to sites used for sampling mussels for scope for growth measurements (see Fig. 8.7); those lettered *A* to *I* refer to stations used during the ICES/IOC Bremerhaven Workshop (see Fig. 8.5 and 8.6, where they are laballed 1 to 9)

ter north of Stavanger, while a period of more than 3 years is required for waters in the German Bight and the western part of the central North Sea (Maier-Reimer 1977). Hydrodynamic models demonstrate that the age of water in the North Sea increases anti-clockwise round the coastlines of Scotland, England, France, Holland, Germany and Denmark (van Pagee and Postma 1986).

The southern North Sea is more vulnerable to pollution, not only because of the longer residence times, but because the water is shallower, providing a lesser volume in which contaminants can be diluted. In addition greater loads of contaminating inputs occur in the more urbanized and industrialized south particularly·from the inputs of rivers Rhine and Thames. These are ameliorated to some extent by the small inputs of Atlantic water via the Dover Straits, which provide a diluting input through the Channel to the north along the coasts of Belgium and the Netherlands, plus additional contaminants originating from along the Channel coast.

The major inputs of contaminating effluents and wastes is principally by way of the major rivers. In addition there are more direct inputs from boating and shipping

(5 000 shipping movements at any one time), from oil exploitation activities and platforms (approx. 150 presently in the North Sea). In recent years there has been an improved awareness of the importance of diffuse atmospheric inputs (Quality Status
Report for the North Sea 1993). The accuracy of riverine load estimates is much greater
(+20–30%) compared to those for atmospheric inputs (+50–100%). Nevertheless, estimates of the proportions of different classes of contaminants indicate that those entering the North Sea by atmospheric wet or dry deposition represent a significant proportion of the total inputs (Nitrate 34–51% (QSR 1993); cadmium 11–108%, mercury
20–42%, copper 13–36%, lead 31–58%, zinc 17–43%, chromium 2–11%, nickel 43–63%
(Kersten et al. 1988); PCBs 96% (Huiskes and Rozema 1988); PAHs 50–1 300% (van Aalst
1988)). There seems to be agreement that elevated lead levels in North Sea sediments
are due to atmospheric inputs (Kersten et al. 1988), probably from motor vehicles, and
that the relationship with PAHs (Preston et al. 1992) implicates a similar route. However at the time of the Quality Status Report (1993) there were no input data for PAHs.
Inputs to the sea surface from the atmosphere tend to be held in the sea surface
microlayer, where concentrations of contaminants (e.g. metals) are elevated by one or
two orders of magnitude, and may be toxic (Hardy and Cleary 1992) to life exposed to
the microlayer.

Contaminants entering the marine environment most often disperse, dilute and
become degraded, such that their toxicity is decreased. Some may be accumulated preferentially by organisms in their food or because they filter large volumes of water
(e.g. bivalve molluscs). Bio-accumulation of contaminants may cause toxic effects due
to bio-magnification of concentrations in higher trophic levels. However, it is also useful in monitoring, since concentrations are not only higher, but tissue levels can be
expected to relate to toxicity and assist in establishing causality (Stebbing et al. 1980;
Chapman 1997). Many other contaminants become partitioned to particles, when it is
the movement of suspended particulate matter that becomes responsible for contaminant flux. South of a line from Hull to the Skagerrak the North Sea is significantly more
turbid (1–10 mg l^{-1}) than to the north (0.1–1 mg l^{-1}) (Eisma and Irion 1988). Fine-grained
suspended matter in the North Sea is the most important phase for heavy metals and
synthetic organic contaminants. It consists of a clay fraction (illite, mectite, kaolinite)
with large surface areas for adsorption relative to volume, and complex ion exchange
properties. The fine fraction is also rich in Particulate Organic Matter (POM) derived
from plankton, such as faecal pellets of copepods and the degradation products of
planktonic production in the water column. The high adsorption capacity of clay minerals and the complexation of organic compounds by POM, scavenges the water column of contaminants. Thus the concentrations of heavy metals and organic contaminants on particles (μg g^{-1} dry weight) is much higher than their concentrations in solution (μg l^{-1}), and the fine fraction accumulates contaminants preferentially. The interpretation of contaminant flux, and the capacity of the North Sea to assimilate wastes,
depends critically on the behaviour and movement of fine-grained, organic-rich suspended particulate matter (Kempe et al. 1988).

The partitioning of contaminants between particulate matter and solution is defined by a partition coefficient (K_d), i.e. the ratio of concentrations at steady state in
particulate and dissolved phases. Since K_ds vary with salinity and the concentrations
of contaminants and particles, their use in estuaries and near-shore waters is complex
and their utility constrained where they could be most useful. K_ds have been widely

used to interpret metal equilibria and behaviour in the marine environment, particularly those of radionuclides (IAEA 1985). For organic chemistry hydrophobic partitioning depends on the organic carbon content of suspended particulate matter; this partitioning can be calculated from the octonol-water partition coefficient (K_{OW}) (Harris et al. 1993). However, recent findings suggest that some polyaromatic hydrocarbons (PAHs) detected on suspended marine and estuarine particles may not be subject to particle-water equilibration and that for those important contaminants (see Sections 8.5.1 and 8.6.4 the partition coefficient is not adequately developed as a means of describing and predicting contaminant behaviour in the environment (Zhou, pers. comm.), let alone their use for management purposes.

This organic-rich fraction is highly mobile, settling to the bottom in calm conditions and low current speeds, providing a superficial covering of a few mm thickness in depositional areas. However this layer is quickly remobilized and carried up into the water column by storms and the turbulence they create, only to settle again through the water column where currents weaken. It is during repeated deposition and resuspension that particulate organic matter scavenges the water column of many contaminants. Deposited organic material provides the basis for production by benthic ecosystems. Many species live by removing fine grained particles a few microns in diameter from suspension (e.g. filter feeding bivalves) or from within the sediment itself (e.g. polychaetes and ophiuroids). Since it is this fine fraction to which many contaminants are adsorbed or bound, even where it represents <1% of the total sediment, these organisms will be exposed to much higher concentrations than those in the sediment as a whole (Kempe et al. 1988). Hence it is likely that organisms will take up and may concentrate the associated contaminants.

The residence time of contaminants adsorbed to sediments in the North Sea is much longer than for those in solution. The average residence time for adsorbed contaminants is likely to be decades (Lohse 1988), if inputs were to cease, it might take centuries for levels to return to those in pre-industrial times (Kempe et al. 1988). Although there are temporary depositional areas on the Dogger Bank, and sedimentation occurs in the Wadden Sea and German Bight, it is predominately in the Norwegian Trench (depth 225–700 m) that suspended sediments are deposited. Some 77–84% of the total net accumulation of fine-grained material in the North Sea occurs here, together with most of the persistent chemical contaminants which are bound to it (Skei 1981). Thus Lohse (1988) speculated that, from sources in the southern North Sea to their sink in the Norwegian Trench, organochlorine contaminants will, over a period of years, be digested and excreted several times before final deposition.

8.2.2
Perceptions of the Health of the North Sea

The North Sea is rich in the variety of its natural habitats and the diversity and productivity of life that inhabit them. Regular "Quality Status of the North Sea" reports review the monitoring effort by those countries that bound the North Sea, providing an overall assessment of its health. The 1987 Quality Status Report concluded "In general deleterious effects, at present, can only be seen in certain regions, in the coastal margins, or near identifiable pollution sources. There is as yet no evidence of pollution away from these areas".

Six years later the 1993 Quality Status Report concluded: "Large areas can be shown to be subject to concentrations of contaminants that are clearly above the North Atlantic background level. Generally, the impact of these enhanced concentrations, which are directly attributable to inputs from around the North Sea, is only clearly identifiable where the concentrations are highest, that is, close to sources e.g., in estuaries or deposition areas such as the Norwegian Trench and parts of the Dogger Bank, or in the Wadden Sea and the sea areas where the pattern of water movements restricts water exchange, e.g., along the Dutch and Danish coasts".

The Report goes on to say that "some detectable effects can be attributed to particular contaminants, as, for example, the effect of high concentrations of PCBs on the reproductive success of seals and the effect of TBT on the shell shape of oysters or the induction of imposex in dogwhelks".

A comparison of the conclusions of successive Quality Status Reports suggest that either the North Sea is becoming more polluted, or our awareness of its polluted state is improving. Ministerial Declarations following North Sea Conferences identify new contaminants for monitoring and new targets for reducing inputs (e.g. endocrine mimics, PAHs, nutrients; Ministerial Declaration 1996).

Given the suggestion of a decline in water quality in the North Sea over the period 1987 to 1993 (QSR 1987, 1993), it must be concluded that the measures introduced by Ministers of the riparian North Sea states have been ineffective in their aim "... to protect and enhance the quality of the North Sea environment" (Ministerial Declaration 1987). Some researchers are unrestrained in their views on the state of the North Sea. Salomons et al. (1988) write of "ecosystem deterioration". Degens (1988) similarly writes that the North Sea is deteriorating at an 'alarming rate'. MacGarvin (1990) extravagantly concludes that any capacity the North Sea might have had to assimilate waste was exceeded decades ago.

We believe any approach to controlling the quality of the North Sea environment by means of chemically-based legislation and monitoring is inappropriate. Among many problems is the changing nature of chemical contamination, which now includes large numbers of synthetic organic contaminants. These compounds may be biologically active at very low levels and available to the biota in various matrices in which they may be preferentially adsorbed and concentrated.

The technical task of analysing large numbers of synthetic compounds in sea water at low concentrations, in several matrices, in all receiving waters, at regular intervals has greatly outstripped the resources available for the task. The problem has been compounded by the number of new chemicals that enter the environment each year and the rate at which legislation can be drawn up to provide a mandate for regulators to monitor them. There is now a tendency to enact legislation more quickly through "Ministerial Orders in Council", but the requirements lag well behind the needs. It is clear that a different approach is now essential, which this chapter attempts to justify.

8.3
Assimilative Capacity

The concept of assimilative capacity has its origins in quantitative environmental toxicology and chemistry. It should also be considered as a valuable and exploitable economic resource. In this section we consider the concept from both disciplinary perspectives.

8.3.1
Definitions

The original definition of assimilative capacity was applied to freshwater pollution and meant the ability of an ecosystem to cope with certain levels of waste discharges, without suffering any significant deleterious biological effects (Cairns 1977). This was broadened later to mean "the amount of material that could be contained within a body of seawater without producing an unacceptable biological impact" (Goldberg 1981). It was then redefined by the GESAMP (Group of Experts on the Scientific Aspects of Marine Environmental Protection) and given a wider meaning as "a property of the environment, defined as its ability to accommodate a particular activity, or rate of activity, without unacceptable impact" (Pravdic 1985). (GESAMP renamed "assimilative capacity" as "environmental capacity", but this change has not been taken up by environmental economists and social scientists, so the original term will be retained here.) The history and semantics of the concept are considered in more detail by Stebbing (1992).

We emphasize two points: First, Cairn's definition referred to an ability to "cope" with discharges, while Goldberg's refers to the amount of material that could be "contained", but Pravdic's wording provides more insight describing assimilative capacity as accommodating a "rate of activity". This is an important distinction, as we will show in Section 8.3.2. Second, throughout the evolution of the concept, the aim has been to permit the use of the assimilative capacity of receiving waters to accept wastes without causing harmful biological effects.

The GESAMP considered the concept in some depth and agreed that it was based on three premises, which give greater meaning to their definition:

1. A certain level of some contaminants may not produce any undesirable effect on the marine environment and its various uses.
2. Each environment has a finite capacity to accommodate some wastes without unacceptable consequences.
3. Such capacity can be quantified, apportioned to a certain activity, and utilized.

GESAMP changed the emphasis of the definition to "levels of contaminants", rather than their "biological effects", and much hangs on the use of the word 'undesirable'. We consider this to have been a retrograde shift of emphasis, since it is as though the presence of chemical contaminants in the marine environment, rather than their deleterious effect, that is of more importance.

The related concept of "pollution" has been defined by GESAMP and their definition generalized by Holdgate (1979) is as follows:

> "The introduction by man into the environment of substances or energy liable to cause hazards to human health, harm to living resources and ecological systems, damage to structures or amenity, or interfere with legitimate uses of the environment." (Authors emphasis)

The concepts of "pollution" and "assimilative capacity" are clearly related terms, although the way in which they have been defined does not link them explicitly (Pravdic 1985). The underlying intent of both "assimilative capacity" and "pollution" is the prevention of deleterious effects on biological systems. Pollution or "harm to living resources and ecological systems" can only be avoided where rates of input are "without unacceptable impact", so it is clear that pollution occurs where assimilative capacity is exceeded. We suggest that the twin concepts of assimilative capacity and pollution need to be more explicitly linked.

Some have considered the concept of assimilative capacity inherently permissive, which it is in relation to the precautionary principle, which is inherently preventative. There has been considerable debate regarding the virtues of the precautionary principle as a way of controlling pollution, rather than the use of assimilative capacity. However, the concepts are not mutually exclusive (Stebbing 1992). It is possible, and environmentally desirable, to be precautionary in the use of assimilative capacity. Both concepts have become incorporated in the UK government's policy in managing pollution. While the assimilative capacity concept is permissive up to a biologically defined threshold, it is important to accept that the concept itself is essentially neutral and the constraints built into the use of assimilative capacity may be as stringent as is necessary to maintain environmental quality in the light of scientific uncertainties.

8.3.2
Quantification

A rigorous approach to quantifying assimilative capacity requires a quantitative knowledge of all significant and relevant processes of geochemical cycling of a pollutant. Some have employed the concept of assimilative capacity in a practical context, using EQS's as a proxy for a biological threshold when assimilative capacity is exceeded (Portmann and Lloyd 1986). Others have used critical pathway analysis approach to identify the most sensitive target species, whose susceptibility is assumed to protect others in the receiving waters (Krom 1986), since knowledge of the processes involved (see Table 8.1) is insufficient for more informed estimates. In an example involving copper pollution of the Krka river estuary (Adriatic Sea) Pravdic and Juracic (1988) identify the key steps in developing a mass balance model to estimate assimilative capacity for copper, recognising that ideally it would need to incorporate the hydrography (flushing times), chemistry (reactivity), sedimentology (deposition/remobilisation) and biological activity.

Here we propose a general model for assimilative capacity

$$\frac{dC_{i,x,n}}{dt} = I_{i,x,n} - D_{i,x,n} \; . \tag{8.1}$$

Equation 8.1 says that the change in the concentration C of contaminant i at place x and compartment n is the difference between those processes I leading to increases in concentration at that place and/or compartment, and other processes D leading to reductions in concentration. Both I and particularly D may themselves be functions of the contaminant concentration, i.e. $I\{c\}$, $D\{c\}$. x may represent a discrete point, a defined area or volume as appropriate. The significant environmental compartment n may be water, sediment, individual fish, shellfishery, or biological community, etc. I and D represent the sum of those individual processes (units concentration per unit time) contributing to increased or decreased contaminant concentrations. So I includes input and transport processes, bio-accumu-

Table 8.1. Summary of some of the processes that may increase (*I*) or decrease (*D*) assimilative capacity by influencing the concentration, availability or biological impact of contaminants in the sea

Mechanism	Increase in concentration, availability and impact (I)	Decrease concentration, availability and impact (D)
Hydrographic		
Distribution in water	Reconcentration (frontal system)	Dilution (tidal mixing)
Distribution of particles	Benthic deposition, settlement (further concentration by gyres and at turbidity maxima)	Remobilisation (fast currents, waves and storms)
Partitioning on particles	Desorption (salinity, pH dependent)	Adsorption (salinity, pH dependent)
Interfacial effects	Accumulation at interfaces (sea bottom, surface, thermocline, pycnocline)	Dispersion from interfaces; Burial
Boundary effects	Importation and accumulation on shores (e.g. litter, oil)	Exportation
Chemical		
Complexation and chelation	Ligand unbinding	Ligand binding (e.g. metals on humic acids)
Transformation	Potentiation	Degradation (e.g. UV photo-oxidation of organic contaminants)
Biological		
Adaptation	Sensitisation	Homeostatic control, acquired tolerance
Joint toxicity	Synergistism	Antagonistism
Bioavailability	Remobilisation	Sequestration (e.g. in shells)
Benthos	Biodeposition	Bioresuspension
Pelagos	Bioresuspension	Biodeposition
Tissue concentration	Bioaccumulation, biotransformation (e.g. methylation of mercury)	Excretion
Ecosystems	Biomagnification	

lation, bio-magnification, adsorption onto, synergy with. *D* is the opposite: processes contributing to contaminant export, dilution, dispersion, transformation from, deadsorption, degradation, detoxification, etc. An appropriate mass-balance equation for a contaminant *i* can be constructed where the size of the compartments (n, x) and flux of contaminant are known. Where the rate *I* exceeds the rate *D*, then concentrations will increase. Some of the principle hydrographic, chemical and biological processes capable of contributing towards assimilative capacity are summarized in Table 8.1.

The unutilized assimilative capacity may be approximated by the difference between *D* and *I* $(D > I)$ for all natural (i.e. non-anthropogenic) inputs of contaminant *i*. If or when concentrations exceed a threshold (i.e. $C > C_{crit}$) at which deleterious biological effects are observed for significant compartments *n* in the area of interest, then pollution has occurred and assimilative capacity will have been exceeded. For sustainable long-term management, it is the regulators role to maintain the rate of input to this point (or area, or whatever) below that for which (over an appropriate timescale)

$$\frac{dC_{i,x,n}}{dt} \leq 0 \quad \text{and} \quad C_{i,x,n} \leq \text{EcoQS}_i = fC_{\text{crit}_{i,x,n}} \cdot \quad (8.2)$$

It is this rate of input that is a measure of the assimilative capacity of the environment of interest. Historically D is simply treated as the capacity of the environment to dilute, disperse (e.g. metals) and degrade (especially organic sewage). C_{crit} is taken to be the toxic threshold determined for a small range of species n, usually from short-term, acute laboratory toxicity tests (i.e. the Ecological Quality Standards or EcoQS for contaminant i), including precautionary extrapolation factor (Zabell, pers. comm.) f (normally in the range 0.01–0.1). We know that C_{crit} also depends on the quality of the environment (i.e. on x). It appears to be a working assumption, largely untested, that for the marine environment D will always exceed I.

We wish to make the following points:

i. Natural processes lead to concentrations of contaminants in certain areas (estuaries, fronts, areas of reduced water movement such as the Norwegian Trench and Dogger Bank) and compartments (sediment, water in estuaries, surface microlayer, biota) that are greater than the relevant EcoQS, and actually have a polluting effect (PCBs, etc.) (Table 8.1).

ii. Assimilative capacity of the system for a particular contaminant is determined by the balance of those particular processes (e.g. bio-accumulation and chemical transformation, etc.; dilution and degradation, etc.) that lead to concentrations at a particular point (or area or volume), that exceed a threshold of effect on a living component. It is not clear whether the assimilative capacity for particular contaminants can be considered independent, or whether certain contaminants with similar properties may act additively (e.g. hydrocarbons may sum dependent on their octonol-water partition coefficients (K_{OW}) (see Donkin et al. 1989).

Those processes that contribute to assimilation capacity, in relation to any species (n), or contaminant (i) or site (x), will tend to increase (I) or decrease (D) assimilation capacity. Such processes are summarized in Table 8.1, indicating the hydrographic, chemical and biological processes that contribute towards greater, or utilize, assimilative capacity. The way in which they do so is self-evident in many cases, but some of the less well known should be described.

Even where contaminants enter the marine environment from diffuse rather than from point sources, e.g. by aerial deposition, they rarely become distributed homogeneously in sea water, but are typically reconcentrated by various processes, often at interfaces. Those that result in benthic accumulation have been discussed, but significant accumulation may also occur at the sea surface (Hardy 1982). Metals (Hardy et al. 1985) and organometals (Clearly and Stebbing 1987) may accumulate to concentrations one or two orders of magnitude higher than those at the immediate subsurface. Accumulation in the sea surface microlayer is of importance as a site of exposure of the permanent and transient members of the neuston, which include developing eggs and larvae of fish and benthic invertebrates. Marine mammals (seals, cetaceans), and sea birds are inevitably exposed to contaminants in the microlayer, as are the littoral fauna and flora with each tidal excursion. While it is not known to what extent the indigenous biota are affected, microlayer samples taken offshore in the German Bight were found to be toxic when tested with bioassay techniques, and TBT concentrations were found to exceed the UK EQS (Hardy and Cleary 1992).

Different mechanisms are involved in the accumulation of contaminants at the frontal systems that are the boundaries between water masses. Work by Tanabe et al. (1991) has shown that persistent organochlorines (PCBs, DDT, HCH isomers) occur at elevated concentrations in frontal regions, due to their affinity for lipids and particles concentrated by fronts. This has been demonstrated for water samples taken at the sea surface, but it is also true for the benthic sediments beneath frontal regions, due to the deposition of particles to which organochlorines become adsorbed. The possible accumulation of contaminants at fronts in the North Sea does not appear to have been investigated.

There is evidence that the thermocline and pycnocline may provide the kind of density discontinuity that results in the accumulation of contaminants. Further reconcentration of sea surface contamination can occur due to Langmuir circulation in open water, or axial estuarine convergences on the flood tide causing slicks (R. G. Uncles, pers. comm.). Suspended sediments and their associated contaminants may be concentrated in estuaries at turbidity maxima due to tidal pumping (Uncles et al. 1988), or by gyres drawing suspended particles to their centres due to centripetal forces and depositing them (Stebbing et al. 1984). Accumulation at all such interfaces is of particular importance as these are often sites of intense biological activity. This is because the same processes that accumulate biogenic material, (organic matter and nutrients) on which organisms feed, also concentrate chemical contaminants. The localized coincidence of contaminants and biota is likely to result in exposure and cause toxic effects.

8.3.3
For Individual Contaminants

Early thinking on the "dispersion capacity" of the environment (Holdgate 1979) was formulated in terms of the relationship between concentrations of an individual contaminant and the capacity to disperse it (Fig. 8.2). The key statutory instrument by which water quality is managed is the environmental quality standard (EQS). EQSs are required under both EC (Dangerous Substances Directive, 76/464/EEC) and UK (Environmental Protection Act 1990) legislation. An EQS is defined as the concentration of a substance which should not be exceeded in the receiving water in order to protect the use of the water; they provide the standards by which the regulatory authorities operate to monitor the aquatic environment and assess the possible biological effects of contaminants.

Critically, EQSs provide the one statistic that links contaminant concentration in the environment to its potential biological impact, so it is important to consider briefly how these aim to protect biological water quality by controlling chemical inputs (Zabell, pers. comm.). First, the acute and chronic toxicity data are reviewed. After considering available data on a contaminant (chemical/physical properties, behaviour pathways and fate, analytical methods, environmental concentrations),

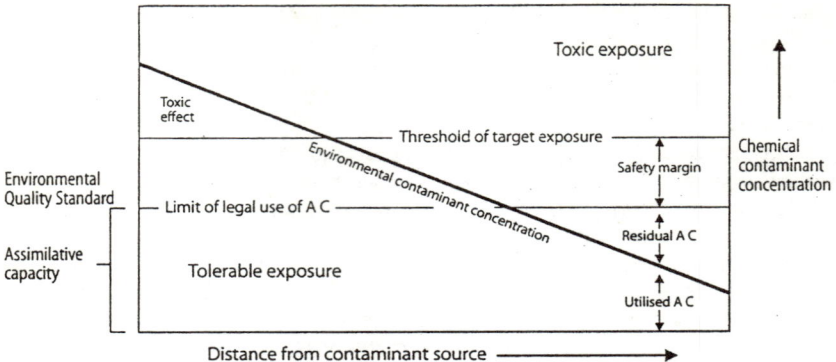

Fig. 8.2. Relationship between contaminant concentration, assimilative capacity and the environmental quality standard for individual contaminants (Derived from Holdgate 1979)

the lowest reliable and relevant concentration is selected from laboratory toxicological data. An extrapolation factor is then applied. Typically the factor is 100 for acute and 10 for chronic/sublethal toxicological thresholds. The factor may be varied depending on the persistence of a chemical contaminant in the environment, its tendency to bioaccumulate, or due to the acute/chronic threshold ratio. The preliminary EQS is then evaluated in relation to field studies before being recommended, reviewed by the regulatory authorities and finally adopted as a standard against which environmental concentrations of each contaminant are monitored. This approach to legislation and monitoring is now not adequate for various reasons.

i. The large number of contaminants in the North Sea that are potentially harmful are now numbered in tens of thousands. Even though EQSs sometimes relate to a group of chemically related compounds, through QSARs for example (Donkin et al. 1989), it is clearly impracticable that there should be an EQS for every contaminant that could potentially have a toxic effect.
ii. The paucity of toxicological threshold data, both chronic and sublethal, limits the appropriateness of EQSs for their purpose. Available toxicological data exist for <5% of chemical contaminants and <1% of species, so extrapolation between compound and species is a necessity.
iii. EQSs based on the toxicity of individual contaminants do not adequately account for their interactions, which may be antagonistic, additive or synergistic.
iv. EQSs depend upon acute or chronic laboratory toxicity experiments under controlled conditions. While protocols aim to achieve reproducibility (e.g. constant temperature, light, feeding regime etc.), environmental relevance is thereby lost. Many environmental factors, such as pH, turbidity and salinity, the test organisms' resistance, and the complexation capacity of the water, affect contaminant toxicity, yet are often not taken into account.
v. Numbers of contaminants in the environment may occur at concentrations less than their various EQSs, yet collectively have a deleterious biological effect.

vi. EQSs can be based on 'annual average' concentrations or 'maximum allowable concentrations'. They may thus allow transient concentrations that are toxic, while the annual average concentration remains less than that permitted by an EQS.

vii. The extrapolation factor used varies between nations, and there seems to be agreed protocol for their specification. In the UK it is arbitrarily determined (Wharfe and Tinsley 1995), and adjusted in the light of experience, rather than having a basis in scientific understanding. It is intended to account for uncertainties and lack of knowledge in extrapolating between species, from acute short term exposure to chronic long term exposure, from one environment ecosystem to another. Extrapolation factors may vary between ranges of 1–5, where confidence is greatest that effects in the field may be avoided, to 200–1 000 for persistent new chemicals, where the toxicological database is sparse. Some use larger extrapolation factors account for possible synergistic effects between contaminants.

Derived in this way, EQSs for listed contaminants provide the means by which statutory limits of environmental contamination are determined and controlled, providing the standards to which the regulatory authorities operate in order to prevent pollution. However, chemical EQSs have led to a waste of monitoring effort, measuring inconsequential contaminants (i) in the wrong place (x) and in the wrong compartment (n) (see Section 8.3.2).

8.3.4
A Resource of Economic Value

The marine environment has a capacity to assimilate wastes and effluents by their dilution, dispersion and degradation, such that they are reduced to harmless concentrations, are transformed into less toxic species, or utilized and recycled by the ecosystem. This capacity is an environmental service of considerable economic value. Globally such services are approximately valued at $1.28 tr p.a., which is 7.1% of the global Gross National Product p.a. (using data from Costanza et al. 1997).

When the basic ecotoxicological and economic variables are considered together (Fig. 8.3), it is clear that a number of factors bear on the management of the rates of contaminant inputs to the system. A low level of loading by nutrient-rich organic wastes (Fig. 8.3a) promotes primary and enhanced heterotrophic production. Increasing nutrient inputs can cause eutrophication. Primary production has increased in the German Bight by 3–4 times since 1962 (Radach et al. 1990). In stratified waters eutrophication and the consequential phytoplankton blooms can lead to hypoxia. The system becomes overloaded at the point at which hypoxia and metabolic by-products have an adverse effect on productivity, although community and other ecosystem changes may occur earlier. Where such events are severe, mortalities of benthic and pelagic organisms occur. The stages of marine eutrophication and its harmful consequences are considered by Gray (1996) and will not be considered further to give greater consideration to toxic contaminants and their control.

With non-degradable toxic contaminants such as metals (Fig. 8.3b), resistance to low concentrations may incur a physiological cost, but typically there is a threshold concentration at which toxic effects occur in an individual organism (Willows 1994), or may occur in a population or community (Warwick and Clarke 1992).

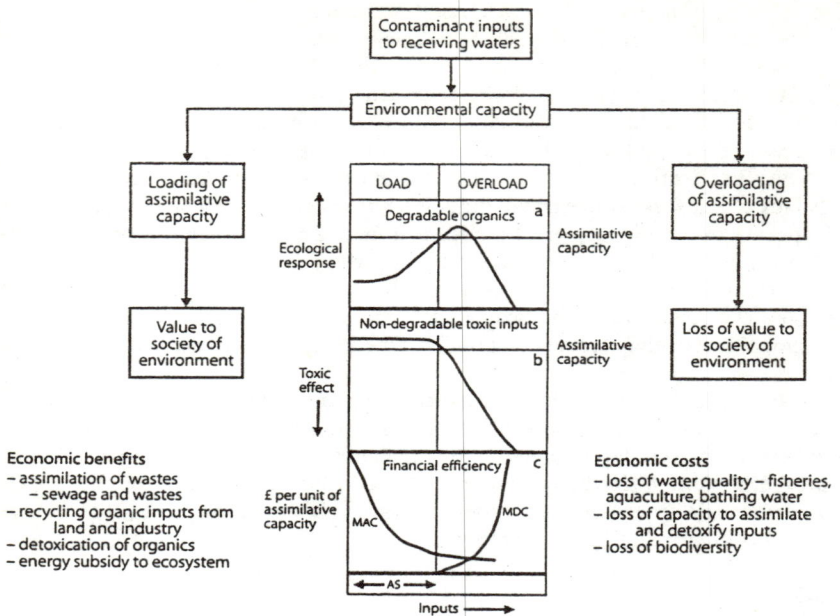

Fig. 8.3. Relationships between the effect of organic inputs **a** and toxic contaminants **b** to the economic efficiency **c** of using assimilative capacity. *MAC* Marginal Abatement Costs; *MDC* Marginal Damage Costs; *AS* Assimilative Capacity

Such relationships represent an over-simplification, yet provide an adequate basis to consider the economic efficiency of utilising assimilative capacity (Fig. 8.3c). With increasing levels of input, Marginal Abatement Costs (MAC) are likely to fall due to efficiencies of scale, while in progressing from load to overload, Marginal Damage Costs (MDC) may increase as toxicological thresholds are exceeded and the ecosystem is damaged. Thus as the assimilative capacity is exceeded, the capacity to assimilate further wastes will be reduced by toxic effects on biota that contribute to the sequestration, degradation and detoxification of chemical contaminants (Table 8.1). The reduction of assimilative capacity may be accentuated by positive feedback. Expressed in this way, the optimal economic use of assimilative capacity occurs where MAC = MDC (Turner et al. 1994). However, such an optimum is not sustainable, since even at that level biological contributors to assimilative capacity will be damaged and their contribution reduced thereby (Pearce 1976).

To protect the ecosystem and its biological contribution to assimilative capacity, a safety margin is desirable (Fig. 8.3), perhaps as great as an order of magnitude less that the lowest toxic threshold, but for clarity is not shown (see Section 8.3.3). Some contaminants may be so toxic that it is assumed there is no capacity to assimilate them. Expressed in this way the assimilative capacity concept can be formulated in a way that allows for precautionary margins to safeguard living resources.

In relating toxicological to an economic analysis of environmental contamination, it is clear that an economic optimum level of contamination (where MAC and MDC intersect) would permit higher levels of inputs than would be desirable on ecological grounds alone. It is evident that, near threshold concentrations, disproportionately large effects may result from small changes in concentration. Since many environmental factors influence the precise threshold, a safety margin is essential. That the utilisation of assimilative capacity then becomes less economically efficient indicates the cost of precaution and the reduction of risk.

8.4
Chemical Versus Biological Monitoring of Assimilative Capacity

Environmental legislation is both enabled and constrained by the monitoring techniques necessary to implement it. Progress in adopting more effective legislation is constrained by advances in research and monitoring techniques. Often the adoption of such techniques by a regulator may also depend on simplicity of application and cost-effectiveness, as well as efficacy.

Chemical analysis presently provides the most important means by which regulators monitor the effectiveness of environmental legislation. The EQS itself is expected to provide the link between chemical contamination and its biological rele-vance, since it is based on laboratory toxicity data, as already discussed (see Section 8.3.3). However, it is significant, when considering the role of particles and sediments as binding sites for many contaminants, that there are no EQS for them in the particulate phase.

We ask whether it is better to monitor "targets" or "factors" (Holdgate 1979), the chemical causes of pollution or their biological effects, or some integration of the two approaches. To present the arguments clearly, we consider first the advantages and disadvantages of an approach that depends on analytical chemistry (Section 8.4.1), before considering one that relies on biological techniques (Section 8.4.2). However, it is evident that monitoring of chemical contaminants that does not relate to their biological consequences or effects is of little benefit in determining the use of assimilative capacity. For those that do, it is only necessary to control contaminants that have biologically harmful consequences. It is essential to establish causality rigorously, to provide an adequate body of evidence to impose regulation and control with the minimum of delay and without excessive cost. Thus after considering the arguments for chemical and biological approaches alone, we go on to advocate an integrated approach (Section 8.4.3), as we will demonstrate that while chemical analyses do not adequately indicate toxic impacts, biological techniques do not adequately identify their chemical causes.

8.4.1
Chemical Monitoring

A chemical approach to the control of pollution was appropriate for an era when contamination of the water system was principally by point source inputs of effluents with relatively few chemical constituents of concern entering rivers; essentially one dimensional systems with unidirectional flow. Contamination of the water course is now more complex because of the numbers of chemicals released into the environment, the many and diffuse routes that they may take, and that our concerns have now extended from rivers to marine systems, where they tend to accumulate. Any system based on controlling chemical contaminants individually is now outmoded and inappropriate for its purpose.

The following points summarize important ways in which we believe a chemically-orientated pollution control system is proving inadequate:

Environmental Quality Standards. While the protocol by which EQSs are determined is rigorous enough, too many assumptions have to be made in their application for them to serve their purpose adequately. EQSs are sometimes based on insufficient data, or are not revised to take account of new data. Others have not been set because the hazard posed was not recognized until the chemical was in widespread use and a common contaminant (e.g. endocrine disrupters). The utility of EQSs depends crucially on the relevance of laboratory-based short term toxicology data to the environment, which for many reasons has always been doubtful, except as a means of determining the relative toxicity of a group of compounds.

Relationship of chemical data to toxicological thresholds. The utility of any EQS to its purpose hinges on the extent to which toxicological thresholds, determined from short-term laboratory experiments, have relevance to the biological impact of the chemical in the environment (see Section 8.3.3) for which in situ data are much more relevant). Laboratory experiments typically do not take into account many of the factors known to be important in the environment in determining the health or susceptibility of the organism (sex, season, breeding condition, etc.), or the bioavailability of the chemical in the experimental medium (turbidity, DOC concentration or complexing capacity, salinity). Sensitivity of biota to toxic effects depends on the environment experienced by the biota, which is at variance to that used in standard toxicity tests. The relevance of an EQS to pollution depends on how well it indicates the likelihood of a biological impact in the environment. The use of short-term lethal thresholds to predict long-term sublethal effects assumes a constancy in the acute/sublethal toxicity ratio. This is unjustified, since the ratio is low for narcotic toxicants and high for others that have a specific mode of action (e.g. TBT).

Weight of numbers. The sheer number of chemical contaminants entering the environment, particularly synthetic organic chemicals, has become too much for a regulatory system based on the control of individual contaminants. The effluent of the river Rhine is now estimated to contain as many as 40 000 individual contaminants. The problem of control is aggravated by the fact that some classes of chemicals are biologically active at concentrations of nanograms per litre, and many are persistent with half-lives of years in the marine environment. The growing burden for those regulatory authorities with responsibilities for monitoring listed chemicals under EC and UK legislation has become overwhelming. Many significant contaminants go unmonitored, and the frequency and spatial definition of monitoring is inadequate for its purpose, and delays between sampling and analysis devalues the data. As it is, the chemically-based legislation motivates monitoring; links to the biological significance of the data are often overlooked.

Redundancy in chemical legislation. Some classes of contaminants for which there is pollution legislation and a requirement to monitor are now known to pose much less of a threat to environmental quality than was assumed. Mechanisms to remove contaminants from monitoring programmes whose environmental threat is now known to be minimal or inconsequential are ineffective. For example, most metals are no longer considered to pose the threat to water quality in the North Sea, yet they are a class of contaminants that have long featured in pollution legislation. The ICES/IOC Bremerhaven Workshop deployed over 50 biological effects techniques on what is perhaps the most marked pollution gradient in the North Sea (Stebbing et al. 1992), but no deleterious effects related to metals were demonstrated. Similarly, the chemical causes of the pollution gradient identified by Widdows et al. (1995) along the UK East Coast do not include metals. There is therefore a need to critically examine the appropriateness of chemical monitoring programmes for their purpose, since present knowledge suggests there is redundancy. Chemical monitoring of contaminant levels that are biologically insignificant, and do not cause pollution, is a waste of resources. The problem has grown considerably with the number of micro-contaminants now present in the marine environment, but which are unmonitored.

Appropriateness of chemical analyses. Regrettably environmental legislation, and the chemical analyses to enforce it, do not necessarily relate to the same fraction, species or phase of a contaminant that is biologically available and potentially toxic. For example, legislation for copper and cadmium relates to the concentration of the metal in sea water, while the ionic activity of the metal determines its bioavailability and toxicity. Since a large fraction of these metals is bound to organic matter in the case of copper (Sunda and Guillard 1976) and for cadmium to inorganic ligands (Sunda et al. 1978), the ionic fraction which is determined is only a proportion of the total present in the environment. The problem of the biological relevance of chemical analysis can be circumvented by monitoring bio-accumulated tissue burdens, which can be used to predict effects (Chapman 1997) for some chemicals. Not only do the enhanced levels ease the analytical problems, but tissue burdens are integrated over time.

Inadequacy of spatial definition in monitoring. There is a growing awareness that contaminant behaviour in the environment does not lead inevitably to their dilution, dispersion and/or degradation. Thus any sampling strategy should recognize those processes that reconcentrate contaminants and to sample accordingly, particularly when those sites are foci of biological activity. One example of importance is the potential of the turbidity maximum in estuaries to accumulate contaminants (Uncles et al. 1988). Similarly coastal fronts have been shown to accumulate organochlorine pesticides at the sea surface and in benthic sediments (Tanabe et al. 1991). The air-sea interface is a well known site of contaminant accumulation (Hardy 1982), besides being one of the accumulation of sensitive early life stages of pelagic and benthic species. Thus monitoring programmes assume homogeneity in distribution, when heterogeneity is the norm.

8.4.2
Biological Monitoring

Pollution as defined (see Section 8.3.1) relates primarily to "harm to living resources and ecological systems" due to contaminants of various kinds. Since concern relates to the biological impact of pollution, and the criteria for environmental quality are bio-logical, the compelling logic is to use biological indices of pollution. The advantages of such an approach are outlined below.

Detection of new and unsuspected pollutants. To depend on a chemically-orien-tated approach presumes that it is known which chemical contaminants are likely to be important, but repeatedly the detection of environmentally significant, new contami-nants has depended upon research that recognized their biological impact. Numerous examples could be cited, such egg shell thinning in birds due to organochlorine pesti-cides, but the best marine example is probably that of TBT (tributyl tin), a biocide used in antifouling paints (see Champ and Seligman 1996). It was first detected due to the failure of oyster larvae to metamorphose and the shell-thickening of adult oysters (see Fig. 8.4). It could not have been detected chemically when it first became a sig-nificant pollutant, because analytical techniques had not been developed to detect it at extremely low concentrations ($ng\ l^{-1}$) at which TBT and its degradation products are biologically active. The inescapable conclusion, from this and other examples, is that biological surveillance monitoring is the most effective way of detecting new and pre-viously unsuspected pollutants.

Integration of the combined effects of chemical contaminants. Many biological indices have been developed and tested over the last ten years which provide an over-all measure of the quality of the environment that organisms inhabit, whether it be the benthic sediments, the water column or the sea surface. Suitable indices using sensitive organisms, reflecting the combined effect of different mechanisms of toxic action, pro-vide an overall measure of environmental quality, integrating the combined effect of the many stresses to which organisms are exposed. More specifically, the combined effect of numerous contaminants, their synergistic or antagonistic interactions, and the many

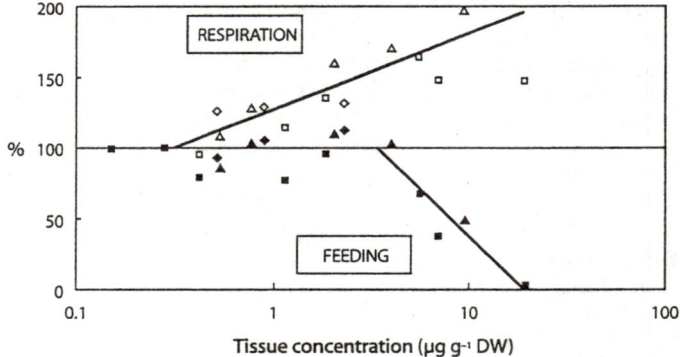

Fig. 8.4. The effect of tissue concentrations of total butyl tins, ($\mu g\ g^{-1}$ dry tissue weight) on two compo-nents of scope for growth: respiration rate and feeding rate expressed as a proportion of control values. Note the difference in toxic thresholds for the two physiological processes and the difference in the rate at which they are affected by the tissue concentration of butyl tins (see Willows 1994, for more details; data redrawn from Widdows and Page 1993)

physical and chemical environmental effects that modify contaminant bioavailability and impact are integrated by such indices. A suite of four techniques have been advocated by the North Sea Task Force, and were tested with many others on the same pollution gradient in the German Bight (Stebbing et al. 1992).

Environmental quality: chemical factors or biological targets? The benefits of measuring the impact of pollutants on biological targets is obvious, since they relate directly to the most commonly used criteria for environmental quality, where biological activity, richness and diversity are the standard. The argument that the detection of toxic effects implies that damage is done, and that their use cannot be preventative, is not the case where techniques utilize adoptive responses to toxic exposure. To monitor chemical factors, rather than biological targets (Holdgate 1979; Stebbing 1996) implies that the causal relationship between each is sufficiently adequate to predict one from the other. Such comprehensive knowledge of environmental processes (Table 8.1) is unlikely to be attained in the medium term. Besides which, the burden of monitoring has grown to the extent that, logistically, target monitoring is the only way to manage the task cost-effectively. In the past, biological techniques lacked the sensitivity and reproducibility required, but this is no longer the case, in that a suite of techniques is now available that potentially fulfil the necessary criteria.

It has never been suggested that biological techniques should be used alone. Their role is as an initial monitor and surveillance of water quality, that can then be used to direct subsequent analytical effort to establish the chemical cause(s). Used alone, biological monitoring techniques are of limited utility because:

a they typically do not identify the specific cause of the toxic effect detected. Specific indicators like imposex in gastropods for TBT are rare,
b only representative species can be used from a limited number of communities to represent whole ecosystems. Extrapolation between species is not straightforward,
c the availability of species and use of biological techniques may be seasonal, or may not cover the geographic range to be investigated,
d few techniques have been used widely enough for rigorous protocols to be established, and problems of reproducibility of results between laboratories resolved,
e only in recent times have a range of biological techniques achieved the necessary sensitivity with reproducibility and precision.

Over the last decade the advantages and disadvantages of chemical versus biological monitoring have been debated within international bodies, such as ICES and IOC. The adoption of research developments in biologically-based indices have been held back by the commitment to legislation expressed predominantly in chemical terms. Nevertheless the rationale for biological monitoring has a logical appeal that is gradually winning support, as concern over the effectiveness of chemical monitoring and its cost-effectiveness have risen, while the rigour of improved biological techniques has been repeatedly demonstrated (Bayne et al. 1988; Addison and Clarke 1990; Stebbing et al. 1992; Wharfe and Tinsley 1995).

8.4.3
An Integrated Approach to Monitoring

While the arguments for and against biological and chemical monitoring have been considered separately, it is clear that an integrated approach is essential. Chemical analysis of contaminants could no more be expected to give a true impression of their biological impact than biological effects techniques alone could pin point their chemical causes. Advances in environmental toxicology have combined the relevance and cost-effectiveness of using biological techniques, closely allied to chemical analyses that enable the identification of the chemical causes of toxic effects. Issues of causality, the generality of biological techniques and their deployment, in an environment where contaminants become heterogeneously distributed, remain key issues and are considered below.

Establishing causality. In any chemically orientated monitoring, potential causality between chemical contaminant and its toxic effect is implied by environmental concentrations that are in excess of the EQS. Historically the limitations of techniques to establish causality have weighed against the wider use of biological techniques for monitoring. Thus operational techniques depending on weight of evidence, rather than rigorously demonstrated causality, have been advocated (Stebbing 1992). Such techniques depend on correlative evidence and the relationships between environmental concentrations, or tissue concentrations, and toxicological threshold concentrations. In the approach advocated here, the need to establish causality becomes an integral part of the method.

Recent developments have improved the rigour of relationships between contaminants and biological effects. Thus the toxicological interpretation of tissue burdens of contaminants using QSARs (quantitative structure activity relationships) is adequate to identify PAHs and organotins as two important classes of contaminants causing the depressed scope for growth in mussels along the UK East Coast (Widdows et al. 1995). Similarly the use of ion exchange resins allied to sensitive water quality bioassays can be used to demonstrate metal pollution (Stebbing 1979a), or to concentrate contaminants and bioassay the eluate to demonstrate the toxicity of organic contaminants (Bening et al. 1992). Such extraction and fractionation techniques have considerable potential for establishing causal relationships between chemical contaminants and their biological effects.

Several biochemical techniques which measure enzyme activity help to identify their chemical causes. The induction of Mixed Function Oxygenases (MFOs) identifies the limited range of organic compounds (including PAHs and PCBs) which induce their production. Thus the induction of EROD in dab indicates pollution by such organic compounds (Fig. 8.6). Similarly induction of the metal-binding protein metallothionein indicates exposure to metals, typically copper, zinc, cadmium and mercury (Hylland et al. 1992). Acetyl cholinesterase activity (AChE) in freshwater fish is used to indicate the impact of some pesticides (including organophosphorous and carbamate pesticides). The occurrence of increased activity of AChE in dab in the German Bight follows the same distribution as MFO activity, suggesting that chemicals which induce AChE have the same distribution as the chlorinated or polynuclear aromatic hydrocarbons (Addison 1992). The induction of imposex in marine gastropods has been found to be specific enough that the degree of imposex is used as a surrogate for chemical analysis to monitor TBT pollution (Gibbs et al. 1966).

Biological indices of water quality allied to the distribution of chemical contaminants may provide correlative evidence of causality, when applied along a relatively simple pollution gradient but such evidence is rarely adequate for management action. Those biological techniques that indicate the class of contaminants responsible have an important role, especially where they are as specific as imposex in gastropods. The technique with greatest scope for application is the toxicological interpretation of tissue burdens with the aid of QSARs. Thus, biological effects techniques that enable chemical effort to be focussed on establishing causality where there is demonstrable pollution, offer the most cost-effective approach to monitoring water quality.

Of those techniques currently available, imposex in gastropods is unique as a specific index of organotin pollution (see Section 8.6.2), while EROD induction indicates exposure to organic contaminants (see Section 8.5.2). The use of QSARs and tissue contaminant concentration-response relationships allied to scope for growth in mussels (Widdows et al. 1990) provides a toxicological interpretation of observed reductions in physiological (energetic) health terms of contaminant tissue burdens (see Section 8.5.4); an approach that could be extended to other species. Similarly the use of liquid-solid extraction technology for the selective extraction and concentration of different classes of contaminants (Bening et al. 1992) has potential, when linked to water quality bioassay techniques, as an aid to establishing causality, but has not yet been used operationally.

The generality of biological techniques. The lower the level of biological organisation at which a technique measures contaminant effects, the more likely it is to have generality of application and comparability, since organisms resemble one another more closely at lower levels. Thus genetic, biochemical and cellular indices, collectively referred to as biomarkers may be used effectively in organisms as diverse as fish and molluscs. Indices at subcellular or cellular levels may usefully indicate specific classes of contaminants due to adaptive metabolic responses to them, but those at the organismal level are more integrated and relevant to management issues. However, the organismal significance of toxic effects or responses at lower levels has rarely been clear. While techniques may be transferred between taxa, there remain difficulties in relating the results of indices at different levels of biological organisation, although advances have been made (Moore 1992; Willows 1994; Goss-Custard and Willows 1996).

More important is whether indices of toxicity in one or a few organisms can be used to indicate the biological quality of different habitats, the health of ecosystems or the North Sea. Clearly, no single species can adequately represent a community of species occupying a single habitat, since the routes by which toxins become biologically available vary, as do the expressions of different mechanisms of toxic actions likely in different taxa (neurotoxin, respiratory inhibitor, genotoxin, endocrine mimic). Thus, a suite of suitable species representative of different taxa from the plant and animal kingdoms is essential in establishing an EQS. Toxicological data from algae and/or macrophytes, arthropod (e.g. crustacean), non-arthropod (e.g. mollusc), fish are considered necessary (Zabell, pers. comm.). However, additional species representative of different habitats, and ecological niches are desirable, since contaminant behaviour will result in the accumulation of contaminants at different sites (air-sea interface, surficial benthic sediments).

Biologically-orientated objectives and standards for marine environmental quality (i.e. EcoQSs) should be based on a taxonomic range of species (microalgae to fish), representatives of functional ecological groups (autotrophs, heterotrophs, symbionts) from different habitats (pelagic, benthic epifauna and infauna, planktonic), modes of toxic action (genotoxic, respiratory inhibition, endocrine mimic) and life phases (egg, embryo, larva, adult).

The heterogeneity of contaminant distribution. The deployment of techniques to assess contamination in the marine environment does not reflect the known heterogeneity in contaminant distribution (see Table 8.1). It has been shown that contaminants bind to particles, especially those which are fine and organic (POC). They are also highly mobile, and both a substrate and food for many species. At the sea surface buoyant organic-rich particles provide the substrate for a specific community of microzooplankton and specialist neuston species, apart from the transient embryonic and larval stages of benthic invertebrates and pelagic fish. Yet here contaminants accumulate to concentrations orders of magnitude greater than the immediate subsurface. Thus contaminants accumulate on more dense but highly mobile organic particulate substrates, which are sites of biological activity. If such sites are where contaminants accumulate, it is here that monitoring effort should be focussed, particularly since it is where biological activity is often greatest.

Cost-effectiveness of an integrated approach. Present legislation, and consequent monitoring effort, requires chemical analysis for those named contaminants for which legislation exists. This strategy leads to considerable redundancy in monitoring effort. Many contaminants are below the detection threshold of available analytical techniques. Other contaminants are only present at inconsequential concentrations. Although required by legislation, such effort is wasted, leading regulators to direct effort in a way that is required by the legislation, but does not adequately serve its purpose, which is to protect the biological quality of the marine environment.

The approach advocated here is to use biological monitoring as the means of overall monitoring and surveillance, using a suite of tested and robust techniques. Such techniques should be widely deployed over the area for which the regulator is responsible at regular intervals. The distribution of sampling efforts needs to be informed by knowledge of the relevant environmental processes. In particular understanding of contaminant transport and behaviour incorporated in simulation models (Taylor 1987) should be used to focus and minimize monitoring (Radford and West 1986)

No chemical technique should be as widely or frequently deployed, since it is assumed failure to detect any biological impact implies that environmental health is satisfactory and sustainable. The chemical monitoring capability should be reserved for deployment alongside biological techniques to establish causality when depressed water quality is detected (Stebbing and Harris 1988). Such a strategy would not only be more effective in detecting new contaminants, but could more economically provide improved awareness of the utilisation of assimilative capacity and the health of the North Sea ecosystems.

8.5
Environmentally Significant Pollutants

The advantage of using biological indices is that not only do they provide an overall, integrated measure of environmental quality, but where techniques identify the contaminants responsible for toxic effects, those classes of chemicals of greatest significance in coastal waters can be clearly identified. Here we consider some of the more significant contaminants that have been detected in this way.

8.5.1
Polyaromatic Hydrocarbons (PAHs)

Wild and Jones (1995), in their budget for PAHs in the UK environment could not quantify marine inputs. PAHs in the fine fraction (<63 µm) of sediments are significant contaminants in the Humber estuary (0.7–2.7 mg kg^{-1} dry weight) and follow the path of the freshwater plume offshore (Klamer and Fomsgaard 1993). Such findings mirror those for the Elbe-Weser estuarine plume, whose contaminant chemistry (including metals, aromatic and chlorinated hydrocarbons) shows declining concentrations over two order of magnitude over a pollution gradient of 200 km to the north east and the Dogger Bank (see Stebbing et al. 1992, Appendix 1). Although the inputs responsible for such gradients have not been quantified, they do demonstrate significant PAH fluxes to the North Sea which can be considered typical of industrialized and urbanized estuaries, having an effect on sediment concentrations of PAHs well offshore.

Wild and Jones (1995) estimate UK inputs of PAHs to the atmosphere amount to 712 t a^{-1}. They estimate a considerable potential export of PAHs to Europe across the North Sea. Approximate estimations of total PAH inputs to the North Sea by atmospheric deposition have been given at 50–1 300 t a^{-1} (reviewed in van Aalst 1988). Preston and Merrett (1991) indicate that the dominant source of hydrocarbons to the North Sea atmosphere is from air which has recently passed over the UK. The aerosol is dominated by hydrocarbons of terrestrial origin. While Brorstroem-Lunden (1996) provide analyses of deposited PAHs from coastal and offshore sites (Swedish west coast) of a similar order of magnitude, Preston and Merrett give evidence of preferential deposition over the sea, which they believe has important consequences for flux calculations, due to the greater proportion of wet deposition over the sea.

It is instructive to consider the distribution of PAHs in relation to lead, whose distribution in the North Sea is partly accounted for by its atmospheric transport and deposition. Markedly elevated lead levels off the UK East Coast have been presented by Kersten et al. (1988) for coastal waters from Flamborough Head to North Norfolk. There is a steep gradient offshore due to scavenging by particles which adsorb strongly and remove fluvial inputs close to the shore (Salomons and Forstner 1984). This has the effect of accentuating the occurrence of elevated concentrations of lead on particulate fines (<20 µm) in the central North Sea to twice that of the surrounding area, implying a different mode of input. Elevated lead concentrations in the sediments from the central North Sea has long been attributed to atmospheric inputs. Since the sediments include a low proportion of contaminant-rich sediment fines (<1%) the shallow waters of the Dogger Bank are considered a temporary deposition site. They are remobilized during the winter gales and likely to be carried further north and finally deposited in the Norwegian Trench (Skei 1981).

Data from Wild and Jones (1995) show that 11% of PAH inputs to the atmosphere are from motor vehicles, and correlations between PAHs and lead in North Sea aerosol samples suggest a source related to road transport (Preston et al. 1992; Chester et al. 1994). They propose the relationship of PAHs with lead, as a tracer of anthropogenic influence, could be used to follow changes due to the adoption of lead-free petrol and the decreasing use of leaded petrol. Airborne contaminant inputs to the northern North Sea are only 50% of those to the southern North Sea (QSR 1993) due to greater urbanisation and industrialisation.

Among the most comprehensive study of estuarine PAHs was that conducted by Readman et al. (1987) for the R. Tamar. They established that PAHs appear to be stable in anoxic sediments, such that PAH concentrations at dated levels in a sediment core reflect inputs to the sediments when they were laid down. He concluded that the exponential increases from 1960–1980 in PAHs and some metals (Cu, Zn, Pb), typical of other UK estuaries, suggests inputs from motor vehicles and road run-off.

Elevated nearshore PAH concentrations also provided the dominant pollutant in the effects observed by Widdows et al. (1995). They conclude that "… at the majority of North Sea coastal (26) sites toxic hydrocarbons are at concentrations capable of inducing significant inhibition of SFG and forming a major component of the overall reduction in SFG". In some instances this accounted for up to 90% of the reduction in SFG.

When considering the effects of North Sea contaminants the QSR (1993) reports that "Information on the occurrence of these compounds in the North Sea is rather scarce and almost no information on inputs is available". Since then preliminary analyses of water samples have been conducted as part of the UK National Monitoring Programme (Law et al. 1997). They show elevated concentrations in industrialized estuaries with levels as high as 10.7 μg l^{-1}, and 15% of samples exceed sublethal toxicity thresholds.

PAHs provide another example (see Section 8.2) where biological techniques are highlighting the importance of a class of compounds which a predominately chemical approach has, until recently, overlooked.

8.5.2
Organotins

The most significant individual contaminant known in recent years to depress coastal water quality is tributyl tin (TBT); the biocide used in antifouling paints. The significance of TBT was first detected by its effect on non-target organisms: shell thickening in oysters and a failure of their larvae to grow to metamorphosis (Alzieu et al. 1982). Subsequently it was found to disrupt the endocrine system in the dog whelk (*Nucella lapillus*), causing infertility by inducing females to become male (imposex) (see review by Gibbs and Bryan 1996). TBT is now known to induce imposex in about 30 species of gastropod, and has been established as a specific index of TBT pollution. TBT constitutes a unique ecotoxicological case study illustrating the complete action cycle, from the problem detected, to causality established, legislation introduced and recovery monitored (Abel 1996; see Champ and Seligman 1996 for a full account).

The impact of TBT is sometimes trivialized as only causing imposex in dog whelks and shell thickening in cultivated Japanese oysters. It is therefore overlooked that TBT was incorporated in antifouling paints as a broad spectrum biocide, to keep vessels free of the full range of sessile communities (algae, barnacles, hydroids etc.). It was also used as an antifoulant on salmon rearing pens. It proved toxic to marine life at such low concentrations that it was probably the first compound for which it was clear that the sea had no assimilative capacity. It is unfortunate therefore that the research programme initiated in the UK was almost exclusively autoecological; the effect of TBT on communities in polluted waters was overlooked, although anecdotal evidence for a wider impact was strong.

Since legislation has effectively controlled the use of TBT on small vessels and the impact of TBT in coastal and estuarine waters has been controlled, attention has turned to the effect of TBT originating from large vessels through studies of imposex in dog whelks around the North Sea. The species was once very widespread and the effects observed during a survey of the entire North Sea coastline in 1992 are likely to prove catastrophic for the species. Offshore, in the German Bight, TBT has been shown to occur in the sea surface microlayer at concentrations in excess of the UK EQS of 2 ng l^{-1} (Hardy and Cleary 1992), while in the southern North Sea high incidences of imposex have been found in benthic whelk (*Buccinum undatum*) populations living in a deepwater offshore shipping lane (Mensink et al. 1996).

The contribution to the depression of scope for growth in mussels due to TBT (Widdows et al. 1990) is significant in increasing the taxonomic breadth of organisms known to be affected in the environment. The results also show the biological impact of TBT in areas (Thames estuary) where dog whelks no longer occur. In the case of TBT, no case needs to be made for the deployment of biological indices with appropriate chemistry, as it has already been established that imposex is a specific biomarker for TBT pollution, to the extent that the degree and frequency of imposex is widely used as a proxy for chemical analyses of TBT.

8.6
Biological Evidence for Pollution Gradients in the North Sea

We now consider examples from the North Sea where the use of biological techniques has demonstrated pollution gradients in terms of the biological impact of contaminants. At the same time these examples demonstrate the extent to which the capacity of the North Sea to assimilate anthropogenic inputs of wastes and effluents is being approached or exceeded.

8.6.1
Gradients Related to Oil Rigs

The North Sea not only accepts inputs of contaminants from the bordering riparian states, it also receives direct inputs from certain offshore activities such as the disposal of sewage sludge, the use of incinerator ships, and as a consequence of the offshore oil and gas industry. While the first two sources of input are now subject to stringent control, oil and gas production and exploration will continue for the foreseeable future. At first it was believed that the major environmental impact of this industry would be a consequence of oil inputs from production platforms. In the event, the largest impacts have resulted from the disposal around the rigs of discharged drill-cuttings (Kingston 1992). Initial drilling activities used water-based drilling muds, but oil-based muds proved more suitable. However, due to evidence of the toxicity of the latter, the oil content of cuttings discharged by platforms in the Norwegian sector was reduced to less than 6% in 1987. In 1993 their disposal was prohibited (Olsgard and Gray 1995). As a consequence there has been a move back to water-based muds. These in turn require larger amounts of weighting agents (up to 10 times more) than oil-based muds, which usually take the form of a barium mineral (barite) which typically contains impurities including a range of other heavy metals (GESAMP 1993). Oil-based drilling muds continue to be used in the UK sector.

Initial evidence from chemical monitoring programmes indicated raised levels of total oil hydrocarbons in sediments sampled in the vicinity of drilling platforms. These can be as much as 10 000 times background levels close to the platforms, but typically decline rapidly at distances greater than 500–1 000 m (Kingston 1992). Oil exploration and exploitation companies are required to conduct both chemical and biological environmental monitoring of their activities. While raised concentrations of contaminants (oil hydrocarbons) from drilling muds can be detected at distances up to 10 km from platforms (Davies and Kingston 1992; GESAMP 1993), evidence for major impacts on benthic animals (reductions in species diversity) were generally limited to 750 to 1 500 m from the platforms. The Norwegian government has a policy of open access to its quality-controlled environmental monitoring data. This has allowed the application of sophisticated multivariate statistical analyses of both the chemical and biological data (Gray et al. 1990; Olsgard and Gray 1995). These analyses have shown that the effects of contamination by drill cuttings of the sea-bed containing a mixture of oil hydrocarbons, barium and other heavy metals, bentonitic clays, together with other bioactive materials, can have a significant impact on the structure of benthic animal communities for areas between 10 and 100 km² around individual platforms. Indeed, these effects can extend beyond the original sampling area, including designated reference sites (Olsgard and Gray 1995). Drilling sites may show significant differences in benthic macrofauna 3 years after drilling activity ended, particularly in the deep burrowing and long-lived species. Where such sediments had been covered by several centimetres of clean and well-mixed sediments on top of the contaminated sediments, the meiofauna communities were remarkably uniform (Heip 1992). In acute toxicity studies barite has not been found to be particularly toxic to marine organisms (see Starczak et al. 1992), and Olsgard and Gray (1995) concluded that the major adverse effects on the biota were probably associated with the hydrocarbons contained in discharged drilling muds, although other components probably make a significant contribution. Given the number and spatial distribution of oil and gas platforms throughout the North Sea, it is likely that benthic communities over a substantial area of the seabed have been affected.

8.6.2
Estuarine Inputs

It is well established that the major route for inputs of contaminants to the North Sea is by way of the river estuaries. Perhaps the most significant pollution gradient in coastal waters is that due to the plumes of the rivers Elbe and Weser which flow into the German Bight. This gradient was intensively studied during an ICES/IOC Bremerhaven Workshop in March 1990 along a transect of stations extending 200 km northwest out to the eastern side of the Dogger Bank (Stebbing et al. 1992). During the workshop,

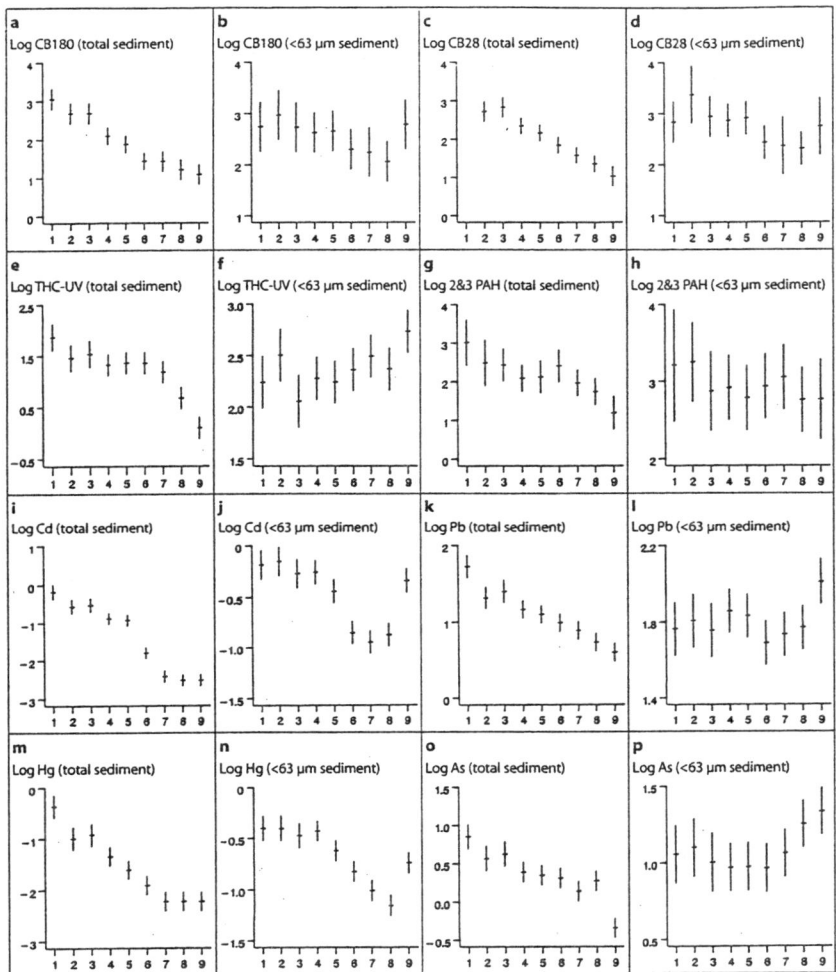

Fig. 8.5. Selected chemical data from the ICES/IOC Bremerhaven Workshop indicating chemical contamination gradients on a transect running northeast across the German Bight from the mouths of Elbe-Weser estuaries (From Stebbing et al. 1992). Units: **a–d, g–h** ng g^{-1} dry wt; **f–g, i–p** µg g^{-1} dry wt. 95% confidence intervals are indicated

numerous biological effects techniques (Fig. 8.5) (including all those adopted by the North Sea Task Force for use in preparing the 1993 QSR), were allied to chemical analyses of water, sediment and tissues (Fig. 8.6).

The biological data relates primarily to the dab (*Limanda limanda*). A wide range of techniques (biochemical, subcellular, organismal responses) indicate that the fish are progressively less stressed with distance offshore (Fig. 8.5). This pattern is reproduced in some of the sediment bioassay data using bivalve larvae. There is a reversal in this trend, at the most offshore station over the Dogger Bank, detected by a number of techniques, whose coincidence suggests that they are significant.

Chemical data showing the concentrations of contaminants in sediments along the transect demonstrate a decrease in concentration with distance offshore. This trend could be seen most clearly in the whole sediment data for PCBs, PAHs, total hydrocarbons and some of the major metals (lead, arsenic, mercury, cadmium). However, when analyses of the fine fraction (<63 μm) are considered, the spatial trends are markedly different, often increasing progressively with distance offshore (total hydrocarbons), increasing only at the most offshore stations (arsenic, lead), or decreasing offshore and then increasing over the Dogger Bank (mercury, cadmium). Such distributions closely match some of the biological data, indicating significant contamination. Inshore the data show marked effects due to pollution of the estuarine plume by the contaminants flowing out of the rivers Elbe and Weser. Effects are reduced progressively with distance offshore as concentrations of contaminants decrease due to dilution. Such data lend support to the interpretation that the association of contaminants with the organic-rich fine fraction is the reason for their disproportionate effect on the biota (Kempe et al. 1988).

Flatfish such as the dab possess a particularly vulnerable lifestyle under these circumstances, since they are likely to be exposed to high concentrations accumulated by their benthic prey, as well as lying cryptically beneath the superficial layer of the fine contaminant-rich floc found in depositional areas (e.g. Dogger Bank). Thus flatfish are exposed to higher concentrations than other demersal fish and, as a consequence, show the adaptive responses of enzyme systems that aid the degradation and excretion of organic contaminants. At higher levels of exposure they are more vulnerable to disease and indeed show increased frequencies of some diseases in parts of the North Sea (Fig. 8.5).

8.6.3
Fish Embryo Abnormalities

As part of the ICES/IOC Bremerhaven Workshop, Cameron and Berg (1992) evaluated the frequency of developmental abnormalities in dab embryos in surface plankton across the German Bight (Fig. 8.5). Subsequently the technique was used for all fish species in a survey of the North Sea. Such teratogenic and chromosomal abnormalities are determined in living embryos at sea. On a transect extending NE across the German Bight frequencies of abnormalities were high at the most inshore stations (32%) decreasing with distance offshore (9%) before increasing again over the Dogger Bank (31%) (Cameron and Berg 1992). Similarly, chromosomal aberrations were highest inshore, lowest offshore, and increased over the Dogger Bank (Fig. 8.5).

It is concluded that the enhanced frequencies of developmental abnormalities and chromosomal aberrations of dab embryos are at least partly due to organochlorines which occurred at elevated concentrations in adult dab livers (see Stebbing et al. 1992, Appendix 1). Experimental work has shown that elevated levels of chlorinated hydrocarbons (PCB, DDE, dieldrin) in flatfish result in malformations during embryonic development which reduce hatching success (Westernhagen 1988). The possibility of abnormalities being caused by the exposure of floating eggs and embryos to contami-

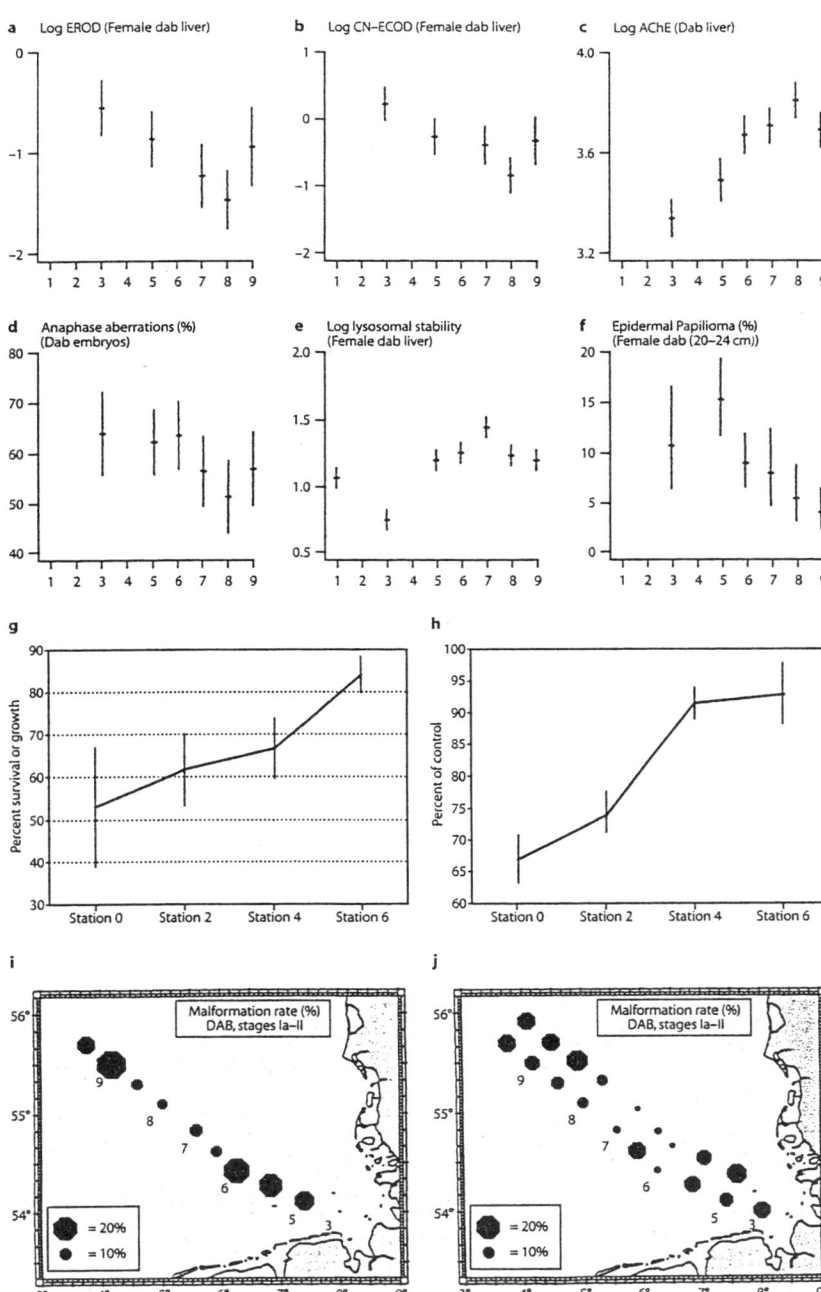

nants concentrated in the sea surface microlayer (Hardy and Cleary 1992) is also a strong possibility, since the toxicity of such contaminants on fish embryos has been demonstrated (Kocan et al. 1987).

A more comprehensive survey of developmental abnormalities for all fish species in the surface plankton (including plaice *Pleuronectes platessa*, sprat *Sprattus sprattus*, whiting *Merlangius merlangius*, rockling *Onos* sp., flounder *Platichthys flesus*) was conducted in 1991 and 1992 (Cameron and von Westernhagen 1996, 1997). The results show malformation rates varying with species between 10% and 80%. Highest malformation rates were found in the plumes of the major industrialized estuaries; Weser 43%, Rhine 29%, Thames 58%, Forth 38%. The authors consider it most likely that the abnormalities are due to anthropogenic contaminants, which have been shown to be capable of inducing such abnormalities (von Westernhagen 1988). Whether these effects are due to parental exposure to contamination in benthic surficial sediments, or eggs exposed at the sea surface microlayer, remains unclear, either or both routes are possible. Up to 85% of such embryonic abnormalities are known to be lethal in 5 days (Cameron and von Westernhagen 1997). There is no evidence yet that the mortality of fish embryos in coastal waters is affecting recruitment of fisheries, nevertheless that must remain a principal concern.

8.6.4
UK East Coast Gradient

Deployments of "scope for growth" in mussels (*Mytilus edulis*) in the North Sea (1990–91) have provided the first measurements of a pollution gradient along the UK East Coast from the Shetland Islands to the Thames estuary (Widdows et al. 1993, 1995).

This technique measures an integration of basic physiological processes in the energy balance of the organism in terms of its potential for growth, and has been used to measure pollution gradients for over 15 years. It has been deployed for intercomparison with a wide range of other techniques in workshops organized by the IOC Group of Experts on the Effects of Pollution (GEEP) in Norway (Widdows and Johnston 1988) and Bermuda (Widdows et al. 1990).

During the 1990 exercise in the North Sea, scope for growth was measured at 26 sites down the UK East Coast, then in 1991 mussels were deployed on 9 lightships off the East Coast from the Humber to the Dover Straits. Scope for growth data from all the sites are plotted here (Fig. 8.7) against the calculated age of North Atlantic water in the North from a winter model simulation (van Pagee and Postma 1986). The results fall into three groups: those for sites under the influence of estuarine water, other coastal

◄ **Fig. 8.6.** Selected biological data from the ICES/IOC Bremerhaven Workshop indicating gradients in pollution on a transect running northeast across the German Bight from the mouths of Elbe-Weser estuaries (From Stebbing et al. 1992). Station numbers *1–9* are indicated in *i* and *j* and in Fig. 8.1b as *A* to *I*. **a** to **f** give ecotoxicological indices of response or effect in the dab (*Limanda limanda*); **a–c** biochemical indices; **d** mitotic abnormalities in developing embryos; **e** index of condition of cellular organelles; **f** frequency of external papillomas indicating skin disease; **g** pooled bioassay data from a variety of techniques following exposure to bulked microlayer and subsurface water samples; **h** growth data for the copepod *Tisbe battaglia* exposed to surface microlayer samples; **i** and **j** morphological malformation frequencies in early developmental stages of dab (*Limanda limanda*). (Papers from which data are drawn are in Stebbing et al. 1992)

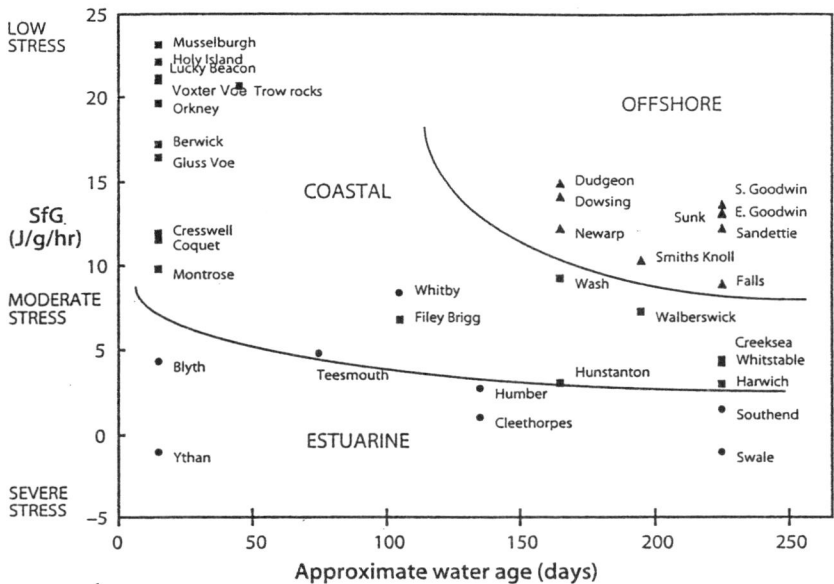

Fig. 8.7. Measured values of scope for growth in mussels (From Widdows et al. 1995) against the simulated age of the water in the North Sea (From van Pagee and Postma 1986) for each site from which mussels were sampled or exposed. *Triangle:* offshore; *square:* coastal; *circle:* estuarine sites whose locations are given in Fig. 8.1b

sites distant from estuaries, and the offshore sites. They show that water quality is poorest in estuaries, improving at coastal sites removed from polluted estuarine inputs, and is best at offshore sites.

More significant are the trends in declining scope for growth from north to south and with age of seawater for all three groups: estuarine, coastal and offshore. The residual currents along the East Coast run from north to south and, as they move south, receive the contaminant inputs from successive major estuaries, the Forth, Tyne, Tees, Humber and Thames. The total burden of the more persistent contaminants is likely to be cumulative and their combined effect on water quality is reflected in declining mussel scope for growth with age of the water and distance south. While less marked in offshore waters, where contaminant concentrations are lower, the same trend is apparent.

Scope for growth, allied to the use of tissue contaminant concentration-response relationships and QSARs, provides a toxicological interpretation of contaminant tissue burdens in mussels. At most of the North Sea coastal sites aromatic and aliphatic hydrocarbons are found in mussel tissues at concentrations high enough to inhibit scope for growth. Such hydrocarbons were the most important class of toxic compounds and are derived from the combustion of oil, coal, wood, petroleum and diesel fuels (see Section 8.5.1). Some details of the mode of toxic action of hydrocarbons, phenolic compounds and TBT are known (Widdows and Donkin 1991; Widdows and Page 1993; Willows 1994). Other biologically significant contaminants detected in this way included TBT and its derivatives.

These data are of especial significant because they show for the first time that pollution in the North Sea is not restricted to the coastal margins (QSR 1987), or localized to areas close to sources or deposition areas (QSR 1993), but affects the North Sea as an entity. They demonstrate that the UK East Coast is a single pollution gradient shown by the decline in water quality from north to south (Fig. 8.7).

At present scope for growth appears to satisfy the various criteria as a biological monitoring technique for general use. It provides in mussels, a sensitive yet robust technique for measuring water quality that is independent of natural stressors, such as salinity and turbidity (Widdows et al. 1995). Combined with QSARs it can provide a toxicological interpretation of contaminant tissue burdens, partitioning the extent of the depression of scope for growth between contaminants. The limiting factor is the availability of QSARs relating tissue concentrations to effects for different classes of contaminants (Donkin et al. 1991, 1997).

8.7
Biological and Ecological Quality Standards

If the regulator's role was set in the context of maintaining assimilative capacity, rather than to serve legislation for individual contaminants, we believe monitoring would be more appropriately directed. Thus EcoQSs should be defined such as to measure and conserve assimilative capacity. That the approach of determining the EQS for individual contaminants is scientifically dubious, and that it would be better to determine and monitor actual effects on the sensitive compartments (i.e., the biota), and use these to direct the need for regulation. The use of biologically defined quality standards (EcoQSs) means that the statutory instrument for regulating pollution can relate directly to the criteria for environmental quality, and the status of biological systems, rather than indirectly through EQSs for specific contaminants based on laboratory toxicity tests.

If the criteria for environmental quality are predominately biological or ecological, then environmental quality objectives should also be expressed in such terms. Such proposals are not new (Elliott 1996), but various factors have prevented their adoption. Defining protocols for biological techniques in a way that yield reproducible results has, in the past, been a problem. There is also the mistaken belief that the inherent variability of biological material prevents its ability to provide precise data. Biological effects techniques typically do not indicate whether they are due to natural factors or to chemical contamination, so the approach must be linked to some means of identifying the chemical causes of significant effects. Their advantage is that they integrate both natural and anthropogenic determinants of environmental quality.

Over the last decade there has been a growing recognition that utilising the integrative capacity of biological techniques is the only way of detecting and controlling pollution by the escalating number of chemical contaminants (Stebbing and Harris 1987). Improved biological techniques are more sensitive and provide more reproducible results. Moreover they now often provide some indication of their chemical causes. We suggest that the evidence presented constitutes a case to develop practical techniques that allow environmental quality to be assessed in biological terms, which should be allied to methods to identify the chemical causes of observed effects. Such techniques will then provide the tools to implement more appropriate legislation, and the regulators with a mandate to apply them.

8.8
Conclusions

1. The perceived state of the North Sea is one of significant local contamination, with evidence of some pollution related to the coastal margins, estuarine inputs, and areas of contaminant accumulation.

2. The definition of pollution is considered from both chemical and biological perspectives; only the latter incorporates a meaningful concept of the environmental effects of pollution.

3. Where biological techniques to measure pollution effects have been deployed on appropriate geographic scale, they demonstrate that significant gradients of pollution exist in the North Sea that have not been defined using chemical monitoring techniques.

4. We propose a definition for assimilative capacity as the balance between those processes that determine the concentration of contaminant in an environmentally meaningful compartment. Assimilative capacity is exceeded when a critical threshold of biological effect is exceeded (i.e. there is pollution), or when the rate of input of a contaminant would lead to such a threshold being exceeded in the future (i.e. there will be pollution). We identify some of the component processes that may act in the North Sea.

5. We argue that the assimilative capacity of the North Sea is a resource of considerable economic value. There may be an optimal economic level of utilisation of this environmental service that does not damage its component processes.

6. The use of biological effects techniques that demonstrate that pollution gradients exist that traverse large areas of the North Sea, and from estuarine plumes to its centre, indicate that assimilative capacity is being exceeded.

7. Existing legislation is orientated to regulating the release of potential pollutants. Further legislation sets chemical EQSs and this determines the use of monitoring effort. This does not provide the most appropriate basis for controlling pollution because links between chemical contamination and its biological impact are often ignored or poorly understood.

8. New contaminants are often not detected as pollutants until they have an impact in the environment; they are recognized through their effects rather than by chemical analysis. Nevertheless investigative chemistry must be included to detect incipient problems.

9. Therefore we advocate the use of biological monitoring techniques that relate to accepted criteria for environmental quality. Such techniques, if strategically deployed, are sufficiently sensitive to provide both detection and early warning of incipient environmental problems. Chemical analyses can then be focussed to establish the causative contaminants.

10. Biologically-based legislation (biological or ecological water quality objectives and standards) that is formulated to protect assimilative capacity, and thereby prevent pollution, would ensure that monitoring effort is more appropriately focussed.

Acknowledgements

We are grateful for critical comments of this chapter from Dr. Peter Donkin, Dr. John Widdows and Miss Nicola Beaumont; to Dr. John Harris and Prof. Geoffrey Millward for advice on sediment/water partitioning and to Mr. Fred Staff and Mr. Martin Carr for help in preparing figures.

References

Aalst RM van (1988) Input from the atmosphere. In: Salomons W, Bayne BL, Duursma EK, Forstner U (eds) Pollution of the North Sea. Springer, p 687

Abel R (1996) European policy and regulatory action for organotin-based antifouling paints In: Champ MA, Seligman P (eds) Organotin: Environmental Fate and Effects. Chapman & Hall, p 623

Addison RF (1992) Biochemical measurements: summary. Mar Ecol Prog Ser 91:61–63

Addison RF, Clarke KR (1990) The IOC/GEEP Bermuda Workshop. J Exp Mar Biol Ecol 138:1–8

Alzieu C, Heral M, Thiband Y, Dardignac MJ, Feuillet M (1982) Influence des peintures antisalissures a base d'organostanniques sur le calcification de la coquille de l'huite Crassostrea gigas. Rev Trav Inst Peches Marit 45:101–116

Bayne BL, Clarke KR, Gray JS (1988) Biological effects of pollutants. Results of a practical workshop. Mar Ecol Prog Ser 46:1–278

Bayne BL, Widdows J, Moore MN, Salkeld P, Worrall CM, Donkin P (1982) Some ecological consequences of the physiological and biochemical effects of petroleum compounds on marine molluscs. Phil Trans R Soc Lond B 297:219–239

Bening J-C, Karbe L, Schupfner G (1992) Liquid/solid phase extraction of water samples used for toxicity testing in the German Bight. Mar Ecol Prog Ser 91:233–236

Brorstroem-Lunden E (1996) Atmospheric deposition of persistent organic compounds to the sea surface J Sea Res 35:81–90

Cairns J (1977) Quantification of biological integrity. In: Ballantine RK, Guarraia LJ (eds) The Integrity of Water. EPA, Office of Water and Hazardous Materials, Washington, DC

Cameron P, Berg J (1992) Morphological and chromosomal aberrations during embryonic development in dab Limanda limanda. Mar Ecol Prog Ser 91:163–169

Cameron P, Westernhagen H von (1997) Malformation rates in embryos of North Sea fishes in 1991 and 1992. Mar Pollut Bull 34:129–134

Champ MA, Seligman PF (1996) Organotins: Environmental Fate and Effects. Chapman & Hall, London, p 623

Chapman PM (1997) Is bio-accumulation useful for predicting impacts? Mar Pollut Bull 34:282–283

Chester R, Bradshaw GF, Ottley CJ, Harrison RM, Merrett JL, Preston MR, Rendell AR, Kane MM, Jickells TD (1994) The atmospheric distributions of trace metals, trace organics and nitrogen species over the North Sea. In: Charnock H, Dyer KR, Huthnance JM, Liss PS, Simpson JH, Tett PB (eds) Understanding the North Sea system, Chapman & Hall, p 222

Cleary J, Stebbing ARD (1987) Organotin in the surface layer and subsurface waters of Southwest England. Mar Pollut Bull 18:238–246

Costanza R, d'Arge R, de Groot R, Farber, S, Grasso M, Hannon B, Limburg K, Naeem S, O'Neill RV, Paruelo J, Raskin RG, Sutton P, van den Belt M (1997). The value of the world's ecosystem services and natural capital. Nature 387:253–260

Davies JM, Kingston PF (1992) Sources of environmental disturbance associated with offshore oil and gas development. In: Cairns WJ (ed) North Sea oil and the environment –developing oil and gas resources, environmental impacts and responses. International Council on Oil and the Environment, Elsevier, London

Deggens ET (1988) Introduction. In: Kempe S, Liebezeit G, Dethlefsen V, Harms U (eds) Biogeochemistry and distribution of suspended matter in the North Sea and implications to fisheries biology. Mitt Geol-Palaeont Inst Univ Hamburg, SCOPE/UNEP Sonderband Heft 65, p 552

Donkin P, Widdows J, Evans SV, Worrall CM, Carr M (1989) Quantitative structure-activity relationships for the effect of hydrophobic organic chemicals on rate of feeding by mussels (*Mytilus edulis*). Aquat Toxicol 14:277–294

Donkin P, Widdows J, Evans SV, Brinsley MD (1991) QSARs for the sublethal responses of mussels (*Mytilus edulis*). Sci Total Environ 109/110:461–476

Donkin P, Widdows J, Evans SV, Staff FJ, Yan T (1997) Effect of neurotoxic pesticides on the feeding rate of marine mussels (*Mytilus edulis*). Pestic Sci 49:196–209

Eisma D, Irion G (1988) Suspended matter and sediment transport. In: Salomons W, Bayne BL, Duursma EK, Forstner U (eds) Pollution of the North Sea. Springer, p 687

Elliott, M (1996) The derivation and value of ecological quality standards and objectives. Mar Pollut Bull 32,762–763

GESAMP (IMO/FAO/UNESCO/WMO/IAEA/UN/UNEP Joint Group of Experts on the Scientific Aspects of Marine Pollution) (1993) Impact of oil and related chemicals and wastes on the marine environment. GESAMP Rep Stud 50:1–180

Gibbs PE, Bryan GW (1996) Reproductive failure in the gastropod *Nucella lapillus* associated with imposex caused by tributyl tin pollution: A review. In: Champ MA, Seligman P (eds) Organotin: Environmental fate and effects. Chapman & Hall, p 623

Goldberg ED (1981) Assimilative capacity of US coastal waters for pollutants. Proceedings of a workshop at Crystal Mountains WA, 29th July–4th August 1979. US Department of Commerce, NOAA Working Paper N° 1

Goss-Custard JG, Willows RI (1996) Modelling the responses of mussel *Mytilus edulis* and oystercatcher *Haematopus ostralegus* populations to environmental change. In: Greenstreet SPR, Tasker ML (eds) Aquatic predators and their prey. Fishing News Books, Oxford p 191

Gray JS (1996) Environmental science and a precautionary approach revisited. Mar Pollut Bull 32:532–534

Gray JS, Clarke KR, Warwick RM, Hobbs G (1990) Detection of initial effects of pollution on marine benthos: An example from the Ekofisk and Eldfisk oilfields, North Sea. Mar Ecol Prog Ser 66: 285–299

Hardy JT (1982) The sea surface microlayer, biology, chemistry and anthropogenic enrichment. Progr Oceanogr 11:307–328

Hardy JT, Cleary J (1992) Surface microlayer contamination and toxicity in the German Bight. Mar Ecol Prog Ser 91:203–210

Hardy JT, Apts CW, Crecelius EA, Fellingham GW (1985) The sea surface microlayer: Fate and residence time of atmospheric metals. Limnol Oceanogr 30:93–101

Harris JRW, Gorley RN, Bartlett CA (1993) ECoS Version 2 – An estuarine modelling shell. Plymouth Marine Laboratory, p 146

Heip C (1992) Benthic studies: Summary and conclusions. Mar Ecol Prog Ser 91:265–268

Holdgate MW (1979) A perspective of environmental pollution. Cambridge University Press, p 278

Huiskes AHL, Rozema J (1988) The impact of anthropogenic activities on the coastal wetlands of the North Sea. In: Salomons W, Bayne BL, Duursma EK, Forstner U (eds) Pollution of the North Sea –an assessment. Springer, p 687

Hylland K, Haux C, Hogstrand C (1992) Hepatic metallothionein and heavy metals in dab *Limanda limanda* from the German Bight. Mar Ecol Prog Ser 91:89–96

IAEA (1985) Sediment K_{d}s and concentration factors for radionuclides in the marine environment. International Atomic Energy Agency Technical Series Report N° 247, p 71

ICES (1983) Flushing times of the North Sea. Coop Res Rep Cons int Explor Mer 123, p 159

Kautsky H (1988) Radioactive substances. In: Salomons W, Bayne BL, Duursma EK, Forstner U (eds) Pollution of the North Sea – an assessment. Springer, p 687

Kempe S, Leibezeit G, Dethlefsen V, Harms U (eds) (1988) Biogeochemistry and distribution of suspended matter in the North Sea and implications to fisheries biology. Mitt Geol-Palaeont Inst Univ Hamburg, SCOPE/UNEP Sonderband Heft 65:11–24

Kersten M, Dicke M, Kriews M, Naumann K, Schulz M, Schwikowski M, Steiger M (1988) Distribution and fate of metals in the North Sea. In: Salomons W, Bayne BL, Duursma EK, Forstner U (eds) Pollution of the North Sea – an Assessment. Springer, p 687

Kingston PF (1992) Impact of offshore oil production installations on the benthos of the North Sea ICES J Mar Sci 49: 45–53

Klamer HJC, Fomsgaard L (1993) Geographical distribution of chlorinated biphenyls (CBs) and polycyclic aromatic hydrocarbons (PAHs) in surface sediments from the Humber plume, North Sea. Mar Pollut Bull 26:201–206

Kocan RM, Westernhagen H von, Landolt ML (1987) Toxicity of sea-surface microlayer: Effects of hexane extracts on Baltic herring (*Clupea harengus*) and Atlantic cod (*Gadus morhua*) embryos. Mar Environ Res 23:291–305

Krom MD (1986) An evaluation of the concept of assimilative capacity as applied to marine waters. Ambio 15:208–214

Law RJ, Dawes VJ, Woodhead RJ, Matthiessen P (1997) Polycyclic aromatic hydrocarbons (PAH) in seawater around England and Wales. Mar Pollut Bull 34 (5):306–322

Lohse J (1988) Distribution of organochlorine pollutants in North Sea sediments. In: Kempe S, Liebezeit G, Dethlefsen V, Harms U (eds) Biogeochemistry and distribution of suspended matter in the North Sea and implications to fisheries biology. Mitt Geol-Palaeont Inst Univ Hamburg, SCOPE/UNEP Sonderband Heft 65:345–365

McGarvin M (1990) The North Sea. Collins & Brown, p 144

Maier-Reimer E (1977) Residual circulation in the North Sea due to the M-2 tide and mean annual wind stress. Dt Hydr Zeitschr 30(3):69–80

Mensink BP, Everaarts JM, Kralt H, Hallers-Tjabbes CC (1996) Tributyl tin exposure in early life stages induces the development of male sexual characteristics in the common whelk, *Buccinum undatum*. Mar Environ Res 42:151–154

Ministerial Declaration (1987) Second International Conference on the Protection of the North Sea, London 24th–26th November 1987

Moore MN (1992) Molecular and cellular pathology: Summary. Mar Ecol Prog Ser 91:117–119

Olsgard F, Gray JS (1995) A comprehensive analysis of the effects of offshore oil and gas exploration and production on the benthic communities of the Norwegian continental shelf. Mar Ecol Prog Ser 122: 277–306

Pagee JL van, Postma L (1986) North Sea pollution. The use of modelling techniques for impact assessment of waste inputs. In: Reasons for concern, Proceedings of the 2nd North Sea Seminar 86, Rotterdam, pp 97–113

Pearce D (1976) The limit of cost-benefit analysis as a guide to environmental policy. Kyklos 29:97–112

Portmann JE, Lloyd R (1986) Safe use of assimilative capacity of the marine environment – is it feasible? Wat Sci Tech 18:233–244

Pravdic V (1985) Environmental capacity – is a new scientific concept acceptable as a strategy to combat marine pollution? Mar Pollut Bull 16:295–296

Pravdic V, Juracic M (1988) The environmental capacity approach to the control of marine pollution: the case of copper in the Krka River estuary. Chem Ecol 3:105–117

Preston MR, Merrett JL (1991) The distribution and origins of the hydrocarbon fraction of particulate material in the North Sea atmosphere. Mar Pollut Bull 22:516–522

Preston MR, Chester R, Bradshaw GF, Merrett JL (1992) PAH/lead relationships. A possible tool for the assessment of anthropogenic influence on marine aerosols. Mar Pollut Bull 24:164–166

Quality Status Report for the North Sea (1987) Second International Conference on the Protection of the North Sea, September 1987, London

Quality Status Report for the North Sea (Interim) (1990) North Sea Conference, March 1990. The Hague

Quality Status Report for the North Sea (1993) Oslo and Paris Commissions, London

Radach G, Berg J, Hagmeier E (1990) Long term changes of the annual cycles of meteorological, hydrographic, nutrient and phytoplankton time series at helgoland and ay LV Elbe 1 in the German Bight. Continental Shelf Res 10(4):305–328

Radford PJ, West J (1986) Models to minimise monitoring. Wat Res 20:1059–1066

Readman JW, Mantoura RFC, Rhead MM (1987) A record of polycyclic aromatic hydrocarbon (PAH) pollution obtained from accreting sediments of the Tamar Estuary, UK: Evidence for non-equilibrium behaviour of PAH. Sci Tot Environ 66:73–94

Salomons W, Forstner U (1984) Metals in the Hydrocycle. Springer, Berlin, p 349

Salomons W, Bayne BL, Duursma EK, Forstner U (1988) Pollution of the North Sea – an Assessment. Springer, Berlin, p 687

Skei J (1981) The entrapment of pollutants in Norwegian fjord sediments – a beneficial situation for the North Sea. Spec Publ Int Assoc Sediment 5:461–468

Starczak VR, Fuller CM, Butman CA (1992) Effects of barite on aspects of ecology of the polychaete *Mediomastus ambiseta*. Mar Ecol Prog Ser 85: 269–282

Stebbing ARD (1979a) An experimental approach to the determinants of biological water quality. Phil Trans R Soc 286:465–481

Stebbing ARD (1996) Foreward: organotins – what help from hindsight? In: Champ MA, Seligman P (eds) Organotin: Environmental fate and effects. Chapman & Hall, p 623

Stebbing ARD, Harris JRW (1988) The role of biological monitoring. In: Salomons W, Bayne BL, Duursma EK, Forstner U (eds) Pollution of the North Sea. Springer, p 687

Stebbing ARD, Akesson B, Calabrese A, Gentile JH, Jensen A, Lloyd R (1980) The role of bioassays in marine pollution monitoring. Rapp P-v Reun Cons int Explor Mer 179:322–332

Stebbing ARD, Cleary JJ, Brown L, Rhead M (1984) The problem of relating toxic effects to their chemical causes in waters receiving wastes and effluents. In: IVth International Ocean Disposal Symposium, Plymouth, April 1983

Stebbing ARD, Dethlefsen V, Carr M (1992) Biological effects of contaminants in the North Sea. Mar Ecol Prog Ser 91(1–3): 1–361

Sunda WG, Guillard RRL (1976) The relationship between cupric ion activity and the toxicity of copper to phytoplankton. J Mar Res 34:511–529

Sunda WG, Engel DW, Thuotte RM (1978) Effect of chemical speciation on toxicity of cadmium to grass shrimp Palaemonetes pugio: importance of free calcium ion. Env Sci Tech 12:409–413

Tanabe S, Nishimura A, Hanoaka S, Yanagi T, Takeoka H, Tatsukawa R (1991) Persistent organochlorines in coastal fronts. Mar Pollut Bull 22:344–351

Taylor, AH (1987) Modelling contaminants in the North Sea. Sci Tot Environ 63:45–67

Turner K, Pearce D, Bateman (1994) Environmental economics. Harvester Wheatsheaf, p 328

Turrell WR (1992) New hypotheses concerning the circulation of the northern North Sea and its relation to North Sea fish stock recruitment. ICES J Mar Sci 49:107–123

Uncles RJ, Stephens JA, Woodrow TY (1988) Seasonal cycling of estuarine sediment and contaminant transport. Estuaries 11:108–116

Warwick RM, Clarke KR (1992) Comparing the severity of disturbance: a meta-analysis of marine macrobenthic community data. Mar Ecol Progr Ser 92:221–231

Westernhagen H von (1988) Sublethal effects of pollutants on fish eggs and larvae. Fish Physiology Vol XIA, Chap 4:253–335

Wharfe JR, Tinsley D (1995) The toxicity-based consent and the wider application of direct toxicity assessment to protect aquatic life. J Ciwem 9:525–530

Widdows J, Johnston D (1988) Physiological energetics of *Mytilus edulis*: Scope for growth. Mar Ecol Prog Ser 46:113–121

Widdows J, Donkin P (1991) Role of physiological energetics in ecotoxicology. Comp Biochem Physiol 100C:69–75

Widdows J, Page DS (1993) Effects of tributyltin and dibutyltin on the physiological energetics of the mussel, *Mytilus edulis*. Mar Environ Res 35:233–249

Widdows J, Burns KA, Menon NR, Page D, S, Soria S (1990) Measurement of physiological energetics (scope for growth) and chemical contaminants in mussels (*Arca zebra*) transplanted along a contamination gradient in Bermuda. J Exp Mar Biol Ecol 138:99–117

Widdows J, Donkin P, Brinsley MD, Evans SV, Salkeld PN (1993) Stress effects and contaminant levels in North Sea mussels. Department of the Environment Final Report, Contract N° PECD 7/7/352 p 45

Widdows J, Donkin P, Brinsley MD, Evans SV, Salkeld PN, Franklin A, Law RJ, Waldock MJ (1995) Scope for growth and contaminant levels in North Sea mussels *Mytilus edulis*. Mar Ecol Prog Ser 127:131–148

Wild SR, Jones KC (1995) Polynuclear aromatic hydrocarbons in the United Kingdom: A preliminary source inventory and budget. Env Pollut 88:91–108

Willows RI (1994) The ecological impact of different mechanisms of chronic toxicity on feeding and respiratory physiology. In: Sutcliffe DW (ed) Water quality and stress indicators in marine and freshwater systems: Linking levels of organisations. Freshwater Biological Association, Ambleside, pp 88–97

[33]

Ocean & Coastal Management, Vol. 31, Nos 2–3, pp. 105–132, 1996
Copyright © 1996 Elsevier Science Ltd
Printed in Northern Ireland. All rights reserved
PII: S0964-5691(96)00037-3   0964-5691/96 $15·00 + 0·00

The Mediterranean: vulnerability to coastal implications of climate change

R. J. Nicholls[a] & F. M. J. Hoozemans[b]

[a]School of Geography and Environmental Management, University of Middlesex, Queensway, Enfield, Middlesex EN3 4SF, UK
[b]Delft Hydraulics, PO Box 152, 8300 AD Emmeloord, The Netherlands

ABSTRACT

The Mediterranean is experiencing a number of immediate coastal problems which are triggering efforts to improve short-term coastal management. This paper shows that coastal management also needs to address long-term problems and, in particular, the likelihood of climate change. Regional scale studies suggest that the Mediterranean is particularly vulnerable to increased flooding by storm surges as sea levels rise—a 1-m rise in sea level would cause at least a six-fold increase in the number of people experiencing such flooding in a typical year, without considering population growth. Protection is quite feasible, however, this would place a greater burden on those Mediterranean countries in the south than those in the north. All coastal wetlands appear threatened. Case studies of coastal cities (Venice and Alexandria), deltas (Nile, Po, Rhone and Ebro), and islands (Cyprus) support the need to consider climate change in coastal planning. However, the critical issues vary from site to site and from setting to setting. In deltaic areas and low-lying coastal plains climate change, particularly sea-level rise, is already considered as an important issue, but elsewhere this is not the case. Therefore, there is a need for coastal management plans to explicitly address long-term issues, including climate change, and integrate this planning with short-term issues. This is entirely consistent with existing guidelines.[1] Given the large uncertainty concerning the future, planning for climate change will involve identifying and implementing low-cost proactive measures, such as appropriate land use planning or improved design standards incorporated within renewal cycles, as well as identifying sectors or activities which may be compromised by likely climate change. In the latter case, any necessary investment can be seen as a prudent 'insurance policy'. Copyright © 1996 Elsevier Science Ltd.

105

1. INTRODUCTION

Humanity is preferentially concentrated in the-coastal zones of the world and these coastal populations are growing more rapidly than the global average.[2,3] This is causing significant changes to the coastal environment, placing increasing demands on coastal resources and increasing exposure to coastal hazards such as erosion, flooding and salinity intrusion. Global climate change will exacerbate all these ongoing problems and its potential implications are causing much concern around the world's coasts.[4,5] The Mediterranean is a good example of a coastal region where human stresses are already significant and continue to grow.[6] Increasing attention is being focused on more effective coastal management within the region, with the overall goal of sustainable development, e.g. the United Nations Environment Programme (UNEP).[1] In addition to the pressing short-term problems, these management activities need to address long-term issues such as global climate change.

This paper examines the vulnerability of the Mediterranean to the coastal implications of climate change, including accelerated sea-level rise. It also considers the range of proactive measures that coastal managers might apply to deal with these problems. A number of studies have already considered this problem, particularly for some of the deltas,[7-9] coastal cities such as Alexandria[10] and the region as a whole.[6] This paper builds on a recent top-down analysis of the possible problems of sea-level rise in the Mediterranean[11] and links this with case studies within the region to provide realistic examples of potential problems and possible approaches to their solution.

2. GLOBAL CLIMATE CHANGE AND ITS COASTAL IMPLICATIONS

The likely changes in most climate variables remain highly uncertain at the local scale relevant to impacts and possible responses (Table 1). However, for future sea levels at least the direction (a rise) and magnitude (faster than today) of change appears certain.

2.1. Sea-level rise

There is a widespread consensus that global sea levels have risen over the last century at a rate of 1–2.5 mm year^{-1}.[12] In addition to global changes local uplift or subsidence of the land surface must also be considered: their sum being termed relative sea-level change. This is by

TABLE 1

Some climate change factors relevant to coastal change and management (adapted from IPCC[12])

Factor	Direction of change	Comments	Potential impacts
Sea level	positive, accelerating	exacerbated by subsidence	numerous (see text)
Sea-water temperature	positive	—	increased algal blooms
Precipitation intensity	positive	—	increased flooding
Wave height	?	—	increased/decreased cross-shore erosion
Wave direction	?	—	increased/decreased longshore transport
Storm frequency	?	—	increased/decreased storm surge occurrence
Riverine sediment supply	?	also sensitive to catchment management	increased/decreased sediment supply to the coast

definition what an observer sees at any particular coastal site. As the sense and magnitude of vertical land movements vary from place to place, so relative rates of sea-level change show similar variability.

In the coming century accelerated global sea-level rise is expected, due to anthropogenic global warming. Given the uncertainties concerning global warming and ocean response, a global rise in sea level ranging from about 0.2–0.9 m by the year 2100 appears possible, with best estimates of a rise of about 0.5 m[12,13] or a two-and-a-half to five-fold acceleration. The large uncertainty concerning future sea levels must be taken into account when considering possible impacts and possible responses. The most serious impacts of sea-level rise are: (1) erosion; (2) inundation; (3) an increased risk of flooding and impeded drainage; (4) salinity intrusion into freshwater supplies; (5) higher water tables which may reduce the safety of foundations.[14] Erosion and inundation both result in land loss. It is important to note that these impacts of sea-level rise can be countered by other natural factors. For instance, a suitable sediment supply would counter beach erosion.

2.2. Other coastal implications of climate change

Many other climate change factors could have significant coastal implications (Table 1). However, the likely changes to most of these climate factors are less certain, with a possibility of both increase or decrease. This uncertainty hinders assessment of the implications of these possible changes and the development of appropriate proactive measures.

However, some climate change factors are more certain. Global climate model simulations indicate that the return period for heavy rainfall events may decrease in many parts of the world, given global warming.[15] This would intensify flooding, particularly in low-lying coastal areas where the base level will be simultaneously increasing due to sea-level rise. It suggests the need for an increased drainage capacity given global warming, particularly in coastal areas.[5] Sea water temperature is also expected to rise. Although adverse coastal implications are not well understood, it can be speculated that many existing problems will be enhanced, such as excessive algal growth. It is worth noting that it has been reported that the deep circulation of part of the eastern Mediterranean has been observed to change recently, although the precise cause remains uncertain.[16]

Climate change may also lead to a change in the frequency and intensity of storms.[17,18] In addition to wind damage coastal storms cause storm surges which flood low-lying coastal areas and allow destructive wave action to penetrate inland. However, it is difficult to construct plausible scenarios and the few studies that have considered this factor have conducted sensitivity analyses.[19] The possibility of an increase in storm frequency causes considerable concern, but a decrease in storm frequency and intensity is also possible with widespread benefits. The development of plausible regional scenarios of future storm characteristics remains an urgent requirement for coastal vulnerability assessment.

2.3. Vulnerability to climate change

To help to provide a better understanding of societal vulnerability to these changes the Intergovernmental Panel on Climate Change (IPCC) Common Methodology (CM) was developed.[20,21,3] The concept of vulnerability embraces: (1) the physical and socio-economic susceptibility to global climate change and (2) the ability to cope with these consequences (i.e. susceptible countries or areas may not be vulnerable). The CM has been applied in a number of countries around the world, including Egypt.[22]

Aggregation of the national results show that the likely impacts of sea-level rise vary from country to country and setting to setting.[5] Certain geomorphic settings are more vulnerable than others, particularly deltas and small islands, most particularly low-lying coral atolls. Coastal wetlands appear to be threatened with loss or significant change in most locations as their present location is intimately linked with present sea level, although their ability to respond dynamically to such changes by sedimentation and biomass production needs to be carefully considered.[23] Developed sandy coasts may also be vulnerable if development is concentrated too close to the shoreline, primarily due to the high costs of maintaining a sandy beach for both recreation and protective purposes.[24] These costs are often highly uncertain.

The CM has also been applied to a few limited factors at a global level—the Global Vulnerability Analysis (GVA)—with the objective of obtaining regional and global results.[25,26] The factors considered include flooding due to storm surge, wetland loss and potential protection costs against flooding. The GVA is a first-order analysis and the global datasets which were available were somewhat limited and a number of assumptions were necessary. Therefore, while national results are provided in appendices only aggregated regional or global scale results are expected to be valid. Overall, considering the world as 20 regions, four appear most vulnerable: the southern Mediterranean, Southeast Asia, Asia Indian Ocean Coast and Africa Indian Ocean coast.[25]

3. THE MEDITERRANEAN

The Mediterranean (Fig. 1) is an enclosed sea characterised by a limited tidal range (often less than 30 cm). It has a long coastline of over 46 000 km, with 75% located in four countries—Greece, former Yugoslavia, Italy and Turkey—and 40% of the coastline located on islands,

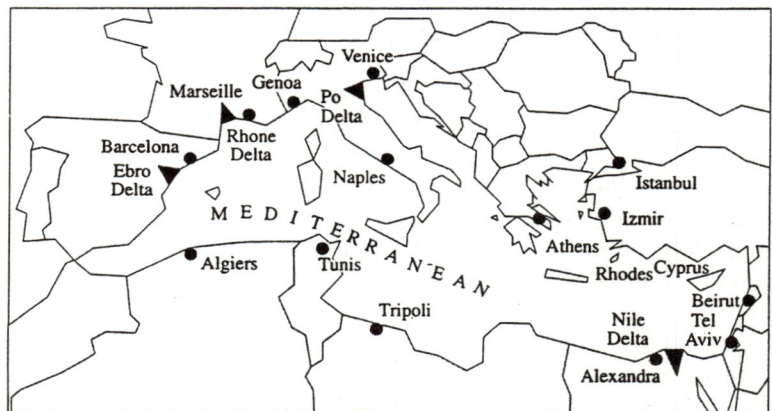

Fig. 1. The Mediterranean, including coastal cities with a population exceeding one million people in 1990, Venice and the major deltas.

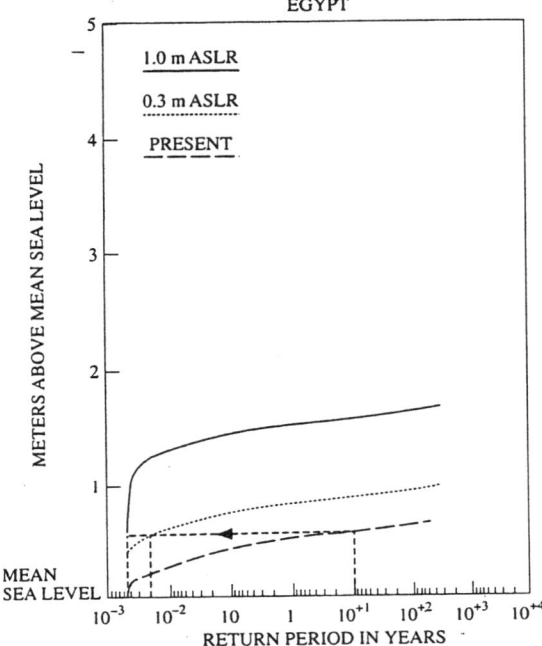

Fig. 2. The flood exceedance curve for the Nile delta, which typifies such curves for the Mediterranean.

mainly in Greece, former Yugoslavia and Italy.[27] About 54% of the coast is rocky and 46% sedimentary: the latter areas are most vulnerable to climate change. Extratropical storms occur in the autumn, winter and spring, and these can cause significant waves and surges given appropriate conditions. Due to the limited tidal range the flood exceedance curves have a low gradient, so a small rise in sea level significantly reduces the return period of a given elevation flood (e.g. Fig. 2). There is a large population located in the coastal zone, exceeding 130 million in 1985, with a number of large cities such as Barcelona, Athens, Istanbul and Tripoli.[27] Many of the cities are growing rapidly. The coastline is also intensively utilized for coastal tourism, with 100 million tourists in 1984 rising to a projected 173–341 million tourists in 2025. This means that significant amounts of tourist infrastructure already exist, or will be built, immediately adjacent to the coast.

The Mediterranean occupies an active geological plate boundary, therefore land uplift or subsidence can be expected to be widespread. Flemming[28] found, using archaeological evidence, that uplift appears to

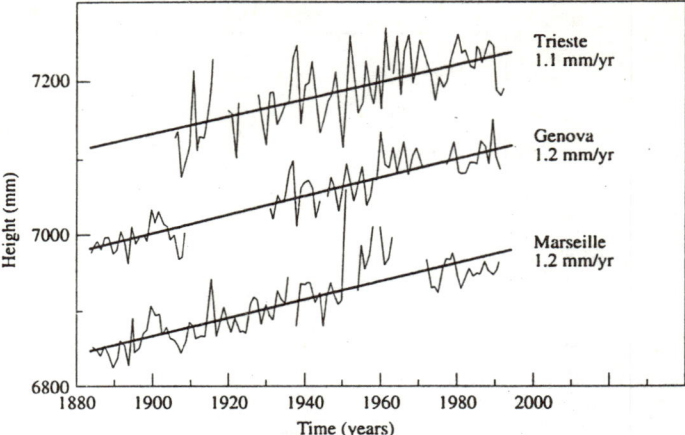

Fig. 3.　Relative sea-level rise data from the Mediterranean for stations exceeding a 50 year duration, including the linear trend. Height is arbitrary. Data provided by the Permanent Service for Mean Sea Level, Bidston.

dominate in the region and hence relative sea-level rise is typically less than global trends, with important exceptions such as subsiding deltaic areas. Tide gauge records longer than 50 years are required to produce robust statistics on rates of sea-level change,[29] which makes some earlier analyses of sea-level trends around the Mediterranean suspect.[30,31] Three suitable records are available from the northwest Mediterranean and show a rise of 1.1–1.2 mm year^{-1} (Fig. 3). This trend in sea-level rise is consistent with IPCC[12] and demonstrates that slow rates of sea-level rise are already a problem within the Mediterranean.

A number of studies suggest that the coastal Mediterranean is vulnerable to a number of aspects of projected climate change, particularly accelerated sea-level rise.[6] The results from the global aggregation provide further insight into the likely magnitude of these problems. While the Mediterranean has no coral atolls, it contains numerous islands and is fringed by a number of deltaic plains, including the Rhone, Po, Ebro and, most particularly, the Nile delta. However, there are few national studies of the vulnerability of this coastline to different scenarios of sea-level rise and climate change in general. The main exception is the Nile delta and environs which has attracted considerable interest given its large size, existing environmental problems, high vulnerability to sea-level rise and its strategic importance to Egypt.[6,7,10,32–36] Similarly, little quantitative regional data is available, except for the GVA. The following discussion includes new calculations using the GVA model, including some impacts for a 0.5 m sea-level rise scenario.

The GVA divides the Mediterranean into a northern (including the Black Sea) and southern region, with a boundary at Istanbul.[1] Therefore, these regions correspond to the European Mediterranean and the western Asian/North African Mediterranean, respectively. Some results are summarized in Table 2. The population living beneath the 1 in 1000 year storm surge level totalled nearly 10 million people in 1990, although many of these people are already protected from flooding by structural measures. Given scenarios of a 0.5-m and a 1-m rise in sea level and no other changes, this population will increase by about 35% and 70% respectively, on both sides of the Mediterranean (i.e. a rise imposed on today's situation). Population growth in the northern Mediterranean is not expected to be significant. In contrast, the southern Mediterranean is expected to experience substantial population growth, increasing the population within the flood zone without any rise in sea level. Taking population projections for 2020 with no rise and a 1-m rise in sea level would produce a cumulative increase of about 65% and 175%, respectively, in the population living beneath the 1,000 year surge compared to 1990. Continued population growth beyond 2020 is expected to further increase the population living in the potential flood zone for the southern Mediterranean, but this has not been quantitatively evaluated.

Based on an estimate of the probability of flooding, one million people presently experience flooding around the Mediterranean in a typical year, nearly all of them living in the southern Mediterranean. People around the northern Mediterranean are generally well protected from flooding with structural measures. However, as sea level rises, so these defences will become less effective and the risk of flooding will increase (Fig. 2). Making first-order estimates of the decline in the level of protection as sea levels rise and the expansion of the hazard zone suggests that the number of people experiencing flooding in a typical year will more than double given a 0.5-m rise and increase more than six times given a 1-m rise, without considering population growth. Globally the corresponding increases are two- and three-fold, respectively,[26] illustrating that the Mediterranean is particularly vulnerable to such changes. The biggest changes would be in the northern Mediterranean, where the increase given a 1-m rise would be more than 100 times present values.

These results suggest that even small rises in sea level cannot be ignored around the Mediterranean and coastal populations in low-lying areas will face a choice between coastal abandonment and

TABLE 2

Some regional consequences of sea-level rise for the Mediterranean region (SLR—global sea-level rise scenario). Risk zone is the area beneath the 1000 year storm surge (see also Hoozemans et al.[25]

Time (years)	Risk zone population (millions) by year and SLR						Protection costs against a 1-m SLR (see text)		Wetlands at loss given a 1-m SLR (see text)	
	1990	2020	1990	2020	1990	2020	US $ (billions)	% GNP (1990)	km²	% Total
SLR (m)	0	0	0.5	0.5	1.0	1.0				
Northern Mediterranean	4.1	4.1	5.7	5.6	7.2	7.1	25.5	2	200	100
Southern Mediterranean	5.6	9.1	7.5	12.2	9.4	15.4	18.1	7	2 600	100
Total	9.7	13.2	13.2	17.8	16.6	22.5	43.6	—	2 800	—

building/upgrading coastal defences. The cost of protection can be evaluated to see how this compares with regional wealth. Assuming that all areas with a present population exceeding 10 people km^{-2} are protected against a 1-m rise in sea level using standard coastal engineering techniques, the absolute costs are higher for the northern Mediterranean, but the costs relative to 1990 Gross National Product (GNP) are significantly higher for the southern Mediterranean. This response would reduce the population which experiences flooding in a typical year below present levels to about 300 000 people (based on 1990 population).

Lastly, the coastal wetlands in the Mediterranean appear highly vulnerable to loss, largely due to the negligible tidal range which restricts their vertical range and ability to accrete vertically in response to sea-level rise.[8,37] Total loss as shown in Table 2 is unlikely, but near total losses should be expected assuming business as usual and a 1-m rise in sea level, with a number of adverse consequences such as a decline in fisheries and coastal biodiversity. Improved management of existing wetlands and allowance for sea-level rise in coastal policy could help to conserve some of these threatened wetlands—improved deltaic management is one area which is being evaluated (see Section 5).[9]

In conclusion, the GVA shows that a rise in sea level could cause some important and often adverse changes to the coast of the Mediterranean. Overall, the southern Mediterranean coast appears more vulnerable to sea-level rise than the northern Mediterranean coast. Some more detailed case studies which consider cities, deltas and islands in the Mediterranean are now examined to better understand these problems.

4. COASTAL CITIES

The protection of cities is expected to be a major cost of accelerated sea-level rise.[38] It would also appear to be one of the more likely responses given the high value of many city areas, although many uncertainties exist.[39] There are a number of historic cities around the Mediterranean such as Venice and 13 coastal cities had a population exceeding one million people in 1990 (shown in Fig. 1), collectively containing over 35 million people.[40] The population of the cities in Europe is growing slowly, with projected rates of increase in the range 3–12% from 1990 to 2010. However, substantial and continuing investment in new and upgraded infrastructure is expected. The cities of western Asia (including Istanbul) and North Africa are growing more rapidly, with projected rates of increase in the range 43%–106% from 1990 to 2010. This means that tremendous investment is required just to stand still.[41] Istanbul is projected to become the first megacity (population >8 million) in the Mediterranean before 2010, with other likely coastal megacities by 2050 being Alexandria, Tripoli and Algiers. Therefore, urban centres around the Mediterranean are likely to place rapidly increasing demands on coastal resources. This is already apparent around many cities in terms of declining water quality and pollution, which are often triggers for improved coastal management efforts, e.g. Izmir.[42]

The likely rate of climate change is much slower than many of the changes occurring in coastal cities, although if poorly managed the impacts of climate change could be serious.[43] Therefore, it is more obvious than in most other settings that climate change will be interacting with a changing landscape. This makes assessment of likely impacts somewhat difficult, but also presents opportunities. When considering proactive response measures for climate change it will be vital to manage other long-term changes to the urban landscape such that vulnerability to climate change is not enhanced and ideally is reduced.

4.1. Venice

One of the most famous and historic cities in the Mediterranean, Venice has always been intimately linked to the sea, being located close to sea level on islands in the Venice lagoon. Therefore, flooding by storm surges has always been a problem. The occurrence of flooding has been exacerbated this century by a relative rise in sea level of about 28 cm:[44] flood frequency in San Marco Square increased from 7 times per year to 40 times per year from 1900 to 1990.[45] These repetitive floods are destroying the fabric of the city, as well as being one contributing factor to population decline in the city. The relative sea-level rise occured due to a combination of subsidence due to groundwater withdrawal[46,47] and global sea-level rise. Given that global sea-level rise has been in the range 10–25 cm century^{-1},[12] global sea-level rise rather than subsidence may be the major contribution to the observed increases in flood frequency this century.

A continued and more rapid rise in global sea levels, as projected, will progressively increase the flood frequency; a 30-cm rise in relative sea level will lead to flooding in San Marco Square 360 times per year (i.e. daily) without other changes.[45] If Venice is to survive some action is essential. Various proposals exist, with flood barrages to separate the lagoon from the sea being most favoured.[44] Other approaches have been suggested, although there appears to be no perfect solution with a clear conflict between the needs of navigation within the lagoon versus flood prevention for Venice.[47–49] Whatever is ultimately decided, it needs to be a long-term solution which mandates considering future rates of sea-level rise as one of the key factors in design.

While no other city on the Mediterranean is as vulnerable to small changes in sea level as Venice, it illustrates that small relative increases in sea level (≤30 cm) can have serious consequences in terms of flood frequency (cf. Fig. 2).

4.2. Alexandria

Alexandria is another historical city on the western edge of the Nile delta. It has a rapidly growing population and stressed environment.[41] The original city developed on a series of shore-parallel ridges up to 12 m above sea level. Human influence on the region has increased and the lagoon south of Alexandria has been largely reclaimed for agriculture, leaving Lake Maiout.[50,51] Lake Maiout is maintained at 2.8 m beneath sea level by continuous pumping to stop these areas being inundated. The prospect of sea-level rise suggested the inundation of this low-lying land leaving Alexandria to the north as a peninsula or an island.[52]

Subsidence in this area appears to be minimal and global sea-level rise scenarios can be applied directly as relative sea-level rise scenarios.[10] Based on a Geographic Information System (GIS) analysis of Alexandria Governorate (the geopolitical unit in which Alexandria is situated), El-Raey et al.[10] estimate that 700 km² of land (35%) is below sea level, with a resident population of 1.2 million (or 37% of the Governorate population) and 50% of industry (by area occupied). Based on the existing population, a 0.5-m rise in sea level would place a further 400 000 people beneath sea level, increasing to 800 000 people given a 1.0-m rise. The low-lying areas are also under development pressure given the lack of alternative sites, so future population and economic growth may tend to occur in the more vulnerable lower-lying locations, substantially increasing the vulnerable population.

Alexandria has a series of pocket beaches which are threatened with near total loss given a 0.5-m rise in sea level.[10] These beaches have a high recreational value and six beaches are already nourished (with land-based sand sources). Therefore, it seems likely that nourishment would be an effective response against sea-level rise, although other factors such as water quality and pollution also need to be considered.

Therefore, proactive responses appear prudent for Alexandria as even small rises in sea level could adversely affect the city. Lake Maiout does not have a direct link to the sea and initially higher sea levels will only increase seepage and decrease drainage capacity. Anecdotal reports of increased waterlogging in low-lying areas may be a manifestation of the existing slow rise in sea level at Alexandria.[10] Failure of the seawall, which allowed the lagoons to be reclaimed, would allow catastrophic flooding of significant areas in Alexandria Governorate and neighbouring areas. Therefore, it would be prudent to evaluate the existing levels of safety for both the drainage system and seawall, and how they would decline with rising sea levels. The possibility of land use planning should also be considered to encourage growth away from the most low-lying (i.e. vulnerable) areas. However, given the present development pressures and rapid increase in population this goal is probably difficult to achieve. All efforts to combat climate change need to be integrated with solutions to existing problems (see Hazma[41]).

4.3. Istanbul

Unlike Alexandria and Venice, Istanbul is not a low-lying city and land loss/increased flooding given sea-level rise are unlikely to be major problems.[53] However, sea-level rise does have implications for water supply and the proper functioning of sewage supply systems. It would be prudent to consider the implications of sea-level rise over the design life of any new systems or upgrades.

4.4. Discussion

These examples demonstrate that global sea-level rise is a concern in coastal cities. While immediate problems require urgent action, their solution should be consistent with longer-term policies to reduce vulnerability to climate change (e.g. UNEP[1]). Land use planning to direct development away from hazardous zones offers significant benefits in rapidly growing cities, but it is these settings where such planning can be most difficult to implement. Further discussion concerning coastal cities and climate change will be found in Nicholls.[43]

5. DELTAS

Deltaic areas are inherently susceptible to sea-level rise, being low-lying plains formed by the deposition of alluvial sediments.[5] Further, deltas naturally subside to varying degrees and hence experience relative sea-level rise without any global rise. Only sedimentation and local biomass production can compensate and hence maintain the deltaic area, but human management of many deltas and their catchments are often reducing this capacity by reducing the availability of sediment and modifying wetland hydrology. This is well-illustrated by the Mississippi delta.[54]

There are at least 31 alluvial–deltaic areas around the Mediterranean[49,55] including the Nile, Po, Rhone and Ebro deltas. These four deltas are experiencing similar problems to the Mississippi delta[8,9,33] and more widely most alluvial–deltaic areas are experiencing a decline in sediment input.[55] Therefore, any global rise in sea level is exacerbating a range of existing problems within deltaic areas which can be related to human management.

5.1. Egypt and the Nile delta

The Nile delta is by far the largest delta within the Mediterranean occupying over 20 000 km². Since ancient times the fortunes of Egypt has been linked to the Nile floods and the crops which this allowed to be grown on the delta. However, during this century increasing regulation of Nile, culminating in the construction of the Aswan high dam in 1964, has completely decoupled the delta from the river system which formed it.[33] Significant shoreline recession has been occurring throughout this century,[32] although this recession is not universal. Based on geological evidence high rates of subsidence of up to 5 mm year⁻¹ are occurring in the northwestern part of the delta.[56]

Together with the Ganges–Brahmaputra delta in Bagladesh, the Nile delta was recognised as highly susceptible to sea-level rise in one of the first vulnerability assessments.[7,32] Given a 1-m rise in sea level these early studies estimated that 8 million people would be displaced from their homes (1985 population). Compounding factors included population growth and socioeconomic developments, which are relocating away from the Nile valley towards the coast,[57] and the possibility of excessive groundwater withdrawals raising the rates of subsidence. These early studies drew attention to the fundamental coastal management problems of deltas and recognised the human role in enhancing or reducing the magnitude of the problems. More recent studies all confirm that planning for sea-level rise must form an important element of coastal management in the Nile delta.

Stanley and Warne[33] analyzed geological/geomorphological changes of the delta evolution over the last few millennia and projected them until 2050. The influence of human activity on the evolution of the delta has been increasing throughout the study period and this trend is expected to continue. Erosion and reshaping of the sandy delta fringe will continue, but with localised accretion. The area of the four remaining lagoons will continue to decline in area due to shoreline recession to the north and reclamation on the inland shores. Combined with increasing levels of pollution, this will lead to loss of fisheries and wildlife areas. The agricultural lower delta plain will also be impacted by relative sea-level rise, with falling crop yields and other adverse impacts.

Delft Hydraulics *et al.*[34] present a rather different analysis based on the IPCC Common Methodology. While this was a national study most results pertain to the Nile delta, including Alexandria (see Section 4.2). Assuming a 1-m rise in sea level by 2100 they came to a number of important conclusions. Firstly, assuming no responses:

- enormous capital values could be lost due to shoreline retreat and, more particularly, increased flooding, while much coastal infrastructure would cease to function as designed;
- about 4.7 million people (1990 population) would be displaced from their homes;
- salt water intrusion and reduced drainage capacity present significant additional problems.

Secondly, assuming appropriate responses are implemented:

- to protect the capital values and people identified above, a national response strategy is necessary, as distinct from a series of local response strategies;
- the problems caused by salt water intrusion are generally insensitive to the possible response measures. The impacts might be alleviated by agricultural responses and this requires further evaluation.
- the costs of measures are large in relation to Egypt's present earning capacity;
- there is a need for institutional and technical capacity building to facilitate planning for climate change and effective and integrated coastal management in general.

Strzepek *et al.*[35] made an integrated study of Egypt using an agricultural economic model to examine the combined effects of changes in crop yields, water supply and arable land on Egypt's agricultural economy. General Circulation Model results were used to develop scenarios for temperature and flow in the Nile river, while a 1-m rise in global sea level was assumed for 2100. Under one scenario the availability of water from the Nile falls significantly and the loss of agricultural areas in the delta due to sea-level rise is of less relevance, as no water would be available for irrigation. This analysis reinforces the importance of the conclusion of Delft Hydraulics *et al.*[34] concerning saltwater intrusion.

Therefore, the Nile delta is highly vulnerable to sea-level rise. Some form of structural solution appears to be the only feasible long-term response. This will require long-term planning. One planning issue that needs to be raised at a strategic level is the continued construction of new development on the coastal fringe and progressive and piecemeal lagoon reclamation.[57]

5.2. Ebro, Rhone and Po deltas

These three deltas are in the northwestern Mediterranean and are being studied as part of the European Union-funded MEDDELT (Impact of Climate Change on Northwestern Mediterranean Deltas) project[9] (see also pp. 737–910 in Ozhan[58]). All three deltas are two orders of magnitude smaller than the Nile delta,[6,58] although both the Po and the Rhone are part of larger areas of coastal lowlands.[51,59] All the deltas have been extensively modified for agricultural and other purposes, and at the same time they are all receiving greatly diminished sediment supplies due to human-induced changes within the catchment. The greatest reduction is in the Ebro delta, which only receives 4% of its former sediment supply.[9] In the Po delta reclamation and fluid withdrawal has lead to significant subsidence in excess of 2 m in some areas and, more broadly, in the northeastern Italian coastal plain over 2300 km² of land is already below sea level and protected from inundation by dikes.[47] In the Ebro delta rice cultivation is important and some of the rice fields are already at or slightly below sea level. In the Rhone delta significant areas of wetland appear vulnerable to inundation.[3] Therefore, continuation with present policies combined with sea-level rise will cause progressive changes and ultimate loss of land, or alternatively require polderisation similar to that already common in the Po delta.

The losses that would result from coastal abandonment are less quantified than in the case of the Nile, but comprise mainly agricultural and aquacultural areas, as well as natural areas of importance to wildlife. Given that some sediment transport is still occurring to these deltas, it is possible to take a natural systems engineering approach. Simply put, it should be possible to maintain some, or possibly all of the existing delta area by managing natural processes.[9,60] Using the self-regulation of natural systems many of the functions of the delta might be maintained into the foreseeable future at minimal cost. Changing catchment management to release more sediment to the deltas, where possible, would further enhance this capability. However, this type of management requires a fundamental change in philosophy, as natural processes must be allowed to occur rather than be excluded or tamed; also, it requires more scientific knowledge of the deltaic processes. In this area the MEDDELT project will provide important new information and management tools.[61] Similar management approaches are being investigated in the context of preserving the Mississippi delta,[54,62] and such approaches might find wide application in response to climate change.

5.3. Discussion

There is some awareness of the possible impacts of climate change in deltaic settings, based largely on the rapid present changes which are already occurring. The possibility of 'working with nature' to counter the impacts of sea-level rise exists in many deltaic settings, including the Ebro, Po and Rhone deltas. However, in the Nile delta the lack of sediment makes this possibility appear unrealistic and diking of the delta appears more likely.

6. ISLANDS

There are 162 islands in the Mediterranean exceeding 10 km^2, with a further 4000 smaller islands, found particularly in the eastern Mediterranean. These areas have a low resource base and have often welcomed significant tourist development.[63] However, this can lead to serious problems and trigger coastal management initiatives.[1] Therefore, the potential problems of climate change have received less attention than deltas. However, they could be serious in a number of ways, including beach erosion (undermining the tourist industry) and more speculatively, a potential decline in rainfall[64] which could cause serious problems for water supply in settings where water shortages already occur.

6.1. Cyprus

Cyprus is the third largest island in the Mediterranean, with a total coastal length of some 735 km, of which about 365 km has been under Turkish occupation since 1974. The 1990 population was about 700 000.

Like many of the islands in the Mediterranean the Cypriot economy increasingly depends on coastal tourism and as a result there is considerable and growing pressure to utilize and exploit the coastal zone.[65] Six main coastal management problems can be recognized: (1) coastal erosion; (2) sea grass and other debris on the beach; (3) growth of algae close to the beach; (4) tourist pressure on the coastal environment; (5) degradation of marine ecosystems; and (6) water pollution.

The shortage of sandy beaches suitable for tourism, together with the widespread occurence of erosion, has led to the construction of many small scale groynes and breakwaters, especially along the popular south coast. However, it is now more widely questioned if the growing number of shore-parallel (detached) breakwaters along many of the newly developed tourist beaches are always an ideal solution to these problems.

6.1.1. *Climate change as a coastal management issue*

Climate change is not presently perceived as a coastal management problem in Cyprus. As a considerable length of the Cypriot coastline is already subject to erosion, and this problem is likely to be exacerbated by climate change, future erosional trends will be considered here to focus on the relative importance of climate change. The major causes of the existing erosion problems are believed to be: (1) the construction of river dams, removing the supply of sediment and (2) extensive beach mining, although other factors may also contribute. Three impacts of climate change (Table 1) might change erosion rates: (1) accelerated sea-level rise; (2) changes in wave characteristics (e.g. increase in storm frequency); and (3) changes in river discharges.

Based on archaeological data,[28] Cyprus appears to be experiencing long-term uplift of between 0 and 1 mm year^{-1}. This uplift will counteract global sea-level rise and given a global rise in sea level of 0.5 m by 2100, relative sea-level rise in Cyprus will be in the range 0.4–0.5 m. This is a significant change which needs to be taken into account when planning sustainable solutions for coastal zone problems. Any rise in sea level is expected to cause beach erosion, unless a supply of sediment is available. Based on a rule of thumb developed from the Bruun Rule, a first-order estimate of the recession is 100–200 times the rise in sea level,[66] or 40–100 m in this case. This estimate takes no account of longshore sediment transport, or other sediment inputs or losses. Therefore, areas which are already eroding will probably erode substantially more than these estimates.

Coastal erosion is also conditioned by wave processes. Given increasing storm frequencies, greater cross-shore sediment transport will occur, generally removing sediment from the upper beach or dunes to the shoreface (>4–5-m depths).[67] This would exacerbate the problems of beach erosion. Conversely, reduced storm frequencies would help to counter the likely erosion due to sea-level rise.

Changes in the amounts and pattern of river run-off are difficult to predict. However, in the case of Cyprus changes in run-off are of little relevance as most rivers (and their sediment yield) have already been separated from the coast by reservoir construction. The net result is a pervasive problem of coastal erosion which is already well apparent.

Therefore, continued erosion is to be expected without any changes in climate. It will be exacerbated by a rise in sea level and this will be conditioned by the uncertain changes in wave climate.

6.1.2. *Possible responses*

Based on these preliminary estimates continued and more rapid beach erosion appears likely, which would have important implications for the long-term sustainability of the tourist industry. A key policy goal would seem to be to maintain the limited beach area and hence to sustain the vital tourist industry. This can be accomplished in a number of distinct ways. On wide sedimentary coastal plains erosion will not lead to beach loss unless a hard structure is built, fixing the position of the shoreline. Therefore, building setbacks for new construction which allow natural dynamics to occur seaward may be appropriate in suitable locations. The precise size of the setback would require more evaluation. In areas of pocket beaches, or locations where existing development is close to the shoreline, protection will probably be appropriate, including beach nourishment. The availability of suitable sand resources may be problematic, but given the large benefits of maintaining the beaches, sources external to Cyprus could be considered if this was necessary. Therefore, with regard to the three potential strategies available to coastal societies (planned retreat, accommodation, or protection[5]), a strategy of protection combined with planned retreat (via setbacks) seems most appropriate. It is likely that the issues raised here will be generic to many of the islands in the Mediterranean.

When considering possible responses institutional considerations are also critical. In Cyprus interest in integrated coastal zone management (ICZM) has been triggered as a process to solve present and near future problems and it is receiving widespread support in this regard.[65,68] Therefore, bringing longer-term issues such as climate change into the Cypriot ICZM process requires reappraisal, including answering the key question: 'What do the Cypriots want to do with their coast in the long-term?' There are many different users (tourism, wetlands, industry, military) with different intentions, with different temporal and spatial claims, with different responsibilities, and hence the potential for a range of conflicts. To facilitate the answering of this central question in an orderly and practical way, two essential requirements should be met:

- the user groups and their representatives should be identified;
- the requirements of each user group should be identified.

These user requirements need to be checked with respect to internal conflicts, national policies and environmental impacts (as triggered by climate change). Then, the necessary institutional and technical infrastructure associated with the users' requirements needs to be defined in a conceptual stage to roughly assess feasibility, cost, performance and side-effects in a preliminary stage. The implications of the inherent uncertainties concerning climate change need to be carefully evaluated. In general, it seems prudent to only implement low-cost strategies at present, such as setbacks in appropriate locations. Other low-cost strategies include allowing for likely rates of sea-level rise when designing breakwaters etc. Most fundamentally the ICZM process should expect the coast to change and be prepared for likely changes in whatever form they might take.

7. DISCUSSION

While this review is not exhaustive, it illustrates that apart from deltaic areas, there is little active concern about sea-level rise and climate change around the Mediterranean. In terms of coastal management immediate problems such as declining water quality and tourist development have been the triggers for coastal management. Within deltaic areas and coastal lowlands generally, interest in sea level and climate change can also be related to existing problems of failing sediment supply and subsidence.

Tide gauge measurements show that relative sea-level rise is already occurring in the Mediterranean (Fig. 3) and in the next century this rise is expected to accelerate. Global rise in sea level by 2100 has a large uncertainty, with a 1-m rise being the likely maximum. Any rise in sea level will have adverse impacts: those impacts being a function of the magnitude of the rise and the human response to that rise. A well-planned response will tend to minimize impacts and *vice versa*. Other implications of climate change are less certain, but many adverse impacts could occur. It is also apparent that climate change will generally exacerbate existing problems, rather than create fundamentally new ones.[6] Therefore, solving existing problems generally increases flexibility in the face of climate change. However, there is also a need for an explicit long-term (50+ years) perspective as part of coastal management and to be most effective, coastal planning will have to simultaneously address the full range of short-term, medium-term and long-term coastal problems. An appropriate institutional mechanism for such planning is integrated coastal zone management. All signatures to Agenda 21 have endorsed this institutional mechanism. Existing regional guidelines for coastal management also address climate change.[1]

An important first step towards incorporating climate change in ICZM would be further local to national scale studies to better evaluate likely impacts and the full range of possible adaptation options. These studies will need to be more comprehensive than many earlier studies and include all pertinent climate change factors, combined with other scenarios, such as population and socio-economic changes. The Common Methodology has formed the basis of many vulnerability assessments, but experience has raised a number of problems and deficiencies when it is applied.[3,5,18] Rather than continue with the Common Methodology this suggests the need for further development of methodologies for vulnerability assessment which are more tailored to local needs. For instance, a methodology designed to assess the vulnerability of Mediterranean deltas is being developed.[9,8,61] The general IPCC guidelines for assessing climate change impacts and adaptations[69] may be useful in guiding such developments.

In the context of the Mediterranean particular attention should be focused on the most vulnerable sectors and areas such as the deltas, developed sandy beaches and coastal wetlands. In addition, a more proactive approach to coastal planning which recognises the possibility of climate change and sea-level rise needs to be fostered. This could take advantage of the opportunity to influence new developments, as well as renewals and redevelopments, such that they are more resilient in the face of climate change. Given the large degree of uncertainty these measures need to be low-cost and/or flexible, such that they will be effective for the full range of likely scenarios. Suitable adaptation options are often limited and have long lead times, so they should be identified and evaluated without delay.[5] In cases where climate change could have serious consequences, additional investment can be seen as a prudent 'insurance' policy. Public consultation and a shared view of the desirable and permitted coastal uses is also important, otherwise technically sound approaches may fail due to the lack of public support. The institutional management structure needs to be considered such that it can evolve and adapt (e.g. National Research Council[70]). Being able to learn from experience is important, as any management plan is inevitably based on incomplete knowledge and can be objectively viewed as an educated experiment. Therefore, constant evaluation and possible tuning and/or redesign of the management efforts are required. This institutional arrangement can also take account of changing user requirements for the coastal zone as they inevitably occur.

Tourist developments on sandy beaches provide one simple example, as already discussed in the context of Cyprus. New tourist resorts could locate all long-life infrastructure a short distance (say 100 m) inland from the coast and only place easily moved structures near the beach. This would maintain a choice between protection (often beach nourishment) and a planned retreat if the beach erodes less than the setback. Tourist reaction to such a policy is poorly understood, but critical to its acceptance to resort operators. In pocket beach settings, which are common in the Mediterranean, erosion may lead to unacceptable beach decline or loss.[10] Therefore, the effectiveness of other measures such as engineering structures and the availability of sand for beach nourishment needs to be assessed. Given the widespread shortage of sand in many parts of the Mediterranean importing external sources of sand may be necessary.

Actual response costs to climate change remain highly uncertain, even for a defined set of climate scenarios, and the determination of optimal responses at sub-national to national levels remains a key research question. The best available analysis is for the Netherlands,[19,71] a country where protection is the only possible response. In most countries national responses to climate change will be more variable and comprise some combination of planned retreat, accommodation and protection.[72,73]

8. CONCLUSIONS

While there is awareness of climate change around the Mediterranean, it is receiving little practical concern except in deltaic areas, where failing sediment supply and subsidence are already problems. While these areas are some of the most threatened given climate change, all coastal activities could be adversely affected. Existing efforts at ICZM are targeted at more immediate problems. These existing management efforts need to be expanded to include longer-term issues, including the implications of climate change. A first step would be more comprehensive vulnerability assessments of some of the more vulnerable settings (deltas, small islands, wetlands, etc.). Given the large uncertainties any measures should be low-cost and highly flexible, such that they are effective for the full range of likely scenarios. In some cases where climate change could cause serious problems aditional investment to counter possible climate change can be seen as a prudent 'insurance policy'. Institutionally the ICZM process needs to be adaptive in nature, so that experience and changing user requirements can be addressed in an evolutionary manner.

REFERENCES

1. UNEP, *Guidelines for Integrated Management of Coastal and Marine Areas—With Special Reference to the Mediterranean Basin*. UNEP Regional Seas Reports and Studies No. 161, Split, Croatia, PAP/RAC (MAP-UNEP), 1995, 80pp.
2. Holligan, P. M. & de Boois, H., ed., *Land–Ocean Interactions in the Coastal zone (LOICZ): Science Plan*. IGBP Report No. 25, International Geosphere–Biosphere Programme, Stockholm, 1993.
3. World Coast Conference, *Preparing to Meet the Coastal Challenges of the 21st Century*, World Coast Conference Report, Noordwijk, November 1993, Rijkswaterstaat, The Hague.
4. Turner, R. K., Subak, S. & Adger, W. N., Pressures, trends and impacts in the coastal zones: interactions between socio-economic and natural systems. *Environmental Management*, **20**, 159–173.
5. Biljsma, L., Ehler, C. N., Klein, R. J. T., Kulshrestha, S. M., McLean, R. F., Mimura, N., Nicholls, R. J., Nurse, L. A., Nieto, H. P., Stakhiv, E. Z., Turner, R. K. & Warrick, R. A., Coastal zones and small islands. In *Impacts, Adaptations and Mitigation of Climate Change: Scientific–Technical Analyses*, ed. R. T. Watson, M. C. Zinyowera & R. H. Moss. Cambridge University Press, Cambridge, 1996, pp. 289–324.
6. Jeftic, L., Milliman, J. D. & Sestini, G., ed., *Climate Change and the Mediterranean*. Edward Arnold, London, 1992, 673pp.
7. Broadus, J., Milliman, J., Edwards, S., Aubrey, D. & Gable, F., Rising sea level and damming of rivers: possible effects in Egypt and Bangladesh. *Effects of Changes in Stratospheric Ozone and Global Climate, Volume 4, Sea Level Rise*. UNEP and US EPA, Washington DC, 1986, pp. 165–189.
8. Day, J. W., Pont, D., Ibanez, C. & Hensel, P. F., Impacts of sea level rise on deltas in the Gulf of Mexico and the Mediterranean: human activities and sustainable management. In *Consequences for Hydrology and Water Management*, UNESCO International Workshop Seachange 1993, Noordwijerhout, 19-23 April 1993, Ministry of Transport, Public Works and Water Mangement, The Hague, 1994, pp. 151–181.
9. Sanchez-Arcilla, A., Jimenez, J.A., Stive, M. J. F., Ibanez, C., Pratt, N.,

128 *R. J. Nicholls, F. M. J. Hoozemans*

Day Jr, J. W. & Capobianco, M., Impacts of sea-level rise on the Ebro Delta: a first approach. *Ocean & Coastal Management*, **30(2-3)** (1996).

10. El-Raey, M., Nasr, S., Frihy, O., Desouki, S. & Dewidar, Kh., Potential impacts of accelerated sea-level rise on Alexandria Governorate, Egypt. *Journal of Coastal Research*, **Special Issue No. 14** (1995) 190–204.

11. Nicholls, R. J. & Hoozemans, F. M. J., Vulnerability to sea-level rise with reference to the Mediterranean region. In *Proceedings of the Second International Conference on the Mediterranean Coastal Environment, MEDCOAST 95*, ed. E. Ozhan. MEDCOAST secretariat, Middle East Technical University, Ankara, Turkey, 1995, pp. 1199-1213.

12. Warrick, R. A., Oerlemans, J., Woodworth, P. L., Meier, M. F. & le Provost, C., Changes in sea level. In *Climate Change 1995: The Science of Climate Change*, ed. J. T. Houghton, L. G. Meira Filho, B. A Callander, N. Harris, A. Kattenberg & K. Maskell. Cambridge University Press, Cambridge, 1996 pp. 359–405.

13. Wigley, T. M. L. & Raper, S. C. B., Implications for climate and sea level of revised IPCC emissions scenarios. *Nature*, **357** (1992) 293–300.

14. National Research Council, *Responding to Changes in Sea Level: Engineering Implications*. National Academy Press, Washington, DC, 1987.

15. Gordon, H. B., Whetton, P. H., Pittock, A. B., Fowler, A. M. & Haylock, M. R., Simulated changes in daily rainfall intensity due to enhanced greenhouse effect: implications for extreme rainfall events. *Climatic Change*, **8** (1992) 83–102.

16. Mackenzie, D., Ocean flip puts modellers on Med alert. *New Scientist*, **147**, No. 1993 (1995) 8.

17. Mitchell, J. K. & Ericksen, J. J., Effects of climate change on weather-related disasters. In *Confronting Climate Change: Risks, Implications and Responses*, ed. I. M. Mintzer. Cambridge University Press, Cambridge, 1992, pp. 141-151.

18. McLean, R. & Mimura, N., ed., *Vulnerability Assessment to Sea-level Rise and Coastal Zone Management*. Proceedings, IPCC/WCC 1993 Eastern Hemisphere Preparatory Workshop, Tsukuba, August 1993. Department of Environment, Sport and Territories, Canberra, 1993.

19. Peerbolte, E. B., de Ronde, J. G., de Vrees, L. P. M., Mann, M. & Baarse, G., *Impact of Sea-level Rise on Society: A Case Study for the Netherlands*. Delft Hydraulics and Rijkswaterstaat, Delft and the Hague, 1991.

20. IPCC CZMS, *Assessment of the Vulnerability of Coastal Areas to Sea-level Rise: A Common Methodology*. Revision No. 1, Report of the Coastal Zone Management Subgroup. IPCC Response Strategies Working Group, Rijkswaterstaat, the Hague, 1991.

21. IPCC CZMS, *Global Climate Change and the Rising Challenge of the Sea*. Report of the Coastal Zone Managment Subgroup. IPCC Response Strategies Working Group, Rijkswaterstaat, the Hague, 1992.

22. Nicholls, R. J., Synthesis of vulnerability analysis studies. In *Preparing to Meet the Coastal Challenges of the 21st Century*, Proceedings of the World Coast Conference, Noordwijk, November 1993, Rijkswaterstaat, The Hague, 1995, pp. 181–216.

23. French, J. R., Spencer, T. & Reed, D. J. ed., Geomorphic response to sea-level rise. *Earth Surface Processes and Landforms*, **20** (1995) 1–103.

24. Nicholls, R. J. & Leatherman, S. P., The implications of accelerated sea-level rise and developing countries: a discussion. *Journal of Coastal Research*, **Special Issue No. 14** (1995) 303–323.

25. Hoozemans, F. M. J., Marchand, M. & Pennekamp, H. A., *A Global Vulnerability Analysis, Vulnerability Assessments for Population, Coastal Wetlands and Rice Production on a Global Scale,* 2nd Edition. Delft Hydraulics and Rijkswaterstaat, Delft and the Hague, 1993.

26. Baarse, G., *Development of an Operational Tool for Global Vulnerability Assessment (GVA): Update of the Number of People at Risk Due to Sea-level Rise and Increased Flooding Probabilities.* CZM-Centre Publication No. 3, Rijkswaterstaat, The Hague, 1995.

27. Baric, A. & Gasparovic, F., Implications of climatic change on the socio-economic activities in the Mediterranean coastal zones. In *Climate Change and the Mediterranean*, ed. L. Jeftic, J. D. Milliman & G. Sestini. Edward Arnold, London, 1992, pp. 129–174.

28. Flemming, N. C., Predictions of relative coastal sea-level change in the Mediterranean based on archaeological, historical and tide-gauge data. In *Climate Change and the Mediterranean*, ed. L. Jeftic, J. D. Milliman & G. Sestini. Edward Arnold, London, 1992, pp. 247–281.

29. Douglas, B. C., Global sea level change: determination and interpretation. *Reviews of Geophysics*, Supplement (July 1995), US National Report to International Union of Geodesy and Geophysics 1991–1994, 1995, pp. 1425–1432.

30. Emery, K. O., Aubrey, D. G. & Goldsmith, V., Coastal neo-tectonics of the Mediterranean from tide-gauge records. *Marine Geology*, **8** (1988) 41–52.

31. Milliman, J. D., Sea-level response to climate change and tectonics in the Mediterranean Sea. In *Climate Change and the Mediterranean*, ed. L. Jeftic, J. D. Milliman & G. Sestini. Edward Arnold, London, 1992, pp. 45–57.

32. Milliman, J. D., Broadus, J. M. & Gable, F., Environmental and economic implications of rising sea level and subsiding deltas: the Nile and Bengal examples. *Ambio*, **18** (1989) 340–345.

33. Stanley, D. J. & Warne, A. G., Nile delta: recent geological evolution and human impact. *Science*, **260** (1993) 628–634.

34. Delft Hydraulics, Resource Analysis, Ministry of Transport, Public Works and Water Management and Coastal Research Institute, *Vulnerability Assessment to Accelerated Sea-level Rise: Case Study Egypt.* Delft Hydraulics, Delft, 1992.

35. Strzepek, K. M., Onyeji, S. C., Saleh, M. & Yates, D., An assessment of integrated climate change impact on Egypt. In *As Climate Changes: International Impacts and Implications*, ed. K. M. Strzepek & J. Smith. Cambridge University Press, Cambridge, 1995, pp. 189–200.

36. Sestini, G., Implications of climatic changes for the Nile delta. In *Climate Change and the Mediterranean*, ed. L. Jeftic, J. D. Milliman & G. Sestini. Edward Arnold, London, 1992, pp. 535–601.

37. Stevenson, J. C., Ward, L. G. & Kearney, M. S., Vertical accretion in marshes with varying rates of sea-level rise. In *Estuarine Variability*, ed. D. A. Wolfe. Academic Press, New York, 1986, pp. 241–259.

130   *R. J. Nicholls, F. M. J. Hoozemans*

38. Turner, R. K., Kelly, P. M. & Kay, R. C., *Cities at Risk*. BNA International Inc., London, 1990.
39. Devine, N. P., Urban vulnerability to sea-level rise in the Third World. MS Thesis, Rutgers, The State University of New Jersey, New Brunswick, NJ, 1992.
40. United Nations Population Division, *World Urbanized Prospects* 1992. United Nations Population Division, New York, 1993.
41. Hazma, A. An appraisal of environmental consequences of urban development in Alexandria, Egypt. *Environment and Urbanization*, 11 (1989) 22–30.
42. Balkas, T., Yetis, U. & Chung, C., The integration of environmental considerations into coastal zone management, Izmir Bay, Turkey. *Coastal Zone Management: Selected Case Studies*, Organization for Economic Co-Operation and Development, Paris, 1993, pp. 85–108.
43. Nicholls, R. J. Coastal megacities and climate change. *GeoJournal*, 37 (1995) 369–379.
44. Zanda, L., The case of Venice. In *Proceedings of First International Meeting (Cities on Water)*, ed. R. Frassetto. Marsilo, Editori, Venice, Italy, 1991, pp. 51–59.
45. Francia, C. & Juhasz, F., The lagoon of Venice, Italy. *Coastal Zone Management: Selected Case Studies*, Organization for Economic Co-Operation and Development, Paris, 1993, pp. 109–134.
46. Holzer, T. L. & Johnson, A. I., Land subsidence caused by ground water withdrawal in urban areas. *GeoJournal*, 11 (1985) 245–255.
47. Bondesan, M., Castiglioni, G. B., Elmi, C., Gabbianelli, G., Marocco, R., Pirazzoli, P. A. & Tomasin, A., Coastal areas at risk from storm surges and sea-level rise in northeastern Italy. *Journal of Coastal Research*, 11 (1995) 1354–1379.
48. Pirazolli, P. A., Recent sea-level changes and related engineering problems in the lagoon of Venice (Italy). *Progress in Oceanography*, 18 (1987) 323–346.
49. Pirazolli, P. A., Possible defenses against a sea-level rise in the Venice area, Italy. *Journal of Coastal Research*, 7 (1991) 231–248.
50. Warne, A.G. & Stanley, D.J. Late Quaternary evolution of the northwest Nile delta coast and the adjacent coast in the Alexandria Region, Egypt. *Journal of Coastal Research*, 9 (1993) 26–64.
51. Sestini, G. The impact of climatic changes and sea-level rise on two deltaic lowlands of the eastern Mediterranean. In *Impacts of Sea-level Rise on European Coastal Lowlands*, ed. M. J. Tooley & S. Jelgersma. Blackwells, Oxford, 1992, pp. 170–203.
52. El-Sayed, M. Kh., Implications of relative sea-level rise on Alexandria. In *Proceedings of First International Meeting (Cities on Water)*, ed. R. Frassetto. Marsilo, Editori, Venice, Italy, 1991, pp. 183–189.
53. Dalfes, H. N., Climatic change and Istanbul: some preliminary results. In *Cities and Global Change*, Proceedings of the International Conference on Cities and Global Change, Toronto, 1991, ed. J. McCulloch. Climate Institute, Washington DC, 1991, pp. 92–107.
54. Boesch, D. F., Josselyn, M. N., Mehta, A. J., Morris, J. T., Nuttle, W. K., Simenstad, C. A. & Swift, D. J. P., Scientific assessment of coastal wetland

loss, restoration and management in Louisiana. *Journal of Coastal Research*, Special Issue No. **20** (1994).

55. Jelgersma, S. & Sestini, G., Implications of a future rise in sea level on the coastal lowlands of the Mediterranean. In *Climate Change and the Mediterranean*, ed. L. Jeftic, J. D. Milliman & G. Sestini. Edward Arnold, London, 1992, pp. 282–303.

56. Stanley, D. J., Recent subsidence and north-east tilt of the Nile delta, Egypt. *Marine Geology*, **94** (1990) 147–154.

57. El-Raey, M., Responses to the impacts of greenhouse-induced sea-level rise on the northern coastal regions of Egypt. In *Changing Climate and the Coast* Volume 2, ed. J. G. Titus. United States Environmental Protection Agency, Washington DC, 1990, pp. 225–238.

58. Ozhan, E. (ed.), *Proceedings of the Second International Conference on the Mediterranean Coastal Environment, MEDCOAST 95*, MEDCOAST secretariat, Middle East Technical University, Ankara, Turkey, 1995, 3 Volumes.

59. Corre, J. J., The coastline of the Gulf of Lions: impact of a warming of the atmosphere in the next few decades. In *Impacts of Sea-level Rise on European Coastal Lowlands*, ed. M. J. Tooley & S. Jelgersma. Blackwells, Oxford, 1992, pp. 153–169.

60. Day, J. W., Pont, D., Hensel, P. F. & Ibanez, C., Pulsing events and sustainability of Mediterranean deltas. In *Proceedings of the Second International Conference on the Mediterranean Coastal Environment, MEDCOAST 95*, ed. E. Ozhan. MEDCOAST secretariat, Middle East Technical University, Ankara, Turkey, 1995, pp. 781–792.

61. Capobianco, M., Jimenez, J. A., Sanchez-Arcilla, A. & Stive, M. J. F., Budget models for the evolution of deltas: Definition of processes and scales. In *Proceedings of the Second International Conference on the Mediterranean Coastal Environment, MEDCOAST 95*, ed. E. Ozhan. MEDCOAST secretariat, Middle East Technical University, Ankara, Turkey, 1995, pp. 737–751.

62. Louisiana Coastal Wetlands Conservation and Restoration Task Force, *Louisiana coastal restoration Plan* (draft). New Orleans District, US Army Corps of Engineers, June 1993, 195 pp.

63. Bandarin. F., The small islands of the Mediterranean: development issues and environmental management. In *Proceedings of the Second International Conference on the Mediterranean Coastal Environment, MEDCOAST 95*, ed. E. Ozhan. MEDCOAST secretariat, Middle East Technical University, Ankara, Turkey, 1995, pp. 537–550.

64. Wigley, T. M. L., Future climate of the Mediterranean Basin with particular emphasis on changes in precipitation. In *Climate Change and the Mediterranean*, ed. L. Jeftic, J. D. Milliman & G. Sestini. Edward Arnold, London, 1992, pp. 15–44.

65. Iacovau, N. G., Loizidou, X. I., Hulsbergen, C. H. & Hoozemans, F. M. J., Coastal zone management for Cyprus. In *Proceedings of the Second International Conference on the Mediterranean Coastal Environment, MEDCOAST 95*, ed. E. Ozhan. MEDCOAST secretariat, Middle East Technical University, Ankara, Turkey, 1995, pp. 491–502.

66. Nicholls, R. J., Assessing beach erosion due to sea-level rise. Paper

132   *R. J. Nicholls, F. M. J. Hoozemans*

presented at Geohazards and Engineering Geology, Coventry, September 1995, The Engineering Group of the Geological Society.

67. Lee, G., Nicholls, R.J., Birkemeier, W: A. & Leatherman, S. P., A conceptual fairweather-storm model of beach nearshore profile evolution at Duck, North Carolina, USA . *Journal of Coastal Research*, **11** (1995) 1157–1166.

68. Loizidou, X. I. & Iacovou, N. G., The Cyprus experience in coastal zone monitoring as a basis for shoreline management and erosion control. In *Proceedings of the Second International Conference on the Mediterranean Coastal Environment, MEDCOAST 95*, ed. E. Ozhan. MEDCOAST secretariat, Middle East Technical University, Ankara, Turkey, 1995, pp. 1019–1024.

69. Carter, T. R., Parry, M. C., Nishioka, S. & Harasawa, H. (ed.), *Technical guidelines for assessing climate change impacts and adaptations*. Working Group II of the Intergovernmental Panel on Climate Change, University College, London and Centre for Global Environmental Research, Tsukuba, 1994.

70. National Research Council, *Science, Policy and the Coast: Improving Decisionmaking*. National Academy Press, Washington DC, 1995.

71. Koster, M. J. & Hillen, R., Combat erosion by law: coastal defence policy for the Netherlands. *Journal of Coastal Research*, **11** (1995) 1221–1228.

72. Turner, R. K., Doktor, P. & Adger, W. N., Assessing the costs of sea-level rise. *Environment and Planning A*, **27** (1995) 1777–1796.

73. Volonte, C. R. & Nicholls, R. J., Uruguay and sea-level rise: potential impacts and responses. *Journal of Coastal Research*, **Special Issue No. 14** (1995) 262–284.

Name Index